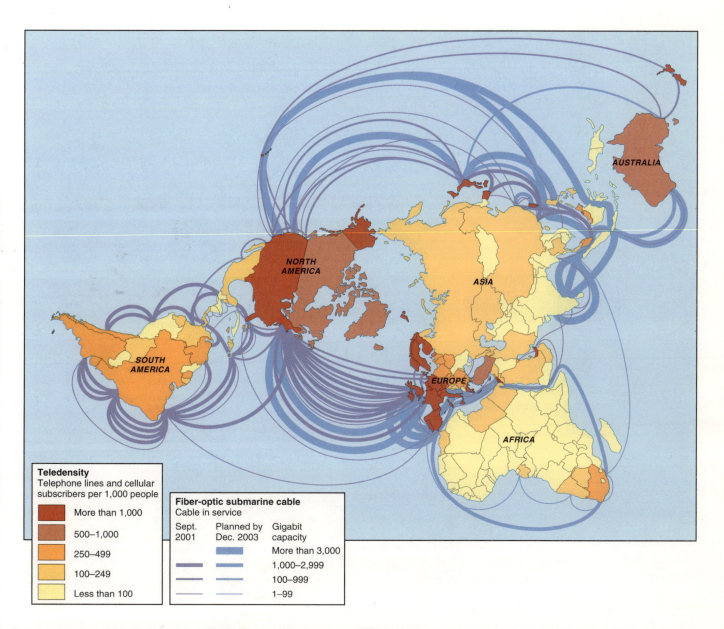

Teledensity
Telephone lines and cellular subscribers per 1,000 people

■	More than 1,000
■	500–1,000
■	250–499
■	100–249
■	Less than 100

Fiber-optic submarine cable
Cable in service

Sept. 2001	Planned by Dec. 2003	Gigabit capacity
		More than 3,000
		1,000–2,999
		100–999
		1–99

Global connections, in place and planned. The first ocean wiring began in 1850 from England to France and across the Atlantic in 1858. In the 1900s competition from radio and satellites made cable almost redundant until in the mid-1990s, when fiber optics provided greater speed and capacity for telephone calls. Satellites continue to be used for video (TV news).

Source: National Geographic Society, Picture ID 759480 Reprinted by permission.

IMPORTANT:

HERE IS YOUR REGISTRATION CODE TO ACCESS

YOUR PREMIUM McGRAW-HILL ONLINE RESOURCES.

For key premium online resources you need THIS CODE to gain access. Once the code is entered, you will be able to use the Web resources for the length of your course.

If your course is using **WebCT** or **Blackboard**, you'll be able to use this code to access the McGraw-Hill content within your instructor's online course.

Access is provided if you have purchased a new book. If the registration code is missing from this book, the registration screen on our Website, and within your WebCT or Blackboard course, will tell you how to obtain your new code.

Registering for McGraw-Hill Online Resources

TO gain access to your McGraw-Hill web resources simply follow the steps below:

1. USE YOUR WEB BROWSER TO GO TO: **http://www.mhhe.com/bradshaw**

2. CLICK ON **FIRST TIME USER**.

3. ENTER THE REGISTRATION CODE* PRINTED ON THE TEAR-OFF BOOKMARK ON THE RIGHT.

4. AFTER YOU HAVE ENTERED YOUR REGISTRATION CODE, CLICK **REGISTER**.

5. FOLLOW THE INSTRUCTIONS TO SET-UP YOUR PERSONAL UserID AND PASSWORD.

6. WRITE YOUR UserID AND PASSWORD DOWN FOR FUTURE REFERENCE. KEEP IT IN A SAFE PLACE.

TO GAIN ACCESS to the McGraw-Hill content in your instructor's **WebCT** or **Blackboard** course simply log in to the course with the UserID and Password provided by your instructor. Enter the registration code exactly as it appears in the box to the right when prompted by the system. You will only need to use the code the first time you click on McGraw-Hill content.

Thank you, and welcome to your McGraw-Hill online resources!

* YOUR REGISTRATION CODE CAN BE USED ONLY ONCE TO ESTABLISH ACCESS. IT IS NOT TRANSFERABLE.

0-07-291982-5 **BRADSHAW: CONTEMPORARY WORLD REGIONAL GEOGRAPHY, 1/E**

MCGRAW-HILL
ONLINE RESOURCES

REGISTRATION CODE

QB8M-CF24-NK4S-U7HA-6EAY

Higher Education

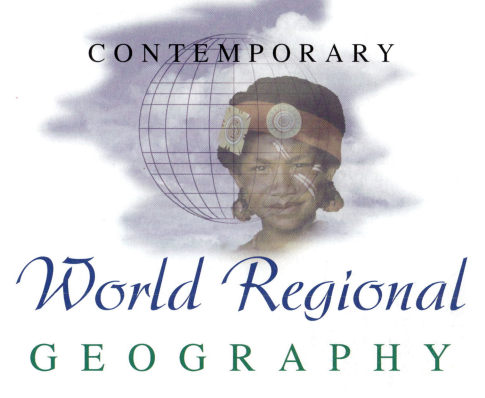

CONTEMPORARY

World Regional

G E O G R A P H Y

Global Connections, Local Voices

Michael Bradshaw George W. White Joseph P. Dymond

College of St. Mark and John   *Frostburg State University*   *Towson University*

with contributions by
Dydia DeLyser
Louisiana State University

 **Higher Education**

Boston   Burr Ridge, IL   Dubuque, IA   Madison, WI   New York   San Francisco   St. Louis
Bangkok   Bogotá   Caracas   Kuala Lumpur   Lisbon   London   Madrid   Mexico City
Milan   Montreal   New Delhi   Santiago   Seoul   Singapore   Sydney   Taipei   Toronto

Higher Education

CONTEMPORARY WORLD REGIONAL GEOGRAPHY: GLOBAL CONNECTIONS,
LOCAL VOICES

Published by McGraw-Hill, a business unit of The McGraw-Hill Companies, Inc., 1221 Avenue of the
Americas, New York, NY 10020. Copyright © 2004 by The McGraw-Hill Companies, Inc.
All rights reserved. No part of this publication may be reproduced or distributed in any form or by
any means, or stored in a database or retrieval system, without the prior written consent of The
McGraw-Hill Companies, Inc., including, but not limited to, in any network or other electronic
storage or transmission, or broadcast for distance learning.

Some ancillaries, including electronic and print components, may not be available to customers outside
the United States.

This book is printed on acid-free paper.

2 3 4 5 6 7 8 9 0 VNH/VNH 0 9 8 7 6 5 4 3

ISBN 0–07–254975–0

Publisher: *Margaret J. Kemp*
Sponsoring editor: *Thomas C. Lyon*
Senior developmental editor: *Donna Nemmers*
Executive marketing manager: *Lisa L. Gottschalk*
Lead project manager: *Joyce M. Berendes*
Senior production supervisor: *Laura Fuller*
Media project manager: *Jodi K. Banowetz*
Media technology producer: *Renee Russian*
Coordinator of freelance design: *Rick D. Noel*
Cover/interior designer: *Jamie E. O'Neal*
Interior design elements created with images from: *Corbis, Digital Stock, and Digital Vision*
Cover images: Front–*©Masterfile/Washing Vegetables Near Taj Mahal by R. Ian Lloyd;*
Rear–*©Masterfile/Open Air Restaurant in Singapore;* Spine–*©Getty Images/Massai*
Warrior on Cell Phone by Joseph Van Os
Lead photo research coordinator: *Carrie K. Burger*
Photo research: *Chris Hammond/PhotoFind, LLC*
Supplement producer: *Brenda A. Ernzen*
Compositor: *Precision Graphics*
Typeface: *10/12 Times Roman*
Printer: *Von Hoffmann Corporation*

All photos, unless otherwise indicated: © Corbis

Library of Congress Cataloging-in-Publication Data

Bradshaw, Michael J. (Michael John), 1935–
    Contemporary world regional geography : global connections, local voices / Michael Bradshaw,
George W. White, Joseph P. Dymond; with contributions by Dydia DeLyser. — 1st ed.
      p. cm.
    Includes index.
    ISBN 0–07–254975–0
    1. Geography.   I. White, George W., 1963–.   II. Dymond, Joseph P.   III. Title.

G116 .B72 2004
910—dc21
                                                        2002044897
                                                         CIP

www.mhhe.com

*to Valerie, Emily,
Maureen and Madison,
and Paul*

Contents in Brief

Contents

 Chapter 10

Latin America 419

 Chapter 11

North America 483

 Chapter 12

Global Connections, Local Voices 537

Meet the Author Team

Michael Bradshaw

Michael Bradshaw and his wife live in Canterbury, England, and have two sons and two grandchildren. Michael taught for 25 years at the College of St. Mark & St. John, Plymouth, as Geography Department chair and dean of the humanities course. He has written texts for British high schools and colleges since the 1960s. In 1985, he was awarded a Ph.D. from Leicester University for his study on the impacts of federal grant-aid in Appalachia. His book, *The Appalachian Regional Commission: Twenty-Five Years of Government Policy,* was published in 1992. Since 1991, he has written for U.S. students and has been responsible for two physical geography texts and the successful world regional geography text, *The New Global Order,* with a second-edition update published in 2002. Michael believes that we should all be better equipped to live in the modern, increasingly global world. Understandings of geographic differences should make us more able to assess crucial issues and value other people who bring varied resources and who face pressures that we find difficult to imagine.

In developing the next-phase text for world regional geography courses, Dr. Bradshaw has extended the experience and expertise of the writing team by adding new coauthors. *Contemporary World Regional Geography: Global Connections, Local Voices* is the outcome of this new collaboration. He is lead author for the first two and the last chapters, and for the regional chapters on East Asia, Southeast Asia and South Pacific, South Asia, Northern Africa and Southwestern Asia, and Africa South of the Sahara.

George White

George W. White grew up in Oakland, California. He pursued graduate work in Eugene, Oregon, completing a Ph.D. at the University of Oregon. He then moved to Frostburg, Maryland, where he met his wife. George is currently an associate professor in the Department of Geography and coordinator of the International Studies Program at Frostburg State University. Political geography and Europe are two of his primary interests. He recently authored a book entitled *Nationalism and Territory: Constructing Group Identity in Southeastern Europe.*

After meeting Michael Bradshaw, George was impressed by Michael's long and distinguished career of teaching, research, and publication. He accepted the opportunity to join Michael in his plans to write a new world regional geography text, taking lead authorship for the chapters on Europe and Russia and Neighboring Countries, as well as contributing to other areas of the text.

George became a geographer because he believes that the field of geography is alive and dynamic, attuned to our ever-changing world and its great diversity. The world regional approach represents the breadth of the field of geography, and world regional geography texts are the epitome of the geographer's art. George White chose to collaborate with Michael Bradshaw on this project because the text combines local practices and global processes, and explains interaction between the two as they shape each other.

Dydia DeLyser, Joe Dymond, George White, and Michael Bradshaw

Joe Dymond

Joseph P. Dymond earned a master of science degree from the Pennsylvania State University in 1994, and a master of natural sciences degree from Louisiana State University in 1999. Joe taught world regional geography courses for the Louisiana State University Department of Geography and Anthropology from 1995 through 2000. During Joe's six years at LSU, he instructed thousands of students and was recognized in the spring of 1997, fall of 1999, and fall of 2000 for superior instruction to freshman students by the Louisiana State University Freshman Honor Society, Alpha Lamba Delta. Joe currently lives in suburban Washington, D.C., with his wife and daughter, and is adjunct faculty in the Department of Geography and Environmental Planning at Towson University. In this new text, Joe is the lead author for the chapters on Latin America and North America.

Joe chose to become involved with this new textbook because he wants to provide students with the geographic tools to better understand the human and environmental patterns in their world. Joe wants to help students understand why certain cultural and physical elements exist where they do, how they got there, how long they have been there, how they have changed over time, and what they might be like in the future. Understanding the relationships of human and physical geographic patterns and relationships creates a strong foundation for a comprehensive and fair perspective on the people and places comprising the regions of the world. The style of this text, including the Point-Counterpoint sections, attempts to tell the regional geographic story from many perspectives. Its structure permits the students to better analyze geographic characteristics around the world and to think critically about issues. It provides students with the opportunity to think on their own and to piece together various data elements so they may establish their own informed opinions.

Dydia DeLyser

Dydia DeLyser is an assistant professor of geography at Louisiana State University, where she has taught world regional geography to thousands of students. She earned her Ph.D. in 1998 from Syracuse University, specializing in cultural geography. Her interests lie in landscapes and the interpretation of the past in the American West, and in flying airplanes.

Dydia became involved to help fashion a textbook that doesn't backpedal on hard topics and introduces students to some of the difficult and complex controversies and challenging issues facing the world. She feels it is important to build students' critical-thinking skills—the skills they will need to work toward constructive resolutions to these issues.

Dydia prefers that a book presents positive elements for every region. She believes that we should never simply give American students the impression that Americans are the best off, but instead reveal some of the world's magnificent complexity and diversity, showing there are wonderful things about every place.

She believes that it is also important to write a text that relates individual places and people with broader cultural/political/economic flows and forces: virtually every "corner" of the globe is now linked to much broader global systems, and yet, as virtually every individual can tell you, local places and local voices remain not only unique but also important in today's world. The study of world regional geography must combine an understanding of broad, large-scale forces with an equally deep understanding of what is local, small-scale, and unique—for it's the combination of those two that makes us who we are.

Preface

The Changing Face of Our World and *Contemporary World Regional Geography*

To contend with the political, economic, demographic, and environmental shifts reverberating throughout the world, the first edition of *Contemporary World Regional Geography* brings together four outstanding geographers. Our team begins with Michael Bradshaw. He is well known for his ability to weave together solid geography content with interesting stories, real-life case studies, and applications to student life experiences.

Joining forces with Michael Bradshaw are coauthors Joe Dymond of Towson University and George White of Frostburg State University. In addition, Dydia DeLyser of Louisiana State University made significant contributions in formulating new ideas and providing guidance for the book. Joe, George, and Dydia are all active instructors who use multimedia approaches to teach hundreds of undergraduates each semester.

Devotion to and passion about teaching unite our team. We thoroughly enjoy telling those interesting stories that help us to understand what is happening—the stories that, when told correctly, mesmerize even the most reluctant students, causing them to perk up and think, "Wow, I never knew that! So that's why …!" We all love to watch students get excited learning about a subject they once viewed as too hard or too intimidating.

Why Focus a World Regional Geography Text on "Global Connections, Local Voices"?

Globalization is a term that is widely used, often with little precision and much controversy. Some globalists even suggest it implies the end of geography because globalization to them means that everyone will eventually become the same, with no variation between geographic places! The authors take a different view—that a focus on globalization is the beginning of a new and particularly satisfying approach to regional geography. Globalization implies an overview of worldwide events and interconnections, together with the contributions from and impacts on world regions, countries, and local areas within countries. We live in a connected world of rapid personal travel and communication of information. What happens in one part increasingly affects all of us in the rest of the world.

Geographic regions within this world can no longer be considered as isolated, unique entities but have to be seen in an interactive and comparative context.

Localization is not merely the opposite of globalization. Local character and local actions often result from, or respond to, globalization. They also give local meaning and outcomes to the global systems. Countries are still the most significant political and economic units in our world and, together with local areas within them, provide the basis of study within our globalizing—but not globalized—world.

Globalization and the Text

Globalization and localization are introduced at the start of Chapter 1 and viewed in the context of a fourfold scale of geographic region: global, world region, country, and (within country) local. The definition of the world regions is carried out in this context, and there is a short discussion of the emergence of increasingly global interconnections through human history. Chapter 2 is a broad overview of the main human and physical geography concepts that will be met in the regional chapters and has an inevitably global view.

In some chapters, there are Global Focus boxes. These highlight specific situations of global significance. For example, "Global Focus: 9/11" in Chapter 1 picks up on the events of September 11, 2001, in New York City, Washington, D.C., and Pennsylvania. In Chapter 5, "Global Focus: Toyota" examines a multinational corporation.

The world region chapter sequence gets the student to embark on a world tour. This tour begins with Europe (Chapter 3), where, although the text avoids a merely Eurocentric view, many of the modern global processes began—from capitalism to modern colonialism and many technologies. While many of the technologies and trading products built on previous Arab, Asian, and African achievements, the European role was often seminal—although not always beneficial to the rest of the world. This provides a significant start to the tour.

Chapter 4 moves eastward to Russia and its neighboring countries, where the European-origin Communist principles applied for most of the 1900s, until the breakup of the Soviet Union in 1991. Culturally, this region is rooted in both European and Asiatic traditions. The region struggles to shift from the self-sufficient central command system to incorporation in the global economic and political systems while retaining the "Russian Empire" motivation.

The tour then turns southward for a study of the contrasting East Asia (Chapter 5) with emphases on Japan and China and their ancient cultures in increasingly global roles. Global considerations also allow the joining of Southeast Asia with the South Pacific (Australia, New Zealand, and Pacific islands) in Chapter 6, despite their differences of physical and human backgrounds. The increasing trade and interactions among Pacific Ocean rim countries make a new center of global development in the 2000s.

South Asia (Chapter 7) brings together the study of countries that attempted to be largely self-sufficient after independence and develop their own non-Western ways of modernization but changed their attitudes toward the global system with varied outcomes in the 1990s. Chapter 8 examines the central (Arab) part of the Islamic world and the internal stresses caused by the presence of Israel, as well as by the uneven distribution of oil and water resources.

In Africa South of the Sahara (Chapter 9), the tour completes the study of the Eurasia, Africa, and Australia continents with a consideration of the impacts of European colonialism and modern independence on many small countries and a few larger ones that face poverty, exploitation, and an HIV/AIDS epidemic.

Skipping across the South Atlantic Ocean, Chapter 10 studies Latin America, its colonization as part of European expansion, and its modern relationships to the United States. The final regional chapter, covering the USA and Canada (Chapter 11), brings together many of the elements of people, movements, and economic and political processes that make the United States the sole world superpower.

We have concluded with a unique Chapter (12) that explores two areas in a geographic context as an endpiece to the book. After studying the global-local interactions in different world regions, a further and deeper consideration of the nature of globalization is relevant. A study of world terrorism combines the global and local effects of a series of processes instigated outside the world's formal political systems. The position of Somalia within this context highlights the strengths of geographic study and the dilemmas in the contemporary world.

This tour could be carried out in varied orders. This text aims to make that possible. Speak with your McGraw-Hill sales representative about customizing this text to best fit your needs or about online content delivery.

A wide variety of print and electronic ancillary components are available to instructors, which are detailed in the following pages. Additionally the following instructor supplements are available through your McGraw-Hill sales representative:

- 125 transparencies
- Computerized Testing Software CD-ROM
- Geosciences Videotape Library

A Text for Students

Pages xiv–xv describe and illustrate some of the ways in which this text provides a source that is well organized, well written, readable, and appropriate to students' interests and abilities. The goals of providing a global view and a personal local view are addressed through the text, maps, and sections such as the Personal View, Global Focus, and Point-Counterpoint boxes. Students are encouraged to think about what it means to be part of a global community.

The authors and publisher have strived to make this a text that students will embrace. To this end, we have taken these steps:

- Written a readable text. Many reviewers commented on the accessible style and the clarity of writing, clear definition of terms, and current examples of events.

- Provided a clear structure to each chapter so that students can more easily compare one world region to another. This consistent structure also helps students become familiar with what to expect in each chapter.

- Included statistical data in straightforward maps and diagrams, without compromising the readability of the text.

- Maintained the same style and types of maps and diagrams throughout, so that students can compare regions with ease.

- Introduced pedagogical aids, such as Point-Counterpoint boxes, which tackle various sides of challenging issues.

- Added study tools to the text, such as the Test Your Understanding sections several times in each chapter to summarize, pose questions, and list key terms. Also, the book's Online Learning Center at www.mhhe.com/bradshaw provides practice test questions, map-based quizzes, and access to a library of related websites, as well as other learning aids.

Art Program

As the authors examined every word of text in the book, they also scrutinized every illustration with a critical look at how it supports the text. The accurate and artistically compelling illustrations and photographs greatly enhance the student's understanding of the processes and concepts in each world region. The illustrations are not only visually spectacular but also pedagogically sound and give *Contemporary World Regional Geography* a consistent look from cover to cover. Over 200 photos have been chosen for their geographic relevance and are often printed in large format across columns. Unusual and interesting photos have been selected to show students glimpses of parts of the world they have likely not seen before.

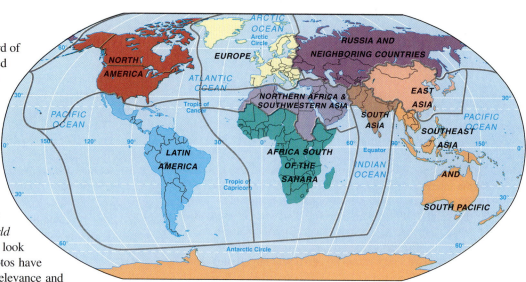

Sets of maps, diagrams, and data figures that are introduced in Chapter 2 follow a consistent theme and scale throughout the regional chapters so that students can more easily assess the similarities and differences among the world regions.

All of the photographs, maps, data figures, and illustrations also are available to adopters in digital format on the Digital Content Manager CD-ROM, provided to instructors by McGraw-Hill.

Notable Comments

"A considerable effort has gone into producing a text that is both readable and precise. Students seeking good characterization of the regional differences present in today's world will find this text most enlightening. It highlights current issues and provides the background explanation to better understand those issues."

—Professor Alice L. Tym,
*University of Tennessee,
Chattanooga*

"The best feature is the readability. The writing is superb. It does not talk down to students, nor is it too challenging. One extremely good aspect of the writing is that it flows very well, holding the attention of the reader, and segues into each of the subjects in a logical order without seeming to start and stop awkwardly. The authors should be commended for their writing."

—Professor Taylor E. Mack,
Mississippi State University

"The chapters follow a coherent structure, which I think is important for both teaching and learning. I commend the authors. The diagrams and tables are consistent throughout."

—Professor Paul Robbins,
Ohio State University

Acknowledgments

This book is the product of the hard work and dedication of dozens of people, all of whom shared a common goal: to produce the very best world regional geography book possible. No single person, no matter how educated, "knows" everything about the world's regions. Even an author team whose collective expertise covers most of the field must rely on an almost unimaginable amount of feedback. We greatly appreciate the help of the many reviewers, consultants, and focus group members—committed teachers who went the extra mile to help make this book what it is.

The manuscript for this text went through three full draft revisions, along with rigorous reviewing by instructors. As we examined every word of text in the book, we also scrutinized every piece of art with a critical look at how it works with the text. An advisory panel was established to help us refine the text and art; a focus group was held to critique the market, text, supplements, and where we're headed for future editions; and other faculty members provided information about course needs and trends.

We offer special thanks to the reviewers who spent hours poring over chapter drafts in meticulous detail, spotting errors and inconsistencies, confirming what works and gently critiquing what didn't, adding new facets, and pointing out sections that we could clarify.

The authors also wish to express their gratitude to Elizabeth Leppman for the skilled editorial services and geographic expertise she brought to this project. They are also thankful for McGraw-Hill's editorial support and assistance provided by Marge Kemp, Tom Lyon, and Donna Nemmers, the marketing expertise of Lisa Gottschalk, and the production knowledge of Joyce Berendes, Rick Noel, and Carrie Burger. In addition, these reviewers and colleagues provided their own photographs to support the printed word: Ian Coles, Jerzy Jemiolo, David C. Johnson, Loren W. Linholm, Mission Aviation Fellowship (MAF), Alexander B. Murphy, Steven M. Schnell, Gerald Selous, Emily A. White, and Ron Wixman. Certainly not least of all, the authors wish to acknowledge the tireless support and patience of their families, who helped in this effort more than they could realize.

Reviewers

Edward Addo
St. Cloud State University

Brad Baltensperger
Michigan Technological University

John Bowen
University of Wisconsin, Oshkosh

John Boyer
Virginia Tech

Robert F. Brinson, Jr.
Santa Fe Community College

Michelle Calvarese
California State University, Fresno

Gabe Cherem
Eastern Michigan University

Jonathan C. Comer
Oklahoma State University

Carlos E. Cordova
Oklahoma State University

Mark W. Corson
Northwest Missouri State University

Marcelo Cruz
University of Wisconsin, Green Bay

Jose A. da Cruz
Ozarks Technical Community College

Robert S. Dilley
Lakehead University

Harold M. Elliott
Weber State University

Allen Finchum
Oklahoma State University

Jay D. Gatrell
Indiana State University

Thomas O. Graff
University of Arkansas

John R. Grimes
Louisiana State University

Mark Guizlo
Lakeland Community College

David C. Johnson
University of Louisiana, Lafayette

Richard L. Kroll
Kean University

James Leonard
Marshall University

Elizabeth J. Leppman
St. Cloud State University

Max Lu
Kansas State University

Taylor E. Mack
Mississippi State University

Susan R. Martin
Michigan Technological University

Calvin O. Masilela
Indiana University of Pennsylvania

Brent McCusker
West Virginia University

Monica Nyamwange
William Paterson University

Bimal K. Paul
Kansas State University

Paul Robbins
Ohio State University

Paul A. Rollinson
Southwest Missouri State University

Steven M. Schnell
Kutztown University of Pennsylvania

Burl Self
Southwest Missouri State University

Jonathan Taylor
California State University, Fullerton

Jeffrey S. Torguson
St. Cloud State University

Alice L. Tym
University of Tennessee, Chattanooga

Kelly Ann Victor
Eastern Michigan University

Charles L. Wax
Mississippi State University

David Welk
Reedley College, Clovis Center

Naim Zeibak
Indiana State University

Course Survey Respondents

Gregory Haddock
Northwest Missouri State University

Doug Heffington
Middle Tennessee State University

Donald Lyons
University of North Texas

Richard Ulack
University of Kentucky

Focus Group Participants

Brad Baltensperger
Michigan Technological University

Michelle Calvarese
California State University, Fresno

Michael A. Camille
The University of Louisiana at Monroe

Francis H. Dillon, III
George Mason University

James D. Engstrom
Georgia Perimeter College

Sherry Goddicksen
California State University, Fullerton

Reuel R. Hanks
Oklahoma State University

Max Lu
Kansas State University

Brent McCusker
West Virginia University

William Preston
California Polytechnic State University

Joel Quam
College of DuPage

Deborah A. Salazar
Texas Tech University

J. Duncan Shaeffer
Arizona State University

Jeffrey S. Torguson
St. Cloud State University

Charles L. Wax
Mississippi State University

Global Connections

Local Voices in the Classroom

For their entire lives, students will be confronted with challenging issues for which there are no easy answers. By exploring such issues from various sides, we seek to develop students' critical-thinking skills to help them understand the complexities that will continue to face them in this increasingly globalizing world. We seek to show not only the interconnections of the world but also the impacts that certain (often powerful) players can have—to understand, for example, how our actions as Americans can affect distant peoples and places. *Contemporary World Regional Geography* emphasizes the relationships between each region of the world and the local issues in each particular region. Cultural, social, historical, and economic topics are examined in a way that provides students with a better appreciation of the world around them.

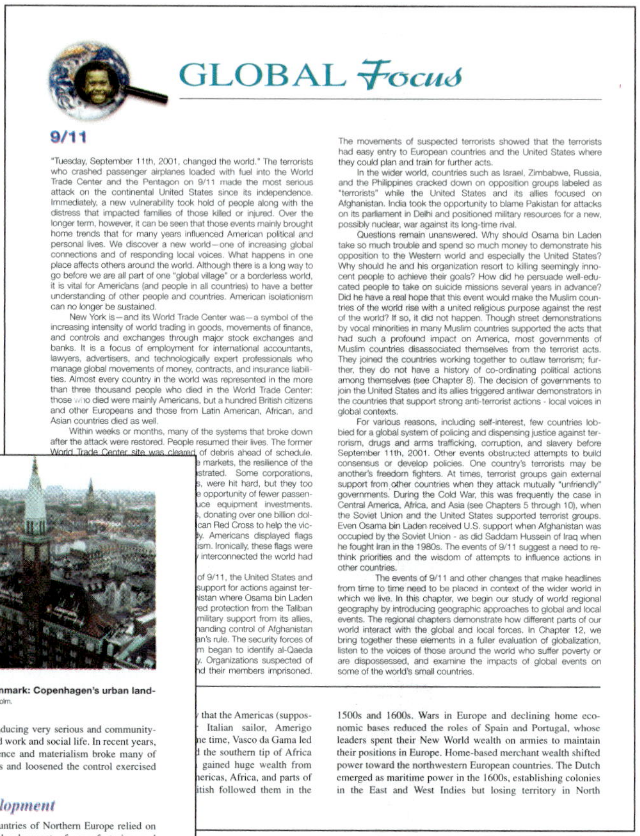

GLOBAL *Focus*

9/11

"Tuesday, September 11th, 2001, changed the world." The terrorists who crashed passenger airplanes loaded with fuel into the World Trade Center and the Pentagon on 9/11 made the most serious attack on the continental United States since its independence. Immediately, a new vulnerability took hold of people along with the distress that impacted families of those killed or injured. Over the longer term, however, it can be seen that those events mainly brought home trends that for many years influenced American political and personal lives. We discover a new world—one of increasing global connections and of responding local voices. What happens in one place affects others around the world. Although there is a long way to go before we are all part of one "global village" or a borderless world, it is vital for Americans (and people in all countries) to have a better understanding of other people and countries. American isolationism can no longer be sustained.

New York is—and its World Trade Center was—a symbol of the increasing intensity of world trading in goods, movements of finance, and controls and exchanges through major stock exchanges and banks. It is a focus of employment for international accountants, lawyers, advertisers, and technologically expert professionals who manage global movements of money, contracts, and insurance liabilities. Almost every country in the world was represented in the more than three thousand people who died in the World Trade Center: those who died were mainly Americans, but a hundred British citizens and other Europeans and those from Latin American, African, and Asian countries died as well.

Within weeks or months, many of the systems that broke down after the attack were restored. People resumed their lives. The former World Trade Center site was cleared of debris ahead of schedule. … markets, the resilience of the … strated. Some corporations, … s, were hit hard, but they too … e opportunity of fewer passen- … uce equipment investments. … , donating over one billion dol- … can Red Cross to help the vic- … y. Americans displayed flags … ism. Ironically, these flags were … interconnected the world had

… of 9/11, the United States and … support for actions against ter- … istan where Osama bin Laden … ed protection from the Taliban … military support from its allies, … nanding control of Afghanistan … an's rule. The security forces of … m began to identify al-Qaeda … y. Organizations suspected of … d their members imprisoned.

The movements of suspected terrorists showed that the terrorists had easy entry to European countries and the United States where they could plan and train for further acts.

In the wider world, countries such as Israel, Zimbabwe, Russia, and the Philippines cracked down on opposition groups labeled as "terrorists" while the United States and its allies focused on Afghanistan. India took the opportunity to blame Pakistan for attacks on its parliament in Delhi and positioned military resources for a new, possibly nuclear, war against its long-time rival.

Questions remain unanswered. Why should Osama bin Laden take so much trouble and spend so much money to demonstrate his opposition to the Western world and especially the United States? Why should he and his organization resort to killing seemingly innocent people to achieve their goals? How did he persuade well-educated people to take on suicide missions several years in advance? Did he have a real hope that this event would make the Muslim countries of the world rise with a united religious purpose against the rest of the world? If so, it did not happen. Though street demonstrations by vocal minorities in many Muslim countries supported the acts that had such a profound impact on America, most governments of Muslim countries disassociated themselves from the terrorist acts. They joined the countries working together to outlaw terrorism; further, they do not have a history of co-ordinating political actions among themselves (see Chapter 8). The decision of governments to join the United States and its allies triggered antiwar demonstrators in the countries that support strong anti-terrorist actions - local voices in global contexts.

For various reasons, including self-interest, few countries lobbied for a global system of policing and dispensing justice against terrorism, drugs and arms trafficking, corruption, and slavery before September 11th, 2001. Other events obstructed attempts to build consensus or develop policies. One country's terrorists may be another's freedom fighters. At times, terrorist groups gain external support from other countries when they attack mutually "unfriendly" governments. During the Cold War, this was frequently the case in Central America, Africa, and Asia (see Chapters 5 through 10), when the Soviet Union and the United States supported terrorist groups. Even Osama bin Laden received U.S. support when Afghanistan was occupied by the Soviet Union - as did Saddam Hussein of Iraq when he fought Iran in the 1980s. The events of 9/11 suggest a need to rethink priorities and the wisdom of attempts to influence actions in other countries.

The events of 9/11 and other changes that make headlines from time to time need to be placed in context of the wider world in which we live. In this chapter, we begin our study of world regional geography by introducing geographic approaches to global and local events. The regional chapters demonstrate how different parts of our world interact with the global and local forces. In Chapter 12, we bring together these elements in a fuller evaluation of globalization, listen to the voices of those around the world who suffer poverty or are dispossessed, and examine the impacts of global events on some of the world's small countries.

… that the Americas (suppos- … Italian sailor, Amerigo … ne time, Vasco da Gama led … the southern tip of Africa … gained huge wealth from … ericas, Africa, and parts of … itish followed them in the

1500s and 1600s. Wars in Europe and declining home economic bases reduced the roles of Spain and Portugal, whose leaders spent their New World wealth on armies to maintain their positions in Europe. Home-based merchant wealth shifted power toward the northwestern European countries. The Dutch emerged as maritime power in the 1600s, establishing colonies in the East and West Indies but losing territory in North

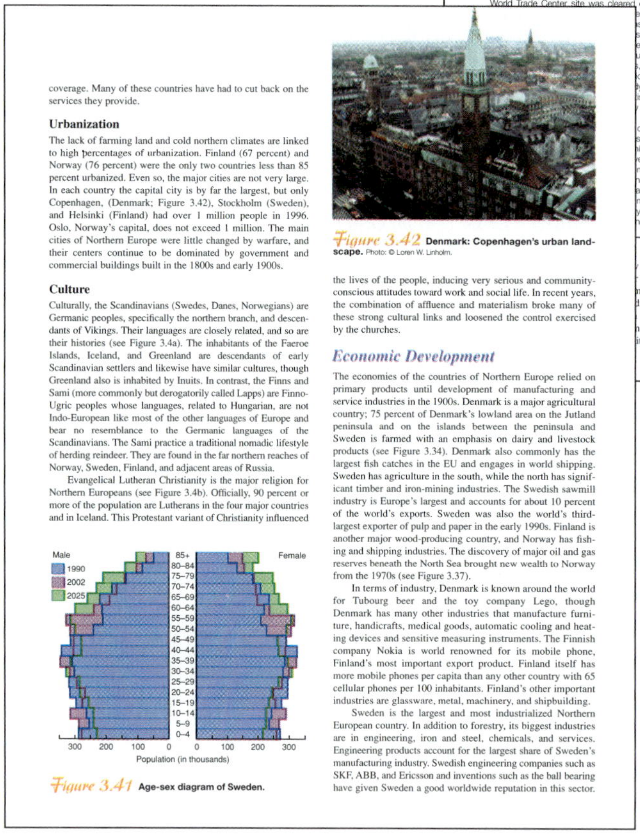

coverage. Many of these countries have had to cut back on the services they provide.

Urbanization

The lack of farming land and cold northern climates are linked to high percentages of urbanization. Finland (67 percent) and Norway (76 percent) were the only two countries less than 85 percent urbanized. Even so, the major cities are not very large. In each country the capital city is by far the largest, but only Copenhagen, (Denmark; Figure 3.42), Stockholm (Sweden), and Helsinki (Finland) had over 1 million people in 1996. Oslo, Norway's capital, does not exceed 1 million. The main cities of Northern Europe were little changed by warfare, and their centers continue to be dominated by government and commercial buildings built in the 1800s and early 1900s.

Culture

Culturally, the Scandinavians (Swedes, Danes, Norwegians) are Germanic peoples, specifically the northern branch, and descendants of Vikings. Their languages are closely related, and so are their histories (see Figure 3.4a). The inhabitants of the Faeroe Islands, Iceland, and Greenland are descendants of early Scandinavian settlers and likewise have similar cultures, though Greenland also is inhabited by Inuits. In contrast, the Finns and Sami (more commonly but derogatorily called Lapps) are Finno-Ugric peoples whose languages, related to Hungarian, are not Indo-European like most of the other languages of Europe and bear no resemblance to the Germanic languages of the Scandinavians. The Sami practice a traditional nomadic lifestyle of herding reindeer. They are found in the far northern reaches of Norway, Sweden, Finland, and adjacent areas of Russia.

Evangelical Lutheran Christianity is the major religion for Northern Europeans (see Figure 3.4b). Officially, 90 percent or more of the population are Lutherans in the four major countries and in Iceland. This Protestant variant of Christianity influenced

Male
1990
2002
2025
Female
85+
80–84
75–79
70–74
65–69
60–64
55–59
50–54
45–49
40–44
35–39
30–34
25–29
20–24
15–19
10–14
5–9
0–4
300 200 100 0 0 100 200 300
Population (in thousands)

Figure 3.41 **Age-sex diagram of Sweden.**

Figure 3.42 **Denmark: Copenhagen's urban landscape.** Photo: © Loren W. Linholm.

the lives of the people, inducing very serious and community-conscious attitudes toward work and social life. In recent years, the combination of affluence and materialism broke many of these strong cultural links and loosened the control exercised by the churches.

Economic Development

The economies of the countries of Northern Europe relied on primary products until development of manufacturing and service industries in the 1900s. Denmark is a major agricultural country; 75 percent of Denmark's lowland area on the Jutland peninsula and on the islands between the peninsula and Sweden is farmed with an emphasis on dairy and livestock products (see Figure 3.34). Denmark also commonly has the largest fish catches in the EU and engages in world shipping. Sweden has agriculture in the south, while the north has significant timber and iron-mining industries. The Swedish sawmill industry is Europe's largest and accounts for about 10 percent of the world's exports. Sweden was also the world's third-largest exporter of pulp and paper in the early 1990s. Finland is another major wood-producing country, and Norway has fishing and shipping industries. The discovery of major oil and gas reserves beneath the North Sea brought new wealth to Norway from the 1970s (see Figure 3.37).

In terms of industry, Denmark is known around the world for Tubourg beer and the toy company Lego, though Denmark has many other industries that manufacture furniture, handicrafts, medical goods, automatic cooling and heating devices and sensitive measuring instruments. The Finnish company Nokia is world renowned for its mobile phone, Finland's most important export product. Finland itself has more mobile phones per capita than any other country with 65 cellular phones per 100 inhabitants. Finland's other important industries are glassware, metal, machinery, and shipbuilding.

Sweden is the largest and most industrialized Northern European country. In addition to forestry, its biggest industries are in engineering, iron and steel, chemicals, and services. Engineering products account for the largest share of Sweden's manufacturing industry. Swedish engineering companies such as SKF, ABB, and Ericsson and inventions such as the ball bearing have given Sweden a good worldwide reputation in this sector.

Geography
in Action

The regional chapters utilize two boxed inserts to portray "geography in action." These readings offer a glimpse of what it's like to live in another part of the world and to view differing perspectives on geographical issues. **Personal View** boxes are vignettes or profiles of individuals who live in a particular country or locality—a local voice with an awareness of global connections. These stories were written following interviews with individuals who are not geographers but who have sensitivity to the experience of their native countries.

The **Point-Counterpoint** boxes encourage analytical thinking as they present significant issues with a historical framework and potential global consequences. Point-Counterpoint boxes tackle challenging issues for which there are no easy answers, and allow students to explore various sides of these difficult topics. Students use "for-and-against" tables to shape and debate their views.

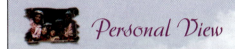

Personal View

BOSNIA-HERZEGOVINA

The wars that ensued in Yugoslavia after the country began disintegration in 1991 were particularly bloody in Bosnia-Herzegovina, lasting from early 1992 until the end of 1995, when the Dayton Peace Accord was brokered (Box Figure 1). Bosnia-Herzegovina was different from other republics in that it was not named after an ethnic group that inhabited it. Instead, it was named partly after a river and partly after a medieval territory. Bosnia derives its name from the Bosna River. Herzegovina stems from the German word *herzog*, which means "duke." Therefore, Herzegovina literally means "dukedom." Bosnia-Herzegovina most likely kept its historical name because no single group occupied the territory to justifiably change its name for that group (see Figure 3.54). For a long time, Bosnia-Herzegovina's multiethnicity worked very well. The people of the republic were even able to showcase their ethnic harmony to the world when they hosted the 1984 Winter Olympics in Sarajevo.

The war destroyed Bosnia's ethnic harmony and caused many to flee and seek refuge elsewhere. Miro Paunović and his family are one such example. Miro was born in Mostar and lived there for most of his life until the beginning of the war in 1992. Mostar is the main city of Herzegovina and is internationally known for its stone bridge. Referred to locally as the "Old Bridge" (*stari most*) (Box Figure 2), it was built by the Ottoman Turks in 1561 and came to symbolize ethnic harmony within all of Bosnia-Herzegovina. Sadly, the bridge was destroyed by the Croatian army on November 9, 1993, the anniversary of *Kristallnacht* (the night when Nazis attacked Jews and Jewish businesses throughout Germany in 1938). Miro has fond memories of the bridge and will never forget how people would travel from all over Yugoslavia every summer to compete in a bridge diving contest.

Miro also misses the old town center where people would come to shop and meet at cafes to socialize. In terms of its site, Miro says that Mostar is not unlike Cumberland, Maryland, where he now lives. Both cities are crowded down along rivers that wind their way through mountains. Mostar's river is the Neretva, which flows south into the Adriatic Sea, some 50 miles away. The short distance to the sea made it easy for Miro and his family to travel to the beaches of the Adriatic, often for just one day. For vacations, Miro's family would typically spend a couple of weeks on the coast, usually traveling to the famous cities of Dubrovnik and Split. Mostar's proximity to the coast means that both the summers and winters are relatively mild. Skiing is popular in this mountainous country, but those interested in doing so must travel farther inland, to the north and east, to places such as Sarajevo before they can find enough snow to ski.

Mostar is a regional center for southeastern Bosnia-Herzegovina. Its factories produce a variety of goods, including aluminum, military airplanes, car and truck parts, and clothing. The mountainous character of Bosnia-Herzegovina means that the country produces far less agri-

Box Figure 1 Map of Croatia and Bosnia.

culturally than many of the other Balkan countries. Nevertheless, some agricultural goods are produced. The area around Mostar is known for wine grapes, nectarines, and plums. Not surprisingly, Mostar has a winery and a distillery that makes a hard alcohol called *Loza*.

When Miro went to school as a young boy in Mostar, the educational system was somewhat different than what his children are now experiencing in the United States. All children went to elementary school, which lasted until the eighth grade. Foreign languages were taught in Miro's school beginning in the fifth grade. English, German, Russian, and French were the choices. Miro selected English, probably the most popular foreign language at the time. From the ninth through twelfth grades, young people went to specialized high schools, of which Mostar had five. The *Gymnasia* was for those bound for university. Another school emphasized automotive training, while a third focused on working with industrial machines. A fourth specialized in food technology, and the fifth centered its curriculum around economics. According to Miro, this latter school went through the greatest transformation in recent years. During Communism, the school emphasized the teachings of Marx and Engels but has since changed to the modern principles of capitalism. Miro states that school was quite rigorous in Mostar. He considers what his children are now learning in school in the

---

...00 people), Warsaw (2.3 million), Katowice (3.5 ...Poland), Prague (Czech Republic, 1.2 million), ...ungary, 2 million), Belgrade (Serbia and ...5 million), Bucharest (Romania, 2 million), and ..., 1.2 million) were the only cities of a million or ...The lack of major world cities in East Central ...from the small size of the countries, the late arrival ...ation, and the focus on manufacturing growth ...um-sized industrial town expansion during Soviet

The city landscapes of East Central Europe suffered great war damage and were subject to post-1945 industrialization and the construction of monolithic buildings (Figure 3.56a) but often retain older townscapes (Figure 3.56b). In Romania, however, the dictator Nicholae Ceausescu demolished villages and whatever remaining historic structures to industrialize. East Central European cities lack the high-rise office buildings containing commercial and professional service facilities and apartments that dominate the central parts of large cities in Western Europe. Many, such as the cities in Polish Silesia, are

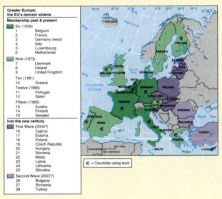

Point COUNTER Point

THE EUROPEAN UNION

The EU is now a major factor in changing the human geography of Europe. Almost all of the countries of Europe either belong to it or have applied for membership, as the map in Box Figure 1 shows. Yet many issues cause continuing debate, including questions asking whether the European Union is a good idea and whether it will succeed or not.

The EU arose from attempts to ensure that the future of war-torn Europe after 1945 would be peaceful. After two world wars in 30 years, political leaders were determined to lock the economies of European countries so closely together that further wars would be difficult, if not impossible. Some anticipated that once economic links were established, political union in a federation like that of the United States might follow. With American aid and protected from the expansionist ambitions of the Soviet Union by the North Atlantic Treaty Organization, the first step was the formation of the Benelux customs union between Belgium, the Netherlands, and Luxembourg in 1949. In 1952, the Benelux countries joined together with France and West Germany to form the European Coal and Steel Community (ECSC). In 1957, the five ECSC countries plus Italy signed the Treaty of Rome, establishing the European Economic Community (EEC). The EEC was expected to create a common market in which goods, capital, people, and services moved freely between countries. The European Commission became the executive arm of the community

and was based in Brussels, Belgium. In 1967, the EEC changed its name to the European Community (EC) to emphasize the move from economic toward political goals. As the organization progressed with integration, other countries were motivated to join. Denmark, the Republic of Ireland, and the U.K. became members in 1973. Greece in 1981, and Portugal and Spain in 1986. By then, the European Parliament, with elected members from all these countries, was created and located in Strasbourg, France. Other EU organs were headquartered in countries such as Luxembourg (Box Figure 2).

The Single European Act, signed in 1986 when Spain and Portugal joined, set out the steps needed to implement a single market. The Treaty of Maastricht (1991) attempted to set a timetable for monetary and political union and changed the name of the organization to European Union in 1993. Austria, Sweden, and Finland joined in 1995. The Swiss and Norwegian referendum votes on joining were close but rejected membership at the time. Many of the countries of East Central Europe, formerly part of the Soviet bloc, along with Turkey and Cyprus hope for a future in the EU, but their admission is a slow process.

European integration has proceeded for almost 50 years. Many Europeans see it as good, while many others do not. Those in favor look at the situation of European countries in the world and note that the countries like the United States, Japan, China, and Russia are much larger than any one European country. This makes it difficult for

Greater Europe:
the EU's domain widens
Membership past & present

Six (1958)
1 Belgium
2 France
3 Germany (west)
4 Italy
5 Luxembourg
6 Netherlands

Nine (1973)
7 Denmark
8 Ireland
9 United Kingdom

Ten (1981)
10 Greece

Twelve (1986)
11 Portugal
12 Spain

Fifteen (1995)
13 Austria
14 Finland
15 Sweden

Into the new century
First Wave (2004?)
16 Cyprus
17 Estonia
18 Poland
19 Czech Republic
20 Hungary
21 Slovenia
22 Malta
23 Latvia
24 Lithuania
25 Slovakia

Second Wave (2007?)
26 Bulgaria
27 Romania
28 Turkey

€ = Countries using euro

Box Figure 1 European Union, its growth to date, and possible further expansion. Reprinted by permission from *Financial Times*, Feb. 2, 1995, and updated.

Online Learning Center

Both students and instructors have access to a wide variety of resources that are only a click away. Use the passcode card available in this textbook to visit the Online Learning Center at www.mhhe.com/bradshaw to utilize this additional content.

Student Center

- Chapter outlines
- Interactive chapter quizzing
- Interactive key term flashcards
- Interactive key places flashcards—including audio pronunciations for difficult-to-pronounce place names
- Interactive maps, with quizzing
- Geography PowerWeb, providing news feeds and related articles
- Printable base maps in a variety of formats
- Data bank related to the tables in the chapters, which provides key data on countries from 2002 and is updated each year
- Related web links for each chapter
- Worldwide embassy resources listing

Instructor Center

- Instructors will have access to all materials within the "Student Center" and also the materials listed below for instructors only.
- Links to professional resources
- Instructor's manual
- *The Power of Place* telecourse video listing
- A collection of 30 printable base maps, available in a variety of formats

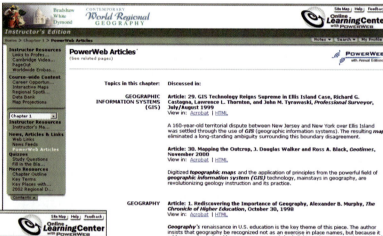

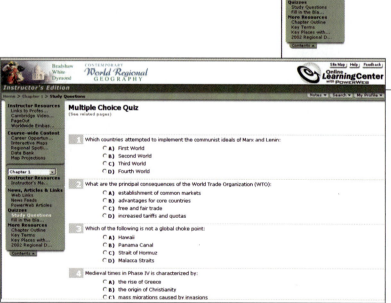

Interactive World Issues
CD-ROM Set

An exciting study tool that captures the power of videos to allow students to visit and closely examine geographic and environmental issues, this two-CD set is FREE with every new copy of *Contemporary World Regional Geography*. The case studies feature high-quality videos that can be enlarged to full screen, animations, maps, and diagrams as activities that students can complete as assignments or as self-paced study. The program visits Oregon, South Africa, Mexico, China, and Chicago to describe events and issues in resource allocation, land reform, water rights, policy, economic conditions, population, and sustainability.

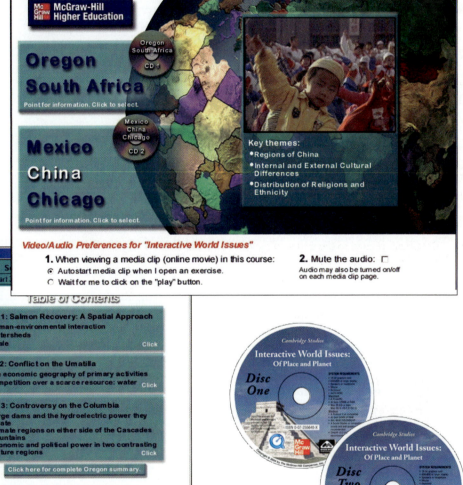

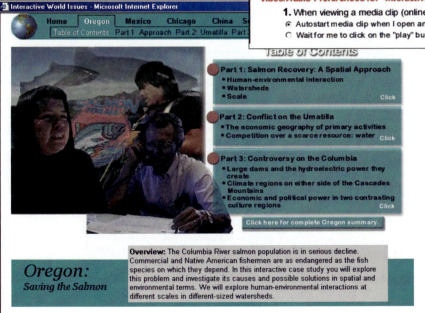

The Power of Place
Video Series

McGraw-Hill is proud to be associated with the well-respected and longstanding *The Power of Place* video series. If you offer a telecourse or have realized the advantage of using videos in your classroom, you'll be pleased to know that *The Power of Place* video series is now correlated to *Contemporary World Regional Geography*.

The Power of Place: Geography for the 21st Century is a revised and updated version of the *World Regional Geography* telecourse funded by the Annenberg/CPB Projects and McGraw-Hill Higher Education. Originally produced by scholars and educational broadcasters from six countries, the update was completed by the U.S. coordinator, Cambridge Studios, Inc. Fifty-two video case studies set in 35 countries around the world involve students in compelling issues of regional and systematic geography. Along with its coordinated McGraw-Hill text, *Contemporary World Regional Geography,* the updated series looks at the new tensions between global connections and local voices in many diverse places.

Through revised case studies, we go to Jerusalem to review the historical geography of territorial conflicts following the breakdown of Jewish-Palestinian relations. In light of the September 11, 2001, attacks, we revisit Lanzhou to consider the role of Islam in China's rugged and isolated northwest. In Tokyo, we observe how mass transportation serves the needs of this world megacity. In Dikhatpura, India, we focus on a small farming village that relies on irrigation to overcome the disadvantages of the local environment. And in Ecuador, we update the work of a geographer monitoring the impending eruption of a deadly volcano. These are just a few of the 52 case studies that offer insight into how place shapes our world.

This newly updated series features commentary by some of the *Contemporary World Regional Geography* authors as well as

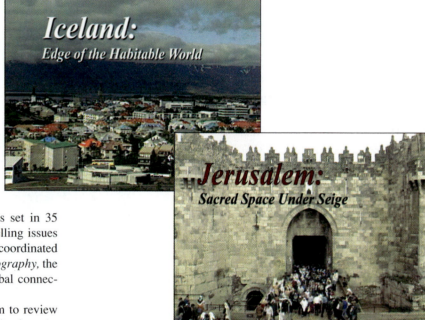

invited faculty guests. These tapes are a great complement to your traditional lecture and a must for institutions providing a world regional geography telecourse.

The Power of Place video series is available for free on the Annenberg/CPB Channel or for purchase on videocassette. For more information, call 1-800-LEARNER, or visit www.learner.org. *The Power of Place* series can also be licensed for distance learning by colleges and universities through the PBS Adult Learning Service (ALS). Call 1-800-257-2578, or visit PBS ALS on the web at www.pbs.org/als for licensing information.

The following supplements are available from McGraw-Hill to accompany *The Power of Place* video series:

Student Study Guide

Organized into sections corresponding to the telecourse's 26 programs, the Study Guide includes unit objectives, key themes, program overviews, discussion of unit themes, study questions, and outline maps with exercises.

Faculty Guide

Offering a week-by-week plan for using the videos, the Faculty Guide includes student assignments, unit overviews, themes/concepts/terms, links to other programs within *The Power of Place,* links to *Contemporary World Regional Geography,* and test questions and answers.

Digital Content Manager
CD-ROM

The Digital Content Manager CD-ROM is a multimedia collection of visual resources that allows instructors to create powerful presentations to help students visualize difficult concepts and view images of faraway places right in the classroom.

The Digital Content Manager CD-ROM contains:

- **All color images, photographs, and illustrations from the text**
- **All data figures from the text**
- **Chapter-specific PowerPoint outlines**
- **All maps from the text**
- **110 additional photographs**
- **Base maps for all world regions as well as major subregions**

The digital assets on this cross-platform CD-ROM are grouped by chapter within easy-to-use folders. The Art Library contains full-color digital files of all of the illustrations from the text, including all data figures. The Photo Library contains digital files of every photograph from the text. Chapter outlines in PowerPoint give an overview of the structure of each textbook chapter.

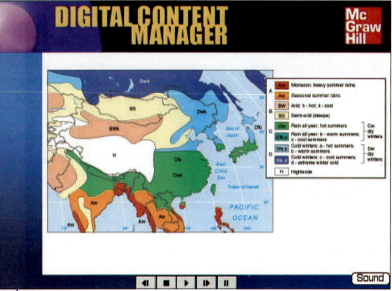

Easily Create:

- **Dynamic classroom presentations using all the art and photographs from the text plus over 100 additional photos from around the world**
- **Visually based tests and quizzes**
- **PowerPoint lecture notes and outlines**
- **Much more**

Course Management Systems
and Packaging Options

Course Management Systems expand the reach of your course. Online discussion and message boards can complement your office hours. Also, a sophisticated tracking system will let you know which students need more attention—even if they don't ask for help. Online testing scores are recorded and automatically placed in your grade book. You can also create special alerts to show up if a student is struggling with coursework. Should you seek advice for your course, you have access to discussion and message boards where you can collaborate and communicate with other professors across campus or around the world. Our own specialists are also ready to answer your questions.

McGraw-Hill's Online Learning Center content is compatible with a variety of course management systems, including **PageOut.** PageOut—McGraw-Hill's exclusive course management system—allows you to create a custom course website with links to book-specific course content. Simply select a template to post course information, and an interactive syllabus links to Online Learning Center content.

Online Learning Center content is also compatible with commercially produced course management systems including:

- WebCT
- Blackboard
- eCollege

Instructors can package *Contemporary World Regional Geography* with one or more of the following outstanding resources at a discount. Contact your McGraw-Hill sales representative for more details.

Global Studies Series

The Global Studies Series is designed to provide comprehensive background information and selected world press articles on the regions and countries of the world. The following are titles in the Global Studies Series:

- Africa
- China
- Europe
- India and South Asia
- Japan and the Pacific Rim
- Latin America
- Middle East
- Russia, the Eurasian Republics, and Central/Eastern Europe

Annual Editions

Completely revised each year, each Annual Editions contains articles addressing such topics as international political economy, peace and security, international trade, and global issues. These readers are complemented by a FREE student website, Dushkin Online (www.dushkin.com/online/), which provides links to related websites and study support tools. The following are titles in the Annual Editions anthology:

- Developing World
- Geography
- Global Issues
- Violence and Terrorism
- Western Civilization Volume 1
- Western Civilization Volume 2
- World Politics

Taking Sides

These editions contain pro-and-con articles that represent the arguments of world leaders, leading political scientists, and commentators on the world political scene. The readings reflect a variety of viewpoints and have been selected for their liveliness, substance, and value in a debate framework. The following are titles in the Taking Sides series:

- Clashing Views on Controversial Global Issues
- Clashing Views on Controversial Issues in World Politics

Student Atlas

The Student Atlas series provides full-color maps and data sets to give your students a clear picture of the recent economic, political, and environmental demographic changes in every world region. The following are titles in the Student Atlas series:

- Environmental Issues
- World Geography
- World Politics

Additional Packaging Options Available

- Fuson's *Fundamental Place-Name Geography,* Ninth Edition
- Getis: *You Can Make a Difference—Be Environmentally Responsible,* Second Edition
- Rand McNally: Atlas of World Geography
- New York Times Subscription
  This 10-week subscription can be packaged with *Contemporary World Regional Geography* for only $20. Each student receives a card containing a special code and URL for a validation site, along with the option of obtaining printed editions or electronic subscription.

Chapter 1

Globalization and World Regions

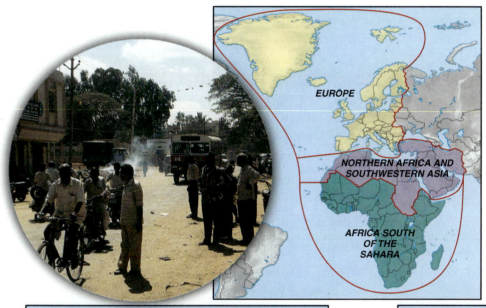

Figure 1.1
The major world regions.

EUROPE

NORTHERN AFRICA AND SOUTHWESTERN ASIA

AFRICA SOUTH OF THE SAHARA

RUSSIA AND NEIGHBORING COUNTRIES

EAST ASIA

SOUTH ASIA

SOUTHEAST ASIA AND SOUTH PACIFIC

NORTH AMERICA

LATIN AMERICA

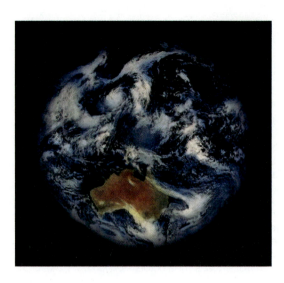

Figure 1.2 **Earth from space.** At the global level, the blue planet is so named because of the predominance (71%) of water on its surface. Australia, the dry continent, stands out, as do the cloud systems of the equatorial region to the north and the midlatitude weather systems to the east.

Global, Regional, and Local Worlds

Different and Changing Worlds

Earth is a planet of great variety (Figure 1.1). Whether we see it as a whole globe or study our own locality in detail, we find diversity. By comparing places, we discover differences and common features in parts of the world. And we live at a time when changes in our political, economic, and social experiences and expectations occur more and more rapidly.

Geographers find the study of our world a challenge because it is such a diverse, complex, and rapidly changing place. While the world stays the same physical size, connections among people bring places closer as cooperation, competition, and conflict with other peoples become more intense. For example, authors living far apart (England, Maryland, and Louisiana) created this text through daily e-mail contact with each other and universal access to the latest data and reports housed in various Internet websites—neither of which was possible ten years ago.

Geographic literacy is essential now more than ever because people are connected and interacting on unprecedented levels. The events of September 11, 2001 (see "Global Focus: 9/11," on the next page), brought home not only the ease with which people could penetrate U.S. security systems but also alerted the United States and the wider world to common concerns and interests as well as alternative views on what the future might hold for different parts of the world.

At the outset, we view Earth at four geographic levels ranging from the whole globe to our local environments. *Global* views from spacecraft (Figure 1.2) show the contrasts between continental land areas and ocean waters. *Major world regions* are whole or large parts of continents and are the divisions used in this text for the regional chapters (see Figure 1.1 and Chapters 3 through 11). *Countries* are the building blocks of major world regions. At the smallest scales, *local regions* are parts of countries and the places where many individuals voice their concerns.

Globalization and Localization

Two apparently opposed geographic trends help us understand what makes places different and causes changes to occur at the present time.

Globalization, in its simplest form, is the increasing level of interconnections among people throughout the world. The speed and intensity of globalization, especially in terms of world trade and the flow of financial investments, increased markedly in the 1990s.

Localization is both a response to and outcome of globalization. On the one hand, global exchanges and flows of information, ideas, people, money, and technology move us toward worldwide political solutions, economic exchanges, cultural attitudes, and environmental concerns. On the other, localization focuses on distinctive identities of places or people in regions, countries, or local areas.

Facets of Globalization

Increasing global connections take place through intensified flows of ideas, goods, and people:

- *Ideas, technologies, and diseases.* Most poorer countries wish to modernize to catch up with the materially wealthier countries. Technology transfer through improved communications and transportation is basic to such progress. The forces that ease the flow of ideas, however, also help terrorism, as well as drug, armament, and slave trafficking and the spread of HIV/AIDS.

- *Goods from many places of manufacture.* Trade in raw materials, food products, and manufactured goods increased rapidly in the late 1900s. Supermarket foods, clothing, electronic products, and autos come from an increasing number of countries, often marketed by multinational corporations of U.S., Asian, or European

9/11

"Tuesday, September 11, 2001, changed the world." The al-Qaeda terrorists who crashed passenger airplanes loaded with fuel into the World Trade Center (Box Figure 1) and the Pentagon on 9/11 made the most serious attack on the continental United States since its independence. Immediately, a new vulnerability took hold of people, along with the distress that impacted families of those killed or injured. Over the longer term, however, it can be seen that those events mainly brought home trends that for many years influenced American political and personal lives. We discover a new world—one of increasing global connections and of responding local voices. What happens in one place affects others around the world. Although there is a long way to go before we are all part of one "global village" or a borderless world, it is vital for Americans (and people in all countries) to have a better understanding of other people and countries. American isolationism can no longer be sustained.

New York is—and its World Trade Center was—a symbol of the increasing intensity of world trading in goods, movements of finance, and controls and exchanges through major stock exchanges and banks. It is a focus of employment for international accountants, lawyers, media and advertising specialists, and technologically expert professionals who manage global movements of money, contracts, and insurance liabilities. Almost every country in the world was represented among the 3,000 people who died in the World Trade Center: those who died were mainly Americans, but a hundred British citizens and other Europeans and people from Latin American, African, and Asian countries were killed as well.

Within weeks or months, many of the systems that broke down after the attack were restored. People resumed their lives. The former World Trade Center site was cleared of debris ahead of schedule. After a period of wild fluctuations in the markets, the resilience of the global economic system was demonstrated. Some corporations, especially airlines and airplane makers, were hit hard, but they too survived and then recovered, taking the opportunity of fewer passengers to trim their staffs and reduce equipment investments. Americans gave generously to charities, donating over $1 billion to organizations such as the American Red Cross to help the victims and their families of the tragedy. Americans displayed flags everywhere with a new sense of patriotism. Ironically, many of these flags were produced in China, illustrating just how interconnected the world had become.

In reaction to the events of 9/11, the United States and its close allies assembled international support for taking action against terrorism. Their forces focused on Afghanistan, where Osama bin Laden and his al-Qaeda terrorist group received protection from the Taliban government. The United States, with military support from its allies, bombed the Taliban into submission, handing control of Afghanistan to local leaders who opposed the Taliban's rule. The security forces of the countries working against terrorism began to identify al-Qaeda support in almost every country in the world. Organizations suspected of links to al-Qaeda had funds seized and their members imprisoned. The movements of suspected terrorists showed that they had easy entry to European countries and the United States, where they could plan and train for further acts.

In the wider world, countries such as Israel, Zimbabwe, Russia, and the Philippines cracked down on opposition groups labeled as "terrorists," while the United States and its allies focused on Afghanistan. India took the opportunity to blame Pakistan for attacks on its parliament in Delhi and positioned military resources for a new, possibly nuclear, war against its longtime rival.

Questions remain unanswered. Why should Osama bin Laden take so much trouble and spend so much money to demonstrate his opposition to the Western world and especially the United States?

Box Figure 1 An airplane full of fuel and passengers crashes into the second World Trade Center tower as the explosion of the first gains momentum, September 11, 2001. Photo: © Cary Conover/Corbis SABA.

Why should he and his organization resort to killing innocent people to achieve their goals? How did he persuade well-educated people to take on suicide missions several years in advance? Did he have a real hope that this event would make the Muslim countries of the world rise with a united religious purpose against the rest of the world? If so, it did not happen. Though street demonstrations by vocal minorities in many Muslim countries supported the acts that had such a profound impact on America, most governments of Muslim countries disassociated themselves from the terrorist acts. They joined the countries working together to outlaw terrorism; further, they do not have a history of coordinating political actions among themselves (see Chapter 8). The decision of governments to join the United States and its allies triggered antiwar demonstrations in many countries by those who support strong antiterrorist actions—local voices in global contexts.

For various reasons, including self-interest, before September 11, 2001, few countries lobbied for a global system of policing and dispensing justice against terrorism, drugs and arms trafficking, corruption, and slavery. Other events obstructed attempts to build consensus or develop policies. One country's terrorists may be another's freedom fighters. At times, terrorist groups gain external support from other countries when they attack mutually "unfriendly" governments. During the Cold War, this was frequently the case in Central America, Africa, and Asia (see Chapters 5 through 10), when the Soviet Union and the United States supported terrorist groups. For example, Osama bin Laden received U.S. support when Afghanistan was occupied by the Soviet Union—as did Saddam Hussein of Iraq when he fought Iran in the 1980s. The events of 9/11 suggest a need to rethink priorities and the wisdom of attempts to influence actions in other countries.

The events of 9/11 and other changes that make headlines from time to time need to be placed in context of the wider world in which we live. In this chapter, we begin our study of world regional geography by introducing geographic approaches to global and local events. The regional chapters demonstrate how different parts of our world interact with the global and local forces. In Chapter 12, we bring together these elements in a fuller evaluation of globalization, study the phenomenon of global and local terrorism, and see how events in Somalia relate to these processes of geographic change.

origin. Many products, such as autos and airplanes, are combinations of parts from a variety of countries assembled in one factory.

- *People migrations for work, political asylum, family consolidation, and long-distance tourism.* Executives of multinational companies "commute" on a worldwide basis. Millions of people displaced by warfare and oppression gather in refugee camps. Wealthier people take vacations in exotic places. Wealthier countries attract people willing to work for low wages, while better-paying jobs attract those with a good education, producing a **brain drain** on the poorer countries they leave.
- *The spread of images and messages through the media of TV, film, the Internet, and print.* Worldwide screening of U.S. movies and TV programs meets challenges from the film industries of India (the world's largest producer of movies) and other countries such as South Korea. The Internet enables international and local firms, political groups, and leisure interests to have exposure among those with access to the World Wide Web. Print media, such as newspapers, magazines, and books, remain significant. While the English language began to dominate through new technologies, the early 2000s see Chinese catching up among users in many of these fields. With Chinese speakers outnumbering English speakers almost 3 to 1 worldwide, Chinese may very well become the leading language of print and computing media.

The rapid and widespread acceptance of globalization terminology by politicians, members of the media, special interest groups, and academics, among many others, created substantial confusion over precise meanings and appropriate definitions. Some people assume globalization is already in place, overriding country boundaries and rendering country governments ineffective; others think the term is overused and unjustified when country governments are still the dominant political entities; a third group is sure that something significant is happening but awaits a fuller understanding of what is going on and where it will lead us. The Point-Counterpoint box, "Facets of Globalization," p. 5 compares these views.

Facets of Localization

Meanwhile, local voices remain loud in our consciousness and ensure that global trends are often far from being fulfilled.

- *Political nationalism* maintains separation of countries and of groups within countries. For example, the countries of Africa, despite criticizing their boundaries as inappropriate, resist amalgamation. Separatist groups such as those of the Basque people on the Spanish-French border and those in Aceh, Indonesia, work for independence rather than incorporation in global linkages.
- Despite globalizing forces, *many local customs and practices* preserve local identities. Western pop music, for example, has a worldwide circulation, but variants in Africa, Asia, and Latin America strongly reflect local attitudes and traditions.
- *Changes and intensification of ideologies, especially religious or political beliefs.* When the breakup of the Soviet

Union in 1991 left many feeling that Communism was discredited and military dictators in other countries fell from power in the early 1990s, more countries had open elections and were increasingly involved in international affairs. Religious and other cultural forces such as language and tradition became more significant, particularly in many materially poorer countries and those in economic transition.

- *Religious differences* among Christian, Muslim, Jewish, Buddhist, and Hindu countries continue to be significant. Increasing migrations place people of different religions together as migrants make their homes among those with a differing religion. This may lead to fragmentation of living areas in cities as each group establishes its places of worship and related cultural symbols.
- *Demonstrators resist the visible economic penetration* of countries around the world by global media and corporations such as CNN, the Murdoch group, McDonald's, Starbucks, Toyota, and Nike that are seen as contributing to poverty or loss of local identity. And yet, greater exposure to media presents these groups with opportunities to emphasize their demands.

Localization commonly involves social and cultural factors—hence, the phrase "global connections, local voices" in this book's title. The Taliban in Afghanistan provided an example of an attempt to establish a culture in isolation from the rest of the world. In the 1990s, its leaders took political control of most of Afghanistan and imposed a strict form of Islam, forcing out other religious groups and Western influences. Its use of the isolation it created to support al-Qaeda terrorists, however, put it into conflict with wider global interests and led to its downfall in 2001.

The rest of this chapter focuses on how regional geography assists the study of globalization and localization. It begins by defining the links among the regional scales and the main issues—political, economic, cultural, and environmental—affecting our lives (Figure 1.3).

🌐 Diverse Worlds

Political Activity: Countries Act

Politics is about power and the ways in which countries are governed, allocate resources, and relate to each other. The global political environment from 1950 was dominated by the **Cold War,** which set the United States and its allies against the Soviet Union and its allies. The Cold War ended in 1991, after never coming to a full "hot war," although many feared nuclear attacks. In the 1990s, political activity shifted from blocs of countries facing off against each other to a fresh emphasis on the roles of individual countries in a global context.

Countries remain self-contained entities with established borders. Within its territory, a country is sovereign: no other being or institution has power to compel it to act. Country governments have powers to tax, provide for defense, negotiate

FACETS OF GLOBALIZATION

Mentions of globalization raise anger in some places; mobs of people demonstrate with violence against it at meetings of international leaders. And yet, others tell us that whatever the growing pains, globalization is good for us in terms of better incomes, better life quality, and better understanding of other peoples. We quote alternative views that have been expressed in the last few years and then summarize them.

SAMUEL P. HUNTINGTON, *Foreign Affairs,* 1993

World politics is entering a new phase. . . . It is my hypothesis that the fundamental source of conflict in this new world will not be primarily ideological or primarily economic. The great divisions among humankind and the dominating source of conflict will be cultural. [Countries] will remain the most powerful actors in world affairs, but the principal conflicts of global politics will occur between them and groups of different civilizations. The clash of civilizations will dominate global politics. The fault lines between civilizations will be the battle lines of the future.

BENJAMIN R. BARBER, *The Atlantic Monthly,* 1992

Just beyond the horizon of current events lie two possible futures—both bleak, neither democratic. The first is a redistribution of large swaths of humankind by war and bloodshed, in which culture is pitted against culture, people against people, tribe against tribe—a *jihad* in the name of a hundred narrowly conceived faiths. . . . The second is being borne in on us by the onrush of economic and ecological forces that demand integration and uniformity and that mesmerize the world with fast music, fast computers, and fast food—with MTV, Macintosh, and McDonald's, pressing nations into one commercially homogeneous global network: one *McWorld* tied together by technology, ecology, communications, and commerce. The planet is falling precipitately apart and coming reluctantly together at the very same moment.

KENICHI OHMAE, *Foreign Affairs,* 1993

The [country] has become an unnatural, even dysfunctional, unit for organizing human activity and managing economic endeavor in a borderless world. It represents no genuine, shared community of economic interests; it defines no meaningful flows of economic activity. In fact, it overlooks the true linkages and synergies that exist among often disparate populations by combining important measures of human activity at the wrong level of analysis.

On a global economic map, the lines that now matter are those defining what may be called 'region states.' The boundaries of the region state are not imposed by political fiat. They are drawn by the deft but invisible hand of the global market for goods and services.

ANNE-MARIE SLAUGHTER, *Foreign Affairs,* 1997

Many thought the new world order proclaimed by George W. Bush [in 1990] was the promise of 1945 fulfilled, a world in which international institutions, led by the United Nations, guaranteed international peace and security and the active support of the world's major powers. That world is a chimera. Even as a liberal

internationalist ideal, it is infeasible at best and dangerous at worst. It requires a centralized rule-making authority, a hierarchy of institutions, and universal membership. Equally to the point, efforts to create such an order have failed. The United Nations cannot function effectively independent of the major powers that compose it, nor will those nations cede their power and sovereignty to an international institution. Efforts to expand supranational authority, whether by the UN secretary-general's office, the European Commission, or the World Trade Organization, have consistently produced a backlash among member states.

A new world order is emerging with less fanfare but more substance than either of these. The [country] is not disappearing, it is disaggregating into its separate, functionally distinct, parts—courts, regulatory agencies, executives, and even legislatures—that are networking with counterparts abroad, creating a new transgovernmental order. Today's international problems—terrorism, organized crime, environmental degradation, money laundering, bank failure, and securities fraud—created and sustained these relations.

CHRISTOPHER BRIGHT, *Foreign Policy,* 1999

World trade has become the primary driver of one of the most dangerous and least visible forms of environmental declines: thousands of foreign, invasive species are hitchhiking through the global trading network aboard ships, planes, and railroad cars, while hundreds of others are traveling as commodities. This 'biological pollution' is degrading ecosystems, threatening public health, and costing billions of dollars annually.

Bioinvasion occurs when a species finds its way into an ecosystem where it did not evolve. In a small percentage of cases, the exotic finds everything it needs—and nothing capable of controlling it. In South and Central America, the growth of specialty export crops—upscale vegetables and fruits—has spurred the spread of whiteflies, which are capable of transmitting at least 60 plant viruses. In southern India, a tropical American shrub is strangling rice paddies and ruining fish habitats throughout the Cauvery River basin.

JAMES N. ROSENEAU, *Current History,* 1997

The mall at Singapore's airport has a food court with 15 food outlets, all but one of which offers menus that cater to local tastes; the lone standout, McDonald's, is also the only one crowded with customers. In New York City, experts in feng shui, an ancient Chinese craft aimed at harmonizing the placement of man-made structures in nature, are sought after by real estate developers in order to attract a growing influx of Asian buyers.

Globalization refers to processes that evolve as people go about their daily tasks: these processes are not hindered or prevented by territorial or jurisdictional barriers, they can readily spread in many directions across national boundaries, and are capable of reaching any community anywhere in the world. Contrarily, localization derives from those pressures that lead individuals, groups, and institutions to narrow their horizons, participate in dissimilar forms of behaviors, and withdraw to less encompassing processes, organizations, or systems. Globalization is boundary-broadening and localization is boundary-heightening; the former allows people, goods, information, and practices to move oblivious to boundaries; the latter inhibits such movements.

PETER D. SUTHERLAND, *Harvard International Review,* 2000

The Seattle Ministerial Conference of the World Trade Organization (WTO) demonstrated with disturbing force the huge confusions that haunt the public mind about . . . the process now known as globalization. The notion that globalization is an international conspiracy on the part of industrial country governments and large firms to marginalize the poorest nations, to exploit low wages and social costs wherever they may be found, to diminish cultures in the interests of an Anglo-Saxon model of lifestyle and language, and even to undermine human rights and cut away democratic processes that stand in the way of ever more open markets is, of course, utter nonsense. The outpouring of misconceived, ill-understood propaganda against a system that has brought vast gains to most nations over the past few decades is extraordinarily dangerous.

We can comment on some of the points made in these quotations:

1. *Globalization is a process or set of processes, not* an established condition leading inevitably to a worldwide society or community. Although European countries move toward economic integration within the European Union (see Chapter 3), political issues remain about the surrender of country sovereignty. Meanwhile the EU is accused of establishing "Fortress Europe" that does not wish to have closer trading relations with other countries except on its own terms. In East Asia (Chapter 5), Japan and China produce many goods for global markets but resist the intrusions of foreign traders in their home markets. In Africa South of the Sahara, countries defend their internal rights and resist external interference.

2. *Global and multinational interconnections involve complex linkages and overlapping networks* among communities, countries, international institutions, nongovernmental organizations, and multinational corporations. There is no necessary complicity among these groupings to achieve globalization based on their own narrow interests. The unevenness of globalization processes and responses of localization reflect partial and selective linkages.

   Research shows that economic integration enables materially poorer countries with about half the world's population to break into global markets for goods and services. Such integration results from both domestic policies and reforms in the poorer countries and from international action to bring access to foreign markets, capital, technology, and aid. Exports of raw materials and other primary products often give way to manufactured goods and services. By such means, some Chinese provinces, Indian states, and countries such as Bangladesh and Vietnam (Chapters 5, 7) reduced poverty in the 1990s by over 100 million people. But many more remain materially poor, providing a challenge to the materially rich countries and the spreading global systems.

3. *Globalization affects all areas of social life*—cultural, economic, political, legal, military, environmental. Interconnectedness is growing among the varied spheres. Understanding the dynamics of globalization requires understanding how each of these spheres contribute to it. The process is both differentiated and multifaceted: it involves lots of different groups of people and institutions that act independently and at different rates.

   This partly explains why there has not been total inclusion among and within countries. While 3 billion people live in countries that are improving their quality of life, around 2 billion are marginalized in the global economy, living in countries where poverty is rising and participation in international trade is falling. Such countries include the failed Afghanistan (Chapter 7) and Democratic Republic of Congo (Chapter 9) that descended into inward-looking anarchy but also others that need help in gaining access to markets in wealthier countries and managing foreign investment and aid. Even within successful countries, there are winners and losers, particularly workers in sectors exposed to a competitive economy and those who have no recourse to social protection against such events.

4. *Globalization cuts through and across political frontiers,* making some social, economic, and political spaces subject to external actions: local, country, or regional patterns form again with or without reference to established political boundaries. New activities emerge at local levels inside countries, reinforcing patterns of localization and nationalist desires.

   While most people hope for improved material wealth, many fear the result will be greater cultural and social homogenization: will everything, everywhere, look the same? Yet the differences among the wealthier countries (e.g., Japan, the United States, France (Chapters 5, 11, and 3) are clearly maintained, while countries such as China, India, Malaysia, and Mexico (Chapters 5, 7, 6, and 10) retain their cultural identities as they become more involved in world trade. Geographical expressions of diversity are robust amid the changes. Country freedoms in aspects such as intellectual property rights, cultural goods, environmental protection, social policies, and labor standards often stand firm against a standardization that may be seen merely as an attempt to safeguard people and organizations in wealthier countries.

5. *Globalization expands the geographic scale of powers* applied. In the increasingly interconnected global system, decisions, actions, and inactions in one place affect others elsewhere at increasing distances. Centers of power may exercise influences increasingly distant among people or places experiencing the effects. In particular, the elite in major metropolitan centers exercise power over many parts of the world from Bangladeshi shirt makers to Burundi farmers (Chapters 7, 9). Multinational corporations also invest in some countries and areas within those countries and not others. However, country governments also influence the choices of locations. Special economic zones in China and Brazil (Chapters 5, 10), among other countries, are country-based decisions to locate jobs and economic growth in designated areas while the surrounding areas experience few or very gradual changes.

with other countries, and develop their own legal system. Within countries, local issues and pressures, as much as international relations, occupy politicians.

In the 1990s, worldwide government through the United Nations achieved only modest results in reducing country rivalries, controlling international problems such as terrorism and the drugs-arms-slave trades or arbitrating internal country strife. The United States and the wealthy European countries, however, acted to stem the destruction caused by war in Kuwait (Gulf War of 1991), Bosnia, and Kosovo.

At the *world regional* level, some countries formed groups for defensive purposes during the Cold War period in the second half of the 1900s. The United States and Canada joined with European countries in the North Atlantic Treaty Organization (NATO) to defend western Europe against potential attacks from the Soviet Union. Indonesia, Malaysia, Singapore, and the Philippines joined the Association of South East Asian Nations (ASEAN) to resist the advance of Communist governments. The countries neighboring South Africa formed the Southern Africa Development Coordination

GEOGRAPHIC SCALE	POLITICAL	LAW AND ORDER	ECONOMIC	CULTURAL	ENVIRONMENTAL
GLOBAL/ WORLDWIDE	United Nations Non-governmental organizations (NGOs) (No united will for a global government)	International Criminal Court. Issues: drugs, nuclear testing, weapons, war criminals, terrorists	Borderless CAPITALIST MARKET SYSTEM: World Bank, IMF, OECD, G8, multinational corporations	Spreading Western and Islamic cultures; Westernization resisted. Olympic Games	Global warming, ozone hole, ocean resources
WORLD REGION/ SUBREGION	European Union, Commonwealth of Independent States, original ASEAN	Hague Court of Human Rights	Regional trading groups: EU, NAFTA, Mercosur, ASEAN, APEC, etc. Regional emphases within the capitalist system.	CULTURAL GROUPINGS: N Africa/SW Asia; Africa S of Sahara; S Asia; China; Japan; Russia; Latin America; North America; etc.	Acid rain, tropical rain forest destruction
COUNTRY (NATION-STATE)	BASIC POLITICAL UNIT: accepted borders; taxing, defense, international relations powers	Each country has its distinctive legal system	Fiscal and monetary policies; rich/poor countries	"Nation-state" concept	COUNTRY-BASED ISSUES: public health; air and water quality; conservation of soils, forests; national parks
LOCAL REGION/ MAJOR CITY	Devolution of administration; world cities; federal states within countries	Variations in application of a country's legal system	Regional product specialization; economic and physical planning; rich/poor areas	Local elements combining with external long-term traditions; distinctive way of life; basis for political pressures	Physical planning impacts

Figure 1.3 **Geographic facets of globalization.** The pink shaded areas highlight the major centers and geographic scale of human activities in each category.

Conference (SADCC) to oppose apartheid (separation of whites and blacks) and resist South African attacks. In the 1990s, NATO gained members from former Soviet Union satellite countries in eastern Europe such as Poland, Hungary, and Romania, while ASEAN and the renamed Southern Africa Development Conference (SADC) focused on new economic objectives.

Continuing political rivalries among and within countries caused some commentators in the 1990s to talk about the "New Global Disorder" rather than a "New Global Order." They pointed to activities outside the jurisdiction of countries, such as terrorism and trade in guns, drugs, and slaves, that increased in the absence of a working system of global governance or law and order. Rivalries among countries and groups within countries prevented worldwide control of illegal activities through the United Nations and allowed both inequalities of opportunity and criminal activities to prosper. Extremist groups grew in significance in all parts of the world, often linking their interests to international groups trained in violence and disruption. The events of September 11, 2001, in New York and Washington, D.C., brought home these trends to many who had ignored them and highlighted the need for a global system of law and order—as opposed to the range of different systems based in countries.

Economic Activities: Global Trends

Economics is the study of how people and groups use resources to meet their material needs through producing, distributing, and consuming goods and services. In this text, we

use the terms *wealthier* and *poorer* people and countries to indicate contrasts of material well-being or poverty.

Wealthier people in wealthier countries enjoy a good life, with security of income, housing, food, health care, and education, together with environmental safety, confidence in the future, and involvement in local and national decision-making. They use a disproportionate share of Earth's resources to maintain such lifestyles, including big houses, gas-guzzling autos, expensive vacations, and high-cost medical services.

Poverty is basically a lack of material wealth that has wider outcomes (Figure 1.4). **Poorer people** do not have security like

Figure 1.4 **Poorer world city: midday street scene in Bangalore, southern India.** Compare this view of a main shopping street in Bangalore with a shopping area or main street with which you are familiar. What ways do people use to get about in each place? What are the signs of poverty? Photo: © Michael Bradshaw.

wealthier people and suffer from malnutrition, poor housing, exhaustion, disease, rejection, and vulnerability. Poor people are deprived of acceptable income, education, health care, housing, clothing, food, and access to jobs. They will not likely have a long life and cannot fully participate in community affairs. Furthermore, some people may not be income-poor but are kept out of the mainstream of society by physical disability, ethnic or racial discrimination, or destructive behaviors associated with alcohol or drug abuse.

As world income increased in the later part of the 1900s, more people enjoyed material well-being as middle classes grew wealthier and in numbers in poorer countries. One indicator of poverty—the numbers of people living on less than $1 per day—increased from 900 million people worldwide in 1820 to 1.4 billion people in 1980 but then fell to 1.2 billion in 2000. Although this is a large number of very poor people, the *proportion* in the whole world population living on less than $1 per day fell (from 85% in 1820 to 30% in 1980 and 20% in 2000). In the 1990s and early 2000s, some areas of the world, including former Soviet Union countries and Africa South of the Sahara, experienced increases in poverty, while East Asia saw major reductions in poverty. Inequalities among households peaked in 1980 and have declined since. By the late 1990s, as poverty fell but continued to affect so many people, the World Bank and United Nations focused economic and related policies on attempts to reduce poverty further in the early decades of the 2000s.

Economic activities underlie increasing global interconnections. The movements of people, driven by the possibility of less oppression or better living conditions, trade, capital investment, and information exchanges, are increasingly international. Even so, parts of the world remain isolated or marginalized by poor links to main economic centers, while others are in transition toward greater global connections. The tensions between increasing interconnections and continuing isolation enhance differences among places.

In the 1990s, the uneven spread of expanding global economic activities caused groups of countries to enter into or revive regional economic agreements, mainly through trade. The best known and farthest advanced is the European Union (Chapter 3), but others include the North American Free Trade Agreement (NAFTA, Chapters 10, 11), Mercosur (southern South America, Chapter 10), the Association of South East Asian Countries (Chapter 6), and the South African Development Conference (Chapter 9). In East and Southeast Asia, the "Asian Way" of free-market economic activity works through family linkages and government-business liaisons more commonly than through the independently verified banking and legal systems regarded as most important in the United States and Europe. The "European Way" is another distinctive regional approach, in which social welfare provides support to people unable to benefit from advanced economic activity.

Within world regions, countries differed in their engagement levels with world trade. The United States, the countries of western Europe, and Japan controlled nearly all the investment, production, and consumption of goods. However, by 2000, China, India, and Brazil increased their contributions. Variations in resource provision, government efficiency, educa-

tion and health levels, and the availability of investments to develop local resources also affected local regions within countries. Urban areas tended to attract most investment, and rural areas were often ignored or subject to the exploitation of people and the environment. Small groups of people living in the Amazon rain forest, large areas of Africa, and in places such as Papua New Guinea engaged little in global economic activities.

Cultural Activities: Major Regions, Local Voices

The **culture** of a group of people results from the learned behavior in the ideas, beliefs, and practices they hold in common and pass on from one generation to the next. Religion, language, the ways in which people do things socially, the design of the items they make, responses to the natural environment, and the level of technology involved are expressions of a shared culture (Figure 1.5). Different cultures often express values through communal and family life, human rights, and the role of women. For example, cultural values are reflected in the number of working hours, the food and meal times, the nature of recreational activities, the design and decoration of clothing, houses, and buildings, and the layout of cities.

Although there are some indications that we may be heading toward "one world" culturally, many of the trends seem to imply that Western cultural norms—with their emphasis on democracy, individualism, and human rights—can (or should) be extended to the rest of the world. However, this is an unlikely scenario because people in many cultures oppose these norms, either in their entirety or the Western forms of them. Many people around the world oppose Western cultures because they see them as emphasizing materialism, consumerism, and other superficial values that lack ethics or spirituality. Western cultural norms are seen to be such a threat that some argue that they must be opposed, even violently if necessary. The events of September 11, 2001, built on such opposition.

Despite opposition, Western cultural norms are spreading. A closer examination, however, shows that the "Cocacola-ization" of eating and drinking habits, the spread of Western TV, motion pictures, and popular music, and the global markets for some consumer goods have not wiped out local cultural differences. On the contrary, local cultural practices and preferences exert themselves and even blend with global culture, adding to the diversity of places. In India, for example, only the wealthier elites can afford to buy imported goods and the new multinational fast food (Chapter 7). The more numerous middle-class groups buy cheaper Indian-made goods and eat at local restaurants. Most Indians prefer their homemade "Bollywood" films.

In the 1990s, cultural activities gained in prominence at the world regional level. One analysis of such differences identified nine major groupings of cultural characteristics called "civilizations" (Figure 1.6). Many of these are linked to specific religious cultural backgrounds: Orthodox (Christianity) in Russia, Buddhist in Southeast Asia, Islamic from Northern Africa and Southwestern Asia through the Indian subcontinent to Indonesia, Hindu in India. The other groupings are defined more broadly and are complex in origins and characteristics. For

Figure 1.5 **City center, Bangalore, southern India.** What aspects of an Indian culture are illustrated, as compared with imported Western culture? Photo: © Michael Bradshaw.

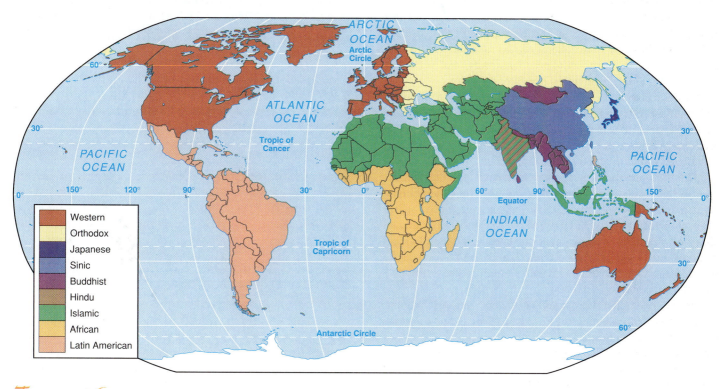

Western
Orthodox
Japanese
Sinic
Buddhist
Hindu
Islamic
African
Latin American

Figure 1.6 **World cultures, or civilizations.** The nine world divisions of Samuel Huntington (1996) gained significance in the 1990s after the end of the Cold War. Source: Reprinted with permission of Simon & Schuster Adult Publishing Group from *The Clash of Civilizations and the Remaking of the World Order* by Samuel P. Huntington. Copyright © 1996 by Samuel P. Huntington. Base map © Hammond World Atlas Corp.

example, the Western category incorporates both European secular and Christian values that spread to North America, the Latin American integrates the Spanish/Portuguese culture with those of indigenous groups, the African includes the outcomes of traditional cultures interacting with European colonial power influences, the Sinic (Chinese) absorbs Confucianism within its administrative structure, and the Japanese combines its national religion with aspects of Buddhism.

The political-with-economic basis of the divisions among blocs of countries highlighted the links between political and economic power and the need to work out differences among Communist, Fascist, and democratic-with-free-market

systems. By 1990, the struggle among these systems largely rejected the Fascist and then the Communist systems, leaving the democratic-with-free-market as the world's only system. Differences among places on a world scale now need to take account of the links between cultural factors and economic processes. For example, people in many countries still adopt aspects of the Western free-market economy and culture they find attractive. Others, especially Muslims and many people in East Asian countries, also wish to improve their material wealth but resist total Westernization, which they see as having many negative characteristics. They seek solutions within their own culture.

Cultural conditions based on ethnicity, class, gender, and political rights are also important at the country and local levels. Few, if any, countries can claim to contain only a single political, religious, linguistic, or pure "national" culture, although some have tried to impose one on their inhabitants. Within countries are often cultural groups that feel underprivileged or discriminated against economically and politically. Examples include the Tamils in northern Sri Lanka, the Basques in northern Spain and southwestern France, the Kurds in parts of Iran, Iraq, and Turkey, and the Native Americans in the United States and Canada (where they are called "First Nations").

Environmental Issues at Varied Scales

Earth is marked by a variety of **natural environments**—our planet as it might be without human modifications. Mountains and rivers, atmosphere and oceans, rocks and soils, and plants and animals are Earth's essential building blocks. The interactive workings of the atmosphere, oceans, and the Earth's surface give rise to varied climates, landforms, and ecologies (Figure 1.7). They create differences among regions.

Natural environments affect human events at global, world regional, country, and local scales. Prediction of hurricanelike storms, effects of acid rain, and damage from river floods and volcanic eruptions present issues at world regional, country, or local scales. At the global scale, effects such as global warming, El Niño, the ozone hole over Antarctica, and the destruction of tropical rain forests are subjects of international debates. In the 1990s, world environment conferences at Rio de Janeiro, Brazil (1992), and Kyoto, Japan (1997), adopted policies to avert environmental crises, but it is left to individual countries to implement the policies. The United States recently decided against following the Kyoto recommendations.

While some natural events are hazardous to human life (volcanic eruptions, earthquakes, violent storms), natural resources such as minerals, fertile soils, and water provide bases for human development. They have often influenced people's decisions to settle in particular areas. Such resources also stimulated later changes in economic activity.

People modify all natural landscapes in which they live, leaving their human (anthropogenic) impacts on the natural world. As they produce cultural landscapes containing buildings, fields, managed woodlands, and transportation links, they, for example, modify local climates by increasing rain runoff from towns that also become "heat islands," change soils by cultivation, and add gases to the atmosphere that enhance or impair its role. Some human modifications improve the productivity, landscape quality, and livability of a place, while others degrade a place's attractions, resources, and future prospects. For thousands of years, people have modified landscapes by removing forests to expand food production and support growing populations. The resulting long-term impacts, however, were not as great as those of the rapid industrialization in the 1800s and 1900s, which was characterized by the digging of huge mine pits, building of larger factories, and extracting of oil and gas that polluted air and water. In places,

(a)

(b)

Figure 1.7 **Contrasting natural environments in California.** (a) The redwood forests that have some human inroads, but where the largest trees require hundreds of years to grow several hundred feet. The flora dominates the landscape. (b) The arid environment of Death Valley with sand dunes, bare rock, and a fan of gravel washed down by occasional rainstorms—but no vegetation.

the buildup of wastes and toxic gases from factories and cities overwhelmed the natural environments' abilities to deal with them, requiring countries to pass laws that penalize polluters.

One impact of the globalization of telecommunications is that people around the world are more aware of the human interactions with the natural environment. For example, those living in the hurricane belt of the U.S. southeast and the Caribbean board up their windows and move inland when TV stations announce that such a storm threatens. Pollution events, such as the "brown cloud" over Asia in the summer of 2002, alert people around the world to the environmental and natural resource issues that are likely to increase at global and regional political, economic, and cultural scales in the 2000s.

The world's growing populations and improved living standards have rising impacts on the resources available and often increase tensions over resource usage. World population rose from 1.6 billion people in 1900 to over 6 billion in 2000, and it could rise to around 9 billion by 2050. The population increase will not be uniform across the globe. Almost all of it will be concentrated in the world's poorer countries, which already find it difficult to cope with their burgeoning populations. World population conferences in the 1990s suggested that population growth should be curbed, but some countries with conservative religious beliefs resist restrictions on family size.

Test Your Understanding 1A

Summary Globalization is the increasing level of interconnections among peoples on Earth. It does not mean that every place is becoming the same. Its impact is varied on places at different levels of geographic scale—global, world regional, country, and local. Localization reflects a set of responses to global trends at smaller geographical scales and intentions of local people to preserve their way of life. The effects of globalization and localization differ relative to political, economic, and cultural activities and with respect to the environment.

Questions to Think About

1A.1 What are the main elements of globalization and localization? Illustrate your answer by referring to recent events.
1A.2 What evidence points to global trends having increased effects on political and cultural activities?

Key Terms

globalization	economics
localization	wealthier people
brain drain	poverty
politics	poorer people
Cold War	culture
country	natural environment

Geography of a Diverse World

What Is Geography About?

One means of understanding our world's complexity and change is through the study of geography. Like other subjects, geography has its special content and methods.

Subject Matter

Geography is the study of where and how human and natural features and events (political, economic, cultural, and environmental) are distributed on Earth's surface, the relationships among them, how their distributions change over time, and how those features and relationships affect human lives. Geographers work to understand where and how people live and apply this knowledge in such specialized fields as planning and resource use. While they begin by identifying and defining the locations where people live, geographers are especially interested in explaining the diversity and distributions of people, political and economic activities, cultural distinctions, and environmental conditions. Although geographers used ideas from other disciplines throughout much of the 1900s, it is now common for those disciplines such as history, sociology, economics, and political science to look for the geographical implications of their work.

The tensions among globalization, localization, and the continuing significance of country governments provide a basis for investigating how people in different parts of the world face changes and move toward either greater interdependence or conflict. Thus, geographers compare places and assess the interactions among them at different levels of geographic scale. For example, people living in coastal towns in China make goods that will be sold in the United States. They work in factories built since the 1980s, often financed by money from Taiwan, Hong Kong, or Japan. These workers still depend on local farms for their food and are subject to government controls determined in Beijing and a cultural heritage from their distant past. The global, world regional, country, and local linkages and their flows of information, people, capital, technology, and ideology give character to each part of the Earth—a character that is voiced in both local and global terms.

Geographic Methods

Geographers use an increasing range of data and methods of analysis to achieve their goals. The method of analysis used often determines what data has to be collected. Modern geographers ask questions about the credibility of the information they use and what it means. This helps them to understand the world and its parts, to bring together understanding from various branches of knowledge, and to contribute to projects involved in planning the future. Geographers adopt the following approaches as they are appropriate to their subject of study:

- Establish scientific "laws" fitting their observations and interpretations after testing in several situations
- Appreciate individual and group perceptions of decisions, for example, in setting up a business or buying a home in a particular place
- Investigate the meanings people give to places
- Investigate and analyze the forces involved in material changes resulting from technological and sociological innovations, including power relations such as social classes that bind people to their areas and societies
- Link the significance of individual decisions and the roles of social institutions
- Assess the complexities of the modern world through the varied "stories" told by individuals and groups, believing that all knowledge is relative to each person's experience
- Assess the complexities of the modern world through a combination of approaches, assuming some common basis

that unites the scientifically observed mechanisms, the actual world based on extensions of precise observations, and how people experience events

We base this text on varied data sources and approaches. The data tables in the chapters report information from the best sources that are updated frequently. In using these sources, however, we allow for variations of accuracy and date among reporting countries. The sections on the natural environment report scientific studies of such phenomena as weather, surface landforms, and soils. The statements made about population, economic events, politics, and cultures derive partly from general overviews and partly from individual stories of people living in specific countries. In reading the book, you can relate it to your own experiences, extending what the text says to places where you live, that you visit, see on TV or in movies, or read about.

In considering two major aspects of geographic study, **physical geography** looks at nonhuman processes and environments across Earth's surface that result in the distributions of climate varieties, plant ecologies, soil types, mountain building, or river action, among other patterns. Physical geographers mainly use quantitative scientific methods. **Human geography** is the study of the distribution of people and their activities (e.g., economies, cultures, politics, and urban changes). Human geographers' studies include scientifically based measurements and testing as well as those that emphasize the experiences and perceptions of individuals: their evidence is qualitative as well as quantitative. Geographers' inquiries include both how the natural environment influences human decisions and also how much people's actions affect the natural world.

First, Geography Is About Places

Geographers study places on Earth's surface as the environments or spaces where people live and through which they make life meaningful. Geography thus provides a place- or space-related, **spatial view** of the human experience.

When we go to visit a **place,** it might be an individual building (convenience store), small town (Freeport, Maine), large city (New York), rural area (Iowa), another state, or another country. Places may be perceived as points on a map or as large areas. However, they all have different relationships to each other in terms of location, direction, distance, and size.

Latitude and Longitude

Absolute location is the precise place on Earth's surface where something is located. For example, many cars are now equipped with global positioning systems (GPS) that define their exact locations on a map.

Latitude and longitude form the framework of the international reference system that pinpoints absolute location (Figure 1.8). **Latitude** describes how far north or south of the equator a place is, measured in degrees. The North Pole is at 90°N and the South Pole at 90°S. The equator is an imaginary line encircling the globe midway between the North and South Poles and

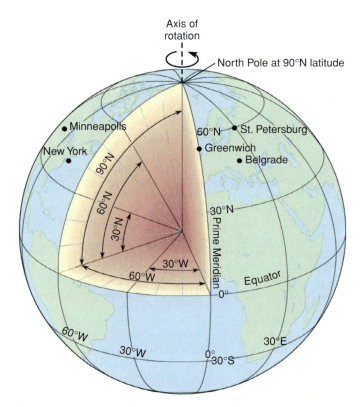

Figure 1.8   **Location: latitude and longitude.** The coordinate system is used for locating place on Earth's surface and the network of parallels of latitude and meridians of longitude. Give the latitude and longitude of each place marked on the globe. The degrees of latitude and longitude result from angles focused at Earth's center. Define the differences between 60°N and 30°N, and between the 0° meridian and 60°W in terms of angles at Earth's center. Source: From *Human and Cultural Geography* by Shelley & Clarke. Copyright © 1994 McGraw-Hill Company.

is the 0° (zero degree) line of latitude. The almost spherical Earth's circumference is around 40,000 km (25,000 mi.) at the equator. A circle that joins places of the same latitude at Earth's surface is called a **parallel of latitude.** The distance from one degree of latitude to the next is approximately 110 km (69 mi.) on Earth's surface. For a long time, latitude was found by measuring the angle of the sun above the horizon at noon.

Longitude measures position east or west of an imaginary line drawn from the North to the South Pole—a half circle—that passes through the former Royal Observatory at Greenwich, London, England. Lines joining places of the same longitude are called **meridians of longitude.** The position of the prime meridian passing through Greenwich (0°) was chosen by an international conference in 1884, when London was the world's most powerful decision-making city. Methods of determining longitude, especially when charting the position of a ship, were more complex and took longer to evolve than latitude measurements. Instead of measuring sun angle at noon, people had to create tables of planetary positions and accurate clocks (chronometers first made in the late 1700s). In the late 1900s, radio beacons and satellites provided standard reference points by emitting radio pulses that could be timed and interpreted rapidly in computerized navigation systems to give accurate position fixes in global positioning system devices.

Distance and Direction

Direction and distance help to define the **relative location** of one place with reference to another. (Relative location is also affected by other factors that slow contacts among people. It involves natural obstacles such as mountains and oceans, political factors such as country boundaries, and cultural factors such as Latino and Anglo differences.) Geographers give **directions** by the points of the compass, such as east, north, southwest. Physical **distance** between places is usually measured in kilometers or miles. However, travel time or travel cost (e.g., cost of gasoline) may be of the greatest significance. The relative measures of time-distance and cost-distance are often substituted for measured distance in geographic studies. The increasing cost and time of distance between places gives rise to the idea of the **friction of distance.** Interaction between people is likely to be less across a distance where costs are higher or journey time is longer. For example, the friction of distance between New York and Chicago was reduced in the nineteenth century, when time for the journey was cut from weeks to days by building the Erie Canal and railroads. Today, air travel between these places takes a couple of hours. The increasing availability of rapid transportation facilities and the "global information highway" (the Internet) bring people into easier contact with each other, making them relatively—but not physically—closer. Reducing the friction of distance makes it cheaper and easier for more people and goods to take that journey.

Maps and Scale

Geographers draw **maps** to represent features on Earth's surface. Maps require a projection, or layout, to convert Earth's curved surface to a flat piece of paper (see website). Maps are drawn to **scale,** relating horizontal ground distance to map distance. Scale is usually quoted as a fraction or ratio (e.g., 1/10,000 or 1:10,000, in which 1 unit on the map represents 10,000 of the same units on the ground). For example, on a map with a scale of 1:10,000, 1 cm represents 10,000 cm, or 100 m, on the ground.

Map scales vary with the size of the area to be mapped and the purpose of the map. *Small-scale* maps usually show areas at fractions of 1:250,000 or smaller (e.g., 1/1 million or a ratio of 1:1 million). They provide much less detail about larger areas (Figure 1.9). The world maps used in this text (see, e.g., Figure 1.6) are examples of small-scale maps, in which the scale along the equator is approximately 1:120 million. Large-scale maps usually have map-to-ground ratios ranging from 1:10,000 to 1:250,000 (e.g., 1:50,000). For the same size of map, they cover smaller areas, as in town maps, so they include more details. Not everything can be drawn to scale on maps or they would be too small to be seen, so roads, rivers, buildings, and other features are replaced by symbols.

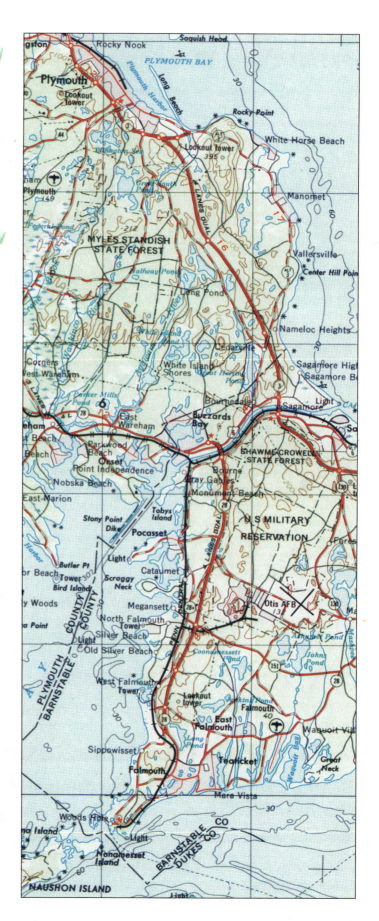

Figure 1.9 **Topographic map: location, distance, direction, and scale.** This is a 1:250,000 map of part of eastern Massachusetts, in which 1 cm on the map represents 2.5 km, or 1 in. represents 4 mi. What other information does the map provide? In what direction does Route 3 travel on leaving Plymouth toward Sagamore, Cape Cod? Source: U.S. Geological Survey.

Next, Geography Is About Explaining the Differences Among Places

The two basic geographic concepts of place and location are combined in three main approaches to geographic information gathering and explanations.

Regional geography evaluates the differences and similarities within and between defined areas, or regions, of the world. A **region** is an area of Earth's surface with similar physical and human characteristics that distinguish it from other regions. Regional geography involves studying the distinctive physical and cultural landscapes within areas. The different scales of region—local regions, countries, and world regions—allow comparisons and understandings of the interactions among the different levels.

Spatial analysis highlights the relationships among places resulting from linkages over the space of Earth's surface. The character and location of places are often considered in terms of geometric points, lines, and areas. Statistical links among places add to the mathematical basis of spatial analysis. Spatial analysis studies may focus on economic or population changes, or on geographic relationships among different sectors of towns or rural areas. Spatial analysis helps to assess linkages among and within regions.

In investigating **human-environment relationships,** geographers consider interactions between physical and human geography. In the early 1900s, the study of the interactions between human activities and the natural environment focused mainly on the impacts of climate, mountains and lowlands, and soil types on human affairs. More recent studies assess the impacts of human activities on the physical environment. The interactions among cultural and natural environments over time are often summed up in visual landscapes. Distinctive elements of townscapes and rural landscapes provide indicators of past interactions that help create and define regional characteristics, and a basis for comparison of regions at different scales.

 # Regions and Globalization

In a world of closer connections among peoples and places and increasingly rapid changes, the interactions and tensions among places and activities at global, world regional, country, and local scales focus attention on regional geography.

Defining Regions and Their Dynamic Features

Each of the geographic scales of region—global, world region, country, local—is a distinctive sector of Earth's surface. Regions are defined by a high degree of uniformity, or limited variability, and by more-or-less lasting boundaries (Figure 1.10). Regional boundaries may include physical features, political boundaries, or economic characteristics. Marketing regions are used by business, while administrative regions are

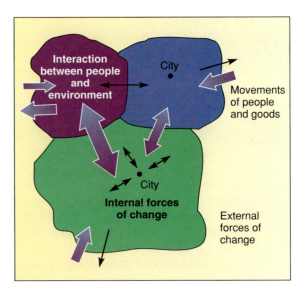

Figure 1.10   **Regions.** Regions are defined by boundaries that include places with similar characteristics but are subject to movements within and across the boundaries, often focused on city nodes.

used by government. Each region is unique in some ways, but those that have related characteristics may be grouped.

Regions are also dynamic geographic entities that have distinctive internal and external flow patterns of such phenomena as people, goods, and ideas. *Nodes* are key features of regions, being specific places from which flows begin or through which they pass; some nodes also direct flows. The area of interaction of a set of nodes may define the boundary of a region. For example, a store or restaurant can be considered a node, and the extent of the node's region is the area from which people come to shop at the store or eat at the restaurant. A country has its political node in the capital city. The counties and municipalities within countries act as local political regions that function in relation to the central government and to the people and activities in their locality. Regions with different sorts of functions and nodes may exist together and overlap.

Flows within and among regions include population migrations; information from the media, Internet, or publications; movements of money; technological innovations in manufacturing processes, information processing, or new transportation modes; and ideology through political and religious beliefs. The dynamic elements of such flows within and among regions affect the prominence of regions within a country, of countries within world regions, and of world regions within the global system. The variety of these flows is generated by characteristics of path, speed, and direction and the different relationships to social structures imposed by governments and other institutions. Breaks or interruptions in the flows may result in social problems such as inequities, injustices, and underresourced livelihoods at the local level. These aspects of regions are some of the most basic features of the regional chapters.

Consider a typical metropolitan region. People living there practice different religions, come from different places, speak different languages, are employed in various jobs, and go to different schools. Yet they use the same transit system and airport, the same hospitals and other emergency services, the same TV stations and newspapers. They go to visit family elsewhere, and the products they make are sold in other regions. There are flows into the metropolitan region, within the region, and out from it.

Global city-regions function as nodes in the global economy. Large cities and ports, established mainly over the last 200 years and linked to outlying areas of resource supply and dependency, often form major centers, or nodes, for today's global exchanges of people, goods, and information. The central cities in global city-regions spread their functions into the surrounding region and form links between both the global and local flows. Both the global and local areas they interact with are part of their functional regions. For example, New York City has flows in and out of the metropolitan area based on transportation links, technology, business services, and media products. These flows link it to the surrounding U.S. states and to countries throughout the world. Some global city-regions, such as New York, London, and Tokyo, have the highest levels of telecommunications links to the rest of the world; other cities link to these cities and to places in the world regions of which they are a part (see map inside front cover).

Changes in Dynamic Regions

Five characteristics of the worldwide mosaic of geographic regions arise from considerations of human-environment relationships, human involvements in change, the significance of political and economic power structures, and the interdependence of regions.

People Create Regions

People living in a region—at local, country, or multicountry scale—determine its characteristics. Human actions at crucial phases of history set a region apart from other regions. Images of a country's or local region's role or identity portray ideas of its significance. Inhabitants may define relations with other regions as friendly or "other."

Human actions are more important than physical environments in defining regions, as is apparent when people live in different ways in similar natural environments. For example, dry-land inhabitants include the wandering herding tribes of the Sahara, farmers in Pakistan using the irrigation canal systems built by British engineers in the late 1800s, and those living today in the urban sprawl of Los Angeles. To see that similar human influences occur across several types of natural environment, one only has to look at the United States, where very similar city landscapes occur across different types of natural environments.

People also "create" regions in a different sense. In deciding what is one region as against another, they bring their own knowledge, feelings, and beliefs—a personal bias. This often means that different people define regions in different ways or create different bases for regional divisions. There are so many possible starting points for defining regions. The student of regional geography is thus not dealing with regional divisions

fixed forever but with entities that may change or be a matter of challenge. For example, many American geographers are wedded to climate divisions derived in 1918 by an Austrian biologist, Vladimir Köppen, from a world vegetation map. The boundaries between the climates identified are often awkward to define and do not relate to any genetic factors such as dominant weather systems. Other climatologists produce different maps.

Regions Shape People's Activities

Each region provides a bound environment for human activities created by the region's dominant political, economic, social, and natural characteristics. People living in arid or very cold environments, or under harsh dictatorships have limited options. People living in New York are constrained by the way in which the city is built and its neighborhoods developed.

So, a two-way process is at work. People are the main forces in creating distinctive regions but are affected by the regional characteristics that others before them established. Over long periods of time, some regional characteristics are perpetuated, causing local cultural traits and social habits to emerge.

People Remake Regions

Regions—and their boundaries—may change over time, instigated by flows of people, information, capital, technology, or the ideas by which political systems are framed. In the 1880s, European colonial powers divided Africa into the country units that, by and large, still exist. These boundaries had nothing to do with traditional African land units (see Chapter 9), but once established, they proved difficult to change. When the Soviet Union and its bloc of satellite countries broke up in 1991, the countries of eastern Europe disassociated themselves from Russia and expressed a wish to join western European countries in the European Union. Regions other than countries (e.g., the Developing World) are less formally defined, are often inventions of geographers or others, and may change more easily. For example, drawing a world regional boundary can place Myanmar (Burma) in either South Asia or Southeast Asia. Wherever the line is drawn today may change in the future as the world changes.

Within a country, individuals and small groups may influence the course of changes. The former Soviet Union arose from a 1917 revolution that brought Vladimir Lenin and later Josef Stalin to power. In the early 1930s, the Soviet Union industrialized and forced independent farmers to work in cooperative state farms. The Moscow government "center" then dictated which crops could be grown. Mikhail Gorbachev and Boris Yeltsin had major roles in the breakup of the Soviet bloc in the 1980s and the development of the new Russia in the 1990s. In the 1990s, the new Russian government tried to persuade those in state farms to return to private enterprise, smaller farm units, and local decisions about crops and livestock.

Regions Interact with Other Regions

No region is an isolated entity. All interact with others through the flows of people, information, capital, technology, and ideology. Events such as news reports, human rights legislation, or action on gender issues spread their influence from major global centers to local regions. Farmers in the U.S. Midwest see their jobs and incomes influenced by world market prices. The world's most remote regions, such as the upper Amazon River basin, are increasingly affected by external demands such as the search for valuable minerals and settlers looking for land. Changes in one region affect other regions: when world markets for peanuts and cotton collapsed and drought struck northern Nigeria, the people abandoned their farms and moved into urban areas in other parts of the country.

Some places affect surrounding regions by their role in funneling trade through narrow ocean-route throughways. The richer countries defend the **global choke points** of the Suez (Egypt) and Panama Canals, the Straits of Hormuz (entrance to Persian Gulf), Dardanelles and Bosporus (entrance to Black Sea in Turkey), and the Malacca Straits (near Singapore) to maintain access to their markets and raw material sources.

Interactions among countries may result in close alliances in trade and defense, complementary exchanges of goods, or conflicts. Different cultural groups within a country, each with territorial claims, may cause minority groups to fight for what they feel are their rights. The Zapatistas in southern Mexico and the Muslims of Chechnya in southern Russia are examples.

Regions Are Used by Those in Power

Regional character may reflect the actions of powerful governments or nongovernmental institutions such as multinational corporations. Central planning under Communist governments in the former Soviet Union industrialized rural areas on a major scale, even north of the Arctic Circle. Groups of people were moved to other parts of the country for state security reasons. In the European Union free-market countries, the outlying ones, such as Portugal and the Republic of Ireland, have their economies stimulated by investments in new roads, port facilities, and airports that help to bring new jobs from American and Asian companies.

In the mid-1900s, despite the end of the political colonialism that mainly European countries instituted, larger and more affluent countries still placed political and economic pressures on poorer ones. The interventions and intrusions of the wealthier countries may affect regional characteristics. For example, the United States went to war with Iraq in 1991 to prevent it from permanently annexing Kuwait and threatening other Persian Gulf countries—as well as to preserve U.S. access to the world's largest oil fields. In 2001, action taken by the United States against the Taliban and al-Qaeda terrorist network in Afghanistan disrupted their bases in that country. The United States obtained support for its actions by assembling a large group of other countries appalled at the events of September 11, 2001, or willing to take political advantage of the situation.

⬤ Major World Regions

This text divides the world into nine major regions on the basis of cultural, political, economic, and physical criteria (Figure 1.11a, b). The geography of each region is described in greater

detail in Chapters 3 through 11. The borders of the major world regions and the countries included in each one are subject to changes over time. However, considerable agreement as to the content of these regions exists among the authors of similar texts.

The main part of this text is a journey through the nine major world regions. Our trip begins in Europe, then moves east through Russia and Neighboring Countries to the Pacific Ocean, where the East Asia region and Southeast Asia and South Pacific region look across the ocean to the Americas. Returning westward, the South Asia region is followed by the Northern Africa and Southwestern Asia region that is limited southward by the Sahara Desert. Africa South of the Sahara is the next destination before a hop across the Atlantic Ocean to Latin America. We finish in North America, taking on board the many influences brought to it from around the world and the many ways in which the United States and Canada impact other world regions.

Europe (Chapter 3) is the source of many Western trends in cultural and economic areas. The region is defined at present as those countries that are members of the European Union or are likely to be in the next decade. Its culture still links to the past dominance of Roman Catholic, Protestant, or Orthodox Christian groups. Europe is marked by the variety of languages spoken, while increased immigration in the late 1900s made it home to a growing number of people from diverse world regions.

Russia and Neighboring Countries (Chapter 4) includes all the countries that emerged from the breakup of the Soviet Union apart from the three small Baltic countries of Estonia, Latvia, and Lithuania. The region extends from easternmost Europe across northern Asia, a huge area that was drawn into the Russian Empire in the last 400 years as an extension of European cultures into Asia and contains a wide range of people, cultures, and languages. Its countries are passing with some difficulty through the transition from Communist to free-market economies. The central Communist government of the former Soviet Union subdued the long-term clashes between Orthodox Christian and Muslim people through much of the 1900s, but such cultures gained significance after the Soviet Union breakup, and conflicts returned to the region.

East Asia (Chapter 5) includes Japan, the Koreas, and a resurgent China, some of the world's successful countries in recent economic growth. The historic and modern influences of the Chinese kingdoms and related cultures were less affected by European colonization than those of other regions. The study of East Asia both holds hope for improving the living quality of the world's poorer people and highlights worries about potential future conflicts over economic differences among China, Japan, the United States, Europe, and India.

Southeast Asia and South Pacific (Chapter 6) includes the independent kingdom of Thailand and the former European colonies of Myanmar (Burma), Malaysia, Singapore, Australia, and New Zealand (British); Indonesia (Dutch); the Philippines (Spanish, taken over by the United States); Vietnam, Laos, Cambodia (French Indochina); and thousands of islands distributed across the South Pacific Ocean (mainly British, French, German, and U.S. colonies, possessions, or trusteeships). Having lost many of their trade links with Europe, Australia and

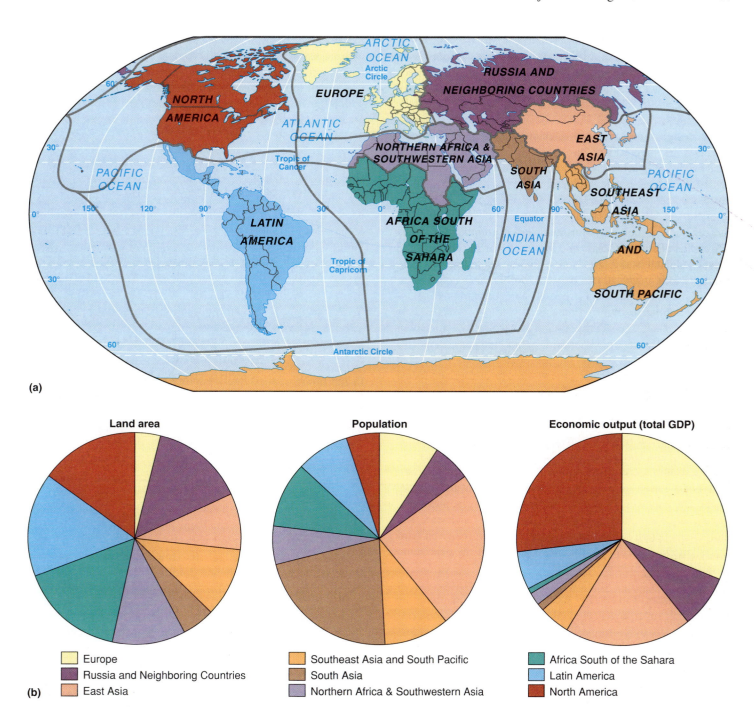

(a)

(b)

Land area

Population

Economic output (total GDP)

Legend:
- Europe
- Russia and Neighboring Countries
- East Asia
- Southeast Asia and South Pacific
- South Asia
- Northern Africa & Southwestern Asia
- Africa South of the Sahara
- Latin America
- North America

Figure 1.11 **Major world regions based mainly on cultural characteristics.** (a) These regions form the subjects of Chapters 3 through 11. World maps in Chapter 2 are divided on this basis so that comparisons may be made. (b) Comparisons of area, populations, and economic output (gross domestic product, GDP). Pie charts show the variations. Which major regions have more of the world's land area than its population? Which major regions have a higher proportion of the world's economic output than its population?

New Zealand increasingly look to Asia for trading opportunities, although they differ in outlook and at times come into conflict with Southeast Asian attitudes. We also include in this region Antarctica, inhabited solely by scientific colonies—the only part of Earth's land surface that is not divided into countries.

South Asia (Chapter 7) is a region where major world religions including Islam, Hinduism, and Buddhism clashed in historic times. Today, Hinduism (the Indian national religion) counts 80 percent of the region's population among its adher-

ents. The British Empire occupation, followed by only partly successful policies of self-sufficiency and neutrality during the Cold War, brought slow economic growth until the adoption of new outlooks in the 1990s. The population in this world region is over 1 billion and includes the largest numbers of poor people. The rivalry between India and Pakistan since their independence in 1947 led to several wars, a continuing dispute over Kashmir, and the development of nuclear arsenals, giving this region the potential for setting off a major conflict.

Northern Africa and Southwestern Asia (Chapter 8), sometimes referred to as the "Middle East," is characterized by its position at the junction of Europe, Africa, and Asia. It includes the birthplaces of the world's three monotheistic (believing in one God) religions, Judaism, Christianity, and Islam. The Islamic religion is the main cultural feature today, partly paralleled in its extent by the Arabic language—although the region's two largest countries, Iran and Turkey, have their own languages. The presence of Jewish Israel in the predominantly Muslim region brings cultural, economic, and political tensions that have already resulted in three wars. The location of the world's largest oil reserves and the largely arid natural environment of much of this region pose problems of uneven resource availability.

Africa South of the Sahara (Chapter 9) was the cradle of the human race and is still largely occupied by the indigenous peoples and their many ethnic groups. The region contains great mineral riches and has other economic potential. Colonial settlement by Europeans occurred much later and on a smaller scale than in the Americas or India. Colonists introduced commercial farming and mining. Decolonization beginning in the 1950s, followed by many poor dictatorship governments, has resulted in the people of this region being the world's most deprived today. The region suffers from political, economic, and health problems that its mainly small countries find difficult to resolve.

Latin America (Chapter 10) has many political and economic issues that relate to the region's position as the nearest neighbor of the United States. The Latin-based languages and Roman Catholic religious culture brought by Spanish and Portuguese settlers in the 1500s influence most of this region. There are enclaves of other European languages, including French, Dutch, and English, particularly in the Caribbean. The indigenous people were reduced in numbers in many parts of Latin America but still form a major portion of the population elsewhere in this region.

North America, comprising the USA and Canada (Chapter 11), is the world's materially wealthiest region, containing the only present superpower (the United States). The two countries are dominated by cultures first brought by European settlers beginning in the 1500s. The settlers had a disastrous effect on the population and cultures of the indigenous peoples. At first, settlers came mainly from western and northern Europe, but from the late 1800s, southern and eastern Europeans, and eventually those from most other parts of the world, became immigrants. French and Spanish are spoken locally, but English is the most widely spoken language in North America. Indigenous peoples formed smaller and smaller proportions of the population as a result of the European influx, but some lands and facilities were restored to them in the later 1900s.

🌐 Development of World Regions

Globalization has a long history, building from the earliest human activities to the current time. A quick look at the way today's diverse world geography developed historically identi-

fies distinct phases in which increased flows of people, information, wealth, technology, and ideas spread from early centers of innovation to incorporate more and more of our world. For a long time, areas of growing material wealth and greater population density occupied small parts of the world, while areas of less intensive production dominated the rest.

Early History

Until around 5000 B.C., most humans lived in small groups, gaining a livelihood by hunting and gathering, and often sheltered in temporary housing that could be rebuilt as they moved to find new resources. The groups had little interaction with each other, and the territory needed to support their small numbers was often huge. Kin alliances linked small local groups across broad territories. The areas needed to support groups varied in size depending on the nature of the local environment and its resources. As a result, some people lived in larger concentrations than others.

Settled Farming

In the midst of the largely hunting and gathering people, the first settled farming began in a few areas of southwestern Asia as long ago as 9000 B.C., followed by small areas of China, other Asian areas, and the Americas. Western Europe and Africa developed settled farming later. Geographic variation was emphasized in the local choice of crops and animals for domestication. Among the crops, wheat and barley were most important in southwestern Asia, rice and millet in China, a wide range of corn, squashes, beans, potatoes, tomatoes, and peppers in the Americas, and sorghums and yams in Africa. Domestic herding groups (nomads) contrasted with the settled farmers and their villages—some of which grew into towns such as Tell es-Sultan (Jericho, c. 8000 B.C.) in the Jordan River valley and Çatal Hüyük in central modern Turkey. The early centers of innovation in agriculture, writing, and other technologies are often referred to as **cultural hearths.** From these areas, people, skills, religions, and languages diffused outward.

City-States and Empires

From 2500 to 1000 B.C., massive surpluses of farm products formed the basis of the first civilizations. Sun and water combined to support irrigation farming in relatively small areas: Tigris-Euphrates Rivers (Mesopotamia), Lower Nile River (Egypt), the Indus River valley (modern Pakistan), the Huang He (or Yellow River) valley in China, coastal Peru, middle America, and western Africa. Political systems varied from city-states to empires.

Each of these centers experienced a huge accumulation of wealth that was often expressed in major buildings and linked to developments of far-reaching technologies—the first writing to inventory trade transactions, the organization of water distribution and land ownership, the foundations of complex mathematics, innovations in metalworking (this was the Bronze Age), pottery, construction materials, and the wheel. Artistic representations became more sophisticated, religious beliefs were linked to mythical prehistories, and laws were created. Recorded history began.

Trading networks linked all these growth regions. Local exchanges of raw materials and craft products complemented longer routes dealing in luxury goods such as precious metals, jewels, and spices. Large areas of the world, where people subsisted on local products, remained outside these areas of major wealth creation.

Trading Empires and "Classical" Civilizations

From 1000 B.C. to A.D. 600, centers of wealth accumulation affected new and larger areas. Increased levels of trade and economic activity spread from Asia along the Mediterranean into western Europe; Chinese dynasties expanded their empire westward and southward. By A.D. 200, trading routes for high-value goods extended from Rome to China along the Silk Road route through Central Asia (Figure 1.12). Slavery remained the main form of labor in the fields, workshops, and domestic realms of life in most of the growing regions. In the Americas, the Mayans developed their own writing system and built cities based around huge public buildings by A.D. 250 (Figure 1.13).

The period from 1000 B.C. to A.D. 600 gave rise to what are known as the "Classical" civilizations. The main centers were in Persia (Zoroaster), Greece and Rome (each with their pantheon of gods), India (Hindu religion developing from more ancient traditions; Buddha), China (Confucius; Lao-Zi and Daoism), southwestern Asia (Judaism, Jesus Christ, Muhammad),

central Europe (Celtic druids), northern Europe (Wotan and Norse gods), Egypt (Isis and other gods), and the Americas (e.g., Mayan nature gods). At the outset, these civilizations were often led by a charismatic leader or linked to a pantheon of gods. Large-scale movements of people resulted in extended warfare, invasions, and changing regional character. Invasions of the Indian subcontinent and of the rich centers of China and the Mediterranean highlighted the spreading knowledge of wealthy locations and their attraction to mobile groups of horse-mounted herders.

Disruptions, Migrations, and Feudalism

Disruptions of established empires occurred from A.D. 600 to around 1450, a period often termed the "Dark Ages." Increased mobility of larger armed groups across greater spaces resulted in new empires. In the Mediterranean, the Roman Empire in the west had succumbed to attacks by this time, but its eastern Byzantine wing, centered on Constantinople (modern Istanbul), lasted longer. After around 650, soon after Muhammad's leadership generated the new religion of Islam, Muslim peoples spread out of Arabia across northern Africa and into southernmost Europe, central Asia, and India.

From 800, hordes of horse-mounted warriors swept out of the Asian grasslands into central Europe, India, and China. The Vikings sailed from northern Europe to western Europe, North America, and down the Volga River valley, where they influenced the early history of the Russian people. China again became the world's most prosperous empire and made huge contributions to

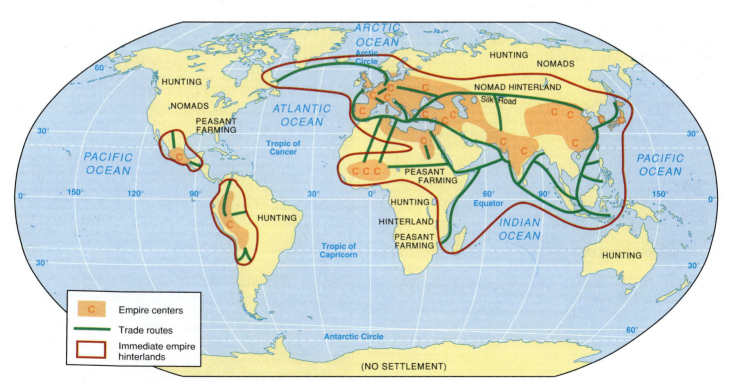

Figure 1.12 **Feudal empires to A.D. 1450.** This period began with disruptions in Europe and included invasions from the Eurasian steppes and Viking homelands. It also saw the development of new empires outside the previous world systems, including in the Americas and western Africa. Some regions experienced important periods of political consolidations and wealth accumulation.

art, philosophy, and technology (such as the first printing). In western Africa, new empires controlled the crossings northward across the Sahara. In the Americas, the Incas and Aztecs dominated large regions—the former in the central Andes Mountains and the latter taking over the Mayan area in Middle America after around A.D. 1300. In many of these places, the growing personal powers of nobles and shortages of people to fulfill production and defense needs allowed the widely effective system of **feudalism** to replace slavery. Under feudalism, people received military protection in exchange for working a proportion of their time for a local overlord. The overlord then gave allegiance to higher orders of knights, kings, and emperors. However, this system resulted in local rather than countrywide control, fostered few new ideas, and slowed the expansion of trade.

The Modern, Globalizing World

Explorations and Colonies

New systems of trade and wealth expansion developed in western Europe around A.D.1450 and spread to the rest of the world. The combination of exploration based on new maritime technology, the zeal to spread the Christian faith, and a new profit motive among merchants formed the basis of European expansion and began the modern phase of globalization.

Spaniards and Portuguese led the way, followed by the French, Dutch, and British. An initial focus on sponsorship by rulers wishing to add new territories and sources of income gave way to investment by groups of speculating merchants. At first, kings supported the Spanish and Portuguese explorers, who claimed lands in the Americas and traded around Africa into the Indian Ocean, reaching China. In northwestern Europe, merchant-funded companies sent out settlers, invested in mines and plantations, and increased trade with eastern North America and India. The slave trade shipped millions of Africans to the Americas. The governments of emerging countries in western Europe claimed as colonies the lands opened up by merchant-funded groups.

Industrialization

From the mid-1700s, new manufacturing technologies in cloth making, metal refining and fashioning, and leather tanning provided investment opportunities in Europe that often brought better returns than gathering produce from the overseas mines and plantations. The growing efficiency and technological breadth of manufacturing processes concentrated increased production in fewer places to make larger ships, steam engines, commercial chemicals, and then airplanes, autos, and a multitude of consumer goods that changed world regional geography. Colonies supplied European and U.S. manufacturers with raw materials, and their elites bought the manufactured goods.

Figure 1.13 **Early civilization achievements.** Among these were huge temple buildings and palaces and advanced decoration and written scripts. The Mayan Kukulkan Pyramid at Chichén-Itzá, built in the 900s A.D., and details of sculpting on its face.

The benefits of these products and the cultural system that enabled them were acknowledged around the world, and most countries wished to modernize, if not to Westernize. Most of North and South America achieved independence from European countries by 1800. By 1900, European countries had colonized much of the rest of the world (Figure 1.14), but by 1960, most of these colonies established their independence.

Globalization, Countries, and Protectionism

Globalization—the increasing interconnections around the world—expanded unevenly in time. In the first phase of European expansion, from around 1450 to the early 1800s, trade and piracy combined on the unpoliced high seas. Much of the wealth from distant lands enhanced differences among European countries. The wealthier ones were able to gain more power as they funded armies to fight for their interests in civil and continentwide wars.

Particularly in Britain, merchants invested much new wealth from overseas ventures in the factory-based concentrations of manufacturing, giving that country advantages in establishing widespread world trade, military power, and colonial dominance. During the late 1800s and early 1900s, globalization spread as world trade expanded further following developments in commu-

nications (telegraph) and transportation (railroads, larger ships), financial investments in newly opened lands, and huge migrations of people—especially from Europe to the Americas.

Toward the late 1800s, countries slowed their levels of international linkages. They defined their own and their colonies' frontiers more strictly, developed nationalist sentiments among their people, established tariffs on trade, began to control the growth and power of large corporations, and restricted levels of immigration. Intense international competition over trade and colonies built political hatreds that led to World War I (1914–1918). After that war, new countries were carved out of old empires in Europe and western Asia. Although international bodies, such as the League of Nations, were established, actions by individual countries weakened any authority they might have had. A major economic crisis in the 1930s (the Great Depression) resulted in restrictions on trade, people migrations, and movements of money, further slowing globalization. Extreme nationalism in Germany and Japan, the closing of the Soviet Union to foreign contacts, and movements for independence by colonial territories diverted attention from international linkages. German and Japanese expansionism led to World War II, causing countries outside Europe and Asia to retreat further inside their borders and build up their own industries.

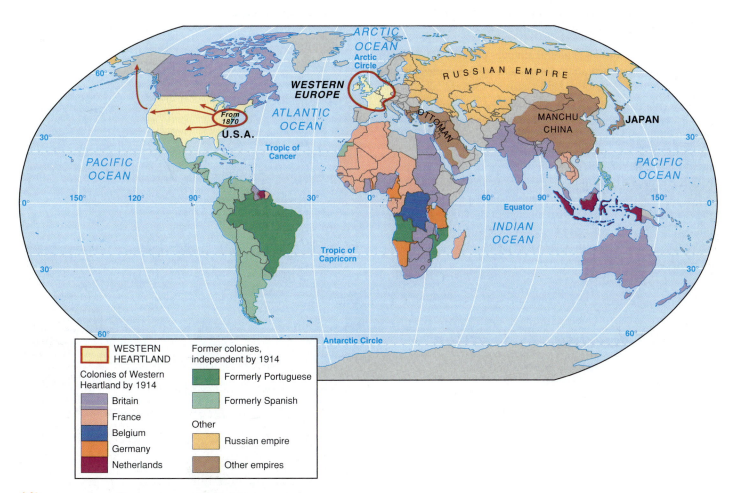

Figure 1.14 **Capitalist world system, about A.D. 1910.** The growth of industrial capitalism, the consolidation of the Western heartland in Europe and North America, and the expansion of European empires.

Following World War II, more international institutions came into being, including the United Nations (UN), the International Monetary Fund (IMF), the World Bank, and the General Agreement on Trade and Tariffs (GATT) with its successor the World Trade Organization (WTO). Their purposes included mediating political issues, assisting the poorer countries, and liberalizing trade. Until 1990, their international, or globalizing, functions were overshadowed by the Cold War divisions. One outcome was the economic growth of the United States, Western or capitalist Europe, and Japan—which developed trade among themselves—and almost no growth in the rest of the world.

After the 1991 breakup of the Soviet Union, the international movement of ideas, capital, goods, and people accelerated, and globalization based on Western-dominated trade and ideas affected all countries. Many of the poorer countries began to grow economically. And yet, international government made slow progress. Some commentators say that the U.S. and European involvements to wrest Kuwait back from Iraq in the Gulf War (1991), to bring order to Rwanda and Somalia, to end Serbian oppression in Kosovo, and to break impasses between Israeli and Palestinian interests demonstrate the ineffectiveness of the United Nations in global governance. Others argue that the United States and European countries took matters into their own hands without giving the United Nations a chance to act. Global political control remains limited and largely unwanted by most countries. Smaller countries fear that the most powerful countries will control any global arrangements. At the same time, social conditions and cultural attitudes created movements resisting external control, both within and from outside current country boundaries. Globalization moves forward slowly in complex and often contradictory trends.

The world regional chapters in this text are based on current situations and enable you to assess the significance of the cultural, economic, and political histories, as well as the natural environments, on contemporary geography. As global connections increase, local voices are heard more clearly supporting alternative views, often linked to traditions and long-established ways of personal and community life. The book's final chapter evaluates trends and views of globalization and studies terrorist activities in that context.

Test Your Understanding 1B

Summary Geography is the study of where and how human beings live in varied ways in different parts of the Earth, comparing places and linkages among them. Location of places is defined by latitude and longitude (absolute location), and by direction and distance (relative location). Geographers use maps to summarize information and a variety of approaches to understanding how and why people live in the places they do.

Regional geography studies the diversity of geographic distributions and human activities through formal and functional regions. People create regions, which then affect the lives of their inhabitants; people recreate regions in times of change; regions interact with each other and are used by those in power. Cultural and physical characteristics define the basis of nine major world regions. The current map of major world regions evolved through the history of human occupation of the Earth. Changes continue.

Questions to Think About

1B.1 Why is regional geography so relevant to understanding our world today?
1B.2 Summarize the differences among the major world regions defined in this text.
1B.3 Where is your nearest global-city region? How does it link to other world regions?

Key Terms

geography	distance
physical geography	friction of distance
human geography	map
spatial view	scale
place	regional geography
absolute location	region
latitude	spatial analysis
parallel of latitude	human-environment relations
longitude	global city-region
meridian of longitude	global choke point
relative location	cultural hearth
direction	feudalism

Online Learning Center   www.mhhe.com/bradshaw

Making Connections

The Online Learning Center accompanying this textbook provides access to a vast range of further information about each chapter and region covered in this text. Go to www.mhhe.com/bradshaw to discover these useful study aids:

• Self-test questions
• Interactive, map-based exercises to identify key places within each region
• PowerWeb readings for further study
• Links to websites relating to topics in this chapter

Chapter 2

Human Development and
World Regional Geography

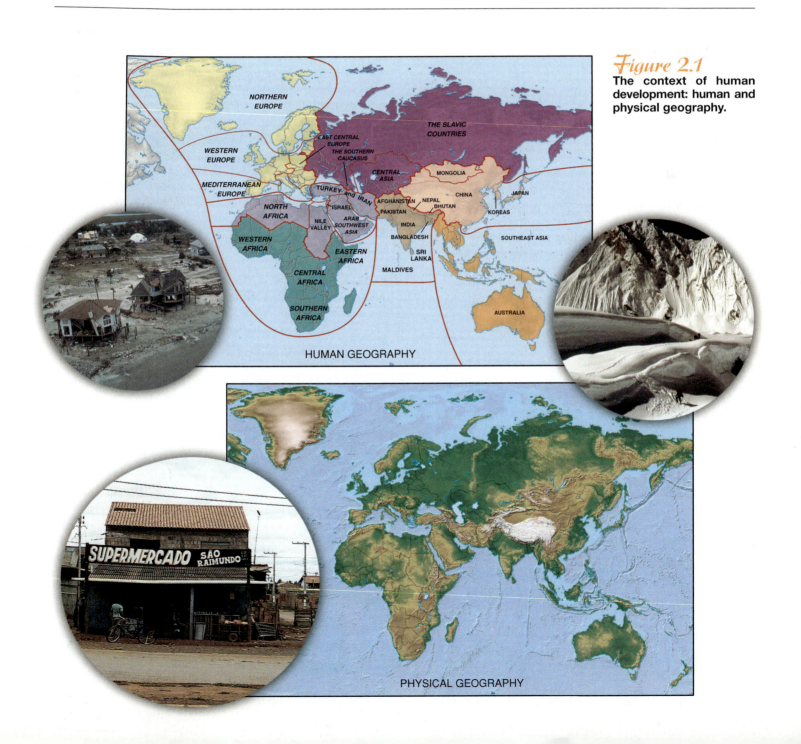

Figure 2.1
The context of human development: human and physical geography.

NORTHERN EUROPE

THE SLAVIC COUNTRIES

EAST CENTRAL EUROPE

WESTERN EUROPE

THE SOUTHERN CAUCASUS

CENTRAL ASIA

MONGOLIA

MEDITERRANEAN EUROPE

TURKEY and IRAN

CHINA

JAPAN

NORTH AFRICA

ISRAEL

AFGHANISTAN

NEPAL

BHUTAN

KOREAS

NILE VALLEY

PAKISTAN

ARAB SOUTHWEST ASIA

INDIA

WESTERN AFRICA

BANGLADESH

EASTERN AFRICA

SRI LANKA

SOUTHEAST ASIA

CENTRAL AFRICA

MALDIVES

SOUTHERN AFRICA

AUSTRALIA

HUMAN GEOGRAPHY

SUPERMERCADO SÃO RAIMUNDO

PHYSICAL GEOGRAPHY

Understanding Global Disparities

Despite the trends toward globalization, the world is full of differences among places (Figure 2.1). The differences between many extremes are as great as they have been in human history. High densities of population in massive cities contrast with empty deserts and polar regions. The material wealth of many Americans contrasts with the extreme poverty of millions in Africa and Asia. Countries range in area from very large (Russia, China, the United States, Australia, India, Canada, and Brazil) to tiny islands. The variety of languages and religions both unites and separates groups of people. Disparities of natural environments and resources result in more or less sun and water and differences in exposure to natural hazards or in access to minerals and good soils.

The differences among places led some countries that were well endowed with natural resources and strong leadership to assume positions of dominance and superiority. For example, the Chinese long believed they were the only civilized people and all others were barbarians. During the 1800s, European countries claimed they were taking civilization and modern ways to parts of the world that they colonized. In the later 1900s, the United States and Soviet Union contested world economic, military, and political dominance, each believing that their systems were superior. After 1991, the breakup of the Soviet Union left the United States as the only superpower with an economic and military might that no other country matched by a long way. And yet, the United States, though dominant, still has to consider the wishes of smaller countries and gain support for its policies through a wide range of political links.

While some countries and peoples assumed dominance or leadership, others had less political or economic power. From the 1950s to 1990, many new countries became independent from colonial governments or the forced incorporation into the Soviet bloc. By the 1990s, increased country independence

added to the variety of challenges to the powerful countries. Greater exposure of voices from local groups of people within countries focused attention on differences in poverty and people's rights to a full human experience. Human development and human rights became significant goals in the globalizing-but-localizing world.

Human Development

Human development is the process of enhancing human capabilities by expanding choices and opportunities to enable each person to live a respected life of value. A human development value can be given to indicate where countries stand relative to each other.

At first, the term **development** was linked to measures of economic performance, such as national income. That called attention to economic disparities among countries, but they are linked inextricably to educational, cultural, political, and environmental conditions. People are central; economic growth is regarded as a necessary, but not the only, process that enables people to enlarge their capabilities and enjoy the richness of being human. **Sustainable human development** involves economic growth that does not deplete renewable resources for the future. It thus relates to both human and natural resources, drawing together studies of human and physical geography.

The United Nations Human Development Program and recent World Bank publications focus on the eradication of poverty as part of this thrust. Human development officially became the province not only of economists but also of a wider range of social scientists and policy-makers. It is encouraging that the last 50 years saw major reductions of income poverty in large parts of the world, improvements in human development indicators—particularly in health and education—and the wider spread of law and fair administration of justice. It remains challenging that so many people in the world are poor.

Human Rights

The concept of **human rights** emerged in part from revolutions in Europe and the United States in the late 1700s, emphasizing, for example, "Liberté, Égalité, Fraternité" in France and enshrined in the U.S. Bill of Rights.

The United Nations' Declaration of Human Rights after World War II and subsequent events increasingly bolstered human rights by international agreements and actions. But some groups state that the imposition of human rights legislation impinges on their traditional rights. Current United Nations literature defines human rights as:

- *Freedom from discrimination* because of gender, race, ethnicity, national origin, or religion
- *Freedom from want* with a decent standard of living
- *Freedom to develop and realize one's human potential*
- *Freedom from fear* of threats to personal security through arbitrary arrest or violence
- *Freedom from injustice*

- *Freedom of thought and speech* to participate in decision-making and form associations
- *Freedom for decent work* without exploitation

In contrast to human development, pursuing human rights has often been a special objective of lawyers, philosophers, and political pressure groups. Women's rights movements are very significant in alerting their governments and others to women's position in society and desire to take up new opportunities, often through community systems of small loans (in which women are more trusted to apply them than men).

During the Cold War, political rhetoric reduced human rights to a propaganda weapon, with the West emphasizing civil and political rights, while the Communist world focused on economic and social rights. Today, human rights activists, particularly in the West, claim all people should have social arrangements to protect them from the worst abuses and deprivations.

Few, if any, countries, however, observe all the rights listed, and there are considerable debates as to what should be included or is feasible in practice. There is also hypocrisy in which human rights issues provide "red herring" arguments. For example, the world's wealthier countries often argue against trading with poorer countries because their factories employ children or ignore environmental protection. In reality, this can be a disguise for protectionist policies that may secure low-wage, low-skill jobs in the wealthier country for a time. Some countries encourage trade with oppressive dictatorships, while others do not. After September 11, 2001, much was made in Western countries of the Taliban government in Afghanistan oppressing women and carrying out amputation punishments for "minor" crimes. And yet, according to Muslim views, many Western governments allow the mistreatment and degradation of women. Different cultural definitions of human rights contribute to differences among places.

Human Development and Human Rights

Although emerging from different sources, the two strands of human development and human rights reinforce each other (see Global Focus: Voices of the Poor, p. 26). Human rights add value to the development agenda, drawing attention to those who are accountable for respecting rights and adding a social justice agenda to economic principles. Priorities shift toward the most deprived and excluded. At the same time, human development brings a long-term perspective to fulfilling the rights by assessing the workings of socioeconomic contexts and institutional constraints. It highlights the resources and policies needed to overcome the remaining gaps.

In the early 2000s, discrimination, poverty, personal insecurity, injustice, and abuses of free speech continue. Internal armed conflicts affect many parts of the world, holding back human development, abusing human rights, and creating increasing numbers of dispossessed refugees. The moves to democracy in Africa and eastern Europe with multiparty elections brought some advances in human development and human rights. These trends, however, also led to new conflicts in some countries over previously suppressed ethnic demands. By contrast with the previous decades of authoritarianism, more open government appeared weak to some political leaders.

The rest of this chapter provides introductions to aspects of the geography of human development that are basic to the regional Chapters 3 through 11.

- Issues of people and land
- Issues of economic inequality
- Issues of cultural freedom and discrimination
- Issues of political freedom
- Environmental issues

Issues of People and Land

People are central to geographic studies: they create, live in the context of, and recreate geographic regions. In late 1999, the world population passed 6 billion, rising at nearly 80 million people per year (Figure 2.2). The rate of population increase is slowing, but several years (or more if the low target is not achieved) of increasing population and pressure on resources will occur before the rise levels off.

Population distribution and growth, together with their impacts on natural resources, are major issues affecting human development and human rights. While no agreement exists on how many people our world can support, present growth rates of over 2 percent per year are judged to be too high. Improvements in education, children's rights, and women's status are basic to making the most of the resources available to control future population growth. World population conferences in the 1990s agreed that population totals should be contained, although some countries resist implementing policies to control birth numbers for cultural or political reasons.

Population Distribution

Population densities—the numbers of people per given area (e.g., square kilometer or square mile)—vary greatly around the world (Figure 2.3). These geographical variations reflect the natural resources available, the historic and present use of those resources, and the type of economy that is dominant. Only 29 percent of the Earth's surface is land, and much of that is uninhabitable (desert, ice cap, mountain) or almost so. The scarce habitable areas vary from low to high densities of people.

- The world's highest densities of population occur in parts of Europe, South Asia, East Asia, and the eastern United States and Canada, as well as in smaller areas elsewhere. These areas have sufficient rain and fertile soils for high-intensity agriculture, resources for industrialization, and growing urban-industrial areas linking into the global economy.

- Medium densities of population occur in regions of more extensive farming, more widely distributed cities, and less favorable natural environments. Such areas include most

GLOBAL *Focus*

VOICES OF THE POOR

"Poverty is a denial of human rights. The Voices of the Poor series provides compelling testimony of the violation of poor people's rights—by agents of the state, by the private sector, and sometimes by civil society in their daily struggle to survive. The last book in the series, *From Many Lands*, underlines the paradox that poor people, who are the most powerless to protect themselves, often receive the least protection from their laws or officials. A human rights approach to development is needed to empower poor people to seek their rights and to hold governments accountable."

Mary Robinson, United Nations Commissioner for Human Rights, 2001

The quotes here are taken from the book mentioned above. The *Voices of the Poor* series records a major attempt to listen to what poor people say instead of wishing untried solutions on them. The first book in the series, *Can Anyone Hear Us?* (2000), reported studies in the 1990s in 50 countries and was based on what over 40,000 people said. The second, *Crying Out for Change* (2000) was based on 1999 fieldwork in 23 countries. The final book, *From Many Lands* (2002), presents a selection of country studies.

The following two stories from the last volume portray two experiences, one in Bangladesh and the other in Russia. As you read them, note the common and different features of these people's lives. Are they related to global trends and connections? How?

Waves of Disaster in Bangladesh

Mariam Bewa, a 40-year-old widow and mother of five, lives in Halkerchar village, five kilometers from the headquarters of the Dewangonj thana (subdistrict) in the northern district of Jamalpur [Bangladesh]. When the banks of the river Jamuna eroded in 1988, Mariam's house was swept away by the river, along with the homes of 450 neighbors. The community rebuilt their village on vacant government land; it has a government-run primary school, a junior high school, and an informal primary school run by an NGO. A small market with a handful of permanent stores stands in the middle of the village. Halkerchar has little transportation infrastructure, but there is a road that leads to the thana center and another that connects the village with the flood control embankment, which is used to access water. Although the land is prone to flooding and erosion, agriculture remains the main source of livelihood. Poor people supplement their incomes by working in fisheries and by picking up day-labor jobs, but the availability of these opportunities fluctuates with the seasons.

Mariam's family used to own a homestead, farmland, and cattle, but when her husband became ill, they sold their assets to cover the costs of his treatment. When he died, Mariam was left destitute except for a house, which washed away in a flood in 1988. Since then, Mariam has raised her family in a small, thatched house. Mariam is excluded from most livelihood opportunities because she is a woman. In her younger years, she supported her family by agricultural work, although it violated local norms. Recently, Mariam took up work as a domestic for wealthy households in the community.

Like many parents in Bangladesh, Mariam raised her four sons hoping that in return for the sacrifices she made for them, they would support her in her old age. Her three oldest sons, however, abandoned her after they married. Her last hope is with her youngest, unmarried son. At present, Mariam's earning from menial jobs and her son's from wage labor are not enough for their household to get by. In addition, Mariam's only daughter will need to marry, which will require that Mariam pay for dowry and other expenses. Anticipating these outlays, she cuts back on household expenses and saves what she can.

Struggling Against the Tide in Russia

Svetlana, a 43-year-old mother of three, lives in Ozerny, a farming village founded in the late 1930s in the Ivanovo [Russia] region. "I used to work at our school as an instruments and equipment keeper," she recalls. "When they started to delay our salary and then stopped paying it, I left." She took a job at the local bakery, but "you had to knead bread two to three times more than the norm, and they paid you only 270 rubles (about US $11), half of what they pay you in Ivanovo," she says. The harsh working conditions and low pay caused her to leave the bakery job. Now Svetlana works as a dishwasher in a retirement home cafeteria. In addition to her work at the cafeteria, Svetlana raises pigs, which provides extra income that helps the family survive. Svetlana's earnings are important because while her husband has a job at a boiler house that supplies the village with heat, his employers do not pay him on time. Svetlana says, "Our life is hard and we watch every penny. I'm lucky my husband doesn't drink. I don't know how we're going to carry on. The only way is to keep farm animals and grow vegetables. We rely on ourselves—no use hoping someone will come and give us something. We are not well off, but we manage. Our only concern is for the kids to make it in life."

Poor children bear the stigma of poverty. Because Svetlana's younger son, who is 15, is embarrassed to wear handmade clothes, Svetlana recently led him to believe that she bought his new jacket at the market, although she had sewn it herself. Having to wear handmade clothing is not the worst effect of poverty on children. Many children from Ozerny's poor families go to school hungry. "What can they understand, what can they learn, if the only thought in their minds is 'How can I get something to eat?'" asks an Ozerny study participant. Svetlana's oldest son has just returned from the army but has yet to find a job. She is afraid her son might take to drinking if he does not soon find something to do.

Last year, many Ozerny residents became officially unemployed when the peat factory finally closed after several years of paying wages very late or not at all. The collective farm remains, but the few people who still work there have not been paid in over a year. Economic problems have made material deprivation a key element of people's daily lives. "It's not life; it's just barely making ends meet," explains one participant when asked to define what it's like to be poor. "We sometimes don't eat bread for five days, and we are so sick of potatoes that we don't know how to ram them down our throats," says another Ozerny resident.

Most housing in Ozerny is dilapidated, mainly due to lack of regular maintenance. A poor man says, "Our apartments badly need regular maintenance. The plumbing is leaking. If the wiring gets short-circuited, there is a danger of fire. There are houses where wiring hasn't been changed in 50 years."

Some key institutions in the village, such as the hospital and the day-care center, have closed. The peat factory used to provide such services at no charge, but a few years ago, the responsibility was transferred to the municipal administration and, lacking secure funding, these services soon disappeared. A discussion group describes the terrible blow of losing the local hospital; "The old hospital was small, but it was there. You could put your mother there. Then they started building a new one, wanted to have a bigger hospital, but didn't have enough resources, and closed it altogether. Now we have to go to Novotalitsy, 50 kilometers away. They seldom agree to admit us to the Ivanovo hospital (20 km away); they say we are from a different district."

For 70-year-old Valentina, who lives alone, medical care is out of reach for several reasons: "Now I have gotten old, my eyes are poor. I have glaucoma and a bad heart but no money to buy medicines. The hospital is in town; it's a long way, and we don't have an eye doctor here. At the hospital, they tell me I need surgery, but my pension will not be enough for that. You have to bring everything with you: clothing, food, and even drugs."

Poor men and women from Ozerny say they now feel insecure about the future and find it impossible to plan ahead. "Now we care only about this day. I live today and don't know if tomorrow will come or not," says a young villager. Like many in Ozerny, old and young alike, a

19-year-old correspondence school student named Sergey sees no future for himself in the village: "This summer, after I pass my exams, I'll go somewhere to look for a job, maybe somewhere on a construction site. . . . In the village it's boring, nothing to do, only a disco on Saturday and drinking. I'm anxious to leave. You can't expect anything good here. The village is dying. But I don't know where to find a job. And I don't want to leave my mother; it'll be difficult for her to be alone."

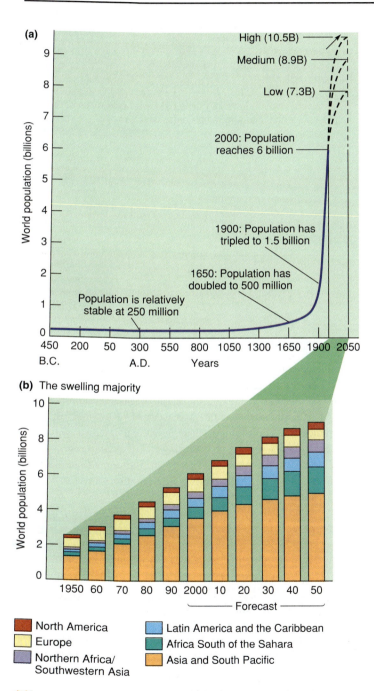

Figure 2.2 **World population growth.** (a) For most of the human occupation of Earth, population growth was slow compared to the last 300 years. The population took 1,300 years to double from 250 to 500 million, then doubled again in 200 years. In the 1900s, world population quadrupled. The high, medium, and low projections are from the United Nations, 1998. (b) When the population increase is broken down by region, all major regions outside North America and Europe show big increases. Sources: (a) United Nations; (b) Data from *The Economist,* 1999.

of Africa South of the Sahara and Latin America, the uplands of the western North America, and hilly areas in Europe and China.

• Very low densities of population reflect the most difficult environmental conditions, such as mountains or the margins of deserts and polar lands.

Other measures of population distribution relate population to resources, but the data required are not so easily available as population numbers and area of land. *Physiological density* is the population numbers per unit of cultivable land. The *population-resource* ratio is an even more sophisticated measure, based on the quantity and quality of each area's material resources and the size and technical competence of its population.

It is estimated that 97 percent of world population growth to 2050 will be in the larger, more densely populated, and often poorer countries: 35 percent of the total will be in China and India—already the world's most populous countries; another 25 percent will be in Pakistan, Indonesia, Nigeria, Brazil, Bangladesh, Mexico, the Philippines, and the USA.

At present, half the world's people live in cities, concentrated at very high densities. This proportion rises to over 70 percent in the wealthier countries. The largest cities in 1990, Tokyo (25 million people) and New York (16 million people), will soon be joined or overtaken in population totals by many cities in the poorer countries. Estimates of total city populations of over 20 million in 2015 raise Tokyo to 29 million, followed by Bombay (India, 27 million), Lagos (Nigeria, 24 million), and Shanghai (China, 23 million), while Jakarta (Indonesia), Karachi (Pakistan), and São Paulo (Brazil) may each have 21 million. New York is unlikely to reach 20 million.

Population Dynamics

Because the human population increases at varied rates and is mobile from place to place, it is necessary to understand the basics of **demography,** the study of population structure and development. Some of the major world regions, such as South Asia and Latin America, have increasing populations; others such as Europe show little change or, like Russia and Neighboring Countries, are declining. Growth or decline in the population of a place is determined by whether births exceed deaths (natural change), whether immigration exceeds emigration, and whether the overall balance between natural change and migration alters. Future population projections are notorious for being subject to unexpected events. For example, the growing threat of HIV/AIDS (see Point-Counterpoint: HIV/AIDS, p. 30) or other deadly diseases not yet known may lower future population totals, while new baby booms may raise them.

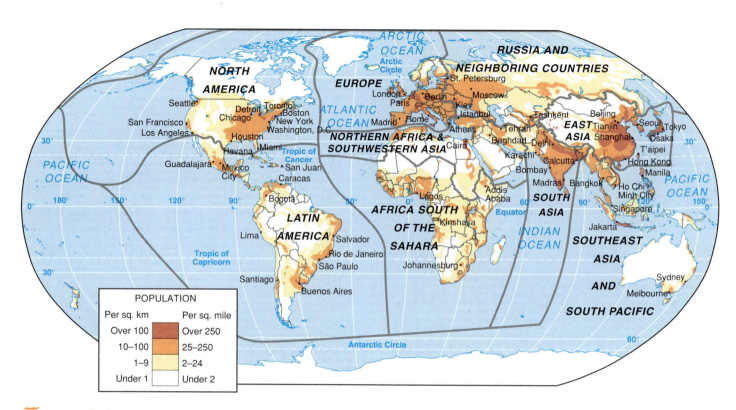

Figure 2.3 **World population distribution.** Which major regions have the highest and lowest densities of population? As you read through this chapter, look for evidence that might explain the differences.

The **birth rate** is the number of births per 1,000 habitants in a year. It is related closely to the **total fertility rate**—the average number of births per woman in her lifetime. Total fertility rates of 6 to 7 are typical of many poorer countries, while many wealthier countries have rates of 2 or below—not enough to maintain a population.

The **death rate** is the number of deaths per 1,000 inhabitants in a year. It is often broken down into age groups. **Infant mortality** (deaths per 1,000 live births in the first year of life) and child mortality (deaths per 1,000 live births in the first five years of life) are examples. Infant mortality rates below 10 (i.e., 10 infants died per 1,000 born) in wealthier countries compare with those above 100 in many poorer countries.

The combination of birth rate and death rate defines the rate of natural population change, whether it is increasing or decreasing. The process of **demographic transition,** summed up in a theoretical model of stages based on the experience of Western countries, relates birth and death rates to economic circumstances (Figure 2.4).

Typically, in today's wealthier countries, the high birth and death rates (Stage 1) for long caused little crucial population change. They gave way over time to the low birth and death rates (Stage 4) that resulted in the current phase of stable population totals. In Stage 2, death rates fell as a result of advances in medical technology, improvements in sanitation, and better education and nutrition. Continued high birth rates, however, caused rapid population growth, placing stresses on the resources needed to feed the extra mouths. In Stage 3, birth rates fell with rising incomes, expanded roles for women,

increasing education, and moves from rural to urban areas. Combined with societies responding to lower death rates by having fewer children, a balance between births and deaths at lower levels was achieved in Stage 4. Some countries have very low birth rates that fall below death rates and declining populations. Since such countries have usually passed through the other stages, this could be Stage 5. It remains to be seen whether poorer countries at present in Stage 3 will go through the later stages or whether those stages merely reflect Western experiences.

Migration is the movement of people into or out of a place. If immigration to a country or region exceeds emigration and natural change is positive, numbers of people increase. Immigration is a major source of population increase in the United States, as natural increase is slow. In the 1990s, net immigration made up at least half of the U.S. population growth.

Periods of globalization link to major migrations of people. In the late 1800s, millions of people moved mainly from Europe to the Americas and Australia. In the late 1900s, migrations occurred more widely. In 2002, 150 million people lived outside the country of their birth, just 2.5 percent of the world population, but many still played important roles in their countries. Large numbers of refugees from war-torn countries in Africa moved into neighboring countries; low- and moderately skilled workers moved from South and Southeast Asia to the Persian Gulf oil countries for work, sending their paychecks home. Highly educated men and women from poorer countries around the world moved to the United States, Europe, and Japan for better-paying jobs.

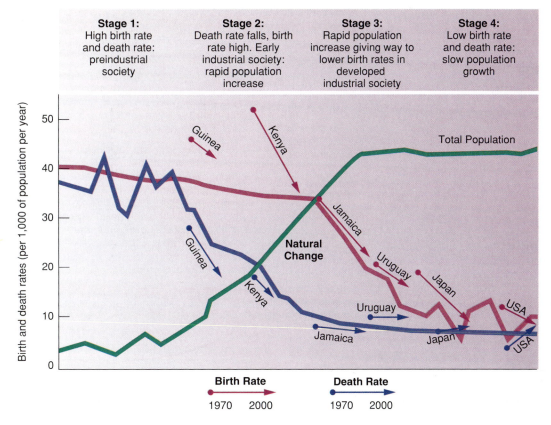

Figure 2.4 **Demographic transition.** Population change is plotted, from high birth and death rates (Stage 1) to low birth and death rates (Stage 4). All countries were in Stage 1 until the 1800s, when Western countries began to move toward Stage 4. Today many poorer countries are in Stages 2 and 3, in which population increases rapidly as death rates fall, but birth rates remain high. Each stage is linked to a phase of economic growth. Similar diagrams are drawn for each world region in this text. Notice where countries depart from this general progression. It has been suggested that there should be a Stage 5, in which birth rates and death rates are both low, but the death rate is greater than the birth rate and the population declines.

Conditions in the sending and receiving countries, the reasons for migration, and the policies adopted by receiving countries all varied and changed over time. In the late 1990s, wealthier countries that previously offered permanent residence and full citizenship rights to immigrants began to examine more closely increasing numbers of asylum claimants and offered limited-term work permits. Political problems arose from the difficulties of integrating immigrant groups with established communities.

When natural and migration changes are combined, an overall population increase of 1 percent will lead to a population doubling in 70 years. An increase of 2 percent means a doubling in 35 years; one of 3 percent means a doubling in 23 years. Wealthier countries today commonly have rates of overall population increase of below 0.5 percent, while poorer countries have rates of 2 to 3 percent that place pressures on economic resources. Countries with high emigration or low birth rates may have population losses.

The composition of a country's population is often summarized in an age-sex diagram, also termed a "population pyramid" (Figure 2.5). These diagrams provide information about the population's recent history and potential future. Migrations into the country or baby booms show up as expansions in particular age and gender groups; deaths in major wars may be reflected in a narrowing of specific age-group numbers.

If there is an expectation of a long life, the older age groups will have more members.

How Many People Can Earth Support?

Growing populations raise concerns about whether Earth and its resources can support them. In the early 1700s, the English economist Thomas Malthus predicted that world population growth would exceed that of food production, leading to widespread famine. Made at a time when rapid population increase was unusual and before industrial processes raised agricultural productivity, his prediction was not fulfilled. Nevertheless, the debate continues today as some scientists and environmentalists place lower limits on the numbers of people Earth can support, while some population experts and economists allow for higher numbers in view of technological, political, and cultural changes. The lack of certainty arises from different assumptions made in calculating Earth's resources, assessing lifestyle requirements, and projecting the numbers of people to be supported.

The ability of world regions and the countries in them to support projected growth in population is not easy to determine. For each region in Chapters 3 through 11, we raise some of the issues. The answer to the question of how many people Earth can support relates to many complex factors in human development. Some of

Point COUNTER Point

HIV/AIDS

HIV/AIDS is a major threat to world health and especially to millions of people in poorer countries, where 90 percent of infections occur. Life expectancy is reduced, and there is a threat of expansion to the rest of the world. First recognized in wealthier countries, HIV/AIDS is now a major plague in southern Africa and is being recognized in the rest of the poorer world.

Only around 1980 was HIV (the human immunodeficiency virus) discovered to cause AIDS (acquired immune deficiency syndrome), which fatally lowers the body's immunity to disease. By 2000, 30 million people had contracted HIV and 6 million had died of AIDS. People contract HIV through unprotected sexual contact with HIV carriers or through contact with HIV-contaminated blood or body fluids (but current medical research suggests not by other contacts with HIV carriers). HIV infections can be passed from mother to baby. Patients become prone to many other sexually transmitted diseases and to other serious illnesses such as tuberculosis (TB). Medical treatments available are complex and expensive, needing close monitoring. They do not cure HIV but can prolong life.

Use the following 2001 newspaper extracts to list the issues and opposing points of view. How do the views conflict? Where in the world is the main problem at present? Is the problem a global or world regional one? Gather further information on the problem and its development since 2001. Have things changed? What can or should be done, and by whom?

MELVIN FOOTE, CHINUA AKUKWE, *Zimbabwe Independent,* 6 April 2001

Imagine the reaction of the developed nations to the grim news that every American who resides in the states of Texas and Tennessee will die before the end of the decade from a deadly disease or that 80 percent of Canadians live with this fatal condition. The developed nations would be up in arms to fight this mortal enemy, with a resolve strengthened by the news that the disease had already claimed so many lives. Panic would set in if it was revealed that 90 percent of all infected individuals are unaware of their status and may unwittingly transmit the infection to other people.

This is the situation in Africa, where more than 25 million individuals live with HIV that causes the fatal disease AIDS. According to the February 2001 report of UN Secretary-General Kofi Annan, Africa has 10 percent of the world's population but is home to 70 percent of adults and 80 percent of children living with HIV infection. By 2010, 40 million AIDS orphans, children less than 15 years old, will live in Africa. In a continent where half the population lives on 65 cents per day, an unchecked AIDS epidemic will shrink its economy by a quarter. And yet, this does not attract great outrage in the West.

BELINDA BERESFORD, *Mail & Guardian,* Johannesburg, South Africa, 2001

The doctor, staring into the bottom of his coffee mug, watched a man and woman decide who was to die first. In their early 30s and

with young children, they have less than one-tenth of a healthy person's immune system between them. From their disability pension and occasional work, they could squeeze $75 per month for basic antiretroviral therapy to enable one of them to recover and raise their children for a few months.

The doctor knew this was futile. If either died, the disability pension would be reduced and there would be no chance of raising the $75 per month. Also, the price depends on drug companies keeping to their low-price promises—that had not yet been put into action by delivering the drugs to doctors. Some offers were for a limited range of countries and not South Africa. The drug price does not, however, cover all the costs of treatment to keep a person safely and healthily on antiretroviral drugs. Multiple tests several times a year and visits to doctors are costly. The treatments often have side effects that require additional treatments.

The uncertainties and complexities are too much for patients close to death, who are tempted to self-medicate. If so, they will waste their money or damage their health further. They will contribute to rising levels of drug resistance that means HIV will become uncontrollable.

ARTICLE IN THE *Toronto Star* (Canada), REPRINTED IN *World Press Review,* June 2001.

Dr. Yusuf Hamied, chair of Cipla (Indian generic pharmaceutical manufacturer), after announcing it would sell a three-drug AIDS cocktail for an at-cost $350 to the group Doctors Without Borders (*Médecins Sans Frontières*) and $600 to African governments, said, "There is a holocaust in Africa, it's my social obligation to society." It was also "a way to break the stranglehold of the multinationals." [Note: The Indian firm has been trying to break through the high prices charged by Western pharmaceutical corporations and sees this as its chance.]

Jeff Tyrwhitt, spokesman for 80-member Pharmaceutical Research and Manufacturers of America: "This is a complex problem with many factors, but that message doesn't seem to have sunk in. Whatever we do, like cutting prices, our critics still say it's too little, too late." The overriding hurdle is the lack of medical infrastructure in much of Africa and therefore the lack of a drug-delivery system. There are too few doctors, with or without AIDS knowledge, not enough clinics, drug-monitoring labs, or storage facilities. In the absence of public education, there is a perpetuation of high-risk behavior and self-medication that may lead to drug resistance (as happened with TB). Although Africa accounts for only 2.5 percent of the industry's $200 billion global sales, Tyrwhitt claims that "half the drugs sold here will never make it to patients."

The pharmaceutical industry contends that its high prices do not reflect production costs but the high costs of research and development, marketing, and lobbying: 95 percent of the drugs developed by North American companies fail and do not recoup costs. Without patent protection, there is no incentive to develop new drugs. Bob Huber, spokesman for Pfizer, Inc., of New York, wonders how several African nations justify buying armaments when they provide so little health care for their people. "Africa's defense ministers just met to see where they could get the cheapest arms," he said. U.S. companies fear that if prices drop dramatically in the developing world, pressure to reduce the cost of AIDS drugs—indeed, all drugs—back home will follow.

Three companies—the U.S.-based Merck & Co. and Bristol-Myers Squibb, and Britain's GlaxoSmithKline—have dropped their previous reduction even further from the May 2000 proposal to cut the price by 80 to 90 percent from $6,000 per person per year to $1,000—still beyond many countries' reach. They propose $600 per drug. But in Merck's case, it wants African governments to guarantee that the drugs will not be reexported to other countries.

A legal case was tried in South Africa in April 2001 in which the major drug companies challenged former President Nelson Mandela's ruling that allowed the import of generic drugs. The companies referred to the World Trade Organization policing of the international treaty on intellectual property, including patents. South Africa could have declared a national AIDS emergency but backed away from this—although it would have meant donation of the drugs rather than the compulsory licensing of cheaper drugs. [President Mbeki denied a link between HIV and AIDS.] The legal case began on 7 March 2001 with a statement by Judge Bernard Ngoepe that it was a landmark case about patent rights and intellectual property but also about life and death. The case became a public relations fiasco and was adjourned until 18 April. On 19 April, the companies withdrew their suit completely but asked the health ministry to consider a law regarding compulsory third-party licensing on patented medicines.

Greg Hartl of the World Health Organization in Geneva says, "We don't vilify the industry here (a point for which WHO has been criticized). There are others who should be in this fight." WHO needs $5 billion a year in new money to combat AIDS and other dangerously resurgent infectious diseases such as malaria and TB. It wants Western countries, foundations, and even private donors to supply it. It seeks a differential pricing system that the industry has long refused. If agreed, mechanisms could be put in place to prevent the reexport of drugs to richer countries. "African governments have to do their part, too. We're trying to go back to the beginning and get them to work on prevention. We're trying to explain that spending money on health care is an investment in their future, a prime factor in economic development."

And yet, there is some optimism and activity at WHO in Geneva, following a lack of progress up to early 2000. Then there was the first UN Security Council debate on the issue and the G-8 summit that outlined the crucial need to resolve the crisis. Now there is a lot of activity behind the scenes.

Dr. Mark Wainberg, former head of the International AIDS Society and now director of McGill University's AIDS Program, thinks the industry has been its own worst enemy in not anticipating the pricing crisis and acting preemptively by lowering costs. But, he adds, many Third World countries have skewed spending priorities. "You can't whitewash anybody in this crisis. Ethiopia says its number-one enemy is Eritrea, India says Pakistan, when in both cases it is AIDS."

Dyann Wirth of Harvard University's School of Public Health is skeptical about drug company motives and their devotion to developing countries. She comments that only 13 of the 1,223 new drugs patented between 1975 and 1997 were for infectious diseases. Yet 6.1 million people died in 1998 of TB, malaria, and respiratory illness. A recent study of 24 multinationals found not one working on a much-needed new malaria drug. Although a drug developed to fight cancer was found to be effective in reviving victims from the terminal coma of sleeping sickness (earning the title "resurrection drug"), it was discontinued in 1999. Aventis, the company, then discovered it could be used profitably as Vaniqua, a female facial-hair remover, and production resumed.

Dr. Michael Schull heads the Canadian branch of Doctors Without Borders, an agency with AIDS projects in 40 developing countries. He dislikes the oversimplified notion of good (his own organization) versus evil (the industry). "There are people out there who want to break down international trade mechanisms, want the companies to give the drugs for free out of altruism. We're not part of that. What I'm concerned about is access to essential drugs." He is clear that if drugs aren't affordable, even the best health system can't tackle the disease. But he agrees that the crisis isn't entirely the drug companies' responsibility. He is disgusted with the idea that Africans with deadly diseases are now to be dependent on drugs having lifestyle benefits in the West. Are AIDS victims to be left in the shadows until the haranguing over intellectual property and patent protection plays itself out?

BENJAMIN STEINBRUCH, *Folha de S. Paulo,* **São Paulo, Brazil, 6 March 2001**

In 1996, 9,600 Brazilians were victims of AIDS; in 2000, this fell to 1,200. That does not mean the disease is under control but that something important is being done.

The Brazilian program to fight AIDS has been considered the best in the world by the international community. It worked because the government made the decision to treat the ill. All people infected with the HIV virus receive the cocktail of drugs and training in how to take it effectively. About 100,000 people take the drug cocktail daily, costing approximately US $163 million per year—a large sum, but it would be US $742 million if Brazil paid the high world market prices.

Brazilian government labs produce the majority of the drugs in the cocktail in Brazil; only two of the 12 drugs are still imported at high prices, costing 36 percent of the total AIDS program budget. It is now planned to produce those two in Brazil as well. Brazilian law authorizes local production of any drug in cases of public advantage or when the lab holding the patent does not produce the medicine in Brazil. Under international law, the country can carry out "compulsory licensing" of those two drugs, which means that the pharmaceutical companies will be paid for temporary use of the formula at prices that the country considers fair.

The United States government petitioned the World Health Organization and World Trade Organization to decide if Brazilian legislation is legal internationally. But victory for the United States would make the Brazilian AIDS program unfeasible and would discourage the entire Third World, which sees Brazil as a model to be copied.

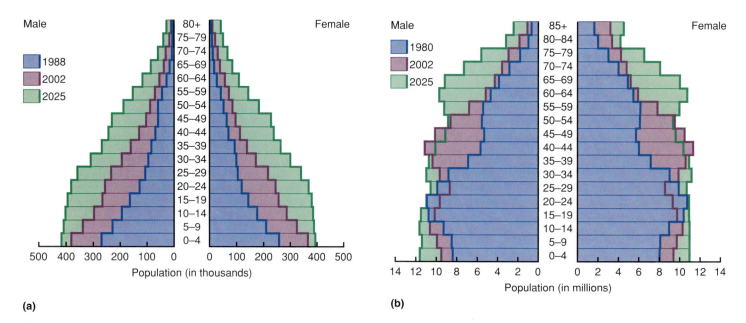

Figure 2.5 **Age-sex diagrams (population pyramids).** Diagrams for three years are overlaid to show changes. In each case, the bars represent a 5-year age group (male and female). Total numbers of people are used, rather than percentages of each age group, to allow comparisons over time and place. (a) Papua New Guinea. This shows a typical poorer country with large numbers of young people and fewer old—with increasing numbers in middle age groups by 2025. (b) United States, a typical wealthier country with a more even spread of numbers in each age group and a baby boom (1950–1965) moving upward through the age groups. These diagrams occur in each subregion of Chapters 3 through 11. Source: U.S. Census Bureau. International Data Bank.

the contributing factors also link to human rights. Asking a set of questions like the following illustrates the difficulties:

- *Economic questions* include:

    What level of well-being is expected? A small proportion of materially very wealthy and many poor, or increasing numbers of people with moderate incomes?

    What levels of technology will be used in growing food, manufacturing goods, and providing services?

- *Cultural questions* include:

    How will average family sizes change?

    What support will be provided for young and old?

    How tightly are people wedded to current habits?

    Are people willing to adopt new lifestyles that might include vegetarian diets, cycling to work, and spending more tax money on schools and health care?

    Can people be forced into adopting new lifestyles, in part by the pressures of globalization?

- *Political questions* include:

    What sort of political system might resolve conflicts among and within countries?

    Will organized violence continue to waste human lives and resources?

    How will domestic and international trade arrangements work out?

- *Natural environment questions* include:

    Do people consider it sufficiently important to maintain a clean environment with conserved wilderness areas that they will alter their demands for cheap and plentiful food?

    How much natural-hazard risk can people accept?

What changes will global warming make?

How long will any predictions last, given uncertainties over the usage of such resources as water and fish stocks?

If all countries consumed resources at the rate of the United States, the world would already be overpopulated. As it is, the United States, with 5 percent of the world's population, consumes half of the world's oil and large proportions of other resources. A cartoon in the *Miami Herald* at the time of the 1992 Rio de Janeiro Environmental Summit showed Uncle Sam telling representatives of poorer countries, "It's a deal. You continue to overpopulate the world, while we squander the natural resources."

We all have choices. Do we belong to the "bigger pie" school, which proposes expanding production through applying more technology in such areas as genetically modified foods, additives, plastics, synthetics, and alternative fuels? Or to the "fewer forks" school, which emphasizes environmental considerations and the slowing, stopping, or reversing of population growth? Or to the "better manners" school, which highlights cultural values as a source of improving the terms on which people interact? Choices of what we eat and wear and of lifestyles will determine the future of humankind on Earth.

🌐 Issues of Economic Inequality

World wealth and poverty are the subjects of economics—how scarce goods and resources are produced, distributed, and consumed. **Economic geographers** study the spatial patterns of production, distribution, and consumption of goods and services.

At present, the distribution of wealth is uneven, as the World Bank map in Figure 2.6 makes clear. For some Americans, perched near the top of the global economic pyramid, the issues of poverty and deprivation may seem quite distant. However, the continuation of American well-being is likely to depend on the encouragement of greater material wealth in the rest of the world. The huge numbers of poor people constitute the greatest economic problem facing our world today. Global connections make them more aware of the differences between their lot and that of wealthier countries. World prospects for expanding people's choices depend first on reducing the extent of poverty.

In contrast to the large numbers of poor people in the world, there are relatively few extremely wealthy people. It is estimated that in 2001, the world had 7.2 million people (0.0012% of the world total population)—up from 5.2 million in 1997—who had investable assets of more than $1 million and controlled one-third of the world's wealth. The 1997 estimate listed 425 billionaires, of which 274 were in the United States. From 1997 to 2000, the numbers of millionaires rose sharply in the United States and Europe, less rapidly in Asia, and hardly at all in Latin America, the Arab world, or Africa. From 2000, however, many of those enriched by high-tech-related and overvalued stocks lost considerable wealth, showing how volatile the system can be.

Those who are poor wish to share the well-being and lifestyles of wealthier peoples but are not impressed by the efforts of wealthier peoples to encourage them to catch up. When the general extent of world poverty first became evident in the 1800s, the European colonizing countries saw it as their mission to "civilize" the rest of the world and so improve the lot of others. From the early 1800s to the mid-1900s, many colonies became independent and had aims to become "developed" according to a Western formula. Some in poorer countries see globalization as a further ruse for drawing the rest of the world into a Western-dominated economic system.

Economic Worlds

In the 1990s, the end of the Cold War and the collapse of the Communist political-economic system left the **free-market system** (capitalism) dominating the world. The free-market system is based on the economic decision-making capacities of individuals, who may choose from a range of products to meet their needs. Those who fulfill these needs—and sometimes create them through advertising—invest financial "capital" with the aim of making profits. They "buy" labor and machinery to produce salable goods at the lowest cost. Competition among small firms, large corporations, and countries is an essential feature of the system. This system has marked the Western countries for over 200 years and involves the private and corporate organization of investment, production, and marketing. Unfortunately, the concept of an "economic man" who considers all possible information and makes soundly judged decisions on investments does not work out in practice. Fallible humans invest, run companies, and generally perform roles to the best of their ability and often take advantage of weaknesses in the system. Even in countries

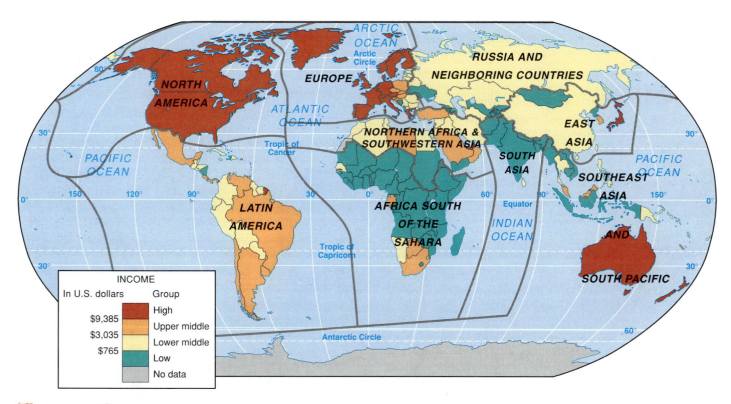

Figure 2.6   **Major income groups: World Bank.** These are based on GNI per head (2000). Relate the distribution of countries in the four categories to the major world regions. Source: Data from *World Bank Atlas*, World Bank, 2002.

in the system. Even in countries with well-regulated economies, major corporations such as Enron and Kmart may crash and create personal catastrophes for employees and customers.

In theory, governments intervene in free-market economies mainly to regulate the terms of trade and ensure "fair play" among producers. In practice, government-based decisions on what trade should happen and what is fair are not always framed or carried out in the public interest. In matters as the taxing of imported goods, political considerations favoring some groups of people influence legislative judgments. Furthermore, the governments of many wealthier countries provide the social services and building infrastructure (roads, airports, harbors, water supplies, waste disposal) that give businesses and people in those countries many cost advantages compared to the poorer countries.

Central Planning

The basis of economic opposition to the free-market Western countries during the Cold War was the Communist centrally planned economic system adopted by the former Soviet Union, its satellite countries, the People's Republic of China, and linked countries such as Cuba. This system places planning and decision-making responsibilities in the central government, on the grounds that the country's interests come first and the central ministries know what is best for the people. Central governments provide medical care and education to support a fit and able workforce, develop strong military defenses, and plan the production of goods considered essential—whatever the cost. For many people, the system provided welcome education, health care, jobs, and housing.

Those in command of centralized policy-making, however, often made large-scale mistakes, handicapped even more than in the free markets by a lack of information or just as much by personal bias or interest. Many leaders feared to change past policies, even if inefficient or oppressive, while regional bureaucrats often obeyed central commands despite knowing the policies would fail. Oppression of political dissidents and restrictions on travel paralleled overproduction of some goods or underproduction of others. These countries failed to produce the consumer goods or possibilities of tourist travel options available in most Western free-market countries. Incomes for most families remained modest, and only the Communist Party hierarchy did better.

Free Market for All

In 1991, the Soviet Union broke up because of the internal and external pressures of the Cold War, comparisons with U.S. affluence, political pressures from the United States, and an internal lack of political freedom. Russia abandoned its integral economic relationships with the countries of the Soviet bloc in eastern Europe and entered the world economic system. Already, from 1978, the People's Republic of China had increased trade with other countries and encouraged investment from them. This brought China high levels of economic growth through the 1980s and 1990s. Most former Communist countries—and those aligned with them—encountered a traumatic transition to the totally different free-market, now global, economic system. Western countries encouraged the transition to friendly democratic governments and increased international trade. The expansion of trade among resource and market areas further strengthened the economies of the Western countries, as well as their multinational corporations.

Global Economic Organizations

The world economic system is bolstered by world organizations that lend money and seek to encourage international trade. The World Bank and International Monetary Fund (IMF), both based in the United States, give loans in exchange for a country agreeing to open its internal economy to external investment and foreign goods, and to reduce its government bureaucracy. Private commercial banks and nongovernmental aid agencies tend to follow the same guidelines for assigning priorities to funding projects in poorer countries.

The role of the World Trade Organization (WTO) is to make trade among countries easier by reducing import and export duties. Despite the fact that all 140 member countries have a veto on decisions, the WTO gives the impression of favoring the wealthier countries and multinational corporations by allowing discrimination against imports of agricultural products and textile manufactures from poorer countries. At the same time, the United States and Europe squabble over such issues as Europe's privileged markets for banana growers in former Caribbean colonies that exclude U.S.-linked producers, the wider adoption of genetically modified foods in the United States and objections to them in Europe, and U.S. concerns over European beef. Some of these issues shift the emphasis from encouraging wider fair trade toward protecting a wealthier country's internal interests.

Regional Emphases

Although the free-market economic system prevailed after 1991, its influence and benefits were not evenly distributed, and the application of its principles varied among countries. The United States, western Europe, and Japan controlled most of the investment, production, and consumption of goods, although by 2000 there were signs that other centers might emerge in China, India, Brazil, and South Africa.

Further, distinctive variants of capitalist economies developed. The "Asian Way" builds on family linkages connected to government-business liaisons, rather than on the independently verified banking and legal systems that are basic to free-market economies in Europe and the United States. The "European Way" makes much of providing social welfare to support those who are not able to benefit from or exploit the capitalist system. Countries in Latin America moved out of self-sufficient economic systems and into the world system with some pain. And yet, despite the growing influence of the world economic system, small groups of peoples living in such isolated areas as the Amazon rain forest and Papua New Guinea, as well as increasing numbers in African rural areas, engaged little, if at all, with it.

Measuring Wealth, Poverty, and Human Development

In attempting to give more precise meanings to wealth and poverty, varied indicators form a common basis for comparing and understanding differences of human development.

The ownership of consumer goods is a vivid indicator of differences in material wealth among countries (Figure 2.7) but focuses on a limited economic sector—the availability of income to spend on poor people's luxuries and wealthier people's normal expectations.

Economic development is commonly measured by two statistics of income that are widely reported. **Gross domestic product** (GDP) is the total value of goods and services produced within a country in a year. Gross national product (GNP), now called **Gross national income** (GNI), adds the role of foreign transactions: it is the total value added from domestic and foreign sources. Per capita figures of the country's total annual income are averages of GDP or GNI per head of the population, not personal incomes. Figure 2.8 compares GDP and GNI with other measures of human development.

The divisions shown on the map in Figure 2.6 are based on GNI per capita, with the World Bank dividing countries into four income groups: low, lower middle, upper middle, and high. The map demonstrates the uneven distribution of the world's GNI among countries.

GDP and GNI values are based on local currencies and converted to U.S. dollars at official exchange rates. Official exchange rates, however, may not reflect the comparable costs of living in a country. The **purchasing power parity** (PPP) estimates of GNI and GDP are more faithful comparisons of living costs among countries. Because prices in India, for example, are much lower for equivalent items you might buy in the United States, US $440 will buy as much in India as US $2,230 does in the United States. Countries with high incomes and high living costs have a lower PPP estimate of income than the GDP or GNI based on exchange rates; poorer countries often have higher esti-

mates. In 1998, Switzerland had a GDP per capita income of $39,980 but a GDP per capita PPP estimate of $25,512; Mexico had comparable values of $3,840 and $6,041. Figure 2.9 shows that over 60 percent of world wealth is produced by 20 percent of the world's population, while just over 20 percent of world wealth is produced by 60 percent of the population.

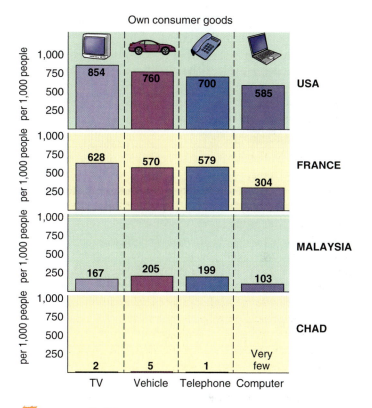

Figure 2.7 **Ownership of consumer goods.** Numbers of goods per 1,000 people in specific countries. Contrasts in affluence occur among wealthier countries (United States, France), countries with middle incomes (Malaysia), and poorer countries (Chad). Similar diagrams occur in each chapter. Source: Data (for 2000) from *World Development Indicators,* World Bank, 2002.

Figure 2.8 **World region comparisons: index data.**

The income per head is based on gross national income purchasing power parity in U.S. dollars averaged across each region. The human development index and gender development index ranks and the human poverty index are also averaged for each region.

World Region	Gross National Income Purchasing Power Parity Per Capita Average 1999 $	World Human Development Index Rank Average (162) 1999	World Gender Development Index Rank Average (146) 1999	Human Poverty Index Average 1999 % of Population
Europe	19,306	26	24	11.7
Russia and Neighboring Countries	4,197	86	72	no data
East Asia	16,648	38	35	15.0
Southeast Asia and South Pacific	8,957	73	61	19.2
South Asia	1,893	127	106	37.4
Northern Africa and Southwestern Asia	5,073	90	79	21.9
Africa South of the Sahara	1,745	150	123	41.1
Latin America	6,200	71	63	14.3
North America	28,600	2	2	15.0

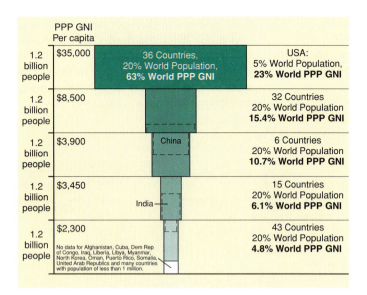

PPP GNI
Per capita

1.2 billion people	$35,000	36 Countries, 20% World Population, **63% World PPP GNI**	USA: 5% World Population, **23% World PPP GNI**
1.2 billion people	$8,500		32 Countries 20% World Population **15.4% World PPP GNI**
1.2 billion people	$3,900	China	6 Countries 20% World Population **10.7% World PPP GNI**
1.2 billion people	$3,450	India	15 Countries 20% World Population **6.1% World PPP GNI**
1.2 billion people	$2,300	No data for Afghanistan, Cuba, Dem Rep of Congo, Iraq, Liberia, Libya, Myanmar, North Korea, Oman, Puerto Rico, Somalia, United Arab Republics and many countries with population of less than 1 million.	43 Countries 20% World Population **4.8% World PPP GNI**

Figure 2.9 **Uneven distribution of world wealth.** The 2000 population was divided into five groups with equal numbers. Countries were linked by purchasing power parity gross national income. What striking information does this diagram provide? Information about each country's position is included in each of the regional Chapters 3 through 11. Source: Data for 2000 from *World Development Indicators*, World Bank, 2002.

The United Nation's **human development index** (HDI) is a broader estimate of human development, incorporating statistics calculated from life expectancy, education attainment, and health, as well as income. Poorer countries investing heavily in education and health care, such as Costa Rica and Sri Lanka, provide a better quality of life for their people and have a higher HDI than GDP (merely based on a country's income) rank. By contrast, many of the oil-rich Persian Gulf countries have high income rankings based on oil exports but lower HDI rankings because of poor provision of schooling, especially for girls.

As the emphasis in thinking about development shifted toward the needs of the poorest people, the United Nations introduced the **human poverty index** (HPI). Linked to the HDI, it focused on levels of deprivation among the least favored groups in society. The indicators used strike balances between individual material poverty and public provision for such needs, and between relevant and available data. For example, the percentage of people expected to die before age 40 indicates vulnerability to death (Figure 2.10); the percentage of illiterate adults indicates restrictions on entry to better jobs and full community life; and a combination of percentages of people without access to health care or safe water and of malnourished children under age 5 indicates a lack of decent living standards.

The HPI records the proportions of populations affected by such deprivations. Values range from around 10 percent in Cuba, Chile, and Costa Rica to over 50 percent in many African countries and Cambodia in Southeast Asia. Contrasts also occur within countries: in China the coastal regions have HPI values of 18 percent, while areas in the remote interior have values of 44 percent.

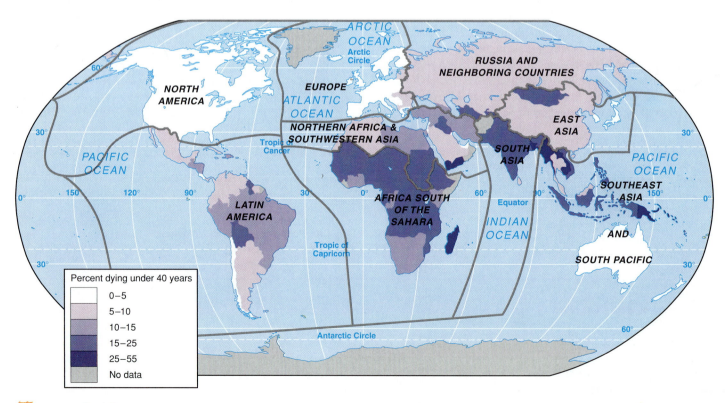

Figure 2.10 **Early death.** This map depicts the proportion of people not reaching 40 years of age for each country. How do the figures relate to major world regions? Source: Data for 1990 from *United Nations Human Development Report*, United Nations, 1997.

Making Development Happen

No simple answers are available to answer the questions of what policies lead to economic or human development and how poor countries become materially wealthier. Several attempts were made in the second half of the 1900s to generate such development but met varied levels of success. Each had its geographical origins and impacts.

Modernization

The Western countries, such as Europe and the USA, experienced economic growth in the later 1800s and through the 1900s by developing wealth from new economic sectors in a process known as **modernization.** The production and trade in **primary sector** goods include outputs from the natural environment (mining, fishing, timber-cutting, and farming). Greater wealth (measured by the value added) comes in the **secondary sector** from converting such products through manufacturing into others of greater value. The **tertiary sector** covers the distribution and consumption of goods produced, including retail and wholesale transactions and the work of business services and government. The **quaternary sector,** emerging from the tertiary sector, includes the growing roles of specialist industries that emphasize information services, including the professions, finance industries, education, media, and government: such industries also create new wealth and high-salaried jobs.

The world's wealthier countries gain more wealth and jobs from the most profitable sectors and have more diversified economies than the poorer countries (Figure 2.11). They have a greater prominence in the new, high-value manufacturing industries such as airplane building, electronics, and pharmaceuticals, as well as in the high-profile business services.

From the 1950s, economic progress policies recommended a passage through the sectors from primary to secondary and tertiary as a formula for modernization. The expansion of manufacturing industries was presented as an essential time of "takeoff" for a country's economy, so countries invested in manufacturing.

The formula worked in a few countries, particularly those, such as Japan and South Korea, that had U.S. help through war recovery investment. Other Asian countries, such as Taiwan, Malaysia, Thailand, and Singapore, also invested in manufacturing and service industries and experienced rapid economic growth. Conditions were different, however, from those that had enabled Western countries to develop the different sectors. In particular, by the mid-1900s, the Western countries themselves dominated the world economy, making it harder for others to emulate them. The East Asian successes in this process used the powers of strong, single-party governments to maintain import taxes while investing in export industries with likely growth. They achieved huge trade surpluses. By contrast, places where traditional cultures predominated and colonial powers had restricted manufacturing, as in much of Africa, appeared unlikely to "take off."

Self-Sufficiency and Import Substitution

Another approach led countries to develop their own manufacturing sector protected from imports and the need to compete in world markets. Some countries adopted this approach in response to being cut off from Western markets in the Great Depression (1930s) and during World War II (1940s). Other countries, such as India, adopted this strategy on becoming independent as a sign of moving away from their past colonial connections into self-sufficiency. Called **import substitution,** the strategy was most successful in larger countries, such as Brazil and India, where the size of the internal market could support a range of local manufactures. Even in Brazil and India, however, protection often resulted in uncompetitive and poorly managed nationalized industries with high costs arising from the political necessity of employing large numbers of workers.

Central Planning Prescription

Other alternatives emerged with the aim of development outside the free-market system. Critics among Western economic development thinkers and in the Communist governments of the Soviet Union and China claimed that the Western free-market

Figure 2.11 World region comparisons: economic sectors.

The role of each sector is assessed by using the value added in each as a proportion of each country's gross domestic product and averaged for the region.

World Region	Agriculture, etc. (Primary)	Industry (Secondary)	Services (Tertiary)
Europe	5	28	67
Russia and Neighboring Countries	23	31	46
East Asia	8	36	55
Southeast Asia and South Pacific	22	33	45
South Asia	29	24	47
Northern Africa and Southwestern Asia	16	34	50
Africa South of the Sahara	32	25	43
Latin America	12	30	57
North America	2	29	39

system made poorer countries dependent on wealthier countries. Trade relations tied them to selling their products to the wealthier countries and, in return, forced them to buy machinery and other sophisticated goods from those countries. This process enabled the wealthier countries to grow at the expense of the poorer: they paid low prices for the raw materials and charged high prices for their own manufactures. And yet, the Soviet bloc, China, and countries aligned with them found that their central planning often failed on a massive scale and their attempts to be self-sufficient outside the Western economy deprived their citizens of the consumer goods they envied.

Marginalization and Dual Economies

Among the poorer countries of Latin America, Africa, and Asia, whole sections of the population were marginalized by poverty, living without hope in huge shantytowns of makeshift homes on the edges of society, often without clean water, electricity, or sanitation. Their experiences provided powerful evidence that it was not possible for all sections of a population to gain from modernization.

In Indonesia, for example, society consisted of dual economies. Traditional rural Asian ways based on farming and crafts conflicted with the new urban-industrial and materially wealthy Westernized lifestyles that many migrants to the towns wished to attain.

In many poorer countries, distinctions emerged between the **formal sectors** of an economy, in which people worked for set wages and paid taxes, and **informal sectors** ranging from unlicensed street trading to small-scale manufacturing and squatter housing. Those engaged in some informal activities, such as trash scavenging, often earn more than formal factory workers. They live in shantytowns, but these can be the rural people's transition into the urban economy. Although the informal sector was often illegal in its activities and use of land, governments began to sponsor it as a means of encouraging self-assistance. They provided building materials in the shantytowns, added infrastructure of power, water, and sewage lines, and introduced some schooling and health care.

Structural Adjustment

In the later 1980s and 1990s, a debt crisis hit many poorer countries, induced by rising interest rates on loans to them. The major world institutions lending to poorer countries, including commercial banks, the IMF, and the World Bank, insisted on new demands, known as **structural adjustment.** Poor governments in debt were encouraged to establish more efficient governments and become involved in world markets. These programs demanded export-based industries, more open trading, privatization, and reduced government involvement in the economy. In the later 1990s, debt forgiveness was linked to such measures. As with other measures, some countries found this approach helpful, but most struggled with political fallout arising from the changes that put many government workers out of jobs.

Technological Innovation

In the 1990s, a new division of the world became apparent—by technological innovation and transfer (Figure 2.12). Around 15

percent of Earth's population living in Canada and the USA, western and northern Europe, Italy, Japan, South Korea, Taiwan, and Australia is responsible for nearly all the technological innovations. They benefit most from the application of new technologies. A further 50 percent of the world's population is able to adopt the technologies in production and consumption. The remaining 35 percent of the world's population lies outside these two zones, is "technologically disconnected," and is often caught in a trap of poverty, disease, low agricultural productivity, and environmental decline: they need technological solutions they cannot afford. These differences indicate a need for new technology-transfer goals to encourage development.

Land Development Banks and Complementary Webs

Frustrated by an increasing knowledge that they are being left behind economically and by the lack of help from their country governments at all levels, many groups of people in poorer countries set up microcredit banks. They built on the experience of the Grameen ("Village") Bank in Bangladesh, which grew out of a program of small individual loans by Professor Mohammad Yunus beginning in 1974 and was formally established in the mid-1980s. Small amounts of money enabled craft workers and others to establish small businesses that helped them emerge from poverty. By 2001, the Bangladeshi Grameen Bank had 1,170 branches with 11,500 staff serving 2.4 million borrowers in over 40,000 villages. Each working day, the bank collects an average of $1.5 million in weekly installments. Of the borrowers, 94 percent are women, and 98 percent of the loans are repaid on time. These methods are now applied in 58 countries, including wealthier countries such as the United States and Canada.

Summary

Figure 2.13 summarizes some of the resources and processes involved in human development. It shows how the nature of government policies may strongly affect the outcome. The influence of internal and external forces will also have important effects on creating and remaking regions within countries.

The results of countries trying to develop for their future are uneven. During the 1990s, East Asian economies grew rapidly but suffered a downturn near the end of the decade, from which most have since recovered. The countries of South Asia, led by India, experienced moderate economic growth through the 1990s. By contrast in this decade, the countries in Africa South of the Sahara, most Arab countries, and the former Soviet bloc of countries in eastern Europe experienced a fall in economic well-being, while Latin American countries slowed in their advance. Even in countries where economic growth was rapid, many people fell below poverty lines.

Several positive results emerge from the experiences of development policies, refuting those who see all previous efforts at development as attempts by the wealthier countries to dominate others. Some voices stress the need to understand local conditions and culture in order to redefine what people and their communities and environments need, can cope with, or hope for. The need to promote grassroots interests and pay greater atten-

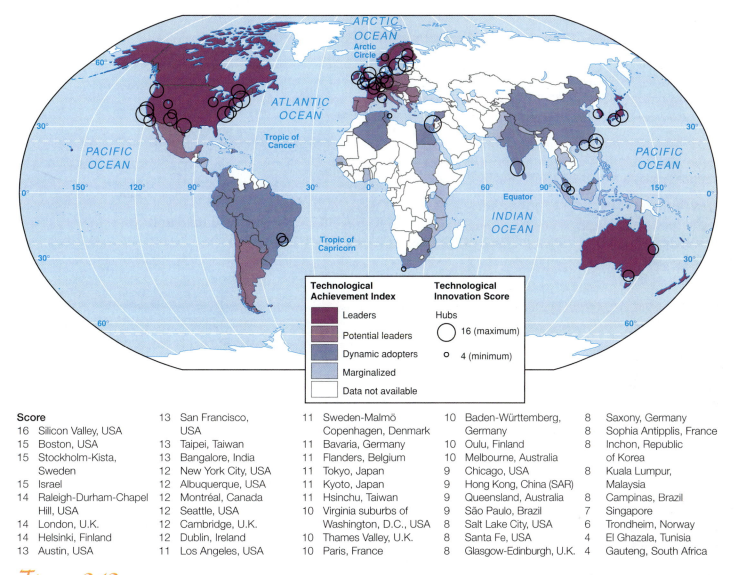

Score				
16 Silicon Valley, USA	13 San Francisco, USA	11 Sweden-Malmö Copenhagen, Denmark	10 Baden-Württemberg, Germany	8 Saxony, Germany
15 Boston, USA	13 Taipei, Taiwan	11 Bavaria, Germany	10 Oulu, Finland	8 Sophia Antipplis, France
15 Stockholm-Kista, Sweden	13 Bangalore, India	11 Flanders, Belgium	10 Melbourne, Australia	8 Inchon, Republic of Korea
15 Israel	12 New York City, USA	11 Tokyo, Japan	9 Chicago, USA	8 Kuala Lumpur, Malaysia
14 Raleigh-Durham-Chapel Hill, USA	12 Albuquerque, USA	11 Kyoto, Japan	9 Hong Kong, China (SAR)	8 Campinas, Brazil
14 London, U.K.	12 Montréal, Canada	11 Hsinchu, Taiwan	9 Queensland, Australia	7 Singapore
14 Helsinki, Finland	12 Seattle, USA	10 Virginia suburbs of Washington, D.C., USA	9 São Paulo, Brazil	6 Trondheim, Norway
13 Austin, USA	12 Cambridge, U.K.	10 Thames Valley, U.K.	8 Salt Lake City, USA	4 El Ghazala, Tunisia
	12 Dublin, Ireland	10 Paris, France	8 Santa Fe, USA	4 Gauteng, South Africa
	11 Los Angeles, USA		8 Glasgow-Edinburgh, U.K.	

Figure 2.12 **Global hubs of technological innovation.** In 2000, *Wired* magazine consulted local sources in government, industry, and the media to find the locations that matter most in the new digital geography. Each was rated from 1 to 4 in four categories, giving maximum scores of 16 and minimum of 4: the ability of area universities and research facilities to train skilled workers or develop new technologies, the presence of established companies and multinational corporations to provide expertise and economic stability, the population's entrepreneurial drive to start new ventures, and the availability of venture capital to ensure that the ideas make it to market. Forty-six locations were identified as technology hubs, shown on the map as black circles. Source: *United Nations Human Development Report,* 2001.

tion to women's issues, and concerns for the natural environment are prominent themes arising out of these experiences.

The Global Economy

While most of the economic aspects affecting human development and human rights operate at the country level, there are increasing trends toward global exchanges and institutions, helped by the distance-reducing roles of improved transportation and communications that speed people, information, and money around the globe. International airlines, courier services, electronic mail, and videoconferences accelerate personal information transfer. As the global connections increased, the world economy began to reduce poverty levels and economic inequalities.

Multinational Corporations

Multinational corporations (MNCs) make goods or provide services for profit in several countries but direct operations from a headquarters in one country. The term *transnational corporation* is often used instead of MNC to refer to corporations that are no longer rooted in a single country. The greater ease of travel and telecommunications contacts, together with the Internet transfer of information, encouraged MNCs to expand in the later 1900s.

Multinational corporations place production facilities in other countries to take advantage of cheaper labor, land, and energy and sometimes of less stringent environmental laws. For example, auto manufacturers spread the manufacture of components across several countries to ensure supplies during labor strikes and to react to local needs (Figure 2.14). The earliest multinational corporations were of United States origin, but

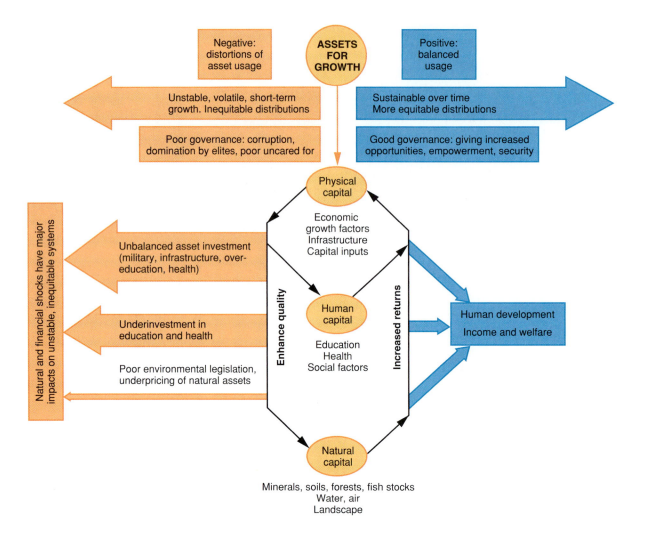

(a)

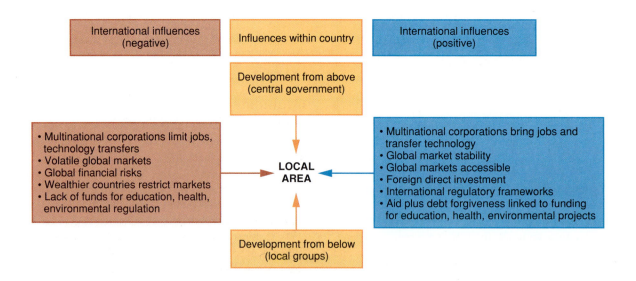

(b)

Figure 2.13 **Aspects of human development.** (a) The importance of government in the wise or wasteful use of assets. (b) The importance of internal and external country influences in the development of regions.

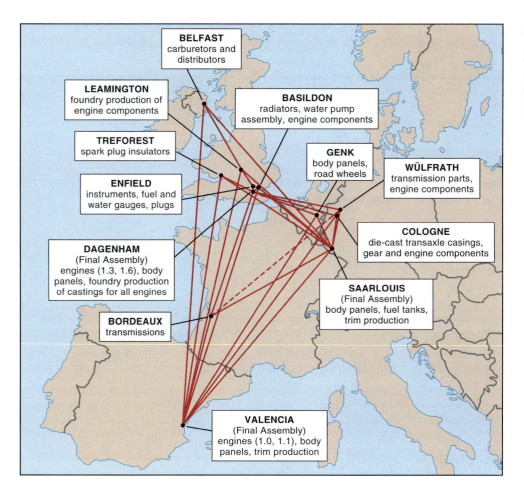

BELFAST
carburetors and
distributors

LEAMINGTON
foundry production of
engine components

TREFOREST
spark plug insulators

ENFIELD
instruments, fuel and
water gauges, plugs

DAGENHAM
(Final Assembly)
engines (1.3, 1.6), body
panels, foundry production
of castings for all engines

BORDEAUX
transmissions

BASILDON
radiators, water pump
assembly, engine components

GENK
body panels,
road wheels

WÜLFRATH
transmission parts,
engine components

COLOGNE
die-cast transaxle casings,
gear and engine components

SAARLOUIS
(Final Assembly)
body panels, fuel tanks,
trim production

VALENCIA
(Final Assembly)
engines (1.0, 1.1), body
panels, trim production

Figure 2.14 **Multinational corporation's international linkages.** The European spread of factories include makers of parts and final assembly locations for the Ford Fiesta in the late 1980s. Source: Peter Dicken, *Global Shift*. Copyright © 1992 Guilford Press, New York, NY.

those of European, Japanese, and South Korean origin are increasingly significant. Of the top 100 MNCs with the highest level of assets outside their home country, 50 are from Europe, 27 from the United States, and 17 from Japan. They produce a huge range of brands sold worldwide, including Coca-Cola, PepsiCo, Ford, General Motors, Volkswagen, Mercedes, Exxon, Shell, Toyota, Sharp, Samsung, Kellogg, Nestlé, Hyundai, and IBM. By the 1990s, multinational corporations accounted for 40 percent of all goods movements among countries. The 600 largest multinational corporations make up a quarter of the total world value added by manufacturing.

Multinational corporations are not only manufacturers. MNCs in service (tertiary sector) industries spread from the 1970s, and by the late 1990s, over 40 percent of foreign direct investments to countries were directed at them. These services include tourism and travel, data processing, advertising, market research, banking, and insurance. Some manufacturing corporations, such as the Ford Motor Company and General Motors, diversified into financial loans and credit cards. Others, such as banking, accountancy, and advertising companies, moved from national to international prominence.

With branch operations in any one country often representing a significant proportion of its labor force, MNCs wield considerable power. In the countries where they operate, some MNCs may strongly influence the local government and act as uncaring monolithic institutions without concern for the best interests of the people they employ in both home and adopted

countries. And yet, not all MNCs adopt that approach, and many transfer wealth and technology to poorer countries, provide jobs where none existed in rural areas, and pay better wages and provide better employee benefits and prospects than local companies.

The clashes between indigenous groups and multinational mining interests in Brazil and Papua New Guinea are examples of the local voices facing up to the globally connected companies. The local people tired of leaving resource development on their lands to outsiders, who may enslave them to work in dangerous conditions and may destroy their environment. Their armed rebellions aimed at making their point by closing mines.

Nongovernmental Organizations

Nongovernmental organizations (NGOs) that are not MNCs include any group of people engaging in collective action of a noncommercial, nonviolent manner and not on behalf of a government. Many have local or country bases, but the largest NGOs engage in international activities. International NGOs generally focus on delivering aid: they are often contracted to do this by governments and international agencies irrespective of country borders. Well-known NGOs, such as the International Red Cross, Oxfam, Save the Children, Greenpeace, and *Médécins Sans Frontières* (Doctors Without Borders) are better known than many smaller countries.

The United Nations gives a consultative status to NGOs at three spatial levels: "general status" is held by the largest NGOs with global influence and extensive memberships; "special status" is given to NGOs with regional or specialist functions; "roster status" is for NGOs with small memberships but highly specialist roles. NGOs working or consulting with the United Nations rose from under 500 in 1970 to over 2,000 today.

Global Financial Services

The expansion of financial services to the global scale in the later 1900s both resulted from global economic activity and acted as an enabler of it. Cross-border movements of funds increased immensely in the late 1900s. A sequence of events contributed to this expansion.

- After 1970, the breakdown of the fixed exchange rates system resulted in more frequent flows among currencies. The opening of borders to foreign investments increased the amounts of capital involved.

- The oil crises of the mid- and late 1970s, caused by the producers raising prices, led to the rapid accumulation of U.S. dollars in foreign banks. Banks holding these dollars created new financial markets, lending some to poorer countries such as Brazil, which used the loans to develop infrastructure such as roads and power dams.

- The hoards of U.S. dollars increased again in the early 1980s, when the United States ran a huge budget deficit financed by borrowing in dollars.

- While money flows in the 1970s and early 1980s were mainly from wealthier to poorer countries, the huge debts incurred by poorer countries and the rising interest rates that resulted from heavy U.S. borrowing prevented the poorer countries from repaying the debts. The late 1980s and early 1990s formed a period in which foreign investment was nearly all supplied by and used in the wealthier countries such as the United States, Japan, and the countries of western Europe (Figure 2.15). Japan, for example, invested heavily in both the United States and Europe.

- Countries such as Taiwan and South Korea avoided the debt problem of many poorer countries by forcing their own peoples to save and invest at home through a form of taxation. By the early 1990s, growing trade surpluses in Japan, South Korea, Taiwan, and Hong Kong added to the funds available for global investments.

- Banks and other financial houses expanded to service this international money explosion, beginning with American banks but followed by others in Europe, Japan, and the Arab countries. Those in New York, Tokyo, and London traded around the clock.

- Danger signs, however, became apparent. Financial markets dealt increasingly in equities (stock and shares) and risky forward contracts based on expectations of commodity production in relation to world prices. Many Asian funds based largely on loans were invested in major building projects around the world and led to overinvestment. A further negative aspect of the global scope of the economic system for many people occurred as corporate mergers increased through buyouts of firms followed by

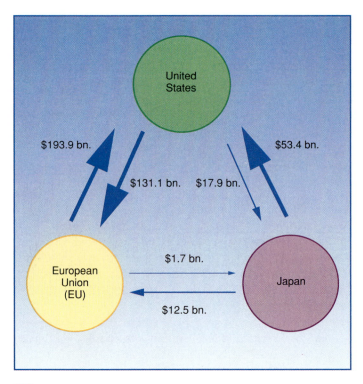

Figure 2.15   **Global investment.** The three-cornered pattern of financial flows in the late 1980s, concentrated on the world's wealthiest areas. Previously, much investment went to materially poor countries, and that began to increase again in the 1990s. Source: Adapted from P. Knox and J. Agnew, *Geography of the World Economy*, 1994. Copyright © 1994 Edward Arnold Publishers, Ltd., London.

the closure of subsidiaries, work forces being paid off, and assets sold.

- In the late 1990s, the momentum set off by increased financial flows slowed for a while when the loans, particularly from Japanese and South Korean banks for construction projects, could not be repaid. In 1997 and 1998, several Asian countries faced economic, political, and social crises as a result; funds used to shore up their economies were not available for other investments, causing countries such as Brazil and Russia to suffer economic slowdowns. Although the Asian country economies soon returned to growth, the social and political impacts remained. Once again, problems raised by the global connections caused local voices to cry out in pain, with the prospects of the poorest being most affected.

Global Information Services

In the 1990s, the rapid expansion of Internet telephone-computer-linked services fueled the growth of the quaternary sector of information services. E-commerce (electronic commerce) is the trade-based sector of such services. Business-to-consumer ("B2C") facilities include retail sales, bidding (e.g., for airline tickets), and auctioning. Success in the initial stages of this area was varied, few companies made rapid trading profits, and many went out of business after initial high share valuations. Although some new car sales, for example, took place over the

Internet, most people with access to the Internet in the United States used it as a source of information before going to their local auto outlet to buy. The largest volumes of e-commerce transactions are business-to-business ("B2B"). Large corporations, such as General Motors, work with their suppliers over the World Wide Web.

Just as some multinational corporations moved manufacturing production facilities to places with lower labor costs, so others moved information handling to such places. In 1983, for example, American Airlines established Caribbean Data Services in Bridgetown, Barbados, to process the paperwork related to its tickets and boarding passes. It became the largest single employer in Barbados. In Montego Bay, Jamaica, the Jamaica Digiport Interstate Communications System links clients in Canada, the United States, and the United Kingdom. U.S. insurance companies process claims in Ireland. India, with its large population of English speakers, grows as one of the world's major call centers for multinational corporations. Some Indian companies even train their staff to respond in American accents.

The global availability of information services also gives people access to data and personal links that are used in local and regional conflicts, including labor disputes and political demonstrations. Offshore banking and organized crime syndicates involved in drugs, armaments, and slave trafficking—and in terrorist attacks to disrupt the world economy—exchange information and make bringing criminals to justice more difficult.

Global City-Regions

The growth of multinational corporations, international financial institutions, dense networks of telecommunications, information processing facilities, and international airline routes, together with the rising significance of quality business services, placed a new focus on some of the world's largest cities. Those cities with an increasing involvement in the global economy often have as much or more contact with foreign centers as with those in their own country. New York, for example, is the center of a region with 18 million people and an economic product greater than countries such as Canada, Brazil, or China; its businesses receive 40 percent of their revenues from foreign sources. Foreign banks with New York offices rose from 47 in 1970 to over 200 in the 1990s; over half of the U.S. law firms with overseas business are based in New York.

Such cities have major impacts on the places immediately surrounding them as well as on cross-border links to other countries. Their geographic scale of size and influence merits the term global city-regions. They have concentrations of high-salaried people, high-end technological and business services, specialized workplaces, hotels, homes, major sports stadia, and concert halls. They have high-rise office and apartment blocks and a wide range of arts and sporting facilities. At the same time, their corporations employ increasing numbers of foreign experts and a growing underclass of poorly paid support workers, often migrants from poorer countries.

One approach to identifying and classifying global city-regions focuses on the importance of four categories of global

corporate services (accountancy, advertising, banking, and law). Prime centers in all categories are New York, London, Paris, and Tokyo, closely followed by Chicago, Los Angeles, Frankfurt, Milan, Hong Kong, and Singapore. Figures 2.16 and 2.17 demonstrate the uneven distribution of these cities—the "control centers" of the global economy: the "top 10" of global cities occur in the United States, western Europe, and East Asia.

Test Your Understanding 2A

Summary Human development and human rights are linked. Much was achieved in the 1900s, but many needs were left unmet. Issues of people and land focus on population distribution and the processes that cause populations to grow or decline. The number of people Earth can support is subject to complex factors.

Economic systems reduced to the free-market basis in the 1990s. Inequality among peoples focuses on the continuing poverty of so many and the huge material wealth of a few. Questions of how development occurs and how poverty may be reduced are foremost concerns of world organizations such as the United Nations and the World Bank, but no recipe is appropriate to all countries or local areas. The global economy spreads world trade through multinational corporations and involves global movements of capital and information. NGOs have local, countrywide, and global roles. Some are well known for bringing aid to needy places across country boundaries and for pursuing human rights and environmental issues. Global city-regions are hubs of international economic activities.

Questions to Think About

2A.1 Why do some countries not follow the demographic transition pattern?
2A.2 How have the varied elements contributing to development emerged from the experiences of different countries?
2A.3 How important is it that one-third of the world's population is poor? What can be done about it?

Key Terms

human development	purchasing power parity
development	human development index
sustainable human development	human poverty index
human rights	modernization
population density	primary sector
demography	secondary sector
birth rate	tertiary sector
total fertility rate	quaternary sector
death rate	import substitution
infant mortality	formal sector
demographic transition	informal sector
migration	structural adjustment
economic geography	multinational corporation (MNC)
free-market system	nongovernmental organization
gross domestic product	(NGO)
gross national income	

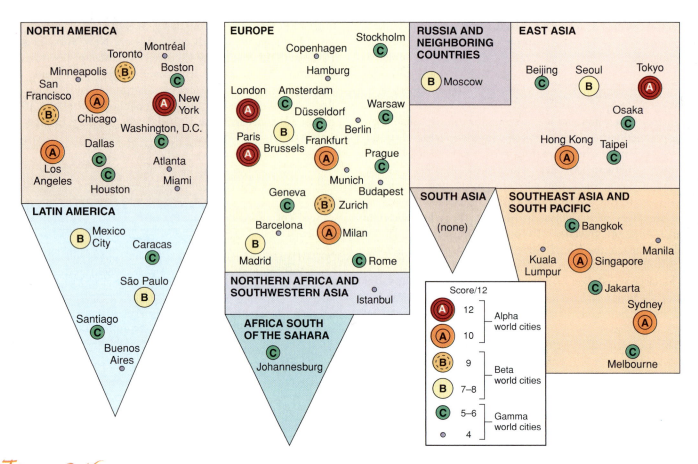

Figure 2.16 **Global city-regions.** This classification is based on activities in four categories of international business services—advertising, accounting, banking, and law—with each given a 1 to 3 ranking for a maximum score of 12. Compare the clustering of global city-regions with densities of submarine cables (inside front cover of this text) and telephone lines. Data source: P. Knox, Research Paper 32, Globalization and World Cities Group, University of Loughborough, U.K.

Issues of Cultural Freedom and Discrimination

Many difficulties facing human development and human rights arise from issues of culture-based discrimination in areas such as religion, ethnicity, class, gender, and age. Worldwide, women's rights movements have been prominent in raising their own and wider-ranging issues, since they are often concerned about education and health issues.

The culture of a group of people is basic to how they create and recreate the regional distinctiveness that in turn affects their lives and those of their descendants. Culture includes:

- A code of acquired beliefs, passed from generation to generation. The resulting cultural traditions produce patterns of meanings, including language and religion, that are copied and updated. These patterns enable people to communicate and develop attitudes to life.

- Material outcomes such as technology, design, art, or music.

- The basis of a system of social relations, worked out in attitudes to ethnicity, class, gender, and age.

Cultural geography affects regions at different scales. Some aspects, such as the Muslim religion, operate at the worldwide

level; others arise from nationalistic pressures within a country or stem from local concerns. For example, a broader European culture is differentiated among and within countries, as in the Dutch and Belgian peoples or northern and southern German groups.

Languages

A **language** is a spoken means of communication among people that grows out of historic experiences and traditions. It may provide a shared identity for a cultural group so that the group chooses to use it as a distinctive feature. Minority groups, such as the French-speaking Québecois people in Québec, Canada, and the Welsh in Britain, are often keen to preserve their language as a means of maintaining their cultural identity. The Québecois and Welsh increased their focus on language when they wanted to demonstrate their identity more forcefully and achieve greater political recognition.

Language is an important factor in geographic diversity. Globalization leads to the marginalization of some languages and the strengthening of others. Regional and internal country variations also occur. In India, for example, each state designates its own official language(s). Many local languages give way to Hindi and the state-designated languages in northern India, while English becomes more significant as a common language alongside the state languages in southern India.

Global City-Regions

Category	Europe	Russia and Neighboring Countries	East Asia	Southeast Asia and South Pacific	South Asia	Northern Africa and Southwestern Asia	Africa South of the Sahara	Latin America	North America
A12	London Paris		Tokyo						New York
A10	Frankfurt Milan		Hong Kong	Singapore					Chicago Los Angeles
B9	Zürich			Sydney					San Francisco Toronto
B8	Brussels Madrid							Mexico City São Paulo	
B7		Moscow	Seoul						
C6	Amsterdam Düsseldorf Geneva Prague		Taipei Osaka	Melbourne Jakarta			Johannesburg	Caracas Santiago	Boston Dallas Houston Washington, D.C.
C5	Rome Stockholm Warsaw		Beijing	Bangkok					
C4	Barcelona Berlin Budapest Copenhagen Hamburg Munich		Shanghai	Kuala Lumpur Manila		Istanbul		Buenos Aires	Atlanta Miami Minneapolis Montreal

Figure 2.17 **Global city-regions.** Each city is scored from 1 to 3 on whether it is a prime, major, or minor center for each of four global corporate services categories: accountancy, advertising, banking, and law. Cities in category A12 are prime centers for all four categories (4 × 3); those in A10 score 10, and so on. Source: Data from University of Loughborough Globalization and World Cities Study Group.

Thousands of languages are spoken around the world, many by small groups of people in isolated environments such as South America's Amazon River basin. Twelve dominant languages each have over 100 million people speaking them. Six of these (English, French, Spanish, Russian, Arabic, and Mandarin Chinese) are official languages of the United Nations, mostly, apart from Arabic, chosen at the end of World War II by the victorious allies.

The many languages can be grouped in families (Figure 2.18), with each family containing a rich variety. For example, the Indo-European family includes most of the languages of the Indian subcontinent, the Slavic languages of eastern Europe and Russia, and the languages of southern, western, and northern Europe. These languages, especially Spanish, Portuguese, English, French, and Dutch, spread worldwide with colonization from around A.D. 1450, dominating the Americas and the South Pacific and becoming important in many parts of Africa and Asia that were European colonies. English is increasingly used in communications among scientists throughout the world, in international air traffic control, in computer software, and in the entertainment industry, where films and TV programs in English are dominant. It remains a common language in former colonies of Britain where rivalries exist among local languages. In the early 2000s, the dominant English computer language met growing competition from Japanese, German, and Chinese.

The Sino-Tibetan language family includes the languages of China (such as Mandarin in the north) and the southeast Asian

land area. The Altaic family includes the languages of Central Asia that diffused westward to Hungary and Finland and eastward to Japan and Korea during medieval conquests. The Afro-Asiatic family of languages dominates northern Africa and the Arabian peninsula, while the Niger-Congo family includes many languages spoken in Africa South of the Sahara. The Malayo-Polynesian family occurs through Malaysia, Indonesia, the Philippines, and the South Pacific. Smaller family groups include those of the pre-European populations of the Americas (Aleut, Inuit, Amerind) and of Cambodia (Mon Khmer).

Religions

Religion generates strong group loyalties and exclusive attitudes. Each **religion** is an organized set of practices, often involving a system of values and faith in and worship of a divine being or beings. Regional emphases, social and legal practices, and visual features, such as building designs, often reflect religious allegiance (Figure 2.19). Religions play a significant role in transferring cultural values and practices from one generation to the next.

Major World Religions

The religions claiming the largest numbers of adherents are Christianity, Islam, Hinduism, and Buddhism. Judaism has a smaller number of adherents but widespread influence, especially in Western countries. Christianity, Islam, and Buddhism

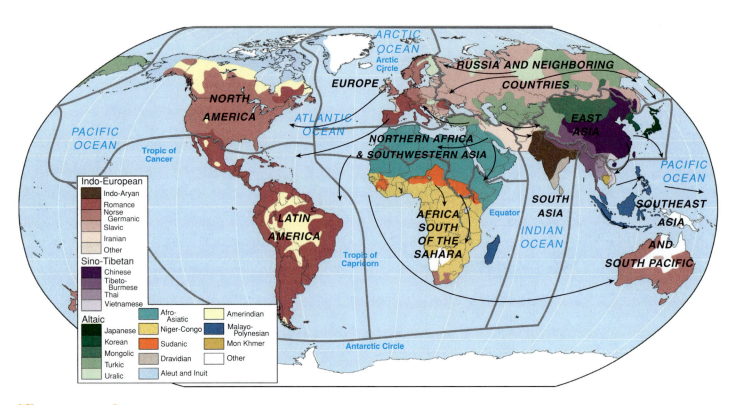

Figure 2.18 **Cultural geography: world language families.** Arrows show how major groups diffused. How do the language families relate to major world regions?

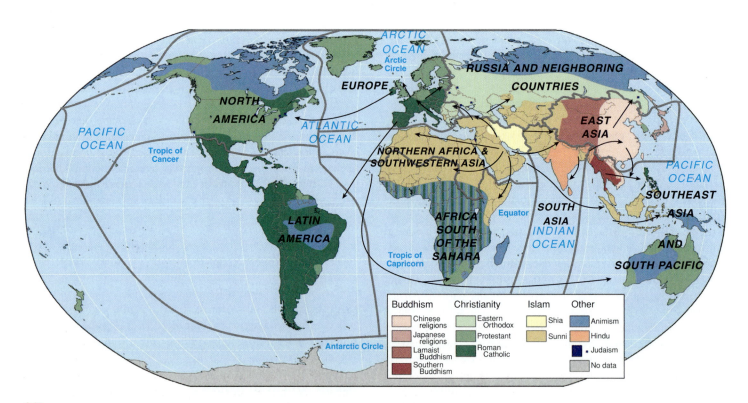

Figure 2.19 **Cultural geography: world religions.** Many religions are practiced, but only Christianity, Islam, Buddhism, and Hinduism have large numbers of adherents. Islam, Christianity, and Buddhism diffused through trade, mission, or conquest. The Jewish religion is spread around the world, mainly in cities. Older and local religious practices remain in small areas, mainly in materially poor countries. Each area mapped reflects the most important religions, and the mixture increases as global interconnections expand. Is there a link between religion and major world regions?

are religions that can be joined by anyone in any country and actively seek to extend their membership: they are **universalizing,** or global, **religions.** Hinduism and Judaism are a matter of birth, closely tied to family and region: they are **ethnic religions.** Although India and Israel are officially secular countries, there are movements to proclaim India as a Hindu country, while religious Jews seek to extend their control in Israel. Buddhism, often combined with older local religious elements, is the main religion of East Asia, while Hinduism is the predominant religion of South Asia.

Close links among religions include the fact that Judaism, Christianity, and Islam are three great monotheistic religions, worshipping one God and building on heritages from Judaism. All three regard Jerusalem as a sacred city. Jews and Muslims both look back to "Father" Abraham, and Muslims recognize Abraham, Moses, and Jesus as prophets—although Islam as defined by Muhammad, their final and greatest prophet, came into existence only in the A.D. 600s. Jesus Christ grew up as a Jew, and Christians use the same sacred writings as Jews in their Old Testament. Connections among Asiatic religions include the fact that Buddha was a Hindu and Buddhism is accepted alongside local religions in China and Japan.

Branches of major religions are linked to specific world regions, although many religions spread out from their earlier centers. In Christianity, for example, the split between Roman Catholicism and the Orthodox churches set apart the cultures of western and eastern Europe. The growth of Protestant churches dominated northern Europe after A.D. 1500 in preference to Roman Catholicism. Overseas explorations from the 1400s led to Christian missionary expansion outside Europe. Today, some 60 percent of world Christians are Roman Catholics, 25 percent belong to Protestant groups, and 10 percent are Orthodox. Christianity has some 1.5 billion adherents and is the main religion of Europe, the Americas, South Pacific, and much of Africa South of the Sahara.

Islam is dominant throughout the Arab world of Northern Africa and Southwestern Asia (see Chapter 8 for a fuller discussion) and extends its influence eastward to Central Asia and through Pakistan and Bangladesh to Indonesia (the most populous Muslim country). At present, one-fifth of the world's population identify themselves as Muslims. The major Islamic groups are the Sunnites, who follow the historical series of caliphs and are in the majority, and the Shiites, who follow a murdered relation of Muhammad and form a majority of the population only in Iran. Ninety percent of the Shiites live in Iran, but Shiites also form large proportions of the populations in the small Persian Gulf countries and southern Iraq (see Chapter 8).

The Jewish faith has a different spatial distribution than other major religions. In A.D. 73, Roman armies destroyed Jerusalem and forced Jews to migrate to all corners of the known world. The dispersed Jewish **diaspora** existed as close-knit groups throughout Europe and extended into other parts of the world with colonization. Laws in Europe, northern Africa, and Asia prevented Jews from holding government positions. Their main means of development was through business. A range of anti-Jewish actions occurred, with the worst excesses during the Nazi holocaust in Europe in the early 1940s. In 1948, the country of Israel was established as a common homeland for many of the dispersed people (see Chapter 8).

Religion and Society

Religious adherence often determines responses to issues affecting society. Roman Catholics and Muslims, for example, officially resist population policies aimed at reducing births—although individual families increasingly make their own choices. Most religions demonstrate a care for the natural environment, although the practice of societies has often not matched these beliefs.

When linked to inequalities of wealth and power, religious differences between peoples may result in conflicts. Such conflicts often arise from exclusive claims that are fanned for their own ends by those with political power. Christians and Jews, Muslims and Hindus, Christians and Muslims fight each other. Conflicts extend to Catholic against Orthodox (former Yugoslavia) or Protestant (Northern Ireland) Christians and to Shia against Sunni Muslims. The Muslim Arabized north of Africa clashes with the partly Christianized black African south, especially in countries crossing the zone between the two, such as Nigeria, Chad, and Sudan.

Recent events refute suggestions that economic globalization and an increasing secularization based on scientific explanations of phenomena will reduce religious influences. Although attendance at religious ceremonies and observance of religious rites are decreasing in some countries, especially in the wealthier West, science does not provide a system of values or guides to the meaning of life. Today, both Western and non-Western societies question the significance of scientific and merely rational approaches as prescriptions for living.

It is likely that religion will continue to influence culture, perhaps in new ways. From the 1990s, the resurgence of Christian, Hindu, Jewish, Islamic, and Buddhist extremists in some countries is one sign that religious motivations are again significant after the Cold War period of largely political conflicts. In Indonesia, Muslims and Christians fight for political control of some of the many islands. The 2001 terrorist attacks on New York and Washington, D.C., brought to light the growing hatred of many Muslims for the West and Christianity, although those events did not succeed in raising a widespread alliance of Muslim countries to resist Western culture. Perhaps the attacks were part of a political movement, cynically attaching their goals to religious beliefs and believers.

Race, Class, and Gender

Status—and the opportunity to carry out or stifle changes in society—varies with culture. In many cultures, status is inherited through the family, as in the caste groups of India or the royal families and linked aristocracies of Europe. In other cultures, status may be achieved by creating wealth, political allegiance, media prominence, or sporting performance. While many of the world's wealthier countries increasingly base status on achievement, some people have a greater chance of achieving higher status than others by virtue of their birth into wealthy families, their gender, their upbringing, or their education. Throughout the world, race, class, and gender influence access to position in society: they vary geographically and are particularly significant in local contexts.

Race and Ethnicity

An **ethnic group** is one with a culture of its own so that the members feel a common origin (real or imaginary) and are set apart by religion, language, national origin, or racial characteristics. Such groups may be called tribes, clans, or segregated minorities. At times, members of certain ethnic groups express contempt for others.

The concept of **race** is assumed by many to be based on essential biological differences, but characteristics such as skin color, eye shape, or hair type vary as much within identified "races" as between them. Although race and racism are often at the center of human conflicts, the most basic human biological features, DNA and blood type, demonstrate little variation across the human species, which is a single reproducing group. More significant is the cultural status given to bodily features as a means of defining what are essentially ethnic "them-us" differences.

In South Africa, the minority white population operated a supremacist apartheid policy until 1993, separating whites, blacks, and other "colored" peoples into segregated neighborhoods. Most African Americans, a minority ethnic group identified mostly by skin color in the United States, still have lower status than those of European origin. Even in Brazil, where European, African, and Asian peoples mix freely, most wealthy people are of European origin and most people of African or Native American origin are poor.

Class Distinctions

Class arises from a stratification of society that is imposed through combinations of religious, economic, and social criteria. Wealth, education, or birth are common bases for class distinctions in all countries. In Britain, the royal family is linked to a hierarchy of hereditary dukes, earls, and knights. In India, the caste system and, increasingly, education define position in society. In Communist countries, the expectation of a classless society is contradicted by the status and privileges given to members of the Communist Party. Most Americans identify general class differences on the basis of possessions and appearances. That basis produces a black underclass and a group identified as "white trash," a lower-middle class of black and white factory and shop workers, a mainly white upper-middle class of managers and professionals, and a group of exceptionally wealthy financiers, property owners, and sports and media stars (both white and black).

Gender Inequalities

Gender—the cultural implications of one's sex—is also responsible for differences and inequalities of opportunity within and among societies. Males dominate most societies and have a history of denying full rights to women. Although major changes occurred in the 1900s, particularly in extending

voting franchises to women, some countries still deny women the human rights as defined by the United Nations.

The Taliban rule in Afghanistan became notorious for taking women out of education, confining them to the home, and selling some of them into slavery. In Africa, female genital mutilation, or "cutting," continues in some societies. In Iran and northern Nigeria, women—but not men—can be stoned to death for adultery. Illiteracy among women is still much higher than among men in most of Africa and Asia because more men than women have been educated. Few European countries allowed women to vote in elections until the 1900s, with Switzerland delaying it until 1971 and one of its cantons until 1990.

Women, even in the world's wealthier countries, commonly receive lower wages than men for the same job and constitute a minority of doctors, engineers, corporate executives, and elected politicians. Some jobs such as nursing, secretarial work, elementary school teaching, and shop clerking have been widely regarded as "women's work" and often have lower status. Further, women commonly take a major share of home management and child care, which affects their career prospects. In some cases, marriage breakups leave women at a financial and social disadvantage.

The United Nations *Human Development Report* for 1995 included a gender-related development index (GDI) for the first time. The GDI focuses on the same criteria as the HDI (see p. 36) but reflects inequalities between men and women (see Figure 2.8). While 162 countries were reported for real GDP (PPP$) per capita and HDI values, only 146 were reported for GDI in the 2001 report (reflecting figures for 1999).

Some aspects of women's inequality are gradually being tackled. The high rates of female illiteracy in poorer countries have major implications for future population-resource issues. Better education gives women confidence, enables them to take jobs, improves their self-esteem, and gains acceptance for them having a role in decision-making. It also impacts such matters as how many children are wanted in a family, since education is vital but expensive. In many places, male attitudes and cultural traditions still prevent women from controlling how many children they have. It is clear that many women prefer smaller families, but fears raised in earlier periods of higher infant and child mortality still encourage families to have more children.

Despite their low status in most societies, women now play major roles in the expanding world economy. During the 1970s and 1980s, the proportion of women in manufacturing jobs in the United States and Canada grew as employment opportunities shifted in their favor. For example, female employment in the electronics industry expanded, but men were laid off when "male" jobs were taken over by machine tools. Expanding numbers of jobs in services such as retailing, health care, teaching, banking, and tourism further increased female employment. In materially poor countries, women comprise around 80 percent of the labor force in export-oriented electronics, apparel, and textile industries.

The wealthier countries and the industrializing countries of East Asia experienced greater household prosperity through dual incomes as females entered the paid labor force. Women's employment prospects in Africa and Latin America, however, deteriorated as the result of measures imposed to solve the debt crises in those countries. In Russia, the wide involvement of female labor in the Soviet Union's Communist regime gave way in the 1990s to a greater focus on male employment.

Cultures and Regions

Cultural differences make the world's major regions, countries, and smaller regions within them distinctive and thus have strong geographic impacts. Many cultural differences can be traced back to early human history and the development of agricultural technology, religion, and language in small regions with specific environmental conditions. Regions that are termed cultural hearths were centers from which techniques, useful materials, and social norms diffused to other regions. Mesopotamia, north-central China, and central Mexico are examples of cultural hearths (see Chapter 1).

Current world geography includes the imprint of many newer cultural overlays on older ones. For example, remnants of the Mayan and Aztec empires, as well as of Spanish colonization, still exist in Mexico. The Mayan remains in Yucatán are being revealed from beneath the forest, while modern urban-industrial buildings are replacing Aztec and Spanish relics around Mexico City. In Southwest Asia, the Muslim culture removed most non-Muslim relics, but some, such as the ruins of medieval Crusader castles, still exist. In Communist China, the palaces of former imperial dignitaries occupy large areas of Beijing, while the new cities of the eastern coastlands impose modern urban-industrial features on the rural landscape that evolved over centuries.

Cultural Fault Lines

Just as cultural hearths identify areas that had formative influences on cultures in wider areas, **cultural fault lines** exist between cultural regions at different geographic scales. In the absence of other constraints, tensions along these lines may lead to conflict. One prominent cultural fault line lies between the Muslim and Christian cultural regions from southern Europe through southern Russia (see Figure 1.6). Although the current disputes are not merely religious, the antagonisms between Muslim and Christian cultures along a line through Bosnia, Kosovo, Macedonia, Bulgaria, Armenia, and the Russian republics such as Chechnya bordering on the Caucasus Mountains are significant.

Issues of Political Freedom

Political geography is the study of how governments and political movements (e.g., NGOs, labor unions, political parties) influence the human geography of the world and its regions. Self-governing **countries** (sometimes confusingly referred to as states or nation-states) are the basic political units: within its borders, a country's government is assumed to have political control, or sovereignty, over the country's inhabitants. Each country is recognized by other countries. The numbers of self-governing countries increased from 62 in 1914 to 74 in 1946 and 193 in 1998.

Not all units shown on maps are fully fledged countries. Some are colonies, while others, such as Taiwan, which the People's Republic of China claims as its territory, lack recognition from other countries.

Despite globalization, country governments still have significant powers to promote and protect their peoples in world affairs. They are central to any understanding of human development or human rights. Countries tax their citizens to provide public services, including military capabilities, and encourage economic and social welfare. On average, during the 1900s, the proportion of a country's wealth taken and used by its government increased from under 10 to over 40 percent and continues to increase. Countries often have systems of regional, state, or local government that carry out some of the governmental responsibilities at different geographic levels. Country governments may also join other country governments in mutual trading or defense agreements. In world regional geography, countries provide the main subunits of study within the world regions.

Political Worlds

Three blocs of countries divided the world during the Cold War but became obsolete in the 1990s (Figure 2.20). The United States led the **First World** bloc—the so-called "free world" countries with Western values, mostly democratic governments, and material wealth based on market economies. The Soviet Union and China led the **Second World** bloc of countries that had Communist governments, marked by centrally planned policies. The **Third World** bloc initially comprised countries searching for political approaches that avoided the pitfalls of the other worlds. The First and Second Worlds competed politically and militarily with each other and vied for influence among the Third World countries, which became "aligned" or "unaligned." Many of the Cold War battles were fought on Third World territory, supported by armaments and funds from the First or Second Worlds. Such actions contributed to undemocratic Third World governments, corruption, and a lack of economic progress. *Third World* became a term to identify the world's poorest countries. During the 1990s, these terms ceased to have meaning as the Second World no longer existed following the breakup of the Soviet bloc. Countries are now grouped into materially wealthier or poorer, or more or less developed, categories.

After the 1991 breakup of the Soviet Union, the United States became the sole world political, economic, and military superpower. Many Americans and many people in other countries, however, resisted the United States exercising its powers as the world's police officers. Neither was there a consensus for global government within a "New World Order." Although its members include almost all the world's countries, the United Nations' weaknesses are shown in its failures to contain civil wars, nuclear testing, or drug, weapons, and slave trafficking. Inaction because of the difficulties in obtaining agreement from all or a majority of its members resulted in the United States, supported by other wealthier countries, intervening in the affairs of other countries such as Caribbean islands, Kuwait, Bosnia,

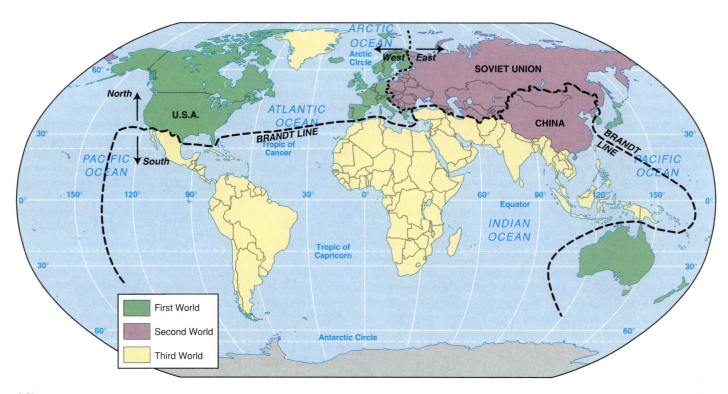

Figure 2.20 **Old global order of the three worlds.** The First World comprised the affluent countries of the West, dominated by the United States and western Europe. The Second World included the Soviet Union, China, and other Communist countries. The Third World made up the rest—the world's materially poor countries. The West-East division marked a Cold War boundary of potential conflict. The Brandt line, proposed in the early 1970s by Willy Brandt, chancellor of West Germany, separated the rich North from the poor South.

Kosovo, and Afghanistan. Worldwide organizations designed to promote the economic development of poorer countries (International Monetary Fund, World Bank) or to liberalize trade among countries (World Trade Organization) took on some governmental roles, but increasingly, nongovernment organizations such as aid bodies (Red Cross, Oxfam, etc.) assumed responsibilities for governmentlike activities, including disaster relief. Any vision of a world government remains a long way off.

The end of the Cold War led to shifting emphases among regional groupings of countries. For example, from the 1950s, the European Union (EU) extended its membership among countries in western Europe (see Chapter 3) as an economic common market paralleled by the defense-oriented North Atlantic Treaty Organization, which included the United States and provided military protection against the Soviet Union. From the 1990s, the EU became both a significant political and also a military force. In the early 2000s, the EU plans to enlarge its membership and include countries in eastern Europe that were members of the Soviet bloc until 1991. The Association of South East Asian Nations (ASEAN) began with political objectives during the Cold War (opposing Communist countries such as Vietnam) but afterward shifted to economic objectives (and admitted Vietnam as a member). As the 1990s progressed, countries refocused their policies and rhetoric. They placed greater emphasis on the reduction of poverty and population growth in the less developed countries, on concerns for the roles of women in society, and on sustaining the natural environment.

Nations and Nationalism

A country is rarely, if ever, the same as a nation. A **nation** is an "imagined community," or ethnic group, in which a group of people believe themselves to share common cultural features, usually linked to a specific area of land. The cultural features may be language, religion, or other characteristics with a historic background. In the country known as the United Kingdom, for example, the English, Scottish, Welsh, and Northern Irish consider themselves distinctive nations. Some foster separatist groups, and they enter separate teams in world sporting events. The Scots, Welsh, and Irish—but not the English—now have separate home parliaments with some limited powers (less than those of a U.S. state), although the people still elect members of the United Kingdom's parliament. Thus, a nation may not have control over a country; it may not have its own sovereign government.

Nationalism—the desire to combine cultural and territorial features and become self-governing—has been, however, basic to the formation of countries in Europe since around A.D. 1800 and has spread to the whole world. At this time, the rise of free-market economies turned allegiances away from previous overriding feudal principles or ecclesiastical loyalties. The increased levels of communication made possible through printed books and newspapers, the telegraph, telecommunications, radio, and television further bolstered nationalism. From the 1800s, universal education supported nationalist themes through selective views of history glorifying the national experience.

Stronger nations often included other peoples in a country they could dominate. Thus, France included Bretons, Basques, and Alsatians. Germany emerged in the later 1800s, when a group of smaller states united under Prussia's leadership and a nationalist linguistic banner. In the 1900s, Germany used the idea of uniting separated German-speaking minorities in neighboring countries as an excuse for talk of national supremacy. This led to world wars, when Germany discovered limits to the expansionism its neighboring countries would tolerate.

Indigenous peoples are the first inhabitants of any given area. With the spread of mainly European colonialism throughout the world from the 1500s to the 1900s, many indigenous groups could not compete, and their numbers fell or they were eliminated as the colonizer took over their lands. Throughout the world, many indigenous groups remain today, generally as minority populations within individual countries. They often have their own cultural and political aspirations as minority "nations." Once incorporated in the modern country, such indigenous groups compete with other interest groups—as when the Native Americans disputed land ownership with European colonists or recently won major rights as the "First Nation" in Canada. Many indigenous groups, however, disappeared. Those that still exist are often regarded as more "primitive" stages of human development rather than as representing an alternative approach to human organization.

Governments

Some countries have a unitary governmental structure, administering all parts from the center for all aspects of government. Other countries, including most of the world's largest (Russia, India, Brazil, Nigeria, the United States, and Canada), have a **federal government** structure, dividing the authority for various activities between a central government and partitions called states or provinces. In the United States, for example, the federal government has responsibility for external relations, defense, and interstate relations, while the states have authority for education, local roads, and physical planning—for everything not stated as "federal" in the Constitution.

Governments differ in the direct influence they impose on internal affairs. Dictatorships and governments where control over production from the center is the dominant economic system intervene much more than democratic governments in economic activities from production to distribution. They may use their powers to coerce and oppress peoples in such actions as forced labor or ethnic cleansing. Examples are the Soviet Union regime from the 1920s to 1991, the holocaust against the Jews under the Nazi Germans in World War II, the Serbian treatment of other ethnic groups in the former Yugoslavia, Chinese actions under Mao Zedong and since, and the actions of African dictators such as Idi Amin and Milton Obote in Uganda.

During the 1900s, however, all governments increased their level of intervention in regulating and promoting economic activity. Increased levels of taxation for spending on rising education, health care, and defense expectations and technology, together with encouragements for foreign investment and the manufacturing of export products, strengthened the role of country governments as global interconnections multiplied.

Government functions are concentrated in **capital cities,** where the head of state lives and the administrative and government offices are situated. Many capital cities are the largest cities in the country, as, for example, London (United Kingdom), Tokyo (Japan), and Nairobi (Kenya). In many federal countries, the establishment of new capital cities avoided jealousies among existing cities. Washington, D.C., Brasilia (Brazil; Figure 2.21), Abuja (Nigeria), and Canberra (Australia) are examples.

Figure 2.21 **Brasilia, the planned capital city of Brazil.** (a) The central avenue lined by public buildings. (b) The cathedral. (c) Unplanned satellite town of Paranoa that sprang up to hold over a quarter of a million people, including the enterprising shopkeeper on the main street. Photos: © Michael Bradshaw.

(a)

(b)

(c)

Country Groupings for Defense or Trade

Countries make agreements with other countries to foster security through common defense and trading interests. The Cold War period generated defense agreements on both sides. The North Atlantic Treaty Organization (NATO) linked North America and western Europe in a common response to a perceived military threat from the Soviet Union (see Chapter 3). The Soviet Union established the Warsaw Pact to unite the countries of eastern Europe in the Soviet bloc. The Warsaw Pact ended after 1991, and NATO extended its agreements to some of the former Soviet bloc countries, despite resistance from Russia.

Governments have a major influence on the conduct of world trade. They may encourage their people to export goods or may control certain imports by charging taxes, or tariffs, on them. During the early 1990s, there was a widespread political will among countries to free world trade from barriers. The General Agreement on Tariffs and Trade (GATT) was established in 1948 to encourage countries to lower tariffs. After 1993, the World Trade Organization took over the role of preventing discrimination among trading partners, but its rules appeared to many to support the wealthier countries against the weaker, leading to riots in Seattle (1999).

Most progress on liberalizing trade has been made at the world regional level in free-trade areas, the members of which have common tariff rates on imports. Opinions differ as to whether the regional free-trade area agreements encourage members to lower their tariffs further or whether they discriminate increasingly between members and nonmembers. The largest trading group at present is the European Union (EU; see Chapter 3). The North American Free Trade Agreement (NAFTA; see Chapter 11) may extend to much of Latin America by 2005 (see Chapter 10). Even more ambitious is the prospect of the Asia-Pacific Economic Cooperation Forum (APEC; see Chapters 5, 6, 10, and 11) that has pledged itself to free internal trade among its wealthier countries by 2010 and among other countries by 2020. The European Union is considering extensions into eastern Europe (see Chapter 3). It has been suggested that NAFTA and the EU could form links. Such groupings of regional interests are considered in each chapter of this text.

Human Rights and Justice Systems

Most human rights relate to personal and group justice and are potentially subject to legislation. Threats to personal security, injustices, and discrimination by gender, race, ethnicity, national origin, religion, or age are problems that supporters of human rights wish to end. Government policies and courts of justice legislate for and enforce some rights, but few countries implement the whole range agreed upon by the United Nations. In some cases, one person's right is another's restriction.

The death penalty is the ultimate act against a human's right to life. Some Asian countries impose it on drug smugglers; some U.S. states provide a death penalty for murderers. Slavery is another restriction of basic human rights and was mostly abandoned in the 1800s, although cases still emerge in countries such as Sudan and Côte d'Ivoire. Long-term imprisonment for political views was common in the Soviet Union and under the South African apartheid regime until the early 1990s and continues in China and Myanmar. Amnesty International is an NGO that monitors such actions. Imprisonment for long periods without accusation or trial occurs in many countries and in 2001 was extended in the United States and United Kingdom as part of antiterrorism policies. Child labor is common in Asian countries, including work in poorly regulated factories and prostitution.

Agreement on the nature of human rights and justice is not global. Muslim countries in particular debate the largely secular basis of the United Nations list that conflicts with their religious beliefs and fierce legal provisions. Saudi Arabia refused to attend UN conferences on such topics on the grounds that it has a God-given right to gender discrimination.

Most justice systems are still based within countries. At the global and world regional levels, new courts are taking up international issues, particularly of human rights and war crimes. The European Union also has a court that considers issues relating to members operating their own systems in accordance with EU rules.

Test Your Understanding 2B

Summary Human rights issues of cultural freedom and discrimination are particularly significant. Languages separate peoples and are a major feature of nationalist claims. Religions are fundamental to world regional divisions and foster both togetherness and conflicts. Race, ethnicity, class, and gender affect status in society, and such issues underlie many local conflicts. Cultural differences trace back to early human history in distinctive cultural hearths.

Political freedoms depend on how countries are governed, the role of nationalism in their policies, and attitudes toward indigenous peoples. Countries where political power is concentrated in the hands of a few rulers often adopt harsh policies toward their peoples.

Questions to Think About

2B.1 In what ways do languages and religions define cultural groups? What effect may differences have on human rights?

2B.2 How are gender issues determined by cultural expectations?

2B.3 Attempt to define *political freedom*.

Key Terms

cultural geography	political geography
language	country
religion	First World
universalizing religion	Second World
ethnic religion	Third World
diaspora	nation
ethnic group	nationalism
race	indigenous people
class	federal government
gender	capital city
cultural fault line	

🌐 Environmental Issues

Physical geography is the study of natural environments and their world distribution. In world regional geography, the outcomes of interactions between human activities and natural environments help to define the nature of each region. Major physical features, such as oceans, seas, mountain ranges, and rivers, often form boundaries between countries. The length of growing season, amount of water available, soil types, and mineral-bearing rocks all influence the type and location of human activities based on natural resources—and the level of human development possible in each region within countries. Natural features and processes not only impact human activities, but human beings also increasingly intervene in and influence the ways in which natural processes function.

Environmental issues focus on whether human activities lower the quality or productivity of natural resources. Lower environmental quality offers a less sustainable prospect for the future. Each chapter in this text discusses the natural environment of the major world region being considered and specific concerns about its management.

Earth's natural environment is a dynamic system of interacting events that combine in various ways to produce regional differences from high forest-clothed mountains (Figure 2.22a) to dried-up desert areas (Figure 2.22b). Natural processes include:

- The workings of the atmospheric and oceanic circulations to produce weather and longer-term climate changes (atmosphere and hydrosphere)

- The workings of Earth's interior that cause huge sections of the crust to collide with each other, producing earthquakes, volcanoes, and mountain systems (lithosphere)

- The interactions of atmosphere, hydrosphere, and lithosphere processes that cause rain, glacier ice, wind, and ocean waves and currents to produce landforms such as hills, valleys, and beaches—the stages on which human activities occur

- The actions of living organisms, plants and animals, in responding to and modifying the local climate, landforms, and soils (biosphere)

Climatic Environments

The **climate** of a place is determined by the transfers of heat and moisture between the linked atmosphere, oceans, and continental surfaces. The transfers are powered by energy from the sun that is filtered as it passes through the atmosphere, making Earth's surface habitable.

Heat Transfers

Visible light rays dominate solar radiation, reaching Earth's surface as other elements of solar radiation (gamma rays, X rays, ultraviolet rays, and infrared rays) are mostly absorbed as they enter Earth's atmosphere. Absorption of these rays causes rock, soil, and ocean water to be heated and to radiate heat rays upward, raising the temperature of the lower atmosphere. Because the sun is more directly overhead for more of the year in tropical regions, its heating impact on the atmosphere is greatest there (Figure 2.23). Tropical areas have an excess of incoming energy over that

(a)

Figure 2.22 **Climates and landforms.** (a) The Rockies: high mountains raised in movements powered by Earth's internal energy and then sculpted by glaciers and rivers. Such environments are favorites of outdoor recreation buffs. (b) Bonneville Salt Flats, Utah. The area was covered by the expanded Great Salt Lake during a period of wetter climate several thousands of years ago. Evaporation of the water during drier conditions left behind the white salt crust. The salt provides a raw material for chemical industries, while its solid, smooth surface is used for world land-speed record attempts. Photo: (b) © Michael Bradshaw.

(b)

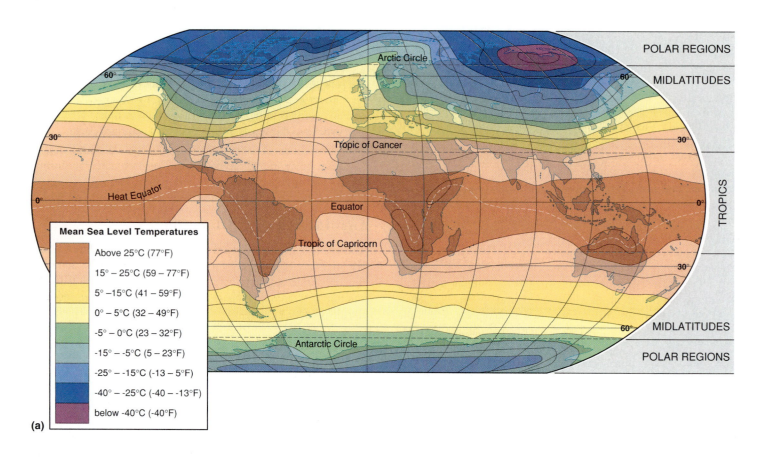

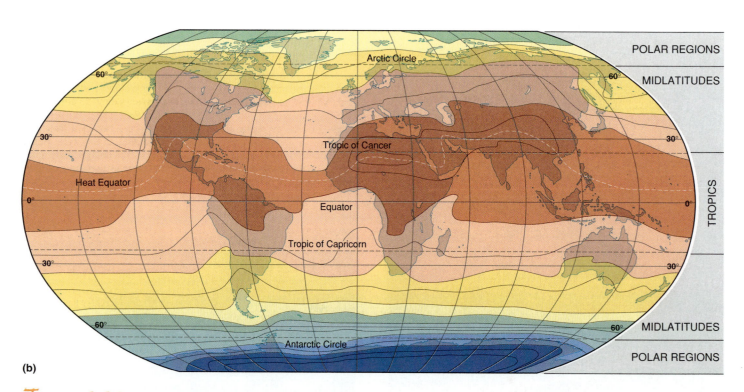

Figure 2.23   **Global temperatures in air near the ground.** Isotherms (joining places of equal temperature) for (a) January and (b) July. The heat equator connects points of highest temperature at each meridian of longitude. Compare the Northern and Southern Hemispheres at these two periods in terms of the extent of very cold temperatures and the position of the warmest band of temperatures. What effect on air temperatures do the oceans have?

lost back to space (Figure 2.24). The polar regions, however, have a deficit of energy. In winter, they have several months of almost complete darkness, losing energy to space without any coming in.

The tropical excess and polar deficit are compensated by flows of air and ocean water between the two regions. Tropical oceans become huge reservoirs of heat that ocean currents move poleward into the atmosphere of middle latitudes. The cooled air and waters return to the tropics, where they are reheated. This system makes human habitation possible into high-latitude regions near coasts.

Earth's Daily Rotation and Annual Solar Orbit

Earth rotates on its axis once a day and revolves in orbit around the sun once a year. The former creates day and night, and the latter seasonal changes. Because of Earth's axial tilt, the sun is directly overhead at noon at the Tropic of Cancer (Northern Hemisphere) between June 19 to 23 and at the Tropic of Capricorn (Southern Hemisphere) between December 19 to 23. Overhead sun brings summers of warmer weather and long polar days to each hemisphere, while the winter hemisphere with low sun angles has cooler weather and long polar nights.

Earth's rotation affects air and water movements—winds and ocean currents—across the surface. The effect increases away from the equator toward the poles, bending winds to form circulating weather systems, including cyclones (counterclockwise circulation in the Northern Hemisphere, clockwise in the Southern Hemisphere) and anticyclones (clockwise circulation in the Northern Hemisphere, counterclockwise in the Southern Hemisphere).

Water Transfers

Oceans are major sources of water that evaporates into the atmosphere, condenses into clouds, and produces rain and snow. Parts of the world with the highest rainfalls lie near the equator (Figure 2.25), where warm, humid airstreams collide, forcing the air to rise, forming clouds, and producing frequent rainstorms. Another zone of high rainfall totals is on the ocean-facing west coasts of midlatitude continents, especially where high mountains (as in Canada and Chile) add to the lifting caused by the meeting of warm tropical and cold polar air in cyclonic systems. Areas between have little, or seasonal, rain.

World Climatic Environments

The receipt and redistribution of solar energy, the atmospheric and oceanic movements, and the circulation of water from oceans to continents all vary in their effects around the world, giving rise to specific weather systems and climatic regions (Figure 2.26).

Tropical climates experience high temperatures throughout the year and have short winters, if any. The main climatic variations in the tropics are seasonal differences of rainfall, as can be seen by referring to Africa on Figure 2.26. Places close to the equator have rain in all months of the year (Af), although a few months may be drier than the rest. Places farther from the equator but still within the tropics have a marked alternation of dry and wet seasons (Aw). Eventually, as distance from the equator increases, the dry season becomes so long that annual water shortages occur, and the climate becomes arid, without appreciable rainfall (BWh). The main tropical climates thus have a north-south distribution, from the equatorial climate with rain at all

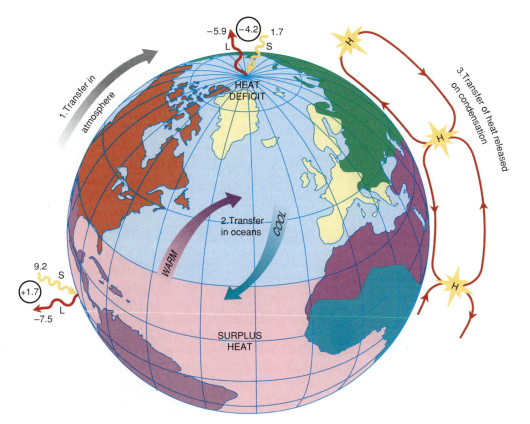

Figure 2.24 **Global heat transfers.** In the tropics, heat losses by longwave radiation from Earth are less than incoming shortwave solar radiation; at the poles, the losses exceed the gains. Three main mechanisms transfer heat from low toward high latitudes, balancing the excess and deficit: winds in the atmosphere, ocean currents, and movements of humid air (heat is trapped during evaporation and released when clouds condense). Flows of warm air and water toward the poles are paralleled by flows of cold air and water toward the equator.

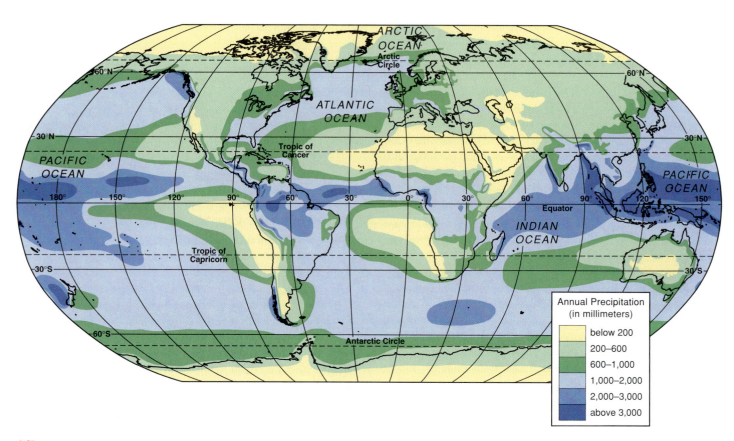

Figure 2.25 **World precipitation (rain, hail, snow).** Locate the main areas of low and high precipitation. Areas with less than 200 mm of precipitation in a year are classed as arid; those with more than 2,000 mm are nearly all in the tropical ocean areas. Warm air (which can hold more moisture than cold air) from the northern and southern tropics converges at the equator, rises, and forms clouds that precipitate rain.

seasons, through the wet-dry seasonal climates, to the very dry climates. The seasonal wet-dry contrasts are greatest in the monsoon climates of the Indian subcontinent (Am).

Tropical climates have distinctive weather systems, including tropical cyclones (also called hurricanes and typhoons), that may cause loss of life and property (Figure 2.27). The frequent rains near the equator, where the effect of Earth's rotation is least and circulating weather systems are rare, come from massive thundercloud developments (Figure 2.28).

Midlatitude climates have marked seasonal temperature contrasts between summer and winter. Such contrasts are greatest in the centers of the North American and Eurasian continents and least where winds blowing from the oceans moderate winter cold and summer warmth over the west-facing coasts of Europe and North America. Most precipitation falls on hills and mountains facing west near these coasts and declines inland to the point where it may become insufficient to support vegetation or agriculture. Midlatitude climates have a west-east distribution, with mild and moist west coasts (Cfb, Cs), dry continental interiors (BS, Dfa, Dfb, Dw), and east coasts with summer-winter temperature contrasts (Cfa, Dfa). Midlatitude weather systems are circular systems around low atmospheric pressures in frontal cyclones—which bring rain and high winds—and calmer high-pressure anticyclones.

Midlatitude climates also have some north-south variants, some as a result of transition from tropical and polar climates.

The humid subtropical variety on the eastern sides of continents forms a transition with tropical seasonal or monsoon climates with short winters and occasional frosts followed by long hot and rainy summers. The humid continental (long rainy summer) variety has noticeable winter seasons with snow cover, while the humid continental (short rainy summer) type in higher latitudes has long, very cold winters.

The marine west coast climates affect extensive latitudinal ranges in North America and Europe because of the ocean currents that warm the air above as it blows toward higher latitudes. On the southern margins, seasonal exchanges with arid climates produce a regime of long, dry summers and mild, wet winters, typical of lands around the Mediterranean Sea and named after that association.

In the large northern continents of Asia and North America, interiors are far from the oceans. They experience drier conditions as well as the greatest summer-winter temperature ranges.

Polar climates are extremely cold throughout the year. Winter conditions dominate the Arctic climates (ET, EF), although short summer spells may melt some of the snows. Truly polar climates are frozen all year. Although expeditions to explore Antarctica and the Arctic Ocean take advantage of the long local summers, temperatures remain low. In winter, there is almost total darkness for several months in polar regions.

Polar regions have no weather systems of their own but are dominated by dense, cold air that sinks and flows outward

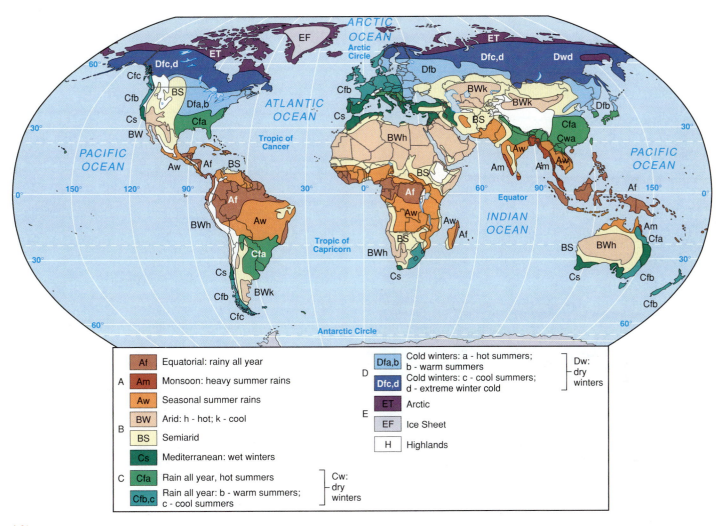

Figure 2.26 **World climate types.** Included are letters referring to the Köppen classification based on the climate characteristics of natural vegetation zones. Source: M. Bradshaw, R. Weaver, *Foundations in Physical Geography,* WCB, 1995.

to deepen the winters of midlatitude regions. Midlatitude cyclones sometimes invade polar regions, bringing high winds and precipitation.

Climate Change

Climates change over time—small amounts over shorter periods and larger over longer periods. Such changes are a normal part of Earth's natural environments and result from changes in Earth's solar orbit shape and the planet's axis angle relative to solar rays.

An intensive freeze, the most recent part of the Pleistocene Ice Age, lasted for most of the last 100,000 years, ending around 10,000 years ago with a period of warming, ice melting, and a sea surface rise of around 100 m (300 ft.) to its present level. The freeze caused huge ice sheets to dominate the northern parts of North America and Europe and lowered sea levels around the world. After the last ice cover retreated, smaller fluctuations of climate brought the warmest conditions around 5,000 years ago. The "Little Ice Age," from approximately A.D. 1430 to 1850, caused upland glaciers to advance several kilometers down valleys and cultivation to retreat from higher areas in midlatitude countries. From the early 1800s, climatic warming resulted in a reversal of those trends.

A present concern is that the carbon gases humans pour into the atmosphere in burning coal, oil, and gasoline will add to the gases that trap solar energy in the lower atmosphere and may lead to enhancements of natural climatic changes through **global warming.** If ice sheet margins melt and the ocean level rises a meter or so within the next 50 years, the low-lying coasts of coral islands, wetlands, and port areas will be at risk. The precise trends and impacts are difficult to predict, however, because natural events are too complex. Even where predictable, the huge scale of size and energy involved in such events as the advances and retreats of ice sheets places them outside the realm of human abilities to alter them.

Shaping Earth's Surface

Earth's surface is 71 percent covered by ocean water and only 29 percent occupied by the continents on which people live. The natural environment of continental surfaces forms the height and slope of the land, or its **relief.** The differences among mountains, valleys, and plains are caused by the interaction of internal Earth-building forces with external Earth-molding influences resulting from weather and the action of the sea (Figure 2.29).

(a)

(b)

Figure 2.28 **Tropical weather systems: thunder-storm groupings over central Africa.** Seen from a space shuttle, several cloud tops amalgamate in fibrous masses of ice that spread outward and downwind. Photo: NASA.

Figure 2.29 **Internal and external Earth forces mold the landscape.** The gash across the Carrizo Plain in southern California was caused by the San Andreas Fault splitting a section of Earth's crustal rocks. Rain and rivers then etched out the line of weakness. Further movement along the fault shifted and offset the line of stream valleys on either side. The San Andreas Fault occurs where two plates move against each other, with the western side moving northward. Photo: © Kevin Schafer/USGS/Tom Stack & Associates.

Figure 2.27 **Tropical weather systems: hurricanes.** (a) Hurricane Elena in the western Atlantic Ocean, as viewed from a space shuttle. Several hundred kilometers across, it has the typical eye of descending air, surrounded by strong winds and swirls of cloud. Hurricanes form over tropical oceans, where the water temperature and evaporation rates are high. (b) Damage caused by Hurricane Hugo when it hit the South Carolina coast in 1989. Photos: (a) NASA, Michael Helfert; (b) S. J. Williams, U.S. Geological Survey.

Earth Interior Forces

Earth is a multilayered planet with a hot molten core. Earth's internal heat provides the energy that forces large blocks of surface rock, thousands of kilometers across and around 100 km (65 mi.) thick, known as **tectonic plates,** to move and crash into each other (Figure 2.30a). The heat from the interior is sufficient to drive the movements of the six major and seven minor plates (Figure 2.30b).

The plate movements produce the major features of Earth's surface, including mountain systems and continents, and cause the opening and closing of ocean basins. The rock that forms the main part of the plate solidifies from the molten state where it erupts along fissures opened up as plates move apart. Such divergent plate margins include the Mid-Atlantic Ridge that erupts the molten lava building Iceland.

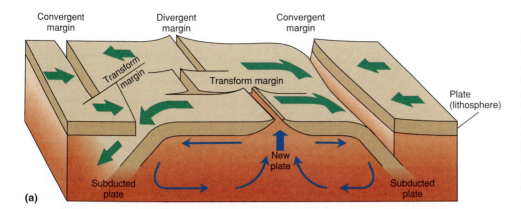

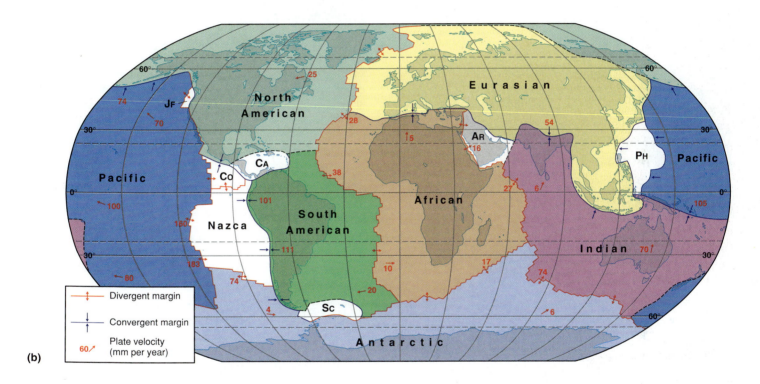

Figure 2.30 **Plate tectonics, continents, ocean basins, and mountain ranges.** (a) The main features of plates and plate margins. Green arrows show plate movement directions, blue arrows depict possible internal Earth convection currents driving movements in the top 100 km of Earth's interior. (b) World map of major and minor plates. The minor plates include Nazca, Cocos (Co), Caribbean (Ca), Juan de Fuca (Jf), Arabian (Ar), Philippine (Ph), and Scotia (Sc). Source: (b) Plate velocity data from NASA.

Plates collide with each other along convergent margins, often forcing one side upward to form mountain systems such as the Andes Mountains of South America. The plate that is forced beneath the raised plate is said to be subducted. As the subducted solid rock melts under high temperatures of burial, the molten rock under pressure rises toward the surface, creating volcanic mountains and piles of lava such as those forming the Columbia Plateau in the northwestern United States.

Earthquakes and volcanic outbursts of molten rock from beneath Earth's surface concentrate along plate boundaries, signaling the effects of plates colliding or moving apart. Where plates move horizontally against each other, transform faults such as the San Andreas Fault in California (see Figure 2.29) create earthquakes, but few have volcanic eruptions.

Surface Changes

Once land emerges above sea level, the details of surface relief result from attacks by the atmosphere and ocean waves. Changes in the temperature and chemical composition of the atmosphere, together with water from rain and snowmelt, react with the rocks exposed at the surface and break them into fragments. Small particles are acted on by chemicals in the water. Such changes are called **weathering.** The broken and dissolved rock material forms the mineral basis for soil. On steep slopes, it moves downhill under the influence of gravity. Such movement may be rapid in slides, flows of mud, or avalanches, or slow in local heaving and downslope creep of the surface. The mobile fraction of this broken rock material often falls into rivers or onto glaciers and is moved toward the ocean. Glaciers formed of ice move slowly, helped by meltwater lubrication in the summer.

The concentrated flows of water or ice and rock particles in rivers and glaciers gouge valleys in the rocks—a process called **erosion** (Figure 2.31). When the flows reach a lake or the ocean, they stop and the rock particles drop to the lake or ocean floor in the process of **deposition.** Wind blows fine (dust-size) rock particles long distances, while sand-sized particles are moved around deserts or across beaches to form dunes. Along the coast, sea waves and tides fashion distinctive eroded cliffs and deposited beach features, often moving the rock particles supplied by river, glacier, or wind.

Physical and Human Landscapes

In a particular region, the relief features of the land surface are determined by the combination of these internal and external

Figure 2.31 **Eroding the land: rivers, glaciers, and the sea.** (a) Stream with boulders up to a meter across, which are used by the stream to wear away its channel bed. (b) Climbers crossing glacier on way to top of Mount Everest, Nepal. The surface snow on the glacier shows the layering caused by successive unmelted snowfalls. The glacier ice beneath opens along tension cracks, or crevasses, making the climb dangerous. (c) Ocean waves attack the cliffs and form beaches in Oregon.

(a)

(b)

(c)

forces, together with modifications added by human occupation, as the two examples in Figure 2.32 show. The landscape outcome depends on whether the region contains a plate margin, which climatic elements fashioned the surface, how long the natural forces operated without catastrophic changes, and the extent to which people affected the surface processes. Over a longer period, changes of climate are important in altering the surface forces at work, such as rivers, glaciers, or wind, as Figure 2.32 shows. In another example, eastern Canada and the northern United States were covered by an ice sheet up to 20,000 years ago, but rivers are now reestablished on the landforms produced beneath the ice.

Plants, Animals, and Soils in Ecosystems

Living organisms may be a unique feature of Earth: we currently know of no other planet where life exists. Plants and animals are sustained by a combination of energy from the sun's rays, water circulating from oceans to the continents and back again, and nutrient chemicals in the soils. Most plants have the ability to capture and store the sun's energy in chemical form, combining with the mineral nutrients drawn from soil to produce the foods that animals require. The atmospheric processes that break up surface rocks, the actions of other organisms, and the water flowing through them make the nutrients available to living organisms.

Ecosystems and Biomes

Plants and animals live in communities in which they share the physical characteristics of heat, light, water availability, and nutrients. An **ecosystem** is the total environment of such linked communities and physical conditions. Ecosystems exist at all geographic scales but for the purposes of this text are discussed in relation to the largest scale, or **biome.** There are five main types of biome—forest, grassland, desert, polar, and ocean. Those on land reflect the major climatic environments, although the natural vegetation and soils of most biome areas have been greatly modified by human activities.

- **Forest biomes** exist in both tropical and midlatitude climatic environments where sufficient water is available. Tropical forest types range from the dense rain forest of equatorial regions, which is characterized by the crowding of great numbers of plant and animal species, to more open forests with fewer species in areas having marked dry seasons. Midlatitude forests include the deciduous forests, typical of the eastern United States, in which the leaves are shed in winter, and the evergreen forests typical of Canada and western North America.

- **Grassland biomes** occur where longer dry seasons or annual burning by humans restrict tree growth. In the tropics, savanna grasslands are characteristic of much of the seasonal climatic regions, as in Africa with its wide range of large animals from lions to elephants (Figure 2.33a). Midlatitude grasslands once occurred in the prairies of North America, the steppes of southern Russia, the pampas of Argentina, and the veldt of Southern Africa. These environments are now dominated by grain farming and grazing.

(a)

(b)

Figure 2.32 **Landscape evolution: two examples explained.** (a) Western Scotland. The rocks in this northwestern part of Britain were formed over 400 million years ago and then crushed and raised into mountain ranges by colliding plates as an earlier Atlantic Ocean closed. Rivers wore down the mountains. The Atlantic Ocean opened again some 50 million years ago, raising this area with others along its margins. Within the last million years, the Ice Age covered the higher parts with thick ice that moved outward, forming deep valleys such as the one in this photo. On melting, the ice poured water back into the oceans, raising its level and drowning the lower part of this valley as a sea loch. Rivers eroded the hills, depositing deltas along the edge of the deep water. (b) Zimbabwe, Southern Africa. The contrast between the rocky hills and the flat plain results from a position far from any plate margin and millions of years of intense chemical decomposition in a tropic environment of seasonal rains. Rock particles weathered from the rocky hills fall to the plain, where rivers at the start of the wet season carry off dry surface particles. Photos: (a) © Michael Bradshaw; (b) © Paul Bradshaw.

With its low shrub vegetation, the Arctic tundra does not qualify as grassland but has some similar characteristics of that form and supports grazing animals. Tundra is present where long cold seasons and strong winds prevent tree growth.

- **Desert biomes** occur in lands dominated by aridity (where evaporation exceeds water supply). Deserts are characterized by little or no vegetation. Any plants that survive in such conditions will have special means for storing water, as in the cactus and short-flowering, seed-producing regimes (Figure 2.33b). Tropical deserts include the zone extending from the Sahara in northern Africa through southwestern Asia to Pakistan. Other tropical deserts occur in central Australia, southwestern North America, and coastal Peru and northern Chile. Midlatitude deserts occur in Central Asia, Patagonia (southern Argentina, South America), and in some areas of western North America.

- **Polar biomes.** In polar lands, ice cover, combined with low inputs of solar energy, provide little sustenance for living organisms. Coastal areas of Antarctica, however, support a variety of animals through the offshore marine resources.

- **Ocean biomes** contrast with those on the land, but differences in water temperature and nutrient availability produce zones of more or less abundant life between tropical, midlatitude, and polar zones. Microscopic plankton plants take advantage of sunlight in the surface waters and form the basis of living systems in the oceans. The richest biomes with the most profuse numbers of organisms occur where rising cold water brings nutrients to the surface, as along the west coasts of Africa and North and South America. Such differences affect the productivity of fisheries, with smaller numbers of a wide range of different species occurring in the tropics and greater numbers of fewer species in midlatitude oceans.

Soils

Soils are a particularly significant aspect of the natural environment for humans. Soil fertility, together with climatic heat and moisture conditions, governs whether a region will produce good crops, support livestock farming, or have little farming potential. Soils form as broken rock matter interacts with weather, plants, and animals. The original rock materials supply or withhold nutrients. Water from rain and snowmelt makes any nutrients present available to plants; decaying plant and animal matter releases the nutrients back to the soil in mineral form.

In the United States, for example, the soils of the eastern half developed beneath deciduous forest, in which the annual leaf fall and ample rainfall returned nutrients to fertile brown soils. Farther north, the long winters restricted growth to evergreen forests and the pine-needle litter made the soils acidic, reducing their nutrients, and making them difficult to cultivate (Figure 2.34a). In the South, rainwater combined with higher temperatures increased the rates of chemical reactions and removed most nutrients from areas of sandy soils through which water passes easily. In the Midwest, development beneath prairie grassland (Figure 2.34b) produced the black soils that proved the most fertile and resilient of all, being less acidic and having a coherent structure that resists erosion. Farther west, the presence of mountains and arid areas resulted in the formation of few well-developed soils; the steep slopes encourage erosion; low rainfall prevents soil cohesion, while high rates of evaporation cause the accumulation of alkaline salts in the upper layers (salinization).

Human Impacts

Natural environments operate largely outside human controls, being powered by energy from the sun or Earth's interior. They are, however, impacted by specific human activities that change rates of erosion, remove natural vegetation cover, or emit pollutants into the atmosphere and hydrosphere.

Early in human history (see Chapter 1), livelihoods depended on the local characteristics of weather, rocks and minerals, landforms, water supply, vegetation, animals, and soils. Even at this early stage of low densities of population, human activities locally changed the workings of the natural environment. Burning of vegetation in dry seasons around forest margins encouraged the extension of grassland and the increase of grazing

(a)

Figure 2.33 **World biome types: savanna grassland and desert.** (a) Tropical savanna grassland, Eastern Africa. The grasses grow in areas of moderate rainfall, supporting large numbers of grazing animals, such as zebras and gazelles. These become food for lions and other carnivores at the end of the food chain. (b) Desert vegetation, western USA, including cactus and other plants that store water and flower briefly after a rare fall of rain. Photo: (a) © Tim Davis/Photo Researchers, Inc.

(b)

animals—such as the bison in North America that provided meat and hides for Native Americans (see Chapter 11).

Farming and Erosion

The first farmers used lighter soils where there was not much vegetation to clear. From around 1000 B.C., new iron implements made it possible to fell trees on a larger scale and extend farmland into heavy clay soil areas. Both phases of woodland clearance increased soil erosion in uplands and deposition in lowlands, together with changes in the species composition of animals and plants. In Asia and elsewhere, the construction of terraces that reduced erosion extended the cultivated area to steep slopes at the expense of forest cover.

Industrial Revolution

The Industrial Revolution after 1750 intensified the scale of human intervention in natural environments (Figure 2.35). The rate of soil erosion from deforested slopes and plowed fields increased, extending to areas of the world that were opened up for agriculture in the 1800s and 1900s by settlers from the colonizing countries with the aim of producing commercial crops for sale to the wealthier countries. Factories in industrializing countries poured their wastes into the atmosphere and rivers. Pollution became a health problem in many industrial cities. City expansion produced surfaces of tile, brick, concrete, and asphalt that caused rain to run off the surface more rapidly, adding to the numbers of floods downstream. Demands for fish by a rising world population led to overfishing of the oceans; increasing demand for timber and land led to the cutting of tropical rain forest. Both overfishing and forest cutting remain major aspects of resource depletion.

Atmospheric Pollution

Environmental problems built up with the increase of the materials entering the atmosphere from industrial processes. The carbon gases emitted by burning coal and gasoline exceeded the amounts that natural systems could absorb , and the carbon dioxide content of the atmosphere rose slowly. In the late 1900s, fears arose that the increasing atmospheric levels of carbon gases would encourage greater heating of the air near the ground and cause global warming.

The growth of the "ozone hole" over Antarctica posed another potential problem for continuing human existence on Earth. Ozone is a gas that is concentrated in the upper atmosphere above the sector in which weather movements occur. The gas forms a protective shield, intercepting incoming ultraviolet rays from the sun. If the ozone layer is destroyed, the sun's rays will destroy plants and animals and increase the risk of skin cancer in humans. In the 1980s, it became clear that chlorine gases, including the human-made chlorofluorocarbons used in refrigeration systems, gradually permeate upward and destroy ozone by chemical reactions. The reactions are most intense during the polar winters over Antarctica: the October measurements (at the end of Antarctic winter) show the greatest depth of ozone depletion, forming the hole. Elsewhere around the globe, the ozone layer thinned slowly. Governments agreed to end the use of the ozone-depleting gases, and the policy is having an effect, although the potential dangers will last for several decades.

While global warming and ozone depletion may have worldwide effects, acid rain affects areas up to several hundred kilometers downwind of major urban-industrial areas. Sulfur and nitrogen gases from power stations and vehicle exhausts react with sunlight in the atmosphere and return to the ground as acids. Soils and lakes that are close to the pollution sources and are already somewhat low in plant nutrients suffer first.

The effects of acid rain may be felt in countries beyond the origin of the pollutants. Canadians claim that emissions from Ohio River valley power plants in the United States affect their eastern forests and lakes; Scandinavians complain about the emissions from western European industrial areas. The phenomenon is a growing menace downwind of new industrial areas in developing countries. Something of the complexity of human interactions with natural process was demonstrated in the early 2000s by the recovery of European forests thought to have been affected by acid rain as global warming brought higher temperatures, more rain, and a new greening of the trees.

(a)

(b)

Figure 2.34 **Soil types: podzol and black earth.** (a) A podzol in France formed by rain removing nutrients from sandy material. Such soils, with their ash-gray layer, are common beneath northern forests in North America and Eurasia. Their low nutrient content makes them difficult to cultivate. (b) A black prairie soil in Texas. The dark surface layers have a high organic and nutrient content, contrasting with the lower white section rich in calcium carbonate. These are among the most fertile soils for farming. Photos: (a) Hari Eswaran, U.S. Department of Agriculture; (b) U.S. Department of Agriculture.

Desertification

Desertification is the destruction of the productive capacity of an area of land. It can occur following natural shifts of climate or because of human activities. Deserts, such as the Sahara, expand in area, but humid areas also experience desertification where vegetation is removed or the soil is stripped away. Environmental contamination, as when mining wastes foul the land or nuclear or other toxic wastes leak, also makes land unusable for many years.

The southern margin of the Sahara experienced desertification from the 1970s after commercial agriculture had extended into subhumid areas and years of drought followed (see Chapter 9). The deaths of cattle and vegetation, and lack of irrigation water caused many communities to move away from their traditional lands.

Resources and Hazards

Natural resources and hazards are distributed unevenly around the world and have important influences on regional geography. Their study unites aspects of physical geography with cultural perceptions of what is useful or presents difficulties to human populations.

Natural Resources

The natural environment provides human societies with **natural resources,** materials that they identify as resources and use to maintain their living systems and built environment. The recognition and availability of resources, their technological and cultural usefulness, and economic viability are basic to economic development and regional differences.

Resources valuable to one society or technology are not always rated highly by others. For example, Stone Age peoples used flint and other hard rocks that flaked with sharp edges to make tools and weapons, but such rocks have few uses today. The clay mineral bauxite was ignored until it was found that refining it produced the strong, lightweight metal aluminum.

Among energy resources, emphasis shifted from wood to wind, running water, coal, oil, gas, and nuclear fuels.

Natural resources include fertile soils, water, and minerals in the rocks. **Renewable resources** are naturally replenished. The best example is solar energy, which provides a constant stream of light and heat to Earth, but the proportion that plants and humans can use is limited. Water is a renewable resource that is recycled from ocean to atmosphere and back to the ground and oceans. Many countries use less than 20 percent of the water falling on their territory, returning much of that to natural systems after use. All renewable resources are, however, ultimately finite in quantity and quality. They may be overused. For example, reaching the limits of water supply

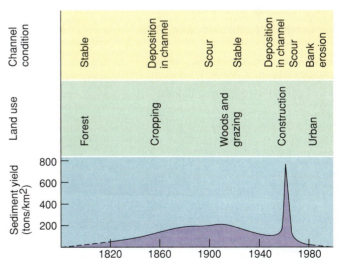

Figure 2.35 **Human impacts on the environment.** The example of a watershed near Washington, D.C., that has been subject to land-use changes over the last 200 years. Describe how clearance for farming, land abandonment, and urban expansion affected the sediment flow and hence changed rates of erosion by the rivers of the area.
Source: Reprinted from "A Cycle of Sedimentation and Erosion in Urban River Channels" by M. G. Wolman from *Geografiska Annaler,* 1967, 49A:385–395, by permission of Scandinavian University Press.

affects irrigation-based development in arid countries, such as along the Nile River valley and in the western United States.

Other resources are nonrenewable: extraction exhausts a nonrenewable resource. **Nonrenewable resources** include the fuels and metallic minerals available in rocks. Technological advances or new and increased demands, however, assist in finding new sources, extracting sources that were once thought to be uneconomic, and recycling metal products. Such technologies extend the lifetime usefulness of nonrenewable resources.

Natural Hazards

The natural environment poses difficulties and challenges for human settlement in **natural hazards** such as volcanic eruptions, earthquakes, hurricanes and other storms, river and coastal floods, and coastal erosion. Hazards interrupt human activities to a greater or lesser extent, although they seldom deter humans from settling or developing a region if its resources are attractive. For example, people are drawn to living in California or the major cities of Japan despite the likely occurrence of earthquakes. Similarly, people continue to live and work in areas prone to hurricane damage or river flooding. In areas such as the Mississippi River valley in the United States and the lower Rhine River valley in the Netherlands, protective walls are built to cope with all but the worst river floods. The highest levels of flooding, as in 1993 along the Mississippi and in 1995 along the lower Rhine, may overtop these walls.

Hazards cause loss of life and destruction of property, but the costs of protection against hazards are also high. Most protection is provided in wealthier countries, where hazards cause the least loss of life and the greatest damage to property. Many poorer countries have few resources available to construct protective measures against natural hazards and often suffer major losses of life after floods, hurricanes, or earthquakes. Comparisons between the few people killed when hurricanes strike the United States and the high number killed in the hurricane devastating Central America in 1998 illustrate this point.

Test Your Understanding 2C

Summary Environmental issues are an outcome of interaction between human beings and the natural world. They are important in sustainable human development. The linked air and water circulations of the atmosphere and oceans, stirred and circulated by solar energy, set off a series of weather events that add up to climatic environments. These can be divided into tropical, midlatitude, and polar variations. Climate change is normal, but at present, human impacts are becoming powerful enough to be part of the causes of such changes.

Earth's surface features are produced by the interactions between internal forces that produce volcanoes, earthquakes, and mountain ranges and climatic forces of temperature changes, wind, water, and iceflow. Plants, animals, and soils are parts of ecosystems fueled by solar energy and with nutrient circulations linked to broken rock materials and atmospheric gases. Forest, grassland, desert, polar, and ocean biomes are distinctive world region–scale ecosystems.

Human interactions with the natural world convert forest and grassland into farmland but also pollute air, water, and soil. Humans designate natural resources and combat natural hazards.

Questions to Think About

2C.1 How do (a) natural environments affect human actions and (b) human activities affect natural environments? Refer to recent events to illustrate your answer.
2C.2 What causes natural environments to change?
2C.3 What are some examples of renewable and nonrenewable resources?

Key Terms

climate	erosion	ocean biome
tropical climates	deposition	soils
midlatitude climates	ecosystem	desertification
polar climates	biome	natural resource
global warming	forest biome	renewable resource
relief	grassland biome	nonrenewable resource
tectonic plates	desert biome	natural hazard
weathering	polar biome	

Online Learning Center

www.mhhe.com/bradshaw

Making Connections

The Online Learning Center accompanying this textbook provides access to a vast range of further information about each chapter and region covered in this text. Go to **www.mhhe.com/bradshaw** to discover these useful study aids:

- Self-test questions
- Interactive, map-based exercises to identify key places within each region
- PowerWeb readings for further study
- Links to websites relating to topics in this chapter

Chapter 3

Europe

Figure 3.1
Europe: the subregions and the physical landscapes.

NORTHERN EUROPE

EAST CENTRAL EUROPE

WESTERN EUROPE

MEDITERRANEAN EUROPE

ATLANTIC OCEAN

North Sea

Baltic Sea

NORTH EUROPEAN PLAIN

ALPS

Mediterranean Sea

This Chapter Is About . . .

- Europe sets the new global order
- Europe's cultural diversity
- Temperate, midlatitude natural environments
- Setting a new global standard in the modern era
- Four subregions:
  - Western Europe: Europe's industrial heartland
  - Northern Europe: high living quality in harsh environments
  - Mediterranean Europe: Europe's oldest civilizations
  - East Central Europe: former Communist countries adopting capitalist practices

Point-Counterpoint: The European Union

Personal View: Bosnia-Herzegovina

European Influences

Many great ideas and inventions originated in other world regions, often long before Europeans made any discoveries. Nevertheless, Europe (Figures 3.1 and 3.2) is a hearth for many contemporary global ideas and practices. Some of these include democracy, Christianity, colonialism, imperialism, capitalism, the Enlightenment, nationalism, fascism, socialism, communism, and genocide. As stated, not all are wholly European. Christianity, for example, began in Southwest Asia, but for centuries, the religion was solely in Europe, where it took on a European character before spreading around the world. Europeans also made such scientific discoveries as the Earth being spherical and revolving around the sun, the laws of motion and gravity, the theory of evolution, genetic science, the causes of disease, radioactivity, and the theory of relativity. European inventions include movable type for the printing press, the microscope, telescope, steam engine, railroad, internal combustion engine, automobile, radio, orbital satellite, and digital computer. The Industrial Revolution also began in Europe, radically altering Europe's economy and society and then making similar dramatic changes in other world regions.

Europeans did not merely use their technologies to explore and gain more material wealth from abroad. They also used their technology to transform Europe's natural environments. Farms and market towns replaced forests, and in time, factories and cities replaced many farms. Rivers were straightened for navigation and dammed to prevent flooding and provide power. The Earth was mined for coal and ore and then drilled for oil. All these activities improved the material standard of living but profoundly affected Earth's ecosystems.

For better or worse, over the last 500 years, Europeans used their local ideas to lay the foundation for the modern global economy, structure the world's political system, and spread their culture around the world. Europeans frequently forced their ways on other cultures, often brutally, as they took control of lands and peoples around the world. Though dominating, they integrated local ideas, technologies, and faiths from cultures in Asia, Africa, and the Americas into their own cultures. After the trauma of World War II, Europeans became less aggressive in their interactions with other cultures, even becoming major proponents of human rights and human development. European cultures continue to interact with and react to cultures in other areas of the world, illustrating that globalization is not destroying geographic differences but is often resulting in different and alternative cultural practices.

Diversity, Conflict, and Technological Innovation

Current attempts to unite European countries into a whole occur against a history of divisions and divergences. Politically, Europe is a very fragmented world region with many small and medium-sized countries that reflect its varied history of conflict and cooperation. The diversity of Europe began centuries ago, when numerous peoples migrated into the world region, bringing with them new cultural practices.

Migrations of Peoples

The current European peoples have not inhabited their region as long as other peoples have occupied their lands. The Greeks, Romans, and Celts were the first to inhabit Europe, followed by Germanic, Slavic, and other peoples.

Figure 3.2 Summary information on Europe's subregions.

Subregions	Land Area (km^2) Total	Population (millions) Mid-2001 Total	2025 est.	GNI 1999 (US $ million) Total	GNI PPP 1999 Per Capita	Percent Urban 2001	Human Development Index Rank of 175 Countries	Human Poverty Index Percent of Total Population
Western Europe	1,421,726	247.2	257.3	6,188,996	26,210	76	12.6	12.2
Northern Europe	1,257,120	24.3	25.8	692,866	25,140	77	9.0	10.6
Mediterranean Europe	1,030,430	118.5	111.4	1,983,815	17,878	65	23.8	12.3
East Central Europe	1,341,140	128.5	125.8	392,581	8,131	60	57.8	no data
Totals or averages	**5,050,446**	**518.5**	**520.3**	**9,258,258**	**19,339**	**70**	**25.8**	**11.7**

Source: Data from *Population Reference Bureau 2001 Data Sheet; World Development Indicators,* World Bank, 2001; *Human Development Report,* United Nations, 2001.

Cultural Groups: Greeks, Romans, Celts

Modern-day **Greeks** only inhabit a small area of Europe in the southeastern corner. Their ancestors, however, were firmly established in Europe by 1000 B.C. The Greek city-states had differing forms of government, but some established democracy, later adopted in other areas of Europe and the world. Early Greeks also made great contributions to the sciences and humanities, and a number of their ideas still influence the way many people view the world.

Around the 100s B.C., Rome's ascendancy eclipsed the Greek city-states. The **Romans** created a large empire that stretched from the eastern Mediterranean to the British Isles. Crucial to keeping their empire together, the Romans built an extensive road system, remnants of which are found today (Figure 3.3). Many of their army camps grew into large cities, such as London and Paris. Aspects of Roman culture were passed down through the ages. The Romans spoke Latin, which persisted in the lands that Romans occupied longest: Gaul (France), Hispania (Spain and Portugal), the Italian peninsula, and Dacia (Romania). Without political unity after the fall of Rome, Latin evolved differentially into the distinct languages of French, Spanish, Portuguese, Italian, and Romanian (Figure 3.4a). The Romans eventually accepted Christianity, making it the empire's official religion in A.D. 381, and thus helping it to become a major world religion. In time, Christianity spread well beyond the former borders of the Roman Empire (Figure 3.4b), but the center of the Roman Catholic Church, the branch of Christianity with the most members, is still located in Rome.

Jews have been in Europe since Roman times, first along the Mediterranean and then eventually everywhere else, especially East Central Europe. Persecution, especially during the Nazi Holocaust in the 1930s and 1940s, sharply reduced their numbers. Nevertheless, Jews have contributed greatly to European culture.

At the same time Greek civilization emerged in Southeastern Europe, **Celtic** peoples occupied large areas of Central and Western Europe. They gave their names to such places as Bohemia, Gaul, the Alps, and the Rhine. Celtic culture began to wane as the Romans and then later peoples took over many of their lands. Remnants of their culture can be found in the language and culture of the modern Welsh, Scots, Irish, and Bretons of western France (see Figure 3.4a).

Cultural Groups: Germanic and Slavic Peoples

During Roman times, **Germanic peoples** of many names arrived from the east, usually Asia, conquering whatever Celtic lands the Romans had not taken, namely the areas just north of the Danube and east of the Rhine. These tribes continually threatened the Roman Empire, sacking Rome itself for the first time in A.D. 410. By the end of the 400s, Gaul was taken over by the Franks, eventually to be renamed for them (France). The Burgundians lent their name to a province (Burgundy) that was eventually absorbed into France. The Visigoths and Lombards moved into the Italian peninsula. The latter name is found in the modern Italian provincial name of Lombardy. The Angles and Saxons moved into the British Isles, pushing the Celtic peoples farther into the fringes of Europe. Even today, the English are considered Anglo-Saxons, often simply called Anglos (Figure 3.5).

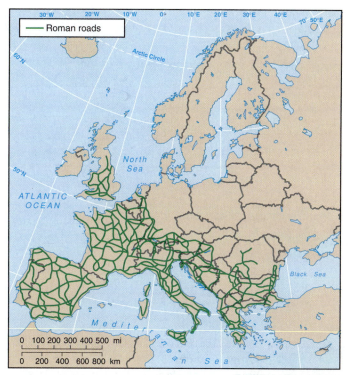

(a)

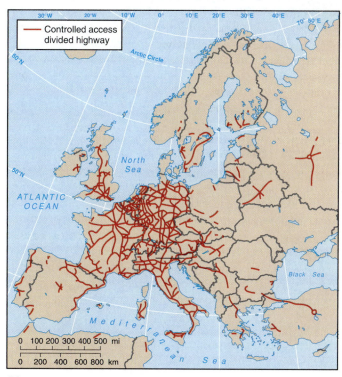

(b)

Figure **3.3 Roman influences.** A comparison of Roman roads (a) with modern, divided highways (b) shows many similarities. Source: Jordon-Bychkov, *The European Culture Area,* 4th ed. Reprinted by permission of Rowman & Littlefield.

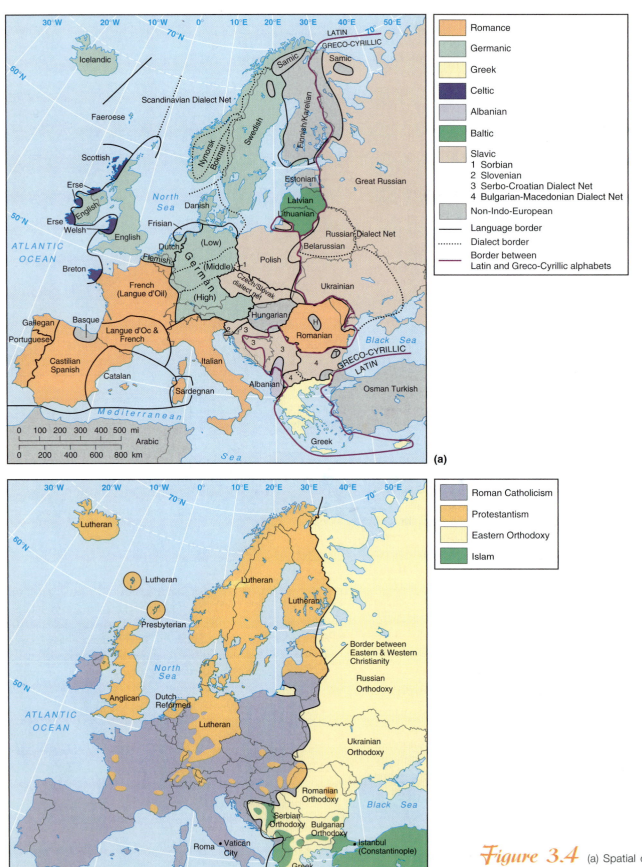

Figure 3.4 (a) Spatial distributions of languages and selected dialects of modern Europe. (b) Spatial distributions of religious groups in modern Europe. Source: Jordon-Bychkov, *The European Culture Area*, 4th ed. Reprinted by permission of Rowman & Littlefield.

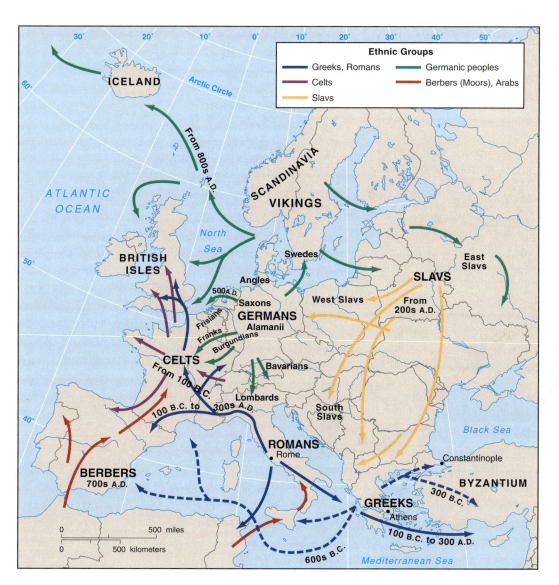

Figure 3.5 **Europe: movements and migrations of people.** Some of the early movements affected later geographic units, from approximately the 300s B.C. to the A.D. 900s. Note how Germanic groups spread outward, displacing the Celts and Romans in the west, and how Slavs moved into eastern Europe, displacing some Germanic groups.

Other Germanic tribes moved north into Scandinavia. By the A.D. 800s, they developed a distinct **Viking** culture. The Vikings were a seafaring and martial people who experienced overpopulation in their home areas. During this time of warm climate, they sailed northward, colonizing the Faeroe Islands, Iceland, and southwestern Greenland, and probably reached North America by the A.D. 900s. From the south, the area of modern Germany, the Vikings felt the pressure of Christianizing forces. In response, they invaded the British Isles and northern France and sacked wealthy monasteries. To the southeast, they had major impacts on the lands around the Baltic Sea and as far south as Kiev by the late 800s, where they played a formative part in founding Kievan Rus', which became one basis for the future Russia. Although they are often caricatured as violent pirates, Vikings mostly settled alongside the local inhabitants of areas that they invaded, organizing wide-ranging trade across Northern Europe. Their influence ended after the 1200s, when cooling climate, volcanic eruptions in Iceland, and political factions caused the Greenland colonies to die out and halved the population of Iceland. Plagues reduced the numbers of people in their homelands.

Germanic culture is still prevalent today. Though the Franks, Burgundians, and Lombards adopted the Romance languages of the Roman provinces they conquered, other Germanic peoples, like the Vikings, maintained their Germanic languages through the centuries and are clearly seen on the map today (see Figure 3.4a). Germans, Austrians, Dutch, and the Scandinavians (e.g., Danes, Norwegians, Swedes) are the most numerous of today's Germanic peoples. The Germanic peoples also converted to Christianity and later became the driving force behind the creation of the branch of Christianity known as Protestantism (see Figure 3.4b).

The last of the major groups to migrate into Europe were the **Slavs,** who began arriving in the A.D. 400s. During the next few centuries, Slavs pushed as far west as the Elbe River in the middle of modern-day Germany and as far south as the Adriatic coast and into the Balkan peninsula, threatening the Greeks. The Slavs are divided into three major groups: western, southern, and eastern. Poles, Czechs, and Slovaks are western Slavs. Slovenes, Croats, Serbs, and Bulgarians are southern Slavs. The eastern Slavs are Russians, Ukrainians, and Belarussians (see Chapter 4).

The groups thus far mentioned account for most of the European peoples and represent considerable diversity. Yet a few other groups migrated into Europe and left their mark. Latvians, Lithuanians, Estonians, Finns, Hungarians, Albanians, Roma (Gypsies), and Basques are the most well known.

The Rise of European Global Power

Our global economy is fundamentally free market, or capitalist, in nature. **Capitalism,** the practice of individuals and corporations owning businesses and keeping profits, traces its origins to Mediterranean and Western Europe. In the late 1400s, mercantile capitalism flourished as merchants invested in trade expeditions that brought profit in the form of precious metals (gold and silver). It began in 1418, when Prince Henry of Portugal established an institute at Sagres, where he brought together scholars to improve and teach the methods of navigation to Portuguese sea captains. He hoped that better skills would lead to the discovery of a sailing route around Africa and on to the Spice Islands in the east, where Europeans could obtain spices and other cherished commodities. Exploration was slow at first, but progress was steadily made. In 1441, a ship finally sailed far enough south to reach wetter parts of Africa south of the Sahara. Portuguese explorers captured men and women for slavery and found gold. News of this event sparked enthusiasm for exploration among many Europeans. Within a short time, many new voyages were launched. The Portuguese and Spanish, later followed by Western Europeans, quickly discovered and conquered new lands, radically altering and frequently destroying many local economies and cultures of indigenous peoples as they began a new era of **colonialism** and **imperialism.**

The best known of the new voyages was made by Christopher Columbus in 1492. Funded by the Spanish crown, Columbus sailed westward in the belief he would get to India by a quicker route. He did not know that the Americas (supposedly later named after another Italian sailor, Amerigo Vespucci) lay in his path. At the same time, Vasco da Gama led the Portuguese explorations around the southern tip of Africa to India. Portugal and Spain both gained huge wealth from trading with and colonizing the Americas, Africa, and parts of Asia. The French, Dutch, and British followed them in the 1500s and 1600s. Wars in Europe and declining home economic bases reduced the roles of Spain and Portugal, whose leaders spent their New World wealth on armies to maintain their positions in Europe. Home-based merchant wealth shifted power toward the northwestern European countries. The Dutch emerged as a maritime power in the 1600s, establishing colonies in the East and West Indies but losing territory in North America to Britain in the 1660s. Over the following century, Britain won its competition with France for supremacy in North America and India, setting the basis for its worldwide empire in the 1800s.

European exploration not only resulted in the spread of European culture and practices. Europeans experienced many things in their travels and brought much back to Europe that they incorporated into their cultures. European diet is one aspect of culture that changed dramatically because of exploration. For example, what would the Irish, German, and Polish cultures be without the potato? Yet the Europeans knew nothing of this tuber until they traveled to the Americas. The potato, along with maize (corn), is significant because it has much greater yields than the grains grown in Europe up until that time. Subsequently, the potato contributed to population growth that fueled further exploration, migration, and even the Industrial Revolution. Through this exploration, the Europeans learned about cotton, tobacco, tea, cocoa, and a wide range of other products that they commonly use today.

Industrial Revolution

The growing overseas trade and merchant wealth of the countries of northwestern Europe led to increasing demands for manufactured goods, stimulating a series of technological innovations and organizational changes. The resources of cottage weavers and blacksmith forges were hard pressed to fill the expanding markets for cloth and metal goods. From the mid-1700s, machinery, at first driven by waterpower, increased productivity in the metal and textile industries. The concentration of machines in factories that were soon powered by steam from coal-burning furnaces required growing amounts of capital and numbers of workers. This led to the expansion of urban-industrial centers on or near coalfields (Figure 3.6) and to systems of banking and investment. Rivers, canals, and the sea were the initial forms of transportation used to assemble raw materials for the new industries and to distribute their products. Huge port facilities grew in the estuaries of major European rivers. During the 1800s, railroads increasingly gained in prominence.

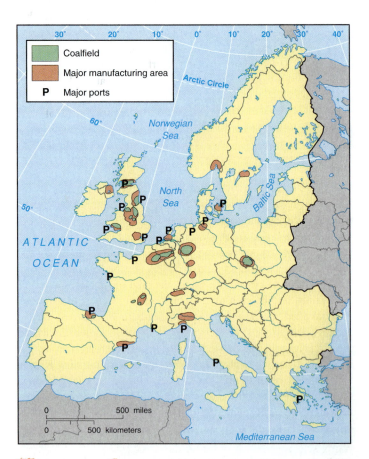

Figure 3.6 **Europe: major manufacturing areas at the start of the 1900s.** Compare the distribution with the late-1900s distribution of population (see Figure 3.21). Note: The contemporary political boundaries are drawn for reference.

Beginning in Great Britain, the **Industrial Revolution** spread across the English Channel to the Netherlands, Belgium, northern France, and the western areas of Germany. In the late 1800s and 1900s, the Industrial Revolution diffused further to Central and Eastern Europe and to other areas of the world. Factories for metal smelting and fashioning, textile machinery, steam engines, and chemical refining were located in areas with plentiful coal resources. The European empires did not transfer their new manufacturing technologies to their colonies. Instead, they used their colonies to produce the raw materials needed in Western European factories. Cotton, wool, indigo, tobacco, and foodstuffs are a few examples. Forced to supply Western Europe with raw materials and having Europe as the only source of finished products, the colonies were pushed into economic dependency. Local economies and traditional ways of life were brought to an end as peoples in the colonies had to change their lives radically to produce exports for Europe.

Modern State System: Nation-States

The current system of international relations and the character of peoples' identities began in Europe. Early in human history, people around the world were loyal to either their tribal leaders or monarchs. Beginning shortly after the first millennium A.D., the identities of European peoples began to change as people began to shift their loyalties to their countries. An important step came with the Treaty of Westphalia in 1648. It marked the end of a bitter religious war between Roman Catholic and Protestant leaders, known as the Thirty Years' War. The treaty laid down rules for religious toleration and ended the arbitrary behavior of monarchs by establishing a legal system of international relations. Monarchs retained considerable power but had to explain to their people why their actions were in the best interests of their country rather than in their own personal interests. Though the rules of the Treaty of Westphalia have been modified, the current system of international relations is still referred to as the Westphalian system.

The Westphalian system led to the development of **nation-states,** first in Europe and then in much of the rest of the world. As noted in Chapter 2, a nation is an "imagined community" of people who believe themselves to share common cultural features, and **states,** often called countries, are politically organized territories with independent governments. Europeans developed the idea of the nation and combined it with that of state to form the **nation-state ideal,** which is the belief that each nation should be free to govern itself and can only do so if it has its own state (i.e., country). Thus, nations become linked with states. The French nation, for example, believes that it is coterminous with the border of France. However, France contains peoples such as the Basques who regard themselves as a separate nation and have tried to establish their own nation-state. Many dominant European nations imposed their national cultures on other nations living within their borders. Germany, for example, tried to incorporate "Germans" living in other countries. This practice resulted in much conflict. To use terms correctly, it should be remembered that nations are peoples, not countries. France, for example, is a nation-state. It is inhabited by the French nation. It is incorrect to refer to France as a nation, though it is acceptable to call France a country.

The grouping of people into nations was not only brought about by the Westphalian system but also by technological innovations such as the printing press, gunpowder, and modern militaries. The scientific ideas of the Enlightenment contributed greatly, too. After Gutenberg invented movable type for the printing press in 1447, books became common. One no longer had to be wealthy to buy and read books. In time, as literacy rose and coupled with Enlightenment ideas of individuality, freedom, and rational thought, common people saw fewer and fewer differences between themselves and their rulers. By the 1700s, many questioned the privileged positions of the aristocracy, turning their loyalties instead to one another and their countries.

As literacy spread, governments had to choose language standards. Regional dialects broke down as people throughout their countries had to conform to the same rules of spelling and writing. Until this time, the dialects of any language were so distinct that people could hardly communicate with one another, though they spoke the same language. Moreover, dialects frequently changed every few miles. The printing press and language standardization made it possible for larger groups of people to communicate with one another. Frequently, the Bible was the main medium for spreading the new language standard.

Gunpowder, too, helped to bind people together. It led to the development of rifles and cannons that required soldiers to drill together constantly so that they could work in unison as a team. Those who preferred the skill and individuality of medieval knighthood were skeptical of armies comprised of soldiers who dressed and acted alike.

Though the new technologies and Enlightenment ideas changed society, people were reluctant to challenge the divinely blessed authority of monarchs. Finally, when people in the 13 British colonies in North America rebelled against the British monarch and defeated the British military, people in Europe saw that their monarchs were not invincible, protected by the hand of God. The American Revolution soon inspired the French Revolution. Napoléon Bonaparte, the leader of France, fed the new nationalist zeal in France, formed a new national army, and defeated the imperial forces of Europe. It required the British and Prussian national armies to defeat Napoléon's national army. Though Napoléon was defeated, he demonstrated that the nation-state, with a population that saw itself as one people and supportive of the state that represented them, was an efficient form of government. The death knell was rung for empires, city-states, and the like, though many hung on for as much as another hundred years before being replaced by nation-states.

Changes in the Modern Era

Nationalism was one of the more significant political ideas that the Europeans brought to the world. As nationalism grew in Europe, it was fanned by the competitive nature of capitalism. Economic competition turned into nationalist competition. Before long, the armies that Western Europeans created to conquer and colonize the rest of the world were turned toward one another. In 1914, war erupted between the European powers, later to be known as World War I. Though Germany and Austria-Hungary were decisively defeated in 1918, the trouble was not over. The war costs and protectionism caused European economies to slump in the postwar 1920s and 1930s, bringing hardship to millions of individuals and families. Discontent and resentment grew in the defeated countries. The nationalist competition became more bitter and fed the more extreme but opposing ideologies of fascism and communism, two other concepts Europe gave to the world. Europe headed down the path of a war that engulfed the rest of the world once again between 1939 and 1945, known as World War II. The intolerant side of nationalism led to fascism and the extermination of millions of people of specific groups such as Jews and Roma (Gypsies), a phenomenon called **genocide.**

After World War II (1939–1945), Europeans seriously reevaluated their role in the world and their relationships with one another. The war had been so devastating that even the winners could be considered losers in some respects, when destruction and financial, political, and cultural aspects were considered. In Western Europe, the U.K., France, and the

Netherlands were confronted with independence movements in their colonies fueled by the ideologies of nationalism and communism that originally came from Europe. Four centuries of building colonial empires on which the "sun would never set" were lost in a couple of short decades.

At the same time, the United States and the Soviet Union emerged as new world powers. The Soviet Union showed its strength when the Red Army moved into most of the East Central European countries at the end of World War II and fostered the establishment of **Communism.** The Communists believed that capitalists used their riches to manipulate their governments in order to protect, even increase, their privileged positions in society and keep the majority of society, especially the working classes, powerless and in relative poverty. Instead, Communists argued for **democratic centralism:** the belief that the Communist Party, the political party of the working class, was the only true representative of the people and, therefore, the only party with the right to govern. To keep capitalists and others from taking advantage of the people, Communists also believed in **state socialism:** governance by the Communist Party, actively running the political, social, and economic activities of the people. The state owned all the businesses and decided what was produced. The capitalist practice of competing companies producing similar products was seen as unnecessary. Rather, large corporations owned by the state made each product. The state, not the free market of consumers, decided what needed to be produced through a **planned economy.**

Natural Environment

The geography of Europe is marked by closeness to the ocean and rapid changes in physical landscape over short distances, factors that influenced human actions over time. The natural environment, particularly the temperate climates, the small scale of geologic provinces, and the long, indented coastline, contributes to the abrupt geographic differences from place to place.

Midlatitude West Coast Climates

Europe's north-south extent is slightly longer than its east-west extent. This is reflected in its range of climatic environments (Figure 3.7). Most are **midlatitude west coast climates** ranging from the icy northern west coasts of Norway near the Arctic Circle and the summer midnight sun to the southern Mediterranean warmth of Portugal.

The far north of Norway, Sweden, and Finland have polar climates, with long, very cold winters and months of snow cover. The coldest weather is somewhat ameliorated along the coasts of these northern lands with air warmed by the North Atlantic Drift current. Most of the western coastal countries from Norway to Portugal have mild winters and warm summers with precipitation throughout the year. Much of their warmth and humidity comes from the winds that bring North Atlantic maritime air. No part of Europe is more than 500 km (320 mi.) from the coast, allowing mild and humid oceanic atmospheric influences to affect the whole region.

Test Your Understanding 3A

Summary Europe set into motion many of the global processes that we experience today. Although individual countries are now less powerful than they were early in the 1900s, many European cultural characteristics are part of other world cultures.

The current human geography of Europe grew out of varied influences, including the movements of Mediterranean, Germanic, Slavic, and Asian peoples. Europe developed global trading links from the 1400s and dominated the Industrial Revolution of the 1700s and 1800s. Europe also developed new political ideas concerning governance.

Questions to Think About

3A.1 What were Europe's historic contributions to the development of the world's present *economic* order? How did they affect Europe itself?

3A.2 What were Europe's historic contributions to the development of the world's present *political* order? How did they affect Europe itself?

3A.3 Which factors promoted European migrations and colonizations from the 1600s to the 1800s?

3A.4 What led to the development of nations and the nation-states?

3A.5 How do government and economy work under Communism?

Key Terms

Greeks	Industrial Revolution
Romans	nation-state
Celts	state
Germanic peoples	nation-state ideal
Vikings	genocide
Slavs	Communism
capitalism	democratic centralism
colonialism	state socialism
imperialism	planned economy

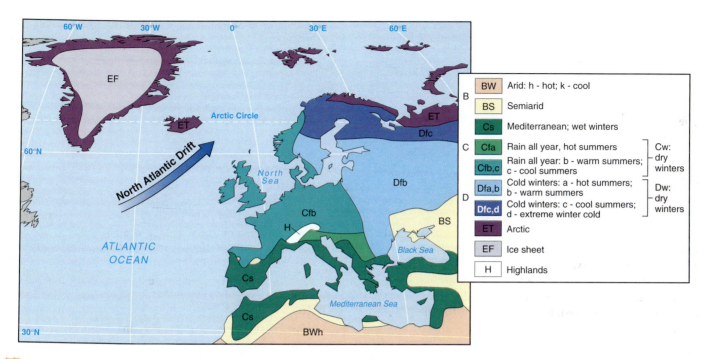

Figure 3.7 **Europe: climates.** Europe is marked by the closeness of all parts of the region to oceanic influences. The North Atlantic Drift brings warmer ocean waters across the North Atlantic Ocean from the Gulf Stream, raising the temperature of the air above and increasing its humidity. In the far north, inland areas become polar in climate, and in the far south, the Mediterranean Sea area has alternating dry, hot summers and mild, wet winters. Toward the east, greater continental heating gives higher summer temperatures, but cold winds blowing from the north and east make winters very cold.

In the far south along the Mediterranean coastlands, there is a seasonal contrast in which the midlatitude belt of cyclones moves southward in winter, bringing rain and wind, while the hot, dry air from the Sahara to the south creates droughtlike conditions in the summer. This southern variant is known as a **mediterranean climate.**

In the **midlatitude continental interior climate** of Central and Eastern Europe, winters can be severe as cold winds bring freezing temperatures from Russia and the northern countries. Summers are warmer than on the coasts and are marked by thunderstorms.

Climate changes affected Europe during the period of human occupation from the later part of the Pleistocene Ice Age. As climatic conditions warmed, the ice sheets, which had extended as far south as the Thames River valley in England and the plains of northern Germany and Poland, melted and retreated northward. Climates throughout Europe got warmer until some 5,000 years ago, when they reached their warmest point. After that time, conditions fluctuated with warmer periods in Roman times, from A.D. 900 to 1200, and around 1850 were punctuated by colder periods. Historians have not fully evaluated the impacts of such climatic changes on economic and political events, but the cooling in the 1300s had much to do with the end of Greenland colonies and with forcing farmers in Europe to lower the upper level of cultivation on hills. Since the mid-1800s, the retreating glaciers in alpine valleys (Figure 3.8a & b) influenced the tourist industry and the generation of hydroelectricity. The retreating glaciers and decreasing extent of snow cover reduced the winter sports season but increased the summer holiday warmth. Inlets to hydroelectricity projects are now sited higher up the valleys, increasing the elevation (and energy) of water falling on the electricity-generating turbines.

Although lands around the Baltic Sea continue to rise gradually above sea level in response to the postglacial melting of the ice sheet, the weight of which caused that area to subside, most of Europe is concerned about the potential impacts of global warming. Rising sea levels would affect the extensive low-lying parts of Europe, including the many areas of former coastal wetland that have been reclaimed and intensively populated. The Netherlands has particular worries in this area, but attention has also focused on the plight of Venice in northern Italy. Venice, a medieval trading city, is one of Europe's greatest architectural treasures and tourist attractions, but it is gradually subsiding (Figure 3.9). Increasingly frequent high tides weaken its foundations. In 1990, St. Mark's Square was flooded 9 times, and in 2000, it was flooded 90 times. In 1966, a south wind raised the tide level by 2 meters (6 ft.), covering St. Mark's Square with oil-polluted water that left stains on all the buildings. As Venice's buildings became less secure, the city's population fell from 175,000 in 1950 to 80,000 in 1990. The industrial development of the lagoon that backs Venice is a major part of the problem, since its industries and oil refineries pollute the waters and make protection of Venice difficult in the face of demands for deepwater channel access.

Geologic Variety

Within its relatively small area, Europe includes almost the world's entire range of geologic features. There are ancient shield areas around the Baltic Sea, the uplands of central Europe, the young folded mountains of the Alps, and the extensive plains

(a)

(b)

Figure 3.8 **Europe: changing climate.** A hundred years has made a difference in Zermatt, Switzerland. (a) The glacier shown on the left side of the 1880 painting disappeared from the modern photo (b), having retreated over a kilometer up the valley, leaving bare rock and icemelt deposits exposed on the valley floor. Locate the church in both views and compare other evidence of the 100-year differences between the two pictures. How might the changing climate have affected tourism? Photos: © Photo Klopfenstein-Adelboden.

in countries around the North and Baltic Seas (Figure 3.10a). Volcanoes erupt and earthquakes occur along the Mediterranean Sea, but much of Europe is almost free of such hazards. The geologic variety in this region, combined with the early development of mining and industrial growth, generated the beginnings of modern geologic science.

The Mediterranean Sea is the remnant of a larger ocean that occupied the area between Africa and Europe but closed as the two continents clashed along a convergent tectonic plate margin (Figure 3.10b). Within this zone, the young folded mountains of the Alps form the highest ranges, with many peaks rising above 4,000 m (13,000 ft.) and the highest point at Mount Blanc (4,807 m, 15,771 ft.) on the French-Italian border. Farther east, the ranges are not so high in Austria, where few peaks exceed 3,000 m (10,000 ft.).

The high Alps are part of a series of ranges that includes the Sierra Nevada in southern Spain, the Pyrenees between France and Spain, the Apennines that form the Italian peninsula, the coastal ranges of the Dinaric Alps in Croatia, Bosnia, Macedonia, and Albania, and the Pindus in the Greek peninsula. The curve of the Carpathian Mountains in Slovakia and northern Romania, continued in the Balkan Mountains of Bulgaria, forms a further extension. Such ranges create the

Figure 3.9 **Venice.** A medieval port city that accumulated great wealth on its unique site in a lagoon. Venice is liable to flooding, as seen here. Photo: © Michael Bradshaw.

dominantly mountainous environment of Alpine Europe and the hilly environments of Mediterranean Europe and the southern Balkans. Lowland areas in these mountainous areas are restricted to river plains and deltas formed by the deposition of rock fragments and particles worn by erosion from the uplands.

To the north of the young folded mountains, most of Europe is lowland, with extensive areas less than 300 m (1,000 ft.) above sea level. The lowlands dominate northern Germany, Poland, and the Baltic countries—part of the North European Plain that continues eastward into Russia. They are formed of more recent layers of rock, covered in the north by glacial deposits.

The lowlands are framed by hilly areas of older rocks that once formed mountain ranges but were worn down by erosion and then raised again by faulting. These include much of Spain and Portugal (the Meseta), France's Massif Central and Brittany areas, the Rhine Highlands of southern Germany, the Bohemian Massif of the Czech Republic, the uplands of western and northern Britain, and those of Norway. As the Atlantic Ocean opened along a divergent plate margin, uplift occurred along continental margins, together with volcanic activity. Rifting extended through the North Sea area, providing a downfaulted block of rocks that became oil reservoirs.

Long Coastlines and Navigable Rivers

Europe is marked by peninsulas such as Scandinavia (Norway and Sweden), Jutland (Denmark), and Brittany (France) in the north, and Iberia (Spain and Portugal), Italy, and Greece in the south. The British Isles formed another peninsula until waves cut the Strait of Dover a few thousand years ago. Arms of the ocean, such as the Baltic, North, Mediterranean, Adriatic, and Aegean Seas, reach far inland. Smaller islands in the Baltic and Mediterranean Seas add to the length of coastline that encouraged many groups to engage in trade and develop ship technology at a time in history when water transportation was easier than land transportation. Skills and technology of that phase were available to expand mercantile capitalism around the globe between A.D. 1450 and 1750.

On land, connections between places and world trade routes were eased by the existence of major river valleys. The Rhine and Elbe Rivers in Germany and the Danube River flowing from Germany through Austria and the Balkans were

ELEVATION

Meters	Feet
4,000	13,120
2,000	6,560
500	1,640
200	656
0	0
Below Sea Level	Below Sea Level

Figure 3.10 **Europe: physical features.** (a) The distribution of mountains and lowlands on a relief map. (b) The relationship of major relief features to plate margins. The southern margin of the continent is part of a convergent plate boundary that was responsible for raising the Alps and other mountain ranges along the Mediterranean Sea. The west-facing coasts result from the opening of the Atlantic Ocean with a divergent boundary in the Mid-Atlantic Ridge that surfaces with volcanic activity in Iceland.

particularly important. The Rhône, Seine, and Loire Rivers in France, the Thames River in England, the Vistula River in Poland, and the Po River in Italy were also significant in movements of people and goods from early times. River valleys large and small also provided good soils. Sites for towns at river crossings attracted and concentrated the settlement of growing populations. Modern investments in the Euroports at the mouths of the Rhine and Rhône Rivers are important for maintaining Europe's continuing role in world trade. Europe is marine in economic outlook and culture as well as in climate.

The Danube River is the longest in this region but is less used for transportation than the Rhine because it flows through less economically developed countries. In the last 50 years, efforts to coordinate the management of Danube waters for transportation and hydroelectricity proved difficult. The Danube flows through more countries (eight of them) than any other major world river, and the Cold War limited traffic eastward from Austria.

The Rhine is Europe's second longest river and the world's busiest waterway (Figure 3.11), used annually by some 10,000 ships that carry 250 million tons of cargo. The watershed includes significant parts of four countries (Switzerland, Germany, France, and the Netherlands). The river waters come from melting snow and ice in Switzerland and from tributaries (Aar, Neckar, Main, and Mosel). The Rhine has major roles in transportation, industrial water supply, and, in the Swiss sector, the generation of hydroelectricity. Human management of the river increased in intensity from the 1700s, but flooding and high levels of pollution remain problems.

Some of the main efforts at managing the Rhine waterway were devoted to making navigation possible for large barges from the international port of Rotterdam at its mouth up to Basel, Switzerland. The river was canalized to that point. As canalization proceeded, rivers were straightened, producing faster flow, channel bed erosion in some sections, and silt deposition in others. Channel deposition partly filled some sectors and caused flooding. Higher levees were constructed to protect from flooding the lands on either side that were used more and more intensively. At the Rhine mouth in the Netherlands, greater efforts were put into keeping out the sea than to maintaining the levees, which were breached in places during the high river levels of early 1995.

The transportation uses of the Rhine River are linked to the growing industrialization of its watershed since the late 1800s. The coalfield and steelmaking areas of the Ruhr and Saar are now less productive and polluting, but some 20 percent of the world's chemical industry output occurs along the river with major centers around Basel, Mannheim, and the Ruhr area. Pollution was at its worst in the 1970s; since then, an international agreement has been reached to reduce the problem, helped by improved water treatment technology. Concerns remain over nondegradable chemicals and metals that are still at high levels in the river.

Forests, Fertile Soils, and Marine Resources

Between 15,000 and 10,000 years ago, Europe was recolonized by forest as the ice sheets retreated northward. Human occupa-

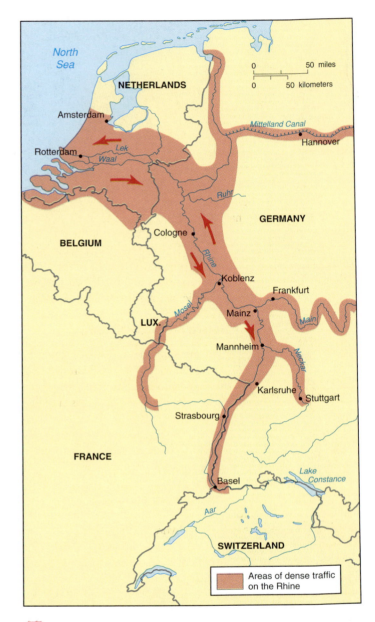

Figure 3.11 **Europe: Rhine River waterway.** Compare the impacts of the industrial areas along the river's length. Rotterdam is Europe's number one port because of its location at the point where maximum Rhine River traffic meets the ocean, requiring change from barge to containers or bulk cargo ship. New canals increased the amount of trade passing through Rotterdam. Source: Reprinted and modified by permission from *Financial Times,* February 2, 1995.

tion began in a forested environment in which the circulation of nutrients through an annual leaf fall raised the quality of soils. The most fertile soils developed on river deposits and on the windblown **loess** covering the common limestone rocks in the northern plains. Fir and pine trees growing on sandy soils and on the thin soils of uplands tended to lower the quality of their soils by acidification.

Because clearance of light woodland was easier, initial settlement favored thinly vegetated uplands on limestone, sandstone, and some granite rocks during the period of warming climate to around 5,000 years ago. Also favored were the easily cultivated loess soils that mantled the southern edge of the North European Plain from Poland to northern France and southern Britain. They provided an easy route of diffusion for farming technology originating around the Mediterranean Sea, via the Danube River valley.

After the introduction of Iron Age tools, human groups rapidly cut into the denser forest on the lowlands. By the time of the Roman occupation of France and lowland Britain, a large proportion of the forest on the lighter lime-rich soils had been cut and some inroads made to the denser forest on heavier soils. Further expansion of the cultivated area on to the clay soils occurred during the Middle Ages.

The marine emphasis of Europe extends to its exploitation of fish resources in the Mediterranean, Baltic, and North Seas and the North Atlantic Ocean. Fishing, related ports, and ships grew in significance in the later medieval period. In the 1900s, overfishing led to decreasing supplies and made sea fishing a source of contention and conflict among European countries.

Natural and Human Resources

Europe contains the natural mineral resources that formed the basis of technological revolutions from the Bronze Age (3500–1000 B.C.) (tin and copper) to modern industrial revolutions (especially coal and iron). Its well-watered lands and moderate temperatures fostered a tree cover and a large proportion of fertile soils that provided the basis for supporting relatively high densities of people at each successive stage of history.

Human ingenuity interacted with the natural environment. By late medieval times, water channels were used widely to generate power in mills and furnaces. Later, the presence of upland areas with high levels of precipitation and snow or ice storage made it possible for engineers to develop hydroelectricity resources there. As populations grew, the development of technologies to support economic growth expanded human resources, making it more likely that the new ideas would spawn further changes.

The rivers and extensive coastline encouraged trade and the exchange of ideas across the region. In the 1800s and 1900s, the human landscapes and historic buildings became resources that made Europe the world's major center of tourism. At the same time, some industrial resources that had been important declined in significance. Coal was replaced by oil and natural gas because of high costs of underground mining and the environmental impacts of burning coal.

Environmental Issues

The occupation of Europe by large numbers of people and their development of technologies for exploiting environmental resources have had major impacts on the natural environment. Today, the wealthier countries are particularly concerned to maintain environmental quality, pouring billions of dollars into such measures.

From Forests to Farms

As soon as humans cut the forest, soil washed down hillsides more rapidly and contributed to the building of lowland river plains and the growth of deltas at their mouths. In the Middle Ages, for example, a combination of growing populations, rigid political systems, close grazing by sheep and goats, and climate change caused intense soil erosion on the hills of the Mediterranean peninsulas, together with the downstream extension of river plains into coastal deltas. Hilly areas of southern Italy and Greece were denuded of their soils, leaving rocky outcrops. In much of Europe outside the alpine area, however, the slopes are less steep, and cultivation methods were adopted that maintained the soils and their productivity over many centuries. In the 1800s, competition from cheap grain imported from newly opened and settled lands in North and South America resulted in once-plowed lands in Europe being sowed with grass for livestock production, further reducing soil erosion and helping to maintain soil quality.

Impacts of Industrialization

The Industrial Revolution and the spread of factory-concentrated production led to widespread pollution of the rivers and air. Many rivers lost their fish stocks. Occasional major pollution incidents still result in fish kills in major rivers such as the Rhine. Great efforts are being made to improve the quality of river water in European countries, and the European Union sets rising standards to be attained by specific dates. In the Thames River of England, the reduction of pollution was so successful that fish stocks revived in the 1990s after decades of absence.

The winter smoke fogs (smog) that blighted the major industrial and urban areas of Europe in the 1950s were controlled by legislated reductions in coal burning in some countries. More recent air pollution occurs where high densities of road traffic pour sulfur, nitrogen oxides, and carbon particles into the air. Such emissions react with sunlight to lower air quality. Under meteorological conditions of slow-moving air, pollutants may accumulate to dangerous levels in broad valleys.

Emissions of sulfur compounds from thermal power stations, especially those burning coal, create **acid deposition**— of dry particles near the source or of wet "acid rain" farther downwind (Figure 3.12). Acid deposition is common around all the main industrial areas of Europe. It affects susceptible areas of coniferous forest, thin soils, and shallow lakes in the Alps and Scandinavia. The highest levels of acid pollution occurred in southern Poland close to the poorly regulated coal mines of the Communist era.

Awareness of environmental problems led to legislation and institutions to monitor and fight air and water pollution, especially in non-Communist Europe. Although road congestion increased with more cars and trucks, new cars in the 1990s emitted 93 percent less carbon dioxide and 85 percent less hydrocarbons and nitrogen gases than in 1970. Lighter vehicles, electronic engine management, more economical engines, and the retiring of older cars will continue to reduce atmospheric pollution. Controlling pollution, however, often increases the costs of industrial production and urban living and causes governments to raise taxes on fuels.

East Central Europe

Environmental degradation was a legacy of Communist governments in East Central Europe. Manufacturing industries did not have to adopt procedures to reduce air and water pollution.

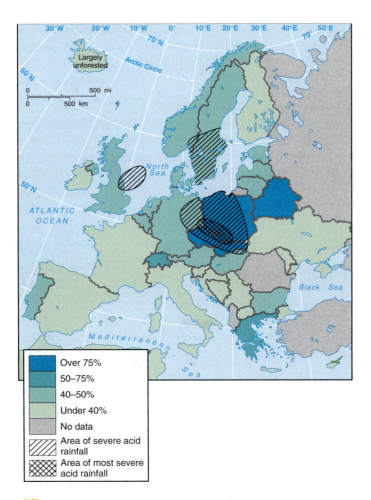

Figure 3.12 **Forest damage as of 1998.** Westerly winds cause much of the damage to the east of major industrial areas. Most severe damage is in the Black Triangle. Compare to Figure 3.6. Source: Jordon-Bychkov, *The European Culture Area,* 4th ed. Reprinted by permission of Rowman & Littlefield.

Industrialized areas were badly affected by the burning of low-quality coal that emitted high proportions of sulfur and particles into the atmosphere (Figure 3.13a, b, c). Chemical works emissions polluted the rivers. When compared to standards of the Environmental Protection Agency in the United States, levels of chemical compounds found in the air and water were hundreds, sometimes even thousands of times higher than what is recommended as safe. One of the worst areas lay along the Czech, Polish, and East German borders. Known as the **Black Triangle,** this area witnessed the death of as much as 90 percent of forests from coal mining and other heavy industries in some locales.

Mediterranean Sea

The growth of population in the 1900s, combined with crowding into coastal locations, industrialization, and the great increase in tourism, led to excessive pollution of the Mediterranean Sea—an important issue for many prospective tourists (Figure 3.14). The Mediterranean Sea is also one of the world's major shipping lanes and has the world's highest level of oil pollution. It is an almost closed sea with hardly any connections to oceans where water

mixing dilutes pollution. Confined within the Mediterranean, the chemicals and other nutrients from pollution decay and thereby deplete the oxygen, killing sea life and creating large algae blooms that thrive in the anaerobic conditions. The Adriatic Sea has some particularly large algae blooms. The most polluted European areas of the sea are those near urban-industrial areas, such as Barcelona, Marseilles, Genoa, Naples, and Athens.

In Spain, for example, the coastal population increased from 12 percent of the national total in 1900 to 35 percent in the 1990s, when the annual influx of tourists added several million visitors. Although the Mediterranean countries of Europe have populations that are not increasing in total, tourism and the movement to the coastal areas continue to grow. Tourists arrive mainly in July and August, when water is short and local sewerage systems find it hard to cope.

The North African and Asian Mediterranean countries—which also pollute the largely enclosed sea—are experiencing major population increases (see Chapter 8). For example, the cities along the African coast are likely to double their populations by 2025. Tourism around the Mediterranean Sea is expected to grow from around 100 million local and foreign visitors in the early 1990s to between 170 and 340 million.

Following a 1975 UN conference, the governments around the Mediterranean Sea tackled their pollution problem through the Mediterranean Action Plan. Sewerage treatment improved in the wealthier countries such as France, but the poorer countries of North Africa cannot afford the necessary investment. Even if pollution is reduced as countries get wealthier, the coastline and its delicate ecosystem are changed irrevocably when coastal wetlands are reclaimed and built over.

Waste Management

One of the major environmental concerns in Europe is the disposal of the rising quantities of trash and industrial wastes. Since 1950, the amount of wastes per head doubled, although it is still less than half that in the United States. Europe, however, has more people on less land, and all its most convenient landfill sites are used up. Further, the nature of trash changed from a majority of ashes and loose dry waste in the 1950s, when coal was burned for domestic heating, to a majority of paper, glass, plastic, and organic waste. Industries produce increasing quantities of hazardous wastes, ranging from poisons to waste oils, heavy metals, and radioactive substances, some of which remain toxic for long periods.

Waste management is becoming a growth industry, requiring major inputs of capital and high technology. As environmental legislation increases, large-scale management corporations will be required. At present, most trash is still buried in landfill sites, the management of which is being improved to exclude gases that might cause combustion or to tap them for use in power generation. Incineration is increasing to reduce the bulk of the large quantities of waste, but it still leaves ashes and emits polluting gases. Recycling is increasingly popular, but the processes are often costly. Recent legislation in Germany obliged companies to take back the packaging used in their products, but the outcome was to overload the German recycling system, creating large "mountains" of paper, glass, and plastic, together with exports of waste to other countries, using up their own capacity.

(a)

(b)

(c)

Figure 3.13 **Environmental degradation in East Central Europe.** (a) Budapest, Hungary. Though the sky is blue and the air apparently clear, the concern for pollution is shown by this electronic board that displays time, temperature, and atmospheric levels of CO (carbon monoxide), SO_2 (sulfur dioxide), and NO_2 (nitrogen dioxide). (b) Kraków-Nowa Huta, Poland. Sędzimir Steel Mill, formerly the Lenin Steel Mill. Communist governments rarely required any kind of pollution controls, and little is different for this factory today. The cemetery in the foreground is a telling commentary. (c) Pollution levels dropped in many areas of East Central Europe with the end of Communism in 1991, after many inefficient factories shut down because they could not compete in the global market. UN funds were provided to shut down this heavily polluting factory in Copşa Mica, Transylvania, Romania. Considered the most polluted place in Europe, local people called it "black town" because everything was black from the air and sky to laundry on clothes' lines and children's faces. Shown a few years after operations ceased, the sky is now blue and the grass green again. Photos: (a) and (c) © George W. White; (b) © Jerzy Jemiolo.

Figure 3.14 **Europe: tourism.** Visitors from northern Europe crowd the sunny resorts of Spain, as here at Cadiz. What are the attractions of this place for so many people? The buildings in the background reflect the occupation by the Muslim Moors up to A.D. 1492. Photo: © Adam Woolfitt/Corbis.

Test Your Understanding 3B

Summary Europe has a mainly coastal environment with midlatitude climates that include cold polar varieties in the north, all-year mild and humid conditions in the west, more continental extreme conditions toward the east, and summer droughts around the Mediterranean Sea.

The physical landscapes of Europe include alpine, lowland, hilly plateau, glaciated, and river plain areas. Lowland and better soils predominate between the northern and alpine mountains. The rocks contain energy and metallic mineral resources. Easy access to ocean transportation was an important factor in Europe's economic growth.

The postglacial forest vegetation that covered the region was largely removed from the areas of better soil and replaced by cultivation and urban-industrial areas. Soil erosion, water pollution, and air pollution resulted from intensive use of the natural environment.

Questions to Think About

3B.1 How does Europe's climate change from north to south? From east to west?

3B.2 Which of Europe's rivers are longest and most navigable?

3B.3 What impacts have industrialization had on Europe's natural environment?

Key Terms

midlatitude west coast climate

mediterranean climate

midlatitude continental interior climate

loess

acid deposition

Black Triangle

Figure **3.15** **Europe: NATO.** The North Atlantic Treaty Organization, showing the countries that are members and those that are applying for membership. In 2002, Russia formed a special partnership with NATO.

Global Changes and Local Responses

With Soviet Communism firmly established in East Central Europe by the early 1950s, many Europeans felt that their nationalist notions and capitalist practices were threatened. Europeans in countries about to lose their colonies knew that their countries were too small to compete individually with the Soviet Union and United States. Moreover, the two world wars revealed the ugly sides of nationalist political competition and economic protectionism, and capitalist competition. Competition could lead to failure as well as success. On the other hand, cooperation in a non-Communist form was thought to result in everyone's success.

Immediate cooperation came about in non-Communist Europe (i.e., Western, Northern, and Mediterranean) with the **North Atlantic Treaty Organization (NATO)** in 1949 (Figure 3.15). NATO included the United States, which was seen as an ally against the Soviet military threat then dominating East Central Europe. After the breakup of the Soviet Union in 1991, many questioned the need for NATO. Others feared that Russia would eventually become a formidable power again, though the Soviet Union no longer existed and Russia was weak at the time. Having emerged from more than 40 years of Soviet domination, East Central European countries were eager to join

NATO to prevent any future domination from the Russians. However, Russian objections and the cost of expansion delayed NATO expansion. Nevertheless, NATO accepted Poland, the Czech Republic, and Hungary as new members and is considering other countries. In 2002, NATO formed a partnership with Russia, the country that NATO was created to defend against! Post-Cold War NATO is actually more focused on resolving or policing disputes within the expanded Europe and its immediate neighbors—as in Bosnia, Kosovo, and Macedonia.

To compete successfully again in the world economy over the long run, Europeans in the non-Communist countries created the European Economic Community (EEC), known today as the **European Union** (EU) (see Point-Counterpoint: The European Union, p. 82). The European Union represents yet another idea emanating from Europe: **supranationalism.** Supranationalism is the idea that differing nations can cooperate so closely for their shared mutual benefit that they can share the same government, economy (including currency), social policies, and even military. Communism could be considered a form of supranationalism, but most Europeans have abandoned the Communist experiment, leaving those in the EU as the primary advocates of supranationalism. Members of the EU are still working out the details of their cooperation, but what they have accomplished is remarkable, considering that the more predominant nationalist idea, subscribed to by most of the world, holds that such cooperation is impossible between nations. Nationalism may still preclude ultimate political union in Europe.

It remains to be seen if supranationalism will work or not. This experiment in new economic and political organization aside, it is clear that despite the destruction of World War II and the loss of economic and political power, by the later 1900s, Europe regained a major place within the global economic and political system and was able to respond to changes in the wider world. European countries have many of the highest GDPs per capita in the world (Figures 3.16 and 3.17).

Devolution Within European Countries

As Europe moves toward expanded economic and possibly political integration, it is simultaneously experiencing **devolution:** the process by which local peoples desire less rule from their national governments and seek greater authority in governing themselves. Devolution occurs in the United Kingdom, for example, where Scots and Welsh seek greater autonomy and have recently obtained their own parliaments. Estonia's, Latvia's, and Lithuania's independence from the Soviet Union and the breakup of Czechoslovakia and Yugoslavia were other forms of devolution. Devolution is also seen in the bloody campaigns fought by the Basque peoples straddling the Spain-France border and the Catholic minority in Northern Ireland.

Devolution pressures are not coming only from ethnic and national minorities. They are also coming from the people of provinces that straddle neighboring countries. For example, Strasbourg is the major city in the mid-Rhine Valley, providing goods and services that the smaller cities do not. Though Strasbourg is in France, it is on the German border, making it a city that is closer to many Germans than German cities of a similar size. Thus, many Germans obtain goods and services in Strasbourg. High unemployment rates on the French side of the Rhine River and high-paying jobs in neighboring Germany lead to many French desiring to work in Germany. In the past, differing government policies of bordering countries—customs, border guards, different currencies—made it difficult for citizens to travel routinely to neighboring countries only a few miles away. Nevertheless, interaction was desired, and peoples of border regions have pressured their national governments to make daily commutes easier. To do this, people living in border areas have formed **Euroregions,** which straddle country boundaries. The people of Euroregions work to make transboundary movement easier within their Euroregions. For example, border controls between France and Germany no longer exist, and passes to museums on the French side of the Rhine boundary are valid in German and Swiss museums on the other side of the Rhine. Euroregional

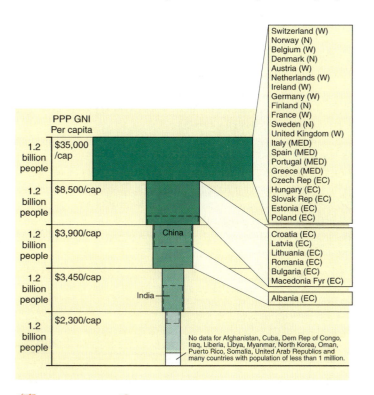

Figure 3.16 **Europe: national incomes compared.** The countries are listed in the order of their PPP GNI per capita. Compare the relative material wealth of the countries in Western (W) and Northern (N) Europe with those in Mediterranean (MED) and East Central (EC) Europe. Source: Data (for 2000) from *World Development Indicators,* World Bank, 2002.

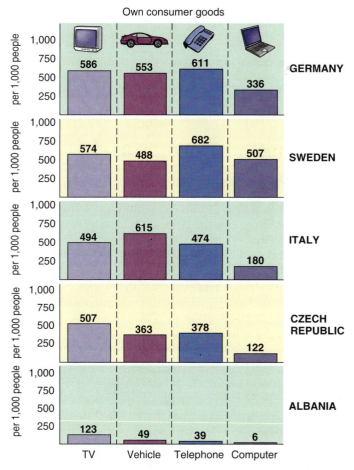

Figure 3.17 **Europe: ownership of consumer goods.** Note the diversity within Europe. Summarize the differences among the countries by subregion. Source: Data (for 2000) from *World Development Indicators,* World Bank, 2002.

✳ ## THE EUROPEAN UNION

The EU is now a major factor in changing the human geography of Europe. Almost all of the countries of Europe either belong to it or have applied for membership, as the map in Box Figure 1 shows. Yet many issues cause continuing debate, including questions asking whether the European Union is a good idea and whether it will succeed or not.

The EU arose from attempts to ensure that the future of war-torn Europe after 1945 would be peaceful. After two world wars in 30 years, political leaders were determined to lock the economies of European countries so closely together that further wars would be difficult, if not impossible. Some anticipated that once economic links were established, political union in a federation like that of the United States might follow. With American aid and protected from the expansionist ambitions of the Soviet Union by the North Atlantic Treaty Organization, the first step was the formation of the Benelux customs union between Belgium, the Netherlands, and Luxembourg in 1949. In 1952, the Benelux countries joined together with France and West Germany to form the European Coal and Steel Community (ECSC). In 1957, the five ECSC countries plus Italy signed the Treaty of Rome, establishing the European Economic Community (EEC). The EEC was expected to create a common market in which goods, capital, people, and services moved freely between countries. The European Commission became the executive arm of the community

and was based in Brussels, Belgium. In 1967, the EEC changed its name to the European Community (EC) to emphasize the move from economic toward political goals. As the organization progressed with integration, other countries were motivated to join. Denmark, the Republic of Ireland, and the U.K. became members in 1973, Greece in 1981, and Portugal and Spain in 1986. By then, the European Parliament, with elected members from all these countries, was created and located in Strasbourg, France. Other EU organs are headquartered in countries such as Luxembourg (Box Figure 2).

The Single European Act, signed in 1986 when Spain and Portugal joined, set out the steps needed to implement a single market. The Treaty of Maastricht (1991) attempted to set a timetable for monetary and political union and changed the name of the organization to European Union in 1993. Austria, Sweden, and Finland joined in 1995. The Swiss and Norwegian referendum votes on joining were close but rejected membership at the time. Many of the countries of East Central Europe, formerly part of the Soviet bloc, along with Turkey and Cyprus hope for a future in the EU, but their admission is a slow process.

European integration has proceeded for almost 50 years. Many Europeans see it as good, while many others do not. Those in favor look at the situation of European countries in the world and note that the countries like the United States, Japan, China, and Russia are much larger than any one European country. This makes it difficult for

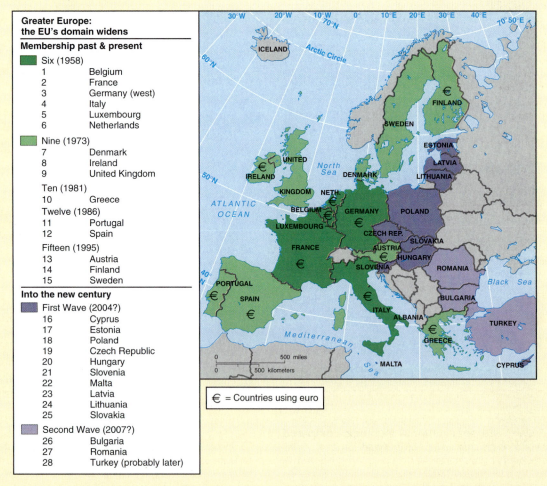

Greater Europe: the EU's domain widens

Membership past & present

■ Six (1958)
1	Belgium
2	France
3	Germany (west)
4	Italy
5	Luxembourg
6	Netherlands

■ Nine (1973)
7	Denmark
8	Ireland
9	United Kingdom

Ten (1981)
| 10 | Greece |

Twelve (1986)
| 11 | Portugal |
| 12 | Spain |

Fifteen (1995)
13	Austria
14	Finland
15	Sweden

Into the new century

■ First Wave (2004?)
16	Cyprus
17	Estonia
18	Poland
19	Czech Republic
20	Hungary
21	Slovenia
22	Malta
23	Latvia
24	Lithuania
25	Slovakia

■ Second Wave (2007?)
26	Bulgaria
27	Romania
28	Turkey (probably later)

€ = Countries using euro

Box Figure 1 European Union, its growth to date, and possible further expansion. Reprinted by permission from *Financial Times*, Feb. 2, 1995, updated.

any single European country to compete with these other countries. An economic and political union gives the Europeans more clout in the global economy and world politics. For example, the United States has 281 million people, Japan 125 million, and Russia 144 million. The largest EU country is Germany, and it has a population of only 82 million. However, the combined population of the 15 members exceeds 370 million people and has considerably more economic and political clout.

Internally, union has other advantages as well. Free movement of labor, capital, and goods strengthens businesses and allows great opportunities to buy the best products at the lowest prices. To facilitate such movements, 12 of the 15 EU members adopted a common currency known as the euro in 2002. It replaced the traditional national currencies such as the German mark, the French franc, and the Italian lira. Imagine if all 50 states in the United States each had their own currencies. Imagine then that if you wanted to buy product from a state other than yours, travel to another state, even for a few hours, or take a trip through several states, you would have to exchange your state's currency for those of other states. In the process, you would have to pay fees for making the exchanges. You probably would not travel to or purchase products from other states as you do now, using the dollar. It would also frequently cost you more to restrict your actions to your own state. This was the situation in the 12 EU countries before they adopted the euro.

As mentioned, not all Europeans are in favor of integration or believe that the EU will work over the long run. For integration to succeed, the member countries have to adopt similar—and in many cases, common—laws and economic policies. For the euro to work, for example, all member countries must limit their spending and keep their annual budget deficits within 3 percent of their GDPs. To do so, many countries may be unable, for example, to pay for their social programs and stimulate their economies as they see fit. They would have to give up a lot of what they value and suffer through economic slumps of high unemployment for the benefit of the other member countries. If they choose to break the rules and spend, then they devalue the euro and damage the economies of the other member countries. In short, many Europeans are opposed to the euro because of the restrictions that come with it. The United Kingdom, Denmark, and Sweden have not adopted the euro and continue to use their own national currencies.

Many Europeans are opposed to more than just the euro. Integration involves the adoption of common laws and the removal of barriers, including border controls between member countries. States will not be able to stop the entry of foreigners, whether from other member countries or from abroad as they enter through other member countries. This means that these foreigners will take local jobs, demand cultural rights, and generally be a visible foreign presence. It also means that countries will not be able to stop the surge of cheap, foreign goods and possibly contaminated food, driving businesses into bankruptcy and creating unemployment. In short, the EU—and with it, integration—is seen as the loss of national sovereignty and the erosion of national identity.

Germany and France, along with the Benelux countries and Italy, have typically been great supporters of the EU. Germany alone contributes 30 percent of the EU's budget. The smaller countries of Ireland, Portugal, and Greece are great proponents of the EU because they receive much more from it than they pay into it. The United Kingdom, Denmark, and Sweden tend to be most opposed to the EU's integration policies. It is even said that the United Kingdom joined the organization to slow down integration and prevent the emergence of a monolith on its doorstep that would be stronger than it. Many of the key hopes and fears of the EU member countries are outlined in Box Figure 3.

One of the biggest concerns of the current EU members involves enlargement. Thirteen countries have applied for membership (Box Figure 1). Most are likely to become members in 2004, though Romania and Bulgaria will likely wait to 2007 and negotiations with Turkey have been postponed indefinitely. The inclusion of these countries could make the EU a more powerful force than it is today, but it could also delay or thwart further integration. Some EU countries have been working together for more than 40 years and are ready to move farther down the road of integration. Current members debate whether it is best to wait for new members to catch up or to allow those ready to move ahead on a new issue to do so. As new members have joined, the EU seems to be adopting the multiple path of continued integration. The adoption of the euro with three members opting not to participate is one example of this tendency.

The other problem with EU enlargement is that the applicant countries are mostly much poorer economically than the current members and would be a financial drain. The 10 East Central European countries that applied for admission have 53 regions similar to provinces. In 1998, 41 out of the 53 regions would be below 50 percent of the EU's GDP. Only the regions around Prague and Bratislava have per capita GDPs close to the EU average. These two regions aside, EU regional development funds would clearly be redirected from Ireland, Portugal, Greece, and southern Italy to the East Central European countries. Germany feels that it already contributes a lot and would have to contribute more. France is worried about protecting its agricultural sector. In addition, worker productivity and wages in the East Central Europe countries were only about 40 percent of the EU average. With freedom of movement, current members fear that unemployment will rise in their countries as East Central European workers would willingly work for less. Thus, even the most supportive countries of the EU have moved to postpone and restrict enlargement in the hopes that the economies of the applicant countries will improve before membership takes effect. Nevertheless, Germany successfully lobbied for a seven-year restriction on the free movement of labor, and Spain, with the support of Italy, Portugal, and Greece, lobbied to ensure that it will continue to receive regional development funds though it will no longer have the poorest regions. Interestingly, the United Kingdom has supported enlargement, probably in the hopes of slowing the integration process.

(Continued)

POINT	COUNTERPOINT
Economic union pools together the resources of member countries and thereby strengthens the economies of every member country.	Economic union undermines the ability of member governments to make economic decisions that are in their national best interests.
Economic union allows for the free flow of capital and labor, permitting capitalist tendencies to strengthen members' economies.	The free flow of capital and labor undermines attempts by member governments to protect their national economies.
The euro, or common currency, further facilitates the movement of capital and labor by doing away with the costs of converting currencies.	The euro forces member governments to have monetary and budgetary policies that may harm their national economies.
Political union increases influence in regional and global politics because it combines the political and military strength of member countries.	Political union forces member countries to adopt foreign policies that go against their national interests (e.g., forcing some, like Ireland, to give up their neutrality).
Political union results in common laws and standards for individuals and the environment in member countries. It makes for better social and natural environments.	Common laws and standards among member countries undermine national needs and traditions, often watering down social and environmental laws.

Box Figure 3 **Hopes and fears of EU countries.**

	Biggest hope	Biggest fear	What they say about themselves	What others say about them
Austria	EU forgets about Haider	Mass immigration from east	Badly misunderstood	Watch those neo-Nazis
Belgium	Brussels as capital of new superpower	Brussels secedes, Belgium disintegrates	We have lots of good ideas	Tiresome zealots
Britain	EU becomes giant free-trade area	Britain submerged into European superstate	We are the only country that obeys the rules	Make up your mind or leave; America's stooges
Denmark	EU integration stops	Welfare state dismantled by EU	Nobody understands us	Please stop having referendums
Finland	EU security arm ends Russian threat	EU-Russian antagonism	We're no trouble, not like those other Nordics	No trouble, nice phones
France	EU as French-led superpower; humiliates USA	EU as German-led superpower; humiliates France	We invented the EU and we are the only people who understand it	Arrogant
Germany	Federal Europe solves "German problem"	Popular backlash in Germany against euro or enlargement	We pay all the bills; we should have more say	Alarmingly ambitious
Greece	EU admits Cyprus, upsets Turkey	Turkey joins EU, upsets Greece	We are Europe's eastern outpost	Levantine rule-breakers
Ireland	Continued boom on back of EU	EU ends Irish neutrality	We are the EU's economic stars	They owe it all to us
Italy	Totally federal Europe	Kicked off top table for being too flaky	We are Europe's true believers	Dodgy politics, dodgy finances
Luxembourg	More EU jobs	EU ends tax-haven status	This is our stage	Too powerful for a microstate
Netherlands	Dutch model adopted by EU	Squeezed between France and Germany	We are founder members; we deserve more respect	Know-it-alls
Portugal	Catches up with EU economically	Small countries marginalized	Don't confuse us with Spain	Poor but honest
Spain	Lots more EU money	No more EU money	We are one of the EU's leading nations	Greedy
Sweden	EU becomes more honest and Swedish	Sweden becomes more corrupt and European	We are the truest democrats	Self-righteous

Source: Modified from *The Economist* Newspaper Group Inc. Reprinted with permission. Further reproduction prohibited. www.economist.com.

organizations also lobby their respective national governments for policies to facilitate movement. Euroregions are appearing in large numbers, now straddling almost every country boundary (Figure 3.18).

Euroregions are challenging national governments everywhere throughout Europe. They are exerting leverage with support from the EU. In addition to the rise of Euroregions, many provinces of countries are exerting their authority. Almost all the provinces in Germany, for example, have offices in Brussels, where they lobby the EU directly for funds. The interests of Euroregions and countries' provinces are not the same as each other's or countries' interests. New kinds of governments and new relationships between levels of government are being worked out in Europe.

Changes in Europe's Economic Geography

Two world wars and intervening economic depressions changed Europe's economy and its economic relationship with the rest of the world. Greater governmental intervention restored **productive capacity** (the amount of goods a country's businesses can produce) and now ensures education, health care, unemployment benefits, and pensions for all. In non-Communist Europe, older,

Figure 3.18 **The Upper Rhine Euroregion.** Notice the close proximity of French, German, and Swiss cities.

"heavy" (**producer goods**) industries, such as steelmaking, heavy engineering, and chemicals located on coalfields, were replaced in value of output and employment by motor vehicles, consumer goods, and light engineering products. The new products could be manufactured wherever there was an electricity supply and plentiful semiskilled labor. The material wealth created and the numbers employed in primary and secondary industries gave way to service industries—based in offices rather than factories. This further increased urbanization. New locations for manufacturing became significant around the expanding consumer markets of major cities and in new regions. The polluted old industrial centers with major investments in factories based on obsolete technologies—often kept going by state subsidies and staffed by members of trade unions that resisted changes—declined and unemployment in them rose.

In the second half of the 1900s, two major features marked the economic geography of Western Europe and now are showing themselves in other areas of Europe and the world. First, the new industrial areas grew at places more suited to the needs of developing technologies and industries, but long-established production continued in the older areas. The major investments in factories, housing, human skills, and infrastructure made it too costly in financial and human terms to undertake sudden locational shifts of production. The advantages of producing goods in an area that has a trained labor force, assembly and distribution systems of transportation, and support by financial and other services, build up and reinforce the original locational advantages as **agglomeration economies.** Keeping production in an area although its costs may be higher than those in possible competing areas creates **geographic inertia.** It is only when a substantial change in costs occurs, new products emerge, or the demand for the original product is reduced that new areas develop and older manufacturing centers decline.

Second, a process known as **deindustrialization** occurred when the numbers of jobs in manufacturing fell rapidly and old factories became derelict as old industrial areas declined. A fall of 20 percent in European manufacturing jobs between 1970 and 1985 was more than balanced by a rise of 40 percent in tertiary sector jobs. The skills of blue-collar miners and factory production-line workers were seldom convertible into the new white-collar office, hospital, or classroom jobs that were taken by younger and better-educated people. In countries where state-run industries had maintained high levels of employees, privatization—with its emphasis on increased productivity—caused rapid increases in unemployment. Those workers unable to retrain for the new jobs faced long-term unemployment. Older production workers, female workers, and young people with a poor education entering the labor force were particularly at risk.

Planning and Privatization

In non-Communist Europe after 1950, the problems of declining old manufacturing areas in particular became political issues and led to additional programs of social welfare, regional policies for siting new industries in the older industrial areas, and retraining programs. Governments attempted to redress the differences in unemployment between new and old industrial areas by forcing manufacturers to locate new factories in materially poorer, rather than wealthier, areas of their country. State industries were located in such areas despite higher operating costs. Italy, France, and Britain introduced such policies. Investment in southern Italy—the Mezzogiorno—is a major example of government-directed industrial location.

Beginning in the 1970s, EU regional policy took over from national policies aimed at redistributing employment opportunities, and the EU invested large sums in infrastructure construction in the lagging regions. The European Regional Development Fund was established in 1975. There was a net flow of funds to the margins of EU countries, to their older industrial areas, and especially to the newer (and materially poorer) member countries in Mediterranean Europe. The main emphasis of EU regional policy was on granting of loans to create jobs directly or indirectly. The loans were used mainly for infrastructure projects, especially roads, telecommunications, water supplies, and waste disposal. Job creation outcomes had modest success, with a few thousand new jobs created for the millions of unemployed. Nevertheless, the EU continually modifies its policies for regional development funds to help the economically poorer areas of Europe. For the years 2000 to 2006, the EU allocates funds according to the following list of priorities (Figure 3.19): (1) development of the most disadvantaged regions (Objective 1), (2) the conversion of regions facing structural difficulties (Objective 2), (3) interregional cooperation (Interreg III), (4) the sustainable development of urban areas in crisis (Urban II), (5) the development of innovative strategies to support regional competitiveness.

Regional development funds have improved the local living situation in the economically depressed areas of the EU. However, such policies produced a modest effect at great cost and, by focusing on internal issues, often made the countries uncompetitive in world markets.

A major shift in Western Europe in the 1980s and early 1990s moved these countries away from a relatively stable economic and social order in which the losers were looked after by government provision of industrial protection and welfare. The other side of this stability had been that the countries of Western Europe began to price their products too high for world markets, resulting in reduced sales and thus income. Management then decided to lay workers off, rather than increase income by lowering the prices of their products to boost sales volume. Moreover, increasing welfare costs put country budgets into deficit and raised internal tax demands. To combat this economic problem, many Western European countries, led by the United Kingdom, moved toward practices followed in the United States, where private enterprise is encouraged, but potential high rewards are balanced by high levels of risk, making for a less stable human environment. During the 1980s, Britain led the way in privatizing state enterprises, many of which made financial losses and were a drain on tax income. The privatized

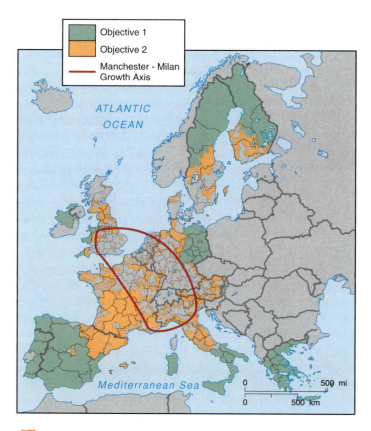

Figure **3.19** **European Union structural funds, 2000–2006, areas eligible under Objectives 1 and 2.** All of Ireland and Portugal, along with most of Scotland, qualified for structural funds until the current round. Note the locations of the most disadvantaged areas. Former East Germany is among them. Source: Copyright © European Communities, 1995–2002.

concerns, including steelmaking and electricity, gas, and water utilities, commonly made considerable profits after this change, especially by cutting their labor forces, leading to more unemployment.

Global City-Regions

One way that Europe plays a leading role in the global economic system is through its cities. Of the global cities identified in Figures 2.13 and 2.14, Europe has by far the most with well over one-third. These European cities are no longer at the top of the list of the most populous cities in the world, but they are among the most globally connected cities. London and Paris, along with only New York and Tokyo, are the most important cities for accounting, advertising, banking, and law. When we think of cities such as London and Paris, we should not simply consider their businesses as only serving Londoners and Parisians, respectively. The businesses within them serve people all over the world, whether in the Americas, Africa, or Asia (Figure 3.20). Many millions of dollars flow in and out of these cities from and to many areas of the world. After London and Paris, cities such as Frankfurt, Milan, Zürich, Brussels, and Madrid rank high in their global importance.

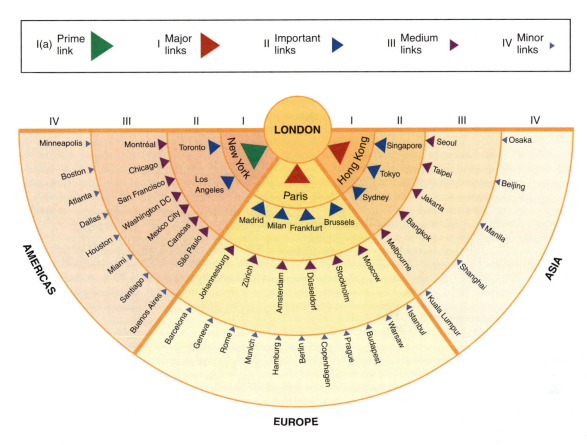

Figure 3.20 **London.** As one of the four most globally connected cities, London is a major center of accounting, advertising, banking, and law.
Source: Taylor, P. J., D. R. F. Walker, and J. V. Beaverstock, *Globalization and World Cities Study Group and Network Research Bulletin 6, Firms and Their Global Service Networks.*

Test Your Understanding 3C

Summary World Wars I and II had a devastating impact on Europe's political and economic systems. Following World War II, Europeans were forced to rethink their relationships with one another and the wider world. At one scale, Europe was divided into a demo-cratic/capitalist West and a Communist East. The West formed the North Atlantic Treaty Organization (NATO) and the European Economic Community (now EU). The East adopted Communism and was tied to the Soviet Union (see Chapter 4). Beginning in the early 1990s, East Central European countries abandoned Communism and applied to join NATO and the EU. All across Europe, devolutionary pressures increased.

Economically, Europe experienced shifting trading links and new locations for manufacturing. European countries still maintain a high standard of living. Europe's cities, particularly those in Western Europe, continued to play a major role in the global economic system. Despite the devastating effects of World War II, Europe regained a major place within the global political and economical system. European countries have many of the highest GDPs per capita in the world.

Questions to Think About

3C.1 What kind of new "supranationalist" organizations did Europeans create after World War II, and how are they supranationalist in their functioning?

3C.2 What are Euroregions and what do they have to do with the concept of devolution?

3C.3 What happened to "heavy" industries in Western Europe in the decades following World War II?

3C.4 What are the spatial patterns of economic development in the European Union? That is, which areas are more economically developed and which areas are less economically developed?

3C.5 Which European cities are Global Cities? Where within Europe are they located? Are they spatially concentrated?

Key Terms

North Atlantic Treaty
   Organization (NATO)
European Union (EU)
supranationalism
devolution
Euroregions

productive capacity
producer goods
agglomeration economies
geographic inertia
deindustrialization

🌐 The Subregions

Though Europe is smaller than many other world regions, it is marked by distinct subregions:

- *Western Europe:* Austria, Belgium, France, Germany, Luxembourg, Netherlands, Republic of Ireland, Switzerland, and United Kingdom
- *Northern Europe:* Denmark, Faeroe Islands, Finland, Greenland, Iceland, Norway, and Sweden
- *Mediterranean Europe:* Greece, Italy, Portugal, and Spain
- *East Central Europe:* Albania, Bosnia-Herzegovina, Bulgaria, Croatia, Czech Republic, Estonia, Macedonia, Hungary, Latvia, Lithuania, Poland, Romania, Serbia and Montenegro, Slovakia, and Slovenia

Population Distribution Patterns

Western Europe has high densities of population, except for its marginal highlands (Figure 3.21). The highest densities are in the urbanized industrial belt that runs southeastward from central Britain, through northern France, Belgium, and the Netherlands, and into Germany. Moderate population densities occur across the French farmlands, eastern and southwestern England, parts of southern Germany, and northern Switzerland. The lowest densities are in the uplands of Scotland, parts of Ireland, the high Alps and Pyrenees in southern France, Switzerland, and Austria.

In Northern Europe, most people live toward the southern part of the subregion. Although (without Greenland and Iceland) they cover an area that is almost as large as Western Europe, these countries, apart from Denmark, have low densities of population because of difficult climates and terrains. (Compare Figures 3.7 and 3.10a with Figure 3.21.)

The distribution of population in the Mediterranean countries of Europe reflects the hilly and mountainous nature of the

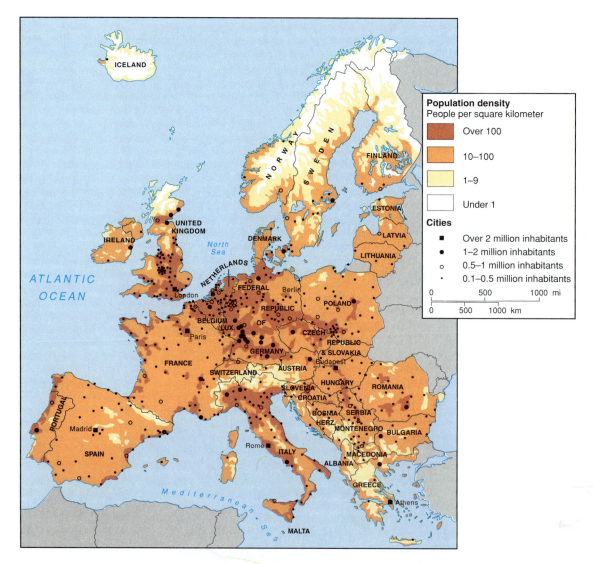

Figure 3.21 **Europe: population distribution.** Explain the heaviest and lightest concentrations of people in each subregion.

Source: Data from *New Oxford School Atlas*, p. 98, Oxford University Press, UK, 1990.

terrain (see Figure 3.21). Most people live in the lower parts of major river valleys and along the coasts. The Po River valley of northern Italy is the largest area of high population density, with others along the Portuguese coast and eastern Spain. The Alps, Apennines (Italy), Greek mountains, and the Pyrenees and parts of central Spain have very low densities.

East Central Europe has a more even population distribution than other European subregions (see Figure 3.21). The main concentrations of people are in the urban-industrial areas on either side of the borders between the Czech and Slovak Republics and southern Poland. Farther south, the Danube River valley has the main population centers in the cities of Belgrade (Serbia and Montenegro), Budapest (Hungary), and Bucharest (Romania) along its length. The predominance of farming elsewhere is reflected in moderate population densities, with low densities on the poor soils near the Baltic Sea and mountainous areas such as the Carpathians and the Dinaric Alps.

Western Europe

The countries of Western Europe include Austria, Belgium, France, Germany, Ireland, Luxembourg, the Netherlands, Switzerland, and the United Kingdom (U.K.) (England, Wales, Scotland, Northern Ireland) (Figure 3.22). This subregion contains three of the four largest populations in Europe (Figure 3.23).

Western European countries have been most significant in creating Europe's image as a global leader, both now and over the course of the last few hundred years. The United Kingdom, France, and the Netherlands were three of Europe's most powerful colonial powers. Along with Germany and Belgium, they were also among the first countries to experience the Industrial Revolution. Colonialism and industrialization gave Western Europe the ability to establish many of the global trade flows that are still in effect today. Though Western European countries are no longer colonial powers, they still import raw materials from countries that were their former colonies and then sell finished

Figure 3.22 Western Europe: countries of the subregion. The western margins include Scotland, Ireland, Wales, southwest England, and Brittany.

Figure 3.23   Western European countries: key data on countries.

Country	Capital City	Land Area (km²) Total	Population (millions) Mid-2001 Total	2025 Est.	GNI 1999 (US $ million) Total	GNI PPP 1999 Per Capita	Percent Urban 2001	Human Development Index Rank of 175 Countries	Human Poverty Index: Percent of Total Population
Belgium, Kingdom of	Brussels	33,100	10.3	10.3	252,051	25,710	97	5	12.4
French Republic	Paris	551,500	59.2	64.2	1,453,211	23,020	74	11	11.9
Germany, Federal Republic of	Berlin	356,910	82.2	80.0	2,103,804	23,510	86	14	10.4
Ireland, Republic of	Dublin	70,280	3.8	4.5	80,559	22,426	58	20	15.3
Luxembourg, Grand Duchy of	Luxembourg City	2,586	0.4	0.6	18,545	41,230	88	17	no data
Netherlands, Kingdom of	The Hague	37,330	16.0	17.7	397,384	24,410	62	8	8.3
United Kingdom of Great Britain and Northern Ireland	London	244,880	60.0	64.1	1,403,843	22,220	90	10	15.1
Austria, Republic of	Vienna	83,850	8.1	8.3	205,743	24,600	65	16	no data
Swiss Confederation	Bern	41,290	7.2	7.6	273,856	28,760	68	12	no data

Source: Data from *Population Reference Bureau 2001 Data Sheet; World Development Indicators,* World Bank, 2001; *Human Development Report,* United Nations, 2001; Microsoft Encarta (ethnic group, language, religion).

products back to them. The economies of Germany, France, and the U.K. rank, respectively, as the third, fourth, and fifth largest in the world. These three countries also account for three of the members of Group of Eight (G8), an informal organization representing the world's most materially wealthy countries. (The other countries in the G8 are the United States, Canada, Japan, Italy, and Russia.) The human environment of this subregion ranks high in health, education, and income standards with all countries ranking within the top 20 on the human development index.

The European Union helped many Western European countries maintain and even increase their economic and political standing in the world. At the same time, these countries helped the EU become the powerful force that it is today. They account for seven of the 15 members and represent five of the six founding members. Beyond the scope of the EU, the political power of Western Europe is underscored by the positions that the U.K. and France hold in the United Nations. They occupy two of the five permanent seats of the UN Security Council, the most powerful organ of the United Nations.

Western Europeans spread many European cultural characteristics around the world. English and French are two of the most commonly spoken world languages. Western Europeans are also the driving forces behind the spread of Protestantism, a branch of Christianity. The British, French, and Dutch, for example, carried Christianity to North America, South Africa, Australia, New Zealand, and a number of other places.

Western Europe is the hearth for international organizations concerned with human rights. Amnesty International, for example, began in London in 1961 and works today to free individuals around the world who are imprisoned for expressing their opinions. The International Movement of the Red Cross and the Red Crescent traces its origins to Switzerland. From the time of its Geneva Convention in 1864, it has helped to set the modern standard for the ethical treatment of wounded soldiers and civilians during wartime. Oxfam (short for the Oxford Committee for Famine Relief) began in Oxford,

England, in 1942 to help starving people in Greece who were caught between Nazi occupation and an Allied blockade. Since that first mission, Oxfam has undertaken numerous operations around the world to bring food and medical supplies to those suffering from war and similar causes. The International Court of Justice, located in The Hague, Netherlands, initially began in 1899. At the time, it was designed to find peaceful resolutions to conflicts between countries. Since then, it has expanded its scope to include many human rights issues. For example, it is involved in assessing the responsibilities of those accused of atrocities in the horrific Balkan wars of the 1990s.

Countries

Located primarily along the maritime edges of Europe, Western European countries exerted their independence early in history, compared to other subregions of Europe. Early independence coupled with technological innovation allowed Western Europe to dominate international relations and the global economy from the 1600s to the early 2000s. Strong local voices within Western Europe abolished the monarchies of the Middle Ages and led to the development of nationalism and the nation-states seen on the map today. Boundaries for these countries continued to change well into the 1900s as West Europeans competed with one another at home and abroad.

France is one of Europe's older and most powerful and influential countries. Beginning with the Capet family in the 900s and its small family holdings around Paris, the country grew over the centuries, becoming a nation-state soon after the French Revolution in 1789. The French Revolutionary expression of "liberty, equality and fraternity" inspired other revolutions in Europe. The French have a strong sense of individuality, but France has a history of having a very centralized form of government with most decisions made in Paris. To satisfy the desire for more local decision-making, the central government created 22 provinces called *regions* in 1982 and

Country	Ethnic Groups (percent)	Languages O=Official	Religions (percent)
Belgium, Kingdom of	Flemish 55%, Walloon 33%	Flemish (Dutch) 56%, French 32%	Protestant 25%, Roman Catholic 75%
French Republic	Celtic/Latin with Teutonic, Slavic, Nordic, African	French, with dialects	Roman Catholic 90%
Germany, Federal Republic of	German 95%, Turkish 2%	German, English	Protestant 45%, Roman Catholic 37%
Ireland, Republic of	Celtic, English	Irish (Gaelic), English	Roman Catholic 93%
Luxembourg, Grand Duchy of	Celtic/French/German, some Portuguese, Italian	Luxemburgisch (O), German, French	Roman Catholic 97%
Netherlands, Kingdom of	Dutch 96%, Moroccan, Indonesian, Turkish	Dutch (O), Frisian, English, German	Protestant 25%, Roman Catholic 34%
United Kingdom of Great Britain and Northern Ireland	English 82%, Scottish (10%), Irish, Welsh	English (O), Welsh, Scottish Gaelic	Protestant 49%, Roman Catholic 16%
Austria, Republic of	German	German (O)	Protestant 6%, Roman Catholic 85%
Swiss Confederation	German 85%, French 18%, Italian 10%	German (O), French (O), Italian (O)	Protestant 44%, Roman Catholic 48%

granted them considerable autonomy in raising taxes and spending funds on local projects (Figure 3.24).

The United Kingdom is one of the other powerful European countries. It adopted its name in 1801 and included the whole of Ireland until 1922. The United Kingdom (U.K.) has a political geography that reflects strong local voices within the country. England, Wales, and Scotland are district entities that together with Northern Ireland comprise the United Kingdom of Great Britain and Northern Ireland (see Figure 3.22). The rise of Scottish and Welsh nationalism over the last few decades resulted in greater autonomy for both Scotland and Wales.

Irish nationalism in Northern Ireland brought about some autonomy at times but also great bloodshed. The Northern Ireland situation is very complex. Independent-minded Irish succeeded in removing all but six counties of Ireland from the United Kingdom in 1922. The freed countries became known as the Irish Free State. Still part of the British Commonwealth, the Irish Free State had to swear allegiance to the British monarch until 1949, when it achieved its independence and adopted the name "Republic of Ireland." The six counties of the north remained part of the United Kingdom after 1922 and became known as Northern Ireland. The violence in Northern Ireland often is depicted as a religious conflict between Roman Catholics and Protestants, but it is really a conflict between Irish nationalists and Unionists, also called Loyalists. Irish nationalists, feeling oppressed by British colonialism, seek to unite the six counties of Northern Ireland with the Irish republic to the south, but others, the Unionists/Loyalists, prefer to maintain the union with Great Britain and have resisted attempts to change the situation. Roman Catholics and Protestants were originally on both sides of the conflict. Because many Irish nationalists are Catholic and have the goal of uniting Northern Ireland with a Catholic country and because Unionists/Loyalists tended to be Protestant and want to maintain a union with a Protestant country, the issue usually is described as a religious conflict. As the conflict became progressively violent, even ambivalent Catholics had to seek protection of other Catholics and Protestants sought refuge with other Protestants. The conflict has led to religiously segregated communities, though almost as many violent acts are committed by Catholics against Catholics and Protestants against Protestants as across religious lines.

The global connections that France and the U.K. had with the rest of the world changed when these countries withdrew from most of their colonies in the 1950s and 1960s. Though these two countries no longer directly control other peoples and their lands, they both are very involved with their former colonies. The U.K. maintains ties, for example, through the British Commonwealth. In 1945, France created a currency known as the CFA franc (franc of the French Colonies of Africa)

Figure 3.24 **Western Europe: regions of France.**
How central is the position of Paris?

for use in its western and central African colonies. These colonies are now independent countries, but the CFA franc still exists and ties their economies closely to France. France also maintains the French Foreign Legion, a military unit of non-French citizens that fights for French interests. France is the main supporter of the Agency for Francophony, an organization that promotes the French language and culture around the world.

The Netherlands, Belgium, and Luxembourg lie along the deltas of the Rhine and Meuse (Maas) Rivers and were once all called "The Netherlands," meaning "Low Countries." The modern country of the Netherlands has one quarter of its land below sea level. Dutch ingenuity and hard work reclaimed much of the country from the sea, mostly in the form of polders (Figure 3.25). In the past, windmills helped to pump the land dry, and more than 1,000 are still in working order today, though modern engines do the pumping.

The people of the Netherlands voiced their desire for independence from Catholic Spain during the Protestant Reformation in the 1500s. The seven northwest provinces, calling themselves the United Provinces of the Netherlands, under the leadership of the province known as Holland, fought for their independence and achieved it in 1648, the year of the Treaty of Westphalia and the end of the Thirty Years' War. In 1815, the southern provinces were joined with the United Principalities to the north. Almost two centuries of separation made union impossible. Most of the southern provinces declared their independence from the Netherlands in 1839 and became the country of Belgium. The province of Luxembourg lost most of its territory to Belgium at that time, but what remained became the country of Luxembourg in 1868.

The Netherlands became one of the great colonial powers, and Belgium had colonies, too. Colonialism helped to make Dutch cities such as Amsterdam and Rotterdam great trading centers, the latter still being the largest port in the world. Over half of the Netherland's GDP comes from international trade. These countries tried to remain neutral in recent wars but were usually overrun because they lay in the path of easy access from Germany to France. After World War II, Belgium, the Netherlands, and Luxembourg worked together, garnering the name **"Benelux"** (**Bel**gium, the **Ne**therlands, and **Lux**embourg). Belgium's capital, Brussels, serves as the seat for the European Commission and headquarters for NATO. The Hague, the Netherlands, is the seat of the World Court, and Luxembourg is the location of agencies of the EU.

The three alpine countries of Austria, Switzerland, and Liechtenstein are centered on the high mountains of the Alps with their glaciated valleys. Austria and Liechtenstein began as dynasties, though Austria grew into one of Europe's largest empires under the Hapsburgs. Modern Austria represents most of the German-speaking regions of that former empire, which disintegrated at the end of World War I. Vienna's grandiose buildings are a clear reminder of earlier, more regal times. Allied agreements after World War II kept Austria from joining the European Economic Community, but in 1995, not long after the breakup of the Soviet Union, Austria joined the European Union.

Switzerland is a country of strong local voices. It began as an attempt not to be incorporated into any dynasty or empire, especially the Hapsburg monarchy. Switzerland began in 1291 when three communes in the Alps swore to help defend one another from outside aggression. The Swiss Confederation is composed

Figure 3.25 **Land reclamation in the Netherlands.** With one quarter of the country's land below sea level, the Dutch have been building dikes to keep the seawater from inundating their land for more than a thousand years. The term *polder* is used to describe land enclosed by dikes. Source: From Hoffman, *A Geography of Europe: Problems and Prospects.* Copyright © 1983. This material is used by permission of John Wiley & Sons, Inc.

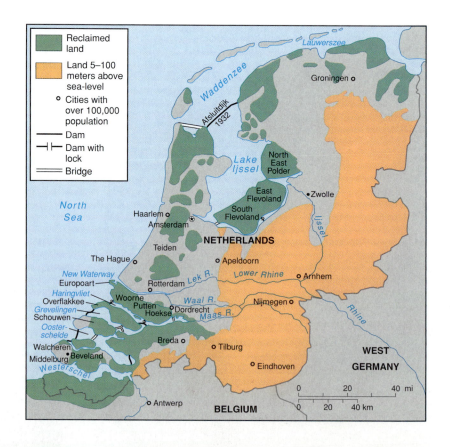

of cantons that have a considerable amount of local decision-making power compared to the federal government. This internal political geography unites a country with a complex geography of languages and religions. Over two-thirds of the country is German-speaking, but French dominates in the west, Italian in the south, and Romansh in the southeast. Roman Catholicism and Protestantism are both common in the country. For external relations, Switzerland has a long history of neutrality in wars and declined to join the EU, though the last referendum on the question came close to achieving the necessary votes.

Germany is one of the younger nation-states in Western Europe, having come into existence only in 1871. Since then, its boundaries have changed. After World War II, the Allies could not agree on a common form of government, so they divided Germany into a Communist country and a democratic/capitalist one. The German Democratic Republic (East Germany) represented the former and the Federal Republic of Germany (West Germany) represented the latter. October 3, 1990, was the day that the two countries reunified (Figures 3.26 and 3.27). Germany's internal political geography is neither as centralized as France nor as devolved as the U.K. With a federal structure like that of the United States, Germany is comprised of 16 provinces called *länder*. Most, such as Bavaria and Saxony, are based on historical territories. Similar to the states of the United States and provinces of Canada, the German *länder* represent local voices and have the ability to craft their own laws and policies concerning such issues as education, transportation, and social welfare.

People

Dynamics

Western European countries are notable for their zero population growth rates (see Figures 3.2 and 3.23). The overall change com-prises a decrease in Germany and small increases in other countries. Rates of natural increase fell from 0.5 percent in the 1970s to around 0.3 percent in 2000. Total fertility rates declined from around 3 births per female in 1965 to around 1.7 in 2000. In the demographic transition, these countries approach or are in the fourth stage (Figure 3.28), where birth rates and death rates are almost equal: after West and East Germany united, their death rate was higher than their birth rate through the 1990s.

Following a baby boom throughout the subregion from around 1950 to 1964, the number of children per family declined with later marriages and the use of family planning methods. The greater number of divorces and later marriages, and increased numbers of widows resulting from the longer life expectancy for women (80 years) created more and smaller households so that more housing units were required. Some of the impacts of such demographic changes are illustrated by the age-sex diagram for Germany (Figure 3.29).

Figure 3.27 **Berlin.** Located near the center of the city, Potsdamer Platz was a thriving business district before World War II. (a) During the Cold War, it was immediately behind the Berlin Wall (foreground) and became a landscape of tank traps, mines, guard dogs, and patrolling soldiers with machine guns. (b) In 1991, with new apartment buildings, a repaved street, and beginnings of a reopened subway station. Use the mound of Hitler's bunker on the left, the East German transmission tower, the church dome, and the tall, almost windowless concrete building as reference points. Photos: © George W. White.

(a)

(b)

Figure 3.26 **Berlin (former East).** "We were the people" sign cynically hangs over a Communist-era East German Trabant, a car with a two-stroke engine. During the Cold War, East Germans were told that they were building an advanced society of equality and economic prosperity for all. After Communism and then unification with West Germany in 1990, East Germans felt conquered and colonized by West Germans with their high-powered Porsches, Mercedes, BMWs, and Audis. Photo: © George W. White.

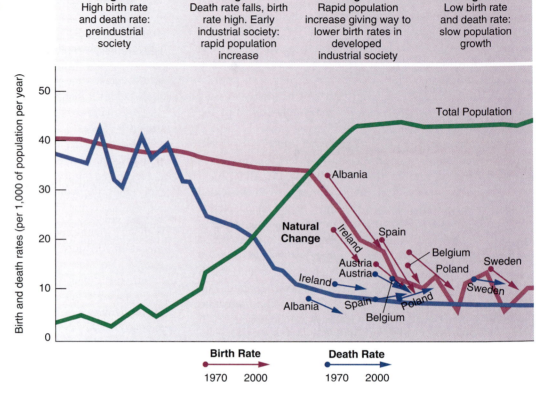

Figure 3.28 **Europe: demographic transitions.** Birth rates and death rates are almost equal in most countries. What does that signify?

Stage 1: High birth rate and death rate: preindustrial society

Stage 2: Death rate falls, birth rate high. Early industrial society: rapid population increase

Stage 3: Rapid population increase giving way to lower birth rates in developed industrial society

Stage 4: Low birth rate and death rate: slow population growth

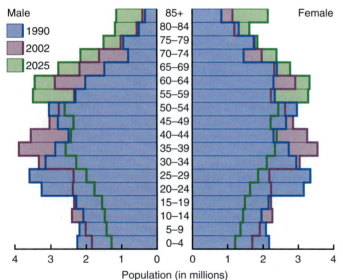

Figure 3.29 **Age-sex diagram of Germany.** Notice the impacts of the 1950–1964 baby boom (ages 38–52 in 2002), the subsequent "baby bust," and the impact of 1900s warfare on the older male populations. Source: U.S. Census Bureau. International Data Bank.

Immigrant Work Forces

World War II destroyed many Western European cities, particularly industrial cities. The advent of the airplane, especially the bomber, resulted in many civilian deaths and the destruction of vast areas far from the front lines. Much rebuilding had to be done, but many young men of prime working age had

been killed during the war. German industry in particular looked for workers in southern Europe: Portugal, Spain, Italy, Yugoslavia, and Greece. When not enough workers were found in these places, Turkey became another source. Over time, Turks came to represent the largest immigrant work force in Germany (Figure 3.30)

The U.K., France, and the Netherlands did not have to search for labor. As decolonization progressed in the 1950s and 1960s, tens of thousands people preferred to migrate to the former ruling countries. Thus, the U.K. received many migrants from the Caribbean, Pakistan, and India, while France accepted large numbers from Algeria and other African countries, as well as Southeast Asia, especially Vietnam. Thousands from Indonesia migrated to the Netherlands.

These foreign workers were frequently granted only limited rights. They obtained the legal right to work but in some cases not citizenship or all the legal rights enjoyed by full citizens. The Germans coined a term for these foreign workers, *gastarbeiter,* or **guest worker.** Though these people were allowed to stay as long as work was available, decades if necessary, the fundamental principle was that they would always be foreigners and, therefore, would eventually go back to their respective homelands. Though more than 7 million immigrants, of which 2 million are Turks, live in Germany, few have more than guest worker status. With so many immigrants, Germany changed its citizenship law in 1999, which previously required German blood, allowing many more immigrants to become German citizens. Of the 4 million foreign-born workers in France, just over 1 million have French citizenship. The United Kingdom granted citizenship to a far

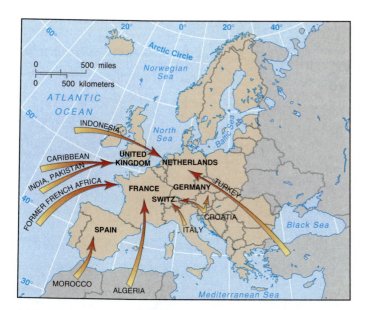

Figure 3.30 **Europe: sources of guest workers, 1970 to 1990s.** The sources vary from the former colonies of the United Kingdom, France, and the Netherlands to the countries around the Mediterranean Sea. What contributions have such workers made to the sending and receiving countries?

greater percentage of its foreign population but tightly controls further immigration.

By the 1970s, Western Europe was rebuilt, but slumping economies hurt further by an international oil crisis meant that less labor was needed. As unemployment began to rise, many Western Europeans began to blame the migrant peoples for taking their jobs. Though many guest workers also lost their jobs, most did not leave. Unemployment benefits allowed them to maintain a standard of living that was higher than any they would have had back in their home countries. Moreover, many had raised families in their host countries, and their children and grandchildren had no desire to relocate to the countries of their parents' birth. Though many guest workers thought fondly of their ancestral homelands, their children viewed these countries as "foreign lands." The situation became apparent in the 1980s, when the German government offered Turkish guest workers cash to go back to Turkey. One stipulation was that they would have to take their entire families with them. As many unemployed Turkish men considered the offer, it was not uncommon for their children to run away in fear that they would have to go live in what they felt was a Third World country where they could not speak the language.

As Western European economies began to improve again in the late 1980s, hostility toward foreign workers subsided for a time. However, following the end of Communism in other areas of Europe in 1989, a flood of new people looking for asylum and work came from East Central Europe. As jobs became scarce, many Western Europeans became hostile toward Eastern Europeans but even more so against non-Europeans. What was often overlooked was that unemployment was not caused simply by foreigners but by changes in the economy and society. As the new service sector replaced

old industry, many could not make the transition from factory worker to computer operator. At the same time, population growth rates among Western Europeans dropped considerably. The current population decline has economic consequences. The number of retired people is growing while the number of workers is declining (Figure 3.31). The declining working generation is increasingly having difficulties supporting pensions for the growing retired generation. Industry also has difficulty finding workers and frequently seeks foreign workers, though it may be unpopular among many citizens.

Refugees

The relatively high level of economic health and political freedom found in Western Europe has been attractive to refugees and asylum seekers from other areas of Europe and the world. With very liberal asylum laws that provide for free housing and education for those under 16 years of age, Germany attracts and receives the largest numbers of refugees. The end of Communism and the onset of wars in the Balkans only increased the number of refugees. People from Bosnia-Herzegovina are the largest group in recent years. Germany has accepted by far the largest share, with over 300,000. Large numbers of refugees in Western Europe come from countries such as Sri Lanka, Turkey, Ghana, Iran, Somalia, Ethiopia, Zaire, Cambodia, and Vietnam. Together with guest workers, they contribute to an ethnic diversity in Western Europe that is much greater than usually acknowledged.

Women, Power, and Social Position

Compared to other areas of the world, women in Western Europe generally have a high degree of political power and a large share of the economic wealth. At a basic level, equality can be measured by life expectancy. For the subregion, men typically live 75 years on average, while women live to 81. Power often begins with literacy and education. In contrast to many other areas of the world, women's literacy levels in Western Europe are equal to men's, at just under 100 percent. Almost an equal percentage of women as men receive university degrees. France and Ireland rank highest with 55 percent of all university degrees received by women. Switzerland is the lowest with 41 percent. In the fields of education, humanities and arts, social and behavioral sciences, and medicine, women commonly receive 60 to 70 percent of the degrees. On the other hand, women receive only 15 to 20 percent of the engineering degrees.

A more direct indicator of women's power is their representation in government. The Netherlands ranked highest among Western European countries by placing fifth in the world in 2001, with women holding over a third of the seats in the lowest house in parliament. Ireland and France were lowest, with Ireland ranked at 55 with 12 percent in the lowest house in parliament and France at 58 with only 11 percent. However, these percentages are still high compared to those in the United States.

Urbanization

Western European countries are among the most highly urbanized countries in the world. In 2000, Belgium had 97 percent of

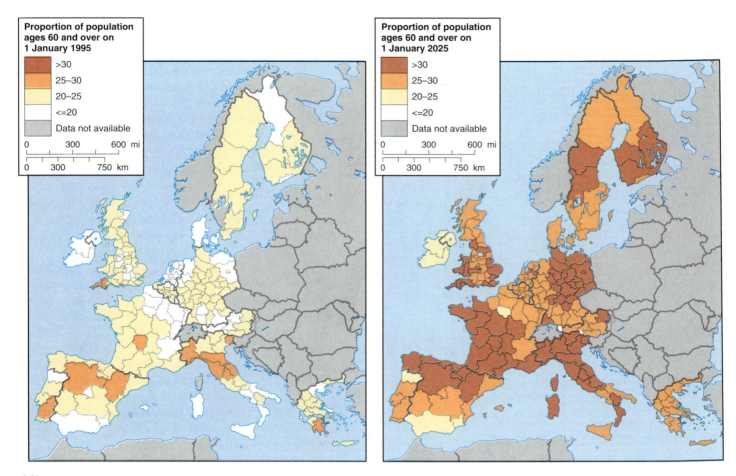

Proportion of population ages 60 and over on 1 January 1995
- >30
- 25–30
- 20–25
- <=20
- Data not available

0   300   600 mi
0   300   750 km

Proportion of population ages 60 and over on 1 January 2025
- >30
- 25–30
- 20–25
- <=20
- Data not available

0   300   600 mi
0   300   750 km

Figure 3.31 **Europe's spatial patterns of aging.** Source: Statistical data: Eurostat, REGIO database; Cartography: Eurostat–Gisco, 03/1999.

its population living in urban areas. For much of the rest of the subregion, many of the countries had 85 to 90 percent of their populations living in cities. Ireland (59 percent) and Austria (65 percent) were the least urbanized but had increased their urban percentages notably over just the previous few years. The high proportions reflect the economic focus on urban-based manufacturing and service industries. There are few extremely large cities, since functions are often spread among several cities. Paris (France), with around 10 million people, and London (U.K.), with over 7 million, are by far the largest urban centers, combining the government and private services of capital cities with diversified manufacturing. Other major cities in France include Lyon, Marseilles, and Lille. Those in the United Kingdom include Birmingham, Manchester, Leeds, and Glasgow. In Germany, Berlin is the reinstated capital. The Ruhr industrial complex around Essen, Düsseldorf, and Cologne forms the largest concentration of people. The major port of Hamburg, the regional manufacturing centers of Munich and Stuttgart, and the financial center of Frankfurt are other major German cities. The largest centers in the low countries are Brussels and Antwerp (Belgium) and the ports of Rotterdam and Amsterdam in the Netherlands. Dublin is the capital of the Republic of Ireland. Vienna, Austria's capital, is by far the largest city in Austria.

In Switzerland, the largest city of Zürich has closer rivals in Geneva, Basel, and the capital, Bern.

Urban Landscapes

The geographic characteristics of the city landscapes of Europe result from centuries of making and remaking built environments. They record the consequences of change (Figure 3.32). A number of Western European cities contain relics of historic cultures from Roman times onward. In medieval times, a network of towns grew with walled defenses and market and administrative functions. Internal spatial differentiation by class was not strong.

With the Industrial Revolution of the 1800s, cities grew tremendously and land uses differentiated into central business districts of shops and offices, industrial areas of large factories, worker housing, and suburbs in which the growing management classes lived. In the early 1900s, the development of public transportation in cities led to further differentiation of land use and the building of suburban housing linked by streetcar or bus to the city centers where retailing, commercial, and manufacturing activities concentrated.

During World War II, bombing and ground fighting reduced parts of many European cities to rubble, destroying older areas, including medieval and industrial buildings. The

Older couples whose children had become independent and younger professionals moved to neighborhoods near city centers. They bought and renovated poor and dilapidated housing, increasing the value of the housing stock and raising neighborhoods from low income to higher income. This process is known as **gentrification.** Allowing only pedestrian traffic in central areas (see Figure 3.33a) and arcades of specialty shops became features of city centers, together with the congregation of high-order services requiring support from people living in a wider area—theaters, opera houses, cinemas,

Figure 3.33 **Europe: urban landscapes in Plymouth, England.** Modern changes in the city center. (a) The main shopping area was totally rebuilt after World War II bombing and remains the largest retail district in the city despite out-of-town supermarket and warehouse shopping developments from the 1980s. Vehicles were excluded from some streets in this shopping area in the early 1990s to encourage pedestrian traffic. (b) The wharves and warehouses around the old harbor—from which Sir Francis Drake (1570–1590) and the *Mayflower* Pilgrims (1620) sailed—are now adapted for tourists with restaurants and gift shops. Photos: © Michael Bradshaw.

(a)

(b)

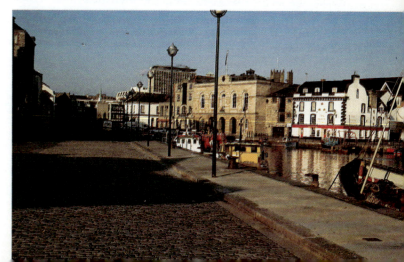

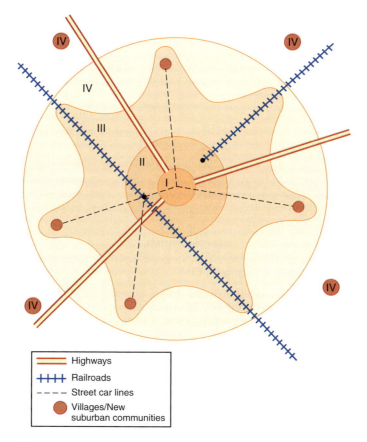

Figure 3.32 **Generalized morphology of non-Communist European cities.** (I) Medieval core. (II) Early growth area later penetrated by railroads and the location of early industrial factories and worker housing. (III) Star-shaped pattern created by suburbanization along street car lines in the early 1900s. (IV) Post–World War II suburbanization with automobiles.

Legend: Highways; Railroads; Street car lines; Villages/New suburban communities

period from 1945 to 1970 was one of rehabilitation, expansion, and restructuring of cities. Nearly all cities expanded their functions and populations. City centers were rebuilt with utilitarian buildings (Figure 3.33a), while large tracts of public housing catered to lower-income groups. New towns were built at a distance from the largest cities, separated by "green belts" of rural land in which new building was seldom allowed. A major result was decentralization—a movement away from the previous focus on the central business district and toward more economic activity in the extensive suburbs built since 1945.

From the 1970s, there was disillusionment over the outcome of postwar reconstruction and the bleak nature of low-cost public housing. Large public housing tracts, especially where high-rise buildings were common, proved unsuccessful and often became centers of unemployment and crime. A significant move began toward public and private cooperation, particularly in efforts to make city centers and old dockland waterfronts more livable (Figure 3.33b). Areas with buildings of historic interest were declared conservation areas, but the finances for pursuing such a policy now came from private investors who saw the potential returns from encouraging tourism and in building new accommodations for those wishing to move back near the city center.

and clubs. Areas around the city center that had become derelict, such as old railroad yards, factories, workshops, and worker housing, became sites for the new facilities.

In the 1980s, the relaxation of planning constraints led to a further dispersal of economic activity from city centers to the suburbs and beyond. Manufacturing had already moved out from cramped inner city sites to suburban industrial estates. Now retail and office facilities followed to suburban warehouse shopping, hypermarkets, shopping malls, and office parks. Nevertheless, densities of European cities are still much higher in the United States and suburbanization is much less pronounced.

Economic Development

Western Europe remains the economic heart of the region, with 60 percent of all manufacturing jobs in the region and 75 percent of the research and development that devises and applies new technology. Germany in particular has the third-largest economy in the world and the largest in Europe. Germany is often referred to as the "motor" of Europe and is responsible for much of Western Europe's economic strength. Total German exports almost meet the combined exports of France and the U.K., and Germany is the major trading partner for almost 20 European countries. By contrast, France and the U.K. each are the main trading partners for only three to four other European countries.

Since 1950, Western Europe's economy and the geographic distribution of economic activity have changed in response to shifting economic sector structures, the use of new energy sources, increasing involvement in the global economy, and the development of the European Union. The continuing high levels of affluence are reflected in the widespread ownership of consumer goods (see Figure 3.17).

Agriculture

Centuries ago, as various European peoples settled lands (see the section on "Migrations of Peoples" in this chapter), they often cleared the forests and established the two- and three-field system. Typically with this system, farmers would divide their land into thirds and rotate what they grew on each third every year, leaving one-third fallow to regenerate. In areas where the soil was poor, farmers alternated their crops between two fields. Wheat was the most common crop, but rye, oats, and barley were cultivated. Livestock played an important role, with much of the grain going to fatten the livestock. Until the Industrial Revolution, most people were farmers who owned small plots of land that primarily fed their own families but frequently produced a small surplus that could be sold for profit.

As the Industrial Revolution progressed, agriculture was transformed. The production of tractors, fertilizers, and pesticides meant that fewer people were needed to work larger farms, otherwise known as **concentration.** Small farms of a few hectares were taken over by more successful farmers, resulting in fewer but much larger farms. It led to a strong flow of people from the countryside to the cities, where they found jobs in factories. Today, agriculture still occupies almost as much land as before (Figure 3.34) but provided only 2 to 7 percent of total employment in 1996.

Tractors, fertilizers, and pesticides also resulted in greater agricultural productivity per hectare, known as **intensification.** Fertilizers led to increased output. Scientific crop rotation, which involved the planting of soil-enriching crops such as clover and turnips every four years, made it unnecessary to leave half or one-third of the land fallow every year.

Modern agriculture also meant that farmers were no longer producing crops primarily for their families' consumption but for sale. The new situation motivated farmers to switch from producing a variety of crops for their families to single crops that brought high profit (Figure 3.35), such as sugar beets or oil seeds. This is known as **specialization.**

Concentration, intensification, and specialization have radically changed agriculture and the rural way of life over the last hundred years. In contrast to the past, modern farms are large and the farmers who run them require management skills. Farms also depend on industries that produce seeds, fertilizers, pesticides, and machinery to produce agricultural goods. They rely on food processing plants and marketing agencies to get their products to the consumer. The term **agribusiness** was coined to describe how businesslike farming has become and the close links that farming has with other industries. Thus, while fewer Europeans work on farms, agriculture is still very significant sector of European economies.

Though agricultural production is high in Europe, the number of farmers continues to decrease. Farmers worry that their declining numbers will result in less political power and will endanger the future of farming. European governments, however, have been sensitive to the need to protect a traditional way of life and maintain a stable food supply. The EU Common Agricultural Policy (CAP) at first provided a set of price supports that made farming almost risk-free at a time of advancing technology. Output increased rapidly in the 1960s and 1970s, although costs to consumers and taxpayers rose. Providing agricultural subsidies to farmers has come to represent the largest expense of the EU's funds. The most productive farming areas of southeastern England, northeastern France, northern Belgium, and the Netherlands were most favored by EU agricultural policy. Farmers in these regions had larger farms, had access to more capital, and adapted easily to new methods. Mountains of grain and butter, and lakes of milk and wine resulted. For example, the cereal crop rose by 40 percent between 1975 and 1985, generating a surplus and huge storage costs. The main fears of farmers in this subregion in the 1990s concerned the possible admission to the EU of countries in East Central Europe with their lower costs, but this change will take years, and at present, those farmers are much less efficient than their counterparts to the west.

In the hilly regions, where more marginal livestock farming replaced crops, farmers received support for social and environmental reasons rather than economic ones. Governments feared rural depopulation and soil erosion if the farms in these areas were abandoned. Governmental support helped the environment and slowed depopulation, but did not reverse it.

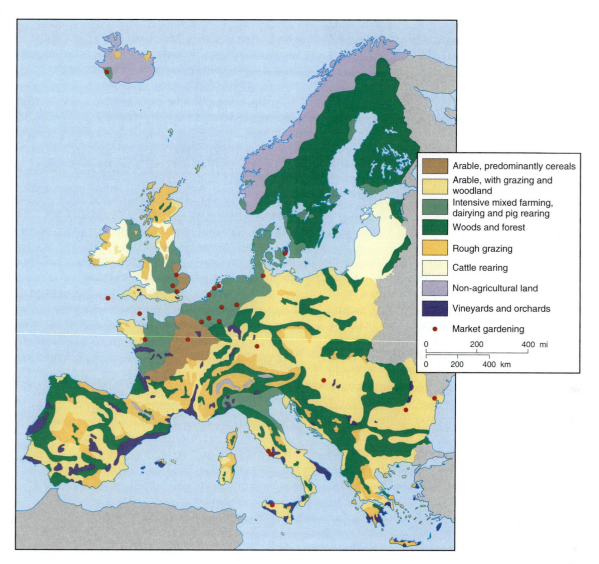

Figure 3.34 **Europe: farming land uses.** Relate the categories on this map to the relief details shown on Figure 3.10.
Source: Data from *New Oxford School Atlas*, p. 98, Oxford University Press, UK, 1990.

During the 1980s, the Common Agricultural Policy changed to reduce the huge surpluses of produce in storage as a result of price guarantees to farmers. It was decided that agriculture should be linked more closely to market prices but that an immediate lowering of guaranteed prices would be devastating to farmers and rural areas alike. More gradual reductions were engineered through downward revisions of price thresholds and milk quotas, and by forcing overproducers to pay part of the storage costs. It was also recognized that less production would reduce stress on the natural environment.

In 1988, further measures encouraged major cutbacks in production. Older farmers were helped to retire early by being given pensions in exchange for ceasing production of surplus capacity crops. Arable land was retired from production in exchange for grants to farmers who would continue to look after the land but not produce certain crops on it. Livestock farmers were encouraged to produce less from the same area, a process known as **extensification.**

By the late 1990s, the results of the attempts to reduce farm output in the EU countries were not clear. Setting aside land does not always reduce productivity in proportion to the area set aside. The poorer land is often retired first. Retired land may continue to be used intensively for nonsurplus crops. The grants are often applied to situations in which farmers would have retired or let land revert to woodland in any case. The geographic impacts are variable, since France in particular resists any move that reduces its agricultural sector. In other countries, there is debate as to whether the poorer hilly lands should be retired to a greater extent than the productive lowlands. Reducing milk quotas and increasing storage charges appear to be the most effective policies for controlling production levels.

As European farmers faced the loss of long-term subsidies and increased competition from new EU members in the late 1990s, they were hit by other events that reduced their incomes and led to reductions in farm acreage. Urban expansion continued, and planning restrictions on the conversion of

Figure 3.35 **Europe: agriculture.** This farm, located east of Dresden, Germany, specializes in growing maize (corn), a crop Europeans learned of during the Age of Discovery when they sailed to the Americas. European farmers most often grow this high-yield crop to fatten animals, usually pigs, which they sell to market for profit.

Photo: © Emily A. White.

farmland were relaxed. The feeding of inappropriate animal matter to cattle in the late 1980s led to the expansion of bovine spongiform encephalopathy (BSE)—a disease similar to Downes cow syndrome in the United States and generally known as "mad cow disease"—in Britain and scares over its transmission to humans through ingestion of the infected beef. The EU prevented the U.K. from exporting its beef, but farmers in other countries were affected by a loss of confidence among meat buyers and the reoccurrence of the disease more widely in Europe. An outbreak of foot-and-mouth disease in the U.K. in 2001 led to the slaughter of thousands of livestock in an attempt to stop the spread of the disease overseas. Though the disease was halted, the outbreak cost millions and further damaged the reputation of the British meat industry.

Major Manufacturing Industries

Two of Western Europe's major industries are automobiles and airplanes. The automobile industry finds it difficult to expand further. It is partly nationalized and partly owned by foreign companies. Renault in France and part of Volkswagen are state corporations, as was the former British Motor Corporation. General Motors and Ford Motor Company of the United States opened their European plants in the 1950s and built a complex set of interrelationships throughout Europe among assembly plants, independent suppliers of components, and specialist factories producing engines and transmission systems (see Figure 2.14). The Japanese carmakers Nissan, Toyota, and Honda all built assembly plants in the U.K. to gain a foothold in the EU market. There are still locally owned private firms, such as Peugeot-Citroën (France) and BMW (Germany).

Germany produces around 4 million cars per year, France around 3 million, and the United Kingdom and Belgium around 1 million each. The total is well above the numbers that can be sold. In the 1980s, only Germany expanded the number of cars produced as overproduction became common—a situation made worse by the opening of Japanese-owned factories in the U.K. Automation was introduced to cut costs, causing tens of thousands of layoffs in the assembly and component factories. Other cost-cutting moves included increased links

between the assemblers and component manufacturers so that inventories could be kept low. This placed a greater emphasis on close geographic links—including location and transportation—between the carmakers and their suppliers.

In 1998, the merger of Daimler-Benz (Mercedes) and Chrysler suggested a new approach toward world markets that involves some of the largest automobile makers in different countries working together. Renault is linked with Nissan (Japan) and Volkswagen with Seat (Spain) and Skoda (Czech Republic).

The aerospace industry is another key sector. Aeronautical research and manufacturing are well developed in Western Europe and have been pursued for decades. Defense was an important reason for the support of this industry, but commercial production receives more attention because of competition with U.S. corporations. The small size of Western European countries makes it difficult for them to compete commercially with a country as large as the United States and its airline manufacturers such as Boeing. To compete, a group of Western European airplane manufacturers formed a consortium in the late 1960s that is now known as Airbus. In the same spirit of the European Union, Airbus pools the resources of several countries. Companies in France, Germany, the U.K., and Spain all manufacture various components that are then shipped to assembly plants in either Toulouse or Hamburg (Figure 3.36). Toulouse is the bigger assembly plant and headquarters for Airbus. The first airplane did not roll off the assembly line until 1972, and it took until 1975 before Airbus was able to capture just 10 percent of market share. Growth has been steady and rapid. Today, Airbus employs 44,000 and was able for the first time in 1999 to capture the biggest share of new passenger-jet orders.

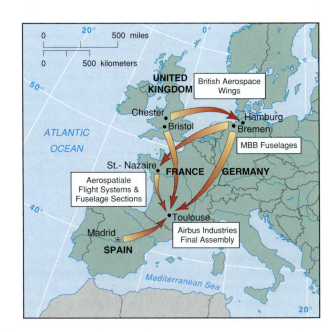

Figure 3.36 **Western Europe: aerospace industries.** Cooperative European manufacture of Airbus aircraft. The high levels of capital inputs and technological complexity required for constructing passenger aircraft make it impossible for the aerospace industry in one country of Europe to support the whole process. What are the political and economic implications of these movements?

Competition between Airbus and Boeing is now fierce. With commercial airline traffic at a high and likewise airport congestion, the two companies are staking their market positions on two different scenarios of what the future may bring. Airbus believes that continued congestion will result in greater difficulties for airlines in obtaining gates at major airports such as London's Heathrow, Tokyo's Narita, and New York's John F. Kennedy. Airbus believes that the best solution is to build a bigger airplane to funnel more passengers through the limited number of gates. Therefore, Airbus is launching the A380, a plane that has 50 percent more floor space and at least one-third more seats than Boeing's 747. The plane is also intended to fly greater distances and offer more comforts such as shower facilities, a piano bar, an exercise room, and a barber shop. In contrast, Boeing believes that greater airport congestion will lead to both new airport construction and the expansion of current facilities, calling for more mid-size and faster jets. Thus, Boeing is developing a modest-sized but very fast subsonic cruiser. It is difficult to predict the future, but European cooperation in aeronautics may make Western Europe the world's leader in the commercial airline industry.

Energy Sources

Domestic coal remained the main source of energy until the 1950s: as late as 1952, it provided 90 percent of energy needs in the U.K. and 92 percent in Germany. After 1955, however, home-produced coal declined rapidly as environmental legislation made burning it more costly and as cheaper imported oil, mainly from Southwest Asia, took over in residential, industrial, petrochemicals, and transportation uses. By 1972, oil supplied 65 percent of the subregion's energy needs, while coal was reduced to 22 percent. Three-fourths of what coal was still produced was used to generate electricity.

Oil-price shocks in 1973 and 1979–1980 raised the oil price ten times above its 1970 level. The growth in oil use slowed, energy conservation became important, and alternative fuels were evaluated. The rising oil prices of the 1970s gave a boost to exploration and production in the North Sea basin, extending northward from the discoveries of natural gas in the Netherlands in 1963 (Figure 3.37). The U.K. and Norway were the main beneficiaries. Early estimates of reserves and production were raised again and again: the Netherlands gas fields have already extracted twice the original reserve estimate and nearly as much again remains available in the rocks. In the North Sea oil and gas fields, production continued to rise in the 1990s, and exploration shifted west of the northern British Isles.

Oil prices in the mid-1980s fell in response to new discoveries that created competition and conservation by users that lowered demand. Natural gas became the cheapest fuel for electricity generators. By discovering its own oil and gas fields, reducing the growth in its demand, and adopting new technologies, Western Europe obtained good prices and a greater security of delivery from competing producers of oil and gas, particularly those in Northern Africa and Southwest Asia. The vast natural gas resources of Russia were brought by pipeline into Germany, France, and Italy. In the future, this source of supply is likely to expand because Russia needs to export vast

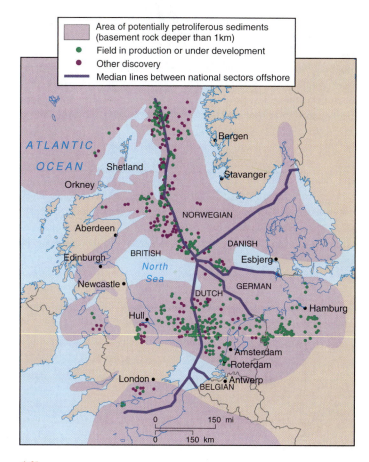

Figure 3.37 **North Sea oil and natural gas fields (1995).**
Source: Pinder, *The New Europe.* Copyright © 1998. Reprinted by permission of John Wiley & Sons, Ltd.

quantities to pay for the consumer, capital, and producer goods that it imports (see Chapter 4).

Of other sources of energy, nuclear-powered electricity generation is the most important, especially in France and Belgium (77 and 57 percent of the national totals, respectively, in 1996). It would have been more important elsewhere if early promises of low cost and safety had materialized and the Chernobyl nuclear power plant explosion in Ukraine in 1986 had not occurred. Hydroelectricity is important in Northern and Alpine Europe. There is some development of wind power in western Britain and of tidal power in northern France, but these and other alternative energy sources provide a tiny fraction of current needs and are often subject to local opposition—against, for example, the visual pollution of wind farms in the landscape.

Service Industries

While manufacturing in Western Europe ventured into new sectors but fell behind the world leaders in the United States and Japan, the development of the service sector almost rivaled that in the United States. In the late 1990s, employment in the service sector made up over 65 percent of the total work force in most Western European countries.

Service jobs grew in importance as the countries increased their populations, became richer, and instituted strong social welfare programs. Jobs in retailing, wholesaling, education, health care, and government employment relate closely to numbers of people and population distribution. The providers range from private corporations, such as food and drink retailing, to state-controlled institutions in health care and education. Such services support and depend on the manufacturing sectors of the economy but play an indirect role in production. For example, better education and health care may provide a more skilled and adaptable work force.

The main growth areas in the service sector are in producer services and tourism. **Producer services** are involved in the output of goods and services, including market research, advertising, accountancy, legal, banking, and insurance. They serve other businesses rather than consumers directly. Greater use of computers and information technology increased **productivity,** which is frequently measured by the amount of product generated or work completed per hour of labor. Yet despite the increase in productivity, producer service jobs have increased by a quarter to a third in the countries of Western Europe since the 1970s and employ around 10 percent of the labor force. Producer services concentrate in the centers of major cities, where agglomeration economies are significant. Agglomeration economies, as stated earlier, involve the clustering of businesses in a location, often near governmental agencies, to produce savings from the sharing of infrastructure, labor pools, market access, and reduced transportation costs. London, Paris, Amsterdam, Frankfurt (Figure 3.38), and Munich have major shares of these industries. While some functions, such as back-office routine processes, are decentralized in suburban and small-town "paper factories"—often in landscaped office parks—major city centers retain the high-order functions in which face-to-face personal contact is important.

Mergers and takeovers after the mid-1980s led to a smaller number of very large producer service firms, each broadening the range of services. Banking, insurance, property services, and accountancy were combined. Centrally located specialist firms took over provincial firms to extend their influence nationwide and internationally. Although financial firms dealing with huge sums of money are liable to occasional scandals, the level of control in Western Europe is high enough to attract many transactions of world financial significance.

Tourism is another growth sector among service industries. Western Europe dominates the international market with over 158 million foreign visitors in 1999. In 1997, European countries held six of the seven top places in terms of international tourist receipts: Italy, France, Spain, U.K., Germany, and Austria (2nd to 7th) had total receipts of $134 billion (compared to the United States, which was 1st with $75 billion).

The numbers of internal and international tourists, both incoming and outgoing, doubled from 1970 to the 1990s. The growth resulted from increased incomes and leisure time, together with new lifestyle expectations. The expansion of package tourism made foreign travel accessible to greater numbers of people. College students, for example, buy rail passes and visit major cities, staying in youth hostels along the way. Working-/middle-class families travel using campgrounds and recreational vehicle (RV) parks. More people take second holidays or have short weekend breaks. Germany and the U.K.

provided the main source of tourists in other European countries. Although Mediterranean countries were the main receiving areas, the mountainous, rural, coastal, and historic urban areas within Western Europe also increased their tourist trade. While tourism is increasingly significant in Europe, tourist industries in other world regions have developed more rapidly. Europe still dominates the world in tourism, though its share of tourism receipts declined from 64 percent of the world total in 1975 to 49 percent in 1997.

Tourism is encouraged by the governments of each country, EU regional investment in infrastructure (roads, airports), and agreements on the sharing of health facilities, currency regulations, and customs. Individual countries promote their facilities through tourist boards. In the mid-1990s, tourism generated one job in eight in the whole European Union, making it the largest industry in Europe. Unlike in many other industries, the growth in tourism has not led to the rise of many large international tourist corporations. Smaller tourist companies remain the norm.

Tourism is often promoted to help in regional development, since many popular rural and coastal areas are located away from urban-industrial economic growth areas and need increased local economic activity to prevent further population loss. Tourist developments provide useful additions to local

Figure 3.38 **Western Europe: financial services.** In one of Europe's major financial centers, Frankfurt, Germany, medieval buildings are preserved in the heart of the city, surrounded by other buildings limited in height, although some were rebuilt after World War II. Beyond these are high-rise office and apartment blocks. Photo: © SuperStock, Inc.

Test Your Understanding 3D

Summary Population in Europe is densest in the industrialized areas of Western Europe. Densities are lowest in the cold, mountainous areas, especially in the north, and more evenly distributed across East Central Europe. Western Europe contains many of Europe's former colonial and imperial powers, and it is where the Industrial Revolution began in the 1700s. Countries of this subregion are still powerful today and rank high on the human development index.

Three of the four largest countries by population in Europe are in Western Europe (France, Germany, United Kingdom). The population of Western Europe is static in overall numbers and aging. Decreasing numbers of people in the working-age groups support increasing numbers of retired people. In the 1960s and 1970s, immigration was encouraged to bring in guest workers.

Western Europe is one of the most urbanized areas of the world. Towns and cities present landscapes that contain relic features from their long history together with massive areas of recent industrial and residential expansion that reflect the growth of urban living in the region.

Agriculture was modernized so that the subregion that was a net importer is now a net exporter of food. The manufacturing industries modernized in type and levels of productivity. Service industries, especially financial and tourism aspects, are the main employers, contributing increasingly to the continuing affluence of the subregion.

Questions to Think About

3D.1 What are the trends in population growth rates in Western Europe? What are the causes of these trends?

3D.2 How have Western Europeans contributed to the struggle to achieve human rights around the world?

3D.3 Why do so many difficulties confront the movement toward peace in Northern Ireland?

3D.4 What happened to urban landscapes in non-Communist Europe in the decades after World War II?

Key Terms

Benelux	specialization
guest worker	agribusiness
gentrification	extensification
concentration	producer services
intensification	productivity

household incomes, often improve the facilities available in an area, and add to amenities that attract other forms of employment. Much tourism employment, however, is seasonal, poorly paid, and female-dominated. The unskilled and low-paid nature of many jobs lead to them being taken up by migrants and thus having less impact on the local economy than anticipated.

Northern Europe

Northern Europe is sometimes called "Norden" and extends from the North European Plain to beyond the Arctic Circle and encompasses the Scandinavian and Jutland peninsulas, Finland, and several islands in the Baltic Sea and North Atlantic Ocean. It includes the countries of Denmark, Finland, Norway, and Sweden, together with the former colonies of Denmark, Greenland, the Faeroe Islands, and Iceland (Figure 3.39).

Northern Europe conjures up images of the Vikings who extended their control to the British Isles, northern France, Iceland, Greenland, much of the Baltic area, and Ukraine from the 800s through 1200s. Later, Sweden, as a modern imperial state, controlled much of northern Europe and areas that are now in Russia, the Baltics, northern Poland, and Germany at various times until the early years of the 1800s. Since then, Sweden and Denmark gave up imperial ambitions and frequently take positions of neutrality in international relations. Northern European countries are great advocates for human rights. Sweden is known for the Nobel Institute, though a committee of the Norwegian parliament awards the Nobel Peace Prize. Finland is known for the International Helsinki Federation for Human Rights. Compared to the rest of the world, women in these countries hold the highest percentage of elected and nonelected positions in government. Though Northern Europe is sparsely populated and contains many resources, some would argue that the high level of women's empowerment has likewise contributed to the subregion's considerable economic prosperity and some of the world's highest GDP per capita and HDI figures (see Figures 3.16 and 3.40).

Countries

Throughout much of their histories, the peoples of Northern Europe remained mostly free of rule from outside the subregion and thus were able to express their local voices in self-governance. Within the subregion, the Danes and Swedes frequently controlled the others, as well as people and lands outside Northern Europe. Norway, Finland, and Iceland have enjoyed independence only in the last hundred years.

Denmark is one of the smallest countries in Europe but has strong global connections since early times. Claiming the oldest capital city and flag, the Danes also claim that their queen, Her Majesty Queen Margrethe II, has the oldest royal lineage in the world, dating from Viking king Gorm in the early 900s. During the Middle Ages, Denmark was a major power in northern Europe. The Danes willingly joined the EC (now EU) in 1973, but they have resisted adopting many EU standards because they have usually been lower than their own. Danes even threatened to unravel the EU in the early 1990s, when they failed in 1992 to ratify the Maastricht Treaty, which called for greater political and economic union among the member countries. In a second vote in 1993, the Danes narrowly ratified the treaty but more recently refused to adopt the euro as their currency.

Though Denmark is not the world power that it once was, it still has close ties with Iceland, the Faeroe Islands, and Greenland, which were all part of it at one time. The populations of the latter two are only 46,000 and 52,000, respectively. Iceland gained some autonomy in 1918 but retained strong links with Denmark until World War II, when U.S. and U.K. forces

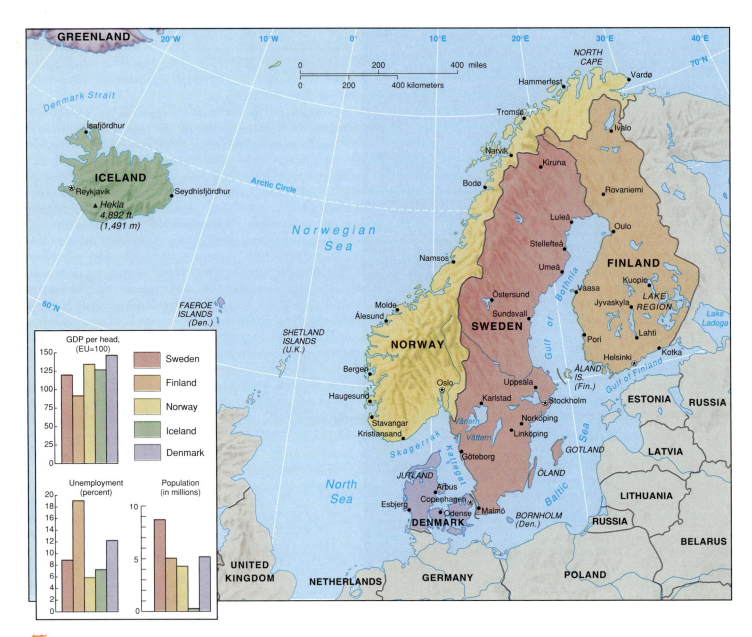

Figure 3.39 **Northern Europe: the countries, cities, and physical features.** Compare the countries in terms of GDP per head, unemployment rate, and population. Norway and Sweden comprise the Scandinavian Peninsula; the Jutland Peninsula is part of Denmark.

occupied it after Germany conquered Denmark. Iceland declared full independence in 1944. The Faeroe Islands and Greenland achieved autonomy in 1948 and 1979, respectively, though both are still part of Denmark. All three places are inhibited by physical difficulties of climate and land and rely on fishing the surrounding waters. In the late 1990s, fish stocks declined as other countries increased their take. Competition for use of the surrounding waters led Iceland and the Faeroe Islands to extend and defend their 200-mile limits against European fishers until international agreement was obtained. Iceland has come under international pressure not to resume whaling.

Norway is a mountainous country with limited farmland primarily around Oslo and Trondheim. During the Ice Age, glaciers rounded river valleys as they cut their way to the sea. After the Ice Age, these river valleys flooded as sea level rose.

These flooded inlets are known as **fjords.** At the very end of these fjords, small patches of good agriculture land formed and were called "Viks." The people who inhabited them were called "Vikings." The Norwegians are proud of their Viking heritage and their seafaring ways. Modern Norway was controlled either by Denmark or Sweden and did not achieve self-governance until 1814. Even then, the Swedish king remained the king of Norway until 1905, meaning that independence was not fully achieved until 1906, when the Norwegians elected their own king. Since then, Norwegians have guarded their independence carefully. Norway is a member of NATO, but Norwegians rejected referendums to join the EU. Until the 1970s, Norway's economy was based on fishing and the smelting of metals using hydroelectricity. It is believed that the Norwegians' desire to protect their fishing grounds and their practice of whaling

Figure 3.40   Northern European countries: key data on countries.

Country	Capital City	Land Area (km²) Total	Population (millions) Mid-2001 Total	2025 Est.	GNI 1999 (US $ million) Total	GNI PPP 1999 Per Capita	Percent Urban 2001	Human Development Index Rank of 175 Countries	Human Poverty Index: Percent of Total Population
Denmark, Kingdom of	Copenhagen	43,090	5.4	5.8	170,685	25,600	72	15	12.2
Finland, Republic of	Helsinki	338,100	5.2	5.3	127,764	22,600	60	13	11.9
Iceland, Republic of	Reykjavik	103,000	0.3	0.3	8,197	27,210	93	9	no data
Norway, Kingdom of	Oslo	323,000	4.5	5.0	149,280	28,140	74	2	11.3
Sweden, Kingdom of	Stockholm	449,960	8.9	9.4	236,940	22,150	84	6	7.0

Country	Ethnic Groups (percent)	Languages O=Official	Religions (percent)
Denmark, Kingdom of	Danish, Inuit (Eskimo), Faroese, Greenlander	Danish, Faroese	Protestant 91%
Finland, Republic of	Finn 90%, Swede 9%	Finnish (O), Swedish	Protestant 89%
Iceland, Republic of	Norwegian/Celtic descendants	Icelandic (O), Danish, English	Protestant 96%
Norway, Kingdom of	Germanic (Nordic, alpine, Baltic)	Norwegian (O), English	Protestant 88%
Sweden, Kingdom of	Swede, Finn, Dane, Norwegian, Greek, Turk	Swedish, English	Protestant 94%

Source: Data from *Population Reference Bureau 2001 Data Sheet; World Development Indicators,* World Bank, 2001; *Human Development Report,* United Nations, 2001; Microsoft Encarta (ethnic group, language, religion).

played a major role in the rejection of EU membership, even after new wealth came with the discovery of North Sea oil and natural gas in the 1970s. The oil was abundant enough to make Norway the world's second largest crude oil exporter after Saudi Arabia in 1995.

Like Norway, Finland achieved its independence late. Finland was within the Swedish kingdom from the 1100s to 1809, when it became a Russian possession. Finnish independence was declared in 1917, but Stalin's armies attacked Finland in 1939 with the intent of reincorporating Finland into the Soviet Union. The Finns fought off the attack but sought a closer alliance with the Axis powers to protect their country's independence. As the Red Army rolled back Hitler's forces, it launched new campaigns against Finland. Vastly outnumbered, the Finns sued for peace, surrendering strips of land adjacent to the Soviet Union and agreeing not to join any political organization, which would include NATO and the EU, or engage in anti-Soviet rhetoric. The Soviets were particularly concerned about Finnish broadcasting, which could be received in nearby Estonia, a Soviet republic where the language was so closely related to Finnish that Finnish was easily understood. After the dissolution of the Soviet Union in 1991, Finland was freed from its forced neutrality, quickly joined the EU, and established close links with the Baltic countries, especially Estonia.

Sweden has been the largest and politically strongest Northern European country throughout history with only Denmark as a rival. At its height from 1610 to 1718, "the Great Power period," Sweden controlled the areas now known as Finland and the Baltics, and northern areas of Poland and Germany. The Swedish army was able to defeat Danish, German, and Russian forces, often simultaneously, before it was permanently weakened in 1721. Sweden lost most of its possessions but retained much of Finland. In 1809, Sweden lost Finland and in 1812 acquired its current boundaries, but a few years later and until 1905, Sweden gained considerable control over Norway.

Though Sweden has not engaged in a war since 1812, the Swedish government cooperated with Nazi Germany by providing it with ball bearings necessary for weapons and by allowing German troops to be transported by rail across Sweden to Norway and Finland. In contrast, Swedes such as Raoul Wallenberg became famous for having rescued thousands of Jews from the Nazis' Holocaust. After the Cold War, Sweden gave up its neutrality and joined the EU in 1995. It has decided, along with Denmark and the U.K., however, that it will not adopt the euro as its currency. Since the end of the Cold War, Sweden has also invested heavily in the Baltic countries.

People

Dynamics

The population of Northern Europe is small, only 24 million people in total in 2000, and likely to rise by just over another million by 2025. Rates of natural population change were between –0.01 and 0.3 percent in 2000, and total fertility rates dropped from around 2.5 in 1965 to under 2 by 2000. Those born in Northern Europe in 2000 expect to live around 80 years. In all countries, birth rates almost equal the low death rate levels (see Figure 3.28). The 2002 age-sex diagram for Sweden (Figure 3.41) shows maximum numbers in the age groups around 30 to 40 years, reflecting the baby boom from the later 1940s into the 1960s and the subsequent fall in births. Those over 65 years make up 15 to 17 percent of the total population, a growing burden on the comprehensive welfare systems of these countries that is already being felt in increased taxes and the reduction of some welfare programs after years of wide-ranging coverage. Many of these countries have had to cut back on the services they provide.

Urbanization

The lack of farming land and cold northern climates are linked to high percentages of urbanization. Finland (67 percent) and Norway (76 percent) were the only two countries less than 85 percent urbanized. Even so, the major cities are not very large. In each country the capital city is by far the largest, but only Copenhagen (Denmark; Figure 3.42), Stockholm (Sweden), and Helsinki (Finland) had over 1 million people in 1996. Oslo, Norway's capital, does not exceed 1 million. The main cities of Northern Europe were little changed by warfare, and their centers continue to be dominated by government and commercial buildings built in the 1800s and early 1900s.

Culture

Culturally, the Scandinavians (Swedes, Danes, Norwegians) are Germanic peoples, specifically the northern branch, and descendants of Vikings. Their languages are closely related, and so are their histories (see Figure 3.4a). The inhabitants of the Faeroe Islands, Iceland, and Greenland are descendants of early Scandinavian settlers and likewise have similar cultures, though Greenland also is inhabited by Inuits. In contrast, the Finns and Sami (more commonly but derogatorily called Lapps) are Finno-Ugric peoples whose languages, related to Hungarian, are not Indo-European like most of the other languages of Europe and bear no resemblance to the Germanic languages of the Scandinavians. The Sami practice a traditional nomadic lifestyle of herding reindeer. They are found in the far northern reaches of Norway, Sweden, Finland, and adjacent areas of Russia.

Evangelical Lutheran Christianity is the major religion for Northern Europeans (see Figure 3.4b). Officially, 90 percent or more of the population are Lutherans in the four major countries

Figure 3.42 **Denmark: Copenhagen's urban landscape.** Photo: © Loren W. Linholm.

and in Iceland. This Protestant variant of Christianity influenced the lives of the people, inducing very serious and community-conscious attitudes toward work and social life. In recent years, the combination of affluence and materialism broke many of these strong cultural links and loosened the control exercised by the churches.

Economic Development

The economies of the countries of Northern Europe relied on primary products until development of manufacturing and service industries in the 1900s. Denmark is a major agricultural country; 75 percent of Denmark's lowland area on the Jutland peninsula and on the islands between the peninsula and Sweden is farmed with an emphasis on dairy and livestock products (see Figure 3.34). Denmark also commonly has the largest fish catches in the EU and engages in world shipping. Sweden has agriculture in the south, while the north has significant timber and iron-mining industries. The Swedish sawmill industry is Europe's largest and accounts for about 10 percent of the world's exports. Sweden was also the world's third-largest exporter of pulp and paper in the early 1990s. Finland is another major wood-producing country, and Norway has fishing and shipping industries. The discovery of major oil and gas reserves beneath the North Sea brought new wealth to Norway from the 1970s (see Figure 3.37).

In terms of industry, Denmark is known around the world for Tubourg beer and the toy company Lego, though Denmark has many other industries that manufacture furniture, handicrafts, medical goods, automatic cooling and heating devices, and sensitive measuring instruments. The Finnish company Nokia is world renowned for its mobile phone, Finland's most important export product. Finland itself has more mobile phones per capita than any other country with 65 cellular phones per 100 inhabitants. Finland's other important industries are glassware, metal, machinery, and shipbuilding.

Sweden is the largest and most industrialized Northern European country. In addition to forestry, its biggest industries are in engineering, iron and steel, chemicals, and services. Engineering products account for the largest share of Sweden's manufacturing industry. Swedish engineering companies such as SKF, ABB, and Ericsson and inventions such as the ball bearing have given Sweden a good worldwide reputation in this sector.

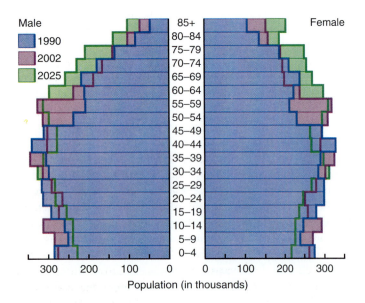

Figure 3.41 **Age-sex diagram of Sweden.** Source: U.S. Census Bureau. International Data Bank.

Volvo and Saab are well-known companies in the automotive field. Saab is also a producer of commercial and military airplanes. Many of Sweden's traditional mechanical manufacturing companies diversified into the electronics industry. ABB is the largest producer of industrial robots in Europe. In telecommunications, Ericsson sells digital exchanges and phone systems. The Swedish chemical industry originally produced matches and explosives, but paint and plastics grew in importance after World War II. In the last few decades, pharmaceuticals became important with companies such as Astra and Pharmacia & Upjohn. The service sector employs the largest number of people in Sweden. Financial, educational, and medical services are the major areas. Much of the employment is in the public sector, but high costs led to increasing privatization.

The four largest Scandinavian countries have some of the highest GDP per capita figures in the world. In recent decades, Sweden maintained its income level, while Finland (new industries such as mobile phones), Norway (oil and natural gas), and Denmark (high-tech industries) increased their affluence. The small, stable populations made it possible to develop societies in which poverty is limited, as reflected in the ownership of consumer goods (see Figure 3.17).

Mediterranean Europe

Mediterranean Europe consists of four large countries—Portugal, Spain, Italy, and Greece—and five small independent countries, Andorra, Monaco, Vatican City, San Marino, and Malta (Figures 3.43 and 3.44). Gibraltar remains a British colony, although discussions between the U.K. and Spain continue over its future. Portugal is not strictly a "Mediterranean" country, because its coast faces the Atlantic Ocean, but it is part of the Iberian Peninsula with Spain, and its history and culture are closely linked to that of its neighbor. Southern France is also oriented to the Mediterranean Sea but its position within France and, therefore, governance from Paris makes it more convenient to include it in Western Europe.

From ancient Greek and Roman ideas to those of the Italian Renaissance, Mediterranean Europe played a major role in directing the course of Western civilization. In the 1400s, Portuguese and Spanish exploration launched the Age of Discovery and soon marked the beginning of European colonization of other lands and peoples. With colonization, the Portuguese and Spanish transplanted their languages and religion—Roman Catholicism—around the world. By the Industrial Revolution of the 1800s, Mediterranean Europe had lost a lot of its political power. In the 1900s, differences grew among the Mediterranean countries. Italy became the most industrialized and Spain followed, while Portugal and Greece still rely on agriculture, fishing, merchant marines, and tourism for much of their overseas income.

Countries

Portugal is one of the oldest countries in Europe, tracing its roots back to the year 1143, but it did not attain its current territory until the 1400s, when the Moors from North Africa were finally expelled from the Iberian Peninsula. Portugal's boundaries also are among the oldest and most stable in all of Europe. Spain, too, traces its roots to the 1100s through the Kingdom of Castile but did not attain its current and familiar shape until the 1600s.

With the riches gained from the sea voyages of the 1400s, Portugal and Spain quickly dominated the world, but competition between them led the pope to divide the world between them by the Treaty of Tordesillas (1494). The treaty designated a line of longitude 370 leagues (approximately 1,500 km, 1,000 mi.) west of the Azores as the eastern boundary of Spanish influence and the western boundary of Portuguese influence. Subsequently, Portugal colonized the coast of what became modern Brazil (see Chapter 10) and parts of Africa and Asia, while Spain established its rule in the rest of Latin America and some Pacific Ocean islands.

The rise of France, the Netherlands, and the U.K. as colonial powers sharply curbed the global influence of Portugal and Spain. By the early 1800s, both countries lost their Latin American colonies, although they maintained African colonies until the 1970s (see Chapter 9). Two dictators, Antonio Salazar, who came to power in Portugal in 1926, and Francisco Franco, who took control of Spain in 1938, kept these two countries inward looking. Following the deaths of these two men in the 1970s, Portugal and Spain became more democratic and outward looking, and joined the European Community (now EU) in 1986.

Spain is much larger than Portugal, but great devolutionary forces are working against it. The Basques, descendants of a people who lived in Europe before the arrival of the Indo-Europeans, seek independence. They live in a northern region along the French border around the city of Bilbao. Some of them use violence to achieve their goal. The Catalans on the northeast coast, where Barcelona (Figure 3.45) is their main city, also press for greater autonomy. Catalans (17 percent of the Spanish population) frequently prefer to fly their own flag and speak their own language that is close to Spanish but still distinctive. The Galicians, north of Portugal, have been less vocal but are culturally different than Castilian Spaniards, having more in common with Portuguese culture.

Though Greece and Italy are hearths for some of Europe's oldest civilizations, the modern nation-states of Greece and Italy came into existence only in modern times. Greece emerged as a nation-state in 1832 but did not take its current shape until the early 1900s. Italy is the youngest of the major Mediterranean nation-states, coming into existence in 1861 and still having experienced some boundary changes as late as the 1940s. Regionalism within Italy is very strong, with most Italians considering themselves Sicilians, Tuscans, Venetians, and so on first and Italians second. A strong north-south differentiation is part of this regionalism. The south is still very rural, Roman Catholic, and under the control of familial organizations such as the Mafia. The Mafia's main reputation is as a crime organization, but it is also a way of doing business through family connections. In contrast, the

Figure 3.43 **Mediterranean Europe: the countries, cities, and physical features.** The countries occupy peninsulas extending southward into the Mediterranean Sea.

north is very urban, modern, and industrialized. The Communists of northern Italy, centered in the industrial towns, long gave Italy the distinction of having the largest Communist party in non-Communist Europe. The Communist Party's influence waned in the 1990s as the Northern League had tremendous growth. The Northern League is a political party and movement that seeks greater autonomy for northern Italy. Unlike the Communists, who draw their support from factory workers, the Northern League is made up of young entrepreneurs engaged in activities tied to the modern global economy.

People

Dynamics

Populations in the Mediterranean countries changed from a period of rapid increase and relatively high fertility and birth rates up to the mid-1900s to one in which they had the world's slowest levels of natural increase and fertility in the 1990s. Total fertility rates dropped from nearly 3 in 1965 to under 1.5 in 2000. Projections suggest that the total population of this subregion, some 118 million in 2000, will fall to 111 million by 2025.

Figure 3.44 **Mediterranean Europe countries: key data on countries.**

Country	Capital City	Land Area (km²) Total	Population (millions) Mid-2001 Total	2025 Est.	GNI 1999 (US $ million) Total	GNI PPP 1999 Per Capita	Percent Urban 2001	Human Development Index Rank of 175 Countries	Human Poverty Index: Percent of Total Population
Helenic Republic (Greece)	Athens	131,990	10.9	10.4	127,648	15,800	59	27	
Italian Republic	Rome	301,270	57.8	55.0	1,162,910	22,000	90	19	11.6
Portuguese Republic	Lisbon	92,390	10.0	9.3	110,175	15,860	48	28	
Spain, Kingdom of	Madrid	504,780	39.8	36.7	583,082	17,850	64	21	13.0

Source: Data from *Population Reference Bureau 2001 Data Sheet; World Development Indicators,* World Bank, 2001; *Human Development Report,* United Nations, 2001; Microsoft Encarta (ethnic group, language, religion).

Urbanization

Apart from Spain, the Mediterranean countries are less urbanized than those of other parts of Western Europe. In 2000, Greece had the lowest percentage of its population living in urban areas with 60 percent, Spain the highest with 78 percent, and the others falling in between. Athens, Greece, and Lisbon, Portugal, are the predominant cities in government, port, and industrial activities in their countries. Italy's capital city of Rome is now exceeded in population by the Milan metropolitan area, Naples, and Turin. In Spain, Madrid and Barcelona (see Figure 3.45) are the largest cities, and Seville and Valencia are important, too.

Mediterranean cities are most distinctive, with their ancient Greek and Roman historic cores surrounded by medieval, 1800s, and modern industrial and residential developments (see Figure 3.32). Athens, Rome, Naples, Florence, Venice, and Milan, Grenada and Barcelona, and Lisbon have distinctive townscapes that include relics of Greek, Roman, Moorish, and medieval contributions. Although their historic sectors gain much attention from the many tourists drawn to them, Mediterranean cities are also often ports, industrial centers, administrative centers, or magnets for beach-based tourism. Port sectors are of great significance to Lisbon, Barcelona, Genoa, Naples, Venice, and Athens. Manufacturing industries created distinctive townscapes in Barcelona and the northern Italian cities. Lisbon, Madrid, Rome, and Athens have growing sectors of government offices. The huge new tourist cities of southern Spain, the French Riviera, eastern Italy, and southern Greece add new urban forms to the old established features.

Culture

The Portuguese, Spanish, and Italians all speak languages in the Romance (also Latinic) branch of the Indo-European language family (see Figure 3.4a) and are Roman Catholic (see Figure 3.4b), having inherited both characteristics from the Romans. Though the Greeks were also in the Roman Empire, their culture predates the Romans and remained distinctive during and after Roman control. The Greeks are Eastern Orthodox Christians and use the Cyrillic alphabet, as do many of the Christians of East Central Europe, Russia, and its neighboring countries. As a language, Greek has its own branch within the Indo-European family.

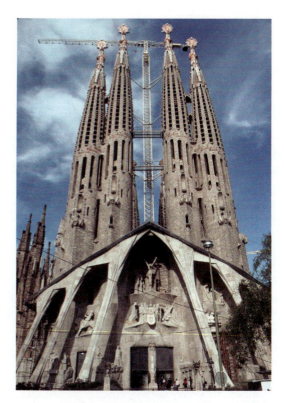

Figure 3.45 **Barcelona, Spain.** Barcelona is world renowned for its Temple de la Sagrada Familia. Begun in 1882 and designed by the famous Catalan architect Antoni Goudí (1852–1926), this church is far from complete. Photo: © Loren W. Linholm.

With levels of life expectancy rising to 75 years (men) and 80 years (women), aging is becoming a major issue (Figure 3.46). All four countries are expected to decrease in population by 2025 (see Figure 3.44). Although three countries—Italy, Portugal, and Spain—are predominantly Roman Catholic in religion, that church's opposition to birth control is clearly having little effect. By 2000, Italy and Spain had the lowest total fertility rate (1.2) in the world. Population growth was only by immigration. In Italy, few babies are born outside marriage, but the access of women to careers, young people continuing to live with parents, and the end of pressures to have children tend to defer marriage and reduce the numbers of children.

Country	Ethnic Groups (percent)	Languages O=Official	Religions (percent)
Helenic Republic (Greece)	Greek 98%	Greek (O), Turkish	Greek Orthodox 98%
Italian Republic	Italian, Sicilian, Sardinian, German, French	Italian (O), German, French, Slovenian	Roman Catholic 98%
Portuguese Republic	Mediterranean	Portuguese (O)	Roman Catholic 97%
Spain, Kingdom of	Mediterranean and Nordic types	Castilian Spanish 74%, Catalan 17%	Roman Catholic 99%

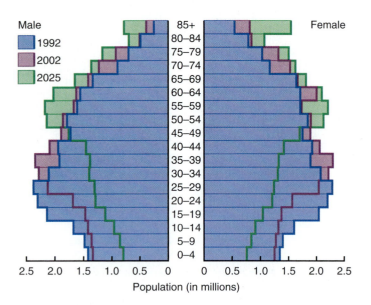

Male
- 1992
- 2002
- 2025

Female

85+
80–84
75–79
70–74
65–69
60–64
55–59
50–54
45–49
40–44
35–39
30–34
25–29
20–24
15–19
10–14
5–9
0–4

2.5 2.0 1.5 1.0 0.5 0   0 0.5 1.0 1.5 2.0 2.5
Population (in millions)

Figure 3.46 **Age-sex diagram of Italy.** Source: U.S. Census Bureau. International Data Bank.

In Italy, there are only tiny areas in the northeast (German) and northwest (French) where Italian is not the dominant language. The modern differences between northern and southern Italy grew out of the medieval occupation of the region south of Rome by Muslims and its later domination by feudal systems under French and Spanish kings. The northern city-states maintained trading links and developed industrial output at an earlier date. It was not until 1861 that Italy achieved unification.

Economic Development

The four main countries of Mediterranean Europe remained as peasant-farming countries with feudal or fragmented types of social organization while Northern Europe industrialized in the 1800s. Industrialization began in the late 1800s in northern Italy and in the Catalonian region around Barcelona in northeastern Spain. Most modernization occurred after World War II and the incorporation of the Mediterranean countries in the EU. Industrialization in Greece and Portugal has been very modest. The two countries remain the poorest of the EU countries and receive large sums of regional development funds (see Figure 3.19).

Italy's GDP is almost twice the total of the other Mediterranean Europe countries and closest to that of the U.K. As one of the world's largest economies, Italy is a member of the Group of Eight (see the "Western Europe" section in this chapter). The country's economic power comes mainly from northern-based and high-tech industries. The Po River valley between the Alps and the Apennines is the largest center of manufacturing in Mediterranean Europe. For example, Fiat manufactures automobiles in Turin. Venice, at the mouth of the Po River, is famous for its glass manufacturing. Milan, the largest city in northern Italy, is a major center of financial and other service industries as well as a producer of a diverse group of manufactured goods including tractors, domestic electronic goods, china, fashion, and pharmaceuticals. Milan

and its surrounding towns produce nearly one-third of the Italian GDP and form one of the major growth areas of Europe.

Agriculture

Agriculture is very important to the Mediterranean countries. The warm dry summers and cool winters that seldom drop below freezing allow the Mediterranean countries to produce crops that are difficult to grow in the other subregions of Europe (see Figure 3.34). Examples of these crops are olives, table and wine grapes, citrus fruits, figs, and specialized cereal grains for pasta. In addition, Portugal produces most of the world's cork for wine bottles, obtained from the bark of the cork oak (Figure 3.47). Furthermore, specialization has led to **market gardening,** the commercial production of these crops.

Tourism

The Mediterranean's warm climate, sunny beaches, and historic centers attract tourists, making tourism a major industry. Each year, Portugal and Greece each receive 10 million tourists, meaning that tourists outnumber the inhabitants of these two small countries. Italy and Spain typically receive three or four times more tourists per year. As a result, English and German are commonly understood in the tourist areas of these countries. In Portugal, the Algarve (along the southern coast) and the Madeira islands are popular. In Greece, many tourists are found on the islands or in the capital, Athens. Barcelona, Madrid, and the cities of southern Spain account for most tourist visits to Spain. In Italy, most tourists go to the historic centers of Rome, Florence, Venice, and Pompeii (near Naples) (Figure 3.48a and b; see Figure 3.9). The east coast beaches, Sicily, many smaller towns with their art treasures, and the winter sports and lake resorts in the Alps attract many others. Venice has become so popular with international tourists that the city has many more tourists than residents. An increasing problem for Venice is rising sea level. While global warming is debated, sea-level rise has resulted in seawater breaching canals and flooding streets more and more every year (see the "Natural Environment" section in this chapter).

Figure 3.47 **A cork oak forest in Portugal.** Bottle corks are made from the bark of these trees. Little processing is required. Once the bark is stripped from the tree, bottle corks are cut. Numbers on the trees refer to a calendar year and indicate when the bark can be stripped from the tree again. Photo: © Alexander B. Murphy.

(a)

(b)

Figure 3.48 **Mediterranean Europe: Italian tourist attractions.** (a) Pompeii, near Naples, is a Roman city that was buried by an eruption from Mount Vesuvius and recently excavated by archaeologists. (b) Florence has its cathedrals and art galleries by the Arno River. Photos: © Michael Bradshaw.

Test Your Understanding 3E

Summary Northern Europe includes Denmark, Finland, Iceland, Norway, and Sweden. These are affluent countries with small populations. Denmark and Sweden were once great empires, but in recent times, Northern Europe is known for its neutrality. The economies of these countries were based on primary products from farm, mine, forest, and ocean, but now income is from manufacturing and services.

Mediterranean Europe comprises the countries of three peninsulas—Portugal and Spain, Italy, and Greece. These countries represent some of the oldest civilizations in Europe. Apart from northern Italy, which is as prosperous as parts of Western Europe, these countries lagged far behind their more affluent neighbors until their incorporation in the European Community (now EU) in the 1980s. They are now developing manufacturing and service industries, and all rely heavily on income from tourism.

Questions to Think About

3E.1 What factors account for the economic well-being of the Northern European countries?

3E.2 What attracts tourists to Mediterranean countries, and what are some of the environmental impacts of tourism (see also the "Natural Environment" section in this chapter)?

Key Terms

fjord   market gardening

East Central Europe

Modern East Central Europe is comprised of the following countries: the Baltics (Estonia, Latvia, and Lithuania), Poland, the Czech and Slovak Republics, Hungary, and what is often called the Balkans: Slovenia, Croatia, Bosnia-Herzegovina, Serbia and Montenegro, Macedonia, Albania, Romania, and Bulgaria (Figures 3.49 and 3.50). These countries are grouped together because Communist forms of government and economies were imposed upon them shortly after World War II ended in 1945 (see the section on "Changes in the Modern Era" in this chapter). Most of these countries were directly controlled by the Soviet Union and called Soviet satellite states. The Baltic countries even were incorporated into the Soviet Union. Though Soviet domination ended by 1991, these countries still share common experiences as they move from Communism to more democratic forms of government and capitalist economies. East Central Europe includes the Baltic countries because they historically belong to the subregion. They also struggled with Communism for the same length of time as other East Central Europeans and not as long as those in the former Soviet Union, though they were part of that country for a time.

Countries

The countries of East Central Europe emerged as nation-states in the late 1900s and early 2000s. For example, Serbia,

Romania, and Bulgaria became independent in the 1870s and 1880s, and Albania was created for the first time in 1912. Others, such as the countries that emerged from the demise of Czechoslovakia and Yugoslavia, achieved independence only as recently as the 1990s. Boundaries also moved considerably over time. Poland, for example, was one of the larger countries in Europe in the 1600s and 1700s. Carved up by Prussia, Austria, and Russia in a short 23-year period, it disappeared in 1795. Poland reemerged after World War I in 1918, only to shift westward after World War II in 1945. Located on the North European Plain, Poles have had difficulty defending their borders through history. The Estonians, Latvians, and Lithuanians had similar problems. Estonia, Latvia, and Lithuania emerged as nation-states after World War I in 1918, only to be overrun by Hitler's armies in 1941 and incorporated into the Soviet Union in 1945, before gaining independence again in 1991 with the dissolution of the Soviet Union.

From the beginning of the nationalist idea in late 1700s to the end of World War I in 1918, most of East Central Europe was dominated by four great empires—the Russian, German, Austro-Hungarian, and Ottoman empires (Figure 3.51a). World War I started in the subregion—specifically in Sarajevo,

Figure 3.49 **Central Europe: countries, cities.** This subregion has been a buffer zone between the major powers of Germany and Russia for centuries, and the countries are in transition from a period of domination by the Soviet Union that ended in 1991 to subsequent orientation westward to the rest of Europe.

Bosnia-Herzegovina—when a Serb nationalist assassinated Archduke Ferdinand of the Austro-Hungarian Empire. In the aftermath of the war, diplomats met in Paris in 1919 to create a lasting peace by redrawing Europe's boundaries. Sir Halford J. MacKinder, a British geographer and Member of Parliament, had long argued that future peace could only be ensured by separating the German and Russian empires. He proposed the creation of a zone of small countries for this purpose. Woodrow Wilson argued for the right of "national self-determination" for the peoples of Europe. With both MacKinder's and Wilson's complementary ideas, a series of new or modified countries were created from Finland to Greece (Figure 3.51b). It was at this time that Yugoslavia and Czechoslovakia came into existence. With them, the term "Eastern Europe" came into use.

After World War II, Stalin's Red Army moved into most of Eastern Europe. Through Stalin's manipulations, including boundary changes (Figure 3.51c), their governments and economies adopted Communist forms. Winston Churchill said that an "Iron Curtain" had fallen across Europe. Politically, Europe was divided into East and West during the Cold War. The term "Central Europe" disappeared with the

Figure 3.50 **East Central European countries: key data on countries.**

Country	Capital City	Land Area (km²) Total	Population (millions) Mid-2001 Total	2025 Est.	GNI 1999 (US $ million) Total	GNI PPP 1999 Per Capita	Percent Urban 2001	Human Development Index Rank of 175 Countries
Albania, Republic of	Tirana	28,750	3.4	4.5	3,146	3,240	46	100
Bosnia-Herzegovina, Republic of	Sarajevo	51,130	3.4	3.6	4,706	no data	40	no data
Bulgaria, Republic of	Sofia	110,910	8.1	6.6	11,572	5,070	68	63
Croatia, Republic of	Zagreb	56,540	4.7	4.4	20,222	7,260	54	55
Czech Republic	Prague	78,860	10.3	10.3	51,623	12,840	77	36
Estonia, Republic of	Tallinn	45,100	1.4	1.2	4,906	8,190	69	54
Hungary, Republic of	Budapest	93,030	10.0	9.2	46,751	11,050	64	47
Latvia, Republic of	Riga	64,500	2.4	2.2	5,913	6,220	69	74
Lithuania, Republic of	Vilnius	65,200	3.7	3.5	9,751	6,490	68	62
Macedonia, Former Yugoslav Rep of	Skopje	25,710	2.0	2.2	3,348	4,590	60	73
Poland, Republic of	Warsaw	312,680	38.6	38.6	157,429	8,390	62	44
Romania	Bucharest	237,500	22.4	21.6	33,034	5,970	55	68
Slovak Republic	Bratislava	49,010	5.4	5.2	20,318	10,340	57	42
Slovenia, Republic of	Ljubljiana	20,050	2.0	2.0	19,862	16,050	64	33
Serbia and Montenegro	Belgrade	102,170	10.7	10.7	no data	no data	52	no data

Source: Data from *Population Reference Bureau 2001 Data Sheet; World Development Indicators,* World Bank, 2001; *Human Development Report,* United Nations, 2001; Microsoft Encarta (ethnic group, language, religion).

Nazis, who used it as a synonym for the empire they wished to create. "Eastern Europe" then referred to the Communist part of Europe between the Soviet Union and Western Europe. The East Germans were added while the Baltic countries were excluded because they were incorporated into the Soviet Union. Finland and Greece were likewise excluded from the concept of Eastern Europe because they did not adopt Communism.

By the dissolution of the Soviet Union in 1991, many of the peoples of this subregion resurrected the old concept of Central Europe and emphasized that they were Central Europeans and

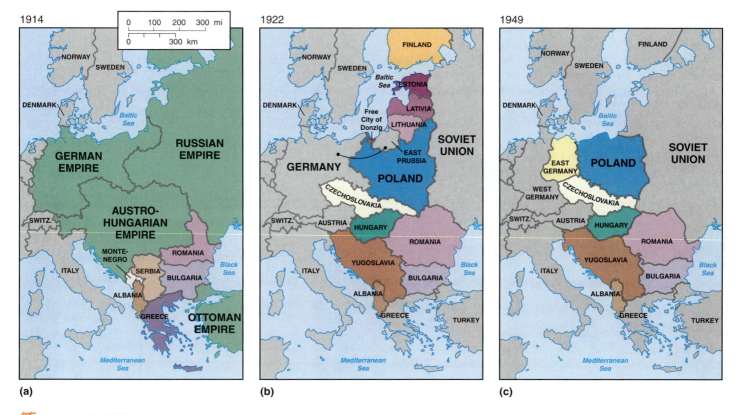

(a)   (b)   (c)

Figure 3.51 **The boundaries of East Central European countries at three different dates in the 1900s.** What are the locations of countries such as Poland, Latvia, and Yugoslavia at each of these times? Source: From Demko & Wood, *Reordering the World*, 1st ed.

Country	Ethnic Groups (percent)	Languages O=Official	Religions (percent)
Albania, Republic of	Albanian 95%, Greek 3%	Albanian (O)	Christian 30%, Muslim 70%
Bosnia-Herzegovina, Republic of	Muslim 44%, Croat 17%, Serbian 31%	Serbo-Croat 99%	Christian 46%, Muslim 40%
Bulgaria, Republic of	Bulgarian 85%, Turkish 9%	Bulgarian	Bulgarian Orthodox 85%, Muslim 13%
Croatia, Republic of	Croat 78%, Serb 12%	Serbo-Croat 99%	Orthodox 11%, Roman Catholic 77%
Czech Republic	Czech 81%, Moravian 13%, Slovak 3%	Czech, Slovak, German	Protestant 5%, Roman Catholic 39%, none 40%
Estonia, Republic of	Estonian 62%, Russian 30%	Estonian, Latvian, Russian	Protestant, Russian Orthodox
Hungary, Republic of	Hungarian (Magyar) 90%	Hungarian 98%	Protestant 25%, Roman Catholic 68%
Latvia, Republic of	Latvian 52%, Russian 34%	Latvian (O), Russian	Protestant, Roman Catholic, Orthodox
Lithuania, Republic of	Lithuanian 80%, Russian 9%, Polish 8%	Lithuanian (O), Russian	Protestant Roman Catholic (most)
Macedonia, Former Yugoslav Rep of	Macedonian 65%, Albanian 21%	Macedonian, Albanian	Macedonian Orthodox 67%, Muslim 30%
Poland, Republic of	Polish 98%	Polish (O), English, German	Roman Catholic 75%
Romania	Romanian 89%, Hungarian 9%	Romanian (O), Hungarian	Romanian Orthodox 70%, other Christian 12%
Slovak Republic	Slovak 86%, Hungarian 11%	Slovak (O), Hungarian	Protestant 8%, Roman Catholic 60%
Slovenia, Republic of	Slovene 91%, Croat 3%	Slovenian, Serbo-Croat	Roman Catholic 96%
Yugoslavia, Federal Republic of	Serb 63%, Albanian 14%, Montenegrin 6%	Serbo-Croat 99%	Orthodox 65%, Muslim 19%

not Eastern Europeans, a term that had taken on very negative connotations. The peoples of the Baltic countries also readopted the term "Central Europe" for themselves. Though many of these countries were part of an older Central Europe, they represented the portion east of the Germans. Thus, we have the term "East Central Europe." During the Cold War, the other Central Europeans—the Germans, Austrians, and Swiss—became so Western oriented that they are rarely labeled Central Europeans anymore, or even "West Central Europeans."

Devolutionary processes occurred in East Central Europe after the end of Soviet domination in 1991. The declarations of independence from the Soviet Union by the Baltic republics are one example. Czechoslovakia divided into separate Czech and Slovak Republics in 1993. In these cases, the movements toward independence advanced peacefully. The Czechoslovak situation was referred to as the "Velvet Divorce," coined from the "Velvet Revolution" of 1989 that peacefully ended Communism in that country. Devolutionary processes also occurred in Yugoslavia but were not so peaceful as the country broke apart. The Yugoslav case is important because it illustrates the ugly aspects of fervent nationalism, and it changed international laws concerning human rights. The Yugoslav situation also took on international significance as it involved other European countries and the United States, including NATO, as well as Russia.

"Yugoslavia"

As of 2002, Yugoslavia ceased to exist as a country. The country's last two republics abandoned the name and now call themselves Serbia and Montenegro. Slovenia, Croatia, Bosnia-Herzegovina, and the Former Yugoslav Republic of Macedonia (FYROM) were republics of Yugoslavia but seceded in 1991 and 1992. For some republics such as Slovenia, independence was achieved relatively quickly and without much bloodshed. For others such as Bosnia-Herzegovina, the opposite was the case. To explain what happened in Yugoslavia, it is necessary to understand the origins of the country (Figure 3.52).

Yugoslavia came into being after World War I. At the time, it was called the Kingdom of the Serbs, Croats, and Slovenes (Figure 3.53a), reflecting the fact that the country was created for a number of nations. In 1929, the name "Yugoslavia" was adopted. "Yugoslavia" translates from the Slavic languages into "the land of the South Slavs."

The relationships of South Slavic peoples are very close and overlapping. Identities are very complex and intertwined. The Slovenes and Croats are Roman Catholic, while the Serbs, Macedonians, and Bulgarians—also South Slavs but not part of Yugoslavia—are Eastern Orthodox Christians. Other South Slavs are Muslims like many Bosnians. Many Bosnians claim that a Bosnian can also be Roman Catholic or Muslim. Despite having differing faiths, all South Slavs speak closely related languages with the Serbs and Croats essentially speaking the same language—Serbo-Croatian—though many Serbs now say they speak Serbian and many Croats say they speak Croatian.

Prior to World War I and the creation of a Yugoslav state, Serbia and Montenegro were independent countries. Both, however, only achieved independence in 1878 and obtained much of their territories during the Balkan wars of 1912–1913. The Slovenes and Croats had lost their independence in the Middle Ages, eventually finding themselves divided among a number of provinces within the Austro-Hungarian Empire before World War I.

In the late 1800s, Serbia's leaders planned to annex to Serbia the territories of their South Slavic brethren in Austria-Hungary. Serb nationalists were upset when Austria-Hungary

Figure 3.52   **Time line of Yugoslavia.**

1878	–	Serbia and Montenegro become independent; Austria-Hungary occupies Bosnia-Herzegovina.
1908	–	Austria-Hungary formally annexes Bosnia-Herzegovnia.
1912/1913	–	First and Second Balkan Wars. Serbia, Bulgaria, and Greece annex portions of Macedonia.
1914	–	Serbian nationalist assassinates Austrian Archduke Ferdinand in Sarajevo. Austria-Hungary declares war on Serbia. World War I begins.
1919	–	Paris Peace Conference. The Kingdom of the Slovenes, Croats, and Serbs is created and includes Montenegro, Bosnia-Herzegovina, and Vojvodina.
1929	–	The Kingdom of the Slovenes, Croats, and Serbs becomes "Yugoslavia."
1941	–	Yugoslavia is dismantled and occupied by Axis powers during World War II.
1945	–	Josip Tito, leader of the Partisans, who are Communists, takes control of a recreated Yugoslavia. Six internal republics are created.
1980	–	Tito dies and nationalists begin expressing themselves. Slobodan Milošević eventually catapults to power by fanning ethnic hatreds.
1991	–	Slovenia and Croatia declare independence; bloodshed ensues as Yugoslav military tries to prevent secession and Serbs in Krajina of Croatia declare their independence from Croatia.
1992	–	Bosnia-Herzegovina and Macedonia declare independence. Bloodshed ensues as many Serbs resist the independence movement.
1995	–	Croatian army retakes Krajina and forcibly expels Serb population. Dayton Peace Accord ends open hostilities in Bosnia-Herzegovina and divides between a Muslim-Croat federation and a Serb republic.
1999	–	Serb military and paramilitary units terrorize ethnic Albanians in the Serb province of Kosovo. NATO begins intensive bombing campaign of Yugoslavia. Millions of ethnic Albanians flee to Albania and Macedonia. Success of the air campaign leads to the stationing of NATO and Russian ground forces in Kosovo.
2000	–	Slobodan Milošević loses a national election and is no longer the leader of Yugoslavia.
2001	–	New Yugoslav government arrests Slobodan Milošević and sends him to the International Court of Justice in The Hague, the Netherlands, where he is the first leader of a country indicted for genocide and crimes against humanity.
2002	–	Yugoslavia changes its name to Serbia and Montenegro.

(a)

(b)

Figure 3.53 **Yugoslavia.** (a) The territorial formation of Yugoslavia in 1919; (b) Yugoslavia's administrative boundaries under Tito (1945–1980). The boundaries lasted until 1991. Source: (a) from Stephen Clissold (ed.), *A Short History of Yugoslavia.* Copyright © Cambridge University Press, 1966; (b) From Fred Singleton, *A Short History of Yugoslav Peoples.* Copyright © Cambridge University Press, 1985.

formally occupied Bosnia-Herzegovina in 1878 and outraged when Austria-Hungary formally annexed the territory in 1908. Not long afterward, a Serb nationalist sought revenge by assassinating Austria-Hungary's Archduke Ferdinand when he visited Sarajevo in 1914. The episode began World War I. Because Austria-Hungary was so much bigger than Serbia, Serbia was always on the defensive and could not use its military to annex the South Slavic lands of Austria-Hungary. Nevertheless, the opportunity came following the defeat of Austria-Hungary and its allies at the end of the war.

Many South Slavs wanted to be free of Austria-Hungary even before the war began, but they also had differing goals for

freedom. Some wanted independence, but not having independent governments of their own, they lacked fully recognized diplomatic corps. Thus, those wanting an independent Slovenia, Croatia, or even Bosnia-Herzegovina had little voice at the Paris Peace Conference in 1919. On the other hand, the Serbs had a diplomatic corps that spoke for the South Slavs. The Western powers were aware that many Slovenes and Croats wanted their own independent countries, yet they felt that these countries would be too small to be economically viable. Thus, they went along with the Serb proposal to unite the South Slavic lands of Austria-Hungary with Serbia and Montenegro to create a single country.

Though the South Slavic movement succeeded in its goal of creating a single South Slavic country, many South Slavs had serious disagreements over the meaning and purpose of this country. Were they to form a common national identity or remain distinct? The answer to this question was required to answer the question of governance. Should the country have a strong centralized government with everyone obeying the same laws, or should the provinces exercise their autonomy? The desired answers to these questions varied not only between the national groups but also within them. In the meantime, Serbia's capital, Belgrade, became the new country's capital, the Serbian king the new country's king, and Serbian symbols such as the flag, the new country's symbols. Resentment toward the Serbs quickly grew before these important questions could be answered. The situation only became worse as Serb nationalists used their power to insist that the Serb way become everyone else's. These tensions led to the disintegration of the country during World War II.

Following the war, the Partisans, who were Communist and under the leadership of Josip Tito, gained control of the country. Tito's parents were Slovene and Croat, yet as a Communist, he did not believe in nationalism, so he suppressed any expressions of national pride. Instead, Tito promoted the "Yugoslav" idea while not using any one group's characteristics for Yugoslav characteristics. He generally allowed everyone to continue with their differing cultural practices while insisting they all consider themselves Yugoslavs. Tito also set up the internal political geography of the country (Figure 3.53b). He created the six republics that we see today, with Serbia having the autonomous regions (later called provinces) of the Vojvodina and Kosovo. Each republic was named after and dominated by one national group except for Bosnia-Herzegovina (Figure 3.54 and see "Personal View: Bosnia-Herzegovina," p. 118). Macedonia became one of these republics. Formerly, it was part of Serbia. Serb nationalists considered the Macedonians South Serbs, but Tito recognized a separate Macedonian identity and thus gave the Macedonians their own republic. In fact, while Tito was Yugoslavia's leader, seven major and 17 smaller groups came to be recognized. The nationalists of various groups were unhappy with Tito's policies and practices, but Tito kept them in check with his charisma and heavy hand.

In 1980, Tito died and the politics within Yugoslavia began to change. Nationalism and old resentments began to surface. Slovenes and Croats felt that the Serbs had always forced them to become more Serbian. In addition, Slovenia and Croatia were the wealthier republics, and the peoples from both felt that the

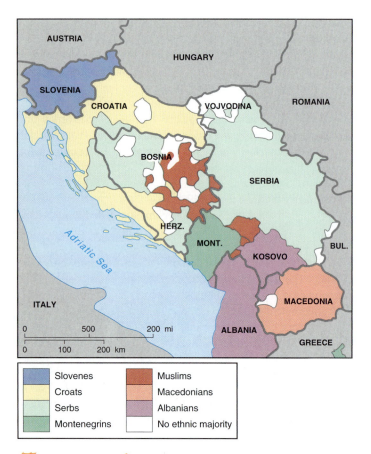

Figure 3.54 **Importance of ethnic differences: The former country of Yugoslavia in the Balkans of Europe.** After 1991, Yugoslavia broke into five independent countries, although Serbia and Montenegro continued to call themselves Yugoslavia until 2002. The independence movements resulted in war in Croatia and Bosnia-Herzegovina. Kosovo (Serbia) experienced violence in the late 1990s, bringing U.S. and NATO involvement in 1999. The colors show areas where particular ethnic groups comprise over 50 percent of the population.

profit from their hard work was going to subsidize the poorer republics. In 1991, Slovenia and Croatia declared independence, and the Yugoslav army attempted to prevent the secession movements. For Slovenia, the fight was brief. Within a short time, Belgrade accepted Slovenia's independence. The situation in Croatia was bloodier. Twelve percent of Croatia's population was Serb. Most of them lived along Bosnia's border in a region known as Krajina ("the borderland") and in eastern Croatia. The Belgrade government, dominated by Serbs, used the Yugoslav army, also dominated by Serbs, to help the Serbs of Croatia establish their own republic. Croatia achieved independence but could not control a significant piece of its territory.

Yugoslavia without Slovenia and Croatia, not to mention Macedonia, which was making plans for secession, left Bosnia-Herzegovina in a Yugoslavia more dominated by Serbs than ever before. Before long, Bosnia-Herzegovina also began moving toward secession. Bosnia-Herzegovina's ethnic complexity, however, made secession a very difficult issue (see Figure 3.54). In the 1991 census, 44 percent of the population was Muslim, 31 percent Serb, 17 percent Croat, and 8 percent

of other groups. Many from all the groups supported the independence of their republic. Serb nationalists were the greatest opponents, and with support from the government in Belgrade and the Yugoslav army, they began an armed insurgency with the intent of keeping Bosnia-Herzegovina within Yugoslavia. Eventually, Croat nationalists turned against government forces with the goal of incorporating territories of the republic into Croatia. Without a neighboring country to support them and under attack from two directions, the Muslims remained loyal to Bosnia-Herzegovina. Serb and Croat nationalists claimed that Muslims were oppressing them in an attempt to make Bosnia a Muslim state. Ironically, Serb and Croat nationalists helped to make Bosnia-Herzegovina more Muslim as they tried to separate its non-Muslim territories.

The international community proposed a number of partition plans to end the conflict. However, the geographic distribution of groups within the republic was complex. Though some groups were associated with particular areas, other patterns existed. For example, Muslims tended to live in the cities, while Serbs were in the surrounding countryside. In 1995, the United States became involved and brokered the Dayton Peace Accord (see Box Figure 1, "Personal View: Bosnia-Herzegovina," p. 118). Though recognizing the integrity of the republic, it temporarily divided the republic into two "entities." The north and east came under the Republika Srpska (Serb Republic) with the capital in Banja Luka. The central and western areas became part of a Muslim-Croat Federation with the capital in Sarajevo. The situation is peaceful, but tensions and animosities remain, as do UN peacekeepers. At the same time, Bosnia-Herzegovina's economy is in shambles. Unemployment has been as high as 70 percent in many communities, and the country has relied heavily on foreign aid. Bosnia's territory is used to funnel prostitutes and illegal immigrants into and across Europe.

As these republics moved toward independence, so did Macedonia. Macedonia was fortunate to escape bloodshed when it negotiated the withdrawal of Yugoslav troops in April 1992. Because war broke out at the same time in Bosnia-Herzegovina, the Macedonians may have benefited from a decision in Belgrade not to engage the Yugoslav army in two conflicts at the same time. Nevertheless, Macedonia was still in a precarious position.

Macedonia is an old region, having existed in ancient times. Alexander the Great, for example, was from Macedonia. Compared to the modern republic, historic Macedonia stretches beyond the boundaries of the former Yugoslav republic into western Bulgaria and northern Greece. Serbia, Bulgaria, and Greece fought the Balkan wars of 1912–1913 to gain control over the territory. Though Serbs, Bulgarians, and Greek nationalists all claimed Macedonia on the grounds that Macedonians were really members of their own respective nations, all three agreed that Macedonians were not their own distinct nation. The situation changed when Tito recognized the Macedonians, created the Macedonian republic, and recognized the Macedonian language. The Bulgarian and Greek governments still do not recognize the Macedonians as a separate people. Moreover, fearing that the Macedonian government would claim historic Macedonia, the Greek government blocked international recognition of the country as long as it insisted on calling itself Macedonia. After international negotiation, a compromise was found for the country's official name: the Former Yugoslav Republic of Macedonia (FYROM). This name prevents the country's government from claiming historic lands in either Greece or Bulgaria.

Though Belgrade allowed Macedonia (i.e., FYROM) to secede without a military struggle, it also left Macedonia landlocked and surrounded by hostile countries. In addition, Macedonia was the former Yugoslavia's poorest republic. Independence has not brought an improvement in the standard of living. Twenty percent of its GDP comes from agriculture, and unemployment is around 30 percent. Macedonia would still most likely have to depend heavily on Yugoslavia, that is, Serbia and Montenegro after Yugoslavia ceased to exist. Twenty-three percent of Macedonia's population is Albanian (see Figure 3.54). In 2001, Albanian nationalists began an armed insurgency. NATO forces are helping local peoples to achieve and maintain peace, as they also are in Bosnia-Herzegovina.

Together, Serbia and Montenegro contain almost half of former Yugoslavia's population with almost 11 million inhabitants. With 62 percent of the population, Serbs are clearly the majority. Montenegro (meaning "black mountains") is by far the smaller of the two republics in the new federation with Montenegrins only representing 5 percent of the country's population. Montenegrins are very closely related to the Serbs. Some simply call them "mountain Serbs." Some Montenegrins have been unhappy with Serb policies and are seeking independence for their republic. Comprising almost 17 percent of the population and found mostly in Kosovo in the south, Albanians are a sizable minority. A Hungarian minority is noteworthy, found mostly in Vojvodina in the north of Serbia.

Though many of the nations/republics of Yugoslavia seceded from the country because they feared or were simply tired of Serb nationalism, the Serbs had a different view of the situation. Like majority groups in other countries, many Serbs believed that their wealth was squandered by Yugoslavia's minorities. They saw their monies invested in the poor areas of Bosnia-Herzegovina, Kosovo, and Macedonia, many of which were Muslim and did not benefit the Serbs. Additional fears were fanned by the fact that these minorities had higher population growth rates. Serbs also felt that they were persecuted by Slovene and Croat nationalists. Slobodan Milosević catapulted to power by playing on these fears. He used the Yugoslav army and paramilitary units of mainly Kosovar Serbs in Croatia and Bosnia-Herzegovina to "protect" Serbs. After losing these wars and bringing ruin to the Yugoslav economy, Milosević began a war against the Albanians of Kosovo. NATO intervention stopped the persecution by a bombing campaign, followed by the stationing of troops in the province. Lost wars, Kosovo occupied by foreign troops, and a severely damaged economy led Serbs to oust Milosević in elections at the end of 2000. Not long after, Milosević was taken into custody and charged with crimes against humanity and genocide at the International War Crimes Tribunal for the Former Yugoslavia in The Hague, the Netherlands. Many Serbs supported Milosević because he was supposedly protecting Serbs throughout the former Yugoslavia. Until recently, only a few had begun to realize the atrocities that he committed in pursuing these goals. Many Serbs will undoubtedly learn more as Milosević's trial unfolds.

People

Dynamics

The populations of the countries in East Central Europe are mostly static to declining (see Figure 3.50), a situation explained by very low fertility rates. Most women have fewer than two children. Birth rates and death rates both fell to low levels, and some countries have death rates that are above birth rates, giving natural decrease (see Figure 3.28). In 2000, most countries had negative natural population change rates of –0.1 to –0.6 percent. The age-sex diagram for Poland, for example, shows that within overall low rates of increase in the later 1900s, Poland experienced alternate periods of more and fewer births (Figure 3.55). The exceptions to population decline are countries (e.g., Albania, Bosnia-Herzegovina, Macedonia) with significant Muslim populations. The traditional way of life still practiced in these countries is just as likely an explanation for population growth as religious belief.

Urbanization

Urbanization levels in the subregion range from a low of 42 percent in Albania to a high of 75 percent in the Czech Republic. Though urbanization levels are relatively low for Europe, Communist policies of industrialization led to most of the urban growth, especially in countries that previously had little industry. Still, no cities are of great size. In 2000, Riga

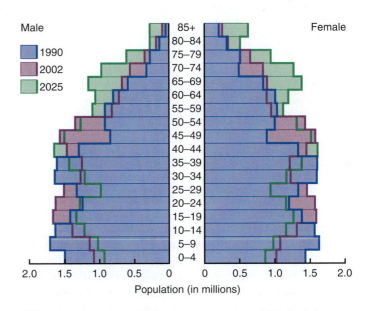

Figure 3.55 **Age-sex diagram of Poland.** Source: U.S. Census Bureau. International Data Bank.

BOSNIA-HERZEGOVINA

The wars that ensued in Yugoslavia after the country began disintegration in 1991 were particularly bloody in Bosnia-Herzegovina, lasting from early 1992 until the end of 1995, when the Dayton Peace Accord was brokered (Box Figure 1). Bosnia-Herzegovina was different from other republics in that it was not named after an ethnic group that inhabited it. Instead, it was named partly after a river and partly after a medieval territory. Bosnia derives its name from the Bosna River. Herzegovina stems from the German word *herzog*, which means "duke." Therefore, Herzegovina literally means "dukedom." Bosnia-Herzegovina most likely kept its historical name because no single group occupied the territory to justifiably change its name for that group (see Figure 3.54). For a long time, Bosnia-Herzegovina's multiethnicity worked very well. The people of the republic were even able to showcase their ethnic harmony to the world when they hosted the 1984 Winter Olympics in Sarajevo.

The war destroyed Bosnia's ethnic harmony and caused many to flee and seek refuge elsewhere. Miro Paunović and his family are one such example. Miro was born in Mostar and lived there for most of his life until the beginning of the war in 1992. Mostar is the main city of Herzegovina and is internationally known for its stone bridge. Referred to locally as the "Old Bridge" (*stari most*) (Box Figure 2), it was built by the Ottoman Turks in 1561 and came to symbolize ethnic harmony within all of Bosnia-Herzegovina. Sadly, the bridge was destroyed by the Croatian army on November 9, 1993, the anniversary of *Kristalnacht* (the night when Nazis attacked Jews and Jewish businesses throughout Germany in 1938). Miro has fond memories of the bridge and will never forget how people would travel from all over Yugoslavia every summer to compete in a bridge diving contest.

Miro also misses the old town center where people would come to shop and meet at cafes to socialize. In terms of its site, Miro says that Mostar is not unlike Cumberland, Maryland, where he now lives. Both cities are crowded down along rivers that wind their way through mountains. Mostar's river is the Neretva, which flows south into the Adriatic Sea, some 50 miles away. The short distance to the sea made it easy for Miro and his family to travel to the beaches of Adriatic, often for just one day. For vacations, Miro's family would typically spend a couple of weeks on the coast, usually traveling to the famous cities of Dubrovnik and Split. Mostar's proximity to the coast means that both the summers and winters are relatively mild. Skiing is popular in this mountainous country, but those interested in doing so must travel farther inland, to the north and east, to places such as Sarajevo before they can find enough snow to ski.

Mostar is a regional center for southeastern Bosnia-Herzegovina. Its factories produce a variety of goods, including aluminum, military airplanes, car and truck parts, and clothing. The mountainous character of Bosnia-Herzegovina means that the country produces far less agri-

Croatians	Bosnian Serb
Serbs in Croatia	Bosnian Croat-Muslim federation

Box Figure 1 **Map of Croatia and Bosnia.**

culturally than many of the other Balkan countries. Nevertheless, some agricultural goods are produced. The area around Mostar is known for wine grapes, nectarines, and plums. Not surprisingly, Mostar has a winery and a distillery that makes a hard alcohol called *Loza*.

When Miro went to school as a young boy in Mostar, the educational system was somewhat different than what his children are now experiencing in the United States. All children went to elementary school, which lasted until the eighth grade. Foreign languages were taught in Miro's school beginning in the fifth grade. English, German, Russian, and French were the choices. Miro selected English, probably the most popular foreign language at the time. From the ninth through twelfth grades, young people went to specialized high schools, of which Mostar had five. The *Gymnasia* was for those bound for university. Another school emphasized automotive training, while a third focused on working with industrial machines. A fourth specialized in food technology, and the fifth centered its curriculum around economics. According to Miro, this latter school went through the greatest transformation in recent years. During Communism, the school emphasized the teachings of Marx and Engels but has since changed to the modern principles of capitalism. Miro states that school was quite rigorous in Mostar. He considers what his children are now learning in school in the

(Latvia, 775,000 people), Warsaw (2.3 million), Katowice (3.5 million) (both Poland), Prague (Czech Republic, 1.2 million), Budapest (Hungary, 2 million), Belgrade (Serbia and Montenegro, 1.5 million), Bucharest (Romania, 2 million), and Sofia (Bulgaria, 1.2 million) were the only cities of a million or more people. The lack of major world cities in East Central Europe arises from the small size of the countries, the late arrival of industrialization, and the focus on manufacturing growth linked to medium-sized industrial town expansion during Soviet times.

The city landscapes of East Central Europe suffered great war damage and were subject to post-1945 industrialization and the construction of monolithic buildings (Figure 3.56a) but often retain older townscapes (Figure 3.56b). In Romania, however, the dictator Nicholae Ceauşescu demolished villages and whatever remaining historic structures to industrialize. East Central European cities lack the high-rise office buildings containing commercial and professional service facilities and apartments that dominate the central parts of large cities in Western Europe. Many, such as the cities in Polish Silesia, are

United States and notes that he learned the same information when he was two to three years younger than American students.

Miro went to the automotive school and received a diploma in auto mechanics. Like all young males at the time in Yugoslavia, Miro had to complete a year of military service before pursuing his career. When he returned to Mostar after his military service, Miro immediately accepted a job in a bar before ever looking for a job as an auto mechanic. He had the position for about three years when the war broke out in April 1992. Like most people in Bosnia-Herzegovina, Miro did not fit easily into any of the ethnic categories of Serb, Croat, or Muslim. Usually people were identified according to their father's status. In Miro's case, his father was a Serb from Serbia, not Bosnia-Herzegovina. Miro's father had come to live in Mostar many years before while serving in the Yugoslav army. While in Mostar, he met and married Miro's mother, a Muslim, and stayed to raise a family. Like many Muslims in Bosnia-Herzegovina, Miro's mother was nonpracticing. With Mostar also having a large Croat population, it meant that Miro did not force himself into any one category, nor did he have to for most of his life. As far as he was concerned, he was a Yugoslav. He, like everyone else, had the right to travel anywhere throughout Yugoslavia and associate with whomever he pleased, regardless of religion.

When the war began in 1992, his family found itself in a difficult situation. Serbs were a minority in Mostar. Moreover, Miro's parents were not categorized together in the same group, although it should be noted that Miro's parents were divorced for a few years by this time. The Yugoslav army offered Serbs where they were minorities in Bosnia-Herzegovina safe passage to Serbia proper. Miro and his sister, Maja, were airlifted to Zaječar in Serbia to join their father, who had moved back to his hometown following his divorce from Miro's mother in 1988. Miro's mother, Melva, stayed in Mostar until some of her family became victims in the war. Her niece Maja, who had the same name as her daughter, was wounded. Maja lost her leg when the shell from a bomb landed in front of her apartment building in Mostar. The war prevented Maja from receiving proper medical care, and her leg had to be crudely amputated. Maja then fell victim to a bone infection. Thanks to the help from Veterans for Peace, Maja was taken to a hospital in Cumberland, Maryland, for proper medical treatment. By the time Maja was ready for transport to the United States, her father was also wounded and could not accompany his daughter. Melva undertook the journey with her. Maja's father was able to join his daughter later.

After Miro's mother was settled in Cumberland, she contacted Miro and his sister in Serbia and asked them to join her, their cousin, and their uncle in Cumberland. By this time, Miro had met a woman named Natasha and married her. They had a son named Milan. Nevertheless, Miro welcomed the opportunity to go to America, so he and his new family moved with his sister to Cumberland.

Miro has a construction job and must drive 70 miles to work every day. Even though Miro never had to commute so far for work in Bosnia, he is happy to be working and living in a place that is peaceful. After some time and with a lot of hard work, Miro and his mother were able to purchase a house in Cumberland. Miro's mother and his sister live in the lower half of the house, while Miro lives with his young family in the upper part. Since arriving in the United States, Miro's wife, Natasha, has had another child, named Nina. Miro is happy to have his relations in close proximity as is the custom in Bosnia-Herzegovina. In many ways, Miro's new family is as complex as the one that he was born into. Miro has Bosnian citizenship while his wife and son are citizens of Serbia and Montenegro. Miro's youngest child, Nina, is an American. However, Miro eagerly awaits for the time in the near future when the whole family will officially become Americans.

dominated by obsolete and rusting industrial facilities built in the 1960s and 1970s.

The largest cities in the southern part of the subregion were often built as expressions of the Austro-Hungarian Empire presence. Budapest, the capital of Hungary, combines two cities on opposite banks of the Danube River (Figure 3.56c) and is a government center with industrial suburbs. Its central areas include spacious 1800s buildings. Bucharest, the capital of Romania, had much of its central area built to resemble Paris. Romania's Communist leader, Nicholae Ceauşescu,

destroyed Bucharest's urban landscape and put up his own grandiose buildings to bolster his image.

Cities in the Balkans often contain historic centers, such as the port of Dubrovnik (Croatia) that was subject to destruction by military action in 1992. An earthquake destroyed much of Skopje, the capital of Macedonia, in the 1960s. Cities such as Sarajevo, the capital of Bosnia, expanded after 1950 with central apartment blocks, suburban industries, and single-family homes but suffered destruction in the early 1990s civil war.

Figure 3.56 **East Central European cities.** (a) Warsaw, Poland: The 38-story Palace of Culture was built with funds from the Soviet Union in 1952–1955. The building, which houses several scientific and cultural institutions, theaters, and Congress Hall, was built on the World War II ruins of the city. (b) Czech Republic: Having escaped much of the destruction of World War II, Prague is one of the best-preserved cities in East Central Europe and subsequently a major tourist destination. Photo shows tourists on the famous Charles Bridge with Prague's castle in the background. (c) Budapest, Hungary: Looking from the castle in Buda across the Danube to the parliament in Pest. Photos: (a) © Jerzy Jemiolo; (b) © Emily A. White; (c) © George W. White.

(a)

(b)

(c)

Culture

Most of the peoples of East Central Europe are Slavs. Poles, Czechs, Slovaks, and Sorbs are Western Slavs. Slovenes, Croats, Serbs, Bulgarians, and Macedonians are South Slavs. Their languages are all closely related, even to those of the East Slavs, such as the Russians (see Figure 3.4a). As Roman Catholics, the West Slavs, Slovenes, and Croats use the Latin alphabet (see Figure 3.4b). The other South Slavs—the Serbs, Macedonians, and Bulgarians—use the Cyrillic alphabet because they are Eastern Orthodox Christians. Cyrillic is a modified version of the Greek alphabet that the Orthodox monks Cyril and Methodus specifically created to convert the Slavs to Christianity. Though all the Slavs are related to their fellow Slavs the Russians, the Eastern Orthodox Slavs of East Central Europe have a closer connection to the Russians, who are also Eastern Orthodox Christians. They have also tended to align themselves politically with the Russians, especially the Bulgarians.

Romanian stands out as a language because it is a Romance language. With Italian as the closest related language, Romanian is the only Romance language spoken in the subregion (see Figure 3.4a). Romanians are primarily Eastern Orthodox, but their language and Western orientation led them to adopt the Latin alphabet. Romanians see themselves as descendants of Dacians, a people who lived 2,000 years ago with an empire centered in Transylvania, and of Romans. Roman ancestry gives Romanians a sense of good pedigree (Figure 3.57).

Lithuanian and Latvian are also together in a category. Having come under the influence of the Poles, the Lithuanians are Roman Catholic. The Latvians accepted Protestantism from

Figure 3.57 **Deva, Transylvania, Romania.** This statue of a Dacian soldier atop a pedestal with the she-wolf nursing Romulus and Remus, the symbol of Rome, illustrates the Romanian belief that their nation descended from a Daco-Roman mix. Located in front of a Hungarian castle, this statue declares the legitimacy of Romanian control over the ethnically mixed territory of Transylvania. Photo: © Emily A. White.

Swedes and Germans. The Estonians, however, do not speak an Indo-European language but rather a Finno-Ugric language like Finnish and Hungarian. The Hungarians are primarily Roman Catholics, but many in the eastern part of Hungary are Protestants. Albanian is an Indo-European language but stands alone in its own category. With 70 percent of its population Muslims, Albania is the most Muslim country in Europe, though a significant number of Albanians are Eastern Orthodox and Roman Catholic.

Jews and Roma, commonly known as Gypsies, also live in East Central Europe. Europe in general was a thriving center for Jewish culture, but both groups experienced intense discrimination throughout history. The Nazi Holocaust in particular resulted in the extermination of more than 6 million Jews, not to mention many from other groups such as Slavs and Roma. Europeans who are not Roma have historically discriminated against the Roma because the Roma have comparatively darker skin and, like Jews, have non-Christian beliefs and practices. Roma still are persecuted intensely and are forced to live as an underclass.

Ethnic Tensions

The breakup of Yugoslavia in the 1990s is Europe's most recent major tragedy involving war and massive human rights violations. Because many of the ethnic groups in the conflict had differing religions, many international diplomats and commentators, as well as governments, tended to label the groups according to religion and even viewed the conflict as a religious one. They rejected recognizing such groups as the Bosnians because Bosnians did not claim to be of a single faith but rather of any faith found in Bosnia-Herzegovina. In doing so, they ignored the fact that many Serbs, Croats, and Muslims fought together to prevent Bosnia-Herzegovina from being carved up by Serb and Croat nationalists.

The fact that many groups of differing faiths have joined together during conflicts, such as the one in Bosnia-Herzegovina, illustrates that conflicts in East Central Europe are caused by more factors than simple religious differences. Historical, political, and economic geography have played great roles. In Yugoslavia, for example, Slobodan Milošević used ethnic politics to gain power. He used thugs in the region of Kosovo to attack Albanians living there. When Albanians fought back, it aroused the fears of many Serbs in Yugoslavia that this ethnic minority, which was economically poor and had high birth rates, was persecuting Serbs and destroying the Serb heartland. Milošević vowed to protect Serbs and Serb culture by calling for the crackdown on all minorities in Yugoslavia. Serb nationalists enthusiastically supported Milošević. In response to Milošević's policies, however, Yugoslavia's ethnic minorities moved toward secession.

The history of Yugoslavia's political geography made it easy for a politician like Milošević to pit the various ethnic groups against one another. Yugoslavia was a country that was created from territories of the Austro-Hungarian and Ottoman empires. Both of these empires had completely differing political and economic traditions and orientations. In Bosnia-Herzegovina, for example, Serbs historically tended to live in the rural areas and Muslims tended to live in the urban areas.

The rural Serbs were comparatively poorer and less educated than the urban Muslims. These rural Serbs resented the urban Muslims, and this resentment expressed itself during the war in the 1990s. Though the two groups each had their own religion, the differing rural and urban traditions played a greater role in the conflict.

Ethnic tensions exist in other countries as well, and the study of geography helps us understand them. Albanians, for example, live beyond the borders of Albania, in the neighboring Yugoslav province of Kosovo, where they account for 95 percent of the population, and in adjacent areas of Macedonia, where they are 23 percent of the population (see Figure 3.54). Hungarians also live in the countries surrounding Hungary. Large numbers of Russians are found in the Baltic countries. The major problem is a disjunction between the distribution of nations and country boundaries. Many nations live in territories that extend beyond the countries designated for them and are seen as minorities in other groups' countries. As a minority, they are a threat to the dominant groups. For example, if Hungarians live in southern Slovakia, the Vojovodina, Transylvania, and other places, should not these territories be given to Hungary? If Albanians live in Kosovo and western Macedonia, should not these territories be awarded to Albania? This solution would certainly bring nations and countries into greater geographical alignment.

The problem with giving minorities their own countries is that it would not end the existence of minorities within countries. As minorities become majorities, new minorities come to the forefront or into being. If Kosovo became an Albanian country, it would have Serb minorities. A similar situation already occurred with the cases of the Hungarians and Russians. When Slovakia, for example, was part of the Hungarian Kingdom, Slovaks were the minority and Hungarians the majority. When Slovakia was detached from Hungary, Slovaks became the majority in a new Slovakia, but the Hungarians living there became a minority. After 1991, Russians found themselves as minorities in new countries (e.g., Estonia, Latvia, Lithuania) that once had been part of the Russian-dominated Soviet Union (see Chapter 4). Because it is not completely possible to give every ethnic group its own country without creating new minorities, many international leaders do not support the breaking up of countries.

Some argue that minorities should leave a country to avoid conflicts. Many minorities, however, occupied their land as long or even longer than the majority. In many cases and as shown in the examples already discussed, boundaries were once drawn differently so that minorities actually formed a majority with others of their group. For example, the boundaries of Hungary once included the Hungarians of Slovakia, Transylvania, and other areas in a Hungary that was much bigger than it is today. Tensions between Hungarians and their neighbors are not just over ethnic differences. Conflict arises over the issue of whether Hungary should be able to regain control of its lost territories. Hungarians who want Hungary to reclaim its lost territories are known as irredentists. **Irredentism** is the desire to gain control over lost territories or territories perceived to belong rightfully to one's group. With boundaries having changed so frequently in East Central European history, irredentism is a major issue and a source of much conflict, more so than conflict arising from simple cultural differences among groups.

Current majority groups, however, rarely want to lose any of their territories, though other groups may live in them. Territories often have great meaning for nations. For example, though Kosovo is more than 90 percent Albanian, Kosovo is the cradle of Serb civilization. It contains the great monasteries and other cultural artifacts of the Serb nation. Many Serbs feel that losing control of Kosovo would result in the loss of much of their culture. A dilemma then emerges. The nation-state ideal implies that a nation may only possess territory occupied by its members. Yet some territories are so important that nations will not give them up, even if their nation is not a majority in them.

To conform to the nation-state ideal, some nations feel that they must eradicate other groups from their territories to justify exercising control over their territories, though the practice certainly violates both the human rights of the victims and international law. Eradicating people from a territory because they are ethnically different is known as **ethnic cleansing.** Serb nationalists coined the term to refer to their own actions during the war in Bosnia-Herzegovina in the 1990s. They attempted to legitimize their claim to the republic by making the territory purely Serbian. Ethnic cleansing is now a term used to refer to similar acts perpetuated in other places around the world, not only currently but also in history. The Holocaust, for example, which refers to the Nazis' attempt to exterminate Jews and other groups during World War II, is now referred to as ethnic cleansing, though the term did not exist at the time.

Ethnic cleansing takes three forms: assimilation, expulsion, and extermination. All three forms were employed in the former Yugoslavia. Bosnia-Herzegovina became infamous for not only concentration camps but also rape camps. Rape is certainly used to terrorize, but it frequently had an added purpose in that war. Despite differing political and economic traditions and differences between individuals, Serbs, Croats, and Muslims were racially and linguistically similar, even sharing many cultural characteristics. The only obvious cultural difference between the groups was religious belief. Extremists thus saw religious conversion as a means of assimilation. Because Balkan peoples typically follow the tradition of raising a child in the father's faith, rape was seen as a means of conversion and thus assimilation. After being raped, many women were told that they were then obligated to raise their children in the rapist's faith, making the child a member of the rapist's nation. Many women were held in the camps so that they could not obtain abortions. Genocide is the attempt to exterminate an entire ethnic group. The attempt by Serb nationalists to end Islam in Bosnia-Herzegovina by exterminating Muslim men and impregnating Muslim women with "Serb" children could be considered **"genocidal rape."**

Widespread atrocities, including rape, in Bosnia-Herzegovina led to changes in international law. To deal with the atrocities, the World Court in The Hague set of a special war crimes tribunal, known as the International Criminal Tribunal for the former Yugoslavia (ICTY). The tribunal tried numerous cases and continues to do so. Previously, international law viewed rape as an unfortunate by-product of war, but the tribunal set the precedent that rape is used as a weapon in war and deserves greater punishment than before. Rape is now officially recognized as a war crime. In one case, the tribunal convicted a man for the crime of not preventing a rape.

Human rights continued to be a major issue after the war as thousands of women were forced into prostitution. The major source countries are Moldova and Ukraine in the Commonwealth of Independent States (CIS; see Chapter 4), both of which have struggling economies, but many are lured from the other poor areas of East Central Europe and the former Soviet Union. In most cases, young women are offered high-paying jobs in the West. Once they depart their countries, however, their passports are taken from them, and they are sold into prostitution. Many women are taken to brothels in Bosnia-Herzegovina. The practice is so widespread that UN soldiers and members of other international organizations have been implicated in prostitution scandals. Many women are also forced to work as prostitutes in Western European cities such as London.

Economic Development

After the Industrial Revolution began in England in the late 1700s, it took considerable time for it to spread to East Central Europe. In addition, because outside empires dominated most of East Central Europe through the 1800s and up to 1918, industrialization was inhibited in many areas. These empires treated their East Central European domains as colonies, preferring to extract resources and agricultural products rather than invest in industrialization. However, many exceptions existed. The Czech and Polish lands were somewhat exceptional in that a number of cities had factories. Hungary and Slovenia also developed industry. Nevertheless, compared to some of the other subregions of Europe, a large number of people still work in agriculture. Albania, for example, ranks as one of the highest with 60 percent of its labor force working in agriculture.

After World War II, Communist economic policies were imposed on most countries in East Central Europe (see the section on "Changes in the Modern Era" in this chapter). During Soviet times, the Communists tried to improve the standard of living by encouraging industrialization, particularly in areas with coalfields, and by providing jobs for everyone, whether it made good economic sense or not. Communism had many failures but it had many successes, too. During Communist times, farming efficiency improved as industrialization provided tractors to replace horse-drawn plows. Great strides also were made in technology, education, health care, and welfare. Literacy increased dramatically and infant mortality plummeted. One of the difficulties in the transition to capitalism is that people in the former Communists countries are not accustomed to paying for health care and higher education. Unemployment is also a new and shocking experience. For many people, capitalism is a step backward from Communism.

In terms of trade, Soviet Communists preferred not to be ensnared by capitalist practices, which they considered to be corrupting. Therefore, they tried to restrict trade among fellow

Communist countries. Stalin set up the Council for Mutual Economic Assistance (CMEA or COMECON) to compete with the European Economic Community (EEC) in Western Europe and even prevent East Central European countries from attempting to join the EEC. COMECON linked the subregions' economies with one another and with the Soviet Union. Stalin also insisted on specialization, with the northern countries focusing on industry and the southern ones producing agricultural products. Such specialization forced greater dependency on the Soviet system and prevented any country from realigning itself with the West. Cheap oil and natural gas from the Soviet Union also increased dependency but left a legacy of contaminated soil and water.

After the end of Soviet control in 1991, the countries of East Central Europe experienced economic crises as they reoriented themselves from incorporation in the Soviet economic system to the world economic system (see Chapter 4). Exchanges with one another and the former Soviet states greatly declined. Countries are challenged with breaking up business monopolies owned by the state, providing productive jobs, and cleaning up the environment. The difficulties in moving to capitalism led to an immediate drop in GDP, but growing trade with Germany in particular and new economic policies caused the GDP to rise again (Figure 3.58). Personal income, which was low during Communist times, remained low and resulted in very low consumer goods ownership compared to the rest of Europe (see Figure 3.17).

The Communist system focused on reorienting the agricultural economies of East Central Europe to industrial economies. While the Communists created factory jobs, similar factory jobs in the West were replaced by service jobs as the world economy continued to develop. In 1997, the service sector accounted for 60 to 70 percent—more in a few countries—of GDP in Western Europe. In contrast, the service sector in East Central Europe accounted for only 50 to 60 percent of GDP on average and was well below those levels a few years earlier. Albania at 23 percent and Romania at 36 percent represented the lowest. Their low rankings illustrate that they have had the greatest difficulties changing their economies and improving the standards of living of their people.

Poland, the Czech Republic, Hungary, Slovenia, and the Baltic countries have had the greatest successes in adopting capitalist practices. These countries were also the most industrialized before and during Communism and have had stronger traditions of democracy. Significantly, they also are closest to countries of the European Union, making them easy beneficiaries of trade and investment from countries in the other subregions of Europe, especially from Germany. For example, 70 percent of Poland's trade was with the EU in 1996.

One of the biggest success stories comes from the Czech Republic. One example is the Czech automotive company Skoda. It was like the automotive companies in the other Communist countries in that it produced poor-quality, two-cylinder vehicles. Volkswagen, however, purchased Skoda, and with German management and huge sums of cash, the company has become the most successful formerly Communist company anywhere. Skoda employs 4 percent of the Czech work force and accounts for 14

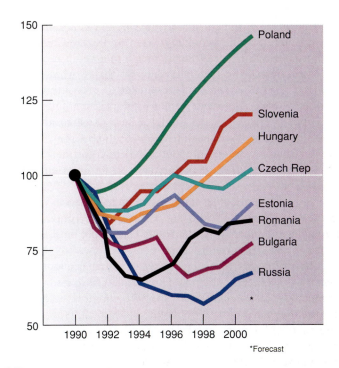

Figure 3.58 **GDP growth in East Central Europe after the end of Communism.** Notice the initial rapid decline in GDP. Source: Copyright © The Economist Newspaper Group, Inc. Reprinted with permission. Further reproduction prohibited. www.economist.com.

percent of Czech exports. Eighty percent of Skoda's vehicles are exported, including to Germany with its competitive auto market.

The Czechs could have great success in earning money through beer exports because they are originators of great beers. For example, Pilsner beers come from the town of Plzeň, and Budweiser traces its roots back to the town of Budweis (now České Budějovice). However, the Czechs have difficulty exporting their beers because the American company Anheuser-Busch uses its financial might, with which it obtained the Budweiser trademark, to keep the Czechs from exporting Czech Budweiser, though the Czech beer was brewed hundreds of years before the American brand existed.

Hungary was able to exert its economic independence during Communist times, probably more than any of the other countries controlled by the Soviet Union. For example, it allowed private businesses that employed fewer than five people. The mixture of large-scale Communist economic policies with small-scale capitalist practices led to the term "goulash Communism," a term derived from the traditional Hungarian stew, which is a mixture of vegetables and meat. Capitalist practices were minimal, but they were greater than in other Soviet Bloc countries and gave Hungary a head start in the transformation process. Though a modest-sized country, Hungary captured over one-third of all 1990s foreign direct investment in the former Communist countries, including the former Soviet Union. In 1989, 65 percent of Hungary's trade was with COMECON. By 1998, 80 percent of Hungary's trade was with the EU.

Before Yugoslavia was engulfed in war beginning in 1991, it had one of the strongest economies and highest standards of living of Communist Europe. Not a Soviet satellite,

Yugoslavia pursued its own course. It employed the Communist idea of centralized planning, but compared to the Communist countries in the Soviet sphere, it also allowed a more genuine practice of the Communist belief that workers should manage their companies. Subsequently, productivity was high in Yugoslavia. Travel to non-Communist countries was not as restricted as in the Soviet sphere. As a result, thousands of Yugoslavs, especially Slovenes and Croats, sought work as guest workers in countries such as Germany. These workers sent millions of dollars back to family members in Yugoslavia.

Slovenia prospered more since independence in 1991 because it further developed its economic contacts with Germany and other EU countries and no longer must send tax money to the other Yugoslav republics. Slovenes now have the highest standard of living and the most modern economy of the former Communist countries. Croatia may have been just as prosperous, but independence in 1991 was followed by the devastation of war and little foreign investment. Now that the wars are over, Croatia is rebuilding and the economy is growing. Serbia and Montenegro's economy faces great difficulties. Economic boycotts during the wars and NATO bombings devastated the country's economy. Though the war is over, the country is receiving little foreign investment. Macedonia escaped most of the ravages of war, but its landlocked position among unfriendly neighbors and its distant location from the wealthier countries of Europe attract little foreign investment to this largely agricultural country.

Test Your Understanding 3F

Summary East Central Europe consists of three small Baltic countries (Estonia, Latvia, and Lithuania), Poland and Hungary with their plains, and the hilly areas of the Czech and Slovak Republics, all with mixtures of farming and manufacturing. Farther to the south in the subregion are former Yugoslavia (Slovenia, Croatia, Bosnia-Herzegovina, Serbia and Montenegro, and Macedonia), Romania, Bulgaria, and Albania, some of which have experienced varying degrees of civil strife from full-scale war in Bosnia and Croatia to ethnic clashes in the other countries, including that in Kosovo.

The countries of East Central Europe struggled through the transition from Communism to democracy and capitalism with many of them desiring to join the European Union. In recent years, many of them experienced economic growth as they established economic trade with the rest of Europe. Others continue to struggle.

Questions to Think About

3F.1 Can you explain the greater economic growth in Poland, the Czech Republic, Hungary, and Slovenia in the 1990s compared to that in the other East Central European countries?

3F.2 What are the causes of, and possible cures for, the strife in Bosnia and Kosovo?

3F.3 What happened to urban landscapes in Communist Europe in the decades after World War II?

Key Terms

irredentism   genocidal rape
ethnic cleansing

Online Learning Center

www.mhhe.com/bradshaw

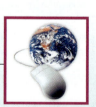

Making Connections

The Online Learning Center accompanying this textbook provides access to a vast range of further information about each chapter and region covered in this text. Go to www.mhhe.com/bradshaw to discover these useful study aids:

- Self-test questions
- Interactive, map-based exercises to identify key places within each region
- PowerWeb readings for further study
- Links to websites relating to topics in this chapter

Chapter 4

Russia and Neighboring Countries

Figure 4.1
Russia and Neighboring Countries.

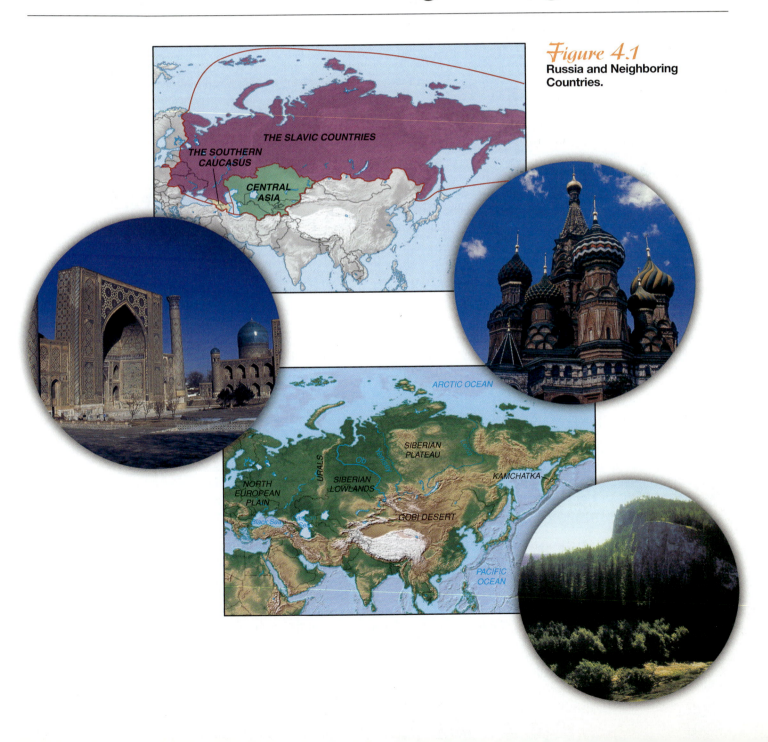

Countries Redefine Their Relationships

Russia (formally called the Russian Federation) and its neighbors comprise 12 countries (Figure 4.1) plus the Baltic states of Estonia, Latvia, and Lithuania (see Chapter 3) that once formed the Union of Soviet Socialist Republics (USSR), more commonly called the Soviet Union. The Soviet Union emerged from the ruins of World War II as one of two world powers. As a Communist country, it competed with the United States, the other world power, for global political and economic influence in what was known as the Cold War (1945–1991).

The Soviet Union was the world's largest country in terms of land area, stretching across Eurasia and inhabited by over 100 national groups (Figure 4.2). The Soviet Union's government tried to keep the country together by integrating every region into a single political and economic system, but it eventually failed, and the Soviet Union dissolved in 1991. Though the 15 republics became independent countries, each found its economy not only weak but also incomplete. Each had been a piece of a larger system, created by the government of a country that did not exist anymore. Within a very short time, the governments of these new countries realized that they would have to cooperate with one another despite any antagonisms or hostilities. In 1991, 11 of these countries formed the **Commonwealth of Independent States** (CIS). Georgia joined in 1993; thus, all the countries discussed in this chapter are members of the CIS. The three Baltic republics (Estonia, Latvia, and Lithuania) decided instead to seek greater cooperation with Europe.

The CIS is an organization of cooperation between independent countries and not a single country itself. It does not undermine the sovereignty of any of the member countries and their peoples. Instead, it is an economic union that recognizes the necessity of maintaining the trade flows established during Soviet times. Member countries meet to discuss and work out economic needs and policies. Beyond economics, the entire membership has very little in common, particularly in terms of culture and natural environment. The Slavic countries (Russia, Ukraine, and Belarus, including neighboring Moldova), the Southern Caucasus (Georgia, Armenia, and Azerbaijan), and Central Asia (Kazakhstan, Turkmenistan, Kyrgyzstan, Tajikistan, and Uzbekistan) are all very different subregions.

In time, these three subregions will likely integrate themselves into other world regions. Some researchers already include some or all the Slavic countries with Europe and Central Asia with Northern Africa and Southwestern Asia (see Chapter 8). In reality, this process is far from complete, and it is premature to redraw the boundaries of the world regions so radically. Despite these countries' differing cultures and orientations, their common Soviet legacy presents them with very similar and recurring problems. These countries have even found it to their mutual benefit to continue working together. Their situation leads some authors to label the region "Post-Soviet Union." This term, however, is a bit dated and does not recognize that the region is changing despite its difficulties. Because the countries continue to work together and Russia remains the dominant political and economic force of the region, we find it appropriate to call the region "Russia and Neighboring Countries."

Crossroads, Imperialism, and Cultural Diversity

The CIS accounts for much more of Earth's land surface than Europe but is divided into only 12 countries, compared to Europe, which has more than 30. The countries also vary in size considerably (Figure 4.3). The Russian Federation is larger than all the others put together, nearly twice the size of the United States. The smallest, Armenia, is only the size of Maryland and Delaware combined.

These differences reflect the cultural and political history of the region. Russia was one of Europe's great empires in modern times, having lasted until the dissolution of the Soviet Union in 1991. Unlike the European empires such as France and the U.K., which conquered lands overseas, the Russian Empire conquered peoples and land adjacent to it, extending its borders as it annexed new territories. As early as the mid-1800s, the Russian Empire had become very multinational, with Russians barely constituting a majority of their empire's population. The countries that recently emerged along Russia's boundaries represent many of those peoples and territories absorbed by the Russian Empire earlier in history. Many of these territories never existed before as separate countries. Russia, however, retained control over most of its conquered lands in 1991. Siberia—a huge, resource-rich, yet sparsely populated area—represents most of the land that the Russians conquered in history.

The other CIS countries, though much smaller in size, represent old cultures. Along with Russia, many of them lay along the crossroads of ancient migration and trade routes. Over the course of time, three major invading influences were Christianity, penetrating the region from the southwest; Islam, entering from the south; and Mongol culture, sweeping in from the east.

Many tribal and ethnic groups moved in from all directions, especially the east. The semiarid steppe grasslands of southern Russia and Ukraine served as the main route for those who invaded Europe throughout much of its history. Many invaders from farther east settled for a time in the steppes before moving onto Europe, while others decided to settle permanently. Central Asia lay along the Great Silk Road (Figure 4.4), the pathway that brought many goods and ideas that eventually helped to build the European empires. The Caucasus Mountains were situated between the ancient civilizations of Southwest

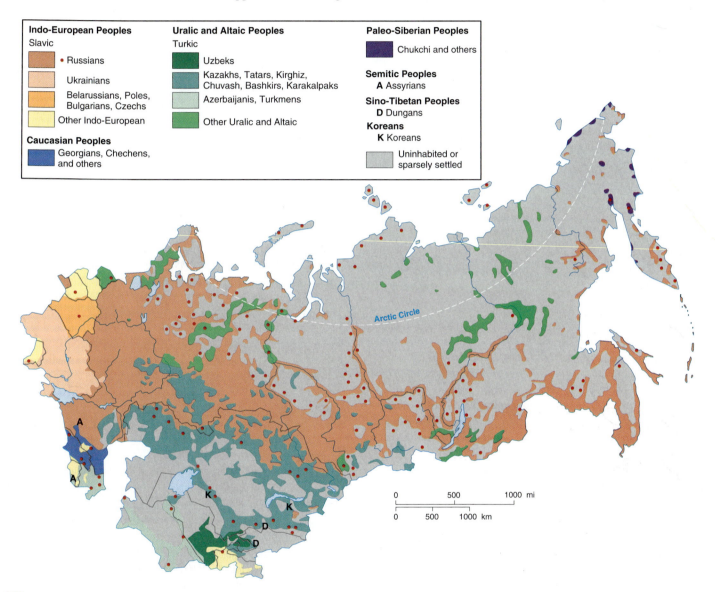

Figure 4.2 **The national groups of the former Soviet Union.** Note how the Russians spread across southern Siberia. Source: © University of Texas Libraries.

Figure 4.3 Summary information on the subregions of Russia and Neighboring Countries.

Subregions	Land Area (km²) Total	Population (Millions) Mid-2001 Total	2025 Est.	GNI 1999 (US $ million) Total	GNI PPP 1999 Per Capita	Percent Urban 2001	Human Development Index 1999 Rank of 174 Countries
Slavic countries	17,920,400	207.8	195.9	363,595	4,833	64.3	81.5
The Southern Caucasus	186,100	17.4	18.7	8,945	2,450	58.0	91.7
Central Asia	3,994,400	56.6	69.5	42,764	3,195	40.0	93.8
Totals or Averages	**22,100,900**	**281.8**	**284.1**	**415,304**	**3,493**	**54.1**	**89.0**

Source: Data from *Population Reference Bureau 2001 Data Sheet; World Development Indicators,* World Bank, 2001; *Human Development Report,* United Nations, 2001.

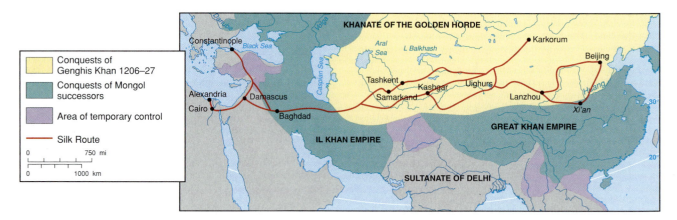

Figure 4.4   **The Great Silk Road.** Note how Central Asia serves as an important link in this old trading route. Source: From Milner-Gullard, *Cultural Atlas of Russia*, © 1998 Andromeda Oxford Limited, www.andromeda.co.uk.

Asia and the cultures of the North European Plain. Trade routes between the Viking homelands in Scandinavia and the Byzantine Empire ran through the Slavic areas located on the North European Plain. Centuries of migrations and imperial conquests have created a region that is culturally very complex today.

Eastern Slavs

As early as 1500 B.C., a Proto-Slavic people appeared between the Vistula and Dnieper Rivers. Little is known about these early peoples. Between A.D. 400 and 900, they divided into three main branches. The western and southern branches moved westward and then settled in East Central Europe (see Chapter 3). The eastern branch moved north and east, developing its own character as it came into contact and exchanged ideas and practices with the Baltic, Finno-Ugric, and Viking peoples. Eventually, the Eastern Slavs subdivided further into Russians, Belarussians, and Ukrainians (see Figure 3.4a).

Rus

The Eastern Slavs adopted many of their unique characteristics beginning in the 800s, when Vikings (called Varangians in Russian) moved into the North European Plain along the Volga River. The Varangians established their rule and began trading the amber, beeswax, fur, and precious metals of their newly acquired lands with Byzantium, the Orthodox Christian realm to the south, with its capital at Constantinople. The Varangians did not settle in great numbers but ruled primarily as overlords by establishing their own dynasties or by marrying into and taking over many local dynasties. As Eastern Slavs mingled with Varangians, Balts, and Finno-Ugric peoples, *Rus* came into being. *Rus* applied to both the people and their land.

The most important of the earliest Rus principalities were Novgorod and Kiev. Novgorod was founded first, but Kiev grew more quickly in power, exacting tribute from the other Rus principalities. Kievan Rus reached its zenith in the 900s and 1000s. In 988, Prince Vladimir brought Eastern Orthodox Christianity to Rus, tying the Eastern Slavs to Constantinople and leading to the adoption of the Cyrillic alphabet, which was derived from the Greek alphabet. Over time, Kievan Rus could

not maintain its authority over the other principalities of Rus. Each began to exert its individual authority. Novgorod and a newer principality called Vladimir-Suzdal, with its capital at Vladimir, were two of the most notable (Figure 4.5).

In the 1200s, the Mongols (or as they were earlier called, Tatars) swept in from the east and attacked the principalities of Rus. Those in the south suffered most. Kiev was sacked in 1240. Novgorod was the only principality to escape invasion, but it had to pay tribute to the Mongols. The Mongols established a state that stretched from the steppes to Central Asia and Siberia. It was known as the Khanate of the Golden Horde. It included the southern Rus principalities and extracted tribute from the others. The Mongol influence on the development of Russian culture was very strong. Centuries after the fall of the Mongols, Napoléon said, "Scratch a Russian, and you will wound a Tatar."

Muscovy

Moscow emerged in the principality of Vladimir-Suzdal as a settlement along the trade routes between the Baltic and Black Seas. It was an insignificant trading post at the time of the Mongol invasions, but its isolated, forested location offered some protection from the Mongols, and it soon grew in importance. When the southern Rus principalities came firmly under control of the Mongols, the Orthodox Christian metropolitan (provincial primate) of Kiev and all Rus moved north from Kiev to Vladimir in 1299, increasing the authority of Vladimir-Suzdal. In the early 1300s, the prince of Vladimir-Suzdal cooperated with the Mongols by collecting tributes from the other Rus principalities on the Mongols' behalf. Around 1318, the prince married the sister of the khan of the Golden Horde and received the title of "Grand Prince." The power of Vladimir-Suzdal shifted to Moscow in 1327, when the metropolitan of the Orthodox Christian church moved from Vladimir to Moscow. During the 1400s, the grand princes of Moscow conquered the other Rus principalities. Novgorod, one of the last and most powerful, fell in 1478. The consolidated territories governed from Moscow came to be known as Muscovy.

By 1480, Muscovy was so powerful and Mongol authority so weak that Grand Prince Ivan III (the Great) simply refused to pay tribute to the Mongols any longer. With this refusal,

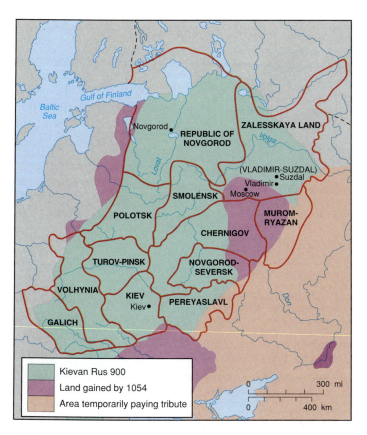

Figure 4.5 **The Rus principalities around 1100.** Kiev was very powerful in the beginning, but Novgorod and then Moscow rose to power. Source: From Milner-Gullard, *Cultural Atlas of Russia*, © 1998 Andromeda Oxford Limited, www.andromeda.co.uk.

Muscovy achieved its independence. Ivan III married Sophie Paleologue, the niece of the last Byzantine emperor. After Constantinople fell to the Ottomans in 1453, Ivan was seen as the true inheritor of the Christian realm. He adopted the title "czar," derived from the Latin *caesar* for emperor. With the succession of Ivan IV (the Terrible), the patriarch of the Eastern Orthodox Church in Constantinople accepted the title "csar." Moscow emerged as the "Third Rome," successor to Constantinople and Rome. Even today, Russians are familiar with the concept that Moscow is the "Third Rome." This, and the link with the Orthodox Church, are features of Russian national identity.

The Russian Empire

In the centuries after Mongol control ended in 1480, Muscovy grew in power and influence. In 1613, Mikhail Romanov ascended the throne. The Romanov dynasty lasted until the last czar was deposed in 1917. From 1480, Muscovy's territory continued to expand, especially to the east, into Siberia and Central Asia but also to the west, bringing the Russians into contact with Central and Western European culture. The greatest territorial expansions were made under Peter the Great (1682–1725), Catherine the Great (1762–1796), and Alexander I (1801–1825).

Peter expanded Muscovy's boundaries to the Baltic Sea, where he founded the city of St. Petersburg and made it his country's new capital in 1712. He moved the capital westward

in an attempt to westernize his country. Peter built the first Russian navy, reorganized the military and government along European lines, and founded modern institutions devoted to the sciences and arts. Peter's efforts transformed Muscovy into the Russian Empire. Catherine continued to westernize Russia. As well as expanding farther east into Siberia, she also conquered lands to the west, incorporating Poland, and to the south, gaining all of Ukraine, where she founded the Black Sea port of Odessa, and also the Crimean Peninsula. She established Russian influence in the Balkans. By the end of Catherine's reign, Russia was clearly one of Europe's great powers in area, population, and military might.

Alexander I was considered the "Savior of Europe" for defeating Napoléon's attack on Moscow in 1812. The victory allowed Alexander to play a major role in redrawing Europe's boundaries at the Congress of Vienna in 1815. Russia's boundaries moved farther west with the acquisition of Finland, more of Poland, and Bessarabia (now Moldova). Around the same time, Russia expanded into the Caucasus, taking the area around Baku. Russia also exerted greater control in Alaska. In the decades following Alexander's reign, Russian czars expanded their empire farther into the Caucasus, Central Asia, and East Asia, taking lands from China.

By the early 1900s, the Russian Empire achieved its greatest extent, stretching from Europe to East Asia and the Pacific Ocean (Figure 4.6). Russia was a world power, but its strength was derived from its sheer size. Internally, it had many problems. Politically, the czars remained in total control, continuing to claim to speak in the name of God, but absolute power inhibited the creation of an efficient government. Technologically, Russia was decades behind Western Europe. The technology of the Industrial Revolution developed very slowly, leaving Russia farther and farther behind as time progressed. Most Russians remained feudal peasants farming with horses or oxen. At the same time, Russia had expanded its boundaries to include more than 100 different peoples. After the rise of nationalism in Europe in the mid-1800s, Russia's nationalities began clamoring for independence. This situation worsened as Russians themselves became nationalistic and began suppressing non-Russians. They developed **Russification** policies to force Russia's minorities to become more Russian ("Russified").

The Soviet Union

World War I (1914–1918) exerted great stress on the Russian Empire. By 1917, starvation and a huge death toll combined with longstanding opposition to czarist rule. Revolutionaries deposed the Czar Nicholas II and set up a provisional government that proved to be weak. A number of national groups staked their claim to territory and declared independence. Finland, the three Baltic countries (Estonia, Latvia, and Lithuania), Belarus, Ukraine, the countries of the Caucasus (Georgia, Armenia, and Azerbaijan), and Central Asia (at the time called Turkestan) all chose independence. The Polish lands of Russia joined with the Polish lands of Germany and Austria-Hungary and declared an independent Poland. Bessarabia (now Moldova) was annexed by Romania.

Figure 4.6   **Russian Federation: history of growth.** The expansion of the Russian Empire from the 1600s to the early 1900s. The Trans-Siberian Railway was built through territories that were regarded as safely Russian.

By late 1917, the Bolsheviks, a group of Communists, also known as "Reds," overthrew the provisional government. Civil war ensued when anti-Communists, called "Whites," tried to dislodge the Bolsheviks from power. The U.K., the United States, and France supported the anti-Communist Whites, even sending troops to Russia. By 1922, the Reds gained the upper hand, expelled their enemies, and reclaimed Belarus, Ukraine, the countries in the Caucasus, and Central Asia.

In 1922, under the leadership of Vladimir I. Lenin, the Bolsheviks established the Union of Soviet Socialist Republics (USSR), commonly called the Soviet Union. This new country was the successor to the Russian Empire, but the Bolsheviks changed the old system of government and economy. They abolished the monarchy and government by a privileged few and replaced them with "soviets." Soviets were workers' and soldiers' councils that drew their members from the common citizens. Economically, the Bolsheviks despised capitalism, viewing it as a system that allowed a select few—the bourgeoisie (middle class)—to grow very wealthy from the exploitation of the masses—the proletariat (the industrial working class). The Bolsheviks set out to turn this system upside down, that is, reverse the results of capitalism, by putting the workers in charge of factories and businesses. They sought to create a "dictatorship of the proletariat" and, thereby, a "workers' paradise." Religion was seen as a tool of the

oppressors because the clergy sanctioned the rule of political leaders and the capitalist exploiters who oppressed the common people. The clergy also encouraged people to accept their suffering on Earth because they would be rewarded in Heaven. Consequently, the Bolsheviks attacked churches and mosques, dynamiting many and turning others into scientific centers such as planetariums to show the error of religious belief.

Five-Year Plans

In the early 1920s, the Soviet Union was economically weak. World War I, its civil war, and wars with countries such as Poland had devastated the country, which had not modernized. Lenin allowed for some freedom of economic action to give the country the opportunity to rebuild before fully implementing Communist practices. The Soviet Union began to recover but only by permitting capitalist practices, especially in agriculture. Lenin died in 1924, and within a few years, Joseph Stalin came to power and radically changed government policy. Stalin's heavy-handed form of governing became known as Stalinism, and some debate whether it was a form of Communism or not.

Stalin believed it necessary to transform the Soviet Union's economy forcefully into the Communist ideal of a "workers' paradise" based on industrialization. This was also necessary to overcome the technology advantages of the West. However, the Soviet economy was largely agricultural, and farmers could not relate to the urban-industrial ideology of Communism. Therefore, Stalin implemented the first **five-year plan** in 1928.

The five-year plan called for collectivization and industrialization. Collectivization was a way of making farmers into factorylike workers and thus more sympathetic to the Communist way. Under collectivization, small family farms were merged together to create large farms, thousands of acres in size. Unlike the small farms, the large farms were better designed to use modern farm machinery, then under production with industrialization. The government became the owner of the collectives and farmers became employees. With collectivization, farmers became more like factory workers, even living in tightly packed housing like urban factory workers.

For industrialization, the plan called for government ownership of all industries. In what is known as the **command economy,** the government ran the economy, setting quotas and favoring heavy industry over production of consumer goods. The five-year plans also established **central planning.** In contrast to capitalist economies, supply, demand, or profit making did not dictate what would be produced. With central planning, the government decided how many goods and services were needed by society, almost without cost considerations (see also the "Changes in the Modern Era" section, p. 71.).

World War II

Rapid and forced industrialization prevented the Soviet Union from experiencing the world economic depression of the 1930s as the government kept investing in the economy and providing jobs. On the other hand, Stalin also purged millions of his enemies, suspected enemies, and potential enemies, including about half of the officers in the military (See the "Human Rights" section in this chapter, p. 150). By the early 1940s, the Soviet Union's defense forces were large but vulnerable, especially to Nazi Germany. In 1939, Stalin entered into a nonaggression pact with Hitler, allowing the Soviet Union to take control of territories that formerly belonged to the Russian Empire: the Baltic states, eastern Poland, and Bessarabia (now Moldova). Finland was also a target, but the Finnish forces successfully defended their country from the Red Army. The victory of the small Finnish military over the huge Red Army underscored the weakness of the Soviet Union and its military.

On June 22, 1941, Hitler launched Operation Barbarossa, the invasion of the Soviet Union. The campaign was lightning quick and devastating. The Soviet Union survived only because of its sheer size. A vast amount of land was lost and hundreds of thousands of people were killed, but even larger areas were left unconquered. The Soviets moved war production farther east, out of the range of the Nazi military, including its air force. With help from the United States and the United Kingdom, production increased. Finally, the harsh Russian winter brought the German Nazi armies to a standstill. In the meantime, Stalin moderated many of his harsh policies, including the persecution of the Russian Orthodox Church. After a fierce battle at Stalingrad in 1943, Soviet forces began rolling back the Nazi invaders. The people of the Soviet Union had turned the tide of war and began winning what they called the "Great Patriotic War." By May 1945, the Red Army had swept through East Central Europe and occupied much of Germany, including the capital, Berlin.

Its victory in World War II allowed the Soviet Union to annex the Baltic countries and Moldova, former territories of the Russian Empire they marched into in 1940, and new territory in East Central Europe never before governed by the Russians (e.g., East Prussia, the northern half of which is now Kaliningrad, and areas of Poland). Victory also allowed Stalin to establish and support Communist governments in East Central European countries. To counter NATO and the Marshall Plan (economic aid from the United States) and later the European Economic Community, Stalin created the Warsaw Pact and Council for Mutual Economic Assistance (CMEA, also known as COMECON). Stalin also began persecutions again, accusing entire ethnic groups and nations of treason during the war. He moved millions from their homes, mostly to Siberia. After Stalin died in

Test Your Understanding 4A

Summary In 1991, 15 new countries replaced the Soviet Union as Communism ended in this world region. Since then, these new countries have worked to establish new relationships with one another and the world. One example is seen in the Commonwealth of Independent States (CIS). During Soviet times, most of the republics were inward-looking and very dependent on Russia. Though old trade links are difficult to change, most seek to break that dependence on the Russian Federation and try to establish stronger ties with other countries, including the United States. Though the Russian Federation prospers by the old dependencies, it too seeks new relationships with countries of other world regions.

The present distribution of peoples and cultures can be traced back to migrations up to around A.D. 1000. The Orthodox form of Christianity, the invasions of Mongol hordes and Ottoman Turks, and the conversions to Islam along the region's southern borders had major influences on subregional cultures. Much of the region remained under feudal and agricultural conditions until the early 1900s.

Russia expanded its territories from the 1600s and began to modernize its productive capacity along Western lines. The Bolshevik Revolution of 1917 led to Communist control of the former Russian Empire as the Soviet Union. After World War II, Soviet influence extended over countries in East Central Europe to form the Soviet bloc with economic and defensive ties.

Questions to Think About

4A.1 What is the relationship between the countries of the Commonwealth of Independent States (CIS)? Do they act as one country? Why or why not?
4A.2 How did the Rus begin as a people, and which of the Rus principalities began the Russian Empire?
4A.3 How did the Soviets design the economy of the Soviet Union?

Key Terms

Commonwealth of Independent States (CIS)
Russification

five-year plans
command economy
central planning

1953, his successors were deliberately much less harsh. However, Stalin's system of government and economy remained intact until the demise of the Soviet Union in 1991.

 # Natural Environment

The variety of natural environments in Russia and Neighboring Countries reflect their huge land area. The continental interior climates, often with long and harsh winters; the vast plains; and the massive areas covered by a variety of natural vegetation types and soils provide a distinctive stage on which the development of human geographies occurred.

Midlatitude Continental Interior Climates

Russian and Neighboring Countries nearly all lie north of 40°N and most of their area is north of 50°N. As such, these countries lie as far north as Alaska and Canada. The region contains the places on Earth that are farthest from major water bodies and their moderating effects. These places become extremely cold during winter and extremely hot during summer, a situation described by the term **continentality.** Continentality was discovered when research on climate was pursued in the region. We now know that it occurs in other regions of the world as well. Central and eastern Canada and the interior United States have similar midlatitude continental interior climates. The warmest climates of the CIS are around the Black and Caspian Seas and in the Central Asian countries (Figure 4.7).

This region contains the world's greatest extent of interior continental conditions. The largest proportion of these lands is more than 500 km (320 mi.) from the ocean, and many parts are over 2,000 km (1,250 mi.) from the ocean. With greater distance inland, the extremes of summer and winter temperatures increase. The greatest contrast is in eastern Siberia, where large areas have January temperatures below –30°C (–22°F) and July averages of 12° to 16°C (56° to 60°F) (annual differences of 45°C, or 80°F). In western Russia, winter temperatures average –5° to –10°C (15° to 25°F), and summer temperatures are 15° to 20°C (59° to 68°F, a difference of 25°C, or 50°F). With greater distance northward, the climate gets colder and the winters longer; much of northernmost Russia lies north of the Arctic Circle. In the far east of Siberia, nearness to the Pacific Ocean results in more humid conditions, although the winters remain long and very cold as Arctic winds sweep out from the continent's interior.

Greater distance inland reduces precipitation. Few parts of the region have over 80 cm (32 in.) a year. Much has moderate precipitation (40–80 cm, 16–22 in.) falling mainly in summer but producing a long-lasting snow and ice cover during the winter. Parts of southern Central Asia are arid because of the high evaporation rates in warmer latitudes and distance from rain-bearing air masses. They rely on snowmelt from the mountains along their southern borders. Parts of the far east are subhumid. Low precipitation and low temperatures combine to make eastern Siberia inhospitable.

Southern Mountain Wall

The southern boundary of the region (Figure 4.8a) is marked by mountain systems. They were pushed up along the convergence of tectonic plates carrying the continent of Eurasia on the north and the land masses of Africa, Arabia, and India to the south (Figure 4.8b). The Caucasus Mountains, between the Black and Caspian Seas, are part of this line of mountain ranges that extends through the Elburz Mountains of northern Iran to the Tien Shan and the Pamir Mountains along the southern borders of the Central Asian countries. These mountain ranges rise to over 7,400 m (24,000 ft.) and include many peaks over 3,000 m (10,000 ft.). They are snowcapped and in spring provide considerable meltwater for streams flowing

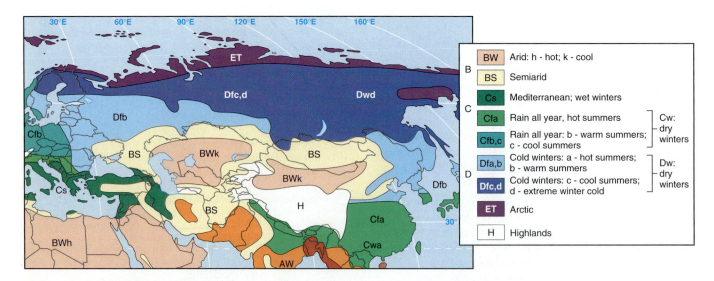

Figure 4.7   **Russia and Neighboring Countries: climates.** The dominant midlatitude continental interior climatic environments are characterized by harsh winter conditions, aridity in the south, and Arctic conditions in the north.

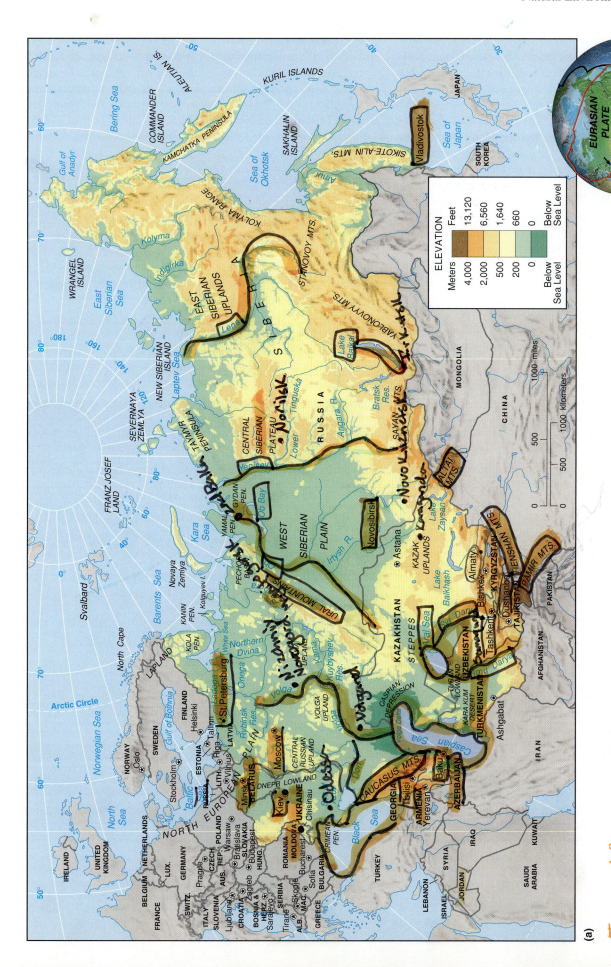

Figure 4.8 **Russia and Neighboring Countries: physical features.** (a) There is predominance of plains and plateaus in the relief, together with a southern rim of mountains that resulted from clashing tectonic plates. (b) The map of tectonic plates and their margins shows that the Eurasian plate clashes with other continental plates from the Caucasus to Central Asia; it clashes with the oceanic Pacific plate in the east.

through the dry areas of southern Russia and Central Asia. In the far east, the East Siberian Uplands and the volcanic peaks on the Kamchatka Peninsula parallel the Pacific coast and its tectonic plate boundary in areas that have few people.

The Earth movements not only raised high mountain ranges but also produced deep basins such as those filled by the Black Sea, Caspian Sea, and Lake Baikal. The Caspian Sea has drainage into it but does not flow into the main oceans. Its sea level and coastline fluctuate according to the inflow of river water. Soviet irrigation projects diverted much water from the rivers that flow into the Caspian Sea, causing the sea level to drop and the coastline to shrink. Between 1922 and 1977, the Caspian Sea level fell to 30 m (65 ft.) below the world ocean level, its lowest level in 500 years. The Gulf of Kara-Bugaz on the eastern shores acted as a natural evaporating pan until the Soviet government blocked the entrance (later reopening it). Since 1980, the Caspian waters rose by just over 2 m (7 ft.) and are likely to rise over 5 m (16 ft.) by 2010. Now, reclaimed lands, tourist centers, and oil installations that were located around the previously shrunken coastlines of the sea are threatened. The rising sea level covered 32,000 km^2 (12,360 mi.^{2}) of former land and now threatens over 100 villages and their farmland. The blocking of Kara-Bugaz accounted for only one-fifth of the rise; the rest is due to increased inflows from the Volga River or decreased evaporation from the Caspian Sea.

Plateaus, Plains, and Major River Valleys

Plains and low plateaus dominate most of the landscapes of this region. The North European Plain widens eastward from Poland into Belarus, Ukraine, and European Russia until it ends against the Ural Mountains (300–1,500 m; 1,000–5,000 ft.) that mark the line between Europe and Asia. Plains around the northern shores of the Black Sea and along the rivers leading to the Caspian and Aral Seas extend into the nearly level, vast West Siberian Plains. Farther east, the relief again becomes more hilly in the Central Siberian Plateau on ancient mineral-bearing rocks. The existence of such extensive areas of plains, interrupted by low hills, was a major factor in easing the invasions from eastern Asia and central Europe into Russia in medieval times.

Across these mainly low-lying landscapes flow some of the world's longest rivers. In the west, the Don River system flows into the Black Sea and the Volga River, draining most of the area between Moscow and the Urals, into the Caspian Sea. The Amu Darya and Syr Darya Rivers flow into the Aral Sea— an interior drainage system. The longest rivers of all, however, are those that rise in the southern mountains of Central Asia and eastern Siberia and flow northward to the Arctic Ocean— the Ob, Yenisey, and Lena. The Ob with the Irtysh is the longest of these and drains the largest watershed, but the Yenisey discharges the largest volume of water into the Arctic Ocean.

While the rivers in western Russia are used intensively for transportation, generating hydroelectricity, and as sources of industrial and domestic water, those farther east, with the

exception of the Angara River, are less exploited. This is partly because the areas through which they flow are sparsely settled but also because the rivers flow northward, are frozen for much of the year, and even in midsummer have frozen mouths along the Arctic Ocean that cause meltwater to back up and flood vast areas of wetland on either bank.

Desert, Grassland, Forest, and Tundra

The natural vegetation, much altered by human occupation, is heavily influenced by its latitudinal location and relationship to climatic factors of temperature and water availability. Hot desert in the south changes to steppe grassland, to deciduous and coniferous forest, and to tundra on the shores of the Arctic Ocean. All are distributed along lines of latitude, interrupted only by mountain ranges (Figures 4.9a and b).

- The deserts of the area east of the Caspian Sea contain few patches of grass and oasis vegetation and were once occupied by nomadic peoples. Trade routes across the region provided long-established links between Europe and

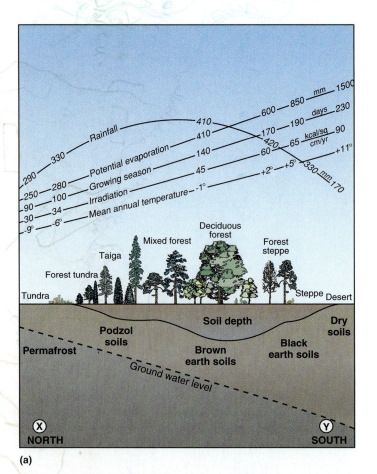

(a)

Figure 4.9   **Russia and Neighboring Countries: natural vegetation and soils.** (a) The section X-Y across the map illustrates the types of vegetation and soils in a north-south succession. (b) Map of natural vegetation and its links to soil types. Source: Data from Archie Brown, et al. (eds.), *The Cambridge Encyclopedia of Russia and the Former Soviet Union,* 1994, Cambridge University Press. Reprinted with permission of Cambridge University Press.

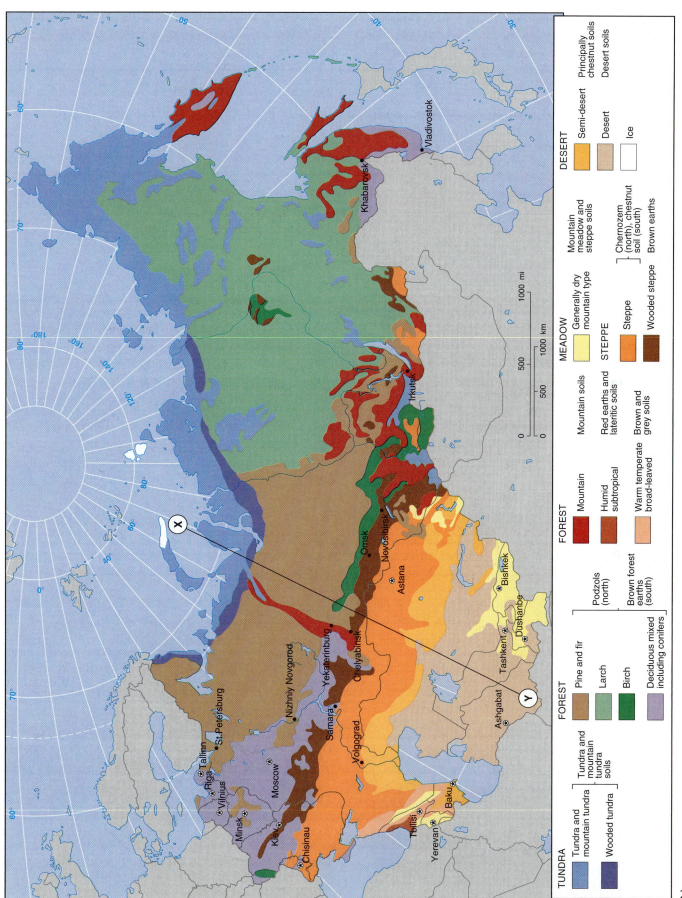

TUNDRA
- Tundra and mountain tundra
- Wooded tundra
- Tundra and mountain tundra soils

FOREST
- Pine and fir
- Larch
- Birch
- Deciduous mixed including conifers
- Podzols (north)
- Brown forest earths (south)

FOREST
- Mountain
- Humid subtropical
- Warm temperate broad-leaved
- Mountain soils
- Red earths and lateritic soils
- Brown and grey soils

MEADOW
- Generally dry mountain type
- Mountain meadow and steppe soils

STEPPE
- Steppe
- Wooded steppe
- Chernozem (north), chestnut soil (south)
- Brown earths

DESERT
- Semi-desert
- Desert
- Ice
- Principally chestnut soils
- Desert soils

1000 mi
1000 km
500
500
0
0

(b)

China through towns sited where water was available along the route.

- North of the desert, the **steppe grasslands** and the fertile **black earth soils (chernozems)** on which they grew are like the North American prairies. They extend from southern Poland eastward through Ukraine into Kazakhstan, becoming one of the world's major arable regions. In the Middle Ages, the steppe grasslands provided the grazing lands through which wave after wave of invaders moved westward into Europe.

- Farther north, where evaporation rates decrease, trees can grow, and a zone of wooded steppe gives way to deciduous forests, linked to fertile **brown earth soils.** The trees have been largely cut for the extension of farmland, except on slopes that are too steep for plowing.

- The deciduous forest gives way northward and eastward—where temperatures and precipitation decline—to forests dominated by hardy birch trees and evergreen pine, fir, and spruce. The underlying poor **podzol soils** have a gray sandy layer under a surface of decaying leaves and above a layer where iron minerals accumulate, and often become very hard, impeding drainage. This **northern coniferous forest,** or taiga, dominates vast areas of land from Moscow northward and across most of Siberia (Figure 4.10). Pine and fir trees in the west give way eastward to larch. The taiga forest occupies areas with very long winters and where glacial action deposited poor material for soils.

- Around the shores of the Arctic Ocean and extending southward into the plateaus farther east is an area where trees will not grow. It is covered by grasses and low-growing shrubs. Such **tundra** vegetation is underlain by permanently frozen ground, or **permafrost,** that also extends southward beneath the taiga forest. Much of the permafrost, which is 3,000 m (10,000 ft.) thick in parts of eastern Siberia, was formed during the glacial phases of the Pleistocene Ice Age. The continuous and broken permafrost areas cover most of central and eastern Siberia. Warmer conditions over the last 10,000 years melted the margins of this area, leaving irregular, small-scale subsidence hollows (i.e., surface depressions) that make farming difficult. Removal of the vegetation cover causes rapid

Figure 4.10   **Siberia.** The taiga. Photo: © Ronald Wixman.

melting of the permafrost and collapse of the underlying soil layers as the released water flows away.

Natural Resources

Russian and Soviet political, military, and economic expansions were driven and enabled by the use of plentiful natural resources in the metal-based manufacturing industries that dominated economies of wealthy countries in the late 1800s and early 1900s. Russia became a superpower at a time when political power was linked to steel and fuel output. The rocks of this region contain huge quantities of a wide variety of mineral resources. Metallic minerals include iron and gold as well as many minerals that are in short supply elsewhere in the world. There are also diamonds and other valuable products. Many of these resources occur in the ancient rocks of the Siberian plateaus or have been washed by erosion into the intervening areas of sedimentary rocks often in the northern, sparsely inhabited areas. Fuels include coal, oil, natural gas, and uranium. The Siberian lowland and the Caspian Basin contain substantial oil reserves, perhaps as much oil as the Persian Gulf area (see Chapter 8).

The countries of this region continue to be world leaders in producing fuels and other minerals. The first Russian Empire and Soviet manufacturing industries built up around the iron ore and coal deposits of Ukraine and southern Russia and extended eastward into Kazakhstan as new resources were discovered. Other sources of minerals and energy occur in and around the southern mountain ranges of the Caucasus and Central Asia. The flanks of the Caspian Sea are particularly rich in oil and natural gas. The main problem for the CIS countries is to build the requisite pipelines to a world oil port, preferably on the Mediterranean Sea (Figure 4.11). This poses many political, economic, and environmental problems at present. By the early 2000s, it appeared that a combination of routes through southern Russia and to the coast in Georgia would be used after the Russians had cleared the way by pacifying Chechnya, although alternative routes through Iran and Turkey remain under consideration. Exports of natural resources assumed great significance in the face of transitions in manufacturing and other economic activities.

The taiga across northern Russia forms the world's largest forest area (see Figure 4.10), covering 7.5 million km^2 (2.9 million $mi.^2$; compared to 5.5 million km^2, or 2.12 million $mi.^2$, of rain forest in Brazil and 4.5 million km^2, or 1.7 million $mi.^2$, of related coniferous forest in Canada). It provides a huge reserve of biodiversity and is important in world climate change, taking up one-third as much carbon dioxide as the tropical rain forest absorbs over a similar period. Owned by the state, this resource has been well regulated and little reduced by logging or fires. State planning located all the Soviet Union paper and pulp mills west of the Urals, making it costly to transport the wood from Siberia. Although large corporations began to cut more of these forests in the 1990s and potential tax revenues from such cutting make it an attractive prospect for Russia, the central bureaucracy continues to slow the process.

Despite its huge area, the extreme climates with extensive subhumid areas and the large areas with arid, podzol, or per-

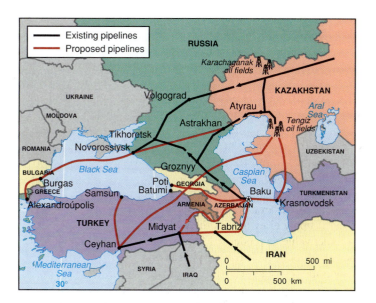

Figure 4.11 **Russia and Neighboring Countries: potential pipeline outlets for oil and gas.** Russia wants oil and gas exports from the countries of the Southern Caucasus and Central Asia to be channeled through its Black Sea port of Novorossiysk via the Groznyy (Chechnya) pipelines. The newly independent countries and their international oil company partners are examining the routes through Turkey to Ceyhan on the Mediterranean Sea or through northern Iran. The Turkish authorities wish to reduce the number of oil tankers passing through the straits at the Black Sea exit, leading to the prospect of a further pipeline linking ports in Bulgaria and northern Greece. How does this example illustrate the interdependence of countries in the world economic system and the importance of individual countries and their actions? Source: Reprinted by permission from *Financial Times*.

mafrost soils leave relatively small portions of the CIS suitable for commercial farming. These are mainly in the west and in parts of arid lands in the south where irrigation is possible.

Environmental Problems

The natural environments of Armenia, Azerbaijan, the Central Asian countries, and the far east of Siberia are affected by earthquake activity along their mountainous southern margins, where tectonic plates converge. The main natural hazards in other parts of this world region include the long winter freezes, land-use problems in areas of permafrost, and the flooding of major river valleys.

Human activities caused serious environmental damage in many parts of the region. For centuries, inhabitants of the region thought that the vast quantities of resources in their lands were inexhaustible. Later, Communism preached that nature should be transformed to serve human needs. Stalin expressed the need for rapid industrialization so that the Soviet Union could catch up to Western European levels of technology and material wealth. Stalin and his successors saw a particular need for industries that would build up the military. Rapid industrialization, as exemplified in the five-year plans, led to the massive exploitation and extraction of minerals and the construction of enormous factories, often built quickly with little thought to environmental protection.

The difficulties of developing the natural resources within this world region and the priorities of production over environmental protection led to increasing environmental degradation of the tundra, forest, and desert areas. The emphasis on heavy industry—especially chemicals and steel, the nuclear industry, the faulty storage of toxic rocket fuel and oil, the testing of weapons, and the excavation of metallic minerals laid waste to huge areas, many of which remain unreported and unreclaimed. A few examples illustrate the range of environmental problems facing Russia and its neighboring countries.

- One of the major legacies of the Soviet Union is frequent oil pipeline breaks and leakages. In the summer of 1994, one occurred on the Pechora River in northern Russia near Usinsk. An earthen dam built in 1988 to contain oil leaking from a pipe collapsed in heavy rain, emptying the oil into the river that drains into the Arctic Ocean. Russian officials played down the scale of the event, but some observers estimated it as being one of the world's largest oil spills. Earlier in 1994, a pipeline carrying oil to Europe burst some 400 km (260 mi.) south of Moscow, creating a 17-acre petroleum lake on frozen ground.

- The city of Norilsk in western Siberia has the most polluted environment in Russia. The region contains some of the world's richest deposits of mineral ore: 35 percent of the world's nickel, 10 percent of the world's copper, significant cobalt reserves, and 40 percent of the platinum group of metals. The Norilsk Metallurgical Combine releases millions of tons of pollutants into the atmosphere each year. In 1991, for example, the amount was nearly 2.4 million tons, with the sulfur dioxide content at 40 to 50 times legal limits. The stinking acidic haze blackens snow, kills trees for miles around, and poisons the river. Many locals, most of whom work in the factories, believe that the noxious fumes inoculate them against disease, but local physicians report high incidents of respiratory illnesses and shortened life expectancy, as low as 50 years.

  The city was built in the 1930s Stalinist era, based on the close occurrence of copper and nickel ores and coal deposits, which were worked at first by political prisoners (see "Human Rights" p. 150). It is the second largest city in the world north of the Arctic Circle but is not connected by road or rail to other parts of Russia except to the port of Dudinka 80 km (50 mi) to the northwest. Life for the 230,000 people of Norilsk is based in functional apartments, with freezing winters and summer mosquito plagues. Little money earned by the mining gets back to the community.

  After privatization in the 1990s, Vladimir Potanin, perhaps the richest man in Russia, used his great wealth and political influence to gain control of the Norilsk Metallurgical Combine for approximately $170 million. The company is now known as Norilsk Nickel Company. With over $1 billion in equipment upgrades and the implementation of capitalist business practices, the $260 million suffered in losses in 1996 turned into profits of $1.5 billion in 2000. Capitalist management, however, also meant job losses. The work force of over 125,000 in 1996 was reduced to about 70,000 by 2000, with more cuts planned. Outside the factories, little work exists in this isolated area of Russia. People want to leave the area but cannot because they lost their savings in the inflation that followed the 1991 breakup of the Soviet Union. To ease the painful economic restructuring, the Russian government recently received an $80 million loan from the World Bank to relocate about 15,000 people to other areas of Russia. However, to keep unemployment down in Norilsk, more people will have to be relocated.

- In addition to industrial pollution, nuclear contamination caused major environmental damage. The 1986 Chernobyl nuclear reactor explosion, which became a symbol for such disasters, affected most of the area immediately around it north of Kiev, but the fallout was worst to the north in Belarus (Figure 4.12). Nuclear dumps on islands, peninsulas, and in seas, especially along the Artic Ocean, have also released radioactive pollution. Well before Chernobyl, the "Kyshtym incident" in 1957 in the Urals exposed almost 500,000 people to harmful doses of nuclear radiation. Coupled with other industrial activities, nuclear contamination puts the Urals high on the list of the most polluted areas of the CIS.

- One of the greatest environmental disasters in the world occurred in the lands east of the Aral Sea (Figure 4.13a). Although arid, Central Asia naturally experiences a major internal circulation of water through the atmosphere and then across its surface in rivers. Water vapor evaporated from the large lakes or inland seas such as the Aral Sea condenses and is precipitated as snow on the high mountains to the east. As the snow melts in spring, the water fills such rivers as the Amu Darya and Syr Darya that flow into the Aral Sea. This water circulation provided the basis of local irrigation farming and small urban settlements through history, but the Soviet Union adopted ambitious plans to use the water on irrigated cotton farms inside and outside the main river basin. Supported by growth-oriented, unbending bureaucrats who often acted against the advice of government scientists, the project was taken too far and so much water was extracted that the rivers stopped flowing into the Aral Sea. The sea is now less than half its previous size (Figure 4.13b), all transportation on it has ceased, and the supply of moisture for the mountain snows has dried up. Glaciers in the mountains are in retreat, groundwater levels have fallen accompanied by ground subsidence, and dust storms now affect areas that had seldom experienced them.

- Sumgait, north of Baku in oil-rich Azerbaijan, is another of the world's most polluted places. Once one of the main suppliers of petrochemicals to the Soviet market, its 30 or so industrial plants are now a huge wasteland of broken concrete, rusting factories, and abandoned railroad tracks. The air smells of chlorine and sulfur. Most factories work at 20 to 30 percent capacity, insufficient to maintain corroding pipes. The town is another legacy of environmental costs resulting from Soviet central planning.

- The Black Sea, which is surrounded by six countries, is also under threat from environmental degradation. Concerns were voiced in late 1994 over a proposal by the Ukraine government to build a large oil terminal and refinery near Odessa to enable the country to reduce its dependence on expensive Russian supplies. The proposal threatened local holiday resort beaches and offshore locations that support up to 70 percent of the sea's fish.

  The practice of putting industry before environment occurs in the other five countries bordering the Black Sea (Russia, Georgia, Turkey, Bulgaria, and Romania) and in the 17 countries whose rivers drain into the sea. Fish stocks and plant and animal life in the Black Sea are killed by oil, toxic waste, ships' ballast water, and nutrients from fertilized fields. Only five of the 26 fish species caught in

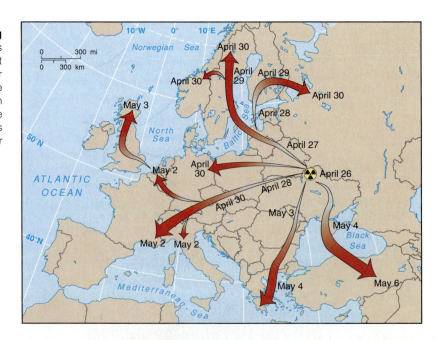

Figure 4.12 **Russia and Neighboring Countries: impact of a nuclear disaster.** The areas affected by the Chernobyl nuclear power plant explosion that released clouds of radioactive gases in 1986. In Ukraine, over 1.5 million people were affected, 10 percent of whom were severely radiated. All 600,000 members of the cleanup team suffered minor illnesses. Local death rates will increase, since the evacuation of the inhabitants after the explosion was delayed. Farther afield, the radiation was detected over Scandinavia and even in the U.K.

the 1960s were still found in 1992, when the total fish catch of the six rim countries dropped to 10 percent of their 1986 haul. Offshore drilling for oil and gas takes place near the Crimean Peninsula, further reducing the prospects of the region's tourist industry.

Figure 4.13 **Central Asia: Aral Sea environmental disaster.** (a) The Aral Sea, Kazakhstan, and its surroundings. The sea is a basin of inland drainage in an arid region. It is supplied by meltwater from the mountains to the southeast. (b) Space shuttle view of the Aral Sea in 1985, taken from the north; the section in the foreground was later cut off from the rest by the falling water levels. Photo: NASA.

(a)

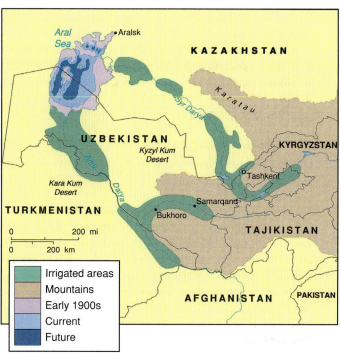

(b)

Test Your Understanding 4B

Summary Russia and Neighboring Countries have midlatitude continental interior climates ranging from arid through humid to Arctic conditions and marked by extreme seasonal fluctuations of temperature. Although mainly a world region of plains, low plateaus, and wide river valleys, its southern margins are marked by high mountain ranges along major junctions of tectonic plates.

The natural vegetation and soils follow the climatic pattern with a south-to-north sequence of desert, steppe grassland, deciduous forest, northern coniferous forest, and tundra.

Natural resources are of vital importance to CIS countries. The rocks of the region contain some the world's most significant reserves of fuels and metallic ores. Human activities, particularly in the Communist era, led to major environmental problems.

Questions to Think About

4B.1 How are the climate and natural vegetation of this region related to each other?
4B.2 How would you characterize the environmental problems of this world region?

Key Terms

continentality	podzol soils
steppe grasslands	northern coniferous forest (taiga)
black earth soils	tundra
brown earth soils	permafrost

Global Changes and Local Responses

In the decades following World War II, Communism thoroughly transformed the political and economic systems of the Soviet Union. With the desire to spread Communist ideology around the world, the Soviet Union led a struggle to overcome the capitalist countries of the West. Opposed to Communist ideology and threatened by the Soviet Union's activities, the United States countered Soviet actions. Both countries, as the world's two superpowers, initiated a "Cold War" as they attempted to destroy each other's systems. To do so, they supported governments, movements, and insurgencies that represented their respective ideologies in other countries, primarily in poorer countries. By supplying arms, they turned local and regional conflicts into international conflicts that sometimes resulted in tremendous death and destruction.

Both superpowers were determined to prove the superiority of their respective systems. In 1957, the Soviet space program launched *Sputnik,* the first artificial Earth satellite, and set a number of other records in outer space. The United States responded with its own space program, setting records such as landing the first man on the moon. At the same time, both countries competed in an arms race, building thousands of nuclear weapons. Though the competition never developed into a hot war in which American and Soviet forces fought one another, both sides were hostile toward each other and engaged in massive war production.

In its antagonism to capitalism, the Soviet Union and its allies closed their economies and societies off from the capitalist world. Refusing to get caught up in the inequities of capitalism, the Soviet Union sought to become economically self-sufficient. The closed Soviet society meant that individuals in the capitalist West knew less and less about life in the Soviet Union as time progressed. From the outside, the perspective of most Americans was that the Soviet Union was a powerful and hostile competitor, one that must be as strong as the United States because it financed Communist movements around the world and could build sophisticated nuclear weapons. Americans and others in the West did not see that the Soviet system had calcified and become extremely inefficient. Inequities also grew within the Soviet Union, especially in relation to ethnic groups. Moreover, support of Communist movements and the arms race were serious economic drains on the country.

Communism at an Economic Standstill

Stalin died in 1953, but his series of five-year plans helped the Soviet Union catch up to the West by the late 1950s, despite the serious setbacks of World War II. The West itself, however, continued to develop economically so that the Soviet Union was still behind despite certain scientific breakthroughs such as the space program. By the end of the 1950s, the Soviet Union had caught up only to the West of the 1920s, the time when Stalin began the five-year plans. Industry in the West now worked with new materials such as plastics and other synthetics and used new fuels such as petroleum and natural gas. The Soviet Union relied on materials such as metal and wood and almost exclusively used coal as a fuel. Realizing this change, Soviet planners implemented a shift in material and fuel use.

However, capitalist economies changed in other ways as well. Consumers in the West were now more demanding. For example, in the early 1920s, Americans were forced to buy black Model Ts because Ford would not produce any other color. In the late 1920s, lagging sales forced Ford to produce Model Ts in other colors. By the 1960s, Americans demanded many choices in the products they bought. American businesses responded, but it required industrial restructuring. Huge factories churning out massive quantities of the same item were replaced by smaller production facilities that were geographically spread out and capable of retooling to respond to constantly changing consumer demand. Production flexibility became the key to success. The Stalinist Soviet economy was so rigid that Soviet planners could not restructure and consumers had little voice to change the system. The military, poor use of natural resources and labor, and centralized planning prevented the Soviet economy from developing.

- *The military:* The Soviet government gave preferential treatment to the military because the defense of the country was the first priority. However, military expenditures produce little profit for society and scarcely raise the standard of living. The Soviet military relied on old-fashioned heavy industry and discouraged the development of smaller, more flexible, and more efficient means of production in the Soviet Union. It also demanded a far greater share of the Soviet Union's economic production than did the American military. Though Americans saw that the Soviet Union's military could match the U.S. military, most Americans did not realize that the Soviet military was a much greater burden to the Soviet economy than the American military was to the U.S. economy.

- *Natural resource use:* Under the Soviet government, the state owned everything, including natural resources. Factory managers paid only minimal costs for the fuels they used, giving them little incentive to use fuels efficiently or find cheaper fuels. Fuel producers, also state agencies, lacking competition or having to worry about paying for the costs of production, saw no need to develop new types of fuels. As fuel sources ran out in the western Soviet Union, new ones were tapped in the north and in Siberia. Exploiting these locations and transporting the fuels cost a lot of money. The Soviet system did not force the companies to carry the burden of increasing costs of production. Instead, the entire economy carried the burden. As the economy wore down, greater pressure was exerted to extract more fuels for export to generate dollar income to compensate for the degenerating economy.

- *Labor:* Communism guaranteed a job for everyone. To accomplish this goal of full employment, the government provided jobs that were not necessary or were even redundant. The government also set wages and kept them at low levels to encourage factory managers to hire many people. With low labor costs, managers also had no incentive to minimize labor costs by keeping their employees trained or installing machinery and robots. Workers, having a guaranteed job, had no incentive to produce quality products. Absenteeism was high, as was labor turnover. Over time, the economy produced less and less using more and more raw materials and labor.

- *Central planning:* Government bureaucrats, not supply and demand, determined what was produced and how much. The bureaucrats proved to be highly inefficient. Along with the military, they wasted natural resources and labor, helping to run the Soviet economy into the ground. As resources became scarcer, the various ministries of the government hoarded them, attempting to become self-sufficient because other agencies could not supply them. This practice created further redundancies and squandering of resources. Bureaucrats played it safe by locating new enterprises where they could best obtain supplies. In most cases, the locations were big cities. This led to excessive migration to cities, creating congestion, high living costs, and environmental problems. Ironically, though government bureaucrats contributed to the ruination of the Soviet economy, they became the unhappiest segment of society and greatly wanted change. As the privileged within the Soviet Union, they were displeased to see their standard of living drop below that of the working class in the capitalist countries.

Perestroika *and* Glasnost

Mikhail Gorbachev became the leader of the Soviet Union in 1985. He immediately set out to reform his country's economic and political system. In doing so, he highlighted two concepts: *perestroika* (economic restructuring) and **glasnost** (informational openness). For *perestroika,* Gorbachev believed that it was necessary to divorce economics from politics, allow more local control, and introduce free market practices. Such policies went against Communist ideals and established interests. The bureaucrats fought Gorbachev and his policies, creating great political turmoil.

As political battles raged, Gorbachev's economic policies ran opposite of what had been practiced since the 1920s. A crisis ensued as the economy attempted to reverse direction. Companies, having existed in a noncompetitive environment, now had to meet their own costs and find their own customers. No longer having the ability to waste resources, they had to cut costs. To generate income, they cut production and raised the prices of their goods, but they soon found that they could not afford to buy supplies from one another. Not only were goods very expensive but cash was also in short supply. For example, the Soviet Union previously sent machinery to Cuba and in return received sugar. With free-market reforms, Soviet factories wanted cash, not sugar. The same problem emerged between companies within the Soviet Union and the COMECON trading bloc of East Central Europe. Furthermore, the lack of a capitalist banking and financial system meant that cash could not flow easily. Poor transportation and communication systems only exacerbated the situation. Production declined. To cut costs, companies laid off workers, increasing unemployment. The Soviet economy continued to spiral downward with frequent labor strikes and rampant crime.

Gorbachev also stressed the concept of *glasnost,* the policy of providing government information to citizens. *Glasnost* was intended to have a positive effect on the country by empowering citizens with knowledge. Freedom of information allowed citizens to learn about corruption, government abuse, forced labor, and many of the other problems of the Soviet government. Non-Russians at last could vent their anger at the Soviet government for its Russification policies. In contrast, Russians complained that the Soviet government suppressed Russian culture, especially the Russian Orthodox Church, and distributed Russia's resources to the country's minorities. To avoid the wrath of citizens, many politicians echoed their anger, championed local and regional causes, and turned against the central government in Moscow. To curry favor with the citizenry, local and regional politicians helped to hoard resources for their people, further contributing to the country's economic crisis. Before long, government leaders helped their republics move toward independence. Even Boris Yeltsin, the elected leader of the Russian Republic in 1990, called for Russia's independence from the Soviet Union.

When Mikhail Gorbachev became the leader of the Soviet Union in 1985, he believed in Communism and tried to preserve the Soviet Union by implementing reforms. Ironically, in 1991, he became the Soviet Union's last leader as his policies led to the unraveling of the country. Gorbachev tried one last time to keep the country together with a new union treaty that relinquished much of Moscow's political power to the republics. On the eve of its signing on August 19, 1991, dissatisfied conservative Communists who wanted to restore the old system attempted to seize power in Moscow while Gorbachev was away. Massive public demonstrations supported Gorbachev. When Boris Yeltsin supported the demonstrators and protected them from a military crackdown, the coup failed. The coup attempt only accelerated the dissolution of the Soviet Union. Led by Lithuania, the Soviet republics all declared independence by December. Economically tied to one another, all the republics, except for Lithuania and the other two Baltic republics of Latvia and Estonia, formed the Commonwealth of Independent States, though Georgia waited a couple of years before joining. However, while Gorbachev's policies led to the rapid demise of his country, his reforms paved the way for further changes, without which the country would have probably disintegrated anyway, perhaps more violently.

After the demise of the Soviet Union in 1991, many people around the world wondered if Russia would remain a world power (see "Point-Counterpoint: Russia: Still a World Power?" p. 142) because it did not have the economic strength of a world power. By 2000, Russia produced only 0.08 percent of the world GDP. The GDPs per capita within Russia and the other CIS countries were far below those of capitalist countries such as those in Europe and the United States (Figure 4.14). Almost 14 percent of Russia's people officially live in poverty, earning less than $31 per month. The ownership of consumer

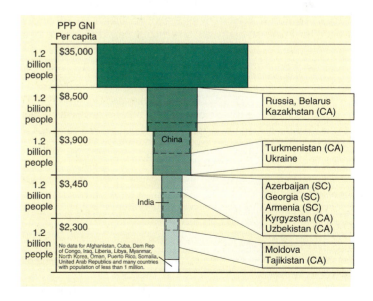

Figure 4.14 **Russia and Neighboring Countries: national incomes compared.** The countries are listed in the order of their PPP GNI per capita. CA=Central Asia, SC=Southern Caucasus. Source: Data (for 2000) *World Development Indicators,* World Bank, 2002.

RUSSIA: STILL A WORLD POWER?

The fall of Communism in East Central Europe and the Soviet Union in 1991 brought about profound change for the world. For most Americans, the demise of Soviet Communism marked the end of a system that threatened the American way of life with both its ideology and nuclear weapons. For many of the non-Russian peoples of the Soviet Union and East Central Europe, the fall of Communism meant the end of Soviet Russian domination. For Russians, the end of Communism was much more complicated. On the positive side, it promised political and economic reform, with greater freedom of expression and a better standard of living. On the negative side, it coincided with the loss of much of their country, the open expression of great anti-Russian feelings from people within their region, and the questioning and potential end of Russia as a world power. These experiences were a tremendous blow to the prestige of the Russian people.

When the 14 non-Russian republics declared their independence from the Soviet Union in 1991, it seemed that the peoples of these republics simply were expressing their right of self-determination. The Russians, however, believed that the Soviet republics were Russian lands regardless of who lived in them. They saw the Soviet Union as rightfully theirs because it was created from the Russian Empire, lands that they struggled for and acquired over the course of centuries. Though these Soviet republics declared independence, Russians do not see these 14 new countries as foreign. Instead, they see them as part of their "Near Abroad" and feel that they have an exclusive voice in both the internal and international relations of these countries.

The unwillingness of Russians to let go of the non-Russian republics is complicated by another factor—the lack of clear boundaries for Russia. Though the Soviet republics, including the Russian Republic, had precise boundaries, their existence was rather artificial for most Russians because Russians controlled all the lands of the Soviet Union, as they had when the country was imperial Russia. For Russians, Russia is much more than the current Russian Republic, but the real boundaries between Russia and those of other lands are not clear in the minds of many Russians. In contrast to the British, the French, and a number of the other European colonizers, who had a clear distinction in their minds between their homelands and their colonies, the Russian colonies were not overseas but adjacent to Russian lands. Over the centuries, the Russians continually expanded the boundaries of Russia to incorporate their newly acquired lands, further making these lands part of Russia in the minds of many Russians. Ukrainian and Belarussian independence particularly is difficult for Russians to accept because much of the land and peoples of these republics were part of *Rus*—the original Russian heartland and people. When Nikita Krushchev, the leader of the Soviet Union, transferred the Crimean Peninsula from the Russian Republic to the Ukrainian Republic in 1954, few Russians objected because the territory remained in the Soviet Union. After Ukraine declared independence in 1991, the Russian government claimed its right to control the Crimean Peninsula, suggesting that the Russians would have never given the Crimean Peninsula to Ukraine if they had known that the Ukrainians would eventually reject Russian rule and keep the Crimean Peninsula.

The spatial distribution of Russians is another reason that Russians have difficulty accepting the dissolution of the Soviet Union. Twenty-five million Russians, or 17 percent of the Russian population of the Soviet Union, live in the "Near Abroad" (Box Figure 1). High concentrations live in Estonia, Latvia, northern and eastern Ukraine, and northern Kazakhstan. In many locales in the Crimean Peninsula, Russians constitute more than 90 percent of the population. The adjustment from majority to minority status has been difficult for Russians, both in and out of Russia. Many of the new countries see their long-established Russian minorities as threats and have insisted that these Russians assimilate (e.g., learn the native language and way of life) or go back to Russia. By 2001, it was estimated that as many as 5 million Russians emigrated from the independent republics to the Russian Federation (Box Figure 2). Many Russians, however, feel that Russians have the right to stay in the former territories of their empire and maintain their Russian culture.

Not only were the countries of the "Near Abroad" part of Russia for centuries and are still the home to millions of Russians, they also are crucial to Russia's status as a world power. The Baltic republics housed key military installations. The Crimean Peninsula is a major Russian naval facility. It is no wonder that the Russian government tried to hold onto the Crimean Peninsula and its naval fleet after Ukraine declared independence. It would take decades and billions of dollars of investment for Russia to re-create a comparable naval facility along the Russian coast of the Black Sea. In the meantime, Russia could lose great influence in world affairs. Kazakhstan was the center of the Soviet space program. The Russians see the Soviet space program as their accomplishment, one that only Americans have matched. Control over the oil in the Caucasus and Central Asia is also key to Russia's role as a world power.

Coping with the loss of the "Near Abroad" is compounded by struggles by non-Russians within the Russian Federation for independence. Chechnya is the most vexing of them all. As the Soviet Union disintegrated in 1991, the Chechens moved toward independence from the Russian Federation and declared it in 1994. The Russian military responded by supporting a Chechen rebel group that sought to overthrow the Chechen government and keep Chechnya within Russia. The military campaign was not as quick as the Russian government had hoped. Military operations have continued, though intervals of negotiations bring relative subsidence of hostilities. The campaign, however, has been brutal. By early 1996, an estimated 40,000 to 100,000 people, mostly civilians, had been killed in a republic of just 1 million. Street-to-street fighting and bombings by the Russian air force have made Chechnya's capital, Groznyy, and other Chechen cities uninhabitable. Accused of gross human rights violations, Russia's international reputation has suffered from the war in Chechnya. However, Russians fear that if they grant Chechen independence, then Russia's other minorities will also move toward independence. In addition, Chechnya, especially Groznyy, is a major transit point for oil leaving the Caspian Sea region (see Figure 4.11). Keeping the oil flowing through this pipeline is crucial for Russia. If proposed pipelines through Georgia and Iran succeed in taking business away from the Russian pipeline, Russia will lose billions in income, along with ability to influence international affairs.

Russia's ability to influence world affairs diminished in other ways in the 1990s. Soldiers in Russia's military are vastly underpaid, unprepared, and demoralized from their experience in Chechnya, which followed the earlier defeat in Afghanistan in the 1980s. Though Russia is a nuclear power, the United States, viewing itself as the sole victor of the Cold War, acted unilaterally in international affairs. For example, the United States resolved the conflicts in the former Yugoslavia in its own way with little consultation from the Russians, who have cultural and historic links with the Yugoslavs. At the same time, the North Atlantic Treaty Organization (NATO) prepared to expand into East Central Europe and include countries formerly dominated by the Soviet Union. For members and potential members of NATO, the expansion was a defensive move. For the Russians, NATO's moves were provocative, particularly since they were taken without Russian consent. NATO was a Cold War institution created to defend Western Europe from Soviet Communism. Because Soviet Communism no longer existed and Russia was very weak, NATO expansion suggested that the organization was not defensive but offensive. It played on Russian fears that NATO was nothing more than another force, not unlike Napoléon or the German armies of the two World Wars, to menace Russia.

By the late 1990s, Russians felt ostracized by the United States and Western European countries. Closer relations with China led to

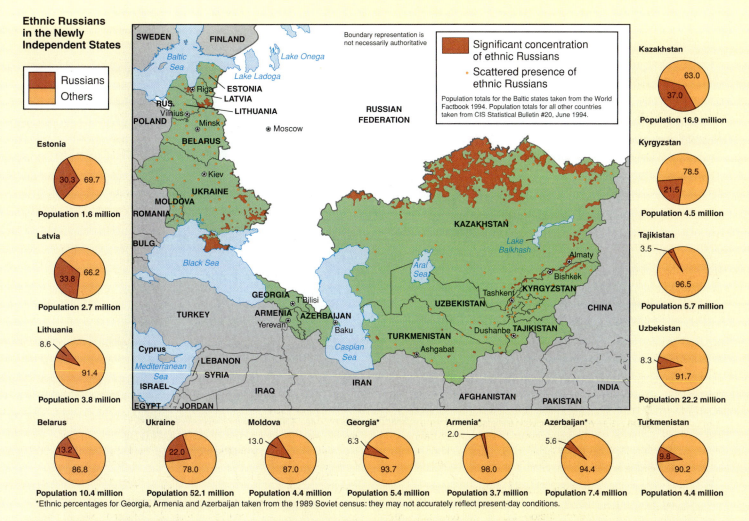

Ethnic Russians in the Newly Independent States

Russians
Others

Estonia
30.3 / 69.7
Population 1.6 million

Latvia
33.8 / 66.2
Population 2.7 million

Lithuania
8.6 / 91.4
Population 3.8 million

Boundary representation is not necessarily authoritative

■ Significant concentration of ethnic Russians
• Scattered presence of ethnic Russians

Population totals for the Baltic states taken from the World Factbook 1994. Population totals for all other countries taken from CIS Statistical Bulletin #20, June 1994.

Kazakhstan
63.0 / 37.0
Population 16.9 million

Kyrgyzstan
78.5 / 21.5
Population 4.5 million

Tajikistan
3.5 / 96.5
Population 5.7 million

Uzbekistan
8.3 / 91.7
Population 22.2 million

Belarus
13.2 / 86.8
Population 10.4 million

Ukraine
22.0 / 78.0
Population 52.1 million

Moldova
13.0 / 87.0
Population 4.4 million

Georgia*
6.3 / 93.7
Population 5.4 million

Armenia*
2.0 / 98.0
Population 3.7 million

Azerbaijan*
5.6 / 94.4
Population 7.4 million

Turkmenistan
9.8 / 90.2
Population 4.4 million

*Ethnic percentages for Georgia, Armenia and Azerbaijan taken from the 1989 Soviet census: they may not accurately reflect present-day conditions.

Box Figure 1 **Ethnic Russians as minorities in the Commonwealth of Independent States and the Baltic countries.** In which countries are Russians the largest minorities? Do Russians outside of Russia enhance or hinder the political power of Russia in the region?

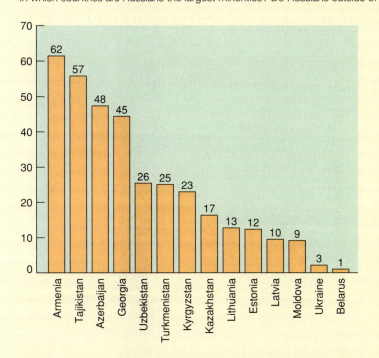

Box Figure 2 **Migration of ethnic Russians from the former Soviet republics from 1989 to 1998.** Each number represents the percentage of ethnic Russians who have emigrated from that country's total population of ethnic Russians.

the signing of a friendship treaty in 2001. The treaty is a significant agreement because it sets aside decades of tension caused mostly by border disputes. The treaty allows for cross-border trade. For example, Russia will send oil and military supplies to China. Both China and Russia agree to crack down on Islamic fundamentalists who straddle their borders and threaten the territorial integrity of their countries. Easier transborder crossings may mean that more Chinese will settle in sparsely inhabited Siberia and someday make territorial claims. Russian high-tech military weapons sold to China may one day be pointed back at Russia if relations between the two countries sour again. However, a Russo-Chinese alliance is a signal to the rest of the world that Russia is still a world power.

Vladimir Putin, the leader of Russia, and the events of September 11, 2001, have changed Russia's relationship with much of the world. Putin has not complained very much about American international diplomacy, even though it offends long-held Russian interests. Instead, he has sought to be more conciliatory to the United States in hopes of gaining U.S. cooperation and increasing Russia's international influence. A warming of relations between the United States and Russia began and then increased after the events of September 11, 2001. Islamic fundamentalism in Afghanistan was a concern that led to the Soviet invasion of that country in the 1980s. The U.S. attempt to suppress Islamic fundamentalism parallels Russian concerns. In fact, the U.S. criticism of the Russian military campaigns in Chechnya turned to sympathy. This occurred when al-Qaeda fighters were taking refuge in Chechnya and aiding the Chechen cause. Russian leaders were able to describe the war in Chechnya as another front on the war on terrorism. Russia was also able to press the Georgian government to do something about Chechen rebels hiding in the Pankisi Gorge. Russians are

(Continued)

sensitive about foreign activity in a "Near Abroad" country such as Georgia, but with Russian-American cooperation in the war on terrorism, Russians voiced little concern about the United States dispatching elite military forces to Georgia to take care of a problem that vexed Russia. The situation is delicate because U.S. military presence in the Caucasus and Central Asia, as it suddenly emerged after September 11, 2001, may play on old Russian fears if the United States does not handle its affairs in the region with Russia carefully.

Since 2000, Russia has become a more full-fledged member of the informal organization representing the world's most wealthy countries, once known as the Group of Seven (G7) but, with Russia,

increasingly known as the G8. In 2002, NATO agreed to give Russia a voice in NATO affairs, though Russia is not a full-fledged member. Russia also has offered to supply the United States and its allies with oil. Showing a preference to buy oil from Russia rather than from many of the countries of Southwestern Asia, such as Iraq and Iran, the United States has become more supportive of the Russian oil pipeline from the Caspian Sea. With an improving economy, more effective political leadership, and a changed climate of international affairs, the Russian Federation has the opportunity to exert itself as a world power.

Debate whether Russia is still a world power:

POINT	COUNTERPOINT
Russia lost possession of 14 republics (now independent countries) and with them sizable populations and resources.	Russia's heartland remains together, and Russia is still the largest country in the world, with a sizable population and considerable resources.
Russia's international power is undermined by the presence of 25 million Russians who now find themselves as ethnic minorities in other countries. Russia must be careful not to offend these other countries and thereby endanger the Russians who live in them. Russia may have to make economic concessions to these countries to protect Russian minorities in them.	Russians living as ethnic minorities in other countries increase Russia's political influence in those countries because they can vote and hold political office. Russia can also pressure those countries on behalf of Russians living in them. The Russian minorities also strengthen Russia's trading links with these countries.
The Russian government has been ineffective in controlling independence movements in its own republics. It has been condemned for human rights violations in Chechnya. Its military is underpaid, unprepared, and demoralized. Russia was unable to control the conflict in the former Yugoslavia and the expansion of NATO.	Russia gained sympathy from the United States for its "war on terrorism" in Chechnya. Russian soldiers are very patriotic, as in the past, and Russia is a nuclear power. With peacekeeping troops in the former Yugoslavia, Russia influences events in that area of the world. Russia also now has a voice in NATO's affairs.
With rampant corruption, Russia's economy is weak and has floundered since the end of Communism in 1991. Russia is unable to capitalize on selling its vast resources on the world market. Its control of Caspian Sea oil is questionable. Russia has little influence in global economic organizations.	Russia's economy is steadily improving, and the government is fighting corruption. Russia has signed agreements to supply oil to other countries and is able to control the flow of the oil from the Caspian Sea. Russia has joined the Group of Seven (G7), making the organization the Group of Eight (G8).

goods reflects this modest income (Figure 4.15). TV set ownership is high, as television provided mass communication from the government to the people during the Soviet era.

Global City-Regions

Russia and its neighboring countries try to better their economies by engaging in the global economy, but their attempts are inhibited by the small role that their cities play as global cities. Of the global cities identified in Figures 2.16 and 2.17, Russia and its neighboring countries have only one—Moscow—and it does not rank high on the list. Almaty (Kazakhstan), Kiev (Ukraine), St. Petersburg (Russia), and Tashkent (Uzbekistan) offer some global city services, but this world region has very few global cities for an area of the world that has such a large number of urban areas. The lack of global connectedness is the direct result of the Communist legacy. Globalization is linked closely to modern capitalism, and Communism precluded the development of the types of accounting, advertising, banking, and law that are so characteristic of globalization. Refusing to engage in global capitalist competition, the Soviet government likewise focused inward, on self-sufficiency, and only engaged in trade with fellow Communist countries. Consequently, earlier global cities of this world region lost many of their global connections. Since the end of Communism in 1991, the capital cities, especially Moscow

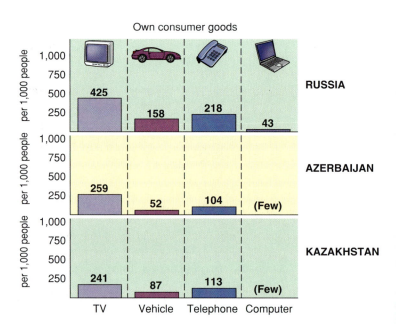

Figure 4.15 **Russia and Neighboring Countries: ownership of consumer goods.** What is the significance of relatively high television ownership? Source: Data (for 1996) from *World Development Indicators,* World Bank, 2002.

The Caucasus Mountains create uneven distributions of people in Georgia, Armenia, and Azerbaijan, with few people in the mountains and most in the plains. The distribution of population in Central Asia is marked by contrasts between areas of few people in the arid and mountainous parts and areas of higher densities in the irrigated lowlands (see Figure 4.16). The main concentrations of people are all separated from other world economic centers by high mountains and the difficulties of having to travel or send goods through Russia.

Test Your Understanding 4C

Summary For a number of years after World War II, the economy of the Soviet Union prospered, but by the late 1960s, it began to languish. Mikhail Gorbachev attempted reforms in the late 1980s with *perestroika* and *glasnost*. The reforms were partially successful, but the Soviet Union disintegrated. *Glasnost* revealed many human rights abuses during Soviet times.

The inward orientation of Soviet Communism kept the world region from developing many global cities. Moscow ranks highest, and it and other cities are likely to rise in the global city rankings as Russia and its neighbors establish new political and economic links with the rest of the world.

Questions to Think About

4C.1 Why did the Soviet economy stall by the 1970s?

4C.2 Do you think that Mikhail Gorbachev's policies destroyed the Soviet Union, or was the country destined to disintegrate anyway? Explain.

Key terms

perestroika   *glasnost*

but also cities such as St. Petersburg, are developing accounting, advertising, banking, and law firms and have received the greatest amounts of international investment. In time, they will likely move up in the global city rankings.

 # Subregions

Russia and its neighboring countries can be divided into subregions that share many similar characteristics:

- *The Slavic countries:* The Russian Federation, Ukraine, and Belarus, including neighboring Moldova
- *The Southern Caucasus:* Georgia, Armenia, and Azerbaijan
- *Central Asia:* Kazakhstan, Turkmenistan, Kyrgyzstan, Tajikistan, and Uzbekistan

Population Distribution and Patterns

In the Russian Federation, the greatest concentration of people is in western Russia. Higher population densities continue eastward along the Trans-Siberian Railway to the southern end of Lake Baikal and Vladivostok on the Pacific coast. The extensive areas of mountains and desert in the south and of permafrost-ridden lands in the north and east contain few people (Figure 4.16). Ukraine, Belarus, and Moldova cluster around the most densely populated western Russia. They have moderate population densities that are evenly spread with greater concentrations of people around industrial cities.

 # The Slavic Countries

The Slavic countries of this subregion are the Russian Federation, Ukraine, and Belarus. Moldova also is included because many Slavs live there, and it is closely tied to the Slavic countries (Figure 4.17). The Russian Federation is by far the largest in land area of any country in the subregion and in the world (Figure 4.18). It has nearly twice as much land area as Canada, the United States, or China. In 2001, it had 77 percent of the CIS area and 52 percent of the CIS population.

Though the Russian Federation and the other countries of the subregion are experiencing economic hardship as their economies move from Communism to capitalism, the Russian Federation still exerts considerable power. The Russian Federation remains a nuclear power and continues to hold one of the five permanent seats of the UN Security Council, the most powerful organ of the United Nations. Because it contains substantial portions of the world's natural resources, the Russian Federation has considerable economic potential. The leaders of the Russian Federation have pressed to have their country added to the Group of Seven (G7), an informal organization representing the world's most wealthy countries. Though no official rules for membership exist, Russian representatives have been regularly invited to meetings, leading many to refer to the organization as the Group of Eight (G8).

Countries

The Slavic countries seen on the map today became independent only in 1991 with the boundaries they had as Soviet republics. This situation is also true for the Russian Federation, though the Russians controlled the Soviet Union and its predecessor, the Russian Empire. Prior to 1991, Ukraine was only independent for a brief period after World War I until it became part of the Soviet Union in 1922. Before that, it was part of the Russian Empire. Belarus was never independent before 1991 and was usually either part of the Russian Empire or the Polish-Lithuanian Kingdom. Of the former Soviet republics, Belarus is the most closely tied to the Russian Federation. Since 1991, Belarussians have considered creating a Russian-Belarussian Federation.

Moldova, too, was never independent and only has been a distinct territory for fewer than 200 years. It was part of the Romanian province of Moldavia until the Russian Empire annexed it in the 1800s and named it Bessarabia. Romania

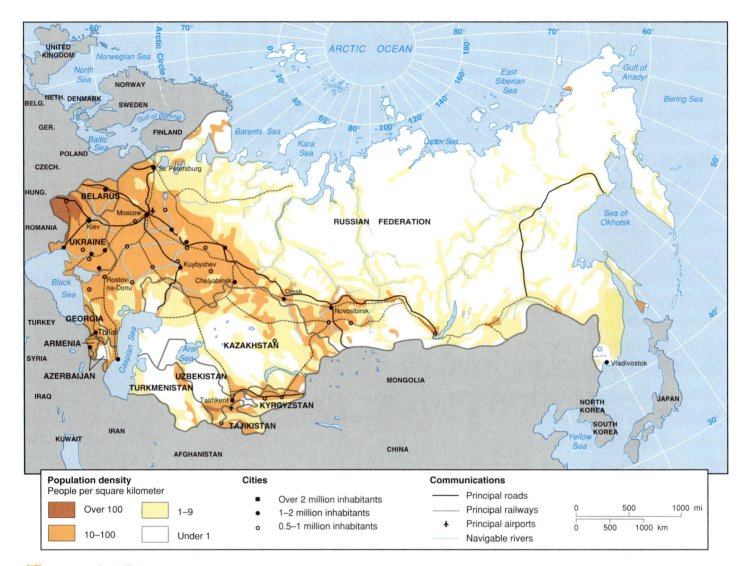

Figure 4.16   **Russia and Neighboring Countries: distribution of population.** For Russia, comment on the location of the highest and lowest densities of population. How do these reflect climatic conditions, initial historic superiority, the line of the Trans-Siberian Railway, and other factors? Source: Data from *New Oxford School Atlas,* p. 75. Oxford University Press, UK, 1990.

annexed the territory after World War I, but the Soviet Union annexed it again after World War II. Many Romanians and Moldovans hoped to unite their two countries after the dissolution of the Soviet Union in 1991, but the Russian military stationed in the country prevented this. Worried about a union of Moldova and Romania, Russians and Ukrainians living in Moldova declared their own republic in the Transnistria region. Other ethnic minorities have made similar proclamations. The government has not been able to suppress these independence movements completely.

People

In 2001, the populations of the four countries in this subregion totaled 207 million. One of the greatest concerns is population decline. Overall, this subregion will decline to 194 million by 2025. Russia, Ukraine, and Belarus had growth rates of –0.5 to –0.7 percent in 2001. Fertility rates were around 1.2, far below

the fertility rate of 2.1 needed to maintain a population at its current number. The age-sex diagram for Russia (Figure 4.19) shows an uneven pattern of age groups as a consequence of World War II and a baby boom from 1950 to 1970, when large families were encouraged by the state. Women dominate the over-60 age groups to a greater extent than in other parts of the world as a result of male deaths in World War II and Stalin's persecutions. In 2001, the birth rates were 8 to 9 per 1,000 of the population and falling, while the death rates were 14 to 15 per 1,000 and rising (Figure 4.20), yielding one of the fastest rates of natural decrease in the world that year.

For the Russian Federation, such figures hide variations within this huge country. In some parts, deaths outnumber births by two to one. The most significant geographic variation is between the western regions of the Russian Federation and the more recently settled regions of the north and Siberia. In western Russia, birth rates are low, death rates high, and the population is aging quickly. Yet the total numbers fall less rapidly

Figure 4.17   **Slavic Countries: major cities and physical features.** Note the distribution of the major cities and their concentration in the western part of the subregion.

Figure 4.18   The Slavic Countries: key data on countries.

Country	Capital City	Land Area (km²) Total	Population (millions) Mid-2001 Total	2025 Est.	GNI 1999 (US $ million) Total	GNI PPP 1999 Per Capita	Percent Urban 2001	Human Development Index Rank of 174 Countries
Russian Federation	Moscow	17,075,400	144.4	136.9	328,995	6,990	73	71
Belarus	Minsk	207,600	10.0	9.4	26,300	6,880	70	60
Moldova	Chisinau	33,700	4.3	4.5	1,500	2,100	46	104
Ukraine	Kiev	603,700	49.1	45.1	6,800	3,360	68	91

Country	Ethnic Groups (percent)	Languages	Religions
Russian Federation	Russian 81%, Tatar 4%, Ukraine 3%	Russian, Tatar, etc.	Russian Orthodox, Muslim
Belarus	Belarussian 78%, Russian 13%, Polish 4%	Belarussian, Russian	Eastern Orthodox
Moldova	Moldovan/Romanian 65%, Russian 13%, Ukraine 14%	Romanian, Russian	Eastern Orthodox
Ukraine	Ukrainian 73%, Russian 22%	Ukrainian, Russian	Ukrainian Orthodox, Roman Catholic

Source: Data from *Population Reference Bureau 2001 Data Sheet*; *World Development Indicators,* World Bank, 2001; *Human Development Report,* United Nations, 2001; Microsoft Encarta (ethnic group, language, religion).

because other Russians are immigrating to western Russia from the former Soviet republics and from the north and Siberia. Emigration is particularly strong from the areas where indigenous peoples have gained control over local government.

Economic decline has been the major cause of population decline in the Slavic countries. Lack of government subsidies for industries in the north and Siberia is one factor. Lack of similar subsidies in the older urban areas, especially the industrial cities, and the inability of industries to adapt to capitalism led to high unemployment. Many who are still working have not been paid for months or years. Abortion rates are high, a legacy of Soviet times when abortion became a common means of birth control. Health care has also suffered greatly with economic decline. Hospitals have frequently run out of basic supplies such as drugs and anesthetics. Economic hardship also has increased the incidence of alcoholism, particularly among men. For many hospitals, 90 percent of emergency cases have been alcohol-related. The entire situation has given women little incentive to have children.

Moldova is also transforming its economic system from Communism to capitalism, yet the population of the country is not declining at the alarming rate of the neighboring Slavic countries. On the one hand, unlike the Slavic countries, Moldova is overwhelmingly agricultural, not industrial. The traditional large farming family has not been greatly disrupted by the transition. On the other hand, growth rates are not nearly so high as they once were. Thus, the transition is having some effect on the population of Moldova.

Urban Population

The highest rates of urbanization of the subregion in 2001 were in areas that lack good agricultural land or emphasize urban-industrial development. For the countries as a whole, the percentage of the population living in urban areas is 73 percent in the Russian Federation, 68 percent in Ukraine, 70 percent in Belarus, and 46 percent in Moldova.

The Russian Federation has the greatest number of cities and the largest cities of the subregion. Moscow (9.3 million in 2000) and St. Petersburg (5 million) in the Russian heartland are by far the largest. The next largest cities, including Nizhniy Novgorod, Novosibirsk, Yekaterinburg, Samara, Omsk, Chelyabinsk, Volgograd, Ufa, Rostov-na-Donu, Perm, and Kazan, all had between 1 million and 1.5 million inhabitants in 1996. Major cities over 1 million in the other countries are Minsk (Belarus, 1.8 million), Kiev (2.7 million), Kharkov (2.7 million), Dnepropetrovsk (1.1 million), Donetsk (1.1 million), and Odessa (Ukraine, 1.0 million). Moldova had no large cities.

Urban Landscapes

In 1917, at the time of the Bolshevik Revolution, 17 percent of Russians lived in mainly small, provincial towns and cities. These prerevolutionary Russian towns were often more like overgrown villages, except for those such as Moscow and St. Petersburg that had grand designs of buildings and roads instigated by the czars (Figure 4.21).

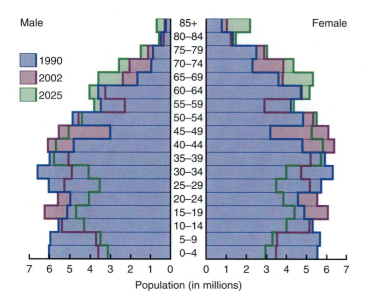

Figure 4.19   **Age-sex diagram of Russia.** Account for the variety of irregular shapes. Do they correspond to historical events experienced by Russia? Source: U.S. Census Bureau. International Data Bank.

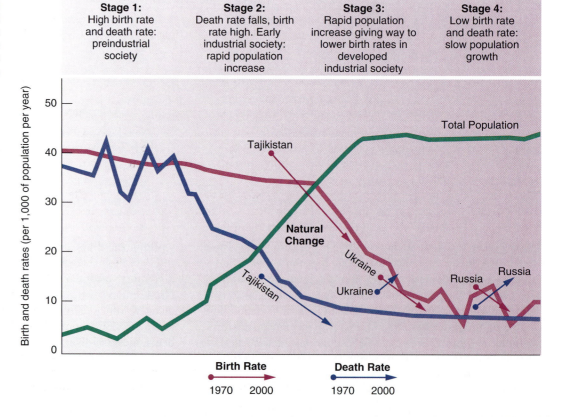

Figure 4.20   **Russia and Neighboring Countries: demographic transition.** Some death rates in the mid-1990s were higher than the birth rates; some countries resembled the world's poor countries.

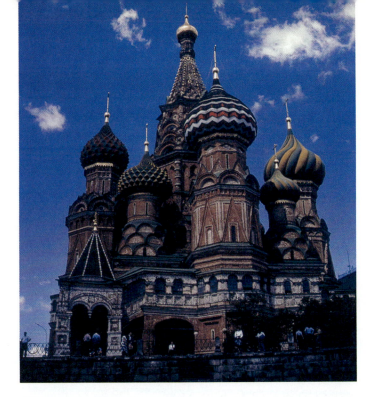

Figure 4.21 **Russia.** The Cathedral of the Intercession, commonly known as St. Basil's Cathedral, is located on Red Square, the center of Moscow. Ivan IV (The Terrible) had the cathedral built to commemorate Russia's victory over the Mongols. Originally intended as eight, smaller clustered cathedrals, each of which to represent Russia's eight victorious battles over the Mongols, the result became a single cathedral with eight domes and chapels signifying the victories. Photo: © Ronald Wixman.

Figure 4.22 **Russia, Moscow.** "Soviet Man and Woman" in the former Park of the Achievement of Communism. The man is holding a hammer, representing industry, and the woman is holding a sickle, symbol of agriculture. In typical Communist fashion, this monument is enormous in size, symbolizing the power and strength of Communism. The car and lampposts in the picture indicate the monument's large size. Photo: © Ronald Wixman.

Joseph Stalin's five-year plans, which emphasized intensive centralization and rapid industrialization, led to fast city growth by the late 1920s, rising to 48 percent of the population by 1959 and more than 70 percent in the 1990s. Urbanization increased as existing industrial centers expanded and new specialist resource centers emerged, often in new towns in remote regions. Stalinist urbanization produced distinctive but standardized cities throughout the Soviet Union. Rapid building was often little coordinated and of poor quality. Housing for the workers consisted of numerous high-rise, concrete buildings with few shopping opportunities in their vicinity because Soviet planners did not foresee the need for much shopping.

Visual symbolism was an important part of establishing the new socialist society. The old czarist memorials, churches, mosques, and even street and city names were replaced with symbols of Communist Party significance. For example, St. Petersburg became Leningrad and Volgograd became Stalingrad. In the 1930s, half the churches in Moscow were demolished. Boulevards and squares were built that could be used for military displays. New statues of Lenin and Stalin, war memorials, and party slogans became a feature of Stalinist cities (Figure 4.22).

One of the major trends to emerge since the breakup of the Soviet Union in 1991 is the building of suburbs, particularly around Moscow. The congestion and pollution of inner-city streets by the growing use of cars and trucks, together with the increase of urban crime, caused richer families to move out. The mayor of Moscow contributed to the movement by ejecting families from apartment blocks designated for renovation and commercial sale. Financial institutions, including banks, fund the new suburban residential developments and assess the customers' ability to pay. Most new house owners are former members of the Communist establishment who transformed their political power into cash. By such means, the market economy began to transform Russian landscapes.

Secret Cities

In addition to the cities marked on official Soviet maps, a number of "secret cities" were revealed after 1991. Some have over 100,000 people. These cities were linked to the nuclear industry, biologic warfare research, or missile and weapon design (Figure 4.23). Built in the Stalinist monolithic style, such cities developed as a combination of scientific research institutes and labor camps; each is surrounded by a zone of cleared land and electrified fences up to 30 km (20 mi.) wide. The cities grew out of the early days of secret nuclear bomb making. Scientists who were enticed to work in them had better living conditions than other citizens. Many such cities were duplicated at great cost in case the United States struck first in a nuclear war. In some, the factories, reactors, and even housing were built underground.

Assembling scientists secretly in specialized centers was successful, turning the Soviet Union into a military superpower, albeit at the expense of consumer needs, long-term economic growth, and environmental damage. There is a concern about the level of environmental damage caused by the dumping of wastes and unreported nuclear accidents on a scale greater than the Chernobyl accident. Leaks from the biological weapons centers gave rise to anthrax outbreaks near Yekaterinburg. The liquid fuel used in Soviet rockets is supertoxic and highly carcinogenic, and the reduction of rocket weapon stocks poses questions about its disposal.

Figure 4.23   **Soviet secret cities and gulag camps: locations of "secret cities" of the former Soviet Union.** Built to isolate special types of military research and with access to them limited, they now have little, if any, role. Also shown on this map are *gulag* camps (slave labor camps) where prisoners were housed and their labor used for economic output. For example, the prisoners in the Kolyma area mined gold. Source: From Hochschild, *The Unquiet Ghost.* Reprinted by permission of George Borchardt.

Entry to these cities is still restricted, but life for the people in them has changed. It is difficult for the cities to get new work apart from contracts to carry out the more dangerous types of nuclear experiments. Privileges and status vanished as salaries shrank and were often not paid after 1991. People stay in the cities, however, since they see them as safe havens from the reported crime and instability elsewhere during the period of transition. New scientists, however, do not volunteer to move there.

Culture

Native Moldovans are very closely akin to Romanians and are not related to the Slavs. Russians, Ukrainians, and Belarussians are Eastern Slavs, offshoots of the broader Rus people and culture that emerged in the 800s and 900s at the time that the Rus adopted Eastern Orthodox Christianity. All three Slavic groups are very closely related as seen, for example, in the name *Belarussian,* which translates as "White Russian." Differences between them developed after Poles, Lithuanians, and Austrians ruled the western lands of the Rus from the 1300s to the 1800s. During this time, Ukrainians and Belarussians developed separate identities from Russians. Polish, Lithuanian, and Austrian influence is seen in the fact that nearly 20 percent of Belarus's population is Roman Catholic. A number of Poles

and Lithuanians live in Belarus today after Poland's boundary was relocated in 1945.

The name *Ukraine,* meaning "borderland," illustrates history's shifting boundaries and influences. To the Russians, Ukraine is Russia's borderland, but to many Ukrainians, who have stronger ties with Europe than the Russians, Ukraine is Europe's borderland. Russian ties with Ukrainians and Belarussians were reestablished in the early 1800s, when the Russian Empire extended west. Russification began soon afterward and was particularly strong during Soviet times. Today, 81 percent of Belarus's population is Belarussian and 11 percent Russian, but 63 percent of the population regularly speaks Russian and not Belarussian. Not surprisingly, Belarus's leaders have expressed interest in uniting Belarus with the Russian Federation. In Ukraine, Russians, living primarily in the industrial areas of eastern Ukraine, account for 22 percent of Ukraine's population. Ukrainians, however, fear Russian domination and cultivate their ties with the West.

Human Rights

In imperial Russia, the czars sent their political opponents to Siberia, far from the political center of the empire. After the Bolsheviks came to power in 1917, they continued the practice of banishing their opponents. Like the czars, the Communists dealt harshly with their opponents. In 1919, for example, several hundred thousand Don Cossacks were killed. Beginning in 1921, the first extermination camp was established for opponents of the Communist regime. Most camps, however, were primarily designed for slave labor but had extremely high death tolls.

Under Joseph Stalin, the camps grew dramatically in size and number. In 1928, they had approximately 30,000 prisoners. By 1930, the numbers grew to 600,000, when Stalin created the Main Directorate for Corrective Labor Camps. The acronym for the Russian name, *Glavnoe upravlenie ispravitel no-trudovykh lagerei* ("Main Directorate for Corrective Labor Camps") is **gulag.** Many of the *gulag* camps were located in Siberia (see Figure 4.23). The first dramatic increase in the number of slave laborers was caused by Stalin's program of collectivization in the agricultural sector. Many of the prisoners were the *kulaks,* successful farmers who engaged in capitalist practices Stalin labeled as "rich." Not allowed to join the collectives after their land was taken away from them, the *kulaks* fell into dire poverty and then were sent to the *gulag.*

Stalin also saw the *gulag* as a useful tool for economic development. Slave labor was essentially free, costing only a few bowls of thin soup and a couple slices of bread a day for each laborer. Slave labor could be used to complete dangerous work that no free worker would agree to do. For example, in 1931, Stalin directed slave labor to build a 227-km (141-mi.) canal across northern Russia from the White Sea to the Baltic Sea. Using pickaxes, shovels, and wheelbarrows, more than 100,000 workers completed the canal in 1933. The cost was tens of thousands of lives. Stalin then used slave labor from the *gulag* to complete such projects as the Moscow-Volga Canal and the Baikal-Amur main railroad line. Slave labor was also used to build hydroelectric stations, thousands of kilometers of roads, and industrial complexes in isolated locations in Siberia (such as the previously mentioned complex in Norilsk) and northern Russia.

The *gulag* was filled with millions more after Stalin began a series of purges in 1934, known as the Great Terror. The purges sent millions of Communist Party members, military officers, scientists, intellectuals, artists, and common citizens arbitrarily accused of treason to the *gulag.* Their free labor further fueled the Soviet economy and helped in the war effort after Hitler declared war on the Soviet Union in 1941. Engineers in the *gulag* system designed the advanced Soviet airplanes and tanks produced during the latter years of the war. Laborers in the *gulag* assembled these weapons. After the war, Stalin sent millions more to the *gulag.* Many were German prisoners of war. Others were accused of collaborating with the enemy, including entire ethnic groups within the Soviet Union. Russians, Ukrainians, and East Central Europeans who lived under German occupation during the war were seen as contaminated with capitalist ideas and were also sent to the *gulag.*

Many *gulag* prisoners were forced to work in coal, copper, and gold mines. The Kolyma gold mining region in far eastern Siberia was one of the most treacherous locations of *gulag* camps (see Figure 4.23). Kolyma was isolated and known for having the coldest recorded temperatures on Earth, sometimes as low as –98°F (–72°C). Thirty percent of the prisoners died in their first year at Kolyma. Very few survived a second year. As many as 3 million perished between 1937 and 1953 in Kolyma alone. About 2 pounds of gold were produced for every person who died.

After Stalin's death in 1953, the number of people sent to the *gulag* dramatically decreased but did not stop. It is difficult to know exactly how many perished in the *gulag,* although estimates are in the tens of millions. People outside the Soviet Union heard of the Soviet *gulag,* but few facts were known. Thanks to Alexander Solzhenitsyn, who highlighted the lack of human rights in the Soviet Union by depicting life in the *gulag* in two books, *One Day in the Life of Ivan Denisovich* and *The Gulag Archipelago,* the world learned more about the horrors of the *gulag.* Solzhenitsyn had personal experience in the *gulag.* After writing a letter containing critical remarks about Stalin, Solzhenitsyn was sent to the *gulag* for eight years, from 1945 to 1953. Solzhenitsyn's later literary efforts earned him the Nobel Prize for literature in 1970. After writing *The Gulag Archipelago,* Solzhenitsyn was sent into exile in 1974. He returned to Russia after the fall of Communism in 1991.

Andrei Sakharov was another major figure who struggled for human rights. A physicist, Sakharov was known as "father of the Soviet hydrogen bomb." As a scientist, he was also concerned about the radioactive hazards of nuclear testing and pressed the Soviet government to use caution in testing nuclear weapons. In time, he also criticized the power of the Soviet government. After he published an essay titled "Reflection on Progress, Coexistence and Intellectual Freedom" in the *New York Times* in 1968, Sakharov was fired from the Soviet weapons program. He continued arguing for peace and human rights and won the Nobel Peace Prize in 1975. In 1980, not long after he criticized his country's invasion of Afghanistan, Soviet authorities sent Sakharov to internal exile in Gorky, 250 miles east of Moscow. After Gorbachev came to power, Sakharov was freed in 1986 and later invited to help in governmental reforms just before his death in 1989.

Since the end of Communism, human rights abuses have decreased but not ended. The system of police practices and prisons that violated human rights did not disappear all at once. Economic hardship has prevented many prison reforms. The court system needs to be reformed. Defendants sit in jail before their cases are heard, and low pay makes many judges prone to accepting bribes. Juries are rare. Instead, judges act with the advice of two citizens who are popularly called "nodders" because they always agree with the judges. Judges frequently serve as prosecuting attorney as well as judge. Not surprisingly, during Boris Yeltsin's administration of the 1990s, the conviction rate was 99.6 percent. The war in Chechnya caused a new series of human rights abuses (see the Point-Counterpoint box, p. 142).

Following the breakup of the Soviet Union in 1991, democratic reforms allowed groups working for human rights to blossom. For example, Amnesty International now has as many as 40 chapters across the CIS. Though some human rights groups are outgrowths of international organizations, some are unique to the concerns of people in the CIS. The Committee of Soldiers' Mothers of Russia, "Mother's Right" Foundation, and the Moscow Center for Prison Reform are a few examples.

Women's Roles

Communist ideology professed that women were equal to men, and women began to achieve this equality following the revolution in 1917, before they were allowed to vote in the United States. During Soviet times, women in the subregion moved into positions traditionally held mostly by men in other societies (e.g., the United States), such as jobs in government, economics, medicine, or engineering. In some professions, women formed the majority. Most physicians, for example, were women. During World War II, some of the most decorated fighter pilots and infantry were women. Though Communism gave women great freedom in their career choices, it also idealized the factory worker, not the physician or engineer. Consequently, women were given the freedom to move into professions that lost much of their prestige compared to the same professions in the capitalist world. At the same time, however, women also were encouraged to do industrial and construction work, holding more than 60 percent of the construction jobs.

Women achieved equality in their careers but usually not at home. After women completed a day's work at their jobs, many arrived home to undertake the traditional responsibilities of housework and child care. Men's attitudes and behaviors in regard to their wives did not change much during Communism. As a result, between home and work, women ended up working long hours. As the subregion is transforming from Communism to capitalism, the roles of women are changing. The idea that women are equal to men still exists, although many women lost their jobs or had jobs that were not paid for long periods. Full and meaningful equality has yet to be realized.

Economic Development

The Slavic countries developed heavy industry during Soviet times (Figure 4.24), yet agriculture remained very significant. The Soviet model strived for self-sufficiency, desiring minimal

Figure 4.24 **Russia.** This cement factory in the city of Voskrenka, east of Moscow, is typical industrial development during Soviet times. Photo: © Ronald Wixman.

contact with the capitalist world. Since 1991, the Slavic countries have worked to update their factories, attract foreign investment, engage in the global economy, and develop more service industries. The Russian Federation is a large country with considerable geographic diversity. Its economic geography deserves special treatment (see "The Russian Federation" p. 154).

Soviet economic policies tied the Slavic countries very closely together, but many Ukrainians, Belarussians, and Moldovans prefer greater independence for their countries. They seek new trading relationships with other countries, but old connections are hard to break and new relationships difficult to form. For example, during Soviet times, eastern Ukraine was the most important iron and steelmaking region of the Soviet Union. Eastern Ukraine can provide for Ukraine's economic independence today, but the area's industries depend on imports of oil and natural gas, primarily from the Russian Federation, to meet 85 percent of their energy needs. Belarus also has much heavy industry that can provide for economic self-sufficiency, but with more than half of its trade with the Russian Federation, it too depends heavily on Russia. Moldova likewise still trades more with the Russian Federation than with Romania, though many Moldovans desire closer ties to Romania.

Agriculture

The Slavic countries' northerly latitudes and distant locations from moderating maritime influences result in low-moisture, low-temperature climates that make farming difficult almost everywhere (Figure 4.25). Even in the best locations, the late arrival of spring or the early appearance of winter can ruin a year's crops. The most productive farming areas are in the west of Russia, Ukraine, and Moldova in the former steppe and deciduous forest areas. Ukraine's vast plains of black soil produce abundant crops of wheat and make the country agriculturally self-sufficient. Moldova also has low relief and good soils like western Ukraine. Agriculture is the main economic activity, with fruits, vegetables, wine, and tobacco as the main products. Belarus, however, is a country with low relief but mostly poor soils. Having been shaped by continental glaciation, the southern part of the country is covered by the Pripyat Marshes and the northern part by glacial moraines. The sandy soils are good for potatoes. Toward the north in the Russian Federation, around Moscow, arable land gives way to livestock farming. East of the Urals, farming is restricted to areas that have sufficient water and length of summer growing season.

When the Soviets initiated collectivization in the 1920s and sent the productive *kulaks* to the *gulag,* agricultural production decreased dramatically as farm families lost their lands and were forced to work on giant collectives. Even after collectives became firmly established, production was low. Farm workers had little incentive to work hard, and faraway bureaucrats, knowing little about local conditions, made poor decisions about farm management.

In the 1950s, Soviet leader Nikita Khrushchev sought to increase agricultural production with his **Virgin Lands Campaign.** This campaign promoted farming in lands where it

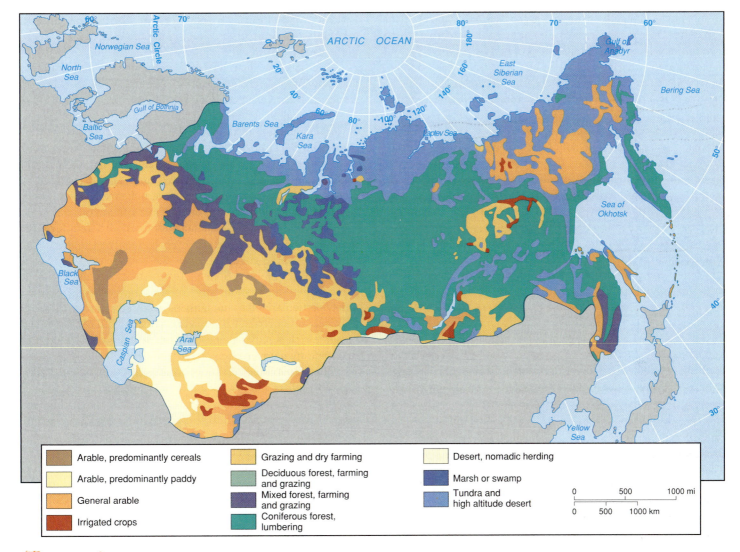

Figure 4.25 **Russia and Neighboring Countries: major land uses.** Relate the different types of farming to the natural regions as designated by climate, natural vegetation, and soils. Source: Data from *New Oxford School Atlas*, p. 98, Oxford University Press, UK, 1990.

had never been done before. Many of these lands were very marginal. They had poor soil quality because they either did not have enough water or heat to grow crops. Much of the land in the Virgin Lands Campaign was in the semidesert and desert areas of Central Asia, especially in the Kazakh Republic. Some of the land was also in the adjacent dry areas of the Russian Republic, much of it stretching across southern Siberia. With the importation of water through irrigation systems, it was believed that these dry areas would be productive. From 1917 to 1987, irrigated land in the Soviet Union increased from 3.5 to 20 million hectares. At the same time, large drainage networks were built in the lands north of Moscow to remove water from the waterlogged soils. From 1956 to 1987, drained land increased from 8.6 to 19.4 million hectares.

The Virgin Lands Campaign proved to be a huge failure over the long run. Production increased, but many problems resulted. Massive amounts of water were diverted from the Caspian and Aral Seas, wreaking havoc with their ecosystems (see Figure 4.13). Pouring water on desert soils only led to soil salinization. Drainage projects were no more successful. Farm

outputs fell on many of the newly established collectives. Eventually, the Soviet government stopped pushing for the expansion of the Virgin Lands Campaign. Over the ensuing decades, much of the land was slowly abandoned, though some of the better lands are still farmed.

Mikhail Gorbachev attempted agricultural reforms in 1988, when he permitted individuals and cooperatives to lease land privately from the government for farming. Within a few years, the government allowed greater privatization. By the end of 1992, all state farms became the property of their members. At the same time, however, the Soviet Union began to disintegrate. Like all other sectors of the economy, the agricultural sector suffered and production fell. For example, the number of cattle fell from 60 million to 36 million from 1985 to 1997. Similarly, pigs fell from 39 million to 20 million and sheep from 65 million to 24 million over the same period.

By 1996, private farms accounted for only 2 percent of agricultural production. Few people own private farms because the infrastructure and economic system do not support them. Many have opted to stay on the collectives, though they are now owners. The collectives still accounted for 52 percent of agricultural production in 1996. Interestingly, much production came from private plots on the collectives. For decades, Soviet authorities allowed individuals to cultivate small plots on the large collectives for their personal use. The total acreage on these personal plots was minuscule compared to the rest of the country's farmland. However, their production was comparatively very high. In 1970, they accounted for 31 percent of production. Their significance only increased during the economic chaos of the 1990s. By 1996, they accounted for 46 percent of production. This method of privately using collectively owned land was more productive than private farms. Individual incentive mixed with collectively owned land and farm machinery that was better integrated into the economic system proved to be more effective than the private farms that lacked machinery and needed to be run as an individual business.

Following the financial crisis in 1998 that affected much of the world economy, the value of the ruble (the Russian currency) dropped tremendously, losing three-quarters of its value against the dollar. Agricultural imports from countries such as the United States became extraordinarily expensive. The situation gave a boost to Russia's farmers and food processors, who could then produce goods cheaper than foreign imports and still earn a profit. Since 1998, private farming began to expand, but it is still a small percentage of Russian farming.

The Russian Federation

The Russian Federation is the modern political state representing the land known as Russia. To many Westerners, Russia is a mysterious land, hidden in cold, dark forests on the eastern and northern fringes of Europe. Europeans have regularly included the Russian heartland within Europe but at the same time have considered Russians too "Asiatic" to be European. For centuries, Europeans struggled to understand Russia. Winston Churchill once remarked that "Russia is a riddle wrapped in a mystery inside an enigma." Russians themselves have frequently alternated between emphasizing their European qualities and their Asiatic characteristics, depending on their relationship with Europe and their desire to be within Europe.

In the 1990s, Russians still wanted their country to be a world power, but economic decline, marked by a 60 percent drop in GDP, challenged that desire. As the Russian Federation struggled to maintain its influence in world politics, its relationships with the former Soviet republics changed considerably. While external relationships evolved in new directions, Russia's internal political geography and its economic and social relationships all dramatically altered.

Political Divisions

The internal political geography of the Russian Federation includes a mixture of political units. To a large extent, the political geography of the country was inherited from the Soviet system, though some changes have been made since 1991. Notably, the Kremlin, the center of government in Moscow, does not have as much power over the political units as before.

Leaders are no longer appointed by the Kremlin simply to carry out the orders of the Kremlin but are elected by the people in each political unit.

The political units fall into two categories: administrative and autonomous. The administrative units consist of 6 federal territories (*krays*), 49 regions (*oblasts*), and 2 federal cities (Moscow and St. Petersburg). Much like the states, counties, and municipalities of the United States, they were created to administer the large country. In contrast, the autonomous units, consisting of 21 republics, 1 autonomous region (*oblast*), and 10 autonomous districts (*okrugs*) (Figure 4.26), are able to craft many of their own laws and govern themselves somewhat differently than in the rest of the Russian Federation.

Together, the 89 political units represent different levels of size, resources, and political power. Unlike the United States, where the federal government has the same relationships with lower levels of government (e.g., states, counties, etc.), Moscow has an asymmetrical relationship with its political units, particularly the autonomous territories, in which each political unit negotiates its own relationship with Moscow. The resource-rich republics tend to exercise the greatest authority over their own governance.

The autonomous territories were established by the Soviet Union to reflect the presence of ethnic minorities, such as the Tatars and Sakha (Yakuts). Soviet law protected minority languages, religions, and cultures. However, only 52 percent of the Russian Federation's approximately 30 million non-Russians live today in the autonomous territories. As a way of controlling their vast country, the Soviets drew boundaries for the republics that deliberately left many members of ethnic groups outside their intended territories and included many Russians within them (Figure 4.27). Also, many recognized nationalities did not receive republic status. Of the 90 numerically significant groups recognized in the 1989 census, only 35 had a homeland. Some ethnic groups without a homeland at all are noteworthy, such as the Volga Germans. In addition to the earlier mentioned policies of Russification, boundary drawing was clearly a means that the Soviets used to divide and dilute non-Russian groups. The situation is particularly true in the North Caucasus, where the native peoples have most fiercely resisted Russian rule through history. For example, the Karbardians were grouped together with the Balkars, though they had more in common with their neighbors the Cherkessians.

Kalmykia illustrates some of the complex factors that produced the current ethnic geography of the Russian Federation. The Kalmyks were Buddhists pushed out of their ancestral lands in western China by Han expansion in the A.D. 1500s. They reached the steppes west of the Volga River mouth in 1608, where Peter the Great later gave them a kingdom, or khanate. From the mid-1700s, Russians and Germans took some of this land, causing a large group of Kalmyks to attempt to return eastward in 1771. Those who returned lost their towns and were resettled on collective farms under Stalin, who then accused them of siding with Germany during World War II and scattered them across Siberia. Allowed back in 1957, the Kalmyks are not a majority in their own republic but claim

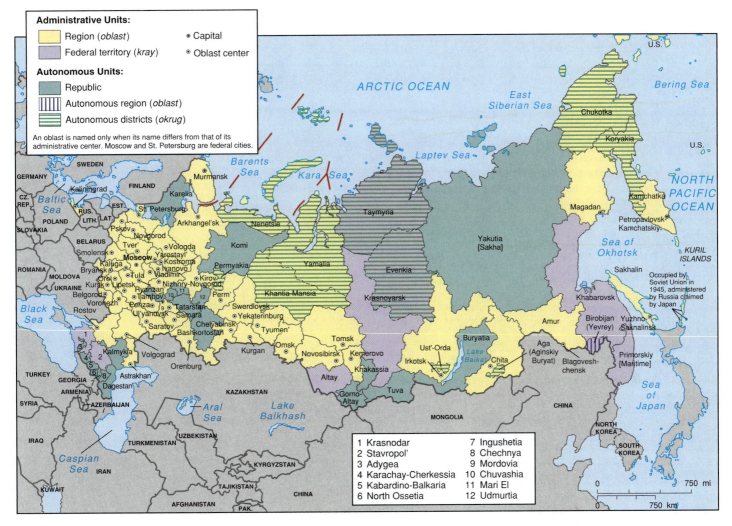

Administrative Units:
- Region (*oblast*)
- Federal territory (*kray*)
- ⊙ Capital
- ⊙ Oblast center

Autonomous Units:
- Republic
- Autonomous region (*oblast*)
- Autonomous districts (*okrug*)

An oblast is named only when its name differs from that of its administrative center. Moscow and St. Petersburg are federal cities.

1 Krasnodar	7 Ingushetia	
2 Stavropol'	8 Chechnya	
3 Adygea	9 Mordovia	
4 Karachay-Cherkessia	10 Chuvashia	
5 Kabardino-Balkaria	11 Mari El	
6 North Ossetia	12 Udmurtia	

Figure 4.26   **Russian Federation and its administrative divisions.**

political power. They rebuilt Buddhist temples, and their old Mongol script is being revived in schools. Soviet agricultural policy, however, turned over half of the land into desert through overgrazing and poor irrigation projects and left the republic in poverty.

The autonomous political units can be summarized into four groups:

- *North European and Middle Volga—Urals:* Both areas have been integrated into the Russian state since the 1500s. The peoples of the North European territories mostly practice Eastern Orthodox Christianity. Their territories are largely in the boreal forests, which are agriculturally poor and have low population densities. The economies are resource-oriented, producing electroenergy, wood products, and ferrous metals. The Middle Volga group represents the meeting ground between Finno-Ugrian, Turkic, and Slavic peoples. The large numbers of Russians are Orthodox Christians, but many of the indigenous peoples are Muslims. Located in the mixed-forest and forest-steppe region, these territories have economies that vary considerably, from traditional

lifestyles centered on agriculture to extractive industries and modern manufacturing.

- *North Caucasus:* Several autonomous territories occupy the foothills on the northern side of the Caucasus Mountains. Some of them extend into the steppe. Most of the peoples are Muslims, but Christians live in the area, too. The Russians were not able to exert full control over the area until the mid-1800s. Ethnically and linguistically, the area is one of the most complex places in the world. The various peoples, however, share certain economic and cultural similarities that facilitate a "mountaineer" identity. Their shared opposition to Russian rule has led to cooperation with one another. The Chechens in particular have resisted Russian rule. Their attempts to make Chechnya independent of Russia since 1991 have resulted in very bloody military conflicts between themselves and the Russian army (see the Point-Counterpoint box, p. 142).

- *Siberia and the Far East:* These autonomous territories stretch east, with many along the Chinese and Mongolian borders. They contain Turkic or Mongolian peoples

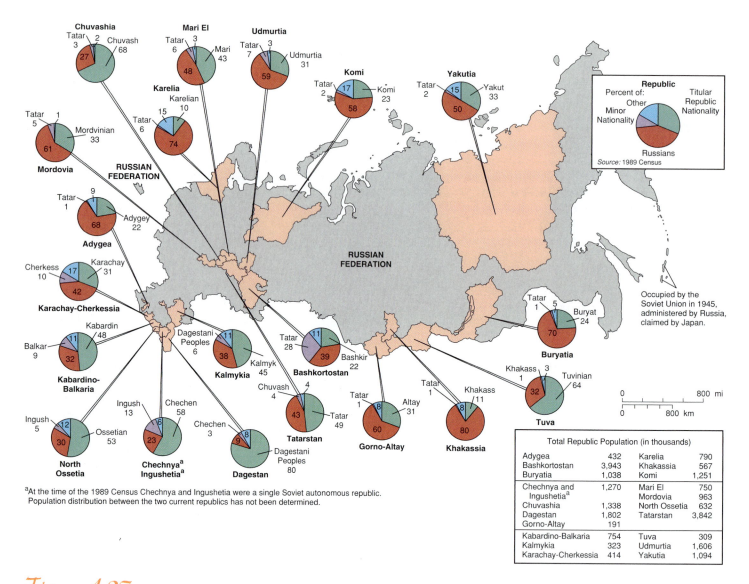

Chuvashia
Tatar 3 | 2 | Chuvash 68 | 27

Mari El
Tatar 6 | 3 | Mari 43 | 48

Udmurtia
Tatar 7 | 3 | Udmurtia 31 | 59

Komi
Tatar 2 | 17 | Komi 23 | 58

Yakutia
Tatar 2 | 15 | Yakut 33 | 50

Mordovia
Tatar 5 | 1 | Mordvinian 33 | 61

Karelia
Karelian 10 | 15 | Tatar 6 | 74

RUSSIAN FEDERATION

Adygea
Tatar 1 | 9 | Adygey 22 | 68

Karachay-Cherkessia
Cherkess 10 | 17 | Karachay 31 | 42

Kabardino-Balkaria
Kabardin 48 | 11 | Balkar 9 | 32

Dagestani Peoples 6

Kalmykia
11 | Tatar 38 | Kalmyk 45

Bashkortostan
Tatar 28 | 11 | Bashkir 22 | 39

North Ossetia
Ingush 5 | 12 | Ossetian 53 | 30

Chechnya-Ingushetia[a]
Ingush 13 | 6 | Chechen 58 | 23

Dagestan
Chechen 3 | 8 | 9 | Dagestani Peoples 80

Tatarstan
Chuvash 4 | 4 | 43 | Tatar 49

Gorno-Altay
Tatar 1 | 8 | Altay 31 | 60

Khakassia
Tatar 1 | 8 | Khakass 11 | 80

Khakassia
Khakass 1 | 3 | Tuvinian 64 | 32

Buryatia
Tatar 1 | 5 | Buryat 24 | 70

Tuva

RUSSIAN FEDERATION

Republic
Percent of:
Minor Nationality | Other | Titular Republic Nationality
Russians
Source: 1989 Census

Occupied by the Soviet Union in 1945, administered by Russia, claimed by Japan.

0 —— 800 mi
0 —— 800 km

Total Republic Population (in thousands)			
Adygea	432	Karelia	790
Bashkortostan	3,943	Khakassia	567
Buryatia	1,038	Komi	1,251
Chechnya and Ingushetia[a]	1,270	Mari El	750
		Mordovia	963
Chuvashia	1,338	North Ossetia	632
Dagestan	1,802	Tatarstan	3,842
Gorno-Altay	191		
Kabardino-Balkaria	754	Tuva	309
Kalmykia	323	Udmurtia	1,606
Karachay-Cherkessia	414	Yakutia	1,094

[a]At the time of the 1989 Census Chechnya and Ingushetia were a single Soviet autonomous republic. Population distribution between the two current republics has not been determined.

Figure 4.27   **The ethnic composition of the Russian Federation's autonomous republics.** Source: © University of Texas Libraries.

(Figure 4.28). Many are Buddhists and practiced nomadic herding until they were incorporated into the Soviet Union. They often practice their traditional ways of life amid large pockets of Russians who engage in extractive industries, and raw material and food processing. Tyva was annexed only as recently as 1944. The 1993 Tyvan constitution claims complete sovereignty for the republic and the right of secession, though this policy violates the constitution of the Russian Federation. Violence directed at Russians caused many Russians to leave.

Stalin created the Jewish (Yevrey) Autonomous Oblast in 1934 and designated it as a homeland for Soviet Jews. It was hoped that it would be a counterattraction to Palestine (now Israel), but its remote location and harsh environment made it unattractive. In 1989, only 9,000 Jews lived there. Eighty percent of the population was Russian.

• *Territories of the Far North:* More than 30 different ethnic groups live in the far north, many only numbering in the thousands. Seven of the groups have *okrugs* (autonomous

Figure 4.28   **Russia.** Yakut (Sakha) woman. Photo: © Ronald Wixman.

districts), meaning that the majority of peoples lack a designated homeland. The territories are rich in diamonds, gold, iron ore, timber, and other natural resources. During Soviet times, a flood of Russians seeking to exploit these resources reduced the percentages of the indigenous peoples to less than 16 percent in the *okrugs*. The Russians concentrated in towns associated with industry. Pipelines, other modern infrastructure, and the resulting pollution made it difficult for the indigenous peoples to practice their traditional ways of life, which include hunting, fishing, and reindeer herding. The Association of Peoples of the North formed in 1989 to voice the concerns of the indigenous peoples.

Heartland and Hinterland in Russia

Another basis of regional differences in Russia is **heartland** and **hinterland.** The heartland lies west of the Urals and includes many of the original territories of Rus and then Muscovy. It contains the greatest concentration of Russian people and accounts for much of the country's economic and political activity. It is also known as the Russian homeland. The Moscow and St. Petersburg urban regions, the Volga River valley, and the Urals contribute to the heartland's prominence. The Moscow region, approximately 400 km (250 mi.) square, is home to 50 million people and was the focus of Soviet central planning and transportation routes linking the entire country. Local manufacturing includes vehicle, textile, and metallurgical industries. St. Petersburg, a major Baltic port north of Moscow, is a smaller manufacturing center but still produces around 10 percent of the total Russian output, including shipbuilding, metal goods, and textiles.

Southeast of Moscow, the Volga River is lined by a series of industrial cities that use the river transport, linked since the 1950s by a canal outlet to the Black Sea. Around 25 million people live in the Volga River region, which was developed for manufacturing during and after World War II—at a distance from advancing German armies and helped by the discovery of major local oil and natural gas fields. Manufactures include specialized engineering and the Togliatti car plant built by Fiat of Italy.

East of the Volga River basin, the Urals contain metal ores and are only a minor barrier to communications. Like the Volga region, the southern Urals were developed during and after World War II, principally as a metals center. Once again, oil and natural gas fields were discovered to the south and east.

The Russian Federation's hinterlands—dependent and tributary to the heartland—include Kaliningrad *oblast* along the Baltic Sea, the Arctic regions around Murmansk, the mining areas east of the Urals, resource-rich Siberia, and the Pacific region inland from Vladivostok. Large expanses of the hinterland are virtually empty of people and economic activity and remain inaccessible to development. Siberia forms a huge area that can be divided into the more developed southern margins along the line of the Trans-Siberian Railway and the northern lands.

Siberia is an essential part of Russia, making up three-fourths of its land and providing a large proportion of its raw materials—a common feature of hinterland regions. In 1990, Siberia produced 73 percent of Russia's oil, 90 percent of its natural gas, 61 percent of its coal, all of its diamonds, and 30 percent of its timber and electricity. It also has a growing fishing industry on the Pacific coast. Farmed lands are mainly in the west of Siberia. Siberia remains a region of promise for enterprising Russians, often based on romantic aspirations such as taming the wilderness or making a fortune.

Along the southern strip of Siberian Russia are a number of centers that were industrialized according to the Soviet planning processes and are separated by large tracts of sparsely settled land. In the Kuzbas region around Kuznetz, 2,000 km (1,200 mi.) east of the Urals, a major coalfield was developed initially to supply coal to the steel manufacturing centers of the Urals. This led to local industrialization, linked to the discovery of iron ore and the return transport of bauxite from the Urals. Aluminum, steel, and products using them are major outputs. The world's largest aluminum plants at Krasnoyarsk, Bratsk, and Sayansk provided metal for the Soviet aircraft and missile industries, but these demands fell away in the 1990s, and the output of aluminum fell by 70 percent from 1991 to 1993. Privatization of the aluminum smelters from 1994 involved foreign capital and led to a major increase in exports.

Farther east, centers close to Lake Baikal, including Irkutsk (see "Personal View: Russia," p. 158), form a narrow belt of industrialization using hydroelectricity generated in the headwaters of streams flowing to the Yenisey and Lena River systems. Mining and lumbering are also important in isolated localities, while larger cities provide services to extensive areas in northern Siberia where population and mining activities are scattered.

The far east, flanking the southern Pacific coast of Russia, has ports such as Vladivostok and Nakhodk, and metal industries along the Amur River. Its development as a major area linking Siberian output to the world economy within the Pacific Rim (see Chapters 5 and 6) is held up by political differences between Russia and Japan over the possession of former Japanese islands. Japanese markets for Siberian minerals and potential investment in eastern Siberia could be a major factor in the future development of these remote areas if the dispute over the islands can be resolved.

Soviet centralized planning often ignored geographic resource variations and sited production facilities for a variety of political, rather than economic, reasons. Special defense-related factories, for example, were built in remote locations where it was expensive to maintain them. Workers were paid more to work in these faraway facilities, and the cities that grew up around them received many subsidies. The Russian Federation stopped giving financial support to these poorly located factories and cities. Without subsidies, many factories were not able to compete economically and had to shut down. In turn, unemployed people emigrated to the Russian heartland. Subsequently, regional differences and local characteristic lifestyles are now greater within the Russian Federation.

The realities of such factors as closeness to consumer markets or ports with world trading connections are causing geographic shifts in regional production patterns. Moscow, for example, has succeeded the best in the transition to capitalism. As much as 20 to 25 percent of Moscow's population is middle class, the highest

Personal View

RUSSIA

Irkutsk is a rather special city in Siberia. It lies five time zones and 88 hours by Trans-Siberian Railway east of Moscow and another three time zones and 72 hours on the train from Vladivostok on the Pacific Ocean. It was one of the earliest Russian settlements in Siberia, dating from 1652, when trappers and traders established a wooden fort at a crossing of the Angara River just downstream from Lake Baikal. Today, it is home to over 700,000 people and has a much greater variety of buildings than other cities along the railway, ranging from the wooden houses with carved window frames and painted shutters that give it a villagelike appearance to the monolithic former Communist Party headquarters. Its people have a strong pride in their city and its cultural history, having produced many poets and artists. The expulsion of Europeanized groups from St. Petersburg to Siberia early in the 1900s added to the ethnic range that includes Slavic Russians and Asian groups, such as the Buddhist Buryats living on the Russian side of the Mongolian border and the Yakuts from northern Siberia who speak a Turkic-derived language.

A major stop and junction on the Trans-Siberian Railway, Irkutsk connects southward into Mongolia and China by rail and northward into Siberia by bus and boat in the summer. It also has good air links to Moscow and other Russian cities, as well as places in northern China, Mongolia, and Japan. Chinese traders regularly appear selling a range of cheap goods. Telephone communications are good, with calls abroad often being clearer and cheaper than local or internal Russian calls. Fax, e-mail, and other telecommunications facilities are available, although many people still use telegrams. The post office remains under bureaucratic control with letters having to be sent separately from parcels and many forms needing to be filled out.

Irkutsk is an industrial and commercial city. Its domestic electricity comes from a local hydroelectric dam. Such power is also used in a nearby aluminum factory, although the future of this and other manufacturing facilities based on metal and engineering products is in doubt. Irkutsk is also a center for coal mining and oil production in the surrounding area.

The city-center shops increased their range of wares rapidly in the mid- and late-1990s: a bakery may also sell cameras, while a pharmacy may sell shoes. The department stores are more like indoor markets, with stalls competing to sell the same products. Fruit and vegetables are commonly sold in street or sidewalk booths. Some sports shops sell Western goods, but the prices are too high for ordinary Russians.

The Siberian climate brings varied hardships to the people of Irkutsk. Its winters are long and very cold. The first snow flurries may come in September, but by October, the ground is covered by snow that lasts until late March, although it is only 15 cm (6 in.) deep and little more falls until spring. Temperatures at their worst plunge to –40°C (–40°F) for a few days and never rise above freezing during winter: the January mean temperature is –21°C (–5°F). In summer, daytime temperatures reach 20°to 25°C (70°–75°F), and the thick winter ice on Lake Baikal melts. There is little rain, but dry, dust-laden winds and mosquitoes make life uncomfortable. Although they are used to these conditions, local people need to wear thick fur coats and hats in winter and prefer to live in the centrally heated apartment blocks with running water rather than in the older wooden houses.

Box Figure 1 **Irkutsk, Siberia.** (a) Unlike many other Russian cities, Irkutsk has many of its prerevolutionary (pre-1917) buildings. (b) Irkutsk was a major monastic center of the Russian Orthodox Church before 1917. The city's major cathedrals were destroyed, but many of its smaller churches, such as this one, still remain. (c) Buildings of the Communist period line the Angara River where people recreate in summer. Photos: © Ronald Wixman.

(a)

(b)

(c)

Although the city-edge apartment blocks are favored for living, they are often difficult to locate because of irregular numbering. The city authorities and large factories built them for their workers, but during the 1990s, the occupiers assumed ownership. Utilities are monopolies and own the household appliances as part of a contract to supply unmetered electricity or gas. Without meters, there is no check on the amounts used, and people tend to be wasteful until supply cuts occur. Water is supplied centrally, as is central heating by water that is pumped around the whole city: warm water is turned on in October and off in April, and there is no other control of this process.

Irkutsk draws its university students mainly from the local area. There are residence halls, but most students live at home. Science, engineering, medical, and language courses, often linked to teacher training, are most popular, but there are also cultural, theater, musical, and historical studies courses. Students get into courses on the basis of their preuniversity qualifications, and they have to pass annual tests to progress and qualify for the small monthly maintenance grant. Tuition is free. Personal achievement measured in tests or longer examinations is a problem, since Russian students are used to working together and may talk to each other across the exam room.

Students have fewer clubs and societies than their counterparts in Western universities, but they enjoy hockey and basketball in the winter and soccer in the summer. Irkutsk won a national hockey competition in 1998 and has an open-air stadium in the middle of the city (where those watching sometimes endure temperatures of −30°C). Russian hockey is played on larger, more open ice-covered fields and has less personal contact than the hockey played in the United States and Canada. Volleyball is popular as an informal sport that requires little equipment. Theater performances, especially of music or the ballet, are low in cost—a dollar or so for a ticket. But Russians generally prefer entertaining at home to going out to be entertained, and there are few night clubs or discos. Birthday parties especially are major social events.

The university halls provide a diet of basic food, often of low quality based on soups and meat snacks. Men living in apartments struggle to feed themselves, since cooking is regarded as a woman's job, as are clothes washing and housecleaning. Some students will work part-time to gain income, but this is not common. Resources in the university are poor because of low funding, and staff and students occasionally strike over pay and conditions. Professors are state employees whose pay is often delayed or withheld; when they retire, their pensions may not be paid for many months.

The issue of low or no pay is important in Irkutsk. People manage to survive, however, by growing their own produce at out-of-town dachas. These are small wooden houses, often near Lake Baikal, on the grounds of which potatoes and other vegetables are grown; they offer some escape from the polluted air of Irkutsk.

Many Russian students entertain strong hopes of moving to a Western country. Their idealistic notions of living elsewhere are fueled by preparations that involve immersing themselves in Western clothes, music, competitive values, and language (English, German, and Japanese courses are popular). Factors such as continuing military service—which all young people have to enter for two years, usually after completing their studies—and the lack of local job opportunities also make them want to emigrate. Even if university graduates obtain a good job in Russia, the pay is poor and they will not be able to live independently with their own apartment. Many continue to live with parents, even after they marry.

People living in Irkutsk, Russia, saw many changes in the 1990s. Shops now contain a greater range of foods and consumer goods, although prices for the most desirable items remain too high for most people. The job outlook is not hopeful, and housing is generally of poor quality compared to Western norms. Easy access to the surrounding countryside, cheap musical events, and the great strength of deep personal relationships—as opposed to the mere acquaintanceships so common in the individualistic West—remain the strengths of Russian society.

percentage in the Russian Federation. In other areas, less than 10 percent of the population has attained middle-class income.

Foreign Investment

In the 1990s, corruption, poor infrastructure, and an unwieldy bureaucracy prevented much foreign investment from flowing into Russia (Figure 4.29). The situation is slowly changing, and Russia is now receiving greater amounts of foreign investment. The greatest amount of investment has come from foreign corporations joining Russian groups to exploit Russia's vast mineral wealth, namely oil, natural gas, and metal ores. The Russian automotive industry has also received investment.

The oil and natural gas reserves are of particular interest to industrialized countries and the main potential source of foreign investment (Figure 4.30). The Lukoil company, which controls one-sixth of Russia's oil reserves and is equivalent in size to major world oil producers, now buys equipment from Western companies and enters joint development ventures. Western oil corporations that hoped to be placed in charge of oil exploration and marketing found that the Russian government expected them to enter joint ventures with companies

such as Lukoil. Gazprom is Russia's biggest company, having a monopoly of producing and distributing natural gas. Some 34 percent of the world's gas reserves are controlled by Gazprom, which supplies 20 percent of Europe's demands.

Russia's mineral wealth extends to a wide variety of metal ores. At present, multinational mining corporations are particularly interested in developing gold mining, as in the world's largest gold deposit at Sukhoi Rog in Siberia. Russia also produces one-fourth of the world's diamonds.

One of the most promising areas for foreign investment is in automobile manufacturing. Russia has only 13 million private autos, with an average age of over 10 years in the late 1990s. The market is growing, however, and all cars made are sold. In 1998, in one of the biggest foreign investments to date, the Gorky firm in Nizhniy Novgorod entered a joint agreement with Fiat of Italy to use idle defense factories for component manufacturers. General Motors has also negotiated similar arrangements with the largest Russian auto producer, Avtovaz.

Trade

Since 1991, trade patterns for the Russian Federation have changed dramatically. During Soviet times, most trade was

Figure 4.29 **Foreign direct investment inflows: dollars per person, 2000.** The Russian Federation receives comparatively little foreign direct investment in terms of total dollars and in terms of dollars per Russian. For example, the Czech Republic, which has a population of 10 million, receives more money then the Russian Federation with a population of 144 million. Source: *The Economist* Newspaper Group, Inc. Reprinted with permission. Further reproduction prohibited. www.economist.com.

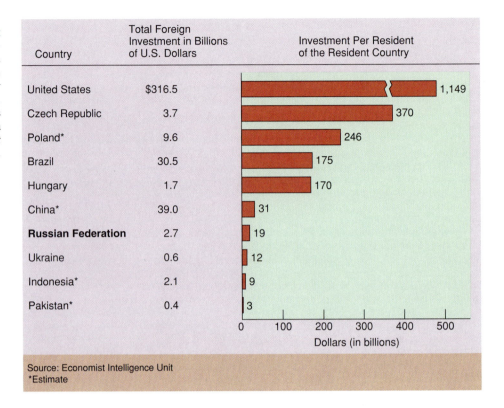

Country	Total Foreign Investment in Billions of U.S. Dollars	Investment Per Resident of the Resident Country
United States	$316.5	1,149
Czech Republic	3.7	370
Poland*	9.6	246
Brazil	30.5	175
Hungary	1.7	170
China*	39.0	31
Russian Federation	2.7	19
Ukraine	0.6	12
Indonesia*	2.1	9
Pakistan*	0.4	3

Dollars (in billions)

Source: Economist Intelligence Unit
*Estimate

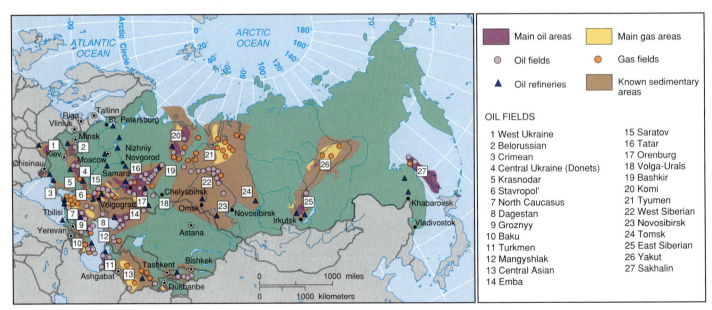

OIL FIELDS

1 West Ukraine	15 Saratov
2 Belorussian	16 Tatar
3 Crimean	17 Orenburg
4 Central Ukraine (Donets)	18 Volga-Urals
5 Krasnodar	19 Bashkir
6 Stavropol'	20 Komi
7 North Caucasus	21 Tyumen
8 Dagestan	22 West Siberian
9 Groznyy	23 Novosibirsk
10 Baku	24 Tomsk
11 Turkmen	25 East Siberian
12 Mangyshlak	26 Yakut
13 Central Asian	27 Sakhalin
14 Emba	

Figure 4.30 **Russia and Neighboring Countries: oil and gas fields.** Assess the availability of oil and natural gas to Russia, the other CIS countries, and the potential for developments in Siberia.

with the republics that would become independent and with the COMECON partners in East Central Europe. In 1990, for example, 70 percent of exports and 47 percent of imports were with the republics that became members of CIS. In 1996, exports and imports to these countries dropped to 21 and 30 percent, respectively. Trade with the former COMECON countries declined similarly. In contrast, trade with the European Union (EU) and the United States increased rapidly. In 1995, the EU accounted for almost 40 percent of Russia's non-CIS

exports and 54.3 percent of its imports. In 1999, the United States and Germany were the Russian Federation's most important trading partners. Italy and the Netherlands played prominent roles, too. The Russian Federation is primarily exporting raw materials such as oil, natural gas, wood products, and metals, and importing finished products such as machinery and equipment. The situation reflects Russia's abundant stocks of natural resources and antiquated industries. The energy exports to countries such as Germany attract high prices. Germany now buys 30 percent of its gas and 12 percent of its crude oil from Russia, and it could be said that Russia is fueling the growth of German industry. The potential for growth in trade between the EU and Russia is great and may eventually equal EU trade with the United States, at present several times greater than that with Russia.

Science, Sports, and Society

Communism strove to excel in science and sports and to improve life in society as never before seen in human history.

Test Your Understanding 4D

Summary Population is not evenly distributed across this world region. Most people live on the fertile soils of the North European Plain west of the Urals. Populations are low in the cold climates of the north, especially in Siberia, and in the mountains and deserts of the south. In the 1990s, the populations of the Slavic countries began to fall as death rates exceeded low birth rates, resulting in slowly decreasing populations.

The Slavic countries of the CIS share a similar history and culture. The Russian Federation is larger than the others combined, and it is difficult for the others to break their dependencies on the Russian Federation. The Slavic countries have changed since the end of Communism, but their urban landscapes still reflect Communist policies, and the people of the subregion continue to wrestle with human rights abuses of the past.

The Russian Federation is well endowed with natural resources and has considerable industry. Belarus and Ukraine also have industry, and Ukraine and Moldova have some of the world's best-quality farmland. All the Slavic countries are changing their economies and trading relationships.

As the world's largest country in land area, the Russian Federation has considerable internal diversity in environment and culture. Since 1991, it has reorganized its internal political geography to give greater power to local voices, for both ethnic minorities and Russians. Economic restructuring occurs as the country tries to compete in the global economy.

Questions to Think About

4D.1 What are the cultural similarities and differences of the peoples of the Slavic countries?
4D.2 What are the population growth trends of the Slavic countries, and what are the causes of these trends?
4D.3 What kind of human rights abuses occurred during Soviet times?
4D.4 Do you believe that women are more equal to men in the Slavic countries compared to women in your own country? Explain your answer.
4D.5 How has the economic geography of the Russian Federation been changing since the end of Communism to compete in the global economy?

Key Words

visual symbolism   heartland
gulag   hinterland
Virgin Lands Campaign

Science and sport achievement was also a way to show that Communism was superior to any other political-economic system. Scientists were pampered in terms of incomes and equipment and concentrated in new cities established for their work—some of which remained secret because they were dedicated to military purposes (see Figure 4.23). Akademgorodsk, outside Novosibirsk in the center of Russia, has around 30 research institutions. The Budker Institute for Nuclear Physics, for example, focused on abstract research on fundamental particle physics. As already stated, the Soviets invested heavily in a space program as well to best the United States.

Like scientists, athletes were pampered and given the best sport facilities in which to train. Athletes were not expected to hold jobs, only train for sports. Though the Soviet Union remained closed to capitalist societies, it sent its athletes to international events, especially the Olympics, where the Soviet Union usually won the most medals of any country.

Everyone in society was guaranteed a free university education and a job. Health care was provided without cost to everyone. Maternity leave was generous. The theater and arts received generous support. The average citizen could go to a play, see a ballet, or visit a museum for almost no cost. The goal of the Soviet government was to improve society to the degree that all citizens received all their basic needs and much more, regardless of their social class. Prior to Soviet times, unemployment and poverty were rampant, but the Soviet government claimed to have eliminated them. Though the Soviet system eventually failed, it brought about great political, economic, and social transformation in its 74 years of existence.

Following the end of Communism in 1991, public institutions and welfare support systems were the main losers as a result of the political and economic changes. Many things that Russians took for granted, such as a job and a free university education, were no longer available to many. After losing support for their efforts, many scientists and athletes migrated to other countries. Not surprisingly, some Russians still support Communist ideals.

The Southern Caucasus

Georgia, Armenia, and Azerbaijan straddle the Caucasus Mountains and are frequently called the Transcaucasus, meaning "across the Caucasus" (Figures 4.31 and 4.32). This term reflects a Russian ethnocentric view of the region as these countries are on the other side of the Caucasus Mountains from Russia and were once Russian colonies. The more neutral term "Southern Caucasus" is used to refer to these countries, and "Northern Caucasus" is applied to the part of the Caucasus in the Russian Federation.

Armenia and Georgia are both very mountainous with many peaks rising above 5,000 m (15,000 ft.). Azerbaijan is mountainous in the west along the borders of Georgia and Armenia, but the eastern areas of the country, formed by the Caspian Sea coast, are flat with areas below sea level. The origins of the word *Azerbaijan* are not clear, but one version derives from Persian words that mean "land of fire." Azerbaijan certainly contains

Figure 4.31   **The Southern Caucasus: countries, cities, and major physical features.**

great quantities of petroleum. "Land of fire" could refer to the surface deposits that burned naturally in the past or to the oil fires in Zoroastrian temples that once dominated the region. Zoroastrianism no longer exists, but it is a religious forerunner to Christianity and Islam, tracing its origins back to Azerbaijan. Zoroastrians believed that the Earth would be consumed in fire following judgment day. Interestingly, this belief developed in an area of the world where the Earth is oil-soaked and burns easily.

Countries

Though the three countries of the Southern Caucasus are relatively new, the peoples and political relationships extend far back into history. Armenians, for example, trace their ancestors back to 6000 B.C. In the 100s B.C., the Armenian empire controlled most of the Southern Caucasus and stretched across what is now northern Iran, Iraq, Syria, and eastern Turkey. Modern-day Armenia is small compared to its predecessors and does not even include a majority of all Armenians. Important historical places, such as the medieval capital Ani and Mount Ararat, the historically accepted landing place of Noah's ark, are now in neighboring Turkey. This situation illustrates that the modern Armenian state only encompasses the eastern portion of historic Armenia.

Georgians have lived in the Southern Caucasus almost as long as Armenians and likewise built empires that stretched across the subregion and beyond, reaching their height of power from the A.D 1000s to 1200s. It was around this time that the Azerbaijanis (also known as Azeris) emerged as a people in the subregion.

Independence and self-rule are features of the distant past. Over the last 800 years, the peoples of the Southern Caucasus were ruled from the outside. The most notable foreign rulers were the Russians, who took control in the 1800s and continued exercising their power until the Soviet Union dissolved in 1991. As former Soviet republics and current members of the CIS, Armenia, Georgia, and Azerbaijan struggle with establishing a viable political and economic existence. Located between Russia, Iran, and Turkey, these countries reflect the differing cultures and political views of the larger neighbors. Past outside political control and continuing outside influence create tension and conflict between the countries of the Southern Caucasus.

People

Significantly, most Armenians and Azerbaijanis do not live in their respective countries. Of the 6.3 million Armenians in 2001, only 3.8 million live in Armenia. Many live in Azerbaijan and Georgia. Of the 19 million Azerbaijanis, less than 8.1 million live in Azerbaijan. Most live in neighboring Iran. Azerbaijan is 90 percent Azerbaijani, but 90 other groups live there as well. In contrast to Armenians and Azerbaijanis, most Georgians live in Georgia, which has 5.5 million inhabitants. However, 30 percent of Georgia's population is composed of other groups such as Armenians, Russians, Azerbaijanis, Ossetians, and Abkhaz (Figure 4.33). The population growth rates for these countries vary slightly from Georgia at 0.0 percent to Azerbaijan at 0.9 percent in 2001 (Figure 4.34). Total fertility rates in Georgia and Armenia were well below replacement levels at 1.2 and 1.1 respectively. Due to this factor combined with emigration, these countries' populations are declining. Emigration is a factor in Azerbaijan as well, but a total fertility rate of 2.2 offsets emigration, leading to slight population growth for Azerbaijan and the subregion as a whole.

Urbanization

Urbanization rates in 2001 were 67 percent in Armenia, 56 percent in Georgia, and 51 percent in Azerbaijan. Major cities

Figure 4.32   **The Southern Caucasus: key data on countries.**

Country	Capital City	Land Area (km²) Total	Population (millions) Mid-2001 Total	2025 Est.	GNI 1999 (US $ million) Total	GNI PPP 1999 Per Capita	Percent Urban 2001	Human Development Index Rank of 174 Countries
Armenia	Yerevan	29,800	3.8	4.1	1,878	2,360	67	87
Azerbaijan	Baku	86,600	8.1	9.8	3,705	2,450	51	103
Georgia	Tbilisi	69,700	5.5	4.8	3,362	2,540	56	85

Source: Data from *Population Reference Bureau 2001 Data Sheet*; *World Development Indicators,* World Bank, 2001; *Human Development Report,* United Nations, 2001; Microsoft Encarta (ethnic group, language, religion).

with over 1 million people in 2000 included the capital cities of Yerevan (Armenia, 1.3 million), Baku (Azerbaijan, 1.9 million), and Tbilisi (Georgia, 1.3 million). Baku is a major oil city and port on the Caspian Sea. Tbilisi is an ancient city that was taken by the Turks in the 1200s and then by the Soviet Union in 1920. Yerevan is economically stagnant following years of warfare against Azerbaijan.

Culture

Though the Southern Caucasus is a small area of the world, the subregion is culturally very diverse (see Figure 4.33), partly because it lies at the historic contact zone between Turkish, Persian, and Russian empires. Many languages are spoken in the subregion. They are of different language families, so most have very little in common. The Georgian language is in the Caucasian language family and has a unique alphabet. Armenian is Indo-European but stands alone in its own branch and has a distinct alphabet of 38 letters, derived mostly from Greek. Azerbaijani is in the Ural-Altaic language family (see Figure 2.18).

The Georgians and Armenians are both Christian, and both accepted Christianity early in history, in the A.D. 300s. The Armenians claim their country was the first in the world to

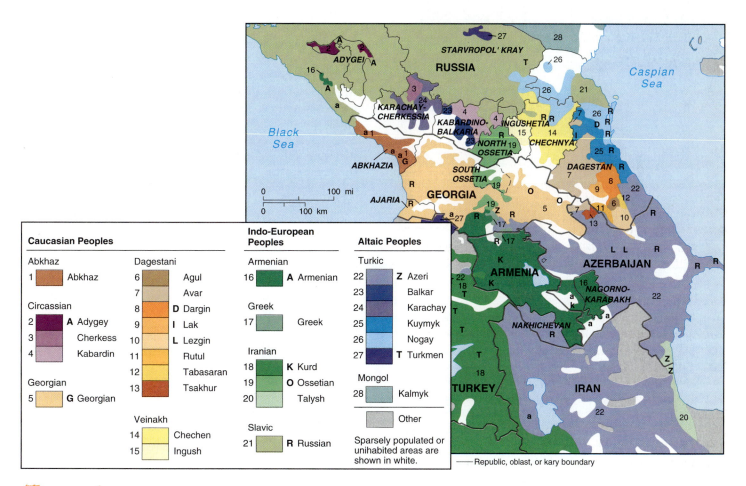

Figure 4.33 **Ethnolinguistic groups in the Caucasus region.** Compare the distribution of groups to the location of political boundaries. Do these comparisons help to explain conflict in the subregion? Source: © University of Texas Libraries.

Ethnic Groups (percent)	Languages	Religions (percent)
Armenian 93%, Azeri 3%, Russian 2%	Armenian 96%	Armenian Orthodox 94%
Azeri 90%	Azeri 89%, Russian	Nominal Muslim 96%
Georgian 70%, Armenian 8%, Russian 6%	Georgian, Russian	Georgian Orthodox 65%, Muslim 11%

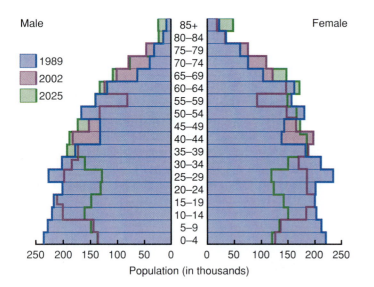

Figure 4.34   **Age-sex diagram of Georgia.**
Source: U.S. Census Bureau. International Data Bank.

adopt Christianity officially. The Armenian Apostolic Church has been independent since the Middle Ages and expresses a unique view of Christianity. The Georgian Church is associated with the Eastern Orthodox Christian churches.

Arabs introduced Islam in the 600s and 700s into Azerbaijan. In the 1500s, the Shia branch of Islam came to the country and now dominates, making Azerbaijan and Iran the only two countries in the world where Shia Islam is practiced by the majority of the population and controls the government. Despite their close ties with Iran, Soviet secular policies influenced Azerbaijanis greatly. For example, Azerbaijani Muslims, unlike those in Iran, drink wine, and women are not veiled or segregated. In 1991, the Azerbaijani government also went against the wishes of Iran and adopted a modified Latin alphabet for Azerbaijani instead of the Arabic alphabet, the original alphabet of the Qu'ran and the alphabet used in many Muslim countries. The Latin alphabet is customarily used in Roman Catholic and Protestant countries but also in nearby Turkey, where the language is similar to Azerbaijani.

Ethnic Peace and Conflict

History has given rise to differing cultural combinations in the Caucasus. Some ethnic groups are closely related and others are not. Many live in peace, while others are locked in conflict. For example, the Adjarians, who have an autonomous republic in Georgia (see Figure 4.31), live peacefully with the Georgians, though they are Muslims and the Georgians are Christians. Despite differing religions, the Adjarians are indistinguishable from Georgians, and most Adjarians consider themselves Georgians.

Relations among other groups have not been so peaceful. The Ossetians and Abkhaz both have their own autonomous republics within Georgia (see Figure 4.31), but they distrust the Georgians and feel no loyalty to the Georgian state. Both groups sought independence for their republics in 1991 and

1992, but the Georgian military intervened and great bloodshed resulted. The Russian army helped to maintain a cease-fire in South Ossetia. Though Russian military personnel aided the Abkhaz cause, the Russian government worked with the UN to establish a cease-fire in Abkhazia.

The persecution of Armenians has had a lasting effect on this part of the world. In 1895, the Ottoman government massacred 300,000 Armenians within its realm. Again in 1915, during World War I, the Ottoman government tortured, exterminated, and deported its Armenian population, claiming that the Armenians were a threat. Somewhere between 600,000 and 2 million Armenians were exterminated out of a prewar population of about 3 million in what can be referred to as the "Armenian genocide." Many Armenians became refugees, migrating across their traditional homeland or leaving it all together. By 1917, fewer than 200,000 Armenians remained in Turkey.

It was not just this one period that inflicted a toll on Armenians. Over 1,000 years, foreign invaders wreaked havoc numerous times and scattered the population. Today, over half of the Armenian population lives outside of Armenia in a diaspora. About half of the diaspora community (e.g., those outside Armenia) lives in other CIS countries. The other half lives in communities from India across to Southwestern Asia, Europe, and North America, with a sizable number in the United States. Interestingly, many Armenians in the diaspora speak the western dialects of Armenian, associated with eastern Turkey, formerly in the Ottoman Empire. Armenians in Armenia speak the eastern dialects. Nevertheless, despite dialectical differences, the Armenian diaspora has close ties with Armenia. Many émigrés now serve in the Armenian government, and the Armenian government considers Armenians abroad to be members of the Armenian nation.

The largest conflict since the dissolution of the Soviet Union in 1991 involved Armenia and Azerbaijan. In 1924, the Soviet government created an autonomous territory within Azerbaijan known as Nagorno-Karabakh. It was 94.4 percent Armenian (see Figures 4.31 and 4.33). By 1979, Armenians represented only 76 percent of the region's population. Armenians began to fear their loss of numbers and objected to Azerbaijani laws that restricted the development of the Armenian language and culture. Clashes between the Armenians of Nagorno-Karabakh and Azerbaijanis began in the 1960s and developed into war by 1992. Armenian forces of Nagorno-Karabakh seized most of the territory and advanced westward to link their territory with Armenia. Afterward, they moved into Azerbaijan proper, but success brought condemnation. The Turkish and Iranian governments warned the Armenians to cease hostilities. Finally, peace talks sponsored by the UN, Russia, Iran, and a number of other countries met with success, and the shooting war ended in 1994. In addition to Nagorno-Karabakh, Armenian forces continue to control approximately 20 percent of Azerbaijan. An official agreement on the governance and the political status of territories has not emerged.

At the time, the Armenian-Azerbaijani dispute took center stage in international relations and seemed to signal the character of a new political order. Until then, the capitalist world opposed the Communist world. In this new situation, the

Armenians were Christian and the Azerbaijanis Muslim. Armenians and Azerbaijanis attempted to gain international support by arguing that they were each fighting for a greater religious cause rather than for national self-interest. As hostilities increased, Christian countries began to line up behind Armenia and Muslim countries behind Azerbaijan. Americans showed a great sympathy for the Armenians. Tensions subsided without much U.S. involvement, but the Armenian-Azerbaijani struggle portended of a new world conflict in which Christian and Islamic countries stand in opposition.

Economic Development

The economies of Georgia, Armenia, and Azerbaijan suffered greatly from the ethnic conflicts, even in Armenia, where little fighting took place. Toward the end of the 1990s, these three countries' economies steadily improved. All three countries have climates warmer than those of the other former Soviet republics and can produce agricultural products not available to the north. Thus, during Soviet times, for example, Georgia supplied over 90 percent of the Soviet Union's tea and citrus fruits. Armenia supplied fruits, especially grapes, and Azerbaijan produced tobacco, cotton, and rice.

While encouraging agricultural production, Moscow limited industrial development in Georgia, Armenia, and Azerbaijan. These policies made these republics heavily dependent on the other Soviet republics, especially Russia, for markets in which to sell their agricultural goods and sources for their industrial goods. Such policies also created a situation in which all three countries still have relatively low ownership levels of consumer goods (see Figure 4.15).

Since 1991, these countries have worked to reduce this dependency on Russia by greatly altering and increasing their industrial and service sectors. Georgia, located on the sunny, warm, eastern shores of the Black Sea, has great tourist potential. Azerbaijan, located on the Caspian Sea, will become one of the world's leading oil producers if it is ever able to exploit its oilfields fully. American oil companies are developing new fields, but continued investment depends on the subregion's political stability. In addition to problems with the Azerbaijani government, foreign investors are concerned about shipping the oil through existing and proposed pipelines (see Figure 4.11). The Russian Federation tries to make Azerbaijan use the existing oil pipeline via Groznyy (Chechnya) to the Russian port of Novorissisk. This is an example of the Russian Federation's continuing attempts to control events in the countries that were once part of the Soviet Union. To avoid continued Russian control, Georgia, Armenia, and Azerbaijan are developing trading relationships with other countries, most notably with Iran, Turkey, the United States, and those in Europe.

 # Central Asia

The former Central Asian republics of the Soviet Union now form five independent countries—Kazakhstan, Tajikistan, Uzbekistan, Turkmenistan, and Kyrgyzstan (Figure 4.35). In 2001, Kazakhstan was the largest of the five countries in area and 9th in the world, but Uzbekistan had the most people (Figure 4.36). These countries have similar landlocked situations, arid or semiarid climates, and Muslim faith of many of their peoples. The subregion's fertile river valleys of the Amu Darya and Syr Darya, are one of the cradles of human civilization. They played important roles in the trading of goods and ideas from the time of the earliest civilizations in Mesopotamia, China, and the Indus River valley. Straddling old trade routes between east and west, most notably the Great Silk Road (see Figure 4.4), great cities such as Bukhoro (Bukhara) and Samarqand (Samarkand) (Figure 4.37) emerged. During the 700s and 800s, Bukhoro became one of the leading centers of learning, culture, and art in the Muslim world. Its grandness rivaled the other Muslim cultural centers of Córdoba (Spain), Baghdad (Iraq), and Cairo (Egypt). Some of Islam's greatest historians, geographers, astronomers, and other scientists came from the area. In the late 1300s, Timur (Tamerlane) emerged as the dominant leader in Central Asia and conquered lands far to the west, south, and east. Within his vast empire, Samarqand served as his capital. A new flowering of culture began as numerous scholars and artisans came to reside here. Timur's grandson was one of the world's first great astronomers. Literature flourished and great religious structures and palaces were built.

Despite a history of great cultural, political, and economic power, Central Asia is often ignored because it appears to be isolated from world trade and has little political or economic power in current world affairs. However, Central Asian countries occupy strategic geopolitical positions with overland access to Russia, Afghanistan, China, Iran, and Pakistan. When the United States overthrew the Taliban as part of the attempt to root out Osama bin Laden and al-Qaeda in 2001, Uzbekistan and Tajikistan played large roles in the military and humanitarian campaign with their airbases and entry points to Afghanistan.

Countries

Central Asia's five countries were once part of a land called Turkestan. Turkestan was not a single, consolidated country but a loose confederation of tribes. Beginning in the 1800s, the Russian Empire expanded forcibly into Turkestan and took firm control of it by the end of the century. Remnant territories of the Central Asian peoples were annexed by neighboring countries. For example, some Uzbeks and Tajiks found themselves in Afghanistan. In the 1920s, not long after the creation of the Soviet Union, Soviet authorities drew boundaries for new republics in Central Asia. These republics became the countries of Central Asia that appeared on the world map after the Soviet Union broke up in 1991.

Accounting for more than 40 percent of the population of the five Central Asian countries and the only one that borders all the others, Uzbekistan is the dominant country in Central Asia. The fact that large numbers of Uzbeks live just across the borders in Kazakhstan, Kyrgystan, and Tajikistan emboldens Uzbekistan's government to declare its right to intervene in its neighbors' affairs to protect Uzbeks living outside its borders. On the constructive side, Uzbekistan has entered into economic agreements with its neighbors and has taken the lead on solving the environmental problems associated with the Aral Sea. During Soviet times, Moscow created the Muslim Board of Central Asia and headquartered it in Tashkent. Tashkent likewise became the center for the religious training of Muslim clerics as the Soviets closed similar Islamic institutions in the other republics. Consequently, Uzbekistan has had a lead in Islamic affairs in the post-Soviet period.

Though Central Asian countries willingly joined the CIS, they are concerned with Russian domination, seek to diminish Russian influence within their countries, and even attempt to reverse past Russification policies. For example, the Russian

Figure 4.35   **Central Asia: countries, cities, and major physical features.** These countries have a mountainous southern border. Under Russian rule, they were carved out of the former Turkestan area. Only Almaty and Tashkent have populations of over a million people.

Figure 4.36   **Central Asia: key data on countries.**

Country	Capital City	Land Area (km²) Total	Population (millions) Mid-2001 Total	2025 Est.	GNI 1999 (US $ million) Total	GNI PPP 1999 Per Capita	Percent Urban 2001	Human Development Index Rank of 174 Countries
Kazakhstan	Astana	2,717,300	14.8	14.7	18,732	4,790	56	76
Kyrgyzstan	Bishkek	198,500	5.0	6.5	1,465	2,420	35	97
Tajikistan	Dushanbe	143,100	6.2	7.7	1,749	no data	27	108
Turkmenistan	Ashkhabad	488,100	5.5	6.5	3,205	3,340	44	96
Uzbekistan	Tashkent	447,400	25.1	34.1	17,613	2,230	38	92

Source: Data from *Population Reference Bureau 2001 Data Sheet*; *World Development Indicators*, World Bank, 2001; *Human Development Report*, United Nations, 2001; Microsoft Encarta (ethnic group, language, religion).

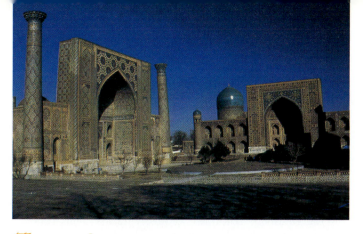

Figure 4.37 **Central Asia: Samarqand, Uzbekistan.** Historic Muslim buildings highlight the importance of Islam in Central Asia. Madrasah (theological seminary) buildings of the 1500s and 1600s border Registan Square. Typical features of Islamic art include arches, tilework, domes, and minarets. Photo: © Ronald Wixman.

and then Soviet government settled many Russians in northern Kazakhstan during the 1800s and then again during the Virgin Lands Campaign of the 1950s and 1960s (see Box Figure 1 in the Point-Counterpoint box, p. 143). Soviet industrialization also integrated the northern part of the country with the rest of the Soviet Union. To diminish the power of the large Russian presence now in Kazakhstan resulting from these earlier policies, the Kazak government moved the capital of the country from Almaty (Alma-Ata) in the Kazak-dominated southeast to Akmola in the Russian-dominated areas of the north on June 10, 1998. Akmola, meaning "white tomb," was renamed Astana, meaning "capital." The government claimed that Almaty had outgrown its location and that Astana (Akmola) was a more central location within the country and thus a better place for the seat of government. The move also means that more Kazaks will likely move to the Russian areas. Feeling unwelcome under many of Kazakhstan's governmental policies, many Russians are leaving the country.

Governmental policies cause concerns about human rights abuses. Amnesty International and Human Rights Watch, for example, have criticized Uzbekistan for having one of the worst human rights records of the former Soviet republics. The government suppresses freedom of expression, frequently arrests members of the opposition, and forbids an independent media. The U.S. government initially sent economic aid to the Uzbekistan but on continued reports of human rights abuses, sharply curtailed the aid. After it initiated a war in Afghanistan in 2001 in the wake of the attacks of September 11, Uzbekistan's government allowed the U.S. military to use Uzbekistan as a staging ground. The United States began sending aid to Uzbekistan again.

People

While most of the other CIS countries have stable or declining populations, the five countries of Central Asia anticipate growth from the 2001 total of around 56 million people to 70 million by 2025. In 2001, natural increases remained below 2 percent in the countries with large Muslim majorities: Kazakhstan (50 percent non-Muslims) had the lowest increases (0.5 percent per year), followed by Kyrgyzstan with 70 percent Muslims (1.3 percent per year). The age-sex diagram for Kazakhstan (Figure 4.38) contrasts with those of the other areas of the CIS. Death rates are very low—a recipe for increasing population (see Figure 4.20). The fertility rates fell from around 6 to around 3 between 1970 and 2000. Such rates contrast with the population decline in the rest of the former Soviet domain.

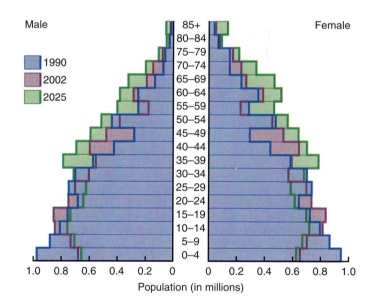

Figure 4.38 **Age-sex diagram of Kazakhstan.** How does its shape compare to those in Figures 4.19 and 4.34. What explains the differences? Source: U.S. Census Bureau. International Data Bank.

Ethnic Groups (percent)	Languages	Religions (percent)
Kazakh 42%, Russian 37%, Ukraine 5%, German 5%	Kazak, Russian	Muslim 47%, Russian Orthodox 44%
Kazakh 52%, Russian 21%, Uzbek 13%	Kirghiz, Russian	Muslim 70%, Russian Orthodox, etc. 30%
Tajik 65%, Uzbek 25%, Russian 4%	Tajik, Russian, Uzbek	Sunni Muslim 80%, other Muslim 7%
Turkmen 73%, Russian 10%, Uzbek 9%	Turkmen, Russian, Uzbek	Muslim 87%, Eastern Orthodox 11%
Uzbek 71%, Russian 8%, Tajik 5%	Uzbek 74%, Russian 14%	Muslim (mostly Sunni) 88%

Urbanization

Urbanization in Central Asia is the lowest in the CIS. Kazakhstan, the most industrialized country, had 56 percent of its population living in towns in 2001, in contrast to other countries, which had 27 to 44 percent living in towns. Only Almaty (former capital of Kazakhstan, 1.2 million) and Tashkent (Uzbekistan, 2.1 million) had a million or more people in 2000. Tashkent was the fourth-largest city in the former Soviet Union, with most of Uzbekistan's Russians living there. Bukhoro (Bukhara), Samarqand (Samarkand) (see Figure 4.37), and Tashkent are historically significant. Their older cores, built with wealth from medieval trade, have architecturally beautiful structures and attract tourists. Soviet additions are typically utilitarian with featureless factories, offices, and apartment blocks.

Culture

Modern Central Asians are settled, but many of their ancestors were nomadic, and their cultures still reflect the traditional ways of life of their ancestors. For example, in traditional Kazakh culture, it is customary to ask about the well-being of someone's livestock before inquiring about the person's health and that of his or her family. The dwelling of these nomadic peoples is the yurt, a circular tent consisting of a willow wood frame covered in wool felt. An opening at the top allows smoke to exit from the fire used for cooking and heating. Modern Central Asians no longer live in yurts, but they use them as decorative motifs for building or erect them in their yards and sleep in them during the summer. The yurt is an important symbol of national identity and appears on the national flag of Kyrgyzstan.

Carpet making and horse breeding are important national traditions for the Turkmens. Five traditional carpet designs are incorporated into Turkmenistan's national flag. Akhalteke, a breed of horses well adapted to the desert, is the breed of national significance, appearing as the central figure in Turkmenistan's national emblem. Many Turkmen still own at least one akhalteke.

The Uzbeks, who prospered greatly from the Great Silk Road (see Figure 4.4), wear clothing made with fine fabrics, color, and ornamentation. All were expensive in earlier times, and the ability to incorporate as much as possible in one's dress showed one's wealth. This earlier culture practice is evident today. On any given day, it is still common to see individuals elegantly dressed, though they have no special function to attend.

Modern Central Asians are by no means homogeneous, as each group has a unique history of ethnic development, though many of them are closely related.

- Kazakhs emerged as a distinct people in the 1400s from a mixture of Turkic and Mongolian nomads of Central Asia.
- The ancestors of the Kyrgyz probably originated in Mongolia. After the 800s, they mixed with Turkic tribes from the south and west and adopted their current name, Krygyz, which means "40 clans" in the Turkic languages. The 40 clans are represented on the national flag with a sun that has 40 rays. During czarist times, both Kazakhs and Kyrgyz were called Kyrgyz, illustrating that the languages of the two peoples are very closely related.
- Turkmens trace their ancestors back to Oghuz tribes that inhabited Mongolia and southern Siberia around Lake Baikal. In the 700s, these tribes migrated into Central Asia and assimilated Turkic and Persian tribes, giving rise to the Turkmen. "Turkmen" probably means "pure Turk" or "most Turklike of the Turks."
- Uzbeks are a Turkic people who moved into the lands now known as Uzbekistan in the 1500s. They are closely related to Turkmens but also have close ties with Tajiks.
- Tajiks probably acquired their name from an old Arab tribe. The Tajik language is a Persian language and was not distinguished from Persian (Farsi) until the Soviets designated Tajik as a unique language. At the same time, Tajiks had not differentiated themselves from Uzbeks, though the Uzbeks are a Turkic people. Both groups commonly spoke each other's languages.

During late Russian imperial and Soviet times, Central Asians were subjected to Russification policies. Soviet policies were more intense, forcing the Central Asians to adopt the Cyrillic alphabet of the Russian language for their native languages. After 1991, many Central Asian countries, with the exception of Kazakhstan, returned to using the Latin alphabet (used for English and Turkish), began expunging Russian words from their languages, and downgraded the status of the Russian language. The policies were problematic because many Central Asians speak only Russian. Moreover, Russians emigrated, and the Russian government was offended and applied political and economic pressure on these countries. By the late 1990s, many Central Asian governments curtailed their attacks on the Russian language, and many of them designated Russian as the second official language of their countries.

Most Central Asians are Muslims. Those living closest to the Islamic heartland, such as the Uzbeks, converted to Islam as early as the 600s, soon after the rise of Islam. Those farther north, such as the Kazaks and Kyrgyz, converted to the religion only in the last 200 years. Kyrgyz also mix Islam with their traditional belief in totemism, the religious idea that humans have spiritual kinship with particular animals (e.g., reindeer, camels, bears). The sun, moon, and stars also have great religious importance for the Kyrgyz.

Within the Russian Empire and later in the Soviet Union, Islamic beliefs were suppressed by deportations, the breakup of nomadic lifestyles by collectivization, and the immigration and resettlement of Russians. Since independence in 1991, the Muslim majorities played significant roles in the political transition of these countries. Some of the countries look to links with Iran (e.g., Azerbaijan and its Shia Muslims) or with other Sunni Muslim countries such as Iraq, Turkey, Kuwait, and Saudi Arabia. These countries finance the building of mosques and other Islamic institutions in Central Asia.

Though Central Asians have a strong faith in Islam, all five countries reduced the impact of Islamic political groups by the late 1990s, holding trials of activists and introducing restrictive laws. In mid-1998, fear of pressure from the victorious Taliban in Afghanistan caused Tajikistan to agree to 30,000 Russian military entering the country to guard the southern border. After the events of September 11, 2001, when the United States formed a coalition against Afghanistan, the Taliban threatened to invade Uzbekistan if the Uzbek government agreed to cooperate with the United States. Though Uzbekistan offered help to the United States, the Taliban in Afghanistan was not able to follow through with its threat.

Ethnic Conflict

The ebb and flow of history created an ethnically complex subregion with significant minority populations in most countries (see Box Figure 1 in the Point-Counterpoint box, p. 143). The ethnic mosaic and many of the tensions that exist between groups can be attributed to Soviet policies. In the 1920s, the Soviets drew the republics' boundaries in a way to help the Communist government in Moscow keep control over Central Asia. For example, the Soviet drew the boundaries of Kyrgystan, Tajikistan, and Uzbekistan so that they arbitrarily zigzag through the fertile and densely settled Fergana Valley, leaving many Kyrgyz, Tajiks, and Uzbeks in their neighbors' countries. These boundaries pitted, and continue to pit, Central Asians against one another, weakening their resistance to Moscow's rule and even resulting in the people looking to Moscow to maintain peace.

Soviet policies also resettled many Russians in the subregion. The presence of Russians was another means for the Soviet government to control Central Asia. During World War II, Stalin built many factories in Central Asia because the subregion was out of reach of the Nazi military. He moved Russians and other peoples from across the Soviet Union to work in these factories. As part of this effort, he also shifted many Central Asians to each other's republics and relocated ethnic groups he did not trust to Central Asia.

Soviet policies eventually split Central Asian societies in two. The groups for whom republics were named (titular groups) tended to maintain traditional ways of life in rural areas, while the imported ethnic minorities, particularly Russians, became the industrial, business, and civil servant class and dominated the urban areas. For example, in Tajikistan's capital city, Dushanbe, only 39 percent of the population was Tajik in 1989. Many Central Asians view the Russians within their countries with resentment and suspicion—as unwanted reminders of Soviet rule. Moreover, while many Central Asians speak Russian, few Russians bother to learn Central Asian languages. Central Asians also resent the privileged economic positions that Russians continue to hold within their countries. New anti-Russian policies and a general anti-Russian attitude of many Central Asians have caused many Russians to emigrate since 1991.

The relationships between Central Asian governments and ethnic groups have not been positive either. For example, anger at the ethnic minorities in Uzbekistan has lead to violence and the emigration of these minorities. Tajik authorities have criticized neighboring Uzbekistan for not recognizing the uniqueness of Tajik culture and forcing many Tajiks in Uzbekistan to register as Uzbeks. At the same time, violence against Uzbeks in neighboring countries caused many Uzbeks to migrate to Uzbekistan. After a civil war in Tajikistan in the 1990s, thousands of Tajiks moved north into Kyrgyzstan in an attempt to escape the conflict, but this only intensified ethnic tensions in Kyrgyzstan.

Economic Development

During the Russian imperial period and especially under Soviet domination, the economic output of these five countries was redirected to supplying the needs of Russia. When the Civil War in the United States in the 1860s prevented Russian industry from receiving its cotton supplies, the Russian government saw a great opportunity in Central Asia to create its own supply of cotton. Later, the Soviets continued developing the cotton industry but also imposed other forms of economic development on the subregion, bringing in its technologists and education systems that drastically altered the human geography of Central Asia. Industrialization was based on producing iron and steel and tractors, while farming focused on growing irrigated cotton. The extraction of the rich mineral resources including coal, iron, chromium, oil, and natural gas turned the subregion into a colony producing raw materials for Russian factories.

Soviet infrastructure still keeps Central Asia dependent on the Russian Federation. Pipelines and transportation lines are of Soviet specifications and run primarily to the Russian Federation. Central Asians are leery of Russia taking advantage of them but do not want to depend on China to the east. Goods could travel through Azerbaijan, Armenia, and Afghanistan, but all three of these countries are very unstable. Turkmenistan is working on a rail line to Iran. Western countries, however, are uneasy about shipping goods through Iran and may not trade with Central Asia if Iran is the only choice for transit. Thus, Central Asia remains heavily dependent on the Russian Federation. Such trade benefits the Russian Federation but not necessarily Central Asia. The Russian Federation buys cheap raw materials from Central Asia and sends back more expensive finished products. Kazakhstan and Turkmenistan need a route to transport oil and natural gas to Western customers and are frustrated by Russia's failure to support new pipelines (see Figure 4.11). The Russian Federation has no incentive to help these two countries in this matter because it also sells oil and natural gas to the west. Kazakhstan barged oil across the Caspian Sea and then through Iran. The practice risked condemnation from Western countries such as the United States that had an economic embargo on Iran.

Because all five countries produce similar commodities (oil, natural gas, and cotton), rivalry rather than cooperation is most common, especially in attempts to attract foreign trade. Kyrgyzstan and Tajikistan, however, do not have large fuel reserves like the other three. Consequently, the other three countries have shut off fuel as a political lever against their neighbors.

The issue of water resources also encourages rivalry. Water is scarce in this dry area of the world but important for agriculture and power generation. Disputes exist over the allocation of water that flows in rivers that pass through a number of the countries. Major rivers originate in the mountains of Kyrgyzstan and Tajikistan and flow down to the lowlands of Turkmenistan, Uzbekistan, and Kazakhstan (see Figure 4.13). Each country frequently complains that the others withdraw an unfair share of the water from the rivers. Serious tensions have arisen between Kyrgyzstan and Uzbekistan over water in the Fergana Valley, where agricultural reform and land privatization programs are endangered by the water disputes.

Test Your Understanding 4E

Summary   The countries of the Southern Caucasus and Central Asia are among the world's most complex in terms of ethnic differences and conflicts. They are tied economically to the Russian Federation but seek new relationships with other countries of the world. Their world position at the heart of Asia makes them strategically significant.

Questions to Think About

4E.1   What are the cultural similarities and differences among the peoples within the countries of the Southern Caucasus and Central Asia?

4E.2   What are the actual and potential sources of political conflict in the Caucasus and Central Asian countries?

4E.3   How do the physical and economic geographies of the Central Asian countries differ from each other?

Online Learning Center   **www.mhhe.com/bradshaw**

Making Connections

The Online Learning Center accompanying this textbook provides access to a vast range of further information about each chapter and region covered in this text. Go to **www.mhhe.com/bradshaw** to discover these useful study aids:

- Self-test questions
- Interactive, map-based exercises to identify key places within each region
- PowerWeb readings for further study
- Links to websites relating to topics in this chapter

Chapter 5

East Asia

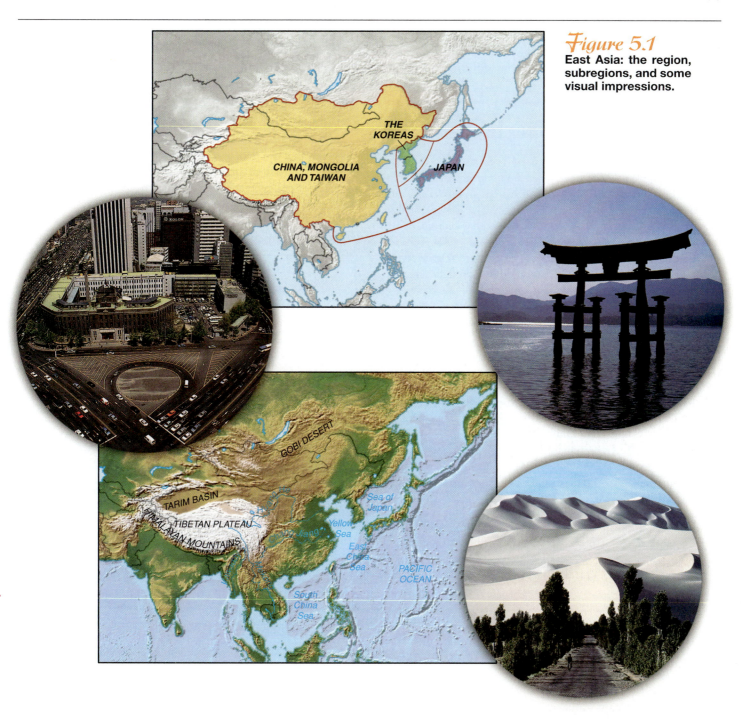

Figure 5.1
East Asia: the region, subregions, and some visual impressions.

THE KOREAS

CHINA, MONGOLIA AND TAIWAN

JAPAN

GOBI DESERT

TARIM BASIN

TIBETAN PLATEAU

HIMALAYAN MOUNTAINS

Sea of Japan

Yellow Sea

East China Sea

PACIFIC OCEAN

South China Sea

East Asian Miracle

The reemergence of East Asia (Figure 5.1) as a world political, economic, and cultural force was a major event of the later 1900s. The changes stemmed from decisions to modernize on European patterns. First, advice from western European and American sources was used by the Japanese government to transform the geography of Japan from the late 1800s. Taiwan and South Korea accepted U.S. advice from the 1950s. Second, Communism backed by Soviet Union advice produced major political and cultural changes—but erratic economic development—in China and North Korea. The subsequent changes have been so great that the early 2000s may see East Asia, led by Japan and China, refocusing global economic growth and political power on the countries situated around the Pacific Ocean. In so doing, this region will not only interact with external globalization trends but will also contribute to those trends.

The region's present character builds on a period of cultural and technological development that is longer than in most other world regions. From ancient times, the Chinese instigated many of the most significant human advances from technology to art. Until well into the 1900s, they insisted that all other people were "barbarians" in relation to their own "civilized" ways. In studying this region, therefore, we must understand the significance of the traditional cultures, the political and economic changes of the last century, and the responses to Western global political and economic dominance.

From Poverty and Defeat to Renewed Eminence

After World War II, poverty, political disruption, and cultural confusion were the rule in East Asia. In the following decades, the region moved to the forefront of the global economy and political status. In 1945:

- Defeated Japan was economically devastated and hated in the region for its wartime acts.

- The Koreas struggled with independence following decades of Japanese occupation and exploitation. Splitting Korea into North and South led to the 1950s Korean War, which destroyed much of both countries.

- China divided into warring Nationalist and Communist factions, which resumed their prewar conflict until the Communists prevailed in 1949 and the remnant of the Nationalist army fled to Taiwan.

Even in 1960, Japan, despite 100 years of modernization, still had a gross national income per capita that was one-eighth that of the United States, while other East Asian countries had GNIs per capita equal to or lower than those of African countries at the time. A major economic and political turnaround occurred over the next 40 years. Japan's export-led growth raised its gross national income one hundredfold between 1960 and 2000 at current dollar values, making it the world's second-richest country (Figure 5.2).

Some 10 years behind Japan, the newly industrializing South Korea, Hong Kong (before and after it became part of China in 1997), and Taiwan increased their incomes six- to sevenfold from 1980 to 2000. Building on the greater national cohesion that the People's Republic of China established over nearly 30 years at tremendous cost in terms of lives and changes of policy, that country experienced its most rapid economic growth from 1980, increasing its total GNI from $202 billion in 1980 to $1,065 billion in 2000. However, neither North Korea with its declining economy nor Mongolia with its tiny economy experienced such economic expansion.

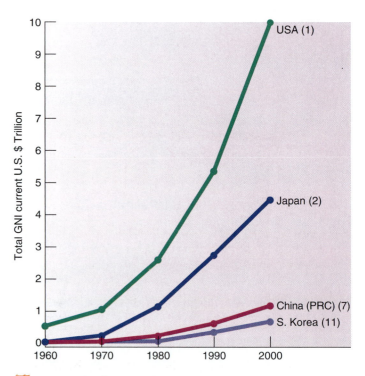

Figure 5.2 **GNI increases in East Asian countries, 1960–2000, compared to the United States, in current U.S. dollars.** World rankings for total GNI are in parentheses.

Will Miracle Growth Continue?

The East Asian economic growth in the later 1900s was geographically uneven and subject to interruptions. For example, in China, the rising incomes of people along the coast, who have access to global trade, contrasted with the continuing extreme poverty of many inland areas.

By the mid-1990s, the rapid economic growth in some countries began to slow, leading to the crisis years of 1997 and 1998. Banks that had accepted too much debt on the expectations of continued economic growth collapsed and the South Korean currency was devalued. That crisis spread from Southeast Asian countries to South Korea and Hong Kong but had less impact on China or Taiwan. The United States and the International Monetary Fund made loans available but demanded an end to protectionist policies that kept Western foreign investments out of East and Southeast Asia. South Korea recovered its economic impetus within a couple of years, but in that time it also suffered unemployment, corporation failures, and an increase in long-term poverty.

In 2001, a World Bank report, "Re-Thinking the East Asian Miracle," reanalyzed the "East Asian Miracle" it had identified in 1993. The 1993 study had argued that distinctive Asian characteristics motivated and fueled economic growth:

- Successful countries had strong governments that managed stable business environments by such measures as keeping inflation low. Their tax structures distributed some of the growth rewards throughout communities. Domestic savings increased.

- Governments encouraged export competitiveness that led to partial global integration.

- Education programs focused on primary and secondary education for both boys and girls as a priority over prestigious higher education for a few.

After the 1997–1998 economic crisis, alternative explanations centered around a combination of global trends and local attitudes:

- Economic growth up to the mid-1990s resulted mainly from increased inputs of capital, labor, machinery, and infrastructure (transportation, power, etc.), rather than from rising productivity (e.g., increasing output per employee) based on technological progress.

- The advantages of activist government industrial policies and liaisons between bankers and industry with government support appeared less convincing when large corporations used government funds wastefully and often corruptly.

- The policy of encouraging export sales and protecting home markets was a poorer motivator of economic growth than openness to total trade, especially in the interdependent world of globalization.

- Increasing global links showed the weakness of such social institutions as strong government, family control of businesses from the smallest to the largest corporations, and regulatory agencies without sufficient powers.

- Many of the profits from export earnings went into real estate speculation rather than further investment in productive facilities. When a 1997 crisis of confidence called for funds to support country economies, many were inaccessible or bound up in such speculation.

Above all, the crisis dealt a blow to assumptions that the "Asian Way" would produce continuing economic expansion. Geographic factors, beginning with the development of distinctive cultures in specific places through history and interactions of people with the natural environment, bring a greater degree of diversity to this region than one simplistic prescription could change in a few years.

Cultural and Political Influences

Chinese, Koreans, and Japanese share many cultural characteristics (Figure 5.3), including Buddhism, the use of Chinese script, close family life and kinship links, and a focus on communal organization. These common elements contributed much to modern attitudes toward the global economy. While the similarities are important, differences emerged between the Chinese, Japanese, Mongols, and Koreans that give character to each of the modern countries. The Chinese-based and other East Asian cultures are very old, and their traditions continue to permeate the lives of people in the region today.

The business phenomenon of the "Asian Way" arose out of Confucian ideals, marked by a hierarchy of relationships that define how people work and live with each other: ruler-subject (boss-employee), father-son, husband-wife, older person-younger person. In each, the former is regarded as protecting, with due consideration, the latter, while the latter respects and obeys the former. There is a parallel emphasis on maintaining poise and proper respect ("face") in public.

Chinese Empires

East Asian cultures developed largely through the interactions of past Chinese civilizations and empires with people in surrounding regions. Through history, Chinese empires added territories or lost them to rivals. At other times, internal anarchy fragmented the area controlled by Chinese emperors. Each phase left its marks on people and landscapes.

Through periods of both unity and disunity, Chinese people produced a series of technological and cultural achievements, including the invention of items as diverse as paper, printing, the compass, the crossbow, paddleboats, clocks, and gunpowder. The Chinese developed a distinctive written script and styles of paintings, pottery (Figure 5.4), and building designs. They took significant steps in astronomy, mathematics, the understanding of weather and minerals, and engineering skills such as deep mining. Many of their achievements were taken up in western Europe at a later stage; for example, Chinese paintings influenced French painter Claude Monet in

(a)

(b)

(c)

Figure 5.3 **East Asia: prominent cultural features.**
(a) Buddha on Eight-Trigram Mountain, Taiwan. (b) Mount Fuji, Japan, sacred to many Japanese people. (c) The Torii Gate, Miyajima, Japan, a water shrine.

the late 1800s. Understandably, the Chinese took for granted their superiority over other people. China was the "Middle Kingdom" at Earth's center, and the Chinese emperor was the link between Heaven and Earth.

Foundations of Chinese Cultures

By 2000 B.C. (see Chapter 1, page 18), the first Chinese civilization developed in the Huang He (Yellow River) valley (Figure 5.5) based on wealth from agricultural surpluses and craft skills. It gained control of much of the lower Huang He

and lower Chang Jiang (Yangtze River) basins, adding a separate center of ancient civilization to those established in southwestern Asia, Egypt, and Pakistan (see Chapters 7 and 8).

The Zhou dynasty (1122–256 B.C.) used Iron Age technology, irrigation, and deeper plowing of the soil to support economic growth. Improving transportation opened wider geographic trading areas but also made the wealthy Chinese lands vulnerable to attacks from outside. During the political and social upheavals of this period, **Confucius** (Kong Fuzi), an administrator, called for a focus on proper relations in all social contexts, proposing a hierarchical system of relationships as a basis for orderly conduct. He wished to weld the growing empire together on the basis of better-trained and more able managers. Hundreds of years later, knowledge of Confucian principles made appointments to public office possible through a system of merit-based written exams. His significance as an educator led to the adoption of his birthday in September as National Teachers' Day.

Daoism, the teaching of Laozi, disdained the Confucian hierarchical system for organizing society, preferring a return to local, village-based communities with little government interference. Daoism is based on the apparently opposing concepts of yin and yang (including the "dark side" and "sunny side" of a hill, negative and positive, male and female, evil and good, Heaven and Earth). Yin is female, linked to the lightweight, a perception that affects Chinese gender values. In fact, yin and yang are complementary, interdependent principles operating in space and time, emblems of the harmonious interplay or balance of all pairs of opposites in the universe. Yin and yang are seen as two "breaths" (*qi*, or cosmic energy, pronounced "chee"). Every person's duty is to strengthen, control, and expand the *qi* received at birth.

Around 220 B.C., the Qin dynasty (spelled "Ch'in" in older forms of Romanization), after which China is named, imposed strict imperial laws to weld the surrounding feudal states into a centralized and culturally uniform empire with a standardized written script. It sited its capital near modern Xi'an. Armies extended the empire southward into the Xi Jiang (West River) basin and westward as far as Lanzhou. The high cost of completing the Great Wall (Figure 5.6) caused internal resentment.

After Southeast Asian Buddhism had been rejected earlier in the south of China, Buddhist monks from India, traveling over the inner Asian Silk Route, reached northern China in the early years of the first millennium A.D. Adding a spiritual dimension to materialistic Confucianism, Buddhism also brought a more systematic framework to Daoist philosophies. After 600, these links resulted in a period of literary and artistic brilliance, religious toleration, and foreign trade. A literate Chinese elite, however, rejected Buddhism as disruptive.

Over the next three centuries, the Tang dynasty spread Chinese influence into the Northeast of modern China, Korea, Japan, and northern Vietnam. Zen Buddhism, a form that is centered in meditation and moments of insight in the course of mundane activities, arrived in Korea and Japan in the early 1200s.

Lamaistic Buddhism was established in Tibet from the A.D. 600s, combining Indian Buddhism with Tibetan rituals. The Dalai Lama leaders had both religious and secular roles. The term *Dalai* ("ocean of wisdom") was conferred by a Mongol ruler in

(a)

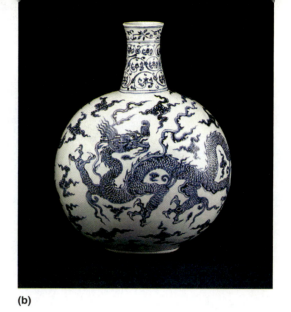
(b)

Figure 5.4 **Chinese art.** (a) Sorting of the cocoons from the *Book of the Silk Industry,* Qing dynasty, early 1800s. (b) Porcelain flask painted in blue underglaze with a running dragon on each side, Ming dynasty, early 1400s. Photos: (a) © Giraudon/Art Resource, NY; (b) © Victoria and Albert Museum, London/Art Resource, NY.

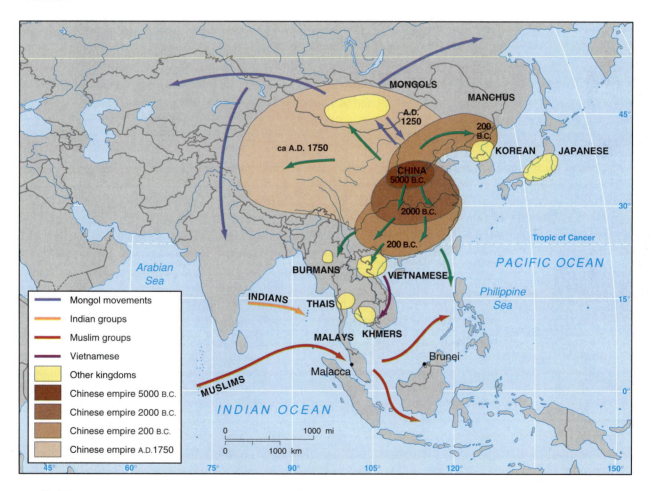

Figure 5.5 **East Asia: the expansion of the Chinese empires and the location of adjacent kingdoms.** The Mongol and Manchu kingdoms were the only external ones that expanded to take over all of China. The green arrows show movements of Chinese people.

the 1200s, and Lamaism had a continuing influence on Chinese dynasties that wished to establish political control over Tibet.

Mongol Invasions

In the 1200s, the Mongols invaded China. Kublai Khan, the Mongol leader, established his capital in northern China at Beijing. He and his descendants ruled as Chinese emperors, driving the center of native Chinese intellectual and economic development southward. In the 1300s, growing Chinese resentment at the new rulers, disruptions, and taxes imposed on them, combined with crop failures and famines, opened the way for rebellion and the proclamation of the Ming dynasty in

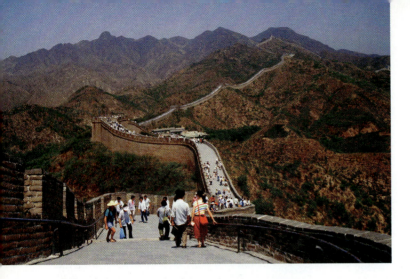

Figure 5.6 **East Asia: Great Wall of China.** Built in stages and connected around 200 B.C., it is nearly 2,400 km (1,500 mi.) long. It was surprisingly effective in keeping out aggressors for many centuries. It is now a major tourist attraction.

1368. By that time, the Mongols had intermarried with Chinese and became sinicized.

Later Empires

Further Chinese expansion occurred under the Ming dynasty (1368–1644), during which the economy, literature, and education advanced further. Chinese naval power under Admiral Cheng-Ho (1405–1433) ventured beyond India to the Persian Gulf, Red Sea, eastern Africa, and Madagascar. Although there was trade, the Chinese did not conquer new lands.

The succeeding Qing (Manchu) dynasty ruled from 1644 to 1911. By the mid-1700s, it had expanded the Chinese realm to its farthest limits in the Northeast, Mongolia, Xinjiang (the Northwest), Tibet, Burma (Myanmar), and Taiwan. The expansion once again brought Lamaistic Buddhists and Muslims into western parts of China. In 1820, China was the world's wealthiest empire, producing nearly one-third of the world GDP. After that time, however, Western influences and conflicts led to a decline of political and economic power.

Japanese Isolationism

The most powerful of the kingdoms, or empires, in the lands surrounding China developed in Japan. Ancient traditions place Japan's foundation in 660 B.C. The Ainu, a non-Mongoloid people, were among the earliest inhabitants but retreated northward as other groups invaded through Korea.

The principle of **Shinto** encapsulated Japanese traditional values in a national religion built on animism, ancient myths, and customs (see Figure 5.3). It is a pantheistic religion of many gods and involves emperor worship and Japanese superiority. It became an instrument of political control. In Shintoism, both living and nonliving objects possess spirits, with Mount Fuji and other mountains being given special reverence. The Japanese avoided living on mountains (in a combination of reverence and fear of volcanic eruptions and earthquakes) and crowded into the lowlands. Japanese Buddhism tolerates Shintoism, and both are followed by most of the population. Chinese Daoism and Confucianism also influenced social practices.

Japanese emperors resided at Kyoto, retiring from public life and delegating administration of the country to leading families of court nobles. From the 1100s, these feudal lords took over the imperial administration, and armed samurai warriors under a shogun (military leader) acted over the heads of the powerless court. The shoguns dominated Japan and largely maintained its isolation from the rest of the world.

When the last shogun resigned in 1867, the Meiji emperor regained his position as titular head of government. The royal capital moved to the shogun center in Edo, later called Tokyo ("eastern capital"). However, in the 1889 constitution, the emperor became a figurehead and political power was taken by leading figures in the parliament.

Korean Origins

The area that is now North Korea was the center of the first Korean kingdom, around which the peninsula was unified in the 600s. After the Mongol invasion and retreat, Confucian principles of government were adopted. The Yi dynasty, founded in Seoul in A.D. 1392, ruled the kingdom of Korea until 1910, although it became subject to the Chinese in 1644. Close proximity to Japan attracted aggressive interest from the growing military power, but Japan was resisted until the late 1800s. In 1894–1895, as rebels opposed the Korean government, the newly strengthened Japanese forces overran Korea, defeating both Chinese and Russian attempts to aid the Koreans. Japan annexed Korea in 1910.

Responses to European Intrusions

The Portuguese led European visits to the region at the start of the 1500s but were soon followed by the Dutch, Spanish, French, and British. Although European colonists never occupied China, Korea, or Japan, the arrival of traders, missionaries, and military forces affected them all. Their partial incorporation into Western global economic and cultural systems began.

China Resists Colonization

In 1557, the Portuguese established Macau as the first trading port on the Chinese coast. China resisted other attempts to set up trade or political links until the global economic system expanded during the 1800s. European countries increasingly pressed China to engage in trade.

The British East India Company raised demand for Chinese tea, exported through the southern port of Guangzhou (Canton). At first, the company paid for tea in silver but replaced silver with opium from India as the trade increased. In 1796, to preserve its privileged trading position at Guangzhou, the company bowed to the Qing emperor's ban on importing opium to China. However, because the company relied for an increasing share of its income on its Indian monopoly over the production and distribution of opium, it sold the drug in open auctions in India. Private merchants then took it to the Chinese coast. By 1819, the opium trade, stimulated by lower prices, increased so much that it exceeded the value of tea exports and required payment in Chinese silver, draining those resources. More merchants, Chinese and European, entered the trade until a Chinese official reasserted the policy of forbidding opium imports. The resulting Opium War of 1841–1842 forced the Qing government to sign the Treaty of Nanjing. This treaty established a framework for continuing British trade with China, the opening of other ports, and the surrender of an island in the Zhu Jiang (Pearl River) estuary upon which was founded the city of Hong Kong. The Kowloon Peninsula opposite Hong Kong Island was ceded to Britain in 1852. In 1898, China leased the New Territories to Britain for 99 years, extending Hong Kong's links to the mainland.

Further local wars during the 1800s ceded rights of Chinese trade and residence through specified ports to Britain, France, and the United States. Although China never became the colony of a single European country, the increasing external control of its economy affected the Chinese deeply. Colonial status was unthinkable for the Chinese because they believed China to be the only civilized country and demanded that all other people should bring them tribute, not vice versa. Foreign trade conflicted with the self-sufficiency of the empire's feudal system. The trading concessions through eastern ports weakened the Qing emperors' authority and internal political structure.

Disrupted Chinese Republic

In 1912, a republic replaced the enfeebled Qing dynasty. Sun Yat-sen, who had experienced some years of exile for rebellion in the 1890s, and his Kuomintang party (Guomindang, or the Chinese Nationalist Party) claimed overall political power. Attempts to reunify China under a common government in the 1920s and 1930s, however, contended with strife among local warlords, a Communist rebellion, and Japanese attacks in the Northeast. The main Japanese invasion of China began with a 1933 expansion out of the Northeast that it had occupied earlier in the 1900s, followed by a general declaration of war in 1937. The Kuomintang and Chinese Communists united in resistance. After World War II, they fought each other again, and the Communist victors established the People's Republic of China in 1949. The defeated Kuomintang leaders fled to Taiwan, protected by U.S. forces, and claimed to continue the true government of China.

Japan Asserts Its Independence

After 1867, the new government of Japan concluded that it should take major actions to prevent world domination by Western countries. The Meiji leaders considered it necessary to follow the Western world's path to international political power through rapid industrialization to catch up with the military capability of Western countries. French advisors helped to modernize the army. Germans gave economic and political counsel. British naval experts, Dutch engineers, and educators from the United States also contributed. At the end of the 1800s, its new military power enabled Japan to take Taiwan and Korea and to win a war against Russia.

After 1911, Japanese governments became more democratic, and social groups established political parties and labor unions. In the worldwide economic depression of the 1930s, however, the economy faltered and military leaders took charge of the government. After invading China in the 1930s, the Japanese military-industrial war machine, invoking the past glories of the shogun leaders and overriding the civil government, went further and occupied much of Southeast Asia in the early 1940s. Japan's antagonism toward the United States, fueled since the early 1900s by the United States slowing the flow of Japanese immigrants into California, came to a head with the surprise attack on the U.S. Naval Base at Pearl Harbor, Hawaii, in December 1941. The Pacific war that followed was a major part of World War II and ended in the 1945 defeat of Japan.

During World War II, the rest of East Asia was made wary of Japanese political imperialism by the harsh manner in which the Japanese military occupied other countries. Japanese military promoted themselves as a master race. Prisoners of war and civilians in occupied countries were used as forced labor, and some women were subjected to being sex slaves for soldiers. After 1945, groups in the formerly occupied countries pressed for redress, and in the 1990s, the Japanese government apologized to them.

Natural Environments

The natural environments of East Asia affected the cultural history and added distinctive local features to the region. East Asia is marked by earthquake and volcanic activity and often-violent weather, especially typhoons. Major rivers cut into the dominant mountainous and hilly landscapes. Climatic environments vary from wet subtropical coastal areas to arid midlatitude continental interiors. The large numbers of people crowd into the small proportions of well-watered lowlands. They are much affected by natural events, especially in rural areas, and themselves have major impacts on the physical environment by modifying natural features and processes.

Subtropical and Midlatitude Climates

East Asia has a variety of climatic environments based around the east coast midlatitude regime of warm, humid summers with air flowing in from the ocean and cold winters as air flows out from the frozen continental interior (Figure 5.7). The combination resembles the eastern parts of North America, from Cuba and Florida in the south to New England and Newfoundland in the north.

Southern China has a subtropical **monsoon climatic environment** with major seasonal wind changes. In summer, southeasterly winds are drawn into central Asia as the continent heats up and rising air occurs in areas of low pressure. These winds bring moisture from the tropical oceans and heavy rains. In winter, the winds blow outward from the high pressure over the cold continent and are cooler and drier. In mid-to-late summer,

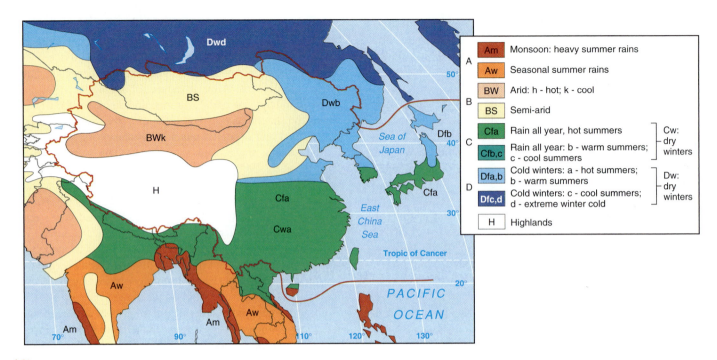

Figure 5.7   **East Asia: climate regions.** Note the contrasts within the region shown by red boundary.

the heating of the Pacific Ocean makes the region subject to typhoons (the Asian equivalent of hurricanes) with strong winds and rain.

Farther north, coastal China, the Koreas, and Japan also have **midlatitude east coast climatic environments** with summer rains and drier winters. Although summers are as warm as those farther south, icy winds blowing from Siberia make winters much colder. Japan receives more winter precipitation than the continental countries after the winds blowing from the continental interior pick up moisture on crossing the Sea of Japan and precipitate snow on the west-facing mountains.

The western parts of China and the whole country of Mongolia have arid or semiarid climatic environments because of their distance from the ocean: humid ocean air loses its moisture by raining before reaching the interior. Winters are extremely cold and little rain falls in summer. In the westernmost parts of China, high altitudes in the mountains and Tibetan plateau cause even greater summer-winter and daily ranges of temperature.

Mountains and Major Rivers

Mountain systems and relatively small areas of lowland (Figure 5.8a) form the diverse relief of East Asia. High mountains extend eastward from the Himalayan Mountain ranges and the Tibetan Plateau, occupying over one-third of the region. On the border with Nepal and India (see Chapter 7), the Tibetan Himalayas are the world's highest mountains, including Mount Everest (8,848 m; 29,028 ft.) and nearly 50 other peaks that rise over 7,500 m (25,000 ft.).

The volcanic and earthquake-prone islands of Japan form active landscapes with major shocks and eruptions every decade or so as tectonic plates clash (Figure 5.8b). Although the high proportion of mountains and destructive events sug-

gest a difficult environment for human occupation, the region supports some of the highest densities of population in the world. In southern China, some of the most distinctive hilly landscapes are in areas of limestone rock, where the almost vertically sided hills are riddled with caves in what is known as a karst landscape (Figure 5.9).

In summer, the meltwater from the high mountain snowfields in the west combines with intense rains toward the coasts, supplying high flows in some of the world's most active rivers. The three major rivers of China are, from north to south, the Huang He (Yellow River), the Chang Jiang (Yangtze or Long River), and the Xi Jiang (West River)—with its wide lower section, the Zhu Jiang (Pearl River). Inland, these major rivers carve deep valleys with steep slopes and boulder-strewn streambeds. The rivers carry large quantities of water toward the sea, together with eroded mud, sand, and rock fragments. As they near the coast, they drop their load of sand, silt, and clay to form wide plains and deltas extending out to sea. The tidal range and wave activity are both low around these coasts, allowing the large river-borne loads of silt and clay to build deltas at river mouths.

Forests, Grasslands, and Desert

The range of climatic and relief environments in East Asia produces a variety of natural vegetation types, although human land uses later replaced or modified these ecosystems. The monsoon forests in southern China have extensive areas that are dominated by a great variety of species, including teak forest. Farther north, midlatitude deciduous and evergreen forests were largely cleared from the lower and more cultivable areas but remain on steep slopes in Northeast China, the Koreas, Taiwan, and Japan. Inland, the increasing aridity causes the

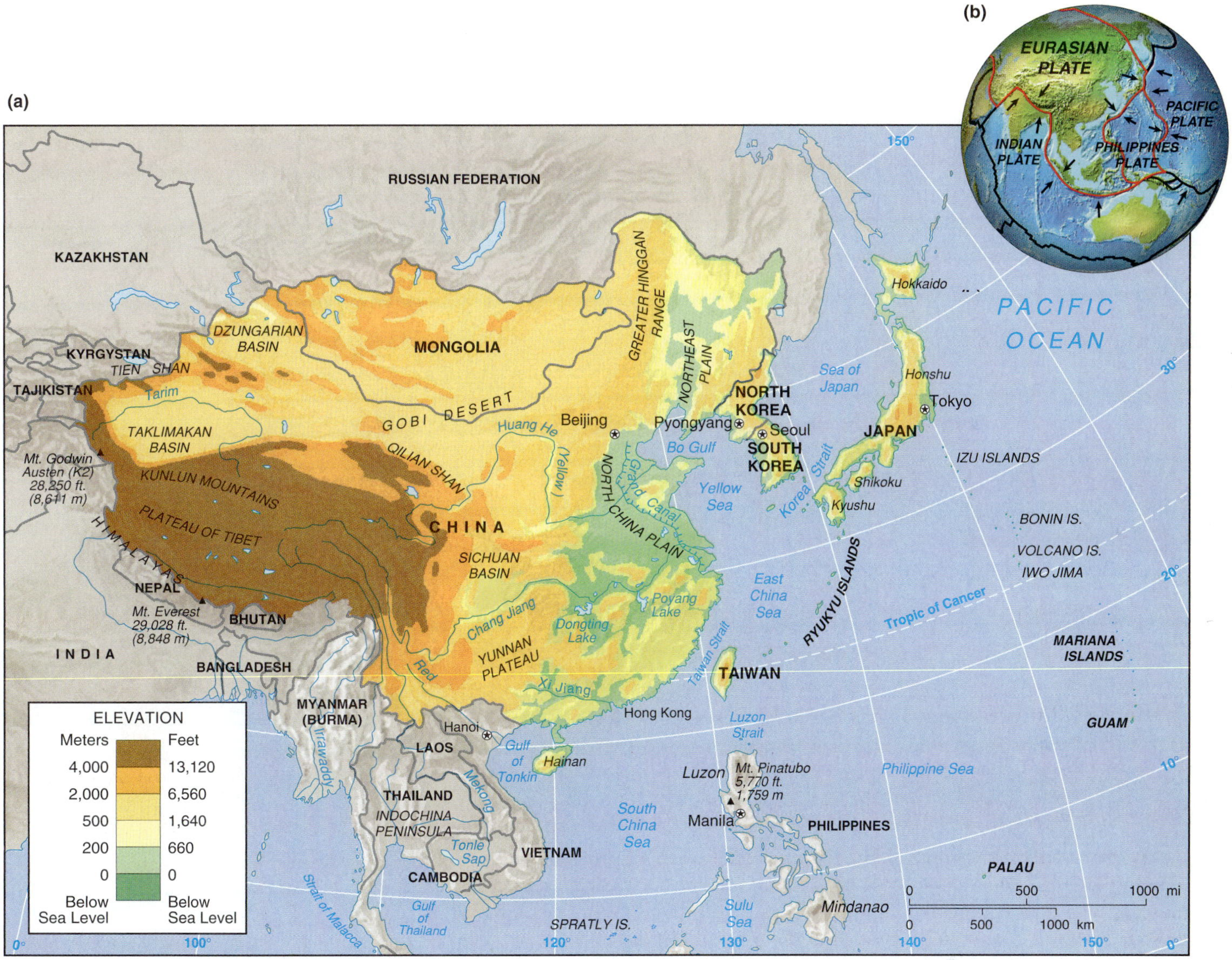

(a)

RUSSIAN FEDERATION

KAZAKHSTAN

KYRGYSTAN

TAJIKISTAN

TIEN SHAN

DZUNGARIAN BASIN

Tarim

TAKLIMAKAN BASIN

Mt. Godwin Austen (K2) 28,250 ft. (8,611 m)

KUNLUN MOUNTAINS

QILIAN SHAN

PLATEAU OF TIBET

HIMALAYAS

NEPAL

Mt. Everest 29,028 ft. (8,848 m)

BHUTAN

INDIA

BANGLADESH

MYANMAR (BURMA)

Irrawaddy

MONGOLIA

GOBI DESERT

Huang He (Yellow)

Beijing

CHINA

SICHUAN BASIN

Chang Jiang

Dongting Lake

YUNNAN PLATEAU

Red

Xi Jiang

Hanoi

Gulf of Tonkin

LAOS

THAILAND

INDOCHINA PENINSULA

Tonle Sap

CAMBODIA

Mekong

Strait of Malacca

Gulf of Thailand

GREATER HINGGAN RANGE

NORTHEAST PLAIN

NORTH CHINA PLAIN

Grand Canal

NORTH KOREA

Pyongyang

Bo Gulf

Yellow Sea

Poyang Lake

SOUTH KOREA

Seoul

Korea Strait

Taiwan Strait

TAIWAN

Hong Kong

Hainan

VIETNAM

South China Sea

Luzon Strait

Luzon

Mt. Pinatubo 5,770 ft. 1,759 m

Manila

PHILIPPINES

SPRATLY IS.

Sulu Sea

Mindanao

Hokkaido

Sea of Japan

Honshu

Tokyo

JAPAN

Shikoku

Kyushu

East China Sea

RYUKYU ISLANDS

PACIFIC OCEAN

IZU ISLANDS

BONIN IS.

VOLCANO IS.
IWO JIMA

Tropic of Cancer

MARIANA ISLANDS

GUAM

Philippine Sea

PALAU

(b)

EURASIAN PLATE

PACIFIC PLATE

INDIAN PLATE

PHILIPPINES PLATE

ELEVATION

Meters	Feet
4,000	13,120
2,000	6,560
500	1,640
200	660
0	0
Below Sea Level	Below Sea Level

0   500   1000 mi

0   500   1000 km

Figure 5.8 **East Asia: landforms.** (a) The relief map shows how the region comprises mountains, lowland plains, and major river basins. (b) Plate margins form the region's boundaries east of Japan and west of China.

forests to give way to grassland and desert in northwest China and Mongolia. The Gobi Desert occupies much of Mongolia and large areas in northern China.

Natural Resources

The major natural resources of East Asia are surface water flows, fertile soils, and minerals in the rocks. The mineral resources of East Asia (Figure 5.10) include coal, oil and gas, iron, gold, and precious stones. Coal deposits are widespread in China, while major deposits of oil and natural gas occur in western China and in many offshore locations that are still being explored. Japan has few mineral resources and relies on imported raw materials for its industries.

Although the initial Chinese civilization near Xi'an was based on millet, the growing of wet rice made it possible to feed much larger numbers of people and maintain high population densities. The water and the annually renewed fertile alluvial soils of the valley floors enabled abundant crops to be harvested.

Figure 5.9 **East Asia: limestone hills near Guilin, Guangxi province, China.** To what extent has human activity affected this landscape? Photo: The McGraw-Hill Companies, Inc./Barry Barker, photographer.

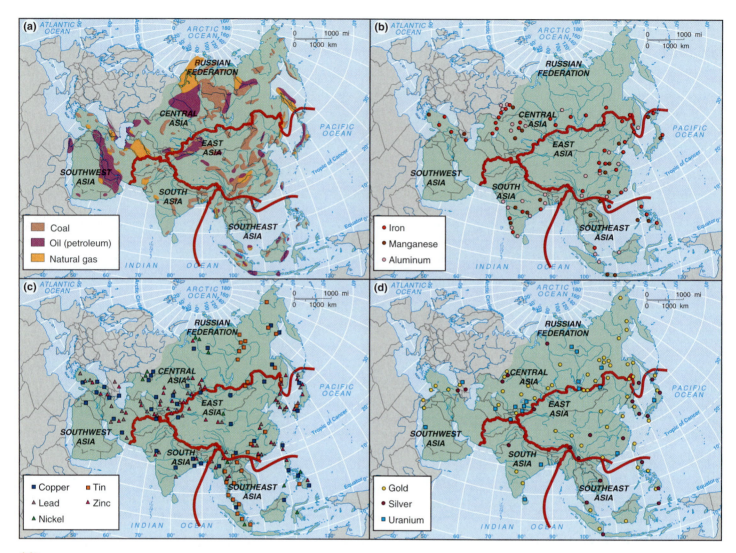

Figure 5.10 **Asia: mineral resources.** Compare the resources of East Asia with other parts of the continent. (a) Fuels: coal, oil (petroleum), natural gas. (b) Metals: iron, manganese ores, bauxite (aluminum ore). (c) Metals: copper, lead, nickel, tin, zinc. (d) Precious metals (gold, silver), uranium.

Country Boundaries

Although most country borders in East Asia follow physical features such as mountain ranges or rivers, their locations reflect cultural and colonial history. Boundary disputes, including those between China and Vietnam, remain central to the political concerns of many countries. Mongolia acts as a **buffer state,** helping to reduce direct conflict across the long boundary between China and Russia.

A dispute where there are no physical boundaries concerns the ownership of the Spratly Islands in the South China Sea (see Figure 5.8a, approximately 8°N, 124°E). All the surrounding countries, including Taiwan, but mainly China and Vietnam, claim the islands. The sea forms a major shipping route for 70 percent of Japan's oil imports, so pressures exist to keep it international. However, in the later 1990s, the discovery of oil in rocks beneath the seas surrounding the Spratlys coincided with disappointing results from Chinese offshore exploration farther north. Both China and Vietnam issued exploration licenses to international oil corporations in adjacent sections of the area.

Environmental Problems

The presence of so many people in the limited natural environments of East Asia causes many problems. Some stem from the natural processes at work; others are outcomes of human interventions and attempts to wrest higher productivity from local resources.

Natural Hazards

The combination of plate boundaries and monsoon rains not only provides resources but also presents hazards to human occupation. Japan lies along a convergent plate boundary and is subject to major earthquakes, such as the one in 1996 that destroyed much of Kobe, and occasional volcanic eruptions. Geologic activity also occurs some distance from the plate boundaries. In 1976, a massive earthquake hit the coal-mining area around Tangshan (near Tianjin) in northeastern China, killing 242,000 people. Rehabilitation took only a few years.

Flooding is an annual hazard on the low-lying major river plains. Chinese flood hazards attract publicity because of the large numbers of people affected when the Huang He or Chang Jiang floods, but similar hazards also occur on a local scale in other river basins and in neighboring countries. The worst floods occur when precipitation and high meltwater flows together enter rivers in the summer months, but human actions often make the flooding worse. The 1981, 1991, and 2002 floods along the Chang Jiang followed abnormal precipitation, but excessive deforestation in the upper parts of the river basin enhanced the rapid runoff. The 1991 floods—the worst in China during the 1900s—affected both the middle basin and the provinces at the river's mouth.

In the mid-1990s, 180 million people in the more northerly Huang He basin suffered the opposite problem—too little water. Building dams in the headwater areas reduced overall flow downstream, while middle-basin irrigation works diverted waters that had been the mainstay of the lower basin. Successive years of drought from 1987 reduced flow in the lower basin to less than half its previous level.

Human Impacts and Responses

River damming for hydroelectricity and irrigation reduces river flow and water resources in rural areas, while urban areas and mining pollute water with toxic chemicals and waste. Deforestation resulting from expanding agriculture and cutting forests for lumber often leads to soil erosion, downstream flooding, and silting. The silt that is swept out to sea blankets and destroys offshore fisheries and coral reefs.

As one example of responses to environmental degradation, China experienced major problems of desertification when it tried to plow the formerly grassed semiarid areas of the Northwest. From 1948 to 1978, soil erosion by wind and water removed over 700,000 hectares (1.7 million acres) per year from productive activities. Government action caused the planting of a "Great Green Wall" of trees between 400 and 1700 km (250–1000 mi.) wide. Dust-storm days in Beijing fell from 30 per year in the 1970s to 12 by the late 1980s, indicating a degree of success, but the incomplete program was ended in the 1980s, after which Beijing experienced an increase in dust storms.

Forest planting became a feature of many parts of China in the 1970s and 1980s. The 1981 flooding in Sichuan highlighted dramatically the effects of poor land-use management, including excessive tree felling. Major tree plantings occurred throughout the coastal region and especially in shelterbelts to reduce wind erosion in the intensive farming areas. By 1989, one-fourth of Chinese local government areas had at least a 10 percent forest cover and shelter belts. The Chinese focus on household economic growth and efficiency since 1978 instead of communal organization, however, reduced funding for the planting program.

Diseases

The countries of East Asia are notable for their improvements in health care provision in the later 1900s. This had major impacts on reducing death rates and infant mortality rates throughout the region.

However, HIV/AIDS is an increasing threat in China. After being a problem mainly of the major cities, it was recognized as a national problem in 2001, when health officials in China admitted to a "very serious epidemic." At the end of 2000, the United Nations AIDS Office estimated there are 1 million HIV sufferers in China and predicted that there might be 20 million by 2010. One of the sources of HIV is in Henan Province, where, from the late 1980s, several counties made an industry of blood collection. The opportunities to earn the equivalent of US $5 per unit at government and army collection stations attracted many people, but failures to follow sterilization guidelines resulted in the transmission of the virus. Local officials covered up the problem, and AIDS sufferers reported harassment by police when applying for help from health agencies. From 2001, Beijing authorities spent heavily on education programs to clean up the blood-collection process, although they could not afford life-prolonging medications for HIV/AIDS sufferers.

Pollution

Pollution of air, water, and soils is a serious problem in this region. As industrial activities increase, more factories emit fumes, more vehicles are used, and farming is made more productive by the use of fertilizers, pesticides, and mechanization. In areas of greater population density and modern industrialization, levels of air and water pollution are high. Large cities regularly exceed the standard levels for concentrations of carbon monoxide, sulfur dioxide, suspended particles, and lead in the atmosphere. Five Chinese cities—Beijing, Shanghai, Shenyang, Xi'an, and Guangzhou—are among the 10 most polluted cities in the world. This is partly because China is the world's largest coal user, depending on coal for 80 percent of its energy needs. Small factories and power plants that burn coal inefficiently produce one-third of the particulates and sulfur dioxide, while domestic users produce another one-third. Acid rain from the particulates and other emissions affects downwind areas, with some Chinese pollution traveling as far as the United States.

Water pollution is particularly serious around the expanding major cities where raw sewage and untreated industrial effluent are common along rivers and in offshore fisheries. In the 1960s, high mercury levels from industrial processes poisoned fish and killed over a thousand Japanese people who ate them. In northeast China, heavy metals from the metallurgical industries around Shenyang polluted local drinking and crop water, causing a surge of birth defects and a life expectancy of 10 years below the national average of 70 years. While access to safe drinking water is increasing in China, water pollution from untreated urban wastes and rural fertilizers caused Shanghai to move its drinking water sources upstream in the mid-1990s. Polluted water and acid rain both reduce crop productivity and make it essential to cook all vegetables by boiling or stir-frying at higher temperatures than the boiling point.

Environmental Policies

Although the countries of East Asia were late in adopting environmental policies, they are now busy setting aside wildlife sanctuaries, establishing forest management procedures, and considering the overall control of water resources. Some countries passed environmental legislation to reduce pollution of air and water but had difficulties enforcing the standards. In China, the government raised coal prices to reduce the inefficient use of this polluting fuel, but it expects that acid rain will spread to new areas in the time it takes to implement the policy. The danger of companies corrupting government officials for short-term gains remains a problem that slows progress in environmental quality in many East Asian countries.

Japan, with so many people and factories on relatively small islands, made the greatest strides in environmental control after the worst years of the 1960s. Tokyo traffic police wearing smog masks became a symbol of pollution, but legislation led to improvements in air quality. One of the major environmental problems Japan faces today, at a time when virtually all its dumping sites are filled, is the disposal of trash, which increased by a third in the 1980s and 1990s. As a result, Japan is becoming a world leader in developing new technologies of recycling materials that do not require landfill areas. These technologies include recycling and the burning of large quantities of trash.

Test Your Understanding 5A

Summary East Asia has 24 percent of the world's population and produces just under 20 percent of the world's economic output. In the early 2000s, the region contained countries that ranged from some of the world's materially wealthiest to its poorest. Countries at different levels of economic development experienced the "East Asian Miracle," although events in the late 1990s led to questions about its causes and longer-term significance.

Chinese culture dominated the region's history, but smaller kingdoms around the margins established separate national entities that formed the basis of modern countries. Buddhism accommodated to Chinese and Japanese religious ideas, while Muslim influences affected the northwestern parts of the region. All the countries in this region resisted European colonization.

The varied physical environments of East Asia include *monsoonal* contrasts of rainfall and drought, the world's highest mountain systems, major river basins, and hazards such as typhoons, floods, earthquakes, and volcanoes. Environmental pollution and degradation increase in countries with huge concentrations of people and industry and high rates of economic development.

Questions to Think About

5A.1 To what extent can East Asia be defined as the "Chinese realm"?
5A.2 What influences worked against the total Chinese or European domination of East Asia before 1950?
5A.3 Does the range of natural environments in East Asia contradict the idea of its character as a major world region? Explain.

Key Terms

Confucius (Kong Fuzi)	midlatitude east coast climatic
Daoism	environment
Shinto	buffer state
monsoon climatic environment	

🌐 Globalization and East Asia

Most countries of East Asia cluster toward the top of the world range of gross national income (Figure 5.11) and ownership of consumer goods (Figure 5.12). They have a major involvement in the global economic system. Two countries, Japan and China, are likely to lead the rest of the world into the next phases of globalization. For U.S. traders, the People's Republic of China, South Korea, and Taiwan were the three main emerging big markets in 2000. North Korea and Mongolia remain largely outside the world economy.

The economic growth of these countries in the late 1900s was at first based on exporting products to American and European markets (see "Global Focus: Toyota," p. 184) but by the 1990s was also bolstered by trade among countries within this region and Southeast Asia. Asian internal trade is now as great as external trade. The gradual evolution of the Asia-Pacific Economic Cooperation Forum (APEC) from the 1990s reflected the global spread of trading links from East Asia to Southeast Asia, the South Pacific, and the Americas.

Global City-Regions

The growth of global and intraregional trade in East Asia led to the emergence of global city-regions (see Figures 2.13 and 2.14) as essential nodes linking this region to the rest of the world through business and political exchanges. Tokyo, Japan, is one of the world's top four global city-regions—with New York, London, and Paris—while Hong Kong is in the second rank. Seoul (South Korea), Osaka (Japan), Taipei (Taiwan), and Beijing and Shanghai (China) are also major cities where employment growth in international business services is significant.

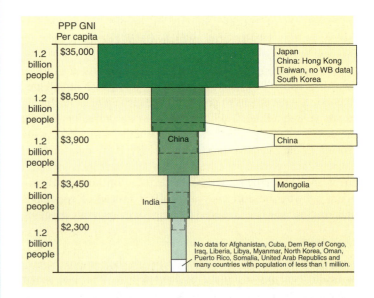

PPP GNI Per capita

1.2 billion people	$35,000 — Japan, China: Hong Kong [Taiwan, no WB data], South Korea
1.2 billion people	$8,500
1.2 billion people	$3,900 — China / China
1.2 billion people	$3,450 — Mongolia
1.2 billion people	$2,300 — India

No data for Afghanistan, Cuba, Dem Rep of Congo, Iraq, Liberia, Libya, Myanmar, North Korea, Oman, Puerto Rico, Somalia, United Arab Republics and many countries with population of less than 1 million.

Figure 5.11 **East Asia: country incomes compared.** The countries are listed in the order of their PPP GNI per capita. Source: Data (for 2000) *World Development Indicators,* World Bank, 2002.

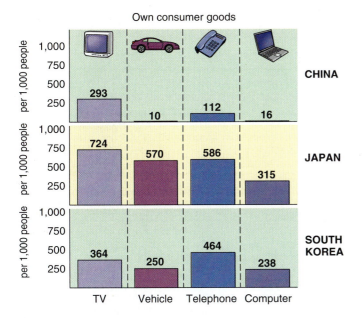

Own consumer goods

CHINA

293   10   112   16

JAPAN

724   570   586   315

SOUTH KOREA

364   250   464   238

TV   Vehicle   Telephone   Computer

Figure 5.12 **East Asia: ownership of consumer goods.** Account for the range of affluence among these countries. Data (for 2000) from *World Development Indicators,* World Bank, 2002.

In the Forefront: Japan, South Korea, and Taiwan

East Asian countries did not come fresh to global trading after 1950. European colonial powers drew them in from the early 1800s. The distinctive feature of the late 1900s was that the independent countries in East Asia made their own decisions to join the global economic system. Japan led the way from the 1960s, followed by South Korea and Taiwan.

Japan

In the early 2000s, despite slower economic growth in the 1990s, Japan remained the world's second-largest economy. It had 8 of the 10 largest world banks, having built its world prominence on huge capital surpluses. By the 1970s, major Japanese industries had sufficient resources to invest in the United States, Europe, and Southeast Asian countries to manufacture goods for their domestic markets. It then invested in China. Japanese companies produced goods in China and Southeast Asia for their brands that sold worldwide.

Multinationals

Japanese, South Korean, and Taiwanese exports of manufactures and foreign investments led to the growth of huge multinational corporations. From Japan, automakers such as Honda, Toyota-Lexus (see the "Global Focus: Toyota" box, p. 184), Nissan, and Mitsubishi, together with electronics and electrical goods makers such as Sony, Sharp, Fujitsu, Panasonic, and Matsushita, built worldwide brands manufactured in many countries. From South Korea, multinationals such as Hyundai, Samsung, Daewoo, and Lucky Goldstar (LG) penetrated markets around the world. Many Japanese and South Korean

employees from managers to technical staff worked abroad in subsidiary companies of the multinationals. In the 1980s and 1990s, around 50,000 Japanese were involved in such situations and the South Korean numbers increased from 150,000 to 250,000. Short business trips abroad went from 600,000 to over 2 million in Japan and from just over 100,000 to nearly 1 million in South Korea. Government employees in overseas embassies also doubled during this period. These figures are similar to those in Europe and North America, demonstrating the role of the two East Asian countries in the global economic system.

International Tourism

Increasing personal wealth and the development of hubs for international airline routes resulted in rising tourist travel, especially for South Koreans, Taiwanese, and Japanese. Tourism became an industry that both brought in visitors and made it possible for people to visit other places around the globe. In 2000, Japan attracted nearly 5 million visitors, while 16 million of its people vacationed abroad—both numbers having risen by 50 percent since 1990. To illustrate the impact of Japanese tourists, direction posts in Switzerland give information in Japanese as well as English and local languages. South Korea received over 5 million tourists in 2000, while nearly 6 million of its people were outbound, almost doubling since 1990.

Recent Entry: China

China has the potential to become a major world power alongside Japan. It is the world's most populous country, and between 1980 and 2000, its economy grew at the fastest sustained rate of any country. During this economic growth, which multiplied the value of its output fivefold, 600 million people escaped absolute poverty. Chinese factories now make increasing numbers of goods bought in Western stores, from electronics to autos and clothing. In November 2001, China became a full member of the World Trade Organization.

Current foreign direct investment to China is fourth in the world, after the United States, Germany, and the United Kingdom, and outstanding among the world's less materially wealthy countries. To support such economic investment, China's huge supply of labor includes college graduates and computer engineers who are paid one-tenth as much as their Taiwanese or Japanese counterparts. However, Chinese entrepreneurs face difficulties at home because, for example, the government is slow to expand the list of companies allowed to go public with share offerings. Those entrepreneurs, therefore, find it difficult to persuade Western financiers to support them after intense scrutiny by external analysts and investors. Most investments in China come from Chinese in Hong Kong, Taiwan, and other Chinese living abroad—but also increasingly from Japan.

Western and Asian multinational corporations trading in China include oil companies, electronics groups, auto manufacturers, and restaurant chains. For example, Kentucky Fried Chicken (KFC) opened a huge outlet in China next to

TOYOTA

The Japanese company Toyota is the world's third-largest automotive group in sales after General Motors and Ford (Box Figure 1). At the end of 2001, its capital resources were double those of either of the other two. Toyota is a model for many other multinationals. Its experiences illustrate some of the ways in which multinational corporations (MNCs) grew to worldwide prominence in the later 1900s and some of the challenges MNCs face in the early 2000s. They also emphasize the continuing significance of the home country's conditions and markets.

Kitchiri Toyota first researched gasoline engines in 1930, switching the family textile business into cars. Family directors still sit on the Toyota board. Toyota's headquarters are in Tokyo, but its Takaoka base is a sprawling collection of 35-year-old factory buildings amid the rice fields of Aichi province to the southwest. Forty years ago, the city of Kolomo was so impressed with the company's local investment that it renamed itself Toyota City. The Toyota City factories produce 700,000 vehicles a year and provide the model "mother factory" for five overseas plants.

At first, Toyota copied—and improved—others' innovations, such as Ford's moving production assembly line. Costs were tightly controlled and profits plowed back. The company continued to be financially flexible and lean. At Takaoka, Toyota developed *jidoka*—automation with human intelligence, combining high quality, short delivery times, and low-cost production. Combined with *kaisen*—

Box Figure 1 **Robots on the production line of Toyota cars, Japan.** Photo: © Michael S. Yamashita/Corbis.

continuous improvement—and just-in-time delivery of components, it led to strong demand for the reliable Toyota vehicles.

In the early 2000s, Toyota developed CCC21 (Construction of Cost Competitiveness for the 21st Century), a program aimed at cutting billions of dollars from its costs. The company is reviewing its design, manufacturing, and other costs with the objective of better using equipment and human resources. It will not, however, fire any of its 36,000 workers at Toyota City but will reduce the 4,000 temporary workers. Toyota also negotiates with component suppliers to reduce costs by up to 30 percent. Denso, its chief supplier (and fourth-largest component maker in the world), employs 85,000 people and is the chief of over 300 suppliers. Toyota holds an average 25 percent interest in most of its component makers. Such policies and links illustrate a typical Japanese conservative attitude that builds in linkages that are not quickly removed when inefficiencies occur because of commitments to lifetime employment and other long-term agreements. Toyota's competitors such as Nissan Motor and Mitsubishi Motor are breaking apart their supplier chains, and this will eventually cause Toyota to rethink its policies.

After 1980, its adaptability enabled Toyota to change from a Japanese auto exporter into a global operator with 41 manufacturing subsidiaries in 25 countries. It has strong market shares in the United States and Europe. Toyota's conservative Japanese outlook applies in these areas. It does not merge with rivals, it does not have foreign managers among its Tokyo vice-presidents, and it has never sent a female executive overseas. The company emphasizes the importance of cash reserves and profitability.

Toyota's successful strategy began with making mid-priced cars for the growing numbers of middle-class Japanese, and it still has 42 percent of its domestic market. The large home sales are a major strength in light of international currency fluctuations. When the company expanded overseas, local production within the United States and Europe minimized problems of currency exchange rate shifts. Toyota also targeted local demands, producing light trucks for the U.S. market that is the source of two-thirds of its current profits. Europeans bought smaller cars. The Lexus luxury brand sells in both regions.

To keep up with market trends, Toyota launches a new model every month. That requires huge investments that are questioned by the increasing numbers of foreign shareholders who replace the more compliant Japanese bank shareholders. The realization that even such large single companies cannot satisfy all the world's market areas led to Toyota joining an alliance with Peugeot-Citroën of France to market small cars in eastern Europe. It is also entering joint research projects with other manufacturers.

Toyota is a Japanese version of multinational corporations. Its policies remained successful as it expanded exports and manufacturing abroad. As globalization becomes more pervasive, Toyota may have to become more transnational by allowing control of parts of its activities to occur outside Japan.

Tiananmen Square in central Beijing in 1987. Despite the 1989 incident in this square, when Chinese army tanks attacked peaceful demonstrators, McDonald's followed in 1992 and by 2001 had nearly 400 outlets throughout China. Both KFC and McDonald's plan to expand further. The first Starbucks outlet in China—serving coffee in a tea-drinking country—opened in 1999, and by 2001, 35 shops held franchises, mainly in Beijing and Shanghai. The most contested site for Starbucks was in a Beijing Forbidden City souvenir shop, with 70 percent of

60,000 people surveyed opposing it. But it continues to serve coffee. Such symbols of an intruding global economy, however, excite little public antagonism. Even at the height of anti-American demonstrations over the bombing of the Chinese Embassy in Belgrade, Yugoslavia, in 1999, the Chinese who stoned the American diplomatic missions left KFC and McDonald's largely untouched. Many Chinese regard the expansion of multinational food franchises as a positive sign that they are involved in the global economy.

In the 1990s, China, including Hong Kong, became the world's fourth-largest center of tourism (after France, the United States, and Spain) as numbers of visitors doubled in Hong Kong (from 6.5 million to 13 million) and trebled in the rest of China (from 10 milion to 31 million) between 1990 and 2000. Outbound tourists made up one-third of the inbound but increased fivefold in the 1990s, indicating the increase of income and adoption of the vacation habit.

The surrounding countries remain wary of China's military and new economic strength, although many Chinese still see their country as poor and struggling to escape centuries of foreign intervention. The neighboring countries fear a Chinese people who continue to perceive a sphere of influence extending to lands it held during its period of greatest expansion in the Ming dynasty.

In the near future, China will expect treatment that reflects its position as a major and growing power in the world. In reciprocation, other countries will take a greater interest in internal Chinese matters such as human rights and democratic institutions. As the People's Republic of China reconsiders its relations with Taiwan, Japan, and South Korea, as well as other Asian countries such as India, each of these countries reorients its outlook, trade, and policies to take account of China's growing role in the region and world.

China and Global Culture

Efforts to be part of the global economic system introduce elements of external cultures to China. Increasingly, however, cultural globalization brings reciprocal changes. The Chinese wish to excel at all they do, especially in front of audiences in other countries. China has long had its Western ballet and orchestral groups and took to soccer and athletics as well as table tennis. In each case, the Chinese contributed to a dominantly Western tradition. In South Korea, young pop stars and record companies adapt Western methods as they work on their links to the Chinese in developing a market among the younger upwardly mobile Chinese. More Western influences appear on CCTV1, the Chinese government monopoly TV station, including children's programs such as the U.S. *Sesame Street* (allowed in as a 1998 "scientific" contribution) and the U.K. *Teletubbies* (*Tianxian Baobao,* "Antenna Babes").

Subregions

Two powerful countries, Japan and China, dominate East Asia. At present, Japan's strength is economic, and it is treated as a separate subregion. China's strength is political-military, cultural, and increasingly economic. Mongolia is linked to China because of its situation between China and Russia. Taiwan is also linked to China, which claims possession of the island. As recent trends bring the two closer to reunification, many Taiwanese bitterly oppose such a move. As separate countries, North and South Korea have very different levels of global involvement but were united until 1950 and may reunite in the future. The rest of this chapter examines the human geography of the subregions:

- *Japan* remains the outstanding country in world economic terms.
- *The Koreas, North and South,* provide contrasts of political and economic development.
- *China—including Hong Kong (Xianggang) since 1997 and Macau since 1999—*is considered with the island of *Taiwan,* which it claims, and the neighboring country, *Mongolia.*

Data for the region and its countries are listed in Figure 5.13.

Population Distribution

East Asia is marked by some the world's highest and lowest population densities (Figure 5.14). The western half of China and Mongolia have densities of fewer than one person per km², except along major routeways. East of a line drawn across China from northeast to southwest, just west of Harbin, Beijing, Lanzhou, and Chengdu, and extending across the Koreas, Taiwan, and Japan, densities are over 10 per km² and exceed 100 in many areas. The densest population areas are along the eastern and southern parts of Japan, the western parts of the Koreas and Taiwan, the north-central part of China, and the coasts and major river valleys of southern China.

 Japan

Japan's (Figure 5.15) present government, economy, and constitution were put in place during the post–World War II U.S. occupation that lasted from 1945 until 1952. While the emperor remains a symbol of national unity, political power rests in networks of politicians, bureaucrats, and business executives. The parliament, or Diet, appoints the prime minister, usually the leader of a majority coalition. Japan's industrial development from the late 1800s, its recovery from defeat and destruction in World War II, and its rapid economic growth since the 1950s constitute the world's greatest economic growth success story of the 1900s. Japan is materially the wealthiest country in East Asia and has the second-highest total annual income in the world.

Following postwar reconstruction from 1945, economic growth was rapid. From 1960 to 1990, Japan's total GDP grew from 9 to 53 percent of the U.S. total and from 3.5 to 14 percent of the world total. In the 1990s, however, as the U.S. economy surged further ahead, the Japanese economy slowed and fell back to less than 50 percent of the U.S. total GDP in 2000 (see Figure 5.2).

Throughout Japan's modernization, its people maintained a traditional culture that encouraged attitudes based on loyalty to the emperor, family, and workplace. The devotion of Japanese workers to their employer's interests in exchange for jobs throughout their working lives, health care, and social support provided an alternative platform to Western

Figure 5.13   East Asia: some population, economic, and cultural data.

Subregion	Land Area (km²) Total	Population (millions) Mid-2001 Total	2025 Est.	GNI 2000 (US $ million) Total	GNI PPP 2000 Per Capita	Percent Urban 2001	Human Development Index Rank of 175 Countries	Human Poverty Index: Percent of Total Population
Japan	377,800	127.1	120.9	4,337,300	25,170	78.0	4.0	12.0
North and South Korea	219,560	70.8	79.0	No data	No data	69.0	No data	No data
China, Mongolia, Taiwan	11,200,260	1,305.1	1,468.0	1,230,549	9,243	67.5	80.0	18.6
Totals or Averages	**11,797,620**	**1,503.0**	**1,667.9**	**5,567,849**	**17,206.7**	**71.5**	**42.0**	**15.3**

Country	Capital City	Land Area (km²) Total	Population (millions) Mid-2001 Total	2025 Est.	GNI 1999 (US $ million) Total	GNI PPP 1999 Per Capita	Percent Urban 2001	Human Development Index Rank of 175 Countries	Human Poverty Index: Percent of Total Population
Japan	Tokyo	377,800	127.1	120.9	4,337,300	26,460	78	4	12.0
Korea, Republic of (South)	Seoul	99,020	48.8	53.3	421,100	17,340	79	30	No data
Korea, People's Dem Republic (North)	Pyongyang	120,540	22.0	25.7	No data	No data	59	No data	No data
China, People's Republic of	Beijing	9,596,960	1,273.3	1,431.0	1,064,500	3,940	36	98	19.0
Hong Kong (to China, 1997)		1,040	6.9	8.4	165,122	22,570	100	24	No data
Mongolia	Ulan Bator	1,566,500	2.4	3.4	927	1,660	57	118	18.2
Taiwan	Taipei	35,760	22.5	25.2			77		

Country	Ethnic Groups (percent)	Languages O=Official, N=National	Religions (percent)
Japan	Japanese 99%, some Koreans, indigenous	Japanese	Shinto/Buddhism 84%
Korea, Republic of (South)	Korean	Korean (O), English	Buddhist 47%, Christian 49%, Confucian
Korea, People's Dem Republic (North)	Korean	Korean	Buddhist, Confucian, mostly atheist
China, People's Republic of	Han Chinese 92%, Zhuang, Mongolian, Tibetan	Mandarin (N); Yue, Min, Wu, Hakka	Buddhist, Confucian, Daoist
Hong Kong (to China, 1997)	Chinese, Europeans	English, Chinese	Buddhism, etc.; Christian
Mongolia	Mongolian 90%, Kazakh, Chinese, Russian	Khalka Mongol 90%, Turkic, Russian	Tibetan Buddhist, some Muslims
Taiwan	Old Chinese migrants 84%, Chinese migrants from 1949 14%	Mandarin Chinese, local dialects	Buddhist 50%, Christian 5%, Confucian/Daoist 25%

Source: Data from *Population Reference Bureau 2001 Data Sheet*; *World Development Indicators*, World Bank, 2001; *Human Development Report*, United Nations, 2001; Microsoft Encarta (ethnic group, language, religion).

motivations for competing in the global economic system. In the late 1990s, however, Japanese employers canceled or modified such agreements in the face of pressures on business costs from newly emerging countries where wages and support costs were less.

People

Ethnicity

Most Japanese people stem from an ancient mixture of Asian continental and Pacific island stocks. From approximately 1650 to 1850, the military shogun rulers isolated Japan from the European influx of traders. This policy had a consolidating effect on its ethnicity and culture. The main Japanese traditions, from Shintoism to Buddhism and Confucianism, agree

on the subordinate status of women. Although women's status is changing with their greater education and involvement in the labor force (see the "Personal View: Japan" box, p. 188), they still work for lower wages, and few attain management positions or lifetime employment status.

Three small Japanese minority groups remain important. The Japanese occupation of the northern island of Hokkaido in the 1800s overwhelmed the Ainu, but they retain ethnic and language differences. On the largest island of Honshu, the Burakumin, numbering around 1.2 million people, were outcasts and allowed to work only in the lowest occupations until their emancipation in 1871. It was not until 1969, however, that the Japanese government put funds into upgrading their settlements to nationwide standards. The third group consists of foreign nationals, three-fourths of whom are Koreans, moved forcibly to Japan in the early 1900s. Until the 1980s, they

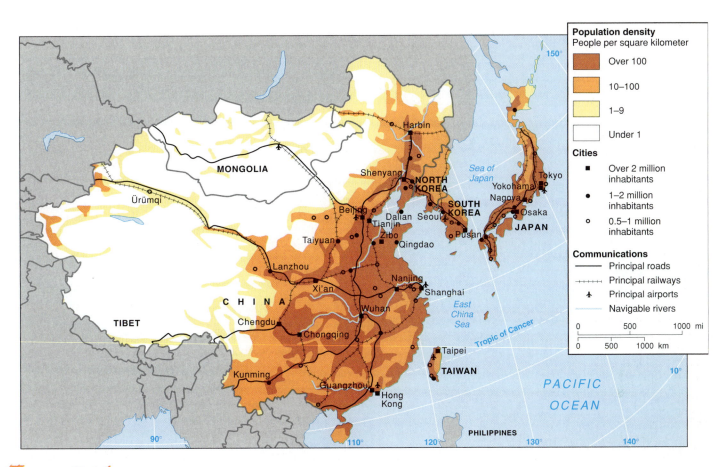

Figure 5.14 **East Asia: distribution of population.** Relate the main areas of high and low population density to physical factors. Is this a satisfactory basis for explaining the differences? Data from *New Oxford School Atlas,* p. 98, Oxford University Press, UK, 1990.

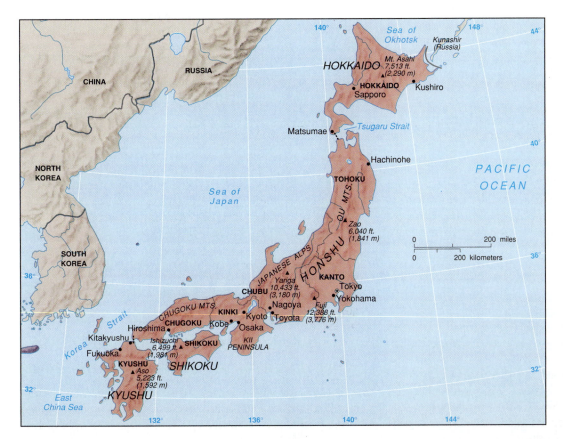

Figure 5.15
Japan. Regions and cities.

Personal View

JAPAN

Shikoku, the fourth largest of Japan's islands, is a pleasant part of the country with a warm climatic environment. One of the southernmost parts of Japan, it has mild winters, early springs with cherry trees blossoming at the end of March, and hot, rainy summers. Typhoons occasionally affect the Pacific coast in summer and fall. It is a hilly to mountainous island with most of the flat land along the northern Inland Sea coast and rather isolated southern coastlands (see the map of Japan in Figure 5.15). Some of the natural subtropical forest is preserved and the higher parts have temperate woodlands. For a long time, Shikoku stood apart from the rush of industrialization and urbanization in Japan.

The building of three bridges to the main island (Honshu), beginning with the Seto Great Bridge that took 10 years to build and was opened in 1988, increased the rate of economic and cultural changes affecting this part of the country (Box Figure 1a). Michiko, now in her early twenties, grew up in one of the ancient fort towns on the northern coast overlooking the Inland Sea (Seto Naikai). These towns were the headquarters of warrior clans during the feudal era up to 1867 but are now industrial and commercial centers. She goes to university there but has traveled to the United States, Australia, and the United Kingdom. Like many Japanese young people, she is attracted to Western influences and wishes to speak English and other European languages, as well as Chinese or Korean. She values greatly the Japanese traditions.

Michiko's family lived for most of her life in a two-story wooden house next to her grandmother but recently moved to a larger, high-tech home built of new materials. Her father is a very busy engineer whose working day typically takes him out of the house from 6 A.M. to 7 P.M., with one week of vacation each year. He has worked for the same company since leaving university. In his small amount of spare time, he enjoys crafts, including making furniture of his own designs. With Michiko's mother also working throughout her lifetime, Michiko's grandmother had a major say in raising her and her sister.

Grandmother clings to the old traditions and attitudes. Born and raised on one of the small islands in the Inland Sea, she married a sailor and spent her whole life as a homemaker. She continues to cook traditional meals with the basic ingredients, care for her home, and maintain social customs and the rules of politeness. For a long time, she continued to sleep on the floor but was eventually persuaded to move into a bed. She finds it difficult to relate to foreigners, such as a U.S. exchange student who stayed with the family for a few weeks. Fearing for her granddaughter's safety, she did not want Michiko to go to other countries. Singing in a choir at the community center provides her relaxation and main social activity: weekly practices lead to annual public performances of traditional Japanese songs. She is not happy with modern technology from vending machines to computers.

Michiko is fairly typical of the generation of Japanese young people who grew up to expect a good education, a consumer society, the use of modern technology, world travel, and a long life. She expects a professional job when her university education is completed, although it may be in a traditional field such as Japanese calligraphy. Teaching in elementary school might be a possibility but is extremely competitive because of its high status and excellent government pension. She likes using a computer, exchanging e-mails with her friends, and looks forward to designing her own website. Michiko enjoys a wide range of Western and Japanese music, preferring relaxing pieces. While friends took up part-time jobs as store assistants or restaurant waiters, Michiko worked with an organization that helps exchange students because she enjoys meeting people from different countries. She goes to movies and karaoki parties, is a member of swimming and tennis clubs, and skis for a week each year.

(a)

(b)

Box Figure 1 **Shikoku Island, Japan: new and old.**
(a) The Seto Great Bridge links Shikoku and Honshu, the largest Japanese island. It took 10 years to build and was opened in 1988. (b) Historic buildings in Matsuyama, Shikoku: entrance to a Japanese Onsen (hot spring). Photos: Permission granted by Japan's Ministry of Foreign Affairs.

A student at her local university, Michiko is used to studying very hard. While at high school, she had no vacations because extra study filled the time. At home, the cable TV was exchanged for the fewer channels on satellite TV so that she would not be distracted from her studies.

Like other Japanese people, Michiko had regular health checks at school, university, and in the workplace, and her family check their blood pressures regularly at home. People in Japan are very concerned with their health. Doctor and hospital costs are met through her parents' employer-supplied insurance coverage. Although the traditional Japanese diet includes a lot of fish, home meals are now more varied, including pasta and Chinese dishes, while meals at fast-food restaurants are very common.

Michiko is glad of her Japanese heritage but enjoys living in other parts of the world and wishes to make the most of new experiences gained from her times in Western countries. She compares the benefits and difficulties of living in Japan with what she found abroad. She is critical of the fact that in Japan, a focus on work and rising incomes means that families are able to meet and talk less. Furthermore, the changing cultural focus is not only replacing Japanese with Western ways but also is resulting in a less polite society. One Japanese habit that has not changed is the refusal to discuss religion with family or friends. Conversation remains an undemonstrative, matter-of-fact activity, focusing on educational and economic achievements and hopes. And yet, Michiko is proud to be Japanese, and her life continues to center around her family.

found it difficult to gain Japanese citizenship and are still subject to discrimination.

Migration into Japan is restricted, and only 1 percent of the population is of foreign origin. Concern over the ethnic background of foreign workers led Japan to recruit 140,000 workers from the Japanese populations of Peru and Brazil—the *Nikkeijin* (ethnic Japanese born overseas). They look Japanese but are culturally Latin American, spoke little Japanese on arrival, and had different behavior patterns. Japan also allows the immigration of foreign brides for men left behind in rural areas by the migration of women to urban jobs. Such "mail-order brides" come from the Philippines, Vietnam, and Thailand and, seldom speaking the local language, take up a life of extreme social loneliness.

Rural to Urban Populations

Overall, Japan is a densely populated country (see Figure 5.14) but has local variations. The main concentrations of people are on the small areas of lowland, and 70 percent of the total population live in the Pacific coastlands from Tokyo to Osaka. This urbanized zone is often known as the "Tokaido **megalopolis**"— a series of almost continuous metropolitan centers with urban functions that exchange flows of people and goods with the surrounding areas.

In the short period of 80 years from 1920 to 2000, Japan changed from a largely rural country, in which industrial towns housed 25 percent of the population, to one in which almost 80 percent live in cities. Up to 1940, Japan's developing industries attracted gradual migration from rural locations to factory towns. At the end of World War II, half of the Japanese population was still rural, but further industrialization brought rapid urbanization. From 1945 to the mid-1970s, jobs in the manufacturing and service sectors multiplied, and migration to urban areas increased rapidly.

After the mid-1970s, improving transportation encouraged people to move out of central cities that were being stifled by congestion and pollution. This was more a process of **suburbanization,** when people move to the outskirts of existing cities, than of **counterurbanization,** when they move away from metropolitan centers to small towns in rural settings.

As urban populations grew, the more remote areas in the mountains and along parts of the western coasts of Japan suffered depopulation. Such regions often have up to 25 percent of their population over 65 years old, and government grants are available to help people stay or settle there.

Population Change

Japan's total population increased from around 90 million in 1960 to 127 million in 2001. The combination of low fertility (1.3) and population growth rates (natural increase of 0.2%) leads to estimates that the total will peak around 128 million in 2010 and fall to 121 million by 2025. Infant mortality at 3.4 per 1,000 live births is the world's lowest, and life expectancy at 81 years is the highest.

In 1948, concern over providing for its large and rising population after World War II led Japan to pass its Eugenics Act, allowing abortions and instigating a series of family planning drives. As the birth rate fell over 50 years, from 34 to 9 per 1,000, the death rate fell from 14 to 8 per 1000. Japan is in the final stage of demographic transition (Figure 5.16). The age-sex diagram for Japan (Figure 5.17) has bulges and troughs. There were more births in the postwar and late 1960s periods. Between those dates, there were fewer births as the economy grew, as there were after 1970, when birth control and affluence combined to lower fertility.

Changes in the population's structure and distribution have important economic and social implications. In particular, Japan's population is aging. In 1970, 7 percent of the population was over age 65; by 2001, the proportion rose to 17 percent. Meanwhile, smaller nuclear families and accommodations in suburban houses became common, reducing the traditional care of older people by eldest sons. The shift to smaller households caused the total number of households to increase from 23 million in 1960, when they averaged 4.5 persons per household, to over 40 million in the 1990s, with 3 persons each. The aging of the population increases pension provision, medical costs, and leisure provision for the retired, but it also adds to firms' wage costs as more employees reach senior positions. Japanese people are now worried that there will not be enough

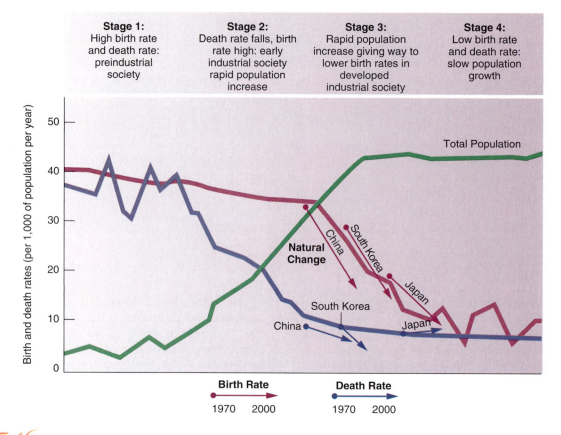

Stage 1: High birth rate and death rate: preindustrial society	Stage 2: Death rate falls, birth rate high: early industrial society rapid population increase	Stage 3: Rapid population increase giving way to lower birth rates in developed industrial society	Stage 4: Low birth rate and death rate: slow population growth

Figure 5.16 **East Asia: demographic transition.** Compare the positions of East Asian countries in this transition with each other and with countries in other major world regions.

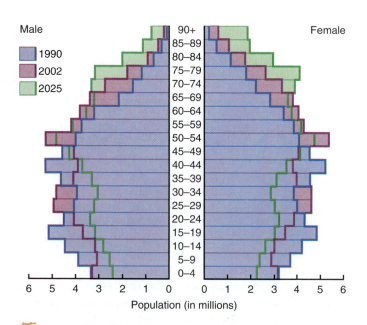

Figure 5.17 **Japan: age-sex diagram for 1990, 2002, and 2025.** Source: U.S. Census Bureau. International Data Bank.

labor for their industries or funds to support the increasing numbers of elderly people. Encouragements for couples to have more children are being considered.

Economic Development

Postwar Changes

The Japanese economy changed massively after World War II. From 1960 to the late 1990s, farming, fishing, and mining—together forming the primary sector of the economy—declined from 33 to 2 percent of the work force, while the service sector rose from 38 to 60 percent.

The postwar economic restructuring began with the breakup of the military-industrial complex and dismantling of factories. It initially produced hardship and uncertainty. Only in the 1950s did American aid, substantial contracts during the Korean War (1951–1953), and a lack of need to invest government funds in defense (because it was forbidden) begin to make a difference to the Japanese economy. Growth was rapid thereafter.

Agriculture

Japanese farms occupy small areas of suitable land within the hilly and mountainous country, often intermingled with housing and industry (Figure 5.18). Urban and industrial users also compete for the land, paying higher prices than farmers could afford without subsidy.

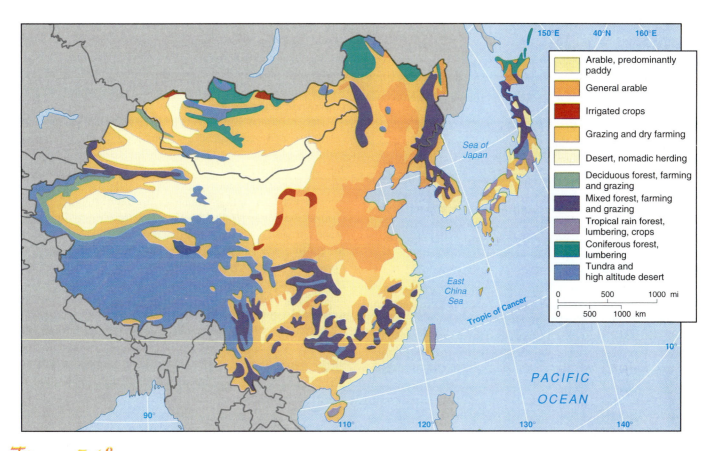

Figure 5.18 **East Asia: farming and other land uses.** Notice the contrast in intensity of production between eastern and western China. Source: Data from *New Oxford School Atlas,* p. 98, Oxford University Press, UK, 1990.

Government support maintains agriculture as a significant sector of the economy. After 1945, the Japanese government bought rice at a price that enabled the farmers to mechanize, and tariffs protected them from world market competition. New tastes for Western foods, however, halved rice demand between 1960 and 1990. The rice that could not be sold formed a stored "mountain" that was too expensive to export. Rice acreage fell as farmers converted land to cultivate vegetables and fruit, reducing the rice store. Rice still costs Japanese consumers six times what Americans pay and 13 times what Thais pay.

Although subsidies made farm incomes secure, migration from rural to urban locations for the better-paid jobs, better education opportunities for children, and a better social life halved the rural population from 1960 to the 1990s. After 1970, only 12 percent of farmers worked full-time without off-farm jobs. The remaining farmers were often aging and found it difficult to compete in a world of declining subsidies and rising foreign competition. The size of most farms remained stable, however, because land increased in value and many owners held on to their land as a family investment.

In the early 2000s, new shocks confronted Japanese farmers. Improved transportation links and refrigeration brought large quantities of Chinese tomatoes, eggplants, onions, and garlic bulbs to Japanese supermarkets. On Tokyo's wholesale markets, Chinese shiitake mushrooms sold for less than one-third the price of the Japanese varieties and quickly took a 40 percent share of the market from Japan's 30,000 shiitake grow-ers. Forest products and the bulrushes used for making tatami mats are also under pressure from cheaper Chinese sources.

Manufactures: Beginnings of Modernization

Manufacturing provided the basis of modern Japanese growth, and its history demonstrates increasing complexity and diversity. From the mid-1800s, modernization in Japan built on the economic stability and government encouragement in the post-shogun era. The economy responded to external trading pressures as well as a national desire to be a military power for self-preservation. New factories built in the coastal zone from Tokyo to Osaka made iron and steel, ships, and constructional engineering products, and textiles.

In the period up to World War II, industrial expansion was based on government-business links, cheap labor, large numbers of small businesses, and improved infrastructure. Overseas, Japan became known for its cheap goods, often of mediocre quality. The 1930s economic depression—when other countries raised tariff charges against cheap Japanese imports—led to an internal pact between Japanese military leaders and the elite families, the *zaibatsu,* who owned the biggest industries. The military leaders took over political power and shifted industrial production to naval ships, air force planes, and other military equipment. This phase culminated in World War II, ending with the destruction of much of Japan's industrial capability by U.S. bombing and postwar dismantling.

Manufactures: Reconstruction

After dismantling Japan's World War II industrial base, the United States poured money into the country to rebuild its industries and infrastructure. The Cold War beginning in the late 1940s, the Communist takeover in China, and the Korean War of the early 1950s resulted in American markets opening to Japanese products. Following the rebuilding of heavy industries such as steelmaking and chemicals in the 1950s, Japan diversified into shipbuilding, automaking, and light manufactures that were demanded by consumers at home and abroad. Japan made timely transportation innovations such as building very large oil tankers and bulk carriers for other commodities in the mid-1950s when the Suez Canal was closed and Western companies sought cheaper ways of transporting oil. Japanese-built bulk tankers and carriers brought home oil from southwestern Asia and coal and iron ore from Australia.

The cheapness of raw materials on world markets and its own low-cost, young labor force enabled Japanese manufacturers to undercut their rivals in the world's wealthier countries and gain overseas markets. The Japanese government assisted and advised industry, with the **Ministry of International Trade and Industry** (MITI) encouraging export sales through a worldwide network of market-intelligence gathering offices.

During the 1960s, total output from the Japanese economy increased by 10 percent per year. At this stage, Japanese workers waited for the gains in prosperity as their firms plowed profits back into new developments. It was not until the 1970s that Japan established a favorable trade balance and paid better wages so that its consumers could buy a range of Japanese-made products. Import restrictions discouraged foreign-made products.

Manufactures: Challenges and Changes

In the 1970s, labor shortages, increased oil prices, a world shipping slump, and reduced demand for some products caused Japanese corporations and their supportive government to rethink policies. For example, as Japan depended on (now more expensive) imported oil for 70 percent of its energy needs, the Japanese government reduced the costs of internal hydroelectricity and nuclear power and subsidized unprofitable coal mines.

By the 1980s, the older industries such as steelmaking, shipbuilding, petrochemicals, and cement-making suffered from overcapacity on world markets and competition from newly industrializing countries such as South Korea. Factory closures led to social problems resulting from unemployment, particularly in the Osaka region.

However, responses to the changes took Japanese industries into a new era. Diversification required greater investment but raised the output of light industries such as those producing cameras and household appliances. After copying others' technologies and designs, Japanese firms became initiators. The increased application of technology brought productivity and international competitiveness gains in industries including electronics, robotics, and new materials. In 2000, Japan had half of the world's industrial robots, four times the numbers installed in the United States. Careful control of inventories by "just-in-time" deliveries cut warehousing costs in assembly industries and freed more factory space for production. In the 1990s, Toyota—a company that led the way in technologies of production—found that the indiscriminate use of robotics and automation was too expensive. The company moved to a system that involved robotics where appropriate but also employed the greater use of people skills in making cars.

Japan's successful exports became known for their quality, reliability, and market-leading technology. In the 1980s, other countries built huge deficits of payments to Japan, causing the Japanese currency, the yen, to double in value compared to other international currencies. This made Japanese exports more costly for other countries to buy. Imports became cheaper, however, so that the Japanese spent more on products made abroad as well as those made in Japan. The major boom in retailing was joined by another in the construction of new houses and apartment blocks. Japan turned from a country of producers into one of consumers. As demand for more luxury and prestige goods rose, Japanese manufacturers moved into these fields, competing with imported goods.

From the 1980s, the Japanese corporations with large yen trading surpluses invested more overseas. They first bought out their foreign mineral suppliers and later established Japanese factories overseas. Building large productive units in other countries during the 1980s and 1990s also helped them penetrate markets that had previously placed quotas on the import of, for example, Japanese-made cars. In the late 1980s, after major investments in the United States and Britain to establish 14 Japanese auto plants, Japan switched investment to Southeast Asia. The economic growth of Thailand, Malaysia, and other countries in Southeast Asia received a major stimulus as Japanese investments poured in.

In the mid-1990s, a further phase of Japanese investment abroad expanded previous investments in cheap-labor, assembly-line factories by moving design and research facilities to Southeast Asian countries. Japan also made increasing investments in China, particularly in the northeastern port of Dalian and other areas close to Japan. By the early 2000s, most Japanese multinational corporations such as Sony, Hitachi, Toshiba, NEC, and the automakers Toyota, Honda, Mazda, and Nissan used parts and whole products manufactured and assembled in China. Japan now has a huge trade deficit with China. Chinese manufactures, often with Japanese corporate badges, began to flood the Japanese markets at home and overseas. Japan's fragile financial system had another economic factor to cope with.

By the early 2000s, Japan again faced a need to change. It suffered from falling prices and business collapses, its established systems of government-business linkages were undemocratic and increasingly inefficient, and its banking systems came under pressure because of huge debts, particularly resulting from investments in Southeast Asia. While Western nations imitated Japanese production processes and competed against Japanese goods in world markets, the Japanese looked more closely again at Western ways, particularly in relation to financial control. However, the irregularities in several large U.S. corporations in the early 2000s resulted in new controls being applied in both Japan and the Western world.

Service Industries

The huge trade surplus from exports of manufactures also resulted in the rapid growth of Japan's financial and business services. In 1986, as well as being the center of government administration, Tokyo became the world's second financial center after New York and in front of London. The additional functions led to further increases of population in the Tokyo region.

Throughout Japan, the increase in service occupations, from education and health care to retailing and tourism, widened the range of employment and contributed to a diversified economy that is not so dependent on the fluctuations of world markets as when manufacturing dominated it.

Japan's Regions and Cities

Japan's cultural and economic prominence requires a fuller understanding of its regional variety. Japanese physical environments—from snowy Hokkaido in the north to subtropical Okinawa in the south and from the volcanic mountain spine to the coastal lowlands—combine with distinctive contrasts in urbanization, manufacturing emphases, and rural ways of life (see the "Personal View: Japan" box, p. 188). Japan consists of four main islands: Hokkaido, Honshu, Shikoku, and Kyushu (see Figure 5.15).

Hokkaido

The northernmost island of Hokkaido has over 20 percent of the land but only 5 percent of the Japanese population and produces just 3.8 percent of the total GDP. The island's previous inaccessibility from the rest of Japan has been reduced by a direct rail link through the Seikan tunnel under the Tsugaru Strait and better air services. The improved access encourages farmers to grow new breeds of rice and produce dairy products and crops linked to the increasing Japanese demand for Western-style foods. Farm products now exceed the value of Hokkaido's traditional natural resource-based industries—fishing, coal mining, and forestry. The 1972 Winter Olympics at Sapporo—now Japan's fifth-largest city with 2 million people—gave a boost to tourism on the island. The recession of the late 1990s, however, affected this region badly, with many bankruptcies and increasing unemployment.

Honshu

The main island of Honshu has three parts. *Northern Honshu* (Tohoku) has a relatively sparse population for Japan—8 percent of the total population on 18 percent of the land—but its density is twice that of Hokkaido. There is better transportation infrastructure, although the central mountains make crossings between the two coasts difficult. This region is Japan's leading rice producer, with 25 percent of the total, much on reclaimed coastal lands.

In the early 2000s, the economic problems facing Japan brought particular hardships to the northern Honshu. Always a vulnerable area with little manufacturing employment, its main sources of income and jobs in fishing and farming suffered from declining stocks of fish and price competition from overseas suppliers. The main fishing port of Hachinohe had newly opened pinball stores that were full of unemployed fishermen

and workers from idle fish-processing plants. The apple orchard owners had to diversify because the price of apples was so low, while Chinese produce undercut the local garlic growers. Public money financed infrastructure projects rather than local industries. Workers in short-term construction jobs had to move elsewhere to continue employment.

In *central Honshu,* the Tokyo region (Kanto) is the heart of Japan in terms of population concentration and economic activity: 31 percent of the population produce 37 percent of Japanese GDP on less than 10 percent of the land. Tokyo, Kawasaki, and Yokohama fuse into a huge urban complex around Tokyo Bay (Figure 5.19). Tokyo itself, having 26 million people in 2000, is at the center of the world's most populous metropolitan area. Tokyo is Japan's center of government, finance, and internal and international service industries, Yokohama is Japan's busiest port, and Kawasaki has heavy industries. Oil refineries, steel mills, chemical plants, and power stations line the waterfront and cover large sectors of reclaimed land. The rest of the plain around this megacity still produces much rice, together with fruit, vegetables, poultry, and pigs for the nearby markets. Volcanic mountains around the northern and western margins of the bay area are used for winter sports facilities and their hot springs attract other tourists.

The Pacific coastal area south of Tokyo is an industrial corridor of connected lowlands where a line of cities along the "bullet train" (*shinkansen*) and major highway routes urbanized former agricultural land. Nagoya is Japan's third-largest city with 3.2 million people and the center of a textile region that now has automaking (Toyota City), oil refining, petrochemicals, and engineering industries. This region's diversified and growing industries suffered least in the late 1990s recession.

West and north of the Tokyo complex, the land in Chubu is more mountainous. The northern coastlands facing the Japan Sea are another rice-growing district. Mountains separate it from the main economic areas of the country, although highway and rail links penetrate these Japanese "Alps." The 1998 Winter Olympic Games were held at Nagano in this area that receives winter snows from winds crossing the Sea of Japan. The mountains include national parks, such as that around Mount Fuji (see Figure 5.3), Japan's highest peak (3,776 m, 12,388 ft.), and have hot springs. Chubu leads the country in generating hydroelectricity.

Southern Honshu contains Japan's second-most prosperous region, Kinki, which includes Osaka and its immediate hinterland, which is called Kansai. Three cities dominate the eastern part of this region—Kyoto is Japan's sixth city (1.8 million people) and was the country's capital from A.D. 794 to 1868. It remains a cultural and tourist center and has craft industries such as silk, pottery, and traditional furniture. Osaka, Japan's second-largest city with 11 million people, and Kobe are industrial ports on Osaka Bay. Following an early specialization in textile manufacturing, the area shifted into iron and steel, chemicals, and shipbuilding in the 1920s, adding oil refining, petrochemicals, automaking, and domestic appliances after 1945. This part of southern Honshu became Japan's "rust belt" of declining industries in the 1980s, and it was hit hard in the late 1990s recession. In 2001, the area around Osaka and

(a)

(b)

(c)

(d)

(e)

Figure 5.19 **Japan: Tokyo-Yokohama-Kawasaki urban area.** (a) Space shuttle view of the harbor and the world's largest metropolis. The extensive port facilities reflect the country's involvement in trade. Although subject to earthquakes at a point where three major, active tectonic plates meet, most of the industrial building is on landfill and artificial islands: such materials are liable to liquefaction in an earthquake, and a major event could cause catastrophic damage. (b) Part of Tokyo harbor, showing detail of industrial building on landfill. (c) Old-fashioned market, Tokyo Asakusa. (d) Sukibayoshi pedestrian crossing, Ginza. (e) Bullet trains. Photos: (a) NASA; (b) © Michael Yamashita/Corbis.

Kyoto had Japan's second-highest unemployment rates of 6 to 7 percent (compared to a national average of 5 percent).

The western peninsula of southern Honshu (Chugoku) has coastal lowlands on either side of a mountainous spine. While a harsh winter climate and poor links with the rest of Japan mark northwest-facing coasts, the southern coast looking to the Inland Sea has a better climate. The latter area responded economically to improved accessibility by train and road after 1945. Reclamation of land from the Inland Sea produced port facilities and heavy industry using imported raw materials. In 1995, the new Kansai airport was opened, built on reclaimed land south of Osaka, but it is the subject of worries about the subsidence of the fill materials and the potential of rising sea level as a result of global warming.

Shikoku

Shikoku is a mountainous, small island with only 5 percent of Japan's land. The Pacific side of the island has a warm, moist climate but is isolated by mountains from the northern coasts that face the Inland Sea. The latter are irrigated for crops and industrialized. Major infrastructure investments in bridges now connect the island with the rest of Japan (see the "Personal View: Japan" box, p. 188).

Kyushu

Kyushu, the westernmost island, is connected by tunnel to Honshu. It was the historic point of entry to Japan for

Buddhism and early Christianity, new writing script, and trade. Two-thirds of the island's population lives in the northern region around Fukuoka and Kitakyushu (2.7 million people), both having long-established metal, chemical, and shipbuilding industries. These began to decline, but in the 1980s, Kyushu attracted automaking and high-tech industries with expanding research and development facilities.

Farming and fishing remain important in the economy. Much of Kyushu is a rice-growing area. It is warm enough for a second crop of vegetables or rushes, while early crops are grown for the large cities farther north. Mandarin oranges, tobacco, and livestock are produced on the hills. There is a growing tourist industry, although the number of shoppers coming on ferries from Korea decreased in the late 1990s.

Ryuku Islands

Offshore, the Ryuku Islands including Okinawa stretch for 1,000 km (650 mi.) southward toward Taiwan. Their economy has not developed with the rest of Japan, and they have high unemployment. Many still have American military bases alongside the traditional fishing and agriculture, and a tourist industry is developing. There are pressures for the removal of the U.S. military personnel, although most of the remaining U.S. military are based on Honshu.

Japanese City Landscapes

While Japanese cities may appear to be formless sprawls, detailed study shows that they reflect irregular historic growth followed by rapid modern changes and expansion. All Japanese cities exhibit the crowding of buildings that results from the shortage of available land.

Castle towns, ports, or centers on main routes to religious shrines now provide the historic cores of huge metropolitan areas. Many modern Japanese central cities retain older street patterns, although the buildings are recent. Only the sturdiest ancient buildings survive. Most of the older buildings in Japanese towns were built of wood and other nondurable materials that fires and earthquakes destroyed. Destruction and rebuilding caused by earthquakes or World War II bombing often interrupt the transitions between different land uses.

After 1970, downtown areas of many Japanese cities were redeveloped with high-rise buildings but retain an older sector that acts as a tourist and specialist shopping area. The modern cities contain sectors for commerce, industry, and residences. The central business districts of the largest Japanese cities contain areas specifically devoted to hotels, department stores, financial businesses, and national or local government offices. High-rise offices, hotels, and apartment blocks are a response to high land costs, as are the underground developments in central Tokyo that include restaurants and car parking lots. Smaller business districts occur in suburban areas of the major cities. Out-of-town shopping only began to take hold in the 1990s.

Japanese industrial development is strongly localized, with heavy industry in coastal locations on reclaimed land because of the needs for space and proximity to ocean transportation links to raw material suppliers and markets. Such

Test Your Understanding 5B

Summary Japan is small, densely populated, and island-based. It became the world's second-wealthiest country following a dramatic economic recovery after 1945.

Japan's agriculture remains highly protected from foreign competition. In the later 1900s, its industrial economy expanded from dependence on importing raw materials and producing goods for export markets. It moved from iron and steel manufacturing to making autos and high-tech goods. In the 1990s, the rising value of the yen caused by selling more exports than buying imports led to competition in world markets from cheaper sources. This caused a slowing of Japan's economy and its making investments in other countries.

Some Questions to Think About

5B.1   In tracing the history of economic growth in Japan from the mid-1800s to the present, how would you relate the events to global connections and local voices?
5B.2   What are the main regional differences within Japan?

Key Terms

megalopolis   Ministry of International Trade
suburbanization   and Industry (MITI)
counterurbanization

zones shut off city centers from the waterfronts. Light industry is more widely distributed and intermixed with housing, with older firms retaining inner-city sites. Location in the inner city takes advantage of the availability of labor, linkages with other firms, and market outlets. Local governments encourage the establishment of light industrial estates on the city margins, including science parks for high-tech industries.

Housing often consists of postwar apartment blocks near the city center, giving way to suburban bedroom areas. Poor, cheaper downtown housing units are refurbished to maintain the labor force in those areas. Transport routes determine the main expansion of suburban housing.

The Koreas

North Korea and South Korea (Figure 5.20)—officially the People's Democratic Republic of Korea and the Republic of Korea, respectively—had a common ethnicity and history to the mid-1900s but after that moved in different directions. They occupy the hilly Korean peninsula that extends southward from the Chinese border along the Yalu River. The peninsula almost touches Japan at its southern end.

At the end of World War II, the United States and Soviet armies defeated Japan and divided the "freed" North and South Korea at the 38th parallel. The Soviet Union supported North Korea, and the United States supported South Korea. Independent republics were created in 1948. After arguments between the two Koreas, North Korean forces invaded South

Figure 5.20 **The Koreas, North and South.** Major geographic features.

Figure 5.21 **North Korea: Pyongyang.** A typical Communist urban landscape, including the National Monument and the 70-ton statue of Kim Il Sung on Mangu Hill. The statue dwarfs people paying homage to the country's former leader. Photo: © Jeffrey Aaronson/Network Aspen.

Korea in 1950. The United Nations sent forces under U.S. commanders to support South Korea, and following destruction over much of the territory as warfare moved back and forth, the two sides agreed on an armistice in 1953. Tensions between the countries remain, causing them to maintain a high level of defensive capabilities. North Korea is rumored to possess the ability to make nuclear and biochemical warheads.

By the late 1990s, there were signs of closer ties developing between the two countries. Presidential visits led to reunions of Korean families separated for 50 years. Talk of reunification of the two countries gathered some momentum in the late 1990s and early 2000s, but both countries had reservations about such a move. North Korea did not wish to lose its independence, while South Korea counted the potential cost.

Countries

With huge support from the United States after the Korean War, South Korea became one of the newly emergent urban-industrial countries linked to the West and the global economic system. Like Japan, South Korea's governments ranged from paternalistic democracies to dictatorships. They used absolute power to promote personal savings and encouraged export industries. By the 1980s and 1990s, such policies led to rising prosperity for many and large trade balances for the country. Freer elections introduced from the mid-1980s led to increases in democratic processes during the 1990s. South Korea's financial sector, together with its mode of industrial organization in large conglomerate firms, or *chaebol,* threatened to weaken much of the country's economy and political stability in the late 1990s.

North Korea maintained a Communist regime under the personal dictatorship of Kim Il Sung (Figure 5.21) and his son Kim Jong-il, who succeeded in 1994. Supported by the Soviet Union and China in the Cold War period, North Korea's isolation from the rest of the world and suspicions of South Korea and the West caused it to invest heavily in military capabilities instead of economic development. After the loss of Soviet Union subsidies in 1991, North Korea depended increasingly on international aid to make up for internal economic collapse and famine. Its main bargaining point is international suspicion that it possesses nuclear and biochemical weapons and ballistic missiles to deliver them. North Korean brinkmanship policies attract international aid to prevent military action.

People

Urban Focus

South Korea is the southern half of the Korean peninsula. It is a hilly land with the highest ridges along the east coast. South Korea's population of 49 million people in 2001 is likely to rise slowly to around 53 million by 2025 because of a low rate of population increase. Life expectancy rose from 47 years in 1955 to over 70 years today. As in Japan, the population in older age groups will increase markedly in the next 20 years (Figure 5.22). These facts and the high proportion living in towns are signs of the modernization of South Korea.

The South Korean urban population increased from 25 percent in 1950 to 50 percent in 1980 and 80 percent in 2001.

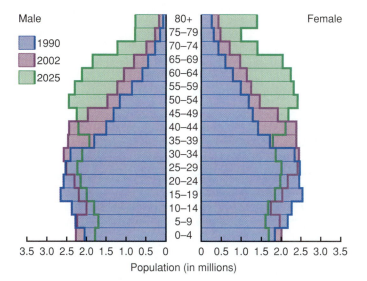

Male   Female

■ 1990
■ 2002
■ 2025

80+
75–79
70–74
65–69
60–64
55–59
50–54
45–49
40–44
35–39
30–34
25–29
20–24
15–19
10–14
5–9
0–4

3.5 3.0 2.5 2.0 1.5 1.0 0.5 0   0 0.5 1.0 1.5 2.0 2.5 3.0 3.5
              Population (in millions)

Figure 5.22 **South Korea: age-sex diagram 1990, 2002, and 2025.** Source: U.S. Census Bureau. International Data Bank.

The main population and industrial facilities are in the northwest (see Figure 5.20), centered on the capital, Seoul (10 million people in 2000), Inch'on (3 million), and Taejŏn (1.6 million), and along the southern coast, including Pusan (4 million), Taegu (2.7 million), and Ulsan (1.6 million). Almost totally destroyed in the 1950s fighting, Seoul was rebuilt rapidly (Figure 5.23).

North Korea had 22 million people in 2001, and this is predicted to rise to nearly 26 million in 2025. Its population is also concentrated around its western coasts with the capital city, Pyongyang (3.2 million people), and Namp'o, the other city with 1 million people. Countrywide, fewer than 60 percent live in towns. Although possessing greater mineral resources than South Korea, particularly coal and metallic ores, North Korea has not developed a major manufacturing capability and drawn people to the towns.

South Korean Economic Development

After World War II and the Korean War (1950–1953), South Korea established import substitution industries that developed manufacturing skills and reduced the need to pay hard currency for foreign goods. Its dictators forced the people to work for low wages while the economy improved. The transportation infrastructure built by the Japanese when they ruled Korea in the early 1900s and rebuilt with external funds after the 1950s war helped develop the Korean economy in its early stages of growth. South Korea built a series of major highways, and its public transport system is one of the world's best.

The devastating Korean War in the early 1950s took place when three-fourths of the South Korean people still gained a living from farming. In postwar land reforms, the government took over large, inefficiently run estates and encouraged a new group of small landowners to apply imported fertilizers. Farm productivity rose, and South Korea became largely self-sufficient in food.

From the early 1960s to the early 1990s, few other countries' economies grew so fast. Exports rose from $33 million in 1960 to $150 billion in 2000. South Korea's total economy is now 66 percent of China's and 10 percent of Japan's. A country of subsistence farmers under largely feudal control was transformed in a generation to the world's main maker of large ships and memory chips, the fifth-largest automaker, and eleventh-largest world economy—with greater output than the whole of Africa South of the Sahara. A farming and rural emphasis gave way to manufacturing, services, and urban living.

The iron and steel, shipbuilding (Figure 5.24), chemicals, automaking, and textiles industries established from the 1950s continue to be important, although in the 1990s, there was a swing toward high-tech industry. South Korea was another major

Figure 5.23 **East Asia: Seoul, capital of South Korea.** Modern buildings overshadow old ones as a result of rapid urbanization. Very few older buildings survived the 1951–1953 war against North Korea. The huge volume of road traffic is accommodated by multilane roads and reflects the growth in economic activity. Photo: © Robin Laurance/Photo Researchers, Inc.

Figure 5.24 **South Korea: Ulsan City shipyard.** The Hyundai *chaebol* is a major builder of large ships. Note the space required for assembling the materials and the range of equipment used in construction. Photo: © Alain Evrard/Photo Researchers, Inc.

investor in modern China in the early 1990s, taking advantage of cheap labor costs in the nearby Chinese province of Shandong. The initial rush of small companies to invest in Chinese locations is giving way to major South Korean corporations such as Daewoo, Samsung, and Hyundai, which link to hundreds of smaller companies producing their components and products.

At an early stage of development, the South Korean government supported the larger family-owned companies that developed into huge conglomerates—the *chaebol,* such as Hyundai, Daewoo, LG (Lucky Goldstar), and Samsung. The *chaebol* became diverse in their products, starting up subsidiaries that often had little chance of success, taking on massive debts, and providing major problems for the country. When Halla, the twelfth-largest *chaebol,* went bankrupt in 1997 with debts of 20 times the company's assets, Hyundai, the largest, was found to have guaranteed around 15 percent of the debts. This highlighted the fact that the founders of the two *chaebol* were brothers and had strong cross-company links. Family loyalty—part of Korean (and Confucian) culture—was exposed as making problems for running large corporate businesses in a free-market context of increasing transparency. The

chaebol also controlled large parts of the South Korean financial markets, providing an environment for building up debts that were shared through many links.

In late 1997, indebtedness reached the point where the South Korean currency unit, the won, and its stock exchange collapsed. Devaluation of its currency increased South Korea's indebtedness to other countries. Bank closures, bankruptcies, and unemployment stirred militant trade unions into political action against the stringent conditions proposed by external loan providers. By 2001, the South Korean economy had largely recovered, helped by the breakup of some family-owned *chaebol,* and several of the new firms merged with U.S. and European corporations.

As an example of *chaebol* history, South Korea was once called the "Republic of Hyundai" after one of its most powerful *chaebol.* After the 1997 crisis exposed its huge debts and damaging grip on the economy, Hyundai was restructured around several independent companies. Hyundai Motor, however, is the only one that makes profits, driven by its U.S. sales.

In 2002, the South Korean government signaled its intention to develop the island of Cheju-do (Jeju) off its southern coast. The island's natural environmental potential includes its pleasant climate and attractive scenery from an extinct volcano to many waterfalls and beaches. It is a traditional honeymoon destination, center of tangerine plantations, and home of professional women divers who pluck seafood from the ocean floor.

The project to be completed by 2010 symbolizes South Koreans' sense of vision, their appreciation of the country's geographic position, and probably their overly-optimistic hopes. South Korean officials managing the project make much of their country's central position between Japan and China and the government's intentions to make business friendlier. Cheju is two hours' flying time from Beijing, Tokyo, and Hong Kong. It is proposed to turn the island into an international business and tourism hub akin to Hong Kong and Singapore, with a free-trade area in which English will be a second language. Overseas universities could set up campuses alongside a science and technology park, duty-free shopping malls, and 20 golf courses. The South Korean government is committed to paying half the projected costs of the project that will require a new airport, better roads, and more hotels. It depends, however, on large investments from the private sector. Moreover, the local population of half a million complains about the lack of consultation and worries about the environmental impacts.

Although mostly noted for its economic growth, South Korea also has an increasing regional contribution in popular music that is both cultural and economic. Young pop stars such as Boa develop alongside other youth culture inputs in movies, TV shows, computer games, and fashion. This trend relates partly to South Korean consumers preferring local products to the invading Western music and movies. It is also part of South Korea's development of knowledge-based service industries. South Korean "K-pop" bands top charts in Taiwan and Cambodia, while Korean TV dramas are highly rated in Vietnam. China, with its huge and growing market for such music, is a prime target for record companies, which envisage setting up an academy for discovering young Chinese pop stars. The main problems for a South Korean venture in China are the rogue bands copying songs and the rampant piracy and copying of CDs. The 2002 World Cup soccer finals, co-hosted in South Korea and Japan, provided a further platform for wider penetration of both Western and South Korean products.

Economic Stagnation in North Korea

When the Japanese ruled the whole of Korea in the early 1900s, the northern part provided iron, coal, and raw materials for a chemical industry and had plenty of hydroelectricity potential. However, from its creation in 1948, North Korea has been an inward-looking country with an economic policy advertised as isolationist self-reliance. This had a historic precedent when Korea was called the "Hermit Kingdom." In fact, the country depended on aid and good prices from China and the Soviet Union. These arrangements ceased in 1989. Factories became idle and people lacked proper nutrition. By the late 1990s, the GDP per capita in North Korea was one-eighth of South Korea's. In 2000, the death rate rose 50 percent higher than in 1994, indicating widespread famine and disease. The North Korean economy collapsed, making the country and its government dependent on international aid, of which the United States is the largest contributor.

North Korea faces a difficult period. After years of a declining economy, the people of North Korea, duped by hype from their leaders or cowed by fear, do not appear likely to revolt. But they have few skills or resources to attract the level of external foreign investment that underlies rapid economic growth in countries such as China. North Korea faced three alternatives in the late 1990s. It could:

- Continue its path to greater poverty and eventual collapse.
- Be reunified with South Korea—although that country views the prospect with unease after Germany's experience incorporating its eastern area (see Chapter 3).
- Follow a path of economic liberalization similar to that of China, take offered aid in exchange for reducing its nuclear program, and encourage investment from foreign countries. Some South Korean multinational corporations are poised to capitalize on low-wage labor and the markets for its goods in North Korea. Many South Korean families with relatives in North Korea wish to help them take advantage of better economic conditions. The North Korean government, however, has avoided following China's lead because it fears losing power.

Following international concerns about North Korea's nuclear weapon capability, South Korea, Japan, and the United States agreed to provide nuclear power stations but on condition that nuclear weapons were outlawed. The first of these power stations was begun at Shupo in mid-1997 and employed 5,000 South Korean engineers. Until these power stations are functioning, North Korea receives half a million tons of free fuel oil each year.

In late 2001 and 2002, North Korea's needs for famine relief—it was the world's largest recipient—competed with the UN World Food Programme's rush of aid to Afghanistan and

Test Your Understanding 5C

Summary South and North Korea have the same ethnic composition of Korean people but contrast in the outcomes of events over the last 50 years following the destructive war of the early 1950s.

South Korea received help with reconstruction after the 1950s war from the United States in particular. The United States opened its markets. South Korea became one of the world's growing urban-industrial centers. Its economic wealth brought increased standing in the world and in East Asia.

North Korea retained its Communist government, backed until 1989 by some economic and political links to China and the Soviet Union. In the 1990s, North Korea became impoverished and under suspicion by other countries of harboring nuclear and biochemical weapons.

Some Questions to Think About

5C.1 Explain the economic growth of South Korea.
5C.2 Imagine a conversation at an occasional meeting of relatives living in North and South Korea. Include comparisons of living conditions and the roles of each country's government.

Pakistan. Fortunately, the harvest in North Korea was not as bad as in the drought year of 2000, but North Korea still needs outside help. The best food produced in North Korea feeds the army and party officials, while other people depend on donated food, often of lower quality. Although China and South Korea reduced aid in 2001, it was likely that the main donors—Japan, the United States, and South Korea—would maintain their contributions to avoid social instability and large numbers of migrants moving out.

China, Mongolia, and Taiwan

The People's Republic of China (PRC) is the world's most populous country, with 1.27 billion people in 2001, and the third-largest by area (Figure 5.25). Hong Kong (Romanized Cantonese for the name, or Xianggang, the pinyin version of Mandarin Chinese) and Macau, previous colonies of the United Kingdom and Portugal, respectively, were returned to PRC control in 1997 (Hong Kong) and at the end of 1999 (Macau). Mongolia (2.4 million) and Taiwan (22.5 million) are tiny by comparison with the PRC but have distinctive roles in relation to China and the rest of the world. These two countries are discussed together after the section on China.

China

From 1949, the Chinese Communist Party's rule had fluctuating impacts, both over time and space. It isolated China economically until the late 1970s, but by the early 2000s, changes introduced in the late 1970s gave China an increasing role in world political and economic affairs.

Figure 5.25 **The People's Republic of China, Mongolia, and Taiwan: major geographic features, including towns and cities, major rivers, and provincial boundaries.** China has 23 provinces (capital letters), four cities with provincial status (Beijing, Shanghai, Tianjin, and Chongqing), and five autonomous regions that have large national minorities (Tibet, Inner Mongolia, Guangxi, Ningxia, and Xinjiang).

China's Ups and Downs Under Communism

At the establishment of the People's Republic of China in 1949, the government's first aim was to turn a backward, feudally divided agricultural country into an advanced and centralized socialist industrial country. From 1820 to 1950, China had declined from producing 33 percent to 5 percent of world economic output. China's new government began the unification and economic revival of a country after 150 years of divisions among competing warlords and ideologies and the disruption of the Japanese invasions in the 1930s.

To restore the country's position, Mao Zedong, the leader and dictator of China from 1949 to his death in 1976, wished to make rapid and massive changes. It is important to understand the ups and downs that his policies caused:

- From 1949 to 1959, Mao established central planning, mainly through the move to the collectivization of agriculture in communes.

- The "Great Leap Forward" of 1959 to 1962 and the "Cultural Revolution" from 1966 to 1976 attempted to change the economy and society in revolutionary ways but caused economic and social devastation.

- After 1976, Mao's successor, Deng Xiaoping, brought in reforms that abolished most of the communes and engaged with the global economic system.

Collectivization and Communes

At first, China modeled its revolutionary changes on the Soviet Union experience, using Soviet advice to establish central control through five-year plans. Large-scale industrialization and the collectivization of agriculture were the first priorities. Two-thirds of available investment went into manufacturing, where output rose rapidly, but hardly any into agriculture despite the fact that most people lived in rural areas and depended on farming.

Although little investment went into improving agriculture, rural social structures were revolutionized. Before 1950, most Chinese were tenants or landless, working tiny parcels of land. The landlords or rich peasants who owned over 70 percent of the cultivated land made up fewer than 10 percent of the population. During the early and mid-1950s, central government control increased and individual decision-making decreased through the following sequence of events:

- In 1952, 300 million landless peasants received their own plots of land, houses, implements, and animals, free from debt.

- By 1954, over half of these peasant farmers joined "mutual aid teams." Each team comprised up to 10 households, and the owners pooled labor, tools, and work animals at planting and harvest. Such teams often became permanent, grouped in larger agricultural producers' cooperatives. By early 1956, over 90 percent of the rural population joined cooperatives.

- Next, land was pooled under central management, with members of the cooperatives receiving income based on their land and labor inputs. In "advanced" cooperatives, the land became the property of the organization apart from tiny garden plots for each family. A typical advanced cooperative included approximately 150 households farming a total of around 400 acres of land. This process, known as **collectivization,** was completed by late 1956.

- Finally, part of the "Great Leap Forward" thinking was to establish communes. A **commune** combined agriculture, industry, trade, education, and the formation of a local militia. The commune leadership planned the agricultural year, other economic activities, and social functions. They guaranteed food, clothing, education, housing, and arrangements for weddings and funerals to members. By September 1958, some 26,425 communes included 98 percent of peasant households. The total number of communes rose to 74,000 in 1965 and then fell to around 54,000 before they were replaced by new policies in the 1980s. Although some communes in favored areas prospered and provided good schooling and health care, few achieved the potential economies of scale based on sharing equipment, buying seed and fertilizer in bulk, and combined production agreements. While some managers attended to community welfare, many became unreasoning dictators. The commune experiment was linked to the Chinese political and economic fluctuations from the 1950s to the 1970s. It achieved a degree of communal working, but local arrangements were often interrupted by political upheavals, and commune life failed to motivate the peasant farmers, for whom political propaganda and punishments never generated the same commitment as individual incentives.

The Great Leap Forward and Cultural Revolution

The "Great Leap Forward" of the late 1950s aimed to increase the rate of industrialization by producing more basic industrial products, such as iron and steel. It attempted to integrate the rural and urban areas by setting up small-scale production units in both. Throughout much of China, community efforts devoted to iron smelting, often in small backyard furnaces, caused a neglect of farm work and made worse the drought-based famine of 1960, in which 25 million to 30 million people died. The plan was a disaster. Both industrial and agricultural production dropped, and China was forced to import wheat from Western countries. Stung by failures and misunderstandings, the Soviet advisers went home, and the Chinese leaders abandoned most Great Leap Forward policies.

After a few years of higher production, Mao Zedong attempted to change the whole basis of Chinese society and its administration, and the country once again turned inward. The "Cultural Revolution" of 1966 to 1976 diverted attention from increasing economic productivity to protests against unchanging local authorities and leadership cliques. Mao Zedong's "Little Red Book" summarized his wish for revolutionary actions to purge society. The Red Guards, composed mainly of young adults, carried out his tenets, destroying management structures at all levels of society. Such fear-based tactics kept Mao Zedong in power until his death. His policies swept aside many of the short-lived economic gains of the early 1960s in a massive persecution of the educated classes, technologists, and former leaders.

China Joins the Wider World

After Mao Zedong died in 1976, his former chief deputy, Deng Xiaoping, redirected China's policies. He reversed Mao's movement toward self-sufficiency and isolation from world markets. Deng's policies were based on China accumulating wealth to diffuse it to the poorer parts of the country.

Changes began on the farm. In the 1980s, commune-based agriculture gave way to the **household responsibility system,** allowing families to decide which crops to grow. At first a local government initiative, central government supported it after 1981 and made it the hub of its reforms. Households could lease collectively owned land for up to 15 years. Small groups entered short-term contracts, leasing land to achieve production quotas set by the government. Farmers could sell any surplus crops or animal products in local markets. By 1984, 99 percent of households participated.

From 1978, Deng encouraged a more open economic policy that attracted investment from outside China and generated the world's highest rate of economic growth from the mid-1980s to the early 2000s. He encouraged foreign investments and the rapid economic growth of many coastal regions of China. The central Communist government, however, stayed in power, and its often oppressive measures restrained personal freedoms.

After Deng Xiaoping died in 1997, his successors developed the policies that brought economic growth and involvement in the global economic system. The acquisitions of Hong Kong and Macau in the late 1990s highlighted the process of reclaiming long-term Chinese territories and the end of foreign ownership of Chinese lands.

Tensions in the 2000s

In the early 2000s, China faces increasing tensions between the government's desire to maintain central control over the economy and the links to the world's free-market system. Many publicly owned industries are no longer cost-effective against international competition and face social consequences from shedding workers. Joining the World Trade Organization in late 2001 brought further challenges to Beijing's centralized control over the economy at a time when the government had to cope with rising urban unemployment, rural unrest, and funding shortages resulting from difficulties in collecting taxes.

The Chinese government handles the tensions in various, often contradictory, ways. For example, in theory, it views the rush to work with new entrepreneurs and international funding and trading companies as a major priority. In practice, however, local entrepreneurs meet difficulties in registering as private businesses with share-based funding. While educated Chinese now have increased opportunities to shape careers, go abroad, or pursue research interests without party interference, the army and security forces continue to respond firmly to dissent and unrest. The justice system is poor and subject to political pressures that prevent magistrates from acting fairly: for most people outside political and business elites, trials and imprisonment remain arbitrary and often long term. The government diverts internal pressures for greater democracy by a renewed emphasis on Chinese nationalism through a historical focus on the pre-Communist triumphs of past civilizations and greater publicity for Chinese cultural and athletic achievements.

Despite cries for reforms, the growing fear of a popular backlash is diminished by a greater fear among ordinary people of acting outside the established system. The reform-blocking local and provincial party leaders are not so much Communists as conservatives hanging on to their powers—very much like their feudal predecessors. Experiments of local democracy in rural areas often lose their impact because central ministries impose restrictions on powers of action and officials continue to demand illegal fees. The endemic corruption requires a new system of independent courts and accountability. Many mid-ranking officials, however, understand "political reform" to mean merely streamlining the bureaucracy. Even among Communist Party members, liberals vie with old-guard Maoists and free-market capitalists with leftists. Each group has its own journals and Internet websites. Although more opinions are aired than was possible in 1989, multiparty democracy is not yet mentioned as even a remote possibility.

The new leaders elected in 2002 will need to deal with widespread corruption, fast-growing crime, social unrest, and encouragement for the aspirations of the growing group of private businesses, once suppressed as "exploiters." In 1991, around 100,000 private businesses employed 1.8 million people; by 2001, 24 million worked in 1.5 million private businesses with over eight employees and around 30 million in smaller, individual concerns. In that decade, private industry grew from a few percent to 40 percent of total Chinese industrial output.

China and the World Trade Organization

China joined the WTO in November 2001. The Chinese leaders expect WTO membership to boost exports and foreign investment, but it will also force its industries to become more competitive in world markets in both quality and price. WTO rules expect China to cut import tariffs and allow foreign businesses to compete in its highly protected areas such as telecommunications, automaking, insurance, and banking industries.

Such trends raise opposition from inward-looking nationalists and the bureaucrats managing the Beijing-based public monopolies. Some Chinese see WTO agreement to Chinese membership as a plot that will enable the United States to gain control of China's economy. They stress the dangers of rising unemployment and fear that foreign multinationals will merely replace current government monopolies or strip the assets of firms they take over in China. Cheaper agricultural imports could mean trouble in rural areas, and openness to integrated global economies could aggravate regional differences within China. Other criticisms focus on the debt and a rising budget deficit caused by increased government spending on housing and other infrastructure construction that external companies require before investing in China.

Neighboring countries also criticize the growing Chinese dominance of international investment and its competitive entry to world markets. Fears of total Chinese economic domination in world markets, however, are unlikely to be realized. Other countries possess different comparative advantages and make moves to develop greater technological expertise. Increased personal wealth inside China expands internal markets for its goods and those from other countries. Perhaps Taiwan will be most affected by China's growth as it invests in production facilities on the mainland but suffers capital and brain drains and a loss of factories to the mainland.

People in China

Although China is now a single country, it includes areas with varied cultures, languages, and religions: it can be thought of as more like Europe, because Chinese languages include a range as great as those in Europe. However, the Chinese writing system is used throughout the country, forming a common element and one that underpins political control.

Ethnicity

The **Han Chinese** make up 92 percent of the Chinese population. They are of similar appearance to most of the national minorities but derive their ethnicity from the emergence of the Chinese administrative culture in the A.D. 200s and 300s. Most non-Han groups did not have their own writing systems and adopted the Han Chinese system. The Han culture diffused by conquest and movements of people to administer new areas. More recently, Han people moved into the Northeast after being excluded until the late 1800s and, from 1949, into westernmost China.

Most minority groups live along the northern borders with Russia and Mongolia and in western China. Central Asian ethnic groups dominate the north and northwest, while Tibetans are the main group in the far west. Tibet was taken under Chinese control in 1950, but, following a 1959 rebellion, the PRC imposed Chinese economic, social, and political values. Despite large subsidies, the material condition of Tibetans did not improve after the Chinese occupation, and the Cultural Revolution was a disaster for Tibet with further bloody repression. The former Tibetan Dalai Lama lives outside Tibet, but his son recently succeeded him under an official Chinese move.

Xinjiang Province, formerly known as East, or Chinese, Turkestan, is linked culturally to Central Asia (see Chapter 4), but China reasserted its control of Xinjiang in 1949. Han Chinese were resettled there, increasing from 5 to 38 percent of the 18 million people in the province. The local Uighurs resist the PRC intrusions by occasional terrorist action. Local autonomy is unlikely, however, as China develops relations with Central Asian countries, extending the rail system to Kashgar and the border, and invests in a pipeline to take oil and natural gas from Xinjiang eastward to Shanghai.

From 1949, the Chinese government encouraged the greater use of the northern Mandarin language that is spoken in north-central China and Beijing alongside the retention of minority languages. At the same time, minorities, such as the Mongols in the north, the Tibetans and Uighur in the west, and the more dispersed Manchu and Muslim Hui, were allowed limited jurisdiction over their affairs, particularly the expression of their culture in music and drama, in autonomous regions and counties within the main provinces. However, former languages were replaced by Mandarin Chinese. Local resentment of the Chinese takeover of these lands, however, causes the eruption from time to time of anti-Chinese demonstrations in Tibet and Xinjiang.

Urban Versus Rural Population in China

Population distribution in China (see Figure 5.14) is marked by the contrast in proportions of growing urban populations and declining rural populations. Until the mid-1900s, most Chinese cities were market and administrative centers, often with walls that restricted expansion. The exceptions were some port cities, of which Shanghai became the largest, growing because of external trade contacts, some of which date back to the 1800s (Figure 5.26).

The largest cities in China (see Figure 5.25) include the capital Beijing with 11 million people in 2000, Shanghai (13 million), and Tianjin (9 million). Chongqing, Shenyang, Wuhan (all 5 million), Guangzhou (4 million), and Changchun, Chengdu (Figure 5.27), Xi'an, and Harbin (all 3 million) each grew from around 1 million or fewer people in 1950. It is likely that further growth of the major cities will take Beijing, Shanghai, and Tianjin to around 15 million people each by A.D. 2015, when China will have over 100 cities with populations of more than 1 million people.

The Communist rule from 1949 alternately favored urbanization and movements back to the countryside. In the Mao Zedong era, the wish to encourage modernization through industrialization led to rapid urbanization from 1949 to 1960. In this period, the urban population more than doubled, from 49 million to 109 million, representing a growth from 9 to 16 percent of the total population. Part of this urbanization resulted from industrialization around previously administrative cities such as Xi'an and Nanjing. Another part came from the development of new inland centers such as Baotou, the iron and steel center in Inner

Figure 5.26 **People's Republic of China: The Bund, Shanghai.** The British built these buildings along the riverfront of the Huang Pu (which flows into the Chang Jiang estuary) in the late 1800s. New building in Shanghai in the 1990s has made major changes in the skyline. Photo: © Kevin Morris/Corbis.

Mongolia. Public rhetoric, however, exhorted the Chinese to avoid the environmental dangers of urbanization and the formation of urban elites that had little interest in the rural population.

From the 1960s to mid-1970s, the rural-urban basis of the registration system (*hukou*), established in the 1950s, prevented rural-born people becoming legal residents of cities. They were denied services, formal-sector jobs, and housing. This policy helped to control urban population growth. During the Cultural Revolution, over 20 million young urbanites, as well as many others, were moved to rural areas to be reeducated in the virtues of rural life. Little investment was made in urban housing or transport infrastructure as any available capital went into expanding heavy industry. By 1978, the Chinese urban population made up around 13 percent of the total, but living conditions in urban areas had declined.

From the late 1970s, Deng's economic reforms led to a further rapid rise of urban population numbers. From 1978 to 2001, the official urban population rose from 13 to 36 percent of the Chinese population, although the *hukou* system still operates. Newly urbanizing areas are classified as rural, including the dispersed, small-town urbanization in the coastal

Figure 5.27 **Modern commercialism and capitalism in the interior Chinese city of Chengdu.** Photo: © David C. Johnson.

provinces, the expanding Beijing-Tianjin area, and the corridors along the lower Huang He, Chang Jiang, and Xi Jiang. These areas raise the total urban population toward half the Chinese total.

From the late 1970s, new industries recruited labor in the countryside on temporary contracts; such workers now make up to 10 percent of urban inhabitants. Many new urbanites do not bother to acquire temporary residence certificates. The new middle-income class in China is almost exclusively urban, working in administration or as managers and technicians. Their numbers rose from 5 million in 1980 to 14 million in 2000. Office, business, and factory workers rose from 23 million to 40 million. The Chinese government believes that 200 million people will be able to afford private cars and housing by the early 2000s. To date, however, it seems that, although the government gives little recognition to them, this group is keener on political security than on initiating changes toward greater democracy.

Over the same period, increased investment in housing and urban infrastructure tripled the housing space built each year, mainly in large apartment blocks. Higher standards of building, sewage services, transportation, and domestic water and gas supplies improved the lot of urban dwellers. The huge new apartment and condominium blocks, however, encroached on farmland that is particularly valuable in a country with so many people and insufficient farmland to grow the necessary food. Although a debate continues over the relative merits of expanding the largest industrial cities or placing more resources in widespread small and medium-sized towns that are integrated with the surrounding rural areas, Chinese people are increasingly making their own decisions and moving to large urban areas.

Chinese Population Dynamics

China's population increased from 583 million in 1953 to 1.27 billion (including the 7 million in Hong Kong and Macau) in 2001 and may grow to nearly 1.6 billion by 2025. A huge population was at first seen as an advantage by Mao Zedong, who attempted to mobilize the great numbers of Chinese people under Marxist principles as a "cohesive force."

When it was realized that around 20 percent of its annual income increase would merely enable the country to keep up with the basic needs of the extra people, China made several attempts to restrict family size (see the "Point-Counterpoint: Population Policies in China" box, p. 206). From the 1970s, the large Chinese population and growth projections resulted in increasingly draconian family planning policies aimed at reducing births. However, the policy of reduced family size gained greater hold in the eastern areas and cities than in the essentially rural west with its minority populations, where the policy is not applied so stringently.

Population growth rates fell from 2 percent per year in 1965 to 1.4 percent in the 1980s and to under 1 percent in 2001. From 1965 to 2001, total fertility fell from 6.4 to 1.8. Even this major achievement merely slowed the continued growth of China's huge population. More children survived early childhood despite fewer babies being born, as infant mor-

tality fell from 200 per 1,000 live births in 1945 to 31 in 2001. The numbers of people living to old age also increased as life expectancy doubled from 35 to 70 years. Investments in education cut adult illiteracy from 80 to 19 percent.

China moved toward the later stages of demographic transition (see Figure 5.16). One projection of the early 1990s showed that, with one child per family, China would reach a maximum population of 1.25 billion in 2000 and would then decline to under 500 million by 2070. With two children, a maximum of over 1.5 billion would be reached in 2050 and subsequent decline would be slow. With three children per family, the population could rise to over 4 billion by 2080. By 2002, it was clear that the Chinese population is still increasing.

The 2002 age-sex diagram for China (Figure 5.28) demonstrates the changing history of population growth. For example, the lower numbers in the 40- to 44-year age group reflect the impact of the major population disaster linked to the Great Leap Forward. The diagram also highlights the changing proportions of men and women. There are more women than men in the older age groups but fewer in the younger groups as a result of the one-child policy, which often means the abortion or murder of girl babies in order to save the family "space" for a boy child.

The future of China's population is linked to potentially political issues such as the problem of caring for the elderly as they grow in numbers. Traditionally, extended families looked after older people, but urbanization often breaks close family ties. In the face of small government provisions for old age and the closure of many publicly owned employers that made better provisions, Chinese families strive to increase their savings for later years.

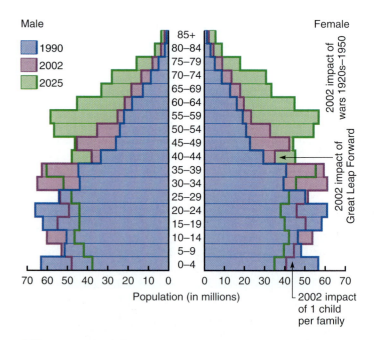

Figure 5.28 **China: age-sex diagram in 1990, 2002, and 2025.** Note the impacts of the Great Leap Forward and the post-1980 one-child policy. Source: U.S. Census Bureau. International Data Bank.

Attacking Poverty in China

Up to the late 1970s, rural life remained harsh. In 1978, China estimated that 33 percent of its rural population (260 million people) lived below its defined poverty line. As the initial impacts of 1980s agricultural reforms doubled output and rural incomes, rural industries created over 100 million new jobs, and the incidence of rural poverty declined from 33 to 9 percent (from 260 million to 97 million people).

As the government's emphasis in the late 1980s shifted to developing the coastal belt, economic disparities between coastal and interior places widened, so that, in the mid-1990s, the human poverty index for interior provinces was 44 percent, compared to 18 percent in the coastal provinces. Some official restrictions remain on migration to urban areas, however, because the urban areas do not have the resources to cope with large additions of population. Migrations of people from rural areas to cities unprepared for their arrival caused urban food subsidies to rise to five times the funding of rural health, education, and relief programs. Even within the cities, Chinese citizens see a few people becoming very wealthy while others face increasing unemployment as publicly owned facilities close.

In 1994, the Chinese government's "8-7 Poverty Reduction Program" (80 million still in poverty to be reduced over seven years) established new antipoverty measures centrally and in the provinces, and committed funds to investment in poor areas. By the late 1990s, the numbers of poor fell to 5 percent of China's population, but this still included a massive 60 million people. Most of the remaining poor live in areas with few natural resources and poor infrastructure—where improvements will be more difficult.

Overseas Chinese

Migration is an important factor in terms of numbers of people leaving China but still represents a tiny proportion of the population. Official international migration out of China in the late 1990s was up to 400,000 persons per year, with 100,000 of those moving to the United States. No figures exist on Chinese migrants smuggled illegally into countries such as the United States, Japan, Europe, and Siberian Russia, but estimates are as high as 200,000 per year. For example, it is known that organized smuggling rings operate from Fujian province. People in Changle county in that province receive a total of around $100 million each year in remittances from relatives working abroad. Nearly 80 percent of the population of Houyu village in Changle county now lives in and around New York.

The latest migrations follow centuries of emigration that made groups of Chinese people—the overseas Chinese—prominent contributors to most East Asian countries. Although their numbers are relatively small compared to the total Chinese population—21 million people in Southeast Asia and 21 million in Taiwan—their influence is much greater. Chinese capital and entrepreneurs dominate the private-sector economies in Indonesia, Malaysia, Singapore, the Philippines, and Thailand. For example, in the late 1990s, they comprised 3.5 percent of the Indonesian population but controlled around 75 percent of the listed companies. In the 1990s, Chinese contract laborers on short-term arrangements began to dominate construction sites around the Pacific Ocean.

Most of the overseas Chinese come from the Shanghai or Guangdong areas and base their business interests on strong family and clan links. Although some anti-Chinese discrimination continues in countries such as Indonesia, Malaysia, Thailand, and Vietnam, their role in economic development makes the Chinese difficult to exclude. Overseas Chinese financiers, particularly in Taiwan and Hong Kong, are also major contributors of foreign investment to new developments within China. The **transnational Chinese economy** is based on these linkages. It became a major factor of economic growth in East Asia and has increasing links to Chinese-based business activities in the United States and the rest of the world.

Economic Development

The Chinese Economy, 1949–1976

From its founding, the Communist government of the People's Republic of China tried to expand output rapidly. However, the Great Leap Forward and Cultural Revolution interrupted and held back the greatest advances. Attempts were also made to distribute economic activities more evenly through the country to increase equality within China and enhance national defense. In 1949, the Northeast together with Shanghai and Tianjin produced 70 percent of national output. The nearness of the Northeast to Japan and the Soviet Union, and the coastal vulnerability of Shanghai caused the Chinese government to move the production of military goods inland.

The policy of wider diffusion had an ideologic, as well as practical, basis. The Communist Party leadership wished to demonstrate that past locations and concentrations of industry depended on the flawed precepts of the capitalist system and foreign intrusions. Factory locations close to raw material sources had better economic foundations. From the mid-1950s, new manufacturing enterprises moved to regions with local coal, hydroelectricity, oil deposits, or strategic factors. However, many of the factories built in the interior proved hugely expensive to run.

Despite the new thrusts of policy and the powers possessed by the central government to make changes, the main centers of manufacturing production remained in the Northeast. In 1976, that region still produced 60 percent of industrial output. Central interior and southern China developed slowly through this period, because of political disruptions and poor transportation linkages to established industrial centers.

New Policies and New Growth After 1976

In 1978, China's total economic output was lower than South Korea's or Taiwan's had been in the 1960s. Deng Xiaoping's major changes in outlook and policy made China increasingly part of the global economic system. Deng began with a new approach to rural life and expanded it into encouraging investments from foreign countries and corporations with a view to increasing manufactured exports. The shift to rapid growth and

POPULATION POLICIES IN CHINA

When an influential nuclear scientist, Song Jian, attended a population symposium in 1979, he made some rough calculations of Chinese population growth on the basis of the poor 1964 census results and formulas used to predict missile trajectories. When asked to flesh out these predictions, he worked with four other scientists and gave results in relation to whether families would have one, two, or three children. It was the first computer-based forecast in China and demonstrated that a rigorously implemented one-child policy would keep China's population to 1 billion by 2000, falling to 700 million by 2050 with advances in medical procedures such as female sterilization, abortion, and other birth control methods. The projections for two- and three-child families suggested continuing increases in the future.

Chinese population change since 1950 was marked by major shifts in policies and attitudes. Although the Maoist government watchword of the early 1950s was "strength in numbers," the census of 1953 registered a total population of 583 million, and the first birth control campaign was launched in 1956 in urban areas. Its impact was minimal, however, in comparison with the 25 million to 30 million who died as a result of the "Great Leap Forward" beginning in 1958. Birth rates increased until China launched another program of birth control in the early 1960s, based on delayed marriages and a wider distribution of birth control pills. It had some effect but was then overtaken by the disrupting Cultural Revolution, which swept aside many of the government-established groups to administer family planning. Once again, birth rates and population totals increased rapidly.

In 1979, Deng Xiaoping approved the one-child policy before further debate, and this key plank of Chinese reforms was implemented without any law being passed by the National People's Congress. Initially a "temporary measure," it has lasted for over 20 years with its complex regulations, harsh punishments, and 80,000 full-time family planning workers to enforce them (Box Figure 1). Despite the new economic freedoms granted to peasant farmers, the policy subordinated individual interests to those of the country; local officials could monitor the most private aspects of individuals' lives.

Previously, some one-child policies had been implemented by urban authorities in Shanghai and other cities in the 1970s. Urban dwellers living in state units could be tracked. But the countrywide one-child policy was directed at rural areas where the communes (and their controls) were being disbanded, food production was increasing, and extra labor was needed.

The policy included "sticks" and "carrots." Families complying with the policy received certificates that enabled them to claim free health care for the mother during pregnancy and delivery, free health care for one child, free education for one child, promotions and better pay for

Box Figure 1 **China's one-child policy.** Large poster at a busy intersection, showing proud, young parents with their well-fed, well-clothed only child. Photo: © Patrick Field; Eye/Ubiquities/Corbis.

parents of one child, free family planning supplies and operations, and sometimes free vacations. However, families not complying lost free health care and free education for the first child, received lower pay or lost jobs, were excluded from social clubs and organizations, and came under pressure and were even forced to have abortions.

The policy was not implemented evenly, however. In some areas, the houses of those resisting the set quotas were burned, and women were imprisoned and subjected to late abortions, with some babies even being killed as they were being born. Crowded urban areas often had stricter enforcements than western rural areas. The southern industrializing area in Guangzhou was not subjected to the policy in order to attract foreign investment to an area of plentiful cheap labor. In Xinjiang province in the far west, the minority people were allowed two and three children. In some rural areas, the peasants rioted, killing local party officials and family planning workers until the policy was softened in 1984 and parents were allowed to have second children under certain circumstances. Elsewhere in rural areas in particular, baby girls were killed, and the act was glossed over before officials knew about it, creating imbalances between males and females in those areas.

The policy did not even meet its target, for the 2000 population was 1.27 billion and continuing to increase instead of slowing markedly. Even that figure is debatable, since peasants mislead officials and officials misreport statistics. In Hubei province, when county officials had insufficient emergency relief food, they discovered that the total population was 10 percent higher than the census figure, with most families having three children, because local officials had reported numbers to match the official quota. Occasional events bring to light families with up to six children. In 2002, a pharmaceutical company revealed that it produced 25 million vaccines for infants, although the estimated baby population was only 20 million, indicating a higher birth rate.

So, would China have been better off without this draconian policy? Comparisons with India suggest it might have been. In India, which has only moderately successful policies on family planning, fertility dropped, and the country claims it has already avoided 230 million extra births and will stabilize its population in 2040, the same year as China hopes for this result. A 1986 study in two counties of Shanxi province allowed peasants to have two well-spaced children. By 1996, the growth rate had fallen, fewer third children were born, fewer abortions were carried out, and less female infanticide occurred. Less coercion went with better relations between the peasants and the authorities. This is an important factor in any country.

The policy continues and is justified by the China Population and Information Center (website: www.cpirc.org.cn) as being effective in some parts of the country and subject to better application of the policies, more consultation, and better advice. Even in the areas of greatest compliance, such as Beijing, only two-thirds of couples held single-child certificates. The center claimed, however, that "family planning has become a social habit."

Debate the value and achievements of the one-child policy and compare Chinese policies with those in other countries.

POINT	COUNTERPOINT
The one-child policy results in fewer births that place claims on national resources.	Family planning is most successful when parents act on their own initiative, perhaps guided by information.
It brings slower population increase and a more balanced population structure in the early stages.	It upsets the population balance between sexes (abortion and murder of baby girls) and age groups (in the later stages when there are small numbers of young people and increasing numbers of older people).
It allows better planning for urban expansion, housing, education, job availability, and transport and utility infrastructure.	It ignores other factors in population growth, including migration, the effect of natural disasters (as during the Great Leap Forward) and diseases, and the impact of political chaos (as in the Cultural Revolution).
It provides control by the government and local officials in the interests of the whole population.	It often requires draconian methods of control that go against human rights.
It brings a rapid realization of the importance of family planning in the face of imminent overpopulation disaster. Other policies had failed.	It ignores cultural and economic factors that favor large families.

international trade multiplied China's total GNP sixfold within 20 years.

After 2000, China experienced a further revolution based on the use of the Internet and mobile phones. The whole telecommunications system was upgraded in the 1990s by a fiber-optic grid laid across the country. That made possible an increase in telephone lines from fewer than 10 million in 1990 to 125 million in 2000, when 2 million were being laid each month. Mobile phone users multiplied from 5 million in 1995 to over 57 million in 2000. However, the lack of a national payments system for buying goods online, the continued state control and policing of website commerce, and sensitivity about website content delay wider use of the Internet.

The Chinese economy does not conform to the pattern followed in other countries, which began economic growth with the production of raw materials and food for export and moved to simple manufactured goods before more sophisticated ones

and a range of service industries. China already produces all types of goods in all economic sectors, from rice, rag diapers, and plastic toys to microchips, and spans the entire value chain. The scale and low costs of production in China affect world prices and other countries that produce similar goods.

Farming in the 2000s

Agriculture remains predominant in the Chinese rural economy, although its role changed after 1978. Through the Mao Zedong years, farming was almost ignored by government and separated from other aspects of the economy, and in 1980, agriculture still employed almost the entire—and growing—rural work force as low levels of mechanization demanded high labor inputs.

By 2000, only half of the rural workers were in farming. From the early 1980s, commune controls were relaxed, and individuals and groups could plan their own program. The personal involvement, longer-term contracts, and lower quotas to be fulfilled before selling on the open market led to growing confidence and higher output. New rural sources of income through the township and village enterprises (TVEs) took underused rural labor and offered higher wages. Over 130 million employees, making up over 30 percent of the rural work force, worked in the TVEs that contributed most to Chinese GDP in the 1980s. Millions of people remained in industrializing rural areas instead of moving to the cities. However, many millions more moved out of agriculture and formed a growing migrant population in the country.

Chinese farming is both land- and labor-intensive as the country attempts to feed 21 percent of the world's population from 7 percent of the world's arable area. Only 10 percent of China is cultivated, and output per unit of land is high. Most agricultural land is cropped, and competition is increasing between land-intensive crops—wheat, corn, soybeans, and cotton in the north and rice and sugarcane in the south—and labor-intensive crops that produce greater value per unit of land, such as vegetables and fruit. The latter are increasing in the coastal provinces. Patterns of farm production moved toward diversification and more complex land use, such as the integrated planting of mulberry, sugarcane, and bananas around fishponds in the Zhu Jiang delta. Meat and milk production continues to be a small-scale, "backyard" enterprise.

The more profitable specialist crops, including industrial crops (cotton, soybeans), fruit, and vegetables, together with livestock products threatened the output of some of the traditional staples, particularly wheat and rice. While the Chinese government needs to ensure the grain supplies, a growing economy requires greater diversification. By encouraging grain growers to increase yields by using more imported potash fertilizer, China produced enough grain to put a surplus into storage. Economists and farming groups inside China advise growing more valuable agricultural products on the scarce agricultural land and importing more grain. In a crisis, the country could return to grain production.

As it faced continuing population increases that made greater demands on food in both quantity and quality, China increased its imports of grain and other foods through the 1990s, funded by increased exports of manufactures. Imports will rise further as the Chinese population increases and tastes change toward Western diets with more meat and other luxury foods that take more land to produce.

While successful in terms of increasing production and raising incomes in many rural areas, the individualistic policy of household responsibility led to the neglect of some aspects of rural land use that communal working had encouraged, including the long-term processes of planting woodland and maintaining road and canal infrastructure. In the 1990s, other problems facing agriculture included long-term soil erosion and the extension of urban-industrial areas that removed good land from production. Increased productivity reached a plateau as more land was taken out of production by urbanization than was added.

Water Resources

Water is as important a constraint on farming as the availability of land. Some 64 percent of arable land is in northern China,

but that is the driest part of the country, with less than 20 percent of the available water; the 36 percent of arable land in southern China has 81 percent of the water but is menaced by floods. In the north, the Huang He often stops flowing before it reaches its mouth at the ocean. The deficit in surface water draws attention to underground sources, but the water table in the rocks underlying the Beijing area is falling rapidly. Forecasts predict that these reserves will run out by 2015.

New projects to move water from south to north will be successful only if they are linked to efficient management of the water. Work on the South-North Transfer Project was due to start in 2002 and will cost three times as much as the Three Gorges Dam project, but such grandiose plans arouse controversy inside and outside China: environments will be damaged and lives disrupted. Even if completed, the transfer project will be only a partial solution to the lack of water in the north. The western route (Figure 5.29) needs tunnels; the central route needs a canal or aqueduct 1,240 km (770 mi.) long; the eastern route will follow established watercourses that are highly polluted. The east and central routes will cost large sums. In the central route, the Danjiangkou reservoir water level will have to be raised 13 m (43 ft.), with major impacts on 200,000 displaced families, deforested hills, and a lowering of water levels below the dam. An alternative approach could begin by coordinating and testing water conservation and antipollution measures in local areas, but governments and their engineers prefer major projects because the proposed scale promises big results from single-focused efforts, the bureaucratic administration is easier to deal with, and more opportunities exist for media attention.

Chinese Energy Policy

Although China ranks third in the world for producing energy (first in coal, fifth in oil, and sixth in hydroelectricity), it suffers from the high costs of distributing the fuels and electrical power over long distances. Oil and natural gas finds are mainly offshore or in the far west, distant from economic centers; coal is produced in the north and hydroelectricity in the south. By the mid-1990s, the shortfall in power supplies was as high as 20 percent in the expanding southern areas. One of the major foreign investment opportunities is to expand power provision. After discussions about the proportion of profits to be allowed to foreign corporations investing in construction projects, building began on the first of around 50 power stations that will use Chinese coal and foreign capital.

China has the world's greatest hydroelectricity potential, but even if it were all developed, it would supply just 6 percent of the country's needs. The building of the Three Gorges Dam beginning in the late 1990s (Figure 5.30) illustrates the environmental and social issues facing major hydroelectricity projects. The outcome should generate enough electricity to save the yearly burning of 45 million tons of polluting high-sulfur coal and should help flood protection and navigation. The dam, however, will create a lake 600 km (450 mi.) long below Chongqing, and over 1 million people will need to be resettled. Farmers on good valley land will be moved to poorer uplands, and some industrial towns will be submerged. The buildup of silt in the slowed river upstream of the reservoir could raise flood levels and submerge additional towns. But this project is Chinese government policy and, although questioned by environmentalists and internal advisers, is pushed ahead.

Coal reserves remain China's chief source of power. They occur mainly in the northern half of the country, resulting in heavy use of the rail network and massive investment in the railcars that carry it southward. Coal transportation takes up 60 percent of the railroad capacity, and plans are in hand to build coal slurry pipelines from interior mining areas to the coast. One of the positive outcomes of the Great Leap Forward and Cultural Revolution was the opening of small local mines run by villages in southern China; these now produce around half of the total output. China's mining economy has two aspects—one with large-scale, capital-intensive, and technologically advanced mines, and another with small-scale, labor-intensive, low-technology mines. The smaller rural mines help with chronic underemployment, but their operation often lacks environmental management.

In the late 1990s, China's oil industry was reorganized into two major companies, Sinopec and Petrochina. They evolved from two government monopolies, one of which was responsible for refining and distribution and the other for exploration.

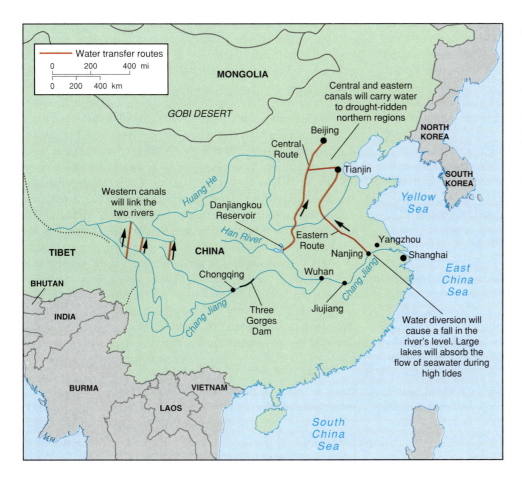

Figure 5.29 **China: potential water diversions.** Northern China and the Huang He basin are increasingly short of water. Three potential diversions have been identified. The western routes are short, but mountainous, requiring tunnels; the central route requires a canal or aqueduct 1,240 km (770 mi.) long direct to Beijing; the eastern route follows existing channels, but their waters are polluted. Compare this map with the climate map, Figure 5.7, and the population map, Figure 5.14, to understand more fully the need for such diversions of water.

Figure 5.30 **People's Republic of China: Three Gorges Dam project.** (a) Chang Jiang Three Gorges Dam, early construction, 2000. (b) Chang Jiang will rise 175 m by 2009, requiring relocation of entire cities to higher ground. Photos: © David C. Johnson.

(a)

(b)

The new companies do both and competed first for a large proportion of the 90,000 gas stations that were mostly owned and run by local governments, rural collectives, or individuals. Moving before foreign corporations could get into the market in the wake of China's entry into the WTO, both companies bought up thousands of the best-situated gas stations, sometimes at ridiculously high prices. They now look to joint ventures with foreign companies, headed by BP, Exxon Mobil, and Shell. The investment presumes more cars and trucks will be on the roads. Although China has fewer cars in proportion to the population than any other country, their numbers increased from 300 to 4,200 per million in the 1990s as incomes rose.

By 2004, China plans to build a pipeline to bring natural gas from the Tarim basin in Xinjiang, in the far northwest of the country, to Shanghai, a distance of 4,200 km (2,600 mi.). This project is likened to the scale of building the ancient Great Wall. Petrochina has the largest stake, along with a consortium of overseas investors from Europe, Russia, and Hong Kong. Points to be resolved in 2002 included testing the output of the Tarim Basin fields, finding the foreign capital, and assessing whether the delivery price would be too high. Other potential problems include the political risks in the production region of unrest among the Tajiks and Uighurs, who resent the Chinese. They see the pipeline as a means of extracting local resources for the benefit of wealthier people in the Shanghai area.

Growth in Chinese Manufacturing

After 1978, the drive to greater efficiency and an open-door policy to encourage foreign investment became more important than self-sufficiency—what has been termed the "Great Leap Outward." Today, China's manufacturing sector comprises a complex and uncertain intermarriage of local enterprise, foreign multinational capital, and public ownership. The facilities operated by largely uncontrolled military groups were banned in the 1990s.

Alongside increases in agricultural productivity from the late 1970s, rural industry boomed. Rural populations invested rising savings in manufacturing industries that had better returns than farming, used surplus labor not required on the farms with higher productivity, and created new markets among rural people. Farm product processing and subcontracting to urban manufacturers provided the basis for the new industries. Initially, this expansion grew out of collectively owned enterprises, rising from 22 to 36 percent of all Chinese industrial output by the late 1980s, but then gave way to individual and private enterprises as the new driving forces in rural industrial expansion.

Government policy gradually relaxed controls to encourage increasing economic relationships with the outside world. At first, this was a response to the need for foreign exchange to buy grain and high-tech equipment. Improving foreign exchange balances stimulated further exports and foreign direct investment. The Chinese government allowed exporters to keep some of the foreign funds their products earned.

Such policies placed pressures on the publicly owned enterprises that formed the heart of Chinese industry up to 1978. From the 1980s, reduced central government subsidies

and the "management responsibility system"—an equivalent to the household responsibility system—placed more decision-making in such firms' hands. By the late 1990s, 95 percent of industrial production sold at market prices. The central government, however, now focuses its own investment on only 1 percent of existing publicly owned enterprises, allowing others to be merged, taken over by workers, or go bankrupt. Much industry continues to be government-run, particularly large sections of the military-industrial complex—a major influence with over 600 factories that produced armaments and other military equipment. Still under military ownership, the sector now produces consumer goods, ships, and aerospace vehicles. Greater legal requirements to provide job security and social benefits make government-owned enterprises less profitable than private firms. Bad loans to failing enterprises and grandiose socialist housing projects amount to around 30 percent of GDP.

China has its own multinational corporations and wishes to generate more with world-known brands. One such company is Qingdao Haier, centered in the coastal city of Qingdao in Shandong province (see Figure 5.25). From being a government-owned firm producing shoddy home appliances, it grew by the late 1990s to sell 15 percent of the Chinese washing machines and 33 percent of the refrigerators. It improved quality and listened to customer needs. For example, when Sichuan peasants used a washing machine to rinse soil off potatoes, the machines were adapted to prevent clogging. A small-scale machine for washing a single change of clothing sells well in crowded urban centers. Haier took over seven other companies in two years, and others wish to join it. As Haier expands, however, it enters the very competitive production of TVs and pharmaceuticals, looking to overseas markets at a time when financial resources and qualified staff are in short supply. Such conditions commonly face expanding Chinese enterprises.

In the late 1990s, bank crises due to overlending affected much of the rest of East Asia, cutting the rapid economic growth rates in the region's countries and reducing the markets for Chinese-made goods. Part of the problem faced by other East Asian countries was the expansion of Chinese output, causing oversupply to common markets.

Increasing Steel Output

As rapid economic growth occurred in southern China, the old industrial area of Northeast China became the country's "rust belt" (Figure 5.31). After China built its heavy industrial capacity in this area in the 1950s, most factories and their equipment were seldom updated. Output of steel products continued, but many were unmarketable because of their price or quality. In the 1990s, when market forces began to take over from government funding, local coal mines (20% of China's total) and oil companies (33% of China's total) could not be paid for their products by the bankrupt industrial users. In the mid-1990s, the central government subsidized many former steelworkers on half pay to prevent social unrest.

Yet, China's steel industry is still the world leader, with increasing production from new coastal and inland mills. The 2001 production was 140 million tons—more than that of

Figure 5.31 **People's Republic of China: Northeast industrial region.** Workers cross railroad tracks at an Anshan steel mill, Liaoning province. Summarize the economic problems that face this part of China. Photo: © Larry Mulvehill/The Image Works.

Regions and Cities in China

With such a large area, China's boundaries enclose many local regional variations. Physically, China is a country of contrasts between the extensive lowlands of the Huang He and Chang Jiang basins, the mountain ranges of the west, and the high plateau of Tibet; between the subtropical monsoon of the southern coastlands, the midlatitude east coast climate with long, cold winters of the Northeast, and the arid inland areas; and between the former forests, grasslands, and deserts, and the different types of human landscapes.

China's human geography is also diverse. Culturally, the regional distributions of languages and dialects, ethnic differences, and historic relics provide a variety of hindrances and resources for human development. Politically, China is divided into 23 provinces and other province-level units (see Figure 5.25), including four great cities (Shanghai, Tianjin, Beijing, Chongqing) and five autonomous regions with large ethnic minorities (Inner Mongolia, Guangxi, Ningxia, Xinjiang, and Tibet). Economic differences between wealthier and poorer areas are linked to regional concentrations—and absences—of resources.

The combination of physical and human characteristics produces a broad regional division of China (Figure 5.32), with

Japan (100 tons) or the United States (90 tons)—after almost doubling in 10 years. The steel from the newer Chinese mills competes with U.S. and Japanese products and often undercuts them in price, causing layoffs and hardships in those countries. The Chinese steel industry, however, is fragmented as the result of central government policy of placing a steel mill in each province. It is also inefficient, with 95 percent of the mills incurring losses, and produces mainly basic steel for which there are few markets. China imports better steel from South Korea and Japan for the sophisticated sheets used in autos and computers.

Figure 5.32 **People's Republic of China: internal geographic regions.** Provincial boundaries define the regions. Locate the space shuttle view of Figure 5.33(a) at X.

major contrasts between the north and south, and between the coastal east and the inland west.

Coastal Regions

The *Northeast* has very cold winters, contrasting with hot and humid summers, somewhat akin to New England in the United States. It was not much settled by Chinese until the early 1900s, after which the Japanese occupied it and developed it into one of the most industrialized regions of China based on the local occurrence of coal and iron. Industrialization of this region continued after the 1949 Communist takeover, making it the center of large-scale industry in China. It produces iron and steel products, machinery, and vehicles, and is supported by a dense railroad network. After 1949, growing corn, soybeans, and rice extended agriculture further across the rich black earth soils.

The *north-central heartland* of China has a long history of occupation and intensive production. Nearly one-third of the Chinese people live in this region. It comprises the plains, formed largely of river delta deposits, into which the Huang He and Chang Jiang drain. After 1949, new drainage and irrigation works improved and regulated the flows of the major rivers. Winter temperatures and rainfall totals both increase southward, and the region has an agricultural base of winter wheat. Its industrial and administrative cities include Beijing and Tianjin. Oil is produced near the mouth of the Huang He.

At the northern end of the southern coastal belt of the *lower Chang Jiang lowlands and central coastlands,* Shanghai is the most advanced and prosperous city of a productive farming region on the Huang Pu River near the mouth of the Chang Jiang. Newly industrialized in the 1990s, this region experienced massive redevelopment and building of new cities after years of slow growth. The cities of the lower Chang Jiang region have traditional manufacturing (textiles) mixed with post-1949 heavy industries (steel, shipbuilding, oil refining) and modern high-tech industries.

Through the 1980s, the policy to encourage growth in the south affected the Shanghai area adversely. While incomes in Guangdong and Fujian doubled, those in Shanghai saw no growth. From 1991, however, the central government made Shanghai a special economic zone, and it became the center of China's main economic growth area.

Shanghai is China's main port after Hong Kong. It is the financial capital and a major center of higher education and research and development. Most of its growth is funded by internal Chinese investment, and many of its products are consumed internally rather than exported. Current work on the construction of bridges, better roads, river port and airport facilities, and an underground train system suggests that much of the city is being rebuilt.

In the late 1990s, the new economic zone in Pudong across the Huang Pu River from Shanghai already had 4,000 enterprises and 400,000 workers. Further developments around Suzhou use the link of the main inland expressway road. The Shanghai authorities initiated Suzhou New District in 1990 to move industry westward out of its central area. Then the Singapore government, in a publicized demonstration of inter-Chinese working, contracted to build the Suzhou Industrial Park and expand supercity facilities east of Suzhou. Competition between the two zones during the late 1990s and decline in foreign investment made rapid returns to Singapore unlikely.

Shanghai is the center of China's growing vehicle-making industries, often based on joint ventures with multinational corporations. Volkswagen of Germany started making cars in Shanghai in 1985 and produces one-third of the national total, expecting to double its output in the early 2000s. Of the parts it uses, 85 percent are locally made. A joint venture between China and General Motors is located in Shanghai's Pudong Development Zone.

Other vehicle manufacturers are based in the Northeast and between Shanghai and Beijing. Brilliance China (luxury vans) and its subsidiary Shenyang Jinbei Passenger Vehicle Company (vans) import engines and components from Japan. In 2002, Brilliance China contracted to produce cars in combination with a U.K. automaker. Those companies and Tianjin Automobile Industry Company (down-market Xiali cars) cut prices in 2002 in the wake of WTO membership and a future of falling import duties.

Peugeot and Citroën (France), Chrysler (U.S.), and Japanese companies also produce cars, trucks, and minivans here. As in other countries in East Asia, the multinational corporations press the government for greater openness to the world economy so that they can establish their own factories and offices inside the country but then encourage the government to protect their investment by excluding potential competitors.

The southernmost part of the Chinese coastal zone, the *maritime south,* had rapid economic growth after the 1978 relaxation of external trading rules. Special economic zones established from 1979 encouraged foreign investment and technological innovation in manufacturing, especially around Guangzhou at the head of the Zhu Jiang estuary. Hong Kong and the offshore economy of Taiwan provide links to the global economy. Environments vary, and rocky coasts alternate with lowland at the river mouths. Summer rainfall and mild winter temperatures increase southward, reflected in the crops that range from winter wheat in the north to rice and tea farther south. Two crops of rice are harvested in the longer summers of the far south.

Interior Behind the Coastal Regions

The regions of the next tier inland are also well populated but experienced less economic growth in the 1990s. The *inner north* region is centered in Shanxi and Shaanxi. Xi'an in Shaanxi was a historic center of Chinese civilization, but this region now forms the northwestern rim of Chinese economic activity. The loess soils of wind-blown origin are easy to work but also easy to erode, attracting major efforts to overcome soil losses. Summers are hotter than on the coast, but winters are colder and rainfall diminishes westward. Spring wheat and millet are basic crops. Local coal mining and hydroelectric potential provide a basis for modern industry, but distance to markets and low levels of technology slow development investments.

The *middle Chang Jiang basin* combines wider plains and hilly areas. It also has cold winters and decreasing rainfall westward. Rice, winter wheat, and cotton grow on the plains, around industrializing cities linked to local hydroelectricity potential and river transportation. Wuhan is the main industrial center of this region.

The *Sichuan basin and Yunnan Mountains* of inland southwest China are less accessible, being cut off by gorges along the Chang Jiang. The highly cultivated rice-growing Sichuan plain contrasts with the surrounding mountains and plateaus. Mineral resources are poorly exploited. The isolation of this area, however, stimulated industrialization for local and national defense needs during World War II and as part of China's strategy after 1949. The Three Gorges Dam project is designed to support industry and transportation (see Figure 5.30).

Deep Interior Regions

The deep interior of China, covering almost half the country, consists of inhospitable mountainous and arid lands with very low densities of population. In the *northern steppes* from Inner Mongolia westward to Xinjiang, plateaus and encircling mountains separate grassy or arid basins (Figure 5.33). The climate is dry with warm summers and very cold winters. Livestock keeping is the main traditional occupation. Ethnic minorities such as Mongols, Uighurs, and Kazaks live there and frequently demonstrate opposition to the central Chinese government. The proximity to the Russian border led China to develop parts of this area, and there are extensive mineral deposits that will be exploited as the transportation links for oil and gas are established.

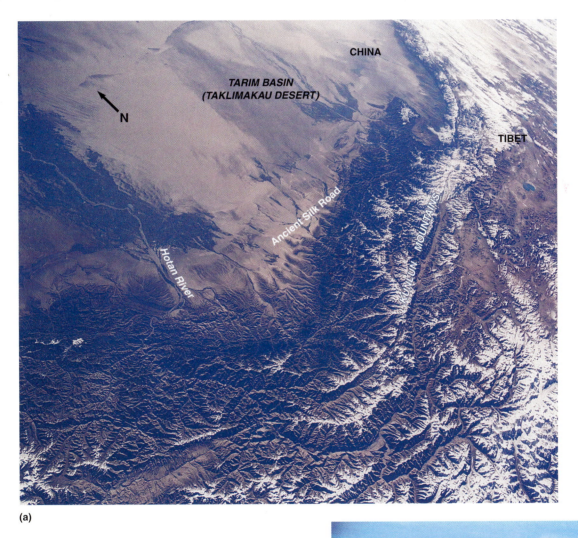

(a)

(b)

Figure 5.33 **People's Republic of China: the arid north and west.** (a) Space shuttle view of western China with the Kunlun Mountains (7,000 m, 20,000 ft) and Tarim basin. The belt of gravel at the mountain foot stores water from snowmelt that is used in irrigated farming—and has been since ancient times. One of the Europe-China Silk Road routes passed through here. (b) The "Singing Sand Dunes" of the Gobi Desert. Photo: (a) NASA.

The *high plateau of Tibet and Qinghai,* which is mostly above 3000 m (10,000 ft.), occupies the southern half of westernmost China. The People's Republic of China reduced the region's isolation and introduced modern industry, but only 6 million people live in a huge area. New buildings in Lhasa, the capital of Tibet, symbolize the Chinese control of the province (Figure 5.34). In 1950, Chinese armies invaded Tibet and, despite subsequent uprisings, maintained rule by oppression.

Chinese Regional Development Policies

In the 1980s, the government divided China into three regions that were expected to have different types of development (see Figure 5.32).

The "Golden Coastline" of the *Eastern Region* includes the capital, Beijing, port cities such as Shanghai and Hong Kong, and other economically developed areas such as the Guangdong region in the south. Though this zone is one province deep along the eastern coast, it contains considerable internal variety. It has the greatest potential for economic development and the easiest links to the global economic system. It specializes in export-oriented goods with the objective of motivating further economic growth and spreading it to the interior. Special concessions were made available to parts of the region, especially to the four **special economic zones** established in the southeast in 1979 and a fifth on the island of Hainan in 1988. Fourteen coastal ports were designated as open coastal cities in 1984. Not only do these places have advantages for access to foreign capital and technology, but they also offer cheap local labor and plants with tax concessions for equipment. Such measures, however, worked at the expense of the rest of the country and shifted much of the economic growth southward and to the coasts.

By 1990, 60 percent of over 12 million township small business enterprises were located in the southern coastal provinces. This area became the Chinese "sun belt." The new industries made textiles and apparel, electronics goods, chemicals, and machine tools. A huge area of rapidly expanding production grew around Shenzhen in Guangdong province inland of Hong Kong. Facilities in this area include an oil refinery, nuclear power plant, airport, deepwater shipping berths, and inland superhighway links.

The *Central Region* is a densely populated interior zone without coastal outlets but with natural resources of minerals, water, soils, and not-too-steep slopes that allow agriculture to flourish. This region specializes in farming and energy production. Such resources provide a basis for local industrialization. Its main problem is the lower prices received for local food and raw materials compared to those paid for manufactured goods from the coastal zone factories. The encouragement of growing interregional linkages, including the Three Gorges project, along the Chang Jiang basin, which is being called "China's Soaring Dragon," partly relieved the tensions between this region and the east coast.

The *Western Region* is sparsely populated with extensive arid and high mountain environments. The region's future is concerned with animal husbandry and mineral exploitation. Relatively little friction occurs across the long international borders because so few people live near them.

Hong Kong

Hong Kong comprises an island (centered on Victoria) and mainland peninsula (Kowloon) that were ceded to the British in the mid-1800s. A more extensive area of mainland, the New Territories, was leased to the United Kingdom for 99 years in 1898. All returned to Chinese rule in 1997, after which Hong Kong did not suffer the expected major capital flight or brain drain but rather experienced influxes of both. The Basic Law, under which Hong Kong operates within China, guarantees retention of its free-market system for 50 years and sees the city as a complementary financial center to Shanghai.

Hong Kong has an excellent natural harbor that is the focus of its economy. It has deepwater access and extensive water frontages (Figure 5.35). In 2000, it was the world's busiest port. Goods received from inland China are sent to worldwide destinations, while others collected from the rest of the world are sent into China. These are the features of an **entrepôt.** As the Chinese economy opened up and expanded after 1976, particularly in the special economic zone around

Figure 5.34 **People's Republic of China: Tibet.** Compare the roles and designs of these buildings, old and new. (a) Potala Palace, Lhasa, Tibet, China, former residence of the Dalai Lama. (b) Telecommunications building, Lhasa, Tibet, China, part of modern Chinese control and modernization. Photos: © David C. Johnson.

(a)

(b)

Shenzhen, the value of Hong Kong's trade multiplied sevenfold. New ports opening along the Zhu Jiang delta in the early 2000s relieved the huge pressure on Hong Kong's port facilities.

Hong Kong's growth as a manufacturing center took off in the 1970s, and by 1990, manufacturing provided a third of all employment. The production of electronic goods and scientific equipment increased, but the textiles and clothing industries declined because of foreign competition and rising labor costs in Hong Kong. In the 1980s, manufacturing moved out of Hong Kong and expanded just over the border in southern China, with its cheap land and labor costs.

At the time of political transfer, Hong Kong was deeply enmeshed with interior China in trade, investment, and personal contacts. In 1996, China accounted for 37 percent of Hong Kong's trade and Hong Kong for 47 percent of China's trade. Over 20,000 trucks crossed the Hong Kong-mainland border each day. As part of China, Hong Kong began with advantages over its internal and external rivals for dealings inside the growing country.

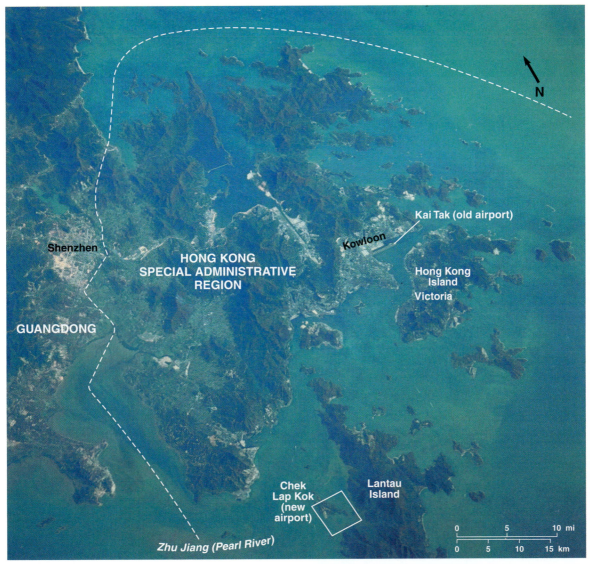

(a)

(b)

Figure 5.35 **People's Republic of China: Hong Kong.** (a) Space shuttle view of Hong Kong and adjacent southern China. The port and other features of the Hong Kong Special Administrative Zone are linked to the mainland by built-up transportation routes. The new city of Shenzhen (left center) is just outside the Hong Kong Special Administrative Region. (b) Hong Kong harbor seen from Victoria Peak on the island and looking across to mainland Kowloon. Photo: (a) NASA.

By 2000, the shift out of manufacturing as factories moved inland led to service industries accounting for 85 percent of Hong Kong's GDP and 80 percent of employment (from around 50% in 1980). Hong Kong is now one of the world's major finance centers and the third-largest gold market, accounting for up to 15 percent of the world total. Hong Kong provides expertise in finance, trade, and public administration and has an important stock market. Its daily turnover of foreign exchange is fifth in the world. It is the regional headquarters for 800 foreign companies. However, some signs suggest a less rosy future. Hong Kong and Chinese regulations reflect different values, with the latter subject to political, rather than market, pressures. Oversubscription for shares in highly valued ("red chip") mainland corporations, heavy borrowing to buy shares, and corrupting measures brought in by mainland corporations could destroy trust in Hong Kong's market.

Tourism became Hong Kong's second major industry with 11.7 million visitors in 1996. Many new hotels catered to this influx of people that added an extra 10 percent to the population at peak times. After Hong Kong's transfer to China in 1997, tourism continued at a high rate. The numbers of visitors reached 13 million in 2000.

The higher wages in factories and service industries force farming labor costs up because farms have to match the wage levels. Even so, Hong Kong grows nearly a third of its food needs on the small area of nonurbanized land in the New Territories.

The population of Hong Kong, nearly 7 million in 2001, is almost totally Chinese. As the main urbanized area became overcrowded from the 1950s, new towns spread into the New Territories. They now accommodate around one-third of the total population. The high population density causes half of government expenditure to go for roads and public transportation. In the late 1970s, electrification upgraded the railroad

route inland to Guangzhou (Canton) and its use increased. Two cross-harbor tunnels and an underground transit system opened in the 1980s. A new Hong Kong International Airport was built offshore north of Lantau Island and linked by bridges to the mainland. It replaced the older airport in Victoria Harbor that could not be expanded on its limited site.

Macau's New Approach to Gaming

The smaller Macau economy remained focused on tourism and gaming after its 1999 incorporation in China. Macau had become a byword for corruption and crime. Over half of all Asian gaming took place in Macau's casinos, and gaming taxes accounted for 60 percent of its government revenue. The Chinese government, instead of closing down the gambling industry in Macau, plans to reform it by ending the monopoly held by Stanley Ho, the Hong Kong tycoon. Three out of four new licenses will go to others, including potential American and Australian casino operators. With a shortage of tax revenues, China now considers combining gaming in Macau with other tourist activities. On the strength of rising Chinese demand, governments in Taiwan, North Korea, as well as in Southeast Asian countries such as the Philippines, Cambodia, and Vietnam, also began to encourage casino gaming with a view to increasing their tax income.

Chinese City Landscapes

Many Chinese cities still retain relics of older cultures (Figure 5.36). Former palaces (Figure 5.37) and temples occupy large sections of central Beijing. In many places, modern roads, squares, and buildings replaced the old city walls and cramped housing. Coastal cities often retain evidence of their 1800s foreign trade, as in Hong Kong, Macao, and Shanghai.

Monolithic and utilitarian buildings typical of centrally planned countries mark areas of Chinese cities built after 1949.

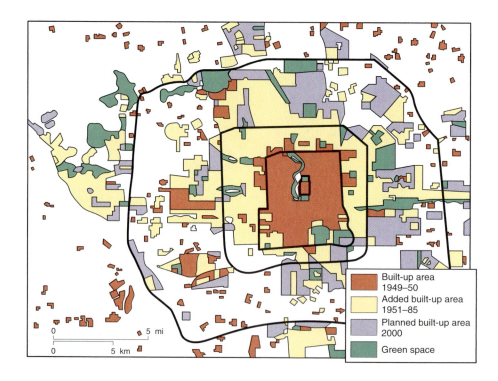

Figure 5.36 **Expansion of built-up Beijing, 1949–2000.** The Imperial Capital/Palace is in the center of the old city, built over a period of 500 years. In 1949, the urban area was a compact walled city with outlying villages. The dark lines are modern ring roads. The lands close to the old city were built on first, and suburban developments later pushed outward. Source: From Sit, Beijing. Copyright © 1996. Reprinted by permission of John Wiley & Sons, Ltd.

Built-up area 1949–50
Added built-up area 1951–85
Planned built-up area 2000
Green space

0   5 mi
0   5 km

(a)

(b)

Figure 5.37 **People's Republic of China: Beijing.** Large areas occupied by former palaces and historic buildings are a feature of Beijing, China. (a) Tiananmen (Gate of Heavenly Peace) is one of these. (b) Tiananmen Square, the scene of revolts in 1989. Photo: (a) © J. Sohm/The Image Works.

The periods of urban neglect and the post-1978 renewal of urban and industrial construction are identifiable in city zones. The production sector and urban infrastructure compete for urban investments. Close control on living accommodations, however, reduced the incidence of **shantytowns**—housing built informally, often without permission, on "spare" land.

The industrial area in southern China's Guangdong province inland of Hong Kong exemplifies the present rapid urbanization. Factories and huge apartment blocks sprang up in the 1990s to form cities of several million people around Shenzhen (Figure 5.38). Many people moved into the new cities from rural areas in southern China, some on temporary work contracts. Shanghai is another center of extremely rapid urbanization.

Mongolian Isolation

Mongolia is landlocked, a country of mountain ranges, the Gobi Desert, and semiarid grassy steppes akin to the northern steppes region of China. It has a small population and struggles to maintain its independence as a buffer state between China and Russia. The capital, Ulan Bator, is connected by rail to both China and Russia.

After being a Chinese province, Mongolia became independent in 1921 with Soviet Union backing, and a Communist government was installed in 1924. The Soviet Union modified Mongolia's East Asian character, replacing the Mongolian alphabet with Cyrillic script and reorganizing the education system. Mongolia became dependent on Soviet aid and trade. Russia's military forces were withdrawn only in 1992. In 1996, a coalition of democratic groups won an election to introduce reforms. After they lost Russian support, the Mongolian people strove to earn a living while attempting modernization in a landlocked, semiarid environment. In 2000, the former Communist Party was overwhelmingly reelected on the basis of policies that combine reform, social welfare, and public order.

Mongolian People

Mongolia has a dispersed, low-density population (see Figure 5.14). Total fertility and natural increase fell in the late 1900s. Life expectancies are now in the 60s for Mongolia. Ethnically,

the Mongol people dominate Mongolia, although it also has 10 percent who are Chinese, Russians, or Kazakhs.

Just over half of Mongolia's population lives in urban places as the result of the pull of growing industrialization and the push of rural poverty. Mongolia's capital, Ulan Bator, houses nearly one-fourth of the country's population, including those living in large tented (yurt) areas (Figure 5.39).

Mongolia's Limited Resources

Mongolia has a strategic location in the heart of Asia but restricted economic potential. Many people still gain a living from herding livestock on the semiarid grasslands. Farmers responded to privatization and decontrol of meat prices in the early 1990s by massively increasing their herds. Cropland area declined as agricultural cooperatives were split and land abandoned because there was no equipment to cultivate it.

For the future, Mongolia increases its output of minerals such as copper, gold, and molybdenum and encourages foreign investment to provide alternatives to the previous Russian links. Private enterprises became the main source of economic growth

Figure 5.38 **People's Republic of China: Shenzhen, Guangdong province.** The downtown area of this city, just inland from Hong Kong, sprang up in the 1980s based on manufacturing goods for export. Many factories moved here from Hong Kong. Photo: © J. Sohm/The Image Works.

Figure 5.39 **Mongolia.** Family in front of their yurt tent home. Photo: © MAF.

in the 1990s, with rising production of gold, although a major fall in world copper prices and weaker prices for cashmere woolens slowed growth. Mongolia suffers from a limited range of resources and the problems of making the transition to a market economy.

Taiwan

Taiwan has 10 times the population on one-fourth the area of Mongolia. It is a well-watered island country, with a mountain "backbone" along its eastern coast. At the end of the 1895 war, China ceded Taiwan, along with Korea, to victorious Japan. After World War II, Taiwan returned to China. In 1949, the Kuomintang forces escaping from China took control of Taiwan. The Kuomintang leaders brought their families to join several thousand military personnel who had been stationed on Taiwan to quell anti-Kuomintang riots. The new rulers governed under U.S. guidance and military protection on the basis of the Chinese 1947 (pre-Communist) constitution, gradually increasing democratic rights for the whole population. Taiwan prospered with high-tech export industries and became a major investor in mainland Chinese industries.

While Taiwan maintains its independence, the People's Republic of China insists it remains a part of the mainland country and intends to reincorporate it. The PRC attitude places Taiwan outside the normal niceties experienced by independent countries such as membership of the United Nations. For 50 years after 1949, the PRC threatened to invade Taiwan and backed off only under U.S. threats. In the early 2000s, "China creep" threatened to replace the threats as Taiwan became more enmeshed in mainland China's economy. Although, Taiwan's government officially limited Taiwanese investments in China, over $60 billion followed indirect routes through Hong Kong, despite few return profits.

China's leadership determines to apply to Taiwan the "one country, two systems" formula adopted for the incorporation of Hong Kong. Around half the Taiwanese population already supports reunification with China based on a Hong Kong type of provision that preserves Taiwan's prosperity and its role as a major investor in Chinese economic growth. Many Taiwanese, however, watch events in Hong Kong before agreeing to become fully part of China again and, in the meantime, continue to strengthen their military capabilities. Supporters of Taiwanese independence argue that it is different from Hong

Kong: Taiwan is 170 km (100 mi.) offshore, has no built-in deadline date for return to China, and has a more mature democracy than Hong Kong and much of the rest of East Asia. The independence position has guarantees of Western support, especially from its strongest ally, the United States. But pressures grow on Taiwan as many of its best workers and managers move to the mainland, unemployment increases in Taiwan, and the World Trade Organization members asks new questions about the Chinese being able to use the Taiwanese ports and airports freely. The relaxation of Taiwanese bans on movements of capital and trade with the mainland will further bring the two countries closer together.

Taiwanese People

Taiwan island is densely peopled. The Taiwanese are nearly all of Chinese origin, and they have demographic characteristics similar to those of other materially wealthy countries, including life expectancies of 75 years. Taipei is the capital and largest city of Taiwan, which has nearly three-fourths of its population living in urban areas as a result of the growing industrial and service-centered economy.

Taiwanese Economy

Until 1945, Japan occupied Taiwan as well as Korea. The Japanese used the Taiwanese as slave labor to build roads and railroads, water projects, coal mines, and factories. They developed farming and a local market for Japanese goods.

After 1949, the Kuomintang government encouraged rapid industrialization, rural change, and urbanization. At first, agriculture provided the financial basis for industrial growth in Taiwan, which remained slow until the late 1960s. The increasing rice surplus gave way to more diversified farm products, including sugar, tea, vegetables, and fruits. The expansion of urban-based industrial jobs and buildings, plus the mechanization of farming, reduced the farm population from 6 million in 1965 to fewer than 4 million in the 1990s, while an initial agricultural trade surplus gave way to increasing food imports. Agriculture's share of Taiwan's GDP fell from 27 percent in 1965 to 5 percent in the 1990s.

Starting in the mid-1960s, export-processing zones focused attention on manufacturing for export. By the 1980s, the emphasis switched to high-technology developments. Private enterprise ownership of firms increased from 52 percent in 1960 to over 80 percent in the 1980s. The government retained control of banks, interest rates, and exchange rates to maintain a consistent financial environment. It cultivated a strong domestic manufacturing industry, at first based on transistor radios, followed in the 1970s by the assembly of Japanese electronic goods, and then promoting links with companies such as Philips (Netherlands) to build its own integrated circuit industry. Taiwan now leads the world in such technologies as integrated circuits, laptop computers, modems, and data communications.

By the 1990s, Taiwan had one of the largest world trade surpluses and was able to export capital, particularly to China and Thailand. As is the case with Japan and South Korea, direct investment in Chinese manufacturing contributed to many "Taiwanese" products. As Chinese costs undercut Taiwanese producers, Taiwan continues to invest in new technologies such as biotechnology and new integrated circuit designs. Such investment, however, demands so much capital that it poses major risks for the Taiwanese government such as overspending and eventually losing in competition with the Chinese. Already, Taiwan's furniture and textile industries moved to mainland China, leaving increased unemployment, and its electronics industries are under a major threat.

Large financial surpluses meant that major financial problems in the late 1990s affected Taiwan less than South Korea and Southeast Asia. Taiwan's other positive features include mainly small corporations and more flexible conditions of operating that enable failing companies to cease trading. Such conditions result in higher productivity compared to the South Korean *chaebol* and Japanese conglomerates.

Test Your Understanding 5D

Summary China has over 20 percent of the world's population. In 1976, still under a Communist regime, it emerged from isolation and instigated a mixture of industrial development, greater agricultural productivity, and greater national unity. It interacted with the global economic system in exporting its products and importing foreign capital.

China's agriculture became more productive after 1976 but is near the limits of areal expansion. The country's industries grow rapidly in southern and coastal China but elsewhere need new plant and infrastructure. China's economy is now a mixture of state-controlled enterprises, military-industrial diversification, and foreign multinational and local investment in manufactured goods for export.

Mongolia remains poor and isolated, has a semiarid climate, and was under Soviet Union domination until 1990. After losing Soviet financial aid, its economy struggled with transition. Taiwan, claimed as an integral part of the People's Republic of China, prospered as an independent developer of high-tech products and invested large sums in mainland China.

Some Questions to Think About

5D.1 How did Chinese government economic policies from 1949 to 1976 and from 1976 to the present differ? What impacts did changes have on China's agricultural, manufacturing, and urban geography?

5D.2 Why do the age-sex diagrams for China and other countries in East Asia show different patterns? What are the consequences of such population structures for education, labor forces, housebuilding policies, and the proportion of older people?

5D.3 What roles might Hong Kong and possibly Taiwan play within China?

Key Terms

collectivization	Han Chinese
commune	transnational Chinese economy
Great Leap Forward	special economic zones
Cultural Revolution	entrepôt
household responsibility system	shantytowns

Online Learning Center

www.mhhe.com/bradshaw

Making Connections

The Online Learning Center accompanying this textbook provides access to a vast range of further information about each chapter and region covered in this text. Go to www.mhhe.com/bradshaw to discover these useful study aids:

- Self-test questions
- Interactive, map-based exercises to identify key places within each region
- PowerWeb readings for further study
- Links to websites relating to topics in this chapter

Chapter 6

Southeast Asia and South Pacific

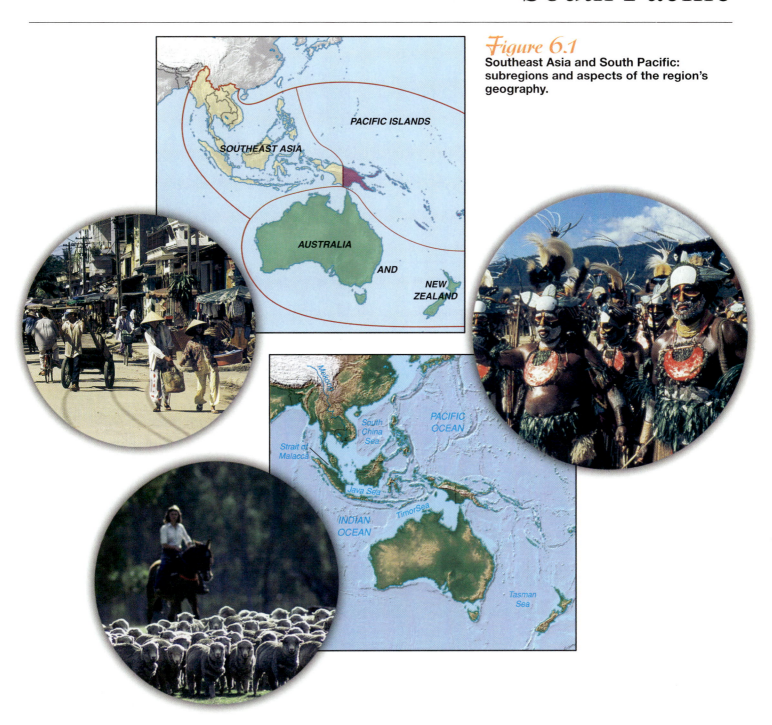

Figure 6.1
Southeast Asia and South Pacific: subregions and aspects of the region's geography.

(a)

 # A World of Influences

Varied Strands Brought Together

The countries of Southeast Asia, Australia, New Zealand, and the South Pacific make up a region of contrasts in natural environments, cultures, and outlooks (Figure 6.1). Situated in the middle of a north-south transition from Asia to the southern oceans and of a west-east transition from the Indian Ocean to the Pacific Ocean, varied and competing influences contributed to the modern geography. Increasing global connections now help to bring the disparate strands together.

Divisions rather than cohesion characterized the region until the late 1900s. Countries had mixed cultural and economic resource bases. The land areas range from continents to many tiny islands, and climates from equatorial to midlatitude, while there are marked differences between the Asian and Australian floras and faunas. Some of the world's longest-established indigenous groups, including the Australian Aborigines and groups in the isolated interiors of Thailand, Indonesia, the Philippines, and Papua New Guinea, still live in traditional ways amid expanding modern cultures.

Before the arrival of European colonists, traders from China, India, and the Arab world exchanged goods in a region of easy maritime access to a large number of productive islands and peninsular locations. The modern Indonesian islands, known as the *Spice Islands* or *East Indies,* became a target for traders from outside the region. Buddhist, Hindu, and Muslim cultures permeated the region (Figure 6.2).

(b)

Figure 6.2 **Southeast Asia: a heritage of Buddhist religious buildings.** (a) Pagoda temples, Lake Inle, Myanmar. (b) Temple at Angkor Wat, Cambodia. (c) Temple on Bali, Indonesia. There are also Muslim, Hindu, and Christian religious buildings in this region. Photos: (a) and (b) © Ian Coles.

(c)

From the 1500s, Portuguese traders, followed by Spanish, Dutch, British, and French merchants and government-sponsored colonizers, took control of most of the kingdoms, sultanates, and sheikdoms of Southeast Asia. Only Thailand succeeded in keeping its independence. Christian missionaries established new religious allegiances. Trading companies used local populations to produce exported raw materials for European industries but had smaller impacts on local cultures than in many parts of the world. They imported Chinese and Indian laborers who worked in mines and on plantations and added to today's multicultural populations and sources of diversity and conflict.

By the 1800s, the European colonial thrusts reached the previously ignored Australian continent, New Zealand, and the Pacific islands. In Australia and New Zealand, the British came to settle and reproduced Western cultures in an "upside-down" world where traditional European winter celebrations took place at the height of summer, often on the beach. They took little account of the indigenous people and kept their distance from Asia. The local impact was total Westernization (Figure 6.3).

In World War II, the Japanese occupied most of Southeast Asia, leaving a legacy of hatred. After that war, Thailand and the island countries of Southeast Asia (Malaysia, Indonesia, and the Philippines) resisted, with U.S. help, the 1950s and 1960s Communist advance that overtook continental Vietnam, Laos, and Cambodia. In the midst of that advance, the **domino theory** was proposed, in which Communists took over the countries one by one, felling the next one from the one that they conquered. In 1967, Thailand, Malaysia, Indonesia, and the Philippines established the **Association of Southeast Asian Nations** (ASEAN) to combat this political threat. Brunei joined in 1984.

Australia and New Zealand remained aloof from their Southeast Asian neighbors, keeping open their main contacts for trading and peopling in Europe but fighting alongside Americans in southern Vietnam. Australians with "Western" views on sustainable resource exploitation and human rights clashed with Malaysian industrialists who were not happy to be accused of "raping the land" and trampling over local people's best interests. Australia disputed submarine oil resources with Indonesia. The Pacific islands attracted little attention or sympathy, apart from exploitative intrusions, from other parts of the region.

Eventually, the external colonial, Japanese military, and expansionist Communist influences gave way to a variety of independent countries placing renewed values on their traditional cultures and coming to terms with their neighboring countries and new roles in the global economic system. One response was to work more closely together. By the 1990s, ASEAN switched from political to economic objectives and embraced its former "opposition" of Vietnam, Laos, Cambodia, and Myanmar as new members. Trade among ASEAN and South Pacific countries and political links among them have been increasing. They now look to the future of the **Pacific Rim** as a new focus of world trade and political leadership in the 2000s (see "Point-Counterpoint: Pacific Rim," p. 224), along with East Asia and the Americas. They perceive common interests in the globally oriented **Asia-Pacific Economic Cooperation Forum** (APEC).

The continent of Antarctica stands apart from the rest of the world and this, the closest region to it. Although Antarctica has never had permanent settlements, many countries claim segments of it and fund scientific research on it. Their findings contribute to global research in such areas as the ozone hole and oceanic fishing resources. We ask the question: Is Antartica a true geographic region?

Varied Economic Achievements

Arising out of these political and cultural histories and human interactions with the natural environments, the region includes some of the world's wealthiest countries, some of those that rose from poorer countries to moderate wealth, and some that remain very poor (Figure 6.4). Australia and New Zealand maintain Western incomes and lifestyles. In 2000, Australia's total gross national income (GNI) was nearly $400 billion, and New Zealand's nearly $50 billion. Thailand, Indonesia, Singapore, Brunei, and Malaysia, followed by the Philippines,

Figure 6.3 **Southeast Asia and South Pacific: contrasts in cultures.** (a) "Long-necked" Karen tribe children, western Thailand. (b) Australian children wave flags at an international cricket match. Photo: (a) © Ian Coles.

(a)

(b)

PACIFIC RIM

The Pacific Rim includes the South Pacific islands, the countries of East Asia and Southeast Asia, Australia, New Zealand, the United States, Canada, and the countries of western Latin America (Box Figure 1). Most of these countries are members of the Asia-Pacific Economic Cooperation Forum (APEC), which seeks to link the growing economies around the Pacific Ocean margins and provide an alternative major world trade grouping to that of the North Atlantic area. The United States became the major power in the Pacific Ocean realm after World War II, although it is now challenged economically by Japan, the newly industrialized countries (NICs) of Southeast Asia, and the emergence of China. Both Australia and New Zealand are members of the APEC forum, but other members of this group resist including their farm products in trade agreements.

There is, however, considerable debate as to whether the Pacific Rim will become the world's most important trading area in the 2000s.

POINT	COUNTERPOINT
Countries bordering the Pacific Ocean are mainly affluent (USA, Canada, Japan, Australia, New Zealand) or growing economically (South Korea, China, Mexico, Southeast Asian countries, Chile). Russia is also a member.	Major differences exist in cultures, languages, attitudes, and demands. Attempts to devise treaties give a false sense of common identity. Most countries still have greater trading links outside this area. Within the area, distances between places are long.
The United States built up economic links to Japan, South Korea, and Taiwan as an outcome of security and defense policies. Now the West Coast of the United States depends on its Asian links. Hong Kong, Singapore, and Los Angeles–Long Beach are the world's three busiest ports.	Such trans-Pacific trade depends on U.S. policy, and the Asian countries are building more trade among themselves.
Australia, New Zealand, and some of the South Pacific islands need to take advantage of the trading opportunities after losing former colonial markets in Europe. They produce wood, agricultural, and mineral raw materials, and are working toward greater penetration of Asian markets through specialist products such as high-quality noodle grain, specialty fruit and nut products, and beef preferences.	The Asian countries wish only to purchase raw materials and not the processed forms that bring greater income to the producers such as Australia. Asian and North American countries protect their agricultural producers, and in 2002, the United States acted to protect its steel producers.
A new global consciousness is overcoming previous isolationist and chauvinist attitudes.	Such attitudes take a long time to change. Australians long gave the impression that they wished to keep out the "yellow peril" of Asians, and their attitudes on environmental issues clash with those of Southeast Asian countries. The Asians act as if Europeans are lower beings and regard Australia and New Zealand (let alone the small Pacific islands) as small markets for their products. Dependency rather than equal roles could result.
Tourism is already bringing people from all these lands together and is a major area of economic growth for East Asia, Southeast Asia, Australia, New Zealand, some Pacific islands—and even Antarctica.	Tourism is too dependent on other aspects of economic growth to be regarded as basic. The area experienced major downturns in visitors following the 1997 financial crisis in Asia and the September 11, 2001, events in the United States.
APEC provides a more inclusive forum than ASEAN.	ASEAN+3 enlarges the scope of the Asian grouping but excludes Australia and New Zealand as well as countries of the Americas. The Asian countries share more interests with each other than they do with other Pacific Rim countries. Australia and New Zealand have involvements with the South Pacific Forum, which questions the fishing and timber felling actions of Asians.
APEC's Shanghai summit in October 2001 called for better cooperation against terrorism. This step was welcomed by Australia and New Zealand, which wish to broaden APEC interests and see security as part of stable economic links.	China is not happy about this because of the U.S. involvement in managing security, while other countries worry about a potential overflow of internal Chinese problems. Muslim countries such as Indonesia, Malaysia, and Brunei were not happy about the U.S. involvement in the 2002 war in Afghanistan.

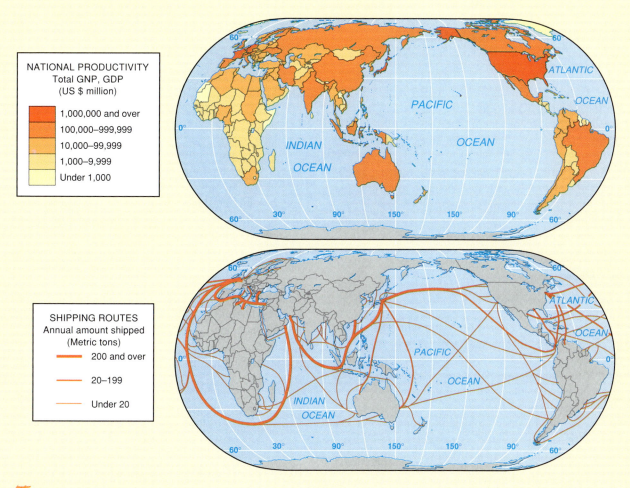

Box Figure 1 The South Pacific countries in the global context. Compare their productivity and distance from main shipping routes with the positions of other wealthier countries.

Figure 6.4 **Southeast Asia and South Pacific: basic data.**

Subregion	Land Area (km²) Total	Population (millions) Mid-2001 Total	2025 Est.	GNI 1999 (US $ million) Total	GNI PPP 1999 Per Capita	Percent Urban 2001	Human Development Index Rank of 175 Countries	Human Poverty Index: Percent of Total Population
Southeast Asia	4,474,170	514	685	529,666	2,648	22.8	116.4	29.7
Australia and New Zealand	7,984,350	23	28	450,644	20,740	81.0	12.5	12.7
South Pacific Islands	527,909	7	12	6,793	3,208	39.8	98.8	18.2
Totals or Averages	**12,986,429**	**545**	**724**	**987,103**	**8,865**	**47.9**	**75.9**	**20.2**

Source: Data from *Population Reference Bureau 2001 Data Sheet*; *World Development Indicators,* World Bank, 2001; *Human Development Report,* United Nations, 2001.

are in the middle group. By 2000, Indonesia ($142 billion), Thailand ($120 billion), Singapore ($98 billion), and Malaysia ($82 billion) all had doubled or trebled their GNIs since 1980. The Philippines ($79 billion) had less growth than the economic leaders.

Vietnam, Cambodia, Laos, Myanmar, Papua New Guinea, and many of the Pacific islands remain poor, and some of the islands depend almost totally on external aid. Vietnam was catching up with the wealthier Southeast Asian countries, having a 2000 economy of $31 billion, up four times since 1990. Cambodia, Laos, and the Pacific island countries had 2000 total GNIs of less than $4 billion each. Myanmar, for which such data are not available from the closed regime, is assumed to remain poor.

🌐 Cultural History and Colonialism

The modern countries and their boundaries were almost entirely the legacy of colonial territories. The varied precolonial and colonial histories contribute to the diversity of people in the modern countries and their economies, cultures, and types of government.

Prehistoric migrants came from the north, filtering southward through the main valleys and moving on across the island archipelagos to Australia, New Zealand, and other nearby islands. They pushed indigenous, black-skinned people into isolated mountain or outback locations.

Khmer, Burmese, Thai, and Vietnamese Empires

Mon and **Khmer** people occupied the Cambodian area from the north (see Figure 5.5), and between the A.D. 800s and 1200s, it was the center of the Khmer Empire. Later, the **Vietnamese, Lao,** and **Burmese** arrived in the territories that now form their national centers. Indian traders brought Hindu and Buddhist religions across the ocean from the west. The distinctive Indian culture and architecture reflected in the temples of Angkor, Cambodia (see Figure 6.2), led to the area that includes Cambodia, Laos, and Vietnam being called "Indochina."

Burma (modern Myanmar) became Buddhist as Sri Lankan monks disseminated Buddhism. The central Pagan kingdom based on Mandalay in a fertile, mountain-rimmed plain in northern Burma fought powerful tribes, including the **Shan** and **Karen,** who lived in the surrounding hills and today continue to resist Burmese attempts to draw them into the country's life. The Mongols (see Chapters 4 and 5) ended the Pagan Empire. After they left, Burma remained divided until the 1700s, when efforts to promote Burmese expansion led to conflicts with the British in India.

At the time of the Mongol advance, a group of **Thai**-speaking people moved from western to southern China and then into the western part of the Indochina peninsula, forming a unified political grouping by the 1300s with its center at Ayutthaya. After conflicts with Burma and Cambodia, the people established their main territory that was never again occupied by other powers and became modern Thailand.

In northern Vietnam, the Red River valley became the center of a kingdom by the first century B.C. The area was later conquered and reconquered by various Chinese dynasties and took on Chinese methods of government and administration. The Vietnamese, however, always fought back and retained their separate identity. In the A.D. 1400s, the Vietnamese not only fought off the Chinese but also extended their lands southward across the older coastal kingdoms to take over some of the Khmer territories at the mouth of the Mekong River.

Cultures Meet on the Southeast Asian Islands

In the island groups that now form Malaysia, Indonesia, and the Philippines, overseas influences brought frequent shifts of power. The basis of present ethnic differences was established before A.D. 1000 on most islands and then overlaid by Islamic ingredients after 1200.

Phases of migration in the Indonesian islands brought together over a hundred ethnic groups and languages. Groups of traders and invaders from India and China came to the islands before a kingdom based on rice growing was established in Java in the A.D. 500s. From the 600s to 800s, ruling dynasties controlled the Malay Peninsula and much of present Indonesia, erecting Buddhist monuments on Java. After around 1000, Buddhism added Hindu elements in some areas, as on Bali (Indonesia), but in most of western Malaysia and Indonesia, both Hinduism and Buddhism gave way to Islam. From mixed Malay, Indonesian, and Chinese influences, a Filipino culture emerged about the 400s.

Little is known of the Malay Peninsula before the arrival of Muslim groups during the 1200s. From that time, the local sultanates established at Malacca (now in Malaysia) and Brunei controlled the islands' trade. Malaysia, Indonesia, and Brunei remain dominantly Muslim in religion, with Indonesia being the world's largest Islamic state.

Australia, New Zealand, and the Pacific Islands

The indigenous people of Australia, New Zealand, and the Pacific Ocean islands included racial and ethnic groups whose origins are still debated. Spread out over vast distances of ocean, the populations of the islands contained mixtures of Melanesian, Polynesian, Micronesian, and Asian types. They have a range of skin colors and body features, but links between them and other racial types are not clear.

When Europeans arrived in the late 1700s, the population of Australia was between 200,000 and 500,000 indigenous, dark-skinned **Aborigines** (Figure 6.5). They were nomadic hunters and gatherers living in groups spread across the continent and speaking 200 different languages. Rock paintings based on their **animistic** (nature worship) religious beliefs and social organization are part of their legacy.

The **Maoris** of New Zealand came from the wider South Pacific around the A.D. 800s, with the final waves of people arriving from Tahiti in 1350. They replaced an earlier dark-skinned people, the Moriori, most of whom they drove out.

The inhabitants of the South Pacific oceanic islands are grouped in three geographic categories—the **Melanesian** ("black islands" because of the dark-skinned people), **Micronesian** ("small islands"), and **Polynesian** ("many islands") people. The Polynesian groups have lighter coloring than the Melanesians. These groups also spread into the eastern islands of Indonesia, where they still form the majority.

(a)

(b)

Figure 6.5 **Australia: Aborigines.** (a) An Aborigine with an ancient rock drawing. (b) Aborigines after shopping at a store in Nangalala, Northern Territory. Tensions arise between traditional ways and Western lifestyles. Photo: (b) © Penny Tweedie/Corbis.

The Colonists

Southeast Asia

The Portuguese established trading centers in the Philippines and other islands, but the Spanish pushed them out of the Philippines in the late 1500s and built their capital at Manila on Luzon (Figure 6.6). Roman Catholic priests poured in, and monastic orders took over large tracts of land and became wealthy. Rapid conversions to this new faith united the disparate groups of Filipinos in what became the only largely Christian country in Southeast Asia. Spain retained its Philippine colonies until the end of the 1800s, when the United States placed them under its jurisdiction after winning a war with Spain over Cuba (the Spanish-American War). Local opposition to colonial rule increased during the 1930s and under Japanese occupation in World War II. The Philippine Republic was established in 1946, and the United States retained its military bases until the 1990s, but reactivated some after the attacks on September 11, 2001.

After 200 years of intermittent fighting among Spain, Portugal, the Netherlands, and Britain, Java and the surrounding islands became the Dutch East Indies colony in 1799. The Dutch made Batavia (modern Jakarta) the capital of the colony after it grew as the trading headquarters of their **Dutch East India Company** from the late 1500s. Local farmers were organized to produce coffee, rubber, and other new crops for export, but neglect of food production resulted in famines. Nationalist movements that began in the early 1900s were further motivated by harsh Japanese treatment during their occupation of World War II. Although the Dutch resisted granting independence after the war, it came in 1950 when the new country of Indonesia was formed.

The Portuguese annexed Malacca on the Malay Peninsula in 1511 to control the spice trade but lost another toehold in the region when the Dutch took over the Malay Peninsula in the

1600s. For over 100 years, the Dutch Batavia-Malacca link gave it a spice-trade monopoly, contested by the (British) East India Company from 1601, which shared facilities at Bantam (west Java) until the Dutch took them over in 1682. Dutch dominance of the East Indies caused the British company to expand trade with and occupy India (see Chapter 7). In the late 1700s, the British founded a port on Penang Island, establishing a rival center linked to their sales of Indian opium in China (see Chapter 5). In 1804, the British and Dutch agreed to take control of Malaya and the East Indies respectively.

The British built a new port at Singapore (1819) and controlled the Malay States by diplomacy. The sultanates on the northern tip of Borneo (Sabah) and the northeastern coastlands (Brunei and Sarawak) became British protectorates administered from Singapore. Tamil Indian and Chinese laborers were brought into sparsely populated areas to work, respectively, in the Malayan rubber plantations and tin mines. In World War II, the Japanese took Malaya, but the British reestablished their rule afterward, fighting against Communist guerrillas until independence was granted to Malaya in 1957. Singapore, Sabah, and Sarawak joined the Federation of Malaysia on their 1963 independence. Brunei decided to stay out of that federation, and Singapore withdrew in 1965 after problems of maintaining its largely Chinese identity among the Malay majority in the rest of Malaysia.

The British colonized Burma from India, making it a province of British India in 1886 after three wars during the 1800s. Developing rice growing in the Irrawaddy River delta, they built the port of Yangon (Rangoon) to export the rice. Over a million Indians moved into this southern part of Burma, becoming the leaders of the country's commercial community. Attempts were made to unify the newly developed Lower Burma with the main Burmese population around Mandalay and the hill tribes. After riots against the British in the 1930s, the main independence group sided with the Japanese in World War II but changed sides toward the end and gained independence in 1948.

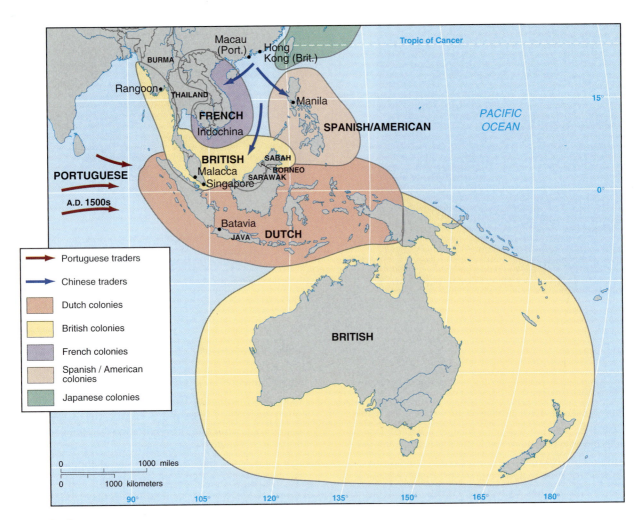

Figure 6.6   **Southeast Asia and South Pacific: colonization.** Portuguese, Dutch, British, French, Americans, and Japanese assumed control of areas easily accessed by sea.

The French occupied Indochina (Cambodia, Laos, and Vietnam) during the 1800s and early 1900s. They built roads and railroads and encouraged manufacturing. During World War II, Nazi Germany occupied France and persuaded the French colonial government to allow the Japanese forces to pass through Indochina. After the war, the French tried to reestablish control of the area, but Communist groups forced them to leave the northern parts of Vietnam in 1954. The divided country, with a Communist north and free-market south, was subject to further warfare in which the United States supported South Vietnam. In 1975, North Vietnam was victorious and reunified the country. Laos went through much strife along its border with Vietnam, slowing economic development. Cambodia became independent in 1953 but suffered 30 years of civil war and invasions by the Vietnamese.

Thailand (Siam)

From 1856 to 1939 and 1945 to 1948, Thailand was known as Siam. The term *Thai* came into use for the first time in the early 1900s. This country never became a colony, maintaining its independence through a strong monarchy. After the 1932 aboli-

tion of absolute monarchy, a rigorous bureaucracy ruled, despite many political changes (19 military coups, 53 governments, and 16 constitutions from 1932 to 1999). The influence of the bureaucracy is reflected in the identical school buildings in all 76 provinces.

Thailand's rulers established commercial treaties with Britain in the 1800s and encouraged modernization. In the 1890s, following a short war when France tried to make it part of Indochina, Thailand yielded land to France (now in Cambodia) and some of the Malay Peninsula to Britain. In World War II, Thailand at first sided with Japan, allowing the latter's armies that easily passed through Indochina to invade Burma and Malaya. Thailand regained territories it claimed in Malaya and Cambodia. Its new government in 1944 wisely supported the victorious allies but had to return the disputed territories at the end of the war. The fighting in Vietnam in the 1960s and 1970s made it seem that Thailand would be the next "domino" to fall to the worldwide Communist advance. However, Thailand stemmed that advance at its borders, attracting considerable financial aid from the United States to help it resist.

Australia and New Zealand

Terra Australis, the "Southland," was the last major inhabited continent unknown to and unexplored by Europeans. The southeast trade winds blew early European traders in the East Indies away from Australia. The Dutch discovered the continent in the 1600s, when their improved ships and occupation of the East Indies led them to explore southward. British ships visited the region, but most early assessments were of dismal lands with little economic or settlement potential for Europeans.

Following Captain James Cook's surveys and more encouraging reports in the 1770s, the British claimed the Australian continent, which they first called New South Wales. From 1787, part of the colony was used as a penal settlement to replace the previous transportation of convicts to the now-independent American colonies. In the early 1800s, settlement increased after initial problems of food supply led to improvements in government administration and the issuance of free land grants that encouraged sheep farming.

Tensions grew between the ex-convicts and other settlers. Most convicts came from the poorer groups within British cities. After serving their sentences, many did well in business or government in Australia. A gold-mining boom in the 1850s drew speculators and new settlers to Australia. By the time the last convict ship arrived in 1868, voluntary migrants outnumbered the convicts and their descendants by 10 to 1.

During the 1800s, British groups established new colonies around the Australian coasts, each with its own main port city and a competitive pride that generated rivalries among the colonies (Figure 6.7). When Canada was granted dominion status

Figure 6.7 **Australia: colonization patterns.** (a) Beginning in the late 1700s and continuing through the 1800s to dominion status in 1901, Australia developed political institutions around the separate colonies (later states). The dates in parentheses indicate when the colonies were established. The straight-line boundaries in many cases were drawn across the interior desert. (b) **South Pacific: mercantile theory of urban patterns in the landscape.** This theory can be applied to Australia and New Zealand, as well as other former colonial lands. The circle size corresponds to the size and range of functions of the urban centers. Imports of manufactures (1) from the home country are paid for by exports of commodities (2). Dashed lines in parts i and ii represent areas of colonizing settlement.

(a)

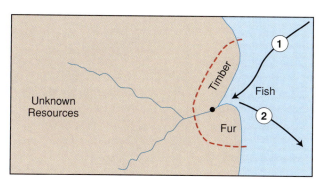

(*i*) New land, Ocean transport

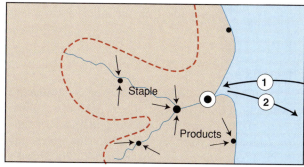

(*ii*) River transport

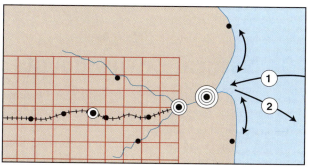

(*iii*) Rail transport

(b)

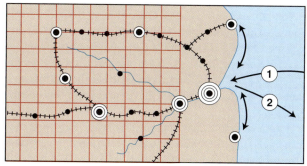

(*iv*) Road/rail transport

within the British Empire in 1867, some Australians began to think in similar terms, but the internal rivalries held back such cooperation. In the late 1800s, the scramble by European powers for colonies in the South Pacific led to worries over common defense, but New South Wales still held aloof from the other Australian colonies that wished to federate. The federal idea finally came to fruition in 1901, when the Commonwealth of Australia within the British Empire was created out of the five colonies that then became states.

The Europeans took little account of the Aborigines, and many died from disease and oppression. The Tasmanian Aborigines were wiped out by 1876, and numbers of Aborigines on the mainland were reduced to fewer than 75,000 by 1933. In the 1900s, attempts to integrate Aborigines into Australian life partly succeeded in health and education terms, and they now number 300,000. Of these, only 10,000 follow traditional ways of life, and the rest live on reservations or in cities. Many are poor, and despite making up only 1.5 percent of Australians, they comprise 29 percent of prison inmates, often being subject to more rigorous judgment for small crimes than people of European heritage.

Colonial settlement came to New Zealand later. Maoris resisted British missionaries and whalers in the 1700s and early 1800s. It was not until after 1840 that the British government encouraged the immigration of farming settlers. As it took sovereignty in New Zealand, it agreed to respect Maori land ownership. After 1860, a short gold rush brought speculators. The technical advance of refrigerator ships made it possible to export fresh meat to Europe from 1882, and more sheep farmers became established. In 1907, New Zealand gained dominion status within the British Commonwealth and a large degree of autonomy. Although subject to some segregation practices, the Maoris are more integrated in New Zealand life than the Aborigines are in Australia.

Pacific Islands

The United Kingdom was the main colonizer of the Pacific Islands, including Fiji, Kiribati (Gilbert Islands), Tonga, Tuvalu (Ellice Islands), some of the Solomon Islands, southeastern New Guinea, and parts of modern Vanuatu, which was shared with the French. Guam and the Marianas were Spanish colonies until taken over by the United States as protectorates just after 1900.

The French colonized New Caledonia and the islands around Tahiti, which, like some Caribbean islands, remain part of France (see Chapter 10). Political decisions are made by the French government in Paris, as recent nuclear tests in French Polynesia emphasized. Like French colonies in the Caribbean, the French South Pacific islands have access to European Union markets.

The Germans were active in the 1880s, when they took the Marshall Islands, together with Nauru, Western Samoa, northeastern New Guinea, and some of the Solomon Islands. All were lost to the United States (Marshalls), Britain, Australia, or New Zealand in World War I.

Test Your Understanding 6A

Summary   The continental and island countries of Southeast Asia, together with Australia, New Zealand, and hundreds of Pacific islands comprise a varied region that had separate historic experiences but is becoming increasingly integrated into global affairs.

Southeast Asia was occupied by people entering from the north, followed by traders from China as well as from South Asia bringing Hinduism and Islam. European colonists claimed much of this territory in the 1800s, but small numbers of Europeans settled permanently, and the present countries achieved independence in the 1960s and 1970s.

Australia and New Zealand act as a local political and economic focus among some of the world's poorest and smallest Pacific island countries. Following colonization by mainly western European countries in the late 1800s, many of the islands remain poor with few products to sell in global markets. Indigenous groups of people in Australia (Aborigines) and New Zealand (Maoris) fared badly under the European settlers, but related groups form the main populations in many South Pacific islands.

Questions to Think About

6A.1 Summarize the variety of external influences that contributed to the present geography of this region.

6A.2 What have been the consequences, good and bad, for Australia and New Zealand of being at a distance from the richer Northern Hemisphere countries?

6A.3 Compare the experiences of Southeast Asian countries with those of Australia and New Zealand in regard to colonialism.

Key Terms

domino theory	Shan
Association of Southeast Asian Nations (ASEAN)	Karen
	Thai
Pacific Rim	Aborigine
Asia-Pacific Economic Cooperation Forum (APEC)	animism
	Maori
Mon	Melanesian people
Khmer	Micronesian people
Vietnamese	Polynesian people
Lao	Dutch East India Company
Burmese	*Terra Australis*

 # Natural Environments

The natural environments of the Southeast Asia and South Pacific region range from continents to thousands of islands, from equatorial to midlatitude climatic environments, and from volcanic islands and coral reefs that are forming now to some of the world's oldest landscapes. The long-term geologic isolation of some lands in this region gave them a unique flora and fauna.

Equatorial, Arid, and Oceanic Climates

Mainly Equatorial Southeast Asia

Malaysia, Indonesia, and the southern parts of the Philippines have hot and rainy weather all year in the equatorial climatic environment (Figure 6.8) that covers up to 10 degrees of latitude on either side of the equator. Air converges and rises, condensing in tall clouds from which rain falls. The islands and narrow peninsular environments surrounded by ocean that occupy much of this area create strong land and sea breezes, giving regular daily weather patterns. Sunny mornings are common, but tall clouds build up in the afternoon as humid air blows in from the ocean and intensive thunderstorm rainfalls occur in the late afternoon and evening. Some places have over 300 days in a year with thundery rain. Nights are mostly clear as air drains back from the cooling land to the ocean.

The northern Philippines, Indochina's mainland, Thailand, Myanmar, and northern Australia experience a monsoon tropical climatic environment in which summer rains are brought by winds from the oceans. These subtropical environments are subject to typhoons (similar to hurricanes) that approach from the ocean, mostly in late summer. Winters are cooler and drier, dominated by winds blowing outward from Asia's interior.

Southeast Asia is at the western end of the oceanic and atmospheric circulations that cause the El Niño fluctuations off western Latin America (see Chapter 10). When Peru is dry, this region has plentiful rains and vice versa. During the major 1997–1998 El Niño, an area stretching from the South Pacific islands into Indonesia and northern Australia suffered intense drought and forest fires while Peru had unusual deluges of rain.

Tropical Ocean Climates

The islands of the South Pacific Ocean are nearly all in the tropical belt. Poleward of 10 degrees north and south, the trade winds blowing from the northeast (Northern Hemisphere) or southeast (Southern Hemisphere) are constant factors. The hilly islands in the path of the trade winds have rainy east-facing flanks. Coral atolls are low-lying, and some that lack hills to cause uplift and rain are often arid with small and uncertain rainfalls.

The tropical oceans have high water temperatures throughout the year, supplying moisture and heat to the air above, which fuel intense tropical storms and typhoons. Tropical disturbances and typhoons occur in late summer in the belt between about 10 and 25 degrees north or south of the equator. In early 1993, tropical cyclones hit the Solomon Islands, Papua New Guinea, and Fiji before passing on to the Philippines and mainland Asia. The rains and raised sea levels swamped low-lying land and made thousands homeless.

Australia and New Zealand

Australia's climates are dominated by its arid continental interior. One-third of Australia is arid, and another one-third is semiarid. Water shortages are permanent characteristics of much of the continent, and even the areas that are normally well watered suffer from lengthy droughts. The aridity of the interior may be broken by storms and flash floods, occasionally filling some of the dried lakes such as Lake Eyre.

The areas with regular and sufficient rains occur around the coasts. Winter rains of the Mediterranean climatic environment, brought by the midlatitude cyclones of the southern oceans, are characteristic of the southwestern corner of the continent and the Adelaide area. Monsoonal summer rains characterize the tropical northern coasts. The southeastern coastlands, where most Australians live, have a warm midlatitude climate with rains all year, most falling in summer.

New Zealand's climate is humid and similar to the British climate that many of its settlers left behind. It has fewer temperature extremes than the British Isles, because there is no continental interior nearby to provide the coldest winter or the hottest summer weather that Britain experiences from time to time because of its closeness to continental Europe. The mountains on South Island create a rain shadow to their east, necessitating the irrigation of some of the farmland.

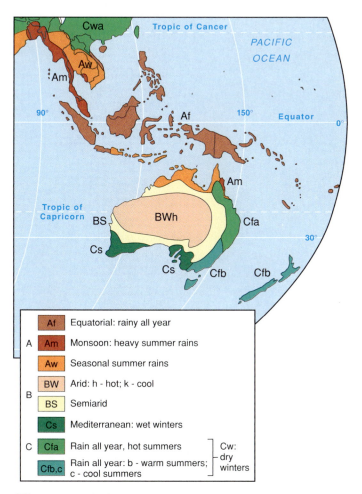

Figure 6.8 **Southeast Asia and South Pacific: climates.** From the equator to the South Pole. Compare the significance of continents and oceans in the range of climates.

(a)

$\mathcal{F}igure$ 6.9 **Southeast Asia and South Pacific: landforms.** (a) Relief map showing the variety of continental and island areas. The Wallace Line marks the division between mainly Asian (west) and Australian (east) species. (b) Plate margins: relate them to relief features. (c) The location of the southern continents 200 million years ago in Gondwanaland.

Continents and Islands

The surface features of Southeast Asia and New Zealand (Figure 6.9a) include mountain ranges and island chains that formed in the recent geologic past. Australia is an ancient continent. The Pacific islands are mainly the product of very recent volcanic activity.

Plate Movements, Mountain Ranges, and Volcanic Activity

The high relief, frequent earthquakes, and many volcanic eruptions result from the clashing of tectonic plates—three major and one minor (Figure 6.9b). Convergent plate boundaries form much of the western and all the southern and eastern boundaries

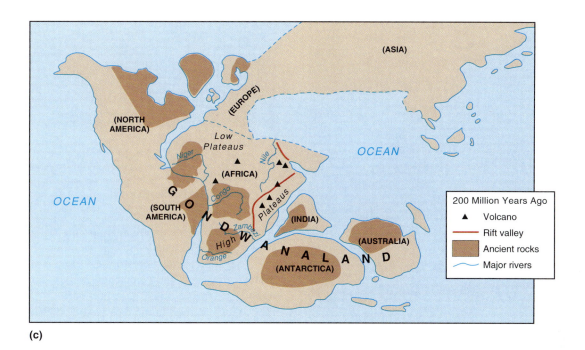

(c)

of Southeast Asia. The boundary between the Indian and Eurasian plates turns southward after following the west-east line of the Himalayan Mountains. Where the Indian plate pushed into Asia from the west, it formed parallel north-south ranges in Myanmar. The fold mountains form Thailand's western border with Myanmar and extend into the Malay Peninsula.

The plate boundary swings eastward south of the line of Indonesian islands including Sumatra and Java but keeps north of Australia. The Indian Ocean floor is subducted beneath the Indonesian islands and melts to erupt volcanic material that forms the islands north of this plate boundary.

On the eastern margins of Southeast Asia, further groups of islands, from eastern Indonesia to the Philippines, formed where the Pacific plate was subducted beneath the eastern margin of the Eurasian and Philippine plates. The volcanic islands of eastern Indonesia and the Philippines provide active landscapes with eruptions every decade or so. The unusual shape of Celebes (Indonesia) marks a front of uplift formed by pressures from the east.

Between the convergence zones along the eastern and southern boundaries of Southeast Asia is a triangular area where the seas are shallow, as in the Sunda Shelf between Malaysia, Vietnam, and Borneo. The Indochina Peninsula and eastern Thailand (Khorat Plateau) are hilly areas formed of ancient rocks brought to the surface by uplift and erosion of the overlying rocks.

The two collision zones of convergence meet in eastern Indonesia and form an active zone between the Indian and Pacific plates running through New Guinea and islands to the east. A transform plate margin (see Figure 2.31a) marks the eastern boundary of the Indian plate and then angles southwestward through New Zealand.

New Zealand is the product of plate margin activity. Its two main islands are part of a fragment of the southern continent that broke away from the main mass of Gondwanaland (Figure 6.9c) around 100 million years ago. The uplift of the

land keeps pace with the glaciers and rivers that wear it down. In the Ice Age, glaciers carved deep valleys in the rising Southern Alps, while rivers flowing from these mountains cut into the sloping eastern plateaus. Rising sea level as the climate warmed at the end of the Ice Age drowned the mouths of the deep glaciated valleys to form fjords (Figure 6.10).

Many volcanic islands also occur farther out in the Pacific Ocean, away from plate margins. These islands form above areas of crustal heating and may have a volcanic core surrounded by coral reefs (fringing reef), a volcanic island with a wide lagoon between it and the coral reef (barrier reef, Figure 6.11), or a central lagoon surrounded by coral reef but no volcanic island (atoll). It is thought that these coral reef forms represent a sequence that begins with active eruptions building a volcanic island and ends with it sinking beneath the waves under its own weight after the volcanic eruptions cease. Coral reefs colonize the islands when tiny animals secrete limy structures at or just below sea level. The coral animals live near the tropical ocean surface, where it is warm enough and where algae on which they feed get access to light and oxygen.

Ancient Continent

Australia was once joined to Africa, Antarctica, South America, and the peninsula of India in the continent of Gondwanaland (Figure 6.9c) but now lies on the eastern portion of the Indian plate at a distance from all plate margins. The relief features of the western half of the continent are low plateaus and plains on the ancient shield rocks. Mountain ranges formed over 600 million years ago were worn down to form landscapes of little relief today, and the lack of vegetation cover in the arid area exposes ancient mountain rock structures.

The Great Dividing Range along the eastern edge of Australia is formed of rocks that were deposited from the erosion of the ancient continent and were then uplifted and

Figure 6.10 **New Zealand: South Island coast.** The area was carved by glaciers and then drowned by rising sea level at the end of the last glacial phase.

Figure 6.11 **South Pacific: coral island.** A typical South Pacific island, with a hilly volcanic core, an outer coral reef, and a shallow lagoon between the reef and mainland. The village is sited where a channel leads from a breach in the reef front. Photo: © Patrick Ward/Corbis.

broken into blocks by faulting. Apart from the steep edges of these eastern uplands, Australia has fewer major relief contrasts than other continents. The world's largest area of coral reefs forms the Great Barrier Reef off the northeastern coast of Australia.

Major Rivers

Long rivers carrying high water and sediment flows in summer are a feature of continental Southeast Asia. They include the Irrawaddy, Salween, Mekong, and Red Rivers (see Figure 6.9a). The Irrawaddy and Salween flow through Myanmar, while the Mekong rises in western China and flows through the countries of Indochina to its delta in southern Vietnam. The Red River lowlands form the northern Vietnam heartland. Western Thailand is drained by a series of rivers that join in the Chao Phraya lowlands near Bangkok, forming the economic and cultural heart of the country. The continent of Australia is mainly dry, but the Murray-Darling River system west of the Great Dividing Range supports farming and hydroelectricity developments.

Distinctive Ecosystems

The range of climatic and relief environments and degrees of long-term isolation from other land areas as the Gondwanaland continent split apart produced a variety of natural vegetation and animal types. The main contrast is between the continental Asian species and the unique groups of species inhabiting Australia, New Zealand, and eastern Indonesia.

The tropical rain forest of the equatorial Indonesian and Malaysian islands and peninsulas contains stands of many different tree species. The monsoon forests, from Myanmar to Indochina and the Philippines, have somewhat fewer species; extensive areas are dominated by teak forest.

Separated by plate movements from the other southern continents some 35 million years ago, Australia's mammals were at an early stage of evolution that did not include a long womb-based gestation of babies. The **marsupials,** such as kangaroos, koalas, wallabies, and possums, raise their young in pouches and compose about half the native animals. They are rare in other parts of the world. The bird life is particularly varied and colorful. Australian vegetation is dominated by species of eucalyptus (Figure 6.12) and acacia, and there are unique desert species in the dwarf mallee community. **Mallee** is formed of drought-resisting eucalyptus shrubs that grow into almost impenetrable thickets of many close-spaced stems rising to 8 or 9 m (25–30 ft.) high.

Some of the distinctive Australasian plant and animal species also occur in the eastern islands that are part of Indonesia today, separated by the **Wallace Line** (see Figure 6.9a) from Asian species. Although this line was drawn by botanist Alfred Russell Wallace in the mid-1800s, it was over 100 years before the origin of the separation was made clear. The line marks the edge of plate tectonics action some 10 million to 15 million years ago that forced the eastern islands fronted by Celebes against the western and brought Australasian plants and animals with them.

During European colonization, species from this region were taken to adorn European gardens, and external species were introduced. Domestic animals from Europe found they had few local predators. For example, wild rabbits introduced in 1859 spread across Australia, destroying large areas of grassland. Efforts to control the rabbits continue following their recovery from the myxomatosis virus introduced in the 1930s.

New Zealand had a unique rain forest-based flora and fauna before European settlement, but the immigrants' domestic animals and crops replaced many of the native plants and animals. Virtually all the original forest was soon cut but was later reforested by quicker-growing Douglas fir trees and pines for commercial uses. Introduced reindeer damaged trees and shrubs but are now contained.

Some of the larger Pacific islands are forested with species closer to those of Indonesia and East Asia than of Australia.

Figure 6.12 **Australia: eucalyptus forest in humid Victoria state.**

Palms are particularly numerous. Few islands have many animals, although bird species are diverse. The surrounding waters contain a wealth of tropical fish varieties, but each is in relatively small numbers, and they are too easily overfished.

Natural Resources

Southeast Asia

The mineral resources of Southeast Asia (see Figure 5.10) include tin, iron, gold, and precious stones. Many of the tin deposits occur where active rivers eroded tin-bearing veins in the rocks and deposited this heavy mineral in concentrated alluvial layers lower down the valley or offshore. Deposits of oil and natural gas occur in Indonesia and in many recently explored offshore locations, such as between Indonesia and Australia.

The recent history of the Mekong River illustrates some of the resource potentials and political difficulties associated with modern water use in the region. The Mekong flows through or along the borders of six countries—China, Myanmar, Thailand, Laos, Cambodia, and Vietnam. It is one of the region's longest rivers, and its development potential includes hydroelectricity, irrigation water, and transportation. Improved flood control would also aid economic development of the lands along the Mekong's length.

Plans for an integrated and multipurpose development of the Mekong waters in the 1940s were abandoned during the subsequent wars in the area. Suspicions among the countries prevented further planning until an initial meeting of the four lower-basin countries in 1994, when an outline agreement was signed to investigate the sharing of waters and their protection from environmental damage. Progress on this accord will be slow. For example, China and Myanmar did not sign the agreement at once; Cambodia and Vietnam are in dispute over transportation on the river; without consultation, Laos interrupts the flow by dams that generate electricity, a major foreign exchange earner; China is already building the large Manwan dam on the upper waters; and Thailand is criticized for taking too much water from the river to irrigate its arid northeast. Furthermore, there are concerns over environmental impacts if the new dams hold back silt that now renews the fertility of soils downstream and if more regular flows in the river affect the freshwater fisheries in Tonle Sap, Cambodia. In 1995, Cambodia, Laos, Myanmar, Thailand, Vietnam, and China's Yunnan province signed an agreement for the future integrated development of the Greater Mekong Sub-Region based on a plan submitted by the Asian Development Bank. This adds formality to previous intentions but does not immediately overcome the difficulties of the cooperation among former enemies and potential economic competitors.

Australian Resources

The ancient rocks of Western Australia—like the shield rocks of similar geologic age in Africa, northern Canada, and Siberia—contain many large deposits of iron ore and other metallic ores such as nickel, gold, platinum, uranium, and copper. The rocks that form the Great Dividing Range in eastern Australia are of more recent origin and contain coal, silver, lead, zinc, and copper ores.

Between the shield rocks and the Great Dividing Range, the lowlands are drained by the Murray-Darling River system in the south and form the Great Artesian Basin in the north. Rain falling on the eastern mountains soaks into the rocks and drains westward and downward, accumulating in the sedimentary rocks of the basin (Figure 6.13). Since the rocks in the mountains remain filled with water, the pressure on water in the rocks beneath the lowlands is so great that wells drilled into the rocks cause water to flow out on the surface without a need for pumping. The northern part of the lowlands was named for these **artesian wells.**

Pacific Islands

Some of the Pacific islands along the plate collision zone have mineral resources. The copper deposits on Bougainville, an offshore island that is part of Papua New Guinea, constitute one of the world's largest reserves, which was intensively mined until terrorist activities stopped this activity. New Caledonia is the world's third-largest producer of nickel ore. Most of the larger islands have a covering of rain forest, but some of the drier and flatter islands have only sparse vegetation.

Environmental Problems

Natural Hazards

Many people in this region live close to plate boundaries and have to contend with earthquakes and volcanoes. The 1995 eruption of Mount Ruapehu in the North Island of New Zealand led to the closure of airports and highways and caused worries over the prospect of a larger eruption. Volcanic and earthquake activity continues along the plate boundaries, with 1990s eruptions in the Philippines and previous events in Sumatra and Java. Threats from typhoons, floods, and droughts also affect various parts of the region.

Human actions, which are especially significant in areas of high population density and modern industrialization, add to the impacts of natural hazards. Expanding agriculture, together with cutting of forests for lumber, caused deforestation. Rapid loss of forest often leads to soil erosion, downstream flooding, and silting, commonly destroying offshore fisheries and coral reefs as the silt is swept out to sea. Warfare in Vietnam in the 1960s and 1970s defoliated forests and left chemical residues

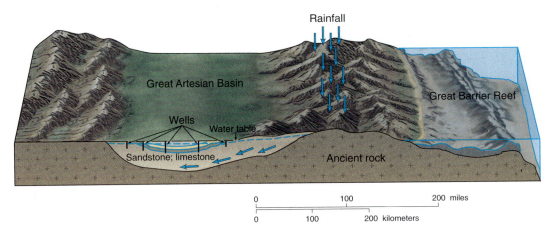

Figure 6.13   Australia: Great Artesian Basin and the Great Barrier Reef. The artesian basin of northeastern Australia is created by the geologic conditions that produce a major groundwater source as rain falls on the Great Dividing Range, seeps into the rocks, and flows downward to replenish the water in the rocks beneath the basin. The water in the deepest rocks flows out under pressure from the water moving down from the hills. The Great Barrier Reef is one of the world's largest developments of coral reef and forms a feature just off the Queensland coast.

and barren hillsides that needed decades to recover. Further examples of environmental problems are discussed in the sub-region sections.

The tropical environments of Southeast Asia harbor many diseases. Although forest clearance, better sanitation, improved diets, and the wider availability of primary medical care improved health since 1960, malaria, cholera, typhoid, and rabies remain serious problems. HIV/AIDS is an increasing threat in Southeast Asia. At first, a problem mainly of the major cities, especially Bangkok (Thailand), it has spread widely.

For the future, the low-lying parts of islands, especially in the South Pacific but also in Southeast Asia and the Great Barrier Reef, face environmental disaster if global warming leads to a rising ocean level. The highest parts of some of the coral atoll islands are only a few meters above that level, and many have a concentration of settlement on low-lying coasts. Many of the coral reefs that form the foundations of the islands, protective barriers, or tourist attractions are under threat from rising sea levels and from predators and pollution that may kill the living corals, leaving the older coral rock to be worn away by the waves.

Pollution, Erosion, and Mining Excavations

Pollution of air, water, and soils is a serious problem in Southeast Asia as concentrations of people and industrial activities increase, more vehicles are used, and farming is made more productive by the use of fertilizers, pesticides, and mechanization. In the late 1990s, the forest fires on Borneo and Sumatra—lit to clear forest for planting oil palms at a time when expected rains did not come to douse the fires—covered much of Malaysia and Indonesia with a choking smog.

Illegal tree felling in Indonesia, Myanmar, Cambodia, and Laos led to erosion and the flooding of bare land. It also occurs in better-controlled Vietnam. Malaysia's claims to manage sustainable logging gave way in face of increased demands for timber, furniture, jobs, taxes, and political influ-

ence. In Sarawak, loggers finance local political parties that allocate areas for felling when in power. In 2001, the Malaysian government agreed to take part in certifying that its timber exports come from well-managed forests. In 2002, Migros, the Swiss supermarket chain, committed itself to buying palm oil from ecologically sound sources in this region, as opposed to the newer plantations that devastate the rain forest.

Australians are increasingly aware of the environmental damage caused to their fragile landscapes by farmers and miners. Farmers in Western Australia, for instance, felled the trees that kept down the water table in soil and rock and slowed the wind's force near the ground. The resulting increase of wind erosion and rise in the water table led to surface salinization and made large areas unusable for farming. Replanting and mixing lighter soils with clays from mining wastes combat the dual menace. Overgrazing and overirrigation in the lowlands between the shield area and the Great Dividing Range have had similar results.

Mining extraction scarred many parts of Australia, causing environmentalists to join their concerns with land claims by Aborigines to delay and impose new conditions on applications for mining licenses. British nuclear tests carried out in the interior in the 1950s left scars on the land and on the Aborigines living in those areas. The United Kingdom still pays compensation for this damage.

When such problems came to public consciousness, the Australian federal government declared the 1990s a "Land Care" decade. Millions of new trees were planted in the early 1990s to make up for the halving of the limited Australian forest area following colonial settlement. It is anticipated that the capital being invested in attempts to restore soil quality may have positive results locally and in technologies that will be extended to poorer countries in arid parts of the world.

In New Zealand, the main forms of environmental degradation include soil erosion and the changes brought about in the fauna and flora by introducing new species. New Zealand takes an increasingly strong stance against potential polluters, and its desire for world peace makes it willing to endanger its relation-

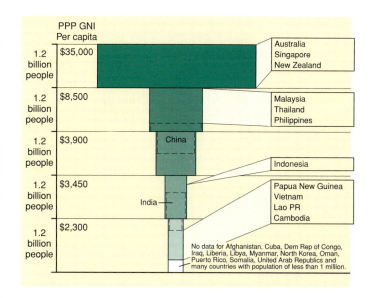

Figure 6.14 **Southeast Asia and South Pacific: country incomes compared.** The countries are listed in the order of their PPP GNI per capita. Source: Data (for 2000) *World Development Indicators*, World Bank, 2002.

Test Your Understanding 6B

Summary Oceanic influences dominate the climatic environments of Southeast Asia and the South Pacific. These environments range from forested equatorial rainy areas through trade wind climates to the stormy seas surrounding Antarctica. Interior Australia remains a desert.

Clashing tectonic plates produce mountains, volcanic islands, and earthquakes from Southeast Asia through Papua New Guinea to New Zealand. They are linked to the deposition of mineral resources such as copper, gold, and nickel. The shield areas of Australia contain mineral resources such as iron, uranium, and gold.

Environmental problems stem from attempts to establish European farming and tropical plantation crops in these new lands. Soil degradation and mining damage are among the main problems faced.

Questions to Think About

6B.1 How can a knowledge of plate tectonics help in understanding this region's relief patterns?
6B.2 Can the small numbers of people living in Australia be related to hostile natural environments? Explain.

Key Terms

marsupial   Wallace Line
mallee   artesian well

ship with the United States by keeping nuclear-powered naval ships out of its waters.

The common perception of the South Pacific islands as a pleasant and untouched part of the world is spoiled by the dumping of oil and other materials from ocean transports. Several of the islands are uninhabitable after being mined for copper, nickel, or phosphate, or after nuclear testing. Nauru was left as an uncultivable skeleton following the extraction of its phosphate deposits.

Globalization

The countries of Southeast Asia, Australia, New Zealand, and the Pacific islands spread across the world range of GNP (Figure 6.14). The countries at the top of the diagram in the figure have the highest incomes—reflected in the ownership of consumer goods (Figure 6.15)—and are most involved in the global economic system. Ever since the European colonial thrust to "civilize" the rest of the world, trade within and outside has connected this region to the global economy.

Southeast Asia experienced a different basis for globalization than Australia and New Zealand or the Pacific islands. In Southeast Asia, the large local population experienced globalization from the 1600s as European merchants and colonists established plantations and mines based on local labor and labor brought in from China and India. Australia and New Zealand had fewer pre-European inhabitants and were settled by Europeans who formed the work force and maintained economic and cultural contacts with Europe. The Pacific islands were also colonized from Europe, but many retained contacts with Asian neighbors and many relied on aid from their colonizers after independence.

The economic growth of the Southeast Asian countries in the late 1900s was at first based on exporting products to American and European markets, but by the 1990s, it was also bolstered by intraregional trade. After 1970, much of the Southeast Asian, Australian, and New Zealand trade was oriented toward East Asia and the United States. The economies of Malaysia and Singapore were particularly linked to the United States and Japanese markets. In both markets, their exports as a percentage of their GDP reached almost 40 percent in 1999, over half of which were in information technology products. In Australia and New Zealand, globalization resulted in less emphasis on a closed self-sufficiency with limited external market dependency and a more open attitude toward neighboring people, markets, and trade. Although many Pacific islands now have better connections to the rest of the world, few are so closely involved in the global economy.

As part of the growth of global and intraregional trade, this region has global city-regions that are centers of international business, bases of multinational corporations' regional offices, and headquarters of some locally based multinational corporations (see Figures 2.16 and 2.17). Among global city-regions, Singapore (see "Global Focus: Singapore," p. 240) is in the second rank (on a par in this sense with Chicago and Los Angeles in the United States). Sydney (Australia) is in the next group, while Jakarta (Indonesia), Melbourne (Australia), Bangkok (Thailand), Kuala Lumpur (Malaysia),

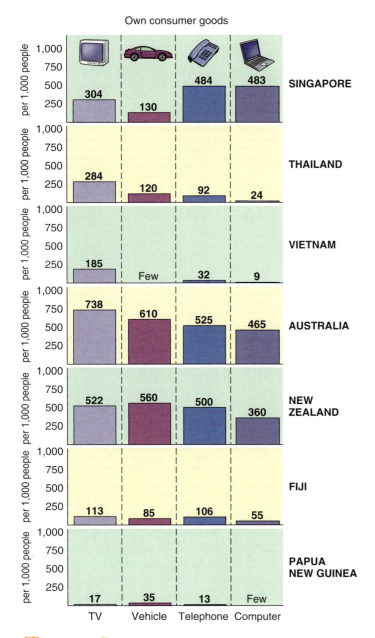

Own consumer goods

SINGAPORE: TV 304, Vehicle 130, Telephone 484, Computer 483

THAILAND: TV 284, Vehicle 120, Telephone 92, Computer 24

VIETNAM: TV 185, Vehicle Few, Telephone 32, Computer 9

AUSTRALIA: TV 738, Vehicle 610, Telephone 525, Computer 465

NEW ZEALAND: TV 522, Vehicle 560, Telephone 500, Computer 360

FIJI: TV 113, Vehicle 85, Telephone 106, Computer 55

PAPUA NEW GUINEA: TV 17, Vehicle 35, Telephone 13, Computer Few

Figure **6.15** **Southeast Asia and South Pacific: ownership of consumer goods.** Note the wide range of affluence among those countries. Source: Data (for 2000) *World Development Indicators,* World Bank, 2002.

Manila (Philippines), and Auckland (New Zealand) also have features of global city-regions. Thus, each of the countries experiencing economic growth has major city-regions linking it to world trade.

Reflecting trade and financial links, the ASEAN group added China, Japan, and South Korea to form a wider Asian trading group with common goals (ASEAN+3). Trade issues are moving from multilateral agreements toward bilateral ones set up by Japan and South Korea in particular with Southeast Asian countries.

As a response to the late 1990s Asian economic crisis, the Asian countries agreed on steps to avoid future crises. Other proposals include an Asian Monetary Fund and a common cur-

rency unit (like the euro), both based on an Asia-wide group of countries that could act like the European Union (see Chapter 3) to resolve historical hatreds through economic links. As yet, internal Asian relationships remain difficult. Most ASEAN countries see fellow members as competitors rather than partners.

While many economic and political events bring the countries of this region closer to each other, others threaten to bring back older conflicts and initiate new ones through cultural clashes, often within country borders. Indonesia, the country with the most people, faces threats from disparate island communities and intensified Islamist pressures after the September 11, 2001, events. Other countries in the region fear that radical and terrorist movements will affect them.

Subregions

Within this world region, subregions are characterized by contrasting population distributions (Figure 6.16). In Southeast Asia, the river lowland-based concentrations of people on the mainland and in island concentrations on Java (Indonesia) and around Manila (Philippines) contrast with upland and less-developed areas. Australia and New Zealand contrast with Southeast Asia, having lower densities of people, in which the huge empty interior of Australia stands out. Only the largest Pacific islands show up on the map at this scale, but densities are often high on small land areas.

Three subregions arise from the interactions of natural environments and cultural and economic histories of the countries:

1. *Southeast Asia:* continental Vietnam, Laos, Cambodia, Thailand, and Myanmar, and island-based Malaysia, Indonesia, East Timor, and the Philippines
2. *Australia and New Zealand:* former British colonies settled mainly by Europeans
3. *South Pacific islands:* many islands spread over great ocean distances

Southeast Asia

Southeast Asia (Figure 6.17) includes part of the Asian continental mainland and a series of islands in the sea-dominated area between Asia and Australia. Although in a global position akin to the West Indies (see Chapter 10)—to the west of a major ocean, between continents to the north and south, and with many islands—the subregion developed very differently within distinctive geographic settings.

The subregion's insular and peninsular character gives it long coastlines. It has a history of increasing interactions among the islands and mainland areas to the north and west, expanding the trade links with South and East Asia. It was incorporated into the global economic system by European trade since the 1500s, colonization in the 1800s, and increasing world trade in the later 1900s. Thailand, Malaysia, Singapore,

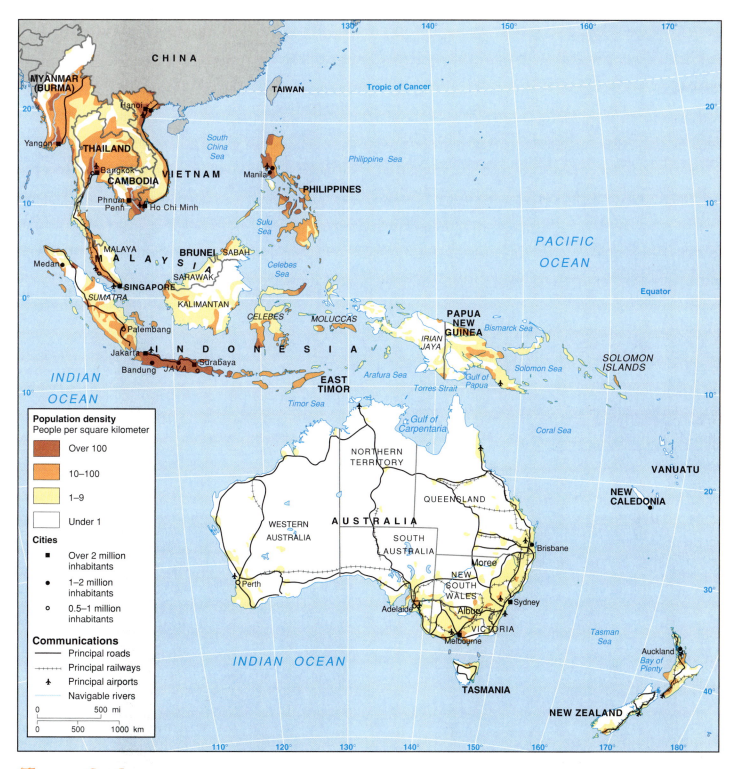

Figure 6.16 **Southeast Asia and South Pacific: distribution of population.** Identify the areas of highest and lowest density.

and Indonesia were among the emerging economies that elicited the term "East Asian Miracle" in 1993 (see Chapter 5), but both within and between them and other Southeast Asian countries, the unevenness of economic development added to diversity.

The subregion's rich cultural heritage left a wealth of historic buildings and traditional ways of life that formed the basis of growing tourist industries in the 1990s (see Figure 6.2). In the early 2000s, the nearness of people with contrasting cultural traditions to each other led to conflicts, as between the central Myanmar government and the Karen and Shan people, and between Muslims and Christians in Malaysia, Indonesia, and Mindanao, the southernmost island of the Philippines. In this subregion, the cultural fault lines are local features and divide many communities.

GLOBAL *Focus*

SINGAPORE

Singapore is an essentially global place with all the right connections, including port and airport facilities, and an enterprising and well-educated population (Box Figure 1). Over 4 million people live on a mere 620 km² (239 mi.²) of an island off the southern end of the Malayan mainland and at the eastern end of the Strait of Malacca that funnels much of the trade between South and East Asia. Despite being the smallest country in Southeast Asia and without natural resources, it is by far the richest in terms of GDP per capita. This is reflected in the comparisons of consumer goods ownership in Southeast Asia (see Figure 6.15).

Like Hong Kong farther north, with which it competes to be the world's busiest port (each has more than double the traffic of any other world port), Singapore has long been an entrepôt, collecting goods from countries in its region and exchanging them on global markets. It differs from Hong Kong in dealing with cargoes to and from many countries and in transferring goods to other ships in its harbor. Hong Kong's trade is nearly all to and from interior China. Singapore continues to refine oil and reship the products, to collect rubber and other agricultural products from Malaysia and Indonesia and tranship them, and to import and distribute autos and machinery.

The first settlement was called Temasek ("people of the sea"), but a visiting prince renamed it Singa-Pura ("city of the lion"). Several settlements were built on the island and later destroyed by marauding groups, but in 1819, Sir Thomas Stamford Raffles established the local headquarters of the British East India Company. In 1824, he purchased the island of Singapore and nearby islands from the sultan of Johore. The appointment of Prince Hussein as ruler gave royal authority to improving the port. Chinese immigrants soon dominated the population.

With Penang and Malacca, Singapore formed the British Straits Settlements colony that joined the Malayan Union in 1946. After Japanese occupation during World War II, Singapore and Malaya were subject to Communist terrorism for some years. Disagreements between Malay and Chinese people held up independence from the United Kingdom until 1963, when the Malayasian Federation was formed. Following major disagreements between Singapore's Chinese majority (77 percent, compared to 15 percent Malays and 6 percent from South Asia) and Malay people, Prime Minister Lee took Singapore out of the federation and declared it to be an independent country, the Republic of Singapore. Some versions of these events state that Singapore was expelled for its ethnic preferences, but Malaysia has different views.

Today, Singapore is a multicultural and multinational global city-state with international business services to rival Hong Kong and major cities in Europe and the United States such as Frankfurt (Germany) and Chicago. Religious groups including Buddhists, Muslims, Hindus, and Christians, with smaller groups of Jews and Sikhs, add to the global culture. The official languages are Malay, English (common language), Chinese (Mandarin), and Tamil (from southern India and Sri Lanka). Various Chinese dialects and other Indian languages (Punjabi, Hindi, Bengali, Tegelu, and Malayalan) are also spoken in specific communities. Occasionally, tensions in race relations come to a head. In early 2002, the government was accused of ignoring the Malay Muslim minority group that makes up 14 percent of the population by cooperating with the United States in the Afghanistan war and arresting 13 Malay Muslims on suspicion of

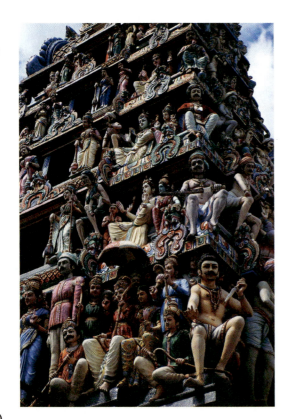

(a)

(b)

Box Figure 1 **Singapore.** (a) Sri Mariamman Hindu Temple with figures of gods on an external wall. (b) Waterfront at night.

planning a bombing campaign. The Malay group is less schooled and earns less than the majority Chinese or many Indians, another minority group. Yet the Singapore government makes special provisions for them, such as Muslim schools.

Singapore diversified into a high-technology manufacturing country within global markets. The strong control exercised by the Singapore government redirected its economy from port-centered activities toward high-tech industry, including information technology and automation. In the 1990s, Singapore firms manufactured 40 percent of world computer disk drives and included some of the leading makers of computer sound cards. As a transportation and communications hub, Singapore became the regional headquarters of many multinational corporations and one of the top 10 global city-regions. By late 2001, however, Singapore's employment was affected by cutbacks in U.S. and Japanese technology and financial services companies, and multinational corporations. It is vulnerable to recession in the global economic system.

Having a largely Chinese population and many links to other Chinese people throughout East Asia, Singapore offered relocation facilities to Hong Kong businesses that feared the Chinese takeover in 1997. A few took up the offer. In the late 1990s, falling economic growth rates in the rest of East Asia and involvement in projects such as building in Shanghai slowed growth in Singapore. Its major thrust in the new century is as a regional business center, promoting investments. Other centers in Indonesia, China (Shanghai), and Vietnam are partially exports of the Singapore business environment.

The global connections, however, have to take account of local pressures. Singapore's existence and prosperity is still resented by Malaysia, which takes opportunities to demonstrate its feelings. In 2001, the opening of a new port, PTP, in southern Malaysia opposite Singapore Island, immediately resulted in the loss of Singapore's major customer. In 2001, a week before the September 11 events, an agreement was reached on water supplies from Malaysia to Singapore, transportation links across the Johore Strait, and development of land on either side of the strait. Singapore depends on Malaysia for 40 percent of its water, but Malaysia has its own water shortage. It was agreed that Singapore should pay more than the 3 Malaysian cents (US $0.008) per thousand gallons of raw water that has been the price since 1961, but although Singapore offered 45 cents when agreements run out in 2011, Malaysia wants 60 cents starting in 2001. The causeway across the narrow Johore Strait also needs to be replaced and is scheduled to be demolished in 2007. The September agreement proposed a new suspension bridge, but Malaysia now wishes to build a high-arch bridge to replace its half of the causeway to allow barges and small vessels to pass through the strait and a swing-bridge rail connection instead of a tunnel. These proposals link to the growth of PTP. There are also rumors that Singapore may reclaim land on its side of the strait to narrow the shipping channel. The tougher Malaysian stance since September 11 may be an outcome of the reduced pressures from its discredited Islamic opposition group to get an agreement with Singapore.

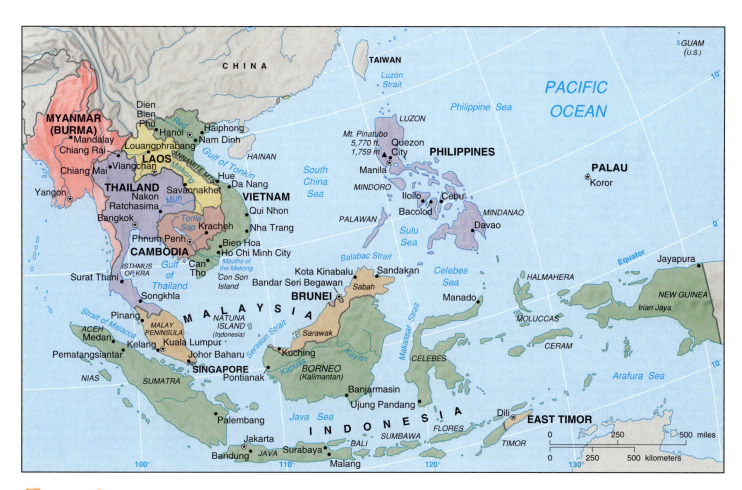

Figure 6.17 **Southeastern Asia: countries and major cities.** How important are ocean links in this region? In 1999, East Timor gained independence from Indonesia after 25 years of military occupation. It had been a Portuguese colony. The weakened Indonesian government is also faced with civil disturbances in other islands.

Along with friction among communities on the land are increasing numbers of pirate attacks at sea. Although incidents extend west to Bangladesh and India, the favorite pirate zone is in Indonesian waters, with over 100 recorded attacks in 2000. In particular, the Malacca Strait between Sumatra, Malaysia, and Singapore is used by around 200 ships each day and accounts for 40 percent of reported attacks. Ships slow down to pass through this busy and narrow waterway, making them vulnerable to attack. Although only one pirate attack occurred in the Malacca Strait in 1998, 75 incidents occurred in 2000 after international controls slipped, linked to Indonesia's deteriorating political situation. New international piracy laws are being prepared, but their implementation requires more policing.

Countries

The countries of Southeast Asia are Brunei, Cambodia, East Timor, Indonesia, Laos, Malaysia, Myanmar, the Philippines, Singapore, Thailand, and Vietnam. In 2001, the populations of Southeast Asian countries ranged from fewer than 1 million people in Brunei and East Timor to fewer than 6 million in Laos and Singapore, 13 million in Cambodia, 23 million in Malaysia, 50 million to 80 million people in each of the larger countries of Myanmar, the Philippines, Thailand, and Vietnam, and almost 210 million people in Indonesia (Figure 6.18). Indonesia is by far the largest in overall area, covering three times the area of the next largest, Myanmar, and having an east-west extent equal to the distance across the United States.

Continental Countries

The continent-based countries of Myanmar (Burma), Thailand, Cambodia, Laos, and Vietnam each center on the ancient kingdoms established in the region (see Figure 5.5). Although those

who established the kingdoms came mainly from the north, later developments owed much to entries from the ocean. Long-term trade with Hindus and Muslims from India and farther west brought religious symbols and buildings to coastal areas. In the late 1800s, the colonization of Burma by the British and of Indochina (Cambodia, Laos, Vietnam) by the French (see Figure 6.6) introduced further external demands.

After being occupied by Japanese military in World War II, Burma (1948) and later Indochina (1954) became independent. Burma was mainly governed by military rulers who changed the country's name to Myanmar in 1989. Vietnam, Laos, and Cambodia divided up Indochina and became Communist-run countries from the 1950s. Northern Vietnam took over the southern half of the country in 1975, when the United States ended its support for South Vietnam after spending US $150 billion, employing nearly 600,000 troops, destroying 70 percent of the northern villages, and leaving much land barren.

The military and Communist governments kept Burma, Cambodia, Laos, and Vietnam aloof from increasing world trade, while internal conflicts slowed economic development. In Burma, the military governments fought persistent hill tribes wanting independence. In Cambodia, the Communist government murdered over a million people in the late 1970s under Pol Pot. Vietnam and Laos were involved in warfare for nearly 30 years, from the 1950s against France and the United States and then with each other to around 1980. Reconstruction since that time has been slow.

In the late 1970s, support from the United States enabled Thailand to resist the Communist advance. With its constitutional monarchy and democratic government, Thailand adopted an open attitude to the expanding world economy, encouraging foreign investment, export manufacturing, and a growing tourist industry. Thailand became one of the economic

Figure 6.18   **Southeast Asia: data for countries.** Note which characteristics are common in wealthier and poorer countries.

Country	Capital City	Land Area (km²) Total	Population (millions) Mid-2001 Total	2025 Est.	GNI 1999 (US $ million) Total	GNI PPP 1999 Per Capita	Percent Urban 2001	Human Development Index Rank of 175 Countries	Human Poverty Index: Percent of Total Population
Cambodia, Kingdom of	Phnum Penh	181,040	13.1	16.4	3,023	1,350	16	137	No data
Laos, People's Democratic Republic of	Viangchan	236,800	5.4	9.0	1,476	1,430	17	140	38.9
Myanmar, Union of (Burma)	Yangon (Rangoon)	676,580	47.8	60.2	No data	No data	27	128	32.3
Thailand, Kingdom of	Bangkok	513,120	62.4	72.1	121,051	5,950	30	67	18.7
Vietnam, Socialist Republic of	Hanoi	331,690	78.7	104.1	28,733	1,860	24	110	28.7
East Timor	Dili	No data	0.8	1.2	No data	No data	8	No data	No data
Indonesia, Republic of	Jakarta	1,904,570	206.1	272.0	125,043	2,660	39	105	27.7
Malaysia, Federation of	Kuala Lumpur	329,750	22.7	33.7	76,944	7,640	57	56	14.2
Philippines, Republic of	Manila	300,000	73.2	107.8	77,967	3,990	47	77	16.3
Singapore, Republic of	Singapore City	620	4.1	8.0	95,429	22,640	100	22	6.6

Source: Data from *Population Reference Bureau 2001 Data Sheet; World Development Indicators,* World Bank, 2001; *Human Development Report,* United Nations, 2001; Microsoft Encarta (ethnic group, language, religion).

leaders in Southeast Asia until the 1997 economic crash, when it was one of the first countries affected.

Just as the 1997 crisis broke, Thailand adopted a new, far-reaching constitution focused on people's rights and challenging corruption and entrenched power at local and central government levels. However, the new measures are contested by rival groups. Thailand, like other countries in Southeast Asia, faces the need to make decisions about its future openness or closure to internal and external pressures.

Island Countries

The island countries of Southeast Asia (see Figure 6.17) are Malaysia, Singapore, Brunei, Indonesia, East Timor, and the Philippines. Malaysia's long Malayan peninsula joins it to Thailand across the narrow Isthmus of Kra, but the country has an island orientation. Since becoming independent, most of the island countries of Southeast Asia have had histories of strong leadership, often alternating with periods of instability.

Indonesia is the most populous country, but it is geographically divided into over 13,000 islands and 300 ethnic groups based on language. Its national motto, "diversity in unity," is appropriate. In the late 1990s, it descended into violence and uncertainty after long periods under the virtual dictatorships of Presidents Sukarno (1949–1966), who emphasized political control of the various islands in the federation but achieved little economic growth, and Suharto (1966–1997), who switched his government's emphasis to economic development. Both periods ended with economic crises and popular uprisings against corruption.

In the Suharto years, annual per capita income rose from US $70 in 1965 to $1,142 in 1996, more social services were made available than in many other Southeast Asian countries, and the number of people living below the poverty level decreased. A rigid power pyramid brought some stability to the diverse country, although it continued inequalities and allowed corruption. At the top were the president, his family, and closely related supporters; below them were the military, top government officials, and business leaders; there was a growing middle class; then there were urban workers, wealthier farmers, and village chiefs; finally came the rest of the people. However, in the mid-1990s, the dynamics changed, and those at the top took more out of the economy, and Suharto's family indulged itself excessively. The middle class who had supported Suharto found upward social mobility capped by corrupt controllers. Parallel to these social changes, foreign debts doubled between 1990 and 1996 as both large and small companies engaged foreign capital. Some large companies owned Indonesian banks and lent to themselves, but when the 1997 Asian economic crisis grew, they could not repay those loans and had to seek foreign financing at a time when worsening exchange rates multiplied the overseas debts. The politicized internal banking system collapsed. The crisis brought down Suharto.

The overturn of the Suharto government in Indonesia was accompanied by social problems, including aggression against the Chinese business community in Indonesian cities, the resettled people in Kalimantan, and the Christian Chinese in Molucca. Several years of hesitant government followed. After Megawati Sukarnoputri became president in 2001, economic conditions began to improve, and the International Monetary Fund was able to make a new agreement to reschedule the debts.

Problems for Indonesia continue with ethnic strife threatening to lead the country—and possibly others in the subregion—into political chaos. Militant and pirate groups flourish on its islands and in neighboring countries. Several Indonesian

Country	Ethnic Groups (percent)	Languages O=Official	Religions (percent)
Cambodia, Kingdom of	Khmer 90%, Vietnamese 5%	Khmer (O), French	Buddhist 95%
Laos, People's Democratic Republic of	Lao 50%, Thai 14%, many tribes	Lao (O), Thai, French, local	Buddhist 60%, local 40%
Myanmar, Union of (Burma)	Burman 68%, Shan, Karen, other tribes	Burmese, local languages	Buddhist 89%, Christian 4%, Muslim 4%
Thailand, Kingdom of	Thai 75%, Chinese 14%, Malays 3%	Thai (O), English, Chinese, Malay	Buddhist 95%, Muslim 4%
Vietnam, Socialist Republic of	Vietnamese 88%, Chinese 2%, tribal groups	Vietnamese (O), French, English	Buddhist 55%, Christian 12%
East Timor	No data	No data	No data
Indonesia, Republic of	Javanese 45%, Sundanese 15%, Mandurese, Malays	Bahasa (Malay), (O), English, Dutch	Christian 9%, Muslim 87%, Buddhist/Hindu
Malaysia, Federation of	Malays 59%, Chinese 32%, Indian 9%	Malay (O), Chinese, English, Tamil	Buddhist, Christian, Hindu, Muslim (main)
Philippines, Republic of	Malays 95%, Chinese 2%	Filipino (O), English (O), local	Christian 92%, Muslim 5%
Singapore, Republic of	Chinese 76%, Malays 15%, Indian 6%	Chinese (O), Malay (O), English (O)	Buddhist, Christian, Muslim

areas, such as Aceh in western Sumatra and the Moluccas and Irian Jaya in eastern Indonesia, seek independence, threatening to break up the country. In 1999, in the northern Moluccas, a dispute over a gold mine turned into fighting between Christians and Muslims in which many died before the Indonesian president imposed a state of civil emergency. Thousands of Muslims and Christians now camp out, too scared to return to homes liable to attack and burning while the military stand aside or are accused of siding with the Muslims. By 2002, groups of militant Islamists from western Indonesia went to support Muslims in the eastern islands. Even on Irian Jaya, where Christian Papuans form a large majority, actions by groups of Muslims increased after the September 11, 2001, attacks in the United States. Laskar Jihad and Laskar Kristus are two paramilitary groups that take sides in these disputes, with the latter linked to the Moluccan Sovereignty Front independence group.

In 1999, with United Nations help, the 800,000 people of *East Timor* began to work toward independence from Indonesia that was achieved in 2002 after 24 years of military occupation. When Portugal left its colony in the eastern part of Timor island in 1975, it was taken over by Indonesian military, who resisted allowing independence until the UN action. In 2001, the East Timor and Australian governments agreed to share the offshore oil and natural gas resources in the Timor Gap between the two countries. These resources should provide East Timor with plentiful funds for its development and defense.

In the *Philippines,* the colonial influences of Spain (1565–1898) and the United States (1898–1946) make it the most westernized country in the subregion. The long-term dictatorship of President Ferdinand Marcos ended in 1986, and since then, a fragile democracy that involves military influences has struggled to maintain order. The economy advanced more slowly than others in the subregion. At present, government forces, assisted by U.S. advisers, seek to restrict an armed Muslim group, the Abu Sayaff, to Basilan Island off southern Mindanao.

Malaysia has a small population relative to its rich natural resources (see "Personal View: Malaysia," p. 245). It has a long-term, strongly led government that seeks democratic mandates under the supreme rule of the sultan, although opposition parties are tightly controlled. The government established policies to promote a competitive economy, enabling the country to experience the most rapid economic growth in the subregion after Singapore.

Singapore, small and urbanized, is the materially wealthiest of the Southeast Asian countries. Although officially a multiparty democracy, Singapore's government is very much single-party in outcomes (see "Global Focus; Singapore," p. 240).

Brunei is a small country on the northern coast of the island of Borneo. Its oil and natural gas riches and population of only 330,000 give it a GDP per capita of around $10,000. Although its geographic position appears to be off the main global routeways, it is central to growing Southeast and East Asia. A British protectorate until independence in 1984, Brunei refused to join the Malaysian Federation that surrounds it. However, the United Kingdom and Singapore made defense agreements with Brunei, which joined ASEAN immediately after independence.

The rule by a Muslim sultanate began in the A.D. 1300s. In 1990, Brunei introduced the concept of a Malay Muslim monarchy to promote Islamic values. Although two-thirds of the population are Muslims, the one-fifth who are Chinese felt alienated by this movement, as did the indigenous groups that inhabit the tropical rain forest interior and expatriates who have chosen to live in Borneo. The 1990 move followed the crushing of prodemocracy movements from 1962 and a state of emergency, after which the sultan ruled by decree.

The oil and natural gas wealth that produces Brunei's huge income was developed largely by the Shell Corporation since 1929 and was a target of Japanese occupiers in World War II. The sultan and his family are the main beneficiaries. Sultan Hassanal Bolkiah is one of the world's richest men and built himself a palace costing $450 million in the capital, Bandar Seri Begawan. Inhabitants of the country do not pay any taxes, government jobs are plentiful, and the provision of health care and education is free and generous. A taxi driver serving a recent visitor had four children at university in Britain, two supported by the Brunei government and two sponsored by Shell.

After September 11, 2001, many Muslims in the subregion were arrested on suspicion of terrorist activities such as the planned (but foiled) attacks on U.S. military and commercial interests in Singapore. Jemaah Islamiah is a regional terrorist organization that seeks to establish an Islamist country, Dauliah Islamiah, to include Indonesia, Malaysia, and the Philippines. Ethnic conflicts in Indonesia following the 1997 Asian financial and social crisis radicalized some Muslims, and links were reported to al-Qaeda training camps on Celebes Island. The Indonesian government does little to control such developments because it fears a backlash from its Muslim majority if it acts strongly against suspected terrorists or Islamist organizations. Singapore, Malaysia, and the Philippines take stronger actions, although a lack of information undermines the confirmation of many accusations.

People

Ethnic Variety

Following centuries of migrations over land and sea, many of which were instigated by Chinese expansion and the European colonial powers, the people of Southeast Asia comprise very diverse groups, including the Burmese, Thais, Cambodians, Vietnamese, Malays, Indonesians, and Filipinos. Chinese communities are significant in all countries, making up 7 percent of the total population. Overseas Indians form important groups in Myanmar, Malaysia, and parts of Indonesia. Groups of indigenous people survive in the hilly and mountainous regions of the mainland, particularly in the uplands of Myanmar (Shan and Karen) and on the islands of Indonesia such as Kalimantan (Indonesian Borneo).

Religious diversity is marked. Animistic religions are significant among indigenous tribes. Islam dominates Indonesia, Malaysia, and Brunei. Buddhism is the chief religion of the mainland countries. The Philippines are mainly Roman Catholic.

MALAYSIA

Malaysia became noticeably more prosperous during and since the mid-1980s. More people moved into the towns, especially the capital, Kuala Lumpur, because of the jobs in government offices, businesses, and banks, the greater range of shopping available in supermarkets and chain stores, and the increasing leisure opportunities from golf courses and swimming pools to badminton courts—the favorite sport. In the 1990s, Kuala Lumpur experienced a huge expansion of high-rise apartment and office construction (Box Figure 1), including investments by corporations moving out of Hong Kong. Housing is more expensive in Kuala Lumpur than in other parts of Malaysia, so many people live in suburbs outside the city and commute by motorcycle, car, bus, train, and the light rail system. In 1995, Malaysia announced plans to build a new administrative capital called Putrajaya 40 km (25 mi.) south of Kuala Lumpur by 2008.

Most Malaysian houses have stone floors to help cool the equatorial air. Urban homes have refrigerators, TVs, and video players. TV programs include many from Australia and the United States, with a few British programs and local programs for Malay, Chinese, or Indian groups. There are radio stations for each of the main languages—Malay, Chinese, Tamil, and English. Most cooking is by gas, bought in cylinders delivered to the house. Water is metered, and most houses have showers rather than bathtubs. Life in the west coast cities still contrasts with life in rural villages and on the east coast, where people live in houses of wood with coconut-leaf or tin roofs instead of brick and concrete.

Family ties remain strong, although family size decreased from about three or four children before 1980 to around two in the 1990s. Parents usually live with the oldest child, who looks after them when they retire. Government policy maintains a high birth rate in order to generate a large domestic market for manufactured goods—hoping to raise the present 20 million population to 70 million by 2100. Large families are encouraged in an attempt to slow the fertility decline linked to increasing affluence and use of contraceptives.

Schools are mostly subsidized by the government, although some are still run by churches; the Chinese and Indian communities run independent schools. Even at the subsidized schools, parents pay fees and buy uniforms and books. Poor families often cannot afford to send their children to school. For most children, schooling lasts from 7 to 17 years of age, with the possibility of another two years for university entrance exams. The school day lasts from 8 A.M. to 12:30 or 1 P.M., with a break during the hot part of the day. Any games take place after 4 P.M.

Most Malaysians use Western medicine, although some use herbal remedies. Doctor visits and hospital treatment have to be paid for. Poor people get free medical care but may have to pay for prescriptions.

Malaysia has a democratic government and is a member of the Commonwealth (of former British colonies). Public support for Malaysia's prime minister for 20 years, Mahathir Mohamad, seemed to wane after the Asian financial crisis of 1997, but he blamed his problems on the opposition Islamist PAS party based mainly in the rural northeast of Malaya. After September 11, 2001, Mahathir denounced extremism and condemned the bombing of Afghanistan by the United States. The opposition PAS appeared to the public to be like the Taliban in Afghanistan, and Dr. Mahathir's opposition melted away.

Malaysia is a multiethnic society, with many people of mixed parentage, such as an Indian father and Thai mother. Malays, who are mainly Muslims, retain control of parliament and often regard themselves as first-class citizens and others as second-class citizens. The Chinese came to work in the tin mines during British rule, when Indians were also brought in for the rubber plantations. Chinese are now mainly middle class with commercial interests, while Hindu Indians include poor laborers, owners of small businesses, and professionals such as lawyers or doctors. Fearing that the Chinese and Indian groups would become too powerful, the Malaysian government passed laws that forced Chinese businesspeople to have Malay partners holding 50 percent control and enforced strict racial quotas in higher education, government jobs, and business ownership. Relationships among the different groups are not so tense as they were following government actions after race riots in the late 1960s and are helped by the increasing well-being of Malaysians.

After 1986, economic growth exposed labor shortages. More women became involved in the labor force for high-tech manufacturing, and their wages contribute to rising family incomes. The Malaysian labor force grew faster than the total population as foreign workers flowed in to take increasing numbers of jobs. Immigrant workers from Indonesia, Bangladesh, and the Philippines extended the multiethnic nature of Malaysian society.

The orientation of Malaysian society changed with its economic focus. British influences remained strong until 1970, but since then, Japanese and U.S. firms established businesses in Malaysia that are seen as forming the foundation of recent economic growth. In addition to the manufactured products of Japanese and other international firms, palm oil, rubber, tin, and iron ore remain important exports, and oil comes from offshore wells in the South China Sea.

In 1997, Malaysians became aware of environmental quality loss as forest fires in Indonesian and Malaysian areas being cleared for oil palm plantings created intense smog over the country. A public health crisis followed in which schools closed. This event, repeated in early 1998, focused attention on lax environmental policies that encouraged timber cutting and mining at the expense of silting estuaries, disappearing coral reefs, and depleted fisheries.

Box Figure 1 **Kuala Lumpur, Malaysia.** The city center, with the historic Sultan Abdul Samad building and gardens, is backed by modern high-rise office blocks. Photo: © David Ball/The Stock Market/Corbis.

Hinduism is locally important, as in southern Myanmar (Bengalis), Malaysia (Tamils), and a remnant of the religion exists in Bali and Lombok (Indonesia). Even where a country's population is mainly of one religion, there are groups with other allegiances. For example, many of the Chinese businesspeople in Indonesia, Malaysia, and Singapore are Christians.

Such diversity often leads to tensions and conflict. Malaysia's 1969 race riots led to new policies to correct the economic "imbalance" in which Chinese owned nearly all the businesses; by 1990, non-Chinese people owned 30 percent of Malaysian shares. However, the policy was relaxed when Chinese investments that could have been made in Malaysia began to be placed in other countries. Such policies to restrict Chinese economic influence, however, tended to make racial differences into rigid and institutionalized social barriers.

In Myanmar, the east remains a battlefield between the hill country people (Shan, Karen, and others) and the central government that tries to crush them, while the southern center around the capital, Yangon, is less troubled (Figure 6.19).

In Thailand, ethnic characteristics give identity to different regions, exemplifying the diversity of Southeast Asia:

- The people of the mountainous northern and western areas are mainly Buddhists with links to those in Myanmar. Minority upland groups, such as the Karen (see Figure 6.3a), migrated from Myanmar, Laos, and China over the centuries.

- In the northeast, people share languages, artistic traditions, and Buddhism with the northerners and Lao people to the east. A time of relative isolation and autonomy until the late 1700s shaped the local identity.

Figure *6.19*   **Southeast Asia: a Burmese street in Yangon.** Photo: © Ian Coles.

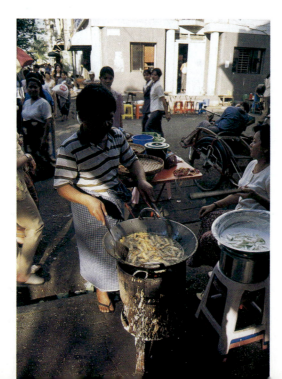

- In the central Chao Phraya River basin, the cultural heartland of Thailand, people speak standard Thai. Their Buddhism is more akin to that of the Khmer people in Cambodia—the more conservative form that is similar to that in Sri Lanka (see Chapter 7). A small proportion of Hindu Brahmins living in the central region act as directors of royal and official ceremonies and use their astrological expertise to prepare the annual calendar. The rapid economic growth of this region adds to the mix by drawing in people from the poorer northeast.

- Along the hilly southern coastal area and extending eastward along the Cambodian border, people of Chinese descent settled in the 1800s to work on sugarcane plantations and now work in timber mills and run small stores.

- The southwest Pak Tai region has people linked closely to those in central Thailand, but Malay-speaking Muslims live in the extreme south.

The people of the Indochinese countries suffered greatly from many conflicts in the latter half of the 1900s. When the Communist Khmer Rouge ruled Cambodia from 1975 to 1979, an estimated 1.7 million people died from starvation, disease, or execution. Many of the killers continued to live prominently in the country without being charged with ethnic crimes until King Sihanouk signed a law in 2001 to create a tribunal to put the remaining Khmer Rouge leaders on trial. The genocidal eradication of hill tribes in Cambodia and Vietnam was a further example of ethnic hatreds being linked to political crusades. The Vietnamese hatred of the Chinese is exposed in occasional border skirmishes, in the expulsion of many Chinese businesspeople from southern Vietnam since 1975, and in competing claims for the Spratly Islands. Central Vietnam has a number of disadvantaged ethnic minority groups, many of whom became Protestant Christians before 1975. Some fled to Cambodia, but although a 2001 United Nations repatriation program expected to send them back to Vietnam, the program was ended in 2002 because the UN was not convinced it would be in the people's interests.

Indonesia's many islands are home to varied ethnic mixes. In western Indonesia, strongly Hindu wet-rice growers on Java and Bali have a sophisticated culture of social and agricultural traditions that evolved over centuries. Islamic coastal groups include Malays on Sumatra and the Makasarese on southern Celebes. Kalimantan has many indigenous tribal groups, dominated by the Dayaks, often still bound to a shifting-cultivation economy.

In eastern Indonesia, the Moluccas have a traditional Melanesian division between the "beach" (coastal) and "bush" (inland) people, although further complexity is added by western island features, as in Ambon and Ceram. Irian Jaya's Papuan groups in the far east of Indonesia also have a coastal-inland differentiation with ethnically mixed coastal people and isolated inland groups that often avoid contact with the outside world.

Population Growth Slows

In 1965, the combined population of the countries of Southeast Asia was 275 million. It doubled to nearly 520 million in 2001

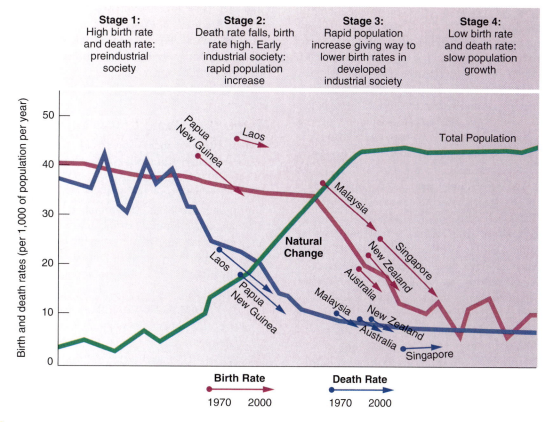

Stage 1: High birth rate and death rate: preindustrial society

Stage 2: Death rate falls, birth rate high. Early industrial society: rapid population increase

Stage 3: Rapid population increase giving way to lower birth rates in developed industrial society

Stage 4: Low birth rate and death rate: slow population growth

Figure 6.20 **Southeast Asia and South Pacific: demographic transition.** Compare the Southeast Asia range with that of the South Pacific countries. Papua New Guinea is typical of many of the islands. The Australian and New Zealand patterns resemble those of other rich countries.

and is expected to be close to 700 million by 2025. From 1965 to 1980, most countries had rapid population growth rates averaging over 2 percent per year, with Thailand and the Philippines close to 3 percent. The exceptions were Cambodia and Laos, with growth rates that reflected the ravages of war at around 1 percent or less. By 2001, annual natural population increase rates in most countries were falling. Thailand, Indonesia, Singapore, and Vietnam were down to around 1 percent, while Laos, Malaysia, and the Philippines remained just above 2 percent.

The slowing population growth rates reflected falls in total fertility rates. All countries were over 6 in 1965, but by 2001, only Laos was over 5 and the other countries were 4 or less. Thailand's family planning policies reduced fertility in that country to under 2. The declines in birth rates led to progress in demographic transition (Figure 6.20). The age-sex diagram for Indonesia (Figure 6.21) demonstrates the impacts of reduced numbers of births. The Vietnamese population profile was affected by wartime losses (those over 10 years old in 1975, over 25 in 2002), but the uncertain economic future after the end of hostilities in 1975 caused population numbers to grow slowly.

Until the 1960s, *Thailand* encouraged births, and its population increased at more than 3 percent per year. A World Bank mission in 1958 recommended a program of birth control, and this became national policy in 1970. The Community Based Family Planning Service, begun in 1974, is a private, nonprofit organization that covers both urban and rural areas. Its program tried to overcome both the ignorance about the

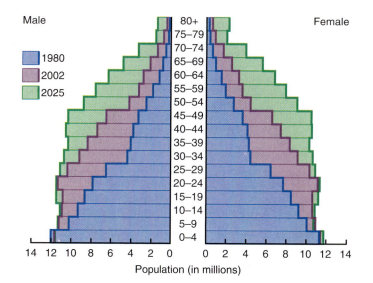

Figure 6.21 **Indonesia: age-sex diagram.** What factors cause Indonesia to have a different pattern than Australia? See Figure 6.30. Source: U.S. Census Bureau. International Data Bank.

mechanisms and value of birth control as well as the cultural constraints that favored large families. A traveling road show was supported by Buddhist monks who used religious texts and blessed contraceptive pills. Bank loans were an incentive for farmers to take part.

Thailand's population natural growth rate fell according to the targets set, from 3 percent in 1972 to 2.5 percent in 1976, 2.1 percent in 1981, 1.5 percent in 1986, and 0.8 percent in 2001.

Indonesia also has family planning policies to reduce family size but placed huge resources into redistributing its population. Sixty percent of Indonesia's population of 200 million in 2001 lived on 7 percent of the land in the four largely metropolitan islands of "Inner Indonesia" (Java, Madura, Bali, and Lombok). Rural population densities on these four islands are often above 1,000 per km^2 (approximately 2,600 per mi.^{2}), whereas they are mainly under 1 per km^2 (approximately 2.6 per mi.^{2}) on the "outer" islands such as Kalimantan (the Indonesian part of Borneo), Sumatra, and the Moluccas—although none of these is "empty."

The **transmigration** program began in 1950 and resettled more than 6.5 million people on the less-populated islands. Practical problems at the outset included the selection processes that placed people of the wrong age groups and skills in inappropriate situations, poor site choices for resettlement, encouragement of unsuitable farming systems, and poor administration. These problems and their local impacts were largely overcome in succeeding years, and some government-sponsored and World Bank-funded projects for volunteers' families were successful. Displaced tribal peoples were given sympathetic treatment. Levels of education and health care provision were much better than in migrants' home areas, and the migrants now owned land and their homes.

Despite this policy's democratic rationale, it was regarded by many as an insidious form of political manipulation and social engineering. People introduced to outlying islands were seen as government supporters and as importers of economic and social norms from Java. Nearly half of the migrants failed to raise their standard of living.

From 1997, there were outbreaks of tribal groups murdering settlers as the settlers cut and burned forest habitats to extend palm oil plantings. Areas of rain forest were felled by illegal logging or mining, causing further social upheavals. Many migrants tried to move back to their home areas, but they no longer had home or land and formed groups of politically unsettled peoples. The Indonesian government ended the program in March 2001.

The Indonesian experience asks whether imbalances of population distribution can be redressed on a large scale by officially sponsored migration. Few good tracts of agricultural land with lower densities of population remain in the islands (or elsewhere in the world), so there is little chance of accommodating major population movements in this way. Indonesia faces the need either to reduce population growth or to increase economic productivity, or both.

Population change in Southeast Asia is increasingly affected by HIV/AIDS. In the 1990s, Thailand and Myanmar became the centers of the expansion of HIV/AIDS in Southeast Asia, despite government denials and optimistic statements about the disease being "self-contained" among drug users, prostitutes and their clients, and (though not mentioned) homosexuals. Asian women are less likely to have extramarital sex than women on other continents, but it is known that drug injectors have sex with prostitutes. There are large sex markets that may involve men who also have sex with their wives. So the disease is not contained, and 10 percent of adult Thais were infected before the government introduced a condom-use program that cut new infections. In 2000, nearly 800,000 Thais had HIV/AIDS and around 400,000 had died. In Myanmar, 500,000 were infected and nearly 200,000 had died.

Impacts of Rapid Urbanization

Urbanization is increasing in all the countries of Southeast Asia, although Cambodia, Laos, and Vietnam still had only around 20 percent of their populations living in urban areas in 2001. Thailand and Indonesia had over 30 percent, and the Philippines and Malaysia around 50 percent urban population. The smallest countries, Brunei and Singapore, had 67 and 100 percent, respectively, living in towns and cities. Most countries had a concentration of people, jobs, infrastructure provision, health care, and education opportunities in the largest, or **primate city.**

- In 2000, Bangkok (7 million), the capital of Thailand, had over 20 times the population of the next largest city, Chiangmai.

- Jakarta (Indonesia, 11 million people), Kuala Lumpur (Malaysia, 1.4 million), Manila (Philippines, 11 million), and Phnum Penh (Cambodia, 1 million) were also capital cities but not quite as dominant in their countries.

- Besides Jakarta, other large Indonesian cities included Surabaya (2.4 million) and Bandung (3.4 million) on Java, Medan (2 million) on Sumatra, and Ujung Padang (1 million) on Celebes.

- Yangon (4 million), the capital of Myanmar, and Mandalay form a dual focus in that country.

- In Vietnam (Figure 6.22), the capital, Hanoi (4 million) with its port, Haiphong (1.7 million), in the north and Ho Chi Minh City (formerly Saigon, nearly 5 million) in the south also form a dual focus.

Modern services, such as hospitals and schools, business services, and transportation, focus on urban centers. For example, individuals and firms in Bangkok, Thailand, have better access to such services than those in the poverty-stricken northeast of the country. Inside cities, governments subsidize urban transportation and provide tax incentives for industrialists to locate factories close to the main work force. Most countries encourage manufacturing to a greater extent than agriculture—a further factor in urban growth at the expense of rural areas. The negative side of growing cities—congestion and pollution—appears to have little effect on their expansion. Bangkok's traffic jams are known as the world's worst. Figure 6.23a shows the features of a typical Southeast Asian city, while Figure 6.23b shows a Southeast Asian city center with religious buildings and commercial traders.

Rapid city growth makes it difficult for governments to expand services and facilities as fast as urban population grows. Government actions are often ineffective in controlling city expansion. For example, the uncontrolled growth of Jakarta, Indonesia, in the 1970s caused the Indonesian authorities to issue housing permits only to those who could prove employment, but this policy only increased the growth of shantytowns.

Urbanization Under Communism

The Communist countries experimented with distinctive urbanization policies aimed at reversing the uncontrollable urban growth and resettling people in rural areas to produce more food. The cities of Cambodia, such as Phnum Penh, were systematically depopulated in the late 1970s under Pol Pot. In early 2002, large areas of squatter settlements along the river in Phnum Penh suffered fires that displaced nearly 20,000 people. Those people were moved to fields outside the city and housed in tents, while the river land was designated for new buildings.

At its end, the Vietnam War focused on the American defense of Saigon. During the war, the frightened and war-torn from surrounding rural areas moved into Saigon and other coastal cities such as Da Nang. Saigon grew from 2.4 million people in 1964 to 4.5 million in 1975.

When the Communist government took over Saigon after the 1975 reunification of North and South Vietnam, they renamed it Ho Chi Minh City and implemented a **deurbanization** policy to reduce the "unhealthy" overcrowding that led to insufficient water, housing, and power. Vietnamese families were directed back to their villages. Economic zones were set up in rural areas. However, ethnic Chinese businesspeople, who mainly lived in the city, were so persecuted that nearly 75,000 left the country between 1977 and 1982. Many fled by boat and were known as "boat people." Eventually, the Vietnamese government realized that the initial target for reducing Ho Chi Minh City's population was too ambitious, and they raised the allowed total to 1.75 million. In the late 1980s, the policy changed again, with a controlled but slower

Figure **6.22** **Southeast Asia: Vietnam.** Street scene in Hoi An. Photo: © Ian Coles.

urban growth alongside a welcome for foreign capital investments in new industries.

Traffic in People

The business of trapping people into laboring jobs or prostitution in other countries is a fast-growing scourge of this subregion, which supplies around half of the women and children trafficked every year across the world. Vietnamese are taken to Cambodia for the sex industry and to Taiwan and China for marriage; Cambodians shipped to Thailand, Taiwan, and Singapore become beggars, sex slaves, or domestic laborers.

People wishing to migrate to escape poverty or trying to get away from the traffickers may be caught by people smugglers who charge large sums (estimated to total over US $7 billion per year) to take them to a new country. Smuggled people often end up in the same occupations as those who are kidnapped. Such illegal smuggling and trafficking creates political tensions, as when Malaysia deports Indonesians or Australia intercepts boatloads of asylum seekers.

Figure **6.23** **Southeast Asian cities.** (a) Generalized plan of a city. Compare this with cities in other regions. (b) Palace and market in central Bangkok. Photo: © Ian Coles.

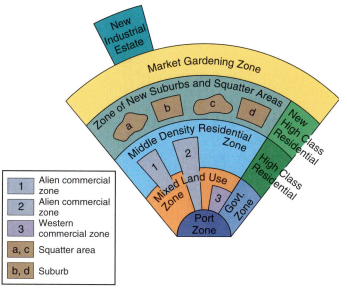

(a)

(b)

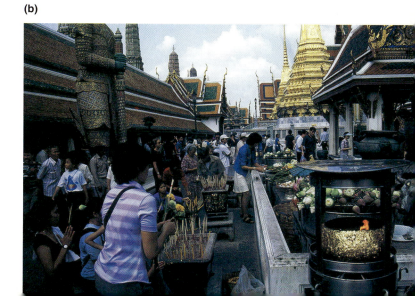

Some 10 percent of Filipinos, especially women, work abroad to support their families and educate their children, resulting in US $6 billion in payments in 2000. In the 1970s, the Philippines was second only to Japan in wealth in this part of the world and employed lots of Chinese refugees escaping through Hong Kong. Today, as a result of poor government, the Philippines is one of the subregion's poorer countries and the situation is reversed: Filipinos are common domestic help, or amahs, in Hong Kong, as well as in the United States, Japan, and Saudi Arabia. Although the women may have college degrees and speak fluent English, the jobs they do often mean virtual slavery.

Economic Development

The countries of Southeast Asia range from the oil-rich Brunei and the prosperous trading center of Singapore, through the economically growing Malaysia, Thailand, Indonesia, and the Philippines, to the military dictatorship of Myanmar and Communist regimes of Cambodia, Laos, and Vietnam with their widespread poverty.

Improvements in health, education, and income from 1960 to 2000 gave these countries increasing levels of gender equality, particularly in adult literacy. In the 1990s, Malaysia made the reduction of poverty a major policy objective, and by its own definition, poverty fell to 21 percent by 1985 and 14 percent by the late 1990s.

ASEAN, Growth, and the Late-1990s Crisis

The ASEAN group was initially a defensive alliance rather than a trading one, but the peace it established formed a basis for economic growth. From the late 1960s, the ASEAN countries experienced a pattern of rapid industrialization stimulated by foreign investment and internal government encouragement. From the late 1980s, the Communist countries and Myanmar changed from isolationist policies to ones fostering greater links to the global economy. They joined ASEAN in the late 1990s, after promoting greater economic freedom without modifying political control, following China rather than the former Soviet Union.

Rapid economic growth in non-Communist ASEAN countries came to a halt in the late 1990s, when overlending by banks in the subregion caused a major financial crisis. Thailand, followed by the Philippines, Malaysia, and Indonesia, had to devalue its currency in 1997. The problems were caused by high levels of foreign borrowing, government budget deficits, major bank loans to glutted property markets, and slower-than-expected economic growth. Huge Japanese investments in the late 1980s and growth in exports from Southeast Asian countries were linked to a high yen value against the U.S. dollar; by the late 1990s, however, the value of the yen fell, forcing the Japanese to slow investment and technical transfer to Southeast Asia. Southeast Asia's problem affected East Asia: too much economic growth, too soon, on too much unsecured credit.

Thailand suffered most. The financial crisis led to a change of government, which imposed austerity measures and distanced itself from the past with a new constitution. Indonesia and Malaysia, ruled by long-term authoritarian governments under President Suharto and Prime Minister Mahathir Mohamad,

respectively, found it more difficult to admit fundamental mistakes in financial regulation. In Indonesia, Suharto was forced to resign, leaving the country's economic and social life in chaos.

By the early 2000s, new challenges came from China's growth, together with slowdowns in the American and Japanese markets for Southeast Asian goods. China provided both increased competition and increased markets. The redirection of foreign direct investment to China in the 1990s slowed Southeast Asia's recovery from the 1997 crisis, while Chinese goods compete in textiles (Vietnam, Indonesia) and higher-value products (Thailand, Malaysia). However, Southeast Asian countries can meet these challenges by placing a greater emphasis on open government to attract investors, lowering trade barriers to open up markets, and upgrading domestic workers' skills.

Economic Changes in Thailand

Thailand's recent experiences are typical of other countries in the subregion. The 1997 financial crisis in Thailand left relicts in the many unfinished and empty skyscrapers in Bangkok. After years of thrift and investment in new manufacturing industries, the returns dwindled and restrictions on foreign investment loosened in the mid-1990s. Better potential returns from property investment made the prices of land and buildings rise as borrowing became more competitive and cheaper, expanding an asset "bubble." The bubble burst when a loss of confidence led to withdrawals of funds from banks, putting pressure on the baht (Thailand's currency), which was allowed to float and moved downward. The cost of foreign debt doubled and banks and businesses collapsed, leaving the building shells as worthless assets.

Five years later, in 2002, the debts still slowed the country's economy, leaving half the industrial capacity idle. The Thai Petrochemical Industries Corporation (TPI) had the largest debts of almost US $3 billion, had not paid interest since 1997, and was slow to impose restructuring and management changes. TPI and other Thai corporations claimed that they would be solvent if the value of the baht had not fallen in 1997, when the government removed the link to the U.S. dollar. In the wider Thai economy, tourism suffered from both the reductions of Asian visitors from 1997 and the events of September 11, 2001.

Despite these difficulties, Thailand's major export industries of oil products and vehicles grew; the country quadrupled exports from its refineries between 1996 and 2000, and welcomed fresh investments from BMW, GM, and Toyota. Worries over China's growth and resultant losses of export markets were calmed by knowledge that exports to China trebled since 1993 and numbers of Chinese tourists to Thailand quadrupled. Thailand possesses comparative advantages that include its membership in ASEAN with a market of over 500 million people, a strong local car parts industry, and tax incentives. Its higher labor costs than Vietnam or Indonesia reduced textile exports, but the food processing industry, including tuna canning, is growing for the local middle-class market.

Thailand's other growing industries are the illegal sex and drugs trades that thrive through boom and bust but generate a frightening public health crisis as HIV/AIDS and drug-related problems spread. *Ya baa,* the methamphetamine "crazy pill," was first produced alongside heroin in the early 1990s, when

Thailand was merely a staging post for movements of heroin. Thais prefer *ya baa,* and countries such as the United States are less concerned about it than about opium. Popularized first by manual workers who take it to remain alert on long shifts, it is now used widely as a recreational drug in colleges and schools. Children as young as 6 become addicts, and hospitals treat more *ya baa* addicts than alcoholics. The worst symptoms after years of addiction are paranoia and hallucinations, producing uncontrolled violence. Cures take several months with no guarantee a relapse will not occur. The number of drug offenders in Thai prisons rose from just over 10,000 in 1992 to 100,000 in 2001, swamping prisons and the courts, with over 80,000 cases per year. Most of the drugs are manufactured in southern Myanmar, but when Thai military close the northern border, traffickers use roundabout routes through Laos and across the Gulf of Thailand.

Farming Changes in Southeast Asia

Most of the people in Southeast Asia still live in rural areas and gain their living from farming, although agricultural employment and agriculture's proportion of each country's income declined with the growth of manufacturing and service industries. As many districts and families changed from local subsistence to a cash-based economy, they linked to and became dependent on growing urban and overseas markets. Such market demands led to new crops and new strains of traditional crops being grown and changing patterns of land ownership. The highest productivity occurs in the major river valleys and island coastal areas (Figure 6.24).

Wet-rice cultivation, or **padi,** remains the chief form of agriculture in Southeast Asia. A reliable water supply comes from rain in Thailand and the Philippines, but irrigation is used in much of Indonesia. From the mid-1960s, the **Green Revolution** was a term applied to the rapid increase in crop yields as a result of a combination of new, high-yielding varieties of rice with the use of fertilizer, pesticides, mechanization, and irrigation to raise production (Figure 6.25a). A major part of the revolution was the **new rice technology** based on agricultural research at Los Banas in the Philippines. It doubled padi yields from 1969 to 1989 at a time when population was increasing faster than food production, and this enabled Indonesia and the Philippines in particular to become almost self-sufficient in grains. The wealthier owners of larger farms in suitable environments with sufficient labor and good management benefited most. They had the capital to invest to obtain higher yields and multiple annual crops. Many poorer farmers could not meet such costs.

The commercialization of farm production changed the use of land and labor. Former communal land sold to a minority of rich landowners forced poorer people to rent land or work as landless laborers. Landlessness increased further with population growth and the expansion of urban areas across former farmland. For example, the amount of rice land per person in Java halved from 1940 to 1980. Communal rights in the rice harvest—vital in supporting many poor families—were replaced by the sale of the standing crop to middlemen, while mechanization displaced wage labor. In Malaysia, for example, the introduction of combine harvesters halved the numbers of workers involved in the rice harvest. As the traditional farming practices of sup-

portive and informal labor exchange during planting and harvest gave way to wages and mechanization, social impacts multiplied. The growing numbers of landless people in the subregion provided support for Communist and Islamist groups, as in some of the islands of the Philippines, and were often associated with ethnic disturbances in Indonesia.

While rice growing is the dominant use of farmland in all Southeast Asian countries, many also produce commercial tree and fruit crops, such as rubber, palm oil, and pineapples in **plantations,** where single species are grown on large estates for commercial purposes. The good management of tree-crop plantations can produce an almost continuous harvest and year-round employment. The plantations were often established under colonial regimes (Figure 6.25b). Although plantations were once owned and run by foreign corporations, Southeast Asian governments nationalized much of the production. Malaysia still produces a large proportion of the world's rubber crop, but a shortage of cheap labor led to falling production as southern Thailand and Indonesia increased their outputs.

Seeking an alternative product, Malaysia became the world's largest producer of palm oil. The crop is less labor-intensive, while the Chinese demand for cooking oil keeps world prices high. Oil palms now constitute the world's biggest vegetable oil crop, despite containing more "unhealthy" fats than midlatitude vegetable oils. Malaysia contributed 58 percent and Indonesia 28 percent of the world total in the late 1990s. Malaysian policies, however, are diverting investment away from agricultural production and into manufacturing.

Indonesian plantation crops include coffee, spices (cloves, nutmegs, cinnamon), tea, rubber, and palm oil products, constituting 13 percent of its exports. In the Philippines, coconut products, sugarcane, pineapples, bananas, and coffee make up 16 percent of exports. In Thailand, plantation crops form 10 percent of exports. Myanmar remains, followed by Afghanistan (see Chapter 7), the world's leading producer of poppies for opium and heroin. It supplies illegal drug traffic through Thailand and other routes.

In the 1990s, Vietnam rose from a marginal coffee producer to rival Brazil and become the world's largest exporter of the cheaper robusta coffee, grown in the central highlands. Farmers from the crowded lowlands migrated into the highlands as world coffee prices rose in the mid-1990s, but Brazil and other coffee producers also boosted production, leading to an oversupply crisis. Vietnam is cutting robusta production but expanding higher-quality arabica coffee trees in the north of the country, as well as encouraging farmers on poorer land to change crops to rubber, peppers, cotton, and hybrid corn.

Southeast Asian Forest Products

Hardwoods from the tropical rain forest that covers much of this subregion are of great economic significance to Southeast Asian countries. Timber products from Malaysia and Indonesia, being greater than tropical timber exports from Africa and Latin America combined, dominate world production. In the 1990s, prices for tropical timber rose by 50 percent at a time when prices for other commodities fell or remained the same, and this stimulated further cutting. Wood-using

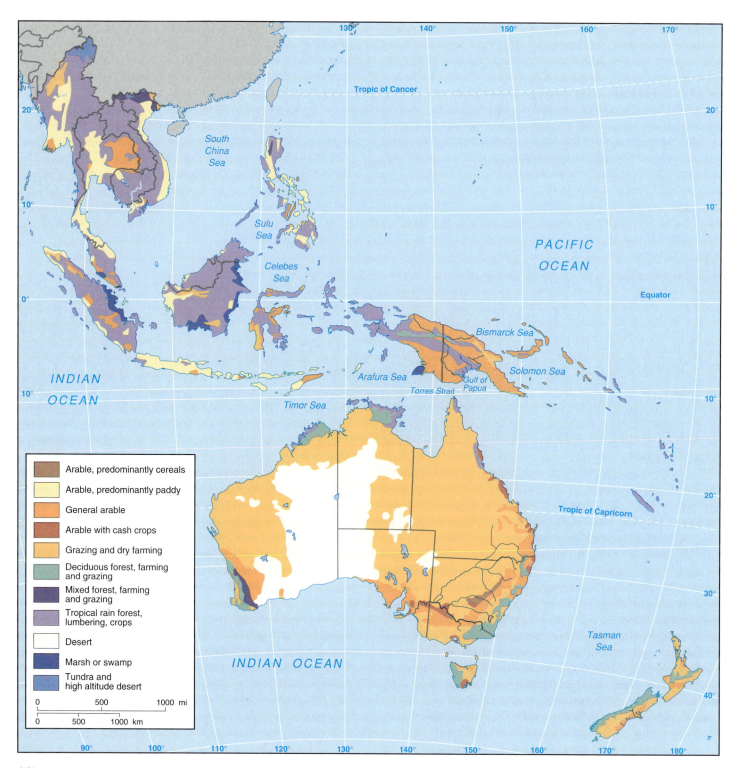

Figure 6.24   **Southeast Asia and South Pacific: farming and other land uses.** What accounts for the many types?

industries became more important in Sumatra and Kalimantan (Indonesian Borneo) as Indonesia increased the income from exploiting its forests. Much of Myanmar remains forest-covered. It has around 250 commercially useful trees, of which teak is a major export. The Myanmar army forces slaves to fell tropical trees destined for world markets.

The limits of cutting without replanting are being reached throughout the subregion. Thailand banned logging in 1989 as

its teak and yang forests diminished. In the Philippines, the forest cover was reduced from 60 percent of the country's area in 1950 to 10 percent in the 1990s. Fears of losing its major resource led Indonesia to curtail clear-cutting of trees and implement reforestation programs. In Malaysia, which claims to be the most responsible country in the subregion for "sustainable logging" and exports more tropical logs and sawn timber than any other country, annual cutting reduces the forest cover by 1 to 2 percent. So far, Malaysia has not adopted international standards of forest management, with loggers disregarding the interests of indigenous forest people in Sarawak, who are losing their livelihood of tree resources, soils, and fish. By 2001, however, Malaysia indicated that it intended to adopt more stringent standards to placate its Western customers. Western objections to unmanaged cutting on environmental grounds are resented by governments in Southeast Asia, which cite their own controls over illegal logging and the historic European and American forest removal.

Mining in Southeast Asia

Mining adds to the primary commodity exports from Southeast Asia (see Figure 5.10). Indonesia is a major oil producer, and much of the world's tin production comes from the subregion. Oil from Sumatra and Kalimantan accounts for 20 percent of Indonesian exports, down from 80 percent in the 1980s because of a fall in new oil discoveries and the development of other income sources in Indonesia's manufacturing base. Growing energy demands with increasing industrialization and without new oil discoveries will soon convert Indonesia into an oil importer.

Malaysia was for long the world's largest tin producer and still exports some, together with oil and natural gas from Sabah and Sarawak on Borneo. Bangka and Billiton (Indonesian islands) and the Malay Peninsula of Thailand are other long-term tin producers. Indonesia's bauxite is mined in the Riau islands and southwest Kalimantan and is processed in the Kualatajuy smelter near Sumatran coal mines. Thailand is one of the world's largest exporters of gems and jewelry, with rubies and sapphires mined along the east coast of the peninsula.

Market-Led Industrialization

While the switch to commercial farming was gradual in the ASEAN countries, moves into manufacturing were more rapid.

By the time the countries of Southeast Asia became independent, most had factory-based industries. At first, import-substitution industries supplied local markets and were protected by tariffs and quotas, but by the 1970s, the small size of domestic markets restricted the scope for further growth.

The combination of small markets, limits of local skills, and restrictions imposed by government licensing offices made the products expensive. In the Philippines, where the government's licensing role led to corruption, the term **"crony capitalism"** described close linkages among bureaucrats and entrepreneurs.

In Indonesia, the government owns large-scale processing facilities for agriculture and mining products. Medium and small-scale factories, often Chinese-owned and based mainly in western Java, dominate consumer goods industries (furniture, household equipment, textiles, and printed matter). The Indonesian textile industry is divided into government- or multinational corporation-owned spinning mills using imported raw materials. Weaving is carried out by smaller firms concentrated in Bandung. Batik, the Indonesian hand-printed textile, is produced in a cottage industry that is being taken over by larger-scale factories.

Beginning in the 1960s, their success in manufacturing goods for export and producing high rates of economic growth stimulated the countries of Southeast Asia to switch to such policies. Many retained some of their less-efficient domestic manufacturing, protected by selective tariffs. The promotion of internationally competitive industries forced local manufacturers to cut costs.

Multinational corporations invested in new factories, employed cheap labor to produce goods for export, and used labor-intensive manufacturing processes. Malaysia and Thailand gained most from this process, although Indonesia and the Philippines also experienced related economic growth. In the 1980s, countries such as Thailand and Malaysia moved from labor-intensive to high-tech industries. However, multinational

(a)

Figure 6.25 **Southeast Asia: farming.** (a) Wet-rice cultivation in Bali. (b) A rubber plantation farmer in Sumatra taps a tree to extract the white sap from just beneath the bark. Photo: (b) © Wolfgang Kaehler/Corbis.

(b)

corporations from the United States, Japan, and South Korea were slow in diffusing innovations to these growing countries and preferred to use them as a cheap source of labor and to open access in their domestic markets.

Much of the export-oriented industrialization relies on imported capital, skills, and management, causing it to be labeled **"ersatz capitalism."** That suggests it is an inferior substitute for development generated by local entrepreneurs based on local savings and investment. The education gap between these countries and the wealthier countries, however, makes it difficult for the Southeast Asian countries to generate and develop their own innovations. Economic growth based on urban manufacturing emphasizes the growing contrasts between moderate-to-high-income urban industrialized areas and low-income rural areas.

Expanding Tourism in Southeast Asia

Some countries in Southeast Asia combine their cultural heritage and new accessibility by airline and telephone in growing tourist industries. Visitors to the main countries—Thailand, Malaysia, Singapore, Indonesia, and the Philippines—increased from 7 million people in 1983 to 20 million in 1990 and 30 million in 2000. Nearly 10 million people visited Thailand in 2000, just over 10 million Malaysia, and 6.3 million Singapore. The highest numbers reflect the political stability of the countries as well as their attractions, which include major cities, precolonial historic places, and beaches and coral reefs. Upland resorts and an increase in **ecotourism** based on natural phenomena (landforms, animals, plants) and historic tourism spread visitors into places once closed by political actions, as in Myanmar and Vietnam. The industry brought in foreign currencies, since fewer local people ventured abroad for vacations; those who did went mainly to Hong Kong and Japan. By contrast, the United States had over 50 million visiting tourists in 2000, while nearly 60 million Americans vacationed abroad.

Communist Economic Stagnation and Change

Economic growth in Cambodia and Laos remains slow. For most people, only bare survival is possible. Vietnam has larger areas of fertile soil, more mineral resources, and a larger population than the other two countries, but its national income grew more slowly than its population, causing per capita incomes to fall in the 1980s.

Much of Vietnam's poverty stemmed from the effects of the fighting on its territory that lasted from the early 1950s until 1975 and disrupted all aspects of the economy. Even after the war to establish the control of the Communist government ended, Vietnam spent heavily on military excursions into Cambodia (1978–1989) and on a border war with China. Internal policies produced little economic growth. The five-year plans for the late 1970s and early 1980s directed a shift into heavy industry and out of agriculture, but the centralized bureaucracy was unwieldy and unresponsive to demands, and it failed to gain the support and further sacrifices of the war-weary population.

The Soviet Union contributed approximately $1.5 billion per year in aid in the 1980s to Vietnam's economy but ceased assistance after 1989. The economic blockade by the United States and the ASEAN countries prevented Vietnam and its neighbors from receiving overseas aid until 1991.

After 1985, however, Vietnam's leaders allowed crops to be sold directly in private markets instead of through the communes and cooperatives selling to government ministries that set prices. The Vietnamese currency was devalued and foreign investment encouraged. By the late 1980s, Vietnam exported rice again, becoming the world's third-largest exporter in 1994—although bad weather in some years caused it to be displaced by India. Vietnam established economic leadership over the two smaller countries of Laos and Cambodia to the west. The shift to greater involvement in the world economy was easier in southern Vietnam—a capitalist country until 1975—than in the northern sector, which had been under Communist control longer. In the mid-1990s, corporations from Japan and South Korea in particular invested in Vietnam. In 1995, the United States "normalized" relations with the country when Vietnam became the first Communist member of ASEAN.

Myanmar's Economy

Myanmar's political isolation results from its lack of democracy. It keeps an unrealistic official exchange rate, 20 times higher than the black-market rate, making its products too

Test Your Understanding 6C

Summary Southeast Asia comprises countries at different levels of political and economic development. Groups of people with different cultures mix within each country, often leading to conflicts. Singapore is a wealthy but small country. Thailand, Malaysia, Indonesia, and the Philippines are emerging economies. Laos, Cambodia, and Vietnam remain extremely poor following years of warfare. Myanmar's exit from military dictator-imposed isolation is slow. All Southeast Asian countries are developing the manufacture of goods for export and are attracting foreign investments. They mostly have governments that take an involved part in industrial and trading policy.

Some Questions to Think About

6C.1 Which are the ASEAN countries? Why was the organization set up, and what is its likely future?

6C.2 Assessing their locations and past and present economies, compare the future prospects of Hong Kong (see Chapter 5) and Singapore.

6C.3 What common issues do North Korea (see Chapter 5), Vietnam, Laos, Cambodia, and Myanmar face?

Key Terms

transmigration	new rice technology
primate city	plantation agriculture
deurbanization	crony capitalism
padi	ersatz capitalism
Green Revolution	ecotourism

expensive to sell on world markets but also making imports cheap. Such subsidies will be removed as the country seeks to join the global economic system. In the 1990s, Myanmar brought in Chinese contractors to build bridges, railroads, airports, and hotels, and entered joint agreements with Japanese and European corporations to develop the economy.

Myanmar creates major problems for its neighbors, especially Thailand. The so-called "Golden Triangle," where Myanmar, Thailand, China, and Laos meet, is a major source of opium and amphetamine drug production. It is estimated that up to 800 million pills passed across the border into Thailand during 2001. The Wa people, a minority group that used to fight the Myanmar army, switched to selling drugs and fought another rebel group, the Shan people. Myanmar's leaders allow the drug trade, but the Thai authorities resist it along with the rising flow of refugees from Myanmar. The Thai resistance is not made easy by the bribing of their officials and the supplies of electricity it sells to eastern Myanmar, some of which is used to produce the pills.

 # Australia and New Zealand

Australia and New Zealand (Figure 6.26) are inhabited almost entirely by European people, who transferred their politics, cultures, and economies to the opposite side of the world (Figure 6.27). From the 1970s, the ties that bound these countries to the United Kingdom in particular were gradually severed and their trade interests reoriented to Asia and the wider Pacific. Australia became a focus of world attention when it hosted the 2000 Olympic Games. Many in more affluent Western countries regard both Australia and New Zealand as pleasant and exciting places to live. Their low densities and population, and high levels of income make them both desirable places for Asians and refugees from many countries.

Australia has a diverse and affluent economy (see Figure 6.18). Although it had only 3.5 percent of the region's 2000 population, it produced 40 percent of its 2000 GNI and has a GNI PPP per capita that is just a little better than that of Singapore. New Zealand is a smaller country but also has a

Figure 6.26 **Australia and New Zealand: major geographic features.** The political divisions, physical features, and major cities.

Figure 6.27   Australia, New Zealand, and Pacific Islands: population, economic, and ethnic data.

Country	Capital City	Land Area (km²) Total	Population (millions) Mid-2001 Total	2025 Est.	GNI 1999 (US $ million) Total	GNI PPP 1999 Per Capita	Percent Urban 2001	Human Development Index Rank of 175 Countries	Human Poverty Index: Percent of Total Population
Australia, Commonwealth of	Canberra	7,713,360	19.4	23.2	397,345	23,850	85	7	12.5
New Zealand	Wellington	270,990	3.9	4.6	53,299	17,630	77	18	12.8
South Pacific Islands									
Fiji, Republic of	Suva	18,270	0.8	1.0	1,848	4,780	46	61	8.6
Kiribati, Republic of	Tarawa	730	0.1	0.2	81	No data	37	No data	No data
Marshall Islands, Republic of	Majuro	179	0.1	0.2	99	No data	65	No data	No data
Micronesia, Federated States of	Palikir	702	0.1	0.2	212	No data	27	No data	No data
Nauru, Republic of	Yaren	20	0.01	0.02	No data	No data	100	No data	No data
Palau, Republic of	Koror	458	0.02	0.03	No data	No data	71	No data	No data
Papua New Guinea, Independent State of	Port Moresby	462,840	5.0	8.3	3,834	2,260	15	129	27.8
Solomon Islands	Honiara	28,900	0.5	0.9	320	2,050	13	118	No data
Tonga, Kingdom of	Nukualota	750	0.1	0.2	172	No data	32	No data	No data
Tuvalu	Fongafele	30	0.01	0.02	No data	No data	18	No data	No data
Vanuatu, Republic of	Port-Vila	12,190	0.2	0.3	227	2,880	21	116	No data
Western Samoa, Independent State of	Apia	2,840	0.2	0.2	No data	4,070	33	70	No data

Source: Data from *Population Reference Bureau 2001 Data Sheet*; *World Development Indicators,* World Bank, 2001; *Human Development Report,* United Nations, 2001; Microsoft Encarta (ethnic group, language, religion).

high GNI PPP per capita; with less than 1 percent of the region's population, it produces 5 percent of its output. The two countries face similar problems, including great distances from their traditional markets, a small local market, and the transition from historic colonial ties with Europe to a closer involvement with Asian and Pacific Ocean countries. Australia is already termed "Asia's farm and quarry."

Countries

Australia

Australia is the only country that is also a continent. Its size produces internal contrasts of physical and human geography.

Political Regions: The States

Separate colonies were established around the initial port cities, giving access to inland areas and connecting them back to the ruling country (see Figures 6.7a and b). The early colonies determined later geographic differences by setting their own patterns of inland expansion, choosing different railroad gauges, and concentrating major developments in their chief cities. In 1901, the Commonwealth of Australia came into being as a federation of the states that emerged from the separate colonies. Sydney and Melbourne competed for the role of national capital, but the question was resolved by building a specially planned city in a new Australian Capital Territory

centered on Canberra, almost midway between them. The federal government moved there in 1928.

Southeast Australia

Most Australians live in the temperate and humid southeastern coastal region from near Brisbane in the north to Adelaide in the south. This area includes three of the five largest cities (Sydney, Melbourne, and Adelaide), as well as the federal capital, Canberra.

For most of its length, the southeastern coastal region is bounded inland by the hills and plateaus of the Great Dividing Range, which formed a barrier to inland movement in the early years of colonization. Three states and their dominant cities occupy this area. Sydney, the largest city, lies at the heart of New South Wales, which also contains industrial cities near the coast and the main concentration of Australia's sheep farming on the western side of the Great Dividing Range. Victoria is focused on Melbourne but is a smaller state, and its inland farming area is concerned mainly with more intensive irrigated farming for grapes, fruit, and grain crops. South Australia has its main city, Adelaide, in a farmed area at the southern edge of the state. Much of the rest is desert.

Northern Australia

From Brisbane northward, the more tropical climate and the offshore Great Barrier Reef—the world's largest continuous coral reef formation—form an area much favored by tourists.

Country	Ethnic Groups (percent)	Languages O=Official	Religions (percent)
Australia, Commonwealth of	European 95%, Asian 4%, Aborigine 1%	English	Protestant 50%, Roman Catholic 26%
New Zealand	Maori 9%, European 88%	English (O), Maori (O)	Christian 67%
South Pacific Islands			
Fiji, Republic of	Fijian 49%, Asian Indian 46%	English (O), Fijian, Hindustani	Christian 52%, Hindu 38%, Muslim 8%
Kiribati, Republic of	Micronesian with Tuvaluan minority	English (O), Gilbertise	Protestant 40%, Roman Catholic 52%
Marshall Islands, Republic of	Micronesian, German, Japanese, U.S., Filipino	English (O), local Malay dialects	Christian (mainly Protestant)
Micronesia, Federated States of	Chuukare 41%, Pohpeian 26%, other groups	English (O), local	Christian 97%
Nauru, Republic of	Naurean 58%, other Pacific 26%, Chinese 8%	Naurean (O), English	Christian (mainly Protestant)
Palau, Republic of	Malay, Melanesian, Filipino, Polynesian	English (O), state languages	Christian, mainly Roman Catholic, local
Papua New Guinea, Independent State of	Melanesian 98%	Melanesian and Papuan languages	Protestant 44%, Roman Catholic 22%, local
Solomon Islands	Melanesian 93%, others	Melanesian, Pidgin, English, local	Christian 95%
Tonga, Kingdom of	Tongan 98%	Tongan (O), English (O)	Christian
Tuvalu	Polynesian 96%	Tuvaluan, English	Church of Tuvalu 97%
Vanuatu, Republic of	Melanesian 94%	English (O), French (O)	Christian, mostly Protestant 84%
Western Samoa, Independent State of	Samoan 93%, Euronesian mixtures 7%	Samoan, English	Christian 99%

Queensland is the second largest state in area with its main city, Brisbane, in the southeast corner. This state has a more tropical climate, is the main cattle state, has major mining operations, and is developing tourist attractions.

Northern Territory is the least populated of the major Australian political divisions. In the far north, the tropical climate brings monsoon rains along the coasts. Most of the few remaining aboriginal people who live in traditional ways are in this territory, but it has fewer inhabitants than Australian Capital Territory.

Interior and Western Australia

Inland of the Great Dividing Range, the Murray-Darling River lowlands have a subhumid climate that supports a major farming region producing cattle, sheep, grain, and fruit in western Queensland, New South Wales, and northern Victoria. Westward of this sparsely populated farming region, the virtually empty Great Australian Desert covers almost all of the rest of the continent. In the southwestern corner, rains come in winter in a Mediterranean climatic regime and provide a basis for agricultural settlements centered on Perth.

Western Australia is the largest state in area but is mostly desert, and its main settlements around Perth are separated by over 1,600 km (1,000 mi.) of uninhabited land from the large cities of southeastern Australia. Perth has farming and mines in its immediate hinterland and large iron ore mines farther north. Parts of South Australia and Northern Territory are also included in the vast Australian interior.

New Zealand

Internal rivalries in New Zealand are less marked than those within Australia. North Island is distinguished by its central upland area that is affected by volcanic activity. It is home to nearly all the Maoris, most of the people of European origin, the largest city, Auckland, and the country's capital, Wellington. South Island has fewer people. Its high western mountains, the Southern Alps, are occasionally rocked by earthquakes. Its eastern plains provide good farming land.

People

Immigrant Populations

In the early 1900s, wild claims were made that "empty" Australia could accommodate over 200 million people, although Griffith Taylor, a British geographer in the 1920s, doubted it could sustain more than 20 million. Despite decades of immigration, Australia's population in 2001 did not reach the smaller figure.

Until the mid-1900s, most Australian immigrants came from the British Isles. In the early days of the Dominion of Australia, an informal **white Australia policy** acted against the pressures of Asians in particular. To increase its population and maintain its links with the United Kingdom, Australia encouraged immigrants from Britain by paying their passage until after World War II. As that source became insufficient, more immigrants came from other parts of Europe. By the 1990s, only 12 percent of immigrants came from the United Kingdom, while 43 percent were from Asia. Australia is becoming a multiracial, multiethnic, and multicultural country—an environment in which there is also greater sympathy for the Aborigines as well as increasing connections to Southeast Asia.

Although the policy to admit only white (European) immigrants ceased in 1972, its legacy still affects some Australian attitudes toward new immigrants and efforts to focus trade in Asia. It has been difficult for many Australians to realize that countries they once looked down on now have as sophisticated economies paying wages as high as their own.

Although high unemployment in the 1990s led to a halving of immigrants allowed into Australia, Australian businesspeople argue for greater numbers in the light of less-than-replacement levels of births and predictions of falling population totals and skills levels. Australia also aims to give 12,000 refugees a year the right to permanent residence, but in 2001 its insistence on preentry interviews offshore earned criticism from human rights groups such as Amnesty International as the Australian navy intercepted boat people and transferred them to the isolation of detention camps on Nauru.

New Zealand's population continues to be more British in origin and allegiance than that of Australia. New Zealand receives fewer immigrants, although it does not discourage nonwhites in the way Australia did until 1972. The indigenous people, the Maoris, take a more significant part in New Zealand life than the Aborigines do in Australia. They total around half a million people, mostly living in North Island. Auckland has over 100,000 Maoris, many of whom have professional jobs.

Urban Populations

Australia is essentially an urban country, with 85 percent of its people living in towns in 2001 and 57 percent in the five largest cities of Sydney (3.7 million in 2000), Melbourne (3.2 million), Brisbane (1.6 million), Perth (1.3 million), and Adelaide (1 million). This reflects the historic primacy of these cities within their states and the growing significance of service industry employment and output. The national capital, Canberra, has one-third of a million people.

The major Australian cities retain their established port functions, although Melbourne grew further in importance by specializing in container trade. They have downtowns that combine state government and financial sectors (Figure 6.28) with major cultural facilities such as the Opera House in Sydney. Industrial suburbs vie with residential areas for space.

Smaller towns are often linked to narrower functions, reflected in their urban landscapes. Thus, Newcastle, New South Wales, is a coal and steel town, while Broken Hill is an inland mining center. Many small mining towns had short lives while the mineral resources or markets lasted and then became ghost towns. Towns along the Queensland coast have tourist functions, while inland towns such as Mildura are farm market centers.

In 2001, 77 percent of New Zealanders lived in towns and cities. New Zealand's main cities on both islands began as ports that acted as "hinges," giving access to the interior and

back to the main overseas markets of the colonial homeland. Processing industries concentrated in the ports, which became market, commercial, financial, and government centers. Auckland, with just over 1 million people in 2000, is the center of an urbanized area that stretches along the peninsula to its north (Figure 6.29). Wellington, with one-third of Auckland's population, is a major port and the center of government. Christchurch and Dunedin are the main towns of South Island. New Zealand towns reflect similar histories to those in Australia, including the most recent growth of manufacturing and service industries. All have port facilities.

Population Dynamics

Australia's population of 19.4 million in 2001 was double that in 1950. As with other materially wealthy countries, Australia's natural rate of growth is slow, with total fertility rates of less than 2. Immigration results in annual population growth rates of over 1 percent, compared to less than 0.5 percent in European countries. Both Australia and New Zealand are well advanced in the demographic transition process (see Figure 6.20). Australia's age-sex diagram (Figure 6.30) reflects a slowing in births after a baby-boom period from the mid-1950s to around 1970 and an aging population.

Figure 6.28 **Australia: the center of Sydney.** Sydney's harbor and Opera House are located on the near promentory.

Figure 6.29 **New Zealand: urban landscape.** Auckland harbor, North Island is the main commercial center of New Zealand. Its waterfront has extensive yacht docking facitilites and oil terminals close to the high-rise city-center offices.

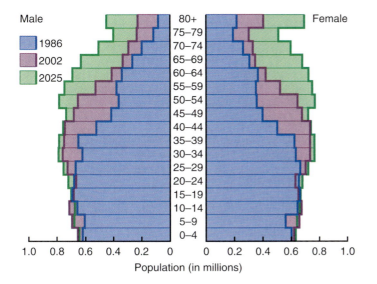

Figure 6.30 **Australia: age-sex diagram.** Source: U.S. Census Bureau, International Data Bank.

New Zealand's total fertility rate remains slightly above that of Australia, but its immigrants keep the annual population rate of increase at just over 1 percent. After gaining dominion status, New Zealand had to survive some years of labor force losses from its loyal support of Britain in two world wars. Today, many younger New Zealanders in particular migrate to Australia for jobs in the professions.

Economic Development

The geographic positions of Australia and New Zealand, so far from their established markets in Europe and with populations dominated by Europeans, influenced their economic development. Until the 1970s, the demands of European markets, including the impact of the European Union and subsequent reduction of trade, determined products and economic structures. The growing Asian markets, however, provided new outlets for Australian and New Zealand products as the older links were severed.

Along with their attempts to develop wider trading links with Asian countries in ASEAN and APEC, Australia and New Zealand are involved in links with other countries inside the South Pacific region. Australia's attempts to form closer economic links with New Zealand in the 1980s, however, slowed in the 1990s, possibly because the attempts to enter Asian markets were seen as more important or because Australians saw New Zealand as benefiting the most from such links. The **South Pacific Forum,** established in 1971, links 13 countries: Australia, New Zealand, Papua New Guinea, the Solomon Islands, the Cook Islands, Fiji, Kiribati, Nauru, Niue, Tonga, Tuvalu, Vanuatu, and Western Samoa. Its aim is to develop regional political cooperation. This agency has had some success in confronting regional problems for the islands in particular. Australia has championed a number of causes, but these often bring it into confrontations with Asian nations, as when it opposed the exploitive Malaysian logging of Papua New Guinea.

Changes in Australia

Australia's production of wool for British factories from the late 1800s led to diversification into fruit, lamb, beef, and dairy products that followed the same routes to markets in Britain and Europe. During the 1900s, mining products grew in significance. The high cost of transporting low-value goods from Europe to Australia enabled the imposition of tariffs to protect the growth of import-substitution manufacturing in the new country, with foundries, textile mills, breweries, and other food and drink processing enterprises. Other industries, such as steel- and automaking and assembling, were later set up under similar or increased protection levels.

By the mid-1900s, Australia had a two-tier economy. The first tier depended on exports from the natural resource base of farm and mine to provide income that paid for imports. The second tier consisted of protected manufacturing and government-owned enterprises such as airlines, railroads, and docks that were often inefficient but had a role in spreading the wealth from the export industries' income over a wider group of people.

In the 1970s and 1980s, the protectionist policies collapsed as exports of agricultural and mineral commodities obtained lower prices on world markets and the expansion of the European Union cost Australia its markets in Britain and Europe. It became clear that Australia could no longer campaign for free markets for its exports while keeping high tariffs for its imports as it shifted its export markets to Asian countries and especially Japan. One implication of the lower tariffs is that many Australian manufacturers will not survive. Of the five automakers that sheltered behind protective tariffs, Nissan closed its Australian operation.

In the late 1980s and through the 1990s, Australia's economy boomed by supplying the growing Asian markets with coal, iron ore, and other minerals and some grain and livestock products. Between 1970 and the 1990s, farm exports fell from 44 to 20 percent, while mine exports rose from 27 to 42 percent of total exports. By the mid-1990s, Japan bought more goods from Australia than those from Europe and the United States combined. A mid-1990s World Bank study that took account of "natural capital" in terms of land value, water, timber, gold, and other minerals, in addition to human resources, ranked Australia as the top world country, just ahead of Canada. Clearly, mineral resources played a major part in this designation.

In 2000, Australia's total GDP placed it fourteenth, after South Korea, among world economies. Its GDP per capita was just below that of the United Kingdom. As in other wealthier countries, the GDP was coming less from agriculture (down from 6% in 1970 to 3%) and manufacturing (down from 24% to 15%) and increasingly from service industries, which rose from 55 to 70 percent.

Problems of Trade Dependence

Australia's economy is increasingly tied to Asian markets. As it accepts more imports of manufactured goods from Asian countries, it needs to export more to them. This is not easy. Australia wishes to add value to its exports by processing its agricultural and mining products in Australia, but Japan and other Asian countries retain import restrictions on processed materials. Certain specialty products created their own growing markets abroad, but Australia is a relatively small market for Asian goods.

The six nations of ASEAN have a regional free-trade agreement among themselves but are reluctant to admit Australia and are more interested in forging agreements with China and Japan. Moreover, the Asian countries have different attitudes than Australians toward the workings of the economy. While Asian governments and businesses work closely together, and environmental and civil rights issues assume less importance, Australians remain fixed to Western values that sometimes upset Asian countries and place political barriers in the way of trade. For example, Australia and Malaysia had major disagreements in the 1990s over Australian criticisms of Malaysian logging practices in the South Pacific.

Australia lobbies in the Asia-Pacific Economic Cooperation Forum (APEC), a wider group of Asian and Pacific countries, including the United States, which is dedicated to trade liberal-

ization. From Australia's point of view, changes need to come sooner than is likely. The increasing links between the United States and Latin American countries further threaten Australian access to markets in the United States.

Australia's Dominant Mining

Mining provides nearly half of the value of goods exported by Australia. It is the world's largest exporter of coal, amounting to three-fourths of its record production of 200 million tons in the 1990s, and the world's largest supplier of high-quality (and low-polluting) coking coal. Plans for expansion could increase this level of production if world demand for coal remains high and as the production of iron and steel increases.

New energy resources include natural gas fields along Australia's west coast, tied to the construction of liquefaction plants that export the gas to Japan. Such projects attract investment by oil companies but compete with producers in nearby Malaysia, Brunei, and Indonesia. Their viability depends on world oil and gas prices.

Australia produces a wide variety of other minerals, but expanding production often faces the complexities of obtaining new mining licenses, particularly in areas that are environmentally sensitive or might have aboriginal sites. For example, Australia produces 10 percent of the Western world's output of uranium and has 30 percent of the Western world's reserves. Australian uranium production is restricted to three mines at present, and there are companies that wish to expand production and opposing groups who wish to curtail it further or see usage restrictions placed on buyers of uranium from Australia. Other minerals of importance to Australia include iron ore, of which it is the fourth largest world producer, bauxite (1st), nickel (4th), and gold (3rd). In the 1990s, Australia also opened the world's largest diamond mine.

Australia's Farm Output

Much of Australia's land is used agriculturally, although only 2 percent is cultivated and the remainder is grazing land (see Figure 6.24). Australia's farmers export 80 percent of their production, selling more in Asia following protectionism in Europe and the United States. Wheat, oilseeds, beef, veal, and wine are the main exports (Figure 6.31). This reliance on world

Figure 6.31   **Australia: farming.** (a) Herding sheep. (b) Harvesting grapes.

(a)

(b)

markets brings problems from time to time. Interruptions of market access to Japan and the United States because of environmental scares have serious consequences for Australian farmers since those two countries between them take three-fourths of all Australian beef exports. Australian farmers suspected that such interruptions were to protect home producers in importing countries. Further challenges from world markets occurred in the early 2000s as other major exporters of farm products, such as Brazil and Argentina, devalued their currencies, making their products cheaper on world markets.

Australia's Underdeveloped Northlands

The 26°S parallel almost cuts the country in half; only 28 percent of Australia's exports are produced north of this line and only 6 percent of Australia's population (slightly more than 1 million people) lives there. The northern part of Australia includes the least developed part of the country because of its combination of arid and tropical monsoon climates and distance from other population centers. The region still lacks good transportation infrastructure, including a rail link south to join up with the country's system at Alice Springs.

Tourism

Numbers of foreign visitors increased from around 1 million in the early 1980s to 5 million in 2000. Over half of these visitors came from Asia, attracted by the beaches, golf courses, and the theme parks along Queensland's Gold Coast (Figure 6.32). It is clear that the industry's prosperity depends on growing affluence in Asia. Many visitors come from Japan and Taiwan partly to purchase goods at lower prices than are available at home.

Tourists from North America and Europe come not so much for mass tourist facilities as for the outdoors experience, including the viewing of wildlife and rock formations. Visits to such features form the basis of ecotourism for more affluent tourists who have the time and money to take the long journey to Australia.

New Zealand

New Zealand's economy, like that of Australia, is based on natural resources and the exports of mainly farm and forest prod-

Figure **6.32** **Australia: tourism.** The Gold Coast area of Queensland attracts many visitors from Asia. The Great Barrier Reef lies offshore. Photo: © Jack Fields/Corbis.

ucts to Europe. New Zealand's main economic products entering world markets are wool, lamb, and dairy products. Over half of New Zealand is in pasture, and over one-third of its exports is from livestock (Figure 6.33). Its income from meat products fell in the 1990s, leading to major reductions in herds of beef cattle and sheep, but dairy product demand and deer farming expanded.

In the development of farming, pastureland replaced much forest. The New Zealand government replanted large areas of forest with Douglas firs in the early 1900s. This afforestation policy provides an increasing harvest that finds markets in Asia as those countries cut their natural forest but do little replanting. Forestry is now as profitable as farming in New Zealand, and the state-owned forests have nearly all been privatized, albeit with strong conservation laws. Processing industries use the capital accrued from foreign sales of timber.

The contributions of agriculture and manufacturing to its economy fell as that of services rose to account for around two-thirds of GDP. In the 1980s, the New Zealand government instituted economic reforms to revive a rather stagnant economy and produce a trading surplus together with lower unemployment and inflation. Tariffs and restrictive port practices were removed and government spending reduced as a proportion

Figure **6.33** **New Zealand: farming.** (a) Sheep show in Rotorua with 20 breeds of sheep, each having advantages and many originating in Scotland (U.K.). (b) Farming area of South Island. The river channels descend from the Southern Alps and form wide valleys across the plateaus of the eastern lowlands. Photo: (a) © David C. Johnson; (b) © James L. Amons/Corbis.

(a)

(b)

Test Your Understanding 6D

Summary   Australia occupies the world's sixth-largest continent but has only 19 million people, largely because half is desert and immigration was restricted for most of the 1900s. Australia's economy ranks it with the world's wealthier countries and is based on exports of a wide range of minerals and farm products, together with growing manufacturing and tourism industries. Its five largest cities hold nearly 60 percent of the total population.

New Zealand is another materially wealthy country. It consists of two main islands. Its 3.9 million people export wool, lamb meat, timber, dairy produce, and fruit.

Some Questions to Think About

6D.1   Why does such a high proportion of Australians live in the major coastal cities of the southeastern quadrant of the country?

6D.2   What factors may slow the further exploitation of Australia's mineral resources?

6D.3   What changes were involved in Australia switching from European to Asian markets?

6D.4   What are some of the environmental impacts faced in Australia and New Zealand?

Key Terms

white Australia policy   South Pacific Forum

of GDP. Lower labor, transportation, and utility costs brought increases in manufacturing productivity. New Zealand's GDP per capita in 2000 fell behind that of Australia and West European countries, although its high quality of life continued.

New Zealand now sells more goods to Japan than to Australia, the United States, and the U.K. combined. Most of its imports still come from Australia and the United States. Like Australia, New Zealand is a member of the South Pacific Forum, and ties with the Pacific islands continue that originated when New Zealand was the United Nations agent in those island countries before their independence.

New Zealand expanded its tourist industry through its outdoors attractions in both North and South Island. Improved and cheaper airline access overcame the great distances from the main sources of richer tourists. In 2000, New Zealand had 1.8 million tourist visitors, double its 1990 number.

South Pacific Islands

The South Pacific islands stretch from the eastern part of New Guinea—the independent country of Papua New Guinea with 4 million people—to tiny islands with a few hundred people thousands of kilometers to the east (Figure 6.34). The smaller islands are grouped in independent countries or colonies. Their small sizes and markets, however, place them at a disad-

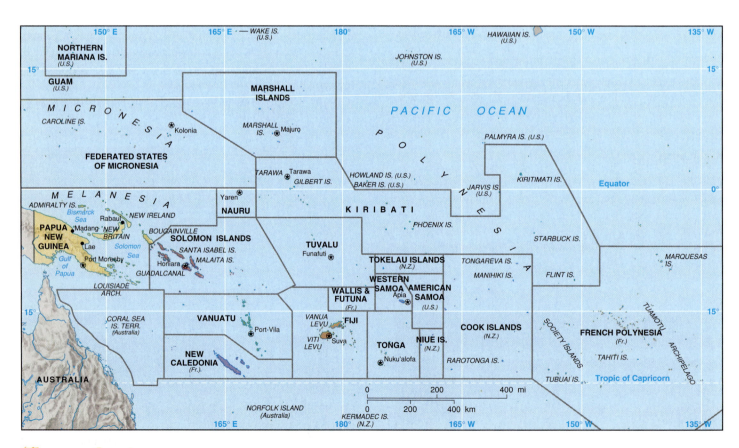

Figure 6.34   **South Pacific islands: major geographic features.** The country groups and main cities. These islands are spread across huge distances.

(a)

(b)

Figure 6.35 **South Pacific: the legend and the naked truth.** (a) An idyllic view of Raiarea, French Polynesia, which is close to the French nuclear testing site. (b) Traditional dress (plus sneakers) in Papua New Guinea. Photo: (b) © MAF.

vantage in arguments with Asian countries or the United States over timber, mineral, agricultural, or fisheries exploitation, and the standards of living of their people suffer accordingly (see Figure 6.27). Only occasionally, as in the violence on the French Polynesian island of Tahiti over the French nuclear testing in mid-1995, do these tensions surface to world news levels. Although these islands were perceived by early travelers as a "Garden of Eden," day-to-day life was not as idyllic as they wrote (Figure 6.35).

The South Pacific Forum, formed in 1971 with Australia and New Zealand as its major members, has its headquarters in Suva, Fiji, and is concerned with the need for new rules to manage the resources in this region. The growing problems faced by the small islands brought closer cooperation on an economic and political agenda that includes achieving independence of remaining colonies, increased investment, and trying to stop French nuclear tests.

Island Countries

Most of the South Pacific islands gained their independence in the 1970s. Western Samoa achieved it in 1962 and the Marshall Islands in 1991. The Micronesian group of islands remains a United Nations trust territory. New Caledonia and French Polynesia are part of France for government purposes.

Although independence appeared desirable to many of the islands, economic difficulties, internal tensions, and dependence on continuing economic aid and protection keep them linked economically to the United States, France, the United Kingdom, Australia, or New Zealand. Kiribati, Tonga, Tuvalu, Western Samoa, and the Solomon Islands remain among the world's poorest countries with few products in world demand.

Modern country groupings of the islands cut across ethnic divisions.

- The southwestern group approximately coincides with former Melanesia and includes Papua New Guinea, the Solomon Islands, Vanuatu, Fiji, and New Caledonia. These are the largest and most populated islands and are the nearest to Southeast Asian markets and Australia-New Zealand links. Papua New Guinea includes the eastern part of the largest of the islands; the western part of this island, Irian Jaya, is in Indonesia.

- The northwestern group, roughly coinciding with former Micronesia, includes Guam, the Northern Mariana and Marshall Island groups, Nauru, Palau, and the Federated States of Micronesia.

- The largest of the three groups in terms of ocean area but the poorest in economic development contains the islands of the central Southern Pacific such as Kiribati, Tuvalu, Wallis and Futuna, Tonga, Tokelau, Western and American Samoa, Niue, Cook Islands, Pitcairn, and French Polynesia.

The islands vary considerably in their well-being, as the consumer goods ownerships in Fiji and New Guinea show (see Figure 6.15). The French colonies of French Polynesia and New Caledonia are heavily subsidized by France and have above-average incomes. Others, such as the Marshall Islands,

Palau, Micronesia, Guam, and the Northern Marianas, have U.S. support.

People

Population Distribution and Dynamics

The South Pacific islands have small total populations, with the largest numbers on those islands nearest to Australia and New Zealand. In 2000, the southwestern group had a total population of some 6 million people, 4.8 million of whom lived in Papua New Guinea and 800,000 on Fiji. The total population of the smaller northwestern islands was under half a million, while that of the widespread islands in the central South Pacific was around 2 million.

Population growth continues on many of the islands due to high total fertility rates (see Figure 6.20). The age-sex diagram for Papua New Guinea resembles those for poorer countries elsewhere and implies a large proportion of young people in the total population (Figure 6.36).

Many islands are on the point of **overpopulation,** where the land and its resources cannot support its people on a subsistence basis. Economic development has not occurred to support larger numbers of people. Although fishing and the cultivation of vegetables and fruit bring good harvests and a healthy diet, after World War II, many islanders switched to imported canned meats and such delicacies as turkey tails that are high in fats and contribute to a high incidence of connected obesity, diabetes, and heart disease. This has been called the "New World Syndrome." Only on Fiji, Guam, New Caledonia, and Palau is life expectancy close to 70 years; for most islands, it is around 55 years. Genetic factors are also thought to influence the trend to obesity.

Small Towns

Few of the South Pacific islands are large enough to have major cities. Most island towns began as ports to give access

for colonial links and continued as political capitals of the independent countries (Figure 6.37). In the South Pacific islands, the towns have populations of a few thousand people and contain a mixture of facilities dating from the colonial era, some processing industries, ports, and shantytown areas. Many islands link by air or sea to the cities of New Zealand and Australia or to the Hawaiian Islands for access to world air routes. Suva on Fiji and Port Moresby in Papua New Guinea, with their international airports, are the main exceptions.

Economic Development

Before colonization, the island communities had complex economies, often linked across the wide ocean expanses. Exchanges took place among the better-watered islands where vegetables and fruit were grown and the arid islands that relied on fishing. Unique maps of direction and distance were constructed, but storms and crop failures brought death and destruction.

After colonization, the interisland trade was commercialized. Some of the islands entered the world economy with sales of coconut palm products such as copra—the dried white meat that lines the inside of the coconut shell and provides an oil used in soaps and candles. Manufactured goods from Europe and the United States and expensive diesel oil for power were imported. Many islands needed subsidies from the colonial countries and remained materially poor. Some islands, including Tuvalu, relied on external aid for nearly 100 percent of their income.

In the 1990s, most South Pacific islands faced economic ruin. Some see external aid as a compensation for past colonial exploitation; others see it as demeaning and sapping self-reliance. With a low economic base, the islands still suffer from increased exploitation rates by foreigners of their few natural resources. Some islanders see tourism as an income prospect, while others view it as a further form of economic colonialism.

Farm, Forest, and Mine Products

Apart from Papua New Guinea (copper and gold), New Caledonia (nickel), and Nauru (phosphate), few islands have natural resources except for a warm climate, nutrient-rich volcanic or coral limestone soils, and the surrounding ocean (Figure 6.38). The small areas available for agriculture allowed few islands to diversify into commercial crops beyond coconuts and copra. Fiji produces sugar and Tonga produces bananas and vanilla in addition to coconuts.

World attention was drawn to the cutting of tropical hardwoods on Papua New Guinea and the Solomon Islands by a statement at the South Pacific Forum in Brisbane in the mid-1990s that the islands were being "ripped off." The imposition of stricter environmental laws in Malaysia and Indonesia turned their logging companies to the islands that had fewer regulations. Cutting is now at levels that could remove all the timber, and little replacement planting is taking place. Local landowners receive US $2.70 for timber that fetches $350 on world markets.

Similar problems face the local fishers as fishing fleets from Japan, the United States, South Korea, and Taiwan take fish from this region to make up nearly 40 percent of the world catch. Although the island countries of the South Pacific have declared

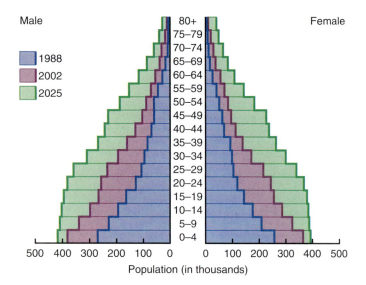

Figure 6.36 **Papua New Guinea: age-sex diagram.**
Source: U.S. Census Bureau, International Data Bank.

200-mile exclusion zones and the foreign fishers often pay access fees, it is impossible for the whole area to be patrolled.

Mining occurs on a few islands but often brings problems. Papua New Guinea suffered an income reduction when a separatist group on Bougainville forced the closure of copper mining that made up 30 percent of Papua New Guinea's exports. After protesting the copper mining and its environmental impacts since 1969, the people of Bougainville declared the island's independence in 1990. Papua New Guinea used military force to prevent secession, but the Australian-owned mining operation in a separatist-controlled area remains closed. The Papua New Guinea government is trying to replace this income by encouraging the development of gold mines in other parts of its territory.

For a few years in the 1970s, Nauru, only 21 km² (8 mi.²) in size, had one of the highest per capita incomes in the region. From the mid-1900s, strip mining for phosphate rock (used in fertilizers) left a barren, jagged surface over much of the island as mining companies cleared the tropical vegetation that was then unable to reestablish itself.

Opting for independence in 1968, the island bought out the phosphate company. New buyers came from Japan and South Korea, injecting income that funded schools and free medical treatment facilities; few Naurians needed to work. The wealthy times attracted Australian managers, Chinese shopkeepers, and miners from other Pacific islands, who now make up one-third of the 12,000 population. Air Nauru bought five Boeing 737s.

Surplus revenue was invested in Pacific Rim properties, but in the 1990s, the value of these assets dropped from over US $1 billion to $130 million. By the 1990s, the depleted phosphate resources and falling world prices also reduced income. The Nauru government made some money by suing former colonial governments for the mining destruction, allowed offshore banking that encouraged the deposit of unrecorded Russian mafia accounts—estimated at US $70 billion in 1998 alone—and in the early 2000s leased land to Australia for a detention camp for asylum seekers. Such sources of income are temporary, and the once-affluent Naurians suffer indignities including rationed electricity, grounding of the last Air Nauru plane, and cancellation of plans to hold the International Weightlifting Federation world championships on the island.

Tourism

Tourism is one of the few ways open to the people of the subregion to increase their incomes. However, the islands have different levels of access to the rest of the world and of interest shown by tourist companies. Tourism is important on some islands that are on international air routes, have particular attractions, or have well-organized tourist facilities, such as Fiji, Guam, the Marshall and Northern Mariana Islands, Toga, Vanuatu, and Western Samoa. It has little or no impact on those that lack such facilities, such as Kiribati, Micronesia, Nauru, New Caledonia, Palau, the Solomon Islands, or Tuvalu. Tourism can, however, destroy or merely exploit the last remnants of traditional culture and often pays low wages to local people. Yet it is often a better industry to work in than the low-wage factories found in other

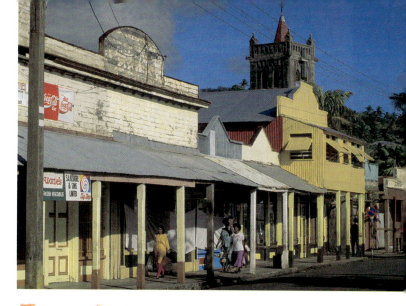

Figure 6.37 **South Pacific: urban landscape.** In Levuka, Fiji, where colonial touches remain in the landscape, a British-style church tower and wooden store buildings line Beach Street. Photo: © Robert Holmes/Corbis.

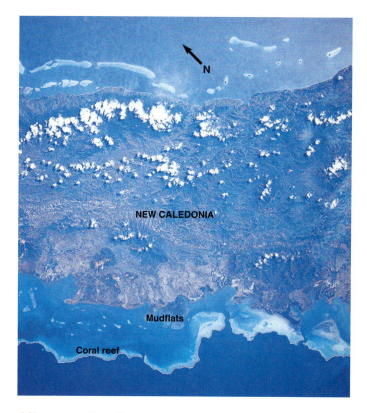

Figure 6.38 **South Pacific: island landscape.** A space shuttle photo of New Caledonia shows evidence of environmental degradation. Several fires are burning what little remains of the forest cover to provide cattle pasture. Much of the land has been degraded by open-pit mining for nickel. Offshore, the coral reef is being killed as mud from eroded land fills the lagoon to produce mudflats off the southern coast. Photo: NASA.

poorer countries. Tourism in this region remains dependent on the retention of an attractive environment and political stability and so is likely to encourage good environmental practices. Where tourism has become important, as in Fiji, it has forced a greater openness to foreign visitors' requirements, creating fresh opportunities for local businesses.

Antarctica: A Region?

Antarctica occupies 10 percent of Earth's land surface but is not settled on a permanent basis (Figure 6.39a). It remains a frozen, empty continent apart with access limited to scientists and occasional small numbers of tourists. Is it a major world region because it is so distinctive? Or does its lack of people and country jurisdictions exclude it from such status? Can it be ignored in a world regions course? Its inclusion as part of another world region is problematic, since it is so different and far removed from the other southern continents.

Antarctica's Global Status

International expeditions to Antarctica from the late 1800s concentrated on the personal physical feat of crossing the frozen continent on foot. They gave way to international claims on sections of the continent during the mid-1900s (Figure 6.39b). Some of the claiming countries established scientific surveys of importance to climate, oceanographic, and glaciological studies.

In 1961, 39 countries signed the Antarctic Treaty as a basis for nonmilitary scientific cooperation, environmental safeguards, and international control. In 1991, the Wellington Agreement banned commercial mining activities and introduced protection regulations. Some countries did not sign this agreement because they did not have an interest in the continent or did not agree with the restrictions imposed.

Antarctica's natural resources are not exploited, and the international agreements stipulate that they should not be. Only the future will tell how long this protective situation will last or whether Antarctica will turn into an open-pit mine for the rest of the world. The main interest in Antarctica in the 1990s concerns its important role as a laboratory for monitoring global climate change and the study of the development of the ozone hole above it, the extent of ice on the continent and the surrounding oceans, and the bounty of the ocean ecosystems. Research carried out in Antarctica finds atmospheric pollutants that indicate a decline in world environmental quality. In their small but increasing manner, however, research stations also degrade the local environment by pouring untreated sewage into the ocean and dumping oil drums on sites that seabirds use for nesting.

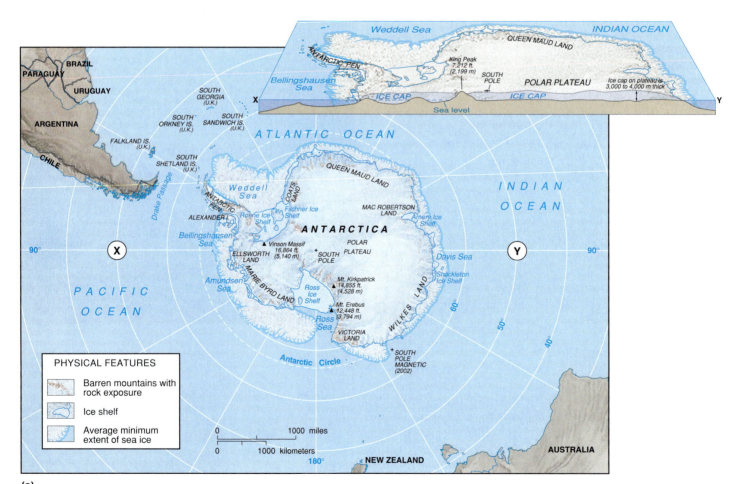

(a)

Figure 6.39 **Antarctica: the ice-covered continent.** (a) The main features and a section through 90°E and 90°W (across the center of the map, X-Y). (b) Political claims: The United States and Russia, among other countries, refuse to make such claims and do not recognize other countries' claims. (c) Penguins: animals depend on resources of the surrounding oceans. Page 267 (d) Tourist cruise ship and Antarctic coast. (e) Tourists from the cruise ship go ashore at a Danish scientific research station. Photos: (d) and (e) © Ian Cole.

Antarctica and the Southern Oceans

Without people, Antarctica's significance is in its natural environment. The divergent plate margin between the Indian and Antarctic plates separates Australia and Antarctica. The Antarctic continent formed the ancient-rock core of Gondwanaland. After the combined continent broke apart, Antarctica remained at the South Pole. As Earth's atmosphere cooled some 20 million years ago, ice accumulated on Antarctica, burying the mountain ranges that continue the line of the Andes from South America (Figure 6.39a). A large part of the continent sank under the weight of ice and is now below sea level. If the ice melted, Antarctica would form several landmasses.

The frozen continent forms its own climate with extreme cold and a heating deficit (see Figure 2.24) throughout the year that built up an ice cover over the last 15 million years. Winters have almost total darkness for several months; summers are all daylight. In winter, the ice-covered area of the oceans around Antarctica doubles in size as the sea surface freezes; in summer, glacier ice calves off icebergs into the surrounding ocean. The waters are very cold in all seasons. Perhaps Antarctica's most significant boundary is between 50 and 60 degrees south,

where cold and warm ocean water and atmosphere converge. The contrast in the atmosphere above creates a sharp boundary between Antarctic air and warmer midlatitude air, generating a frontal zone and a succession of midlatitude cyclonic storms with high winds. The southern oceans surrounding Antarctica have little land to interrupt these strong airflows and cyclonic storms. This zone of confrontation of contrasting air types effectively cuts off Antarctica from warming influences.

Despite the barrier to climatic elements entering Antarctica's atmosphere, small quantities of chlorine gases penetrated from lower latitudes to produce the ozone hole above the continent at the end of the Antarctic winter in October. This "hole" is a thinning of the protective ozone layer high in Earth's atmosphere that filters out harmful ultraviolet radiation from the sun. The hole enlarged in the 1980s and 1990s as the chlorine gases reacted with the ozone but appeared to stabilize around 2000.

During the late 1990s and early 2000s, some of Antarctica's ice shelves, where ice flowed off the continents to cover the ocean surface, melted and broke up—seen by many as a sign of global warming. But a report in 2001 showed that temperatures in the heart of Antarctica are falling rather than rising.

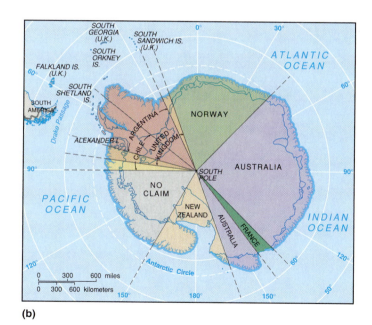

(b)

(d)

(c)

(e)

Antarctica's Resources

Antarctica is not a country and so does not have an economy of its own. Although there is no authority to regulate resource usage, international agreements prevent the known deposits of coal and other minerals from being exploited.

While the natural resources in its rocks are off-limits to exploitation, Antarctica's surrounding oceans draw fishers from all over the world. The oceans surrounding Antarctica are among the world's most profuse life zones. Antarctica's living organisms are dominated by a huge variety of sea birds including penguins that rely on the rich ocean life of plankton, fish, seals, and whales (Figure 6.39c). The Antarctic oceans are an important basis for wider ocean food chains and are being studied to gain an understanding of the sustainable levels of fishing, sealing, and whaling.

In 1982, as commercial fishers from around the world increased their exploitation of the marine resources, it was agreed internationally to regulate such fishing. Fish stocks, such as cod, together with some groups of whales, were declining. Some fishing fleets, however, claimed that they fish outside the Antarctic convergence zone that forms the northern boundary of the partially protected area or professed ignorance of the quotas that were established. Additionally, it is almost impossible to monitor fishing in this extensive area of ocean that has few ships passing through and is not the responsibility of a particular country. Related problems that cause worries over the future of the total ocean ecology include, for example, the long-line hooks used to catch Patagonian toothfish in the mid-1990s that also snared albatross and petrel birds.

Tourism

Tourists, usually based on cruise ships (Figure 6.39d), visit Antarctica to view the scenery, wildlife, and some research stations (Figure 6.39e). In 1994–1995, around 8,500 tourists visited the Antarctic Peninsula. The numbers increased with the arrival of sturdy ice-breaking ships that add landing possibilities and as global warming reduced the extent of ocean ice cover. As more tourists arrive, however, the dangers of environmental damage increase. At present, the Antarctic Treaty system does not have a code regulating the tourism industry.

Test Your Understanding 6E

Summary The South Pacific islands include large numbers of small islands grouped in mainly independent but poor countries. Papua New Guinea is the largest, having over half the total population of all the islands. Many South Pacific islands have no commercial products; others produce coconuts and copra; some export a particular mineral; and some have economies maintained by French colonial support or U.S. grants. Antarctica is a continent without a permanent population. Its resources are not exploited, and its main role is as a scientific laboratory for weather and marine ecosystem study.

Some Questions to Think About

6E.1 What attractions do the countries of this subregion offer to tourists? What are the advantages and disadvantages of the tourism industry for the inhabitants?

6E.2 Should Antarctica be included as a subregion of a world region, be a world region on its own, or be ignored by regional geographers?

Key Terms

overpopulation

Online Learning Center   www.mhhe.com/bradshaw

Making Connections

The Online Learning Center accompanying this textbook provides access to a vast range of further information about each chapter and region covered in this text. Go to www.mhhe.com/bradshaw to discover these useful study aids:

- Self-test questions
- Interactive, map-based exercises to identify key places within each region
- PowerWeb readings for further study
- Links to websites relating to topics in this chapter

Chapter 7

South Asia

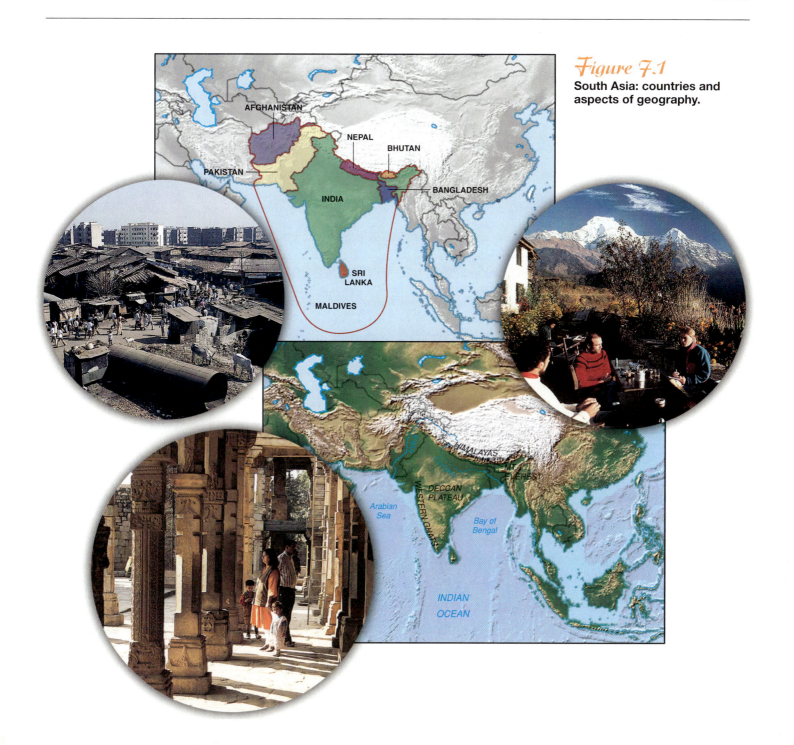

Figure 7.1
South Asia: countries and aspects of geography.

AFGHANISTAN

NEPAL

BHUTAN

PAKISTAN

BANGLADESH

INDIA

SRI LANKA

MALDIVES

HIMALAYAS

MT. EVEREST

DECCAN PLATEAU

WESTERN GHATS

Arabian Sea

Bay of Bengal

INDIAN OCEAN

Past and Present

South Asia consists of the land from the Himalayan peaks to the Indian Ocean—the part of Asia often referred to as the Indian subcontinent (Figure 7.1). Although clearly defined by its physical geography, South Asia has great internal cultural diversity that often translates into political conflicts. Furthermore, the region is home to 1.4 billion people, a total that may rise to nearly 2 billion by 2025, more than any other world region (Figure 7.2).

Today's diversity and internal conflicts flowed from a varied history that left the region with a widely observed culture but also uneven political and economic geographies. Possession of nuclear weapons by India, Pakistan, and China heightens worries that rivalries focused on the province of Kashmir may lead to war. Political instability in the wake of the war on terror-

ism that led to military intervention in Afghanistan added to tensions in the region (see the "Point-Counterpoint: Afghanistan-Pakistan-Kashmir-India," box, p. 274). Such tensions prevent the South Asian Association for Regional Cooperation (SAARC), founded in 1985, from meeting.

The human diversity of the region stems from the waves of people who entered first through the northern mountain passes and later from the ocean. Each left their marks on the land, people, and culture. Historically, the material and cultural wealth of the region vied with that of China and, later, Europe. The complexity of cultures, particularly in response to religious developments, is reflected in a distinctive literature, music, and art (Figure 7.3). The Sanskrit Vedas rank among the world's greatest religious literature.

At each stage of its history, the people of the region responded to external influences, including Muslims and the British invaders, but largely maintained the region's identity and character. India's media industry illustrates how the ancient cultures respond to modern globalization. By the 1990s, satellite delivery of TV programs enabled the globalizing media moguls to replace the single-channel Indian government programs and provide a diet of U.S. and Western films, soap operas, and sports, mainly in languages spoken by fewer than half the total population. In response, new Indian TV stations, such as ZeeTV, brought a mainly Indian emphasis and now provide programs in 15 languages. The Indian film industry that already produced more films than the U.S. industry diversified into TV soap operas with Indian venues.

The new range of TV channels were consumerist but soon focused on Indian interest groups and especially the large number of young adults. Soap operas dealt with issues of local concern such as violence against females. One result of the wider media coverage is a new pan-Indian awareness within an extremely diverse country and an increased confidence in

Figure 7.2 South Asia: comparative data by country.

Country	Capital City	Land Area (km²) Total	Population (millions) Mid-2001 Total	2025 Est.	GNI 1999 (US $ million) Total	GNI PPP 1999 Per Capita	Percent Urban 2001	Human Development Index Rank of 175 Countries	Human Poverty Index: Percent of Total Population
Afghanistan, Islamic State of	Kabul	652,090	26.4	45.9	No data	No data	22	No data	No data
Bangladesh, People's Republic of	Dhaka	144,000	133.5	180.5	47,071	1,530	21	150	44.4
Bhutan, Kingdom of	Thimphu	47,000	0.9	1.4	399	1,260	15	145	41.8
India, Republic of	New Delhi	3,287,590	1,033.0	1,363.0	441,834	2,230	28	132	35.9
Maldives, Republic of	Malé	600	0.3	0.5	322	No data	25	93	25.4
Nepal, Kingdom of	Kathmandu	140,800	23.5	37.0	5,173	1,280	11	144	51.9
Pakistan, Islamic Republic of	Islamabad	796,100	145.0	251.9	62,915	1,860	33	138	42.1
Sri Lanka, Democratic Socialist Republic of	Colombo	65,610	19.5	22.7	15,578	3,230	22	90	20.4
Total or average		**5,133,790**	**1,382.1**	**1,903**	**573,292**	**1,898**	**22**	**127.4**	**37.4**

Source: Data from *Population Reference Bureau 2001 Data Sheet*; *World Development Indicators*, World Bank, 2001; *Human Development Report*, United Nations, 2001; Microsoft Encarta (ethnic group, language, religion).

(a)

(b)

Figure 7.3 **Indian cultural expressions.** (a) Painted sculptures of the Hindu deities Kali and Ganesh adorn the exterior of the 1500s Shree Meenakshi Temple at Madurai, southern India. (b) Posters for movies, Mumbai (Bombay). The Indian movie industry is the world's most prolific and its base is given the name "Bollywood" (Hollywood in Bombay). Photos: (a) © Chloe Johnson/Corbis; (b) © Catherine Karnow/Corbis.

being Indian in a West-dominated world. Moreover, conflicts in Afghanistan, Sri Lanka, and Nepal and between India and Pakistan in the early 2000s brought the countries of this region to the greater attention of people in other parts of the world.

In this chapter, the essential understanding of South Asian cultural and political history is followed by a study of the region's distinctive natural environments that define some possibilities and limitations for human activities. The subregional studies that form the main part of the chapter focus on the modern countries in a global context, including their political status, population characteristics, and economic development.

 # Diverse Cultures

Recurring themes through the history of South Asia include attempts to unify the people of the subcontinent and protect them from external invaders. The outcome was a strong social culture that endured many political and economic changes and remains significant today. Cultural differences related to religious allegiance, including gender issues, later become enshrined in political attitudes.

Precolonial Cultures

Little is known about the earliest events in the evolution of South Asian societies. The dark-skinned Dravidian people (Figure 7.4) are thought to be the modern descendants of some of the oldest inhabitants of the region. By 3000 B.C., the Dravidians most likely established irrigation farming in the Indus Valley, developing the Harappan civilization that rivaled those in the Tigris-Euphrates and Nile Valleys (see Chapter 1). The Harappans left ruined cities and irrigation works but only a small number of seals, which have not been translated.

Country	Ethnic Groups (percent)	Languages O=Official	Religions (percent)
Afghanistan, Islamic State of	Pashtun 38%, Tajik 25%, Hazara 19%	Afghan-Persian 50%, Pashtun 35%, Turkish, 15%	Muslim 99%
Bangladesh, People's Republic of	Bengali 98%	Bangla (O), Urdu, English	Hindu 16%, Muslim 83%
Bhutan, Kingdom of	Bhote 50%, Nepalese 35%	Dzongkha (O), Tibetan/Nepali dialects	Lamaistic Buddhist 75%, Hindu 25%
India, Republic of	Indo-Aryan 72%, Dravidian 25%	Hindi 30%, 14 official regional, English	Christian 2%, Hindu 80%, Muslim 14%
Maldives, Republic of	Sinhalese, Dravidian, Arab, African	Divehi (Sinhala), English	Muslim dominant
Nepal, Kingdom of	Nawar, Indian, Tibetan, Gurungi, Sherpa	Nepali (O), 20 others, English	Buddhist 5%, Hindu 90%, Muslim 3%
Pakistan, Islamic Republic of	Punjabi, Sindhi, Pashtun, Pathan, Muhajir, Baluchi	Urdu (O), English (O), others	Muslim 97%
Sri Lanka, Democratic Socialist Republic of	Sinhalese 74%, Tamil 18%, Moor 7%	Sinhala, Tamil, English	Buddhist 69%, Christian 8%, Hindu 15%

Figure 7.4 **South Asia: Dravidian peoples.** Indian people engaged in a workshop industry in Bangalore, southern India. The dark skins often originated from the Dravidians and other possibly pre-Dravidians such as the Veddas from Sri Lanka. Dravidians occupied India before the invasions of lighter-skinned Aryan groups. Photo: © Michael Bradshaw.

Hindus and Caste

By 1500 B.C., and possibly many years earlier, lighter-skinned and taller Indo-Aryan people invaded through the northwestern passes and drove Dravidians southward into the Indian peninsula. As a result of Aryan influences, Hinduism crystallized in the Indus River valley around 1200 B.C. The Aryans brought together traditional myths and gods to form Hinduism as an inclusive nationalistic religion. **Hinduism** can most accurately be described as "the religion of the people." This makes it an ethnic religion that people are born into—although there were Hindu missions to Southeast Asia 2000 years ago and the modern Hare Krishna groups recruit worldwide. Today, most of the world's 800 million Hindus live in India. Hinduism is not a single centrally organized religion with a single sacred text but has a series of "Great Traditions" as set out by religious experts and "Little Traditions" about local gods, beliefs, and practices. Hinduism recognizes millions of gods, although local groups often see one as dominant, and incorporates beliefs of reincarnation. Many classified as Hindus (82% in the 1992 Census of India), particularly in rural areas, do not recognize that Hinduism is the title given to their local devotions: it is their way of life.

Although caste is a Hindu concept, it may have permeated all South Asian society before the arrival of the Aryans and was adopted by later invading groups. The **caste order** is similar to ethnic or class divisions elsewhere. It arose out of the Aryan wish to control non-Aryan peoples. They imposed their own social groupings: the priests (Brahmins—the top group, who decided the membership of other groups); warriors and rulers (Kshatriyas); and commoner merchants and artisans (Vaishayas). Those outside these divisions, mostly non-Aryans, were the lowest caste of menials and servants (Shudras). Outcasts were believed to be by nature "disgusting, polluting, and unworthy" and labeled "untouchables." Both the menials and outcasts were believed to pollute all they touched, and they had to live separately and use separate wells. Groups not integrated into the caste order are known as tribal peoples. Perceived as "primitive" forest dwellers or descendants of the original population, they have similar physical features to Hindu Indians and may include many who were rejected or not absorbed by Hindu groups.

The religious base of the four major caste divisions (Varna) was paralleled by the common interests of groups of people (Jatis, or subcastes) who worked in similar craft industries of roughly equal status. These economic and social divisions are thought to have evolved from tribal kinship groups into specialist feudal roles. Membership became hereditary, based on intermarriage within the group: birth determined status in society. In the typical South Asian village of the past, landowners, tillers, carpenters, potters, barbers, priests, and other groups worked within local hierarchies for the common (or landowner's) good.

Buddhism and Jainism

Buddhism and Jainism were founded in the Ganges River valley in the 500s B.C. in reaction to aspects of Hinduism. Buddha began life as a Hindu. A major early attraction of his teachings was the rejection of the millions of Hindu gods. Buddha also questioned the rigid caste system and believed in the possibility of greater social mobility. When the Mauryan Empire expanded to incorporate most of the Indian subcontinent from 321 to 181 B.C., its most celebrated emperor, Asoka, adopted Buddhism and propagated the *dharma* policy, a code of conduct focusing on social responsibility, human dignity, and socioreligious harmony. This led to a decentralization of activities, but after Asoka's death, weak successors allowed the empire to fall apart. Brahman Hinduism took over, envisaging a country as a divine creation rather than a Buddhist concept of a contract between ruler and people to control the chaotic and demoralizing effects of mortal possessions. The renewed dominance of Hinduism forced Buddhists to move out of the Indian mainland. The expelled Buddhists converted much of Sri Lanka and spread their faith eastward and northward into Southeast and East Asia.

Jainism originated at the same time as Buddhism, following a nonviolent code that was later taken up by Mahatma Gandhi in his early-1900s campaign for India's independence from the British. Forbidden to farm (and kill worms and insects), Jains became an exclusive group. Many, especially in the Mumbai (Bombay) area, are wealthy traders. They make up a small proportion of the Indian population today, but that is still 3.25 million people.

Many Invasions

Invasions kept historical Indian societies in a state of flux, especially in the northern plains, where a mosaic of distinctive peoples, landscapes, and anarchic governments resulted. For example, the short-lived invasion by the Greeks under Alexander the Great in 326 B.C. at the end of his conquests in western and central Asia left behind military and legal influences. Although the Gupta Empire brought peace, economic growth, and a new flowering of art, music, and literature from A.D. 319 to 950, invading Mongol groups, Muslims, and Turks were unable to establish lasting rule over the next 600 years.

Muslim Presence

After the Muslim invasion in the A.D. 1100s (Figure 7.5a), Muslim influences pushed southward from the Indus and Ganges Plains. However, it was not until the 1500s that the **Mughal (Mogul) dynasty** extended Muslim beliefs to the rest of the region. The Mughals were Persian Turks led by Babar, a descendant of the *Mongols* (see Chapters 4 and 5). Under the Mughals, India experienced developments in such fields as architecture: the Taj Mahal and many examples of Muslim buildings built in the 1600s survive throughout the region (Figure 7.5b).

Sikhism emerged in the early 1500s as one of several responses by some Hindus in the Punjab to the Muslim Mughal dynasties. It later resisted absorption into mainline Hinduism despite pressures from dominant Muslims and Hindus. Sikhism is marked by a strict code of conduct and many temples with linked kitchens providing food for all. The Golden Temple at Amritsar is the holiest temple, constructed by Guru Arjan in the late 1500s, and contains the Sikh scriptures that he compiled. Although small in number, many Sikhs became wealthy from their bumper crops in Punjab and continue to lobby for a separate state or even an independent country (Sikh Kalistan, the "land of the pure").

Mountain Isolation and Island Openness

While such changes affected the greater part of South Asia, some of the places at the mountainous northern margin of the region—the present countries of Afghanistan, Nepal, and Bhutan—mixed isolation with wider influences. Tribes in the Himalayas established their own ways of living. Nepal and Bhutan were occasionally invaded from Tibet (see Chapter 5), and fortified monastic schools built in the 1600s mark Bhutan landscapes.

The area that is now Afghanistan formed a mountain refuge for many ethnic groups. The Pashtuns formed the majority in the southern parts, but Tajiks and Kyrgyz people dominated the north (Figure 7.6). In medieval times, Herat near the present Iranian border, was home to a fusion of Buddhist and Persian culture that left buildings and statues. Until the 1980s, the rich cultural heritage from this time was on display in the Kabul National Museum: Islamic art, Roman bronzes, Egyptian glass, Chinese lacquerware, Indian ivories,

Figure 7.5 **South Asia: historic buildings in Delhi, India.** (a) Quwwatu'l Islam (Might of Islam) mosque inside Tomar fortress. The mosque formed the heart of a late 1100s city, and the buildings were constructed out of ruined Hindu temples. The buildings include a courtyard, cloisters, prayer hall, and the Qutb Minar, a giant prayer tower rising 72.5 m (230 ft.). (b) A 1500s mausoleum built by a Mughal emperor to commemorate the life of one of his ministers and set in a large, square-walled garden. Such buildings, similar to the Taj Mahal at Agra, which was built in the same period, reflect a period of great riches. Photos: © Michael Bradshaw.

(a)

(b)

Point COUNTER Point

AFGHANISTAN–PAKISTAN–KASHMIR–INDIA

The events of September 11, 2001, and the subsequent attacks on Afghanistan brought to light many links of agreement and conflict across sets of countries. One of these was the long-lasting dispute between India and Pakistan over the Kashmir Province at the northern end of their common border. By early 2002, India and Pakistan were drawing military forces to the border and war looked possible. As in its relationships with the former Taliban government in Afghanistan, however, the "War on Terrorism" caused the Pakistan government to hold back and even repudiate groups of Islamic radicals it supported and trained for many years. The situation along the Indian-Pakistani border remained tense—as it had since the partition adopted at independence in 1947.

This is a situation where geographic principles may assist in deeper understanding. First, it is important to know where Kashmir is and how it relates to the surrounding countries (Box Figure 1). Second, the nature of the countries, including those surrounding India and Pakistan, and of their often conflicting central beliefs and political aims is very significant. Third, the influence of external pressures, such as that from the United Nations or from major powers such as the United States, Russia, and China, is pertinent. After September 11, 2001, and its aftermath, it is increasingly difficult for any country to operate without reference to others—be it Afghanistan, Pakistan, India, or the United States. Like many others throughout the world, the relationships among the countries in this region are complex and debated. It is unlikely that there will be a rapid solution.

Some history of the Kashmir situation is basic to present events. In the mid-1800s, the British sold the rulership of the Muslim state of Kashmir to a Hindu maharajah for around US $50 million at today's prices. In 1947 at independence, Kashmir opted for being part of India. Although the maharajah tried to avoid the need to join India or Pakistan, an invasion by Pakistani tribesmen forced his hand, and he chose India in return for military help. The Indian prime minister, Jawaharlal Nehru, in his eagerness to get popular ratification for the accession, brought Pakistan's 1947 invasion to the notice of the United Nations and urged Pakistan forces to withdraw until a referendum allowed Kashmiris to choose between India and Pakistan. After the 1948 cease-fire, Pakistan held onto the one-third it conquered, much of it close to the city of Rawalpindi and later the new capital, Islamabad. At that stage, Kashmir's main Muslim leader, Sheikh Muhammad Abdullah, who became prime minister in 1948, preferred India's secular socialism to the Muslim ideology of Pakistan. The Pakistanis, however, never withdrew, and Nehru never put the matter to a vote.

This is often where the debates begin. Did the maharajah persecute Muslim subjects? Did the tribesmen attack spontaneously for their defense, or were they pushed by Pakistan's new government? Did the British support India's new government in acquiring Kashmir?

At first, the maharajah, who only handed over control of defense, foreign affairs, and communications to India, leaving Kashmir with its own prime minister until 1965, encouraged local politics. In 1953, suspicions (disputed) that he plotted with Pakistan and the United States earned Sheikh Abdullah 11 years' imprisonment.

India progressively integrated Kashmir, moving many of its Hindu population to Delhi and setting them up with craft industry employment making rugs, woolen clothes, and decorated plates and selling them in tourist store outlets. In 1964, India extended to Kashmir the right of the federal government to dismiss state governments. Each decision, however, seemed to thwart the popular will of Kashmiris. India and Pakistan fought over Kashmir in 1965–1966, when India accused Pakistan of backing insurgents.

War broke out again in 1971, when Bangladesh (East Pakistan) became independent, backed by India. Kashmir was another source of fighting in that war, after which the 1972 Simla Agreement divided Kashmir into the Pakistan state of Kashmir and the Indian state of Jammu and Kashmir. This agreement established today's line of control.

Any goodwill generated by the first genuinely free Jammu-Kashmir elections in 1977 that brought back Sheikh Abdullah collapsed with the ousting of his son from power in 1984. Antigovernment Muslim groups contested the 1987 state elections, but Delhi annulled some of their successes. From 1989, separatist violence that India alleges is backed by Pakistan was matched by Indian military repression.

By the late 1980s, Pakistan looked friendlier to Kashmiris than India despite having a very different culture based on Islamic rules and military dictatorships. Indians believe their democratic constitution can provide liberation for Kashmiris. In fact, most Kashmiris see themselves as prisoners of either country and are weary of continuous fighting. They prefer independence.

It is difficult to see who speaks for Kashmiris. The state government based in Srinagar is very unpopular. Separatist politicians based around the All-Party Hurriyat Conference have yet to test their popularity in elections but do not refute suspicions that they repeat Pakistan's policies. Militant groups formed as groups of local young men went to Pakistan for training and arming, and returned with other Pakistanis linked to Afghanistan's Taliban.

Some Kashmiris look to the United States and its allies as potential saviors that can push India into a settlement. But India and Pakistan are the real arbiters. India shifted from inaction to attempts to talk with Pakistan or separatists. The current Pakistani crackdown on terrorists causes Indian cynicism about Pakistan's promises when banned groups such as Lashkar-e-Taiba continue to act after merely

Box Figure 1 **Kashmir.** Map shows the provincial area and the line of control between Indian and Pakistani occupation.

changing their names. India, however, keeps a security force of 400,000 men in Kashmir and hopes the crackdown will moderate the militancy. There is speculation that Pakistan will return support to more moderate Kashmiri groups such as Hizbul Mujahideen.

Any election or referendum faces major problems. If separatist groups, which deeply distrust India, participate, they have to declare themselves Indian citizens to qualify to vote. Without the separatists, however, Kashmiri Muslims will mock the result as worthless. The Indian government rejects proposals for foreign election observers or outside mediation as questioning its authority. This attitude continues a high-handed and superior approach based on the original partition decision, Indian democratic institutions, and resistance to terrorist attacks but avoids discussion of how India alienated the Kashmiri people.

Partition of Kashmir between India and Pakistan with a degree of autonomy for both parts of Kashmir might then occur. Pakistan conceded much after September 11, 2001, withdrawing support for Taliban in Afghanistan and shutting down a potential war against India after terrorists attacked the Indian parliament in Delhi. It remains to be seen whether Pakistan can accept a more democratic and humane version of the status quo Pakistan fought against for over 50 years. President Pervez Musharraf of Pakistan is a military dictator who gained international disapproval when he dismissed the previous democratic (but corrupt) government in 1997, ordered a nuclear bomb test in 1998, and oversaw training camps for Islamic extremists. Against this, he ended ties with the Afghanistan Taliban and

banned Islamist groups likely to annoy India, showing that he is at least preparing the ground to act on his stated wishes to modernize Pakistan and bring in a democratic government again. Most Pakistanis see the winning of Kashmir and supporting the "freedom fighters" there as of much greater importance than backing the former Taliban government and its terrorist al-Qaeda in Afghanistan.

Both India and Pakistan took advantage of U.S. policies arising from the war on terrorism from late 2001, but the Kashmir question threatened to undermine the uneasy peace. The buildup of Indian troops along its border with Pakistan took Pakistani units from their patrols along the Afghanistan border, allowing Taliban and al-Qaeda elements to escape more freely. India used the situation to defuse Islamic militant terrorism in Kashmir and Delhi, getting U.S. President George W. Bush to freeze assets of Pakistan-based groups. The war on terrorism broadened to include cooling antagonisms between two nuclear powers who became allies. Already seeking a new relationship with India as a counterweight to China, the U.S. administration took the opportunities after September 11, 2001, to reshape relations with Pakistan as well. The complexity of having to deal with terrorist groups operating in Kashmir is matched by international support for Pakistan and longings inside Pakistan for a period of peace and development.

This account is written in the midst of the continuing tensions centered on Kashmir and Afghanistan. Add new information, gleaned from newspaper websites, to debate the following Point-Counterpoint issues.

POINT	COUNTERPOINT
Pakistan: Kashmir should be ours; India took it unfairly.	India: We assumed control of Kashmir after a request by the Muslim maharajah following a precipitate Pakistani invasion.
Kashmiris: We prefer independence to either secular Indian rule or Islamic Pakistani rule.	India and Pakistan: Kashmiri independence is not feasible or offered by either country.
India: Pakistan trains terrorists to disrupt society in Kashmir.	Pakistan: Those who resist Indian domination within Kashmir are local tribespeople fighting for their rights of self-determination.
India: Pakistan held onto one-third of Kashmir despite a United Nations ruling that a referendum should decide Kashmir's future.	Pakistan: India delays a referendum for no good reason except it knows it will lose. It even annulled the 1977 elections that would have given a good indication of support for Pakistan control of Kashmir.
Kashmiris: We feel that we live in a prison under Indian military guard and do not like the alternative of Pakistan military dictatorship control. If we take part in elections, India will declare we are their citizens, and we have little confidence that India will moderate its high-handed attitude.	India: The Kashmiris would prefer a democratic constitution and have already experienced good results of Indian provisions. Pakistan: The years of Indian suppression makes the people feel that we are friendlier.
India: After September 11, 2001, Pakistan sent terrorists into India and nearly precipitated war.	Pakistan: The government is against terrorism and did not send those who attacked the Indian parliament in Delhi. We demonstrated this by withdrawing support for the Afghan Taliban against the wishes of many Pakistanis.

Figure 7.6 **Afghanistan: ethnic groups.** The complexity of ethnic groups by language in the country and links across borders.

Ethnolinguistic Groups: Baluch, Pashtun, Hazara, Nuristani, Ismaeliens, Turkmen, Uzbek, Tajik, Kyrgyz, Other

and Buddhist collections. Soviet military, looting and export through Pakistan, and vandalism under the Taliban government from 1996 damaged this collection.

Its closeness to the Indian peninsula and openness to ocean-borne trade and conquest had major impacts on Ceylon (modern Sri Lanka). A migration of Indo-Aryan peoples from northern India formed the majority Singhalese people with Buddhist beliefs. They developed a civilization based on irrigated rice agriculture in the north-center and southeast of the island that lasted from around 100 B.C. to the A.D. 1200s. During the later part of this period, Tamils—Dravidian peoples from southern India—established kingdoms in northern Sri Lanka based around Jaffna. In the A.D. 900s, Tamil mercenaries helped settle disputes over leadership among the Singhalese. As Tamils expanded their influence, Singhalese peoples drifted southward, abandoning the irrigation systems and relying on rain-fed agriculture. They grew spices such as cinnamon. Arab traders settled and controlled the overseas spice trade.

The growing Arab (Muslim) control of medieval Indian Ocean trade routes included the Maldive Islands, which form a large and widespread group stretching southward into the Indian Ocean. Chinese and Arab merchants first set up trading posts to exchange their goods for local fish, coconuts, and shells. The islands became solidly Muslim by the A.D. 1100s and remain so.

Colonial Impacts

Trading Expansion

The great wealth of this region that attracted invaders over many centuries became a goal for European adventurers in the phase of mercantile expansion from the mid-1400s. They

searched for an alternative route to get Indian products and riches to Europe without traveling through the increasingly hostile Muslim countries of southwestern Asia. While most earlier invasions came through the mountain passes from the north, the Europeans arrived by sea and found few defenses. In 1498, the Portuguese explorer Vasco da Gama reached India after sailing around southernmost Africa. The Portuguese and Dutch established trading stations, such as Goa, around the coasts of South Asia. The first British trading post followed in 1612. During the 1600s, the Dutch forced the **(British) East India Company** out of the East Indies (see Chapter 6), and the British switched to India, ousting their Portuguese and Dutch rivals from most trading centers. The British East India Company took the Portuguese port of Bombay ("good bay," now Mumbai), built a new port at Madras (Chennai), and began trading with the most populous area of Bengal around modern Calcutta (Kolkata).

In the early 1700s, internal factions in the Mughal Empire led to political chaos. The region splintered into small and large kingdoms, often established and ruled by Muslim or Hindu leaders who styled themselves as maharajas, or princes. Foreigners took control of the increasing trade in cotton, cloth, rice, and opium. When the French contested British control, they were defeated, and the Treaty of Paris in 1763 left them with only a few small trading ports. Over the next 100 years, the British East India Company, backed by the British Indian army that employed Indians as foot soldiers, increased its hold on South Asia, taking a particularly strong position in Bengal and the peninsula. Many of the company directors and employees became extremely rich, buying huge homes and honors back in England. British ports expanded to handle the cargoes from the new trade. The company's area of influence was extended southward into Ceylon in 1798 and westward into Punjab in the mid-1800s.

British Indian Empire

Following several uncoordinated revolts, a major mutiny of its Indian Sepoy troops in 1857, which many Indians see as "The First Independence War," led the British government to take full political control of South Asia. It abolished the British East India Company and established the **British Indian Empire,** often referred to as the "British Raj" (*raj* means rule or government). Britain did not need to conquer many smaller princely states, since those states were run in harmony with British policies. At the time of independence in 1947, 40 percent of South Asia, with almost one-fourth of the total population, remained under "independent" princely family governments. The 600 princely states ranged from Hyderabad, with 14 million people in 1947, to Jathiamar, with only 200 (Figure 7.7). Almost the whole subcontinent was part of the empire.

The British saw their role as "civilizing" India through Western education, new technology, public works, and a new system of law. However, British interests redirected India's farms toward producing raw materials for British industries. The Indian textile industry was suppressed in favor of British cotton goods. The region entered the expanding global economy of the later 1800s as the export of Indian cotton to other

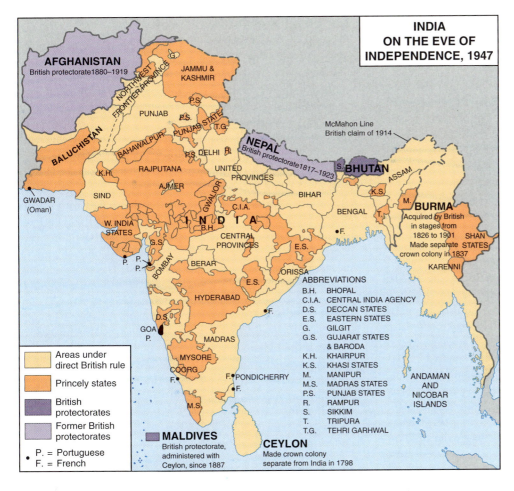

INDIA
ON THE EVE OF
INDEPENDENCE, 1947

Figure 7.7 **South Asia: the British Indian Empire just before independence in 1947.** The princely states had working arrangements with Britain but were not ruled as part of the Indian Empire. At independence, they agreed to be part of the new countries of India and Pakistan and subject to their governments. Kashmir is still disputed between India and Pakistan.

countries compensated for negative trade balances that Britain had with those countries. Furthermore, the British Indian army became a tool of attempted imperial expansion into Afghanistan and Burma in the 1880s and into Tibet in 1903–1904. Burma became part of the British Indian empire in 1885 (see Chapter 6), but Britain established Afghanistan as a separate country.

From the 1850s, Nepal had a close relationship with Britain, and many Nepalese served in the Indian army. Following British attacks in the 1800s, Nepal remained independent with pro-British rulers who hired out Gurkha soldiers (from different Nepalese ethnic groups) to the British. Bhutan was annexed by Britain in 1826 and given autonomy in 1907.

British imperial rule had massive effects on the geography of the peninsula. Its mercantile economy focused on primary products to be exported to Europe, defining the resources that could be developed; determined what was produced; altered patterns of land control; selected areas for development and cities for growth; and controlled external trade. British engineers irrigated land in the Indus and Upper Ganges River basins to produce cotton for export. They built railroads from the main ports to move troops and exports. Former communal land was reallocated to larger and smaller landowners, forming interest groups that were expected, in return, to support the colonial administration. The cities of Calcutta, Bombay, Madras, and Delhi grew faster than others after the British East India Company designated them as focuses where lines of

overseas trade and internal communications met. The English language became the unifying language among the traders and elite families.

Ceylon and the Maldives

Ceylon and the Maldives were also incorporated as British colonies into the world economic system that focused on the demands of the colonizing country and other wealthier markets rather than on local needs. Wealth was transferred to Britain, and the food needs of the local people were largely ignored in favor of exported beverages and spices.

In Ceylon, the Portuguese and Dutch were the early traders and colonists. The Portuguese left a strong legacy of Roman Catholic missions and of Portuguese as a trading language. In 1795, the British took Ceylon, which remained a crown colony until its 1948 independence. British companies set up plantations for growing tea, rubber, and coconuts, instead of the traditional spices. More Indian Tamil labor was brought in to work on the plantations, increasing local ethnic tensions. Their better education in Christian schools in the north enabled the Tamils to take most of the colonial clerical jobs. A new Western-educated elite of local plantation owners attempted to emulate the social life of the colonials. The emphasis on export crops, however, was accompanied by the neglect of traditional agriculture, leading to a decline in the rice crop and a need to import half the island's needs.

Paths to Independence

Afghanistan

After a century of attempted dominance over the Northwest Frontier area of India, the British created the country of Afghanistan in the late 1800s (Figure 7.8). The northern border defined at that time established a limit, or buffer, to Russian expansion. The new country brought together groups fragmented by language, creed, geography, and historic cultural traditions.

King Abdur Rahman (1880–1901), backed by British money and weapons, tried to create the basis of a centralized Afghanistan. His and later attempts to build and unify the new country were often ruthless, resented by the intensely independent groups, and generally unsuccessful. The widespread alienation brought down the monarchy by the 1970s, returning the country to its tribal divisions. In 1978, a Communist revolution attempted to restart a program of modernization but returned to Abdur Rahman's savage methods to achieve those ends—with similar results. Dependence on Soviet Union subsidies and military support sparked resistance against the government from many mujahideen groups, supported by the United States, and they eventually overthrew the Communists in 1992. The victorious mujahideen, however, could not provide any unified authority. After a period of violent chaos, the Taliban, based on young militant Islamists (*taliban* is Arabic for "students"), conquered most other groups by 1996 and imposed a further phase of repressive government, separating girls and women from society. Even after the Taliban defeat in late 2001, many women continued to cover themselves with the pale blue *burka* as a means of protection from male attention.

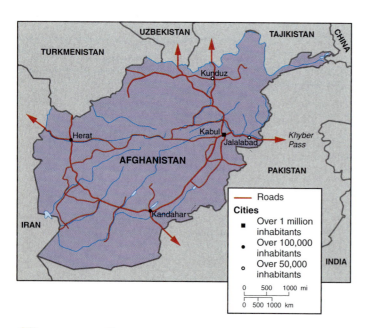

Figure 7.8 **South Asia: Afghanistan and its frontiers.**
Note the few border crossings shown by arrows.

The Origins of India and Pakistan

The extension of British rule during the late 1800s made Indian leaders aware of an emerging national identity across the politically fragmented subcontinent. In 1885, a combination of cultural background and personal interest led a mainly Hindu elite (Brahmins) to form the **Indian National Congress Party** with a secular and multicultural basis. Leaders of the minority Muslim population at first thought it safer to build links with the colonial order, but the growth of National Congress Party influence led them to form the **Muslim League** in 1906 with the aim of establishing an Islamic country.

During World War I (1914–1918), the Indian army fought alongside Europeans in a European war, leaving many in India to suffer. After the war, India became less vital to Britain's economic interests of trade, investment, and political empire. Pressures increased for independence and received sympathy. In the 1920s, Mahatma Gandhi popularized political issues among a deeply religious people and welded together a coalition of interests. His policy of nonviolent resistance brought the country to a standstill over his wish that the British agree to consult more Indians.

By 1940, Muslim fears strengthened that they would be an underprivileged, even persecuted, minority if a single country covered the whole subcontinent. Britain and the (largely Hindu) National Congress Party favored a single country in South Asia with a political structure based on a weak central government and strong provinces.

The Muslim League, however, requested a separate country to be called Pakistan and formed of provinces in which there was a majority of Muslims, including the whole of Punjab and Bengal. These two large provinces, however, had Muslim majorities of only some 56 percent and large numbers of Hindus and Sikhs. The British government proposed dividing the provinces of Bengal and Punjab (see Figure 7.7) between India and Pakistan, placing the great majority of territory and people in a new India and leaving the Muslims with two separate areas to the west and east of India.

At the end of World War II, pressures for a rapid handover of government resulted in a grudging Muslim acceptance of the smaller East and West Pakistan. Few people in these new countries were happy. Many Muslims and Hindus caught in the wrong country fled across the borders to avoid ethnic bloodshed. Some 12 million people were displaced, but over a million died in clashes. Kashmir (see Figure 7.7) remains a major border issue between India and Pakistan (see the "Point-Counterpoint: Afghanistan-Pakistan-Kashmir-India" box, p. 274).

Ceylon Becomes Sri Lanka

The majority Buddhists in Ceylon pressed for independence early in the 1900s. Buddhist revivalism among the Singhalese alerted the Tamil population to the need to preserve their separate identity, which had been protected in colonial times. Ethnic differences and resentments surfaced before independence in 1948. Although Ceylon was not partitioned at independence, tensions were increased by Singhalese nationalists in the new government and their change of the country's name to Sri Lanka.

Natural Environment

The natural environments of South Asia contain great contrasts from the high mountains to broad plains and from arid to rainy tropical climates that have major impacts on the human geography. Long dry seasons, uncertain rainfall, and large areas of arid conditions make water basic and crucial in sustaining the livelihoods of the large population of the region, where some form of farming is still the occupation of two-thirds of the people. Such a large population also places stresses on the natural environments.

Monsoon Climates

The monsoon climatic environment of South Asia (Am and Aw in Figure 7.9) brings heavy summer downpours of rain over much of the Indian subcontinent but little rain at other times of the year. Cut off from contact with Central Asia and its freezing winter air by the Himalayan Mountains, South Asia remains warm and very dry in the winter. In this season of the dry monsoon, winds flow outward from high atmospheric pressure over the northwest of this region. Only northern Sri Lanka and parts of southernmost India receive rain at this season—from winds that blow out from the continent, over the Bay of Bengal, become moist by evaporation from the ocean surface, and turn toward the land.

As the land warms in early summer, temperatures in South Asia become unbearably hot and air rises, lowering atmospheric pressure. When the wet monsoon breaks in June or July, the winds change direction, being drawn into the low atmospheric pressure area over the continent. Southwesterly winds from the Indian Ocean bring moisture, causing the highest rain totals to fall on the western coastal mountains of the peninsula. Precipitation at high elevations, often as snow, occurs on the Himalayas. Some of the world's largest annual rainfall totals, over 10,000 mm (400 in.) per year, are recorded in the Assam hills (northeastern India; see Figure 7.7). In all these cases, the lift forced by humid air flowing up and over mountains (the orographic effect) adds to the amount of condensation and precipitation. Each year during the summer, a small number of tropical cyclones occur in the Bay of Bengal, bringing flooding and death to the Ganges-Brahmaputra Delta in Bangladesh.

The eastern part of the peninsula and the Ganges Plains receive lower and more irregular rainfalls that vary sharply from year to year. Most of the monsoon rains fall on the western mountains, leaving the eastern lands in a rain shadow. Since most people live in regions of lower rainfall, the amount of variability has great human significance.

The monsoon rains also miss most of the northwestern parts of India and nearly all of Pakistan, which remain dry throughout the year, forming one of the world's major arid regions. The main source of water in this part of the region is the rivers, such as the Indus and Ganges that are fed in early summer by the melted snows of the Himalayas.

World's Highest Mountains and Deep Valleys

The Himalayan Mountains include Mount Everest (8,848 m; 29,028 ft.), the world's highest mountain, and nearly 50 other peaks that rise over 7,500 m (25,000 ft.). They form a northern wall-like boundary to the region (Figure 7.10a). The mountain wall is the southern margin of the high and broad plateau of Tibet (China), forming a wide barrier to both climatic influences and incursions of all but the most determined people.

The northern mountain wall of South Asia formed when the continental plate fragment that is peninsular India crashed into the Asian continental plate (Figure 7.10b), thrusting itself beneath the main mass of Asian crustal rocks. The collision of two continent-carrying plates caused thickening of the low-density crustal rocks to produce the world's highest mountains. The continuing thrust of peninsular India beneath Asia, signaled by earthquakes, raises the mountains by 6 cm per year. While they are rising, however, glaciers and rivers cut into them and wear them down. At present, an approximate balance exists between the rate of rise and the rate at which these mountains are being eroded. Many of the plentiful minerals of Asia (see Figure 5.10) are related to such geologic activity along these ancient plate margins.

The headwater streams of the Indus and Ganges River systems breach the Himalayan wall in the far northwest. Deep

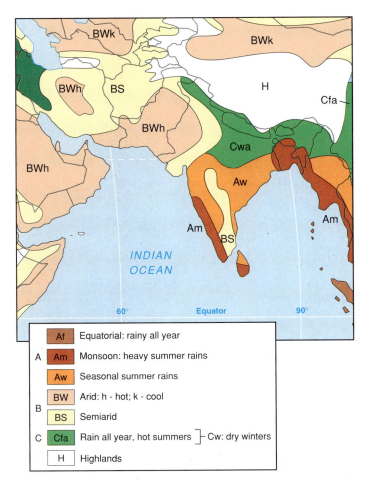

Figure 7.9 **South Asia: climates.** Climate map. Note the extent of full monsoon, lesser summer rains, arid climates, and highland climates.

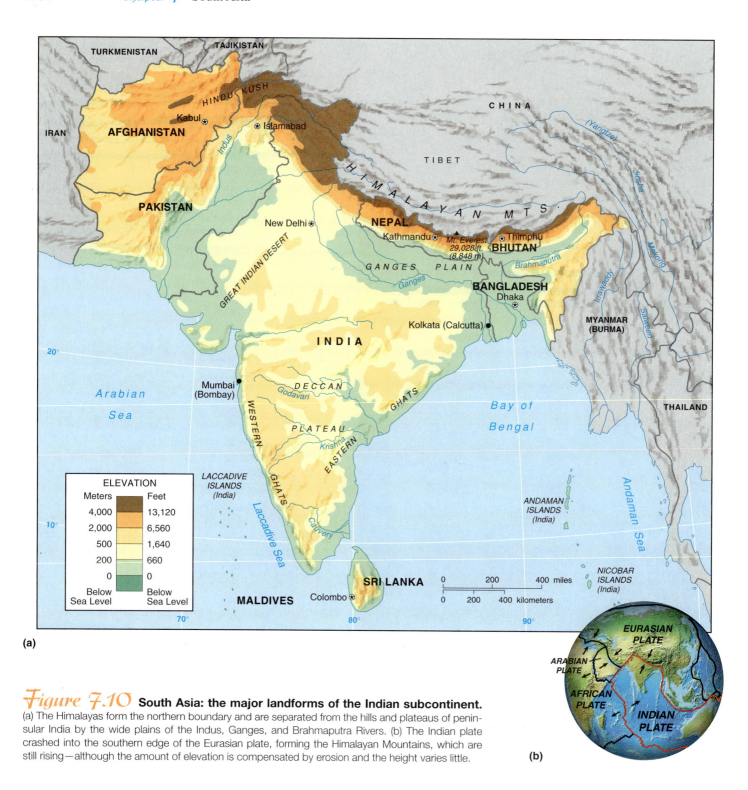

Figure 7.10 South Asia: the major landforms of the Indian subcontinent.
(a) The Himalayas form the northern boundary and are separated from the hills and plateaus of peninsular India by the wide plains of the Indus, Ganges, and Brahmaputra Rivers. (b) The Indian plate crashed into the southern edge of the Eurasian plate, forming the Himalayan Mountains, which are still rising—although the amount of elevation is compensated by erosion and the height varies little.

valleys, carved by glaciers and rivers, lead to lower routes through the Hindu Kush range, such as the Khyber Pass, that give access to Afghanistan and Central Asia and formed historic points of entry for Aryan and Central Asian groups. The passes remain a focus of tensions between India and Pakistan, which vie for control of the Kashmir region (see Figure 7.7). In the northeast, the ranges are lower, as they turn southward into Myanmar, and include breaks eroded by the Brahmaputra River system that afford access from China.

Peninsular Hills and Plateaus

Peninsular India is a continental plate fragment of very ancient rocks that was tilted with the highest points along the western coast—the Western Ghats—rising to just over 2,500 m (8,000 ft.). Much of the peninsula consists of plateaus sloping eastward to the Bay of Bengal—a gradient followed by many rivers. Along the east coast, a broken line of hills, the Eastern Ghats, rises in places to 1,500 m (5,000 ft.). A coastal plain of varied width encircles the peninsula. It is widest where rivers such as the Krishna and Godavari drain off the plateau, creating deltas at their mouths on the eastern coast.

As the continental plate fragment forming the peninsula of South Asia broke away from the other southern continents over 150 million years ago (see Figure 6.9c), layers of volcanic lava poured out through cracks in Earth's crust, covering large areas of older land. The lava flows are most prominent in the northwestern part of the peninsula in the Deccan Plateau.

Major River Basins

Between the Himalayas and the peninsular plateaus is a wide lowland zone crossed by three of the world's major river systems—the Indus, Ganges, and Brahmaputra Rivers—after they flow down from their Himalayan sources. These plains are now densely populated and intensively farmed, with the river water being used for irrigation. The Ganges River has great religious significance for Hindus (Figure 7.11).

The combination of high mountains and heavy precipitation produces some of the most rapid physical landscape modifications in the world. The melting Himalayan snows and the monsoon rains combine in powerful flood flows in the Indus, Ganges, and Brahmaputra River systems. The strength of flow can be gauged from the fact that the Ganges and Brahmaputra sweep debris 3,000 km (nearly 2,000 mi.) out to sea in the Bay of Bengal. The flows in the Indus system bring irrigation water to the Thar Desert.

Rock particles and fragments worn from the Himalayas and carried by these rivers built deposits of **alluvium** up to 3,000 m (10,000 ft.) thick. The surface of the deposits is generally flat but marked by small-scale relief of a few or tens of meters where the rivers cut new channels in older material. Along the northern margins of this plain, large **alluvial fans** of gravel mark the junction of the steep mountain courses with the lowlands. The sudden lowering of gradient as they enter the lowlands causes the rivers to drop the gravel and sand they can carry along their steeper-gradient mountain valleys. In the drier areas of the northwest are sections of **badlands topography,** where occasional rainfall runoff cuts dense networks of steep-sided gullies into unvegetated alluvial deposits.

Forests and Soils

Much of South Asia was originally forested, including the peninsula, northern river plains, and Himalayan foothills. After centuries of clearance by expanding populations, the teak forests of southern India and the forests on the Himalayan slopes are almost all that remain. These are being reduced further in size by increased cutting. Much of the former woodland was cut to expand the cultivated area and for fuel. The northwest with its arid climate forms the Thar Desert.

The most fertile soils formed beneath the forests, in areas subject to annual flooding, or around the lava plateaus. Long use for farming, however, reduced the soil quality, while the drier areas experienced soil erosion, waterlogging, and salinization.

Natural Resources

The ancient rocks of the South Asian peninsula contain mineral ores including iron and uranium, while the newer rocks on top

Figure 7.11 **South Asia: the sacred, life-giving Ganges River.** The river at Varanasi draws great numbers of Hindu pilgrims to bathe in its waters. The river is regarded as a god. It is supplied by Himalayan meltwaters and is liable to flooding. Pollution from agricultural chemicals, industrial wastes, and domestic wastes is an increasing problem in this river. Photo: © Ian Coles.

contain one of the world's largest coal reserves. Deposits of oil and natural gas occur in the thick sediments beneath the major valley areas and just offshore in India and Pakistan.

For most people in this region, the critical natural resource is water. In the many thousands of villages across South Asia, access to water is a constant focus of efforts to manage the environment. Early in history, the monsoon rains made it possible to support large populations at subsistence level. The Ganges and Brahmaputra brought economic life to the lowlands south of the Himalayas, especially after irrigation and water storage techniques were developed. The Indus tributaries brought the water to its lower arid valley, where one of the world's earliest civilizations developed (see Chapter 1). The waters flowing northward from Sri Lanka's central hills also fostered a civilization based on irrigation. Large hydroelectricity projects continue to harness the water power of the mountains.

Water shortages are major problems for both urban and rural areas of India. While the urban middle classes can install storage tanks against times of shortage, the slum dwellers wait in line with buckets. While members of the village upper castes use good wells, the untouchables have to find their water elsewhere.

Irrigation in South Asia is of many types. Large dams and canal systems, irrigating over 4,000 hectares (10,000 acres) each, are supplemented by medium-sized dams that have been built in planned and coordinated projects over the last 100 years. Smaller traditional earth dams, also known as tanks, were built hundreds of years ago by local communities and are most common in the peninsula. Over a thousand large or medium-sized projects were constructed after 1947. Most of the remaining sites involve grandiose construction plans for the upper Ganges and Indus in the Himalayas.

Natural Environmental Problems

People living in South Asia face a number of environmental hazards. Some are an integral part of the dynamic natural environment, but others are the outcomes of human clearance of land, population increase, and the context of the global economic system.

The dynamic natural environment is brought about partly by the clashing plates that produced the Himalayas and continue to set off earthquakes, such as the one in Hyderabad in 1993 and the one that centered on Bhuj, Gujarat, in January 2001. Both caused loss of life and destroyed poorly constructed housing. More frequent earthquakes along the base of the Himalayas are less devastating, although very large ones occurred in 1905 and 1934.

In the future, the Maldive Islands and much of the Ganges-Brahmaputra Delta that are only a few feet above sea level face drowning by a rising ocean level as global warming proceeds. The high costs of raising dikes and building sea defenses make such measures unlikely.

Human-Induced Environmental Problems

In India, Hinduism, Jainism, and Buddhism proclaim the importance of care for the environment. High-caste Hindus avoid killing animals for food, and Jains abhor any killing. Mahatma Gandhi's philosophy was built around a lifestyle that had a low impact on nature and an acceptance that humans have a place within nature instead of over it. Despite such views, the huge rise in India's population and the necessary growth in its economy leave landscapes of exploitation, degraded resources, and pollution. The union (federal) and state governments are concerned first with providing subsistence and jobs for their people.

Pollution of air and water is common. To make the herbicides and pesticides used in the Green Revolution, chemical factories pollute land and water, while concentrations of toxic chemical manufactures cause deaths. In 1984, leaking gases killed 3,000 people and seriously injured 50,000 at the Union Carbide pesticide plant in Bhopal, India (Figure 7.12a). This was a well-publicized, but not isolated, instance of exporting pollution to poorer countries. The multinational corporation had resited "dirty" manufacturing facilities in India after its home country increased its restrictions.

Increasing damage of the Taj Mahal because of air pollution led to the closure of metal workshops in the surrounding area to reduce the sulfur gases that damage building stones. However, it is only possible to take such expensive action on a local scale (Figure 7.12b). The agglomerations of industrial development in the "Golden Corridor" of Gujarat state north of

(a)

Figure 7.12 **South Asia: Indian manufacturing contrasts.** (a) The Union Carbide chemical plant and nearby housing in Bhopal. High-tech industries exist alongside heavy industries, such as large-scale iron and steel and chemical works and thousands of small-scale and craft industries. (b) Air pollution from local industry causes the stonework of the Taj Mahal in Agra, India, to crumble, although local regulation reduced pollution to help save this monument of great tourist interest.

Photos: (a) © Jagdish Agrawal; The Image Works/ (b) © Sheldon Collins/Corbis.

(b)

Mumbai and along the road leading north from Delhi through Uttar Pradesh produce massive pollution from sugar mills that destroys farming livelihoods.

Environmental pressures are high on the Ganges River basin, where nearly 40 percent of the Indian population lives. The land is intensively cultivated (Figure 7.13), and there are many factories and urban areas. For example, heavily polluted effluent from a concentration of leather tanning works enters the Ganges River close to some of the main Hindu ritual bathing sites. The faithful believe Mother Ganga will preserve them from harm, but health dangers from various sources increase. In recent years, an upsurge of trash dumping in the Ganges further worsened the water quality. The Ganga Action Plan formulated and passed by the Indian government in the 1980s envisaged a cleanup of the river. A few projects were implemented, but lack of funding, continuing industrial malpractice, illegal passing of wastes into the river, and illiteracy brought the program few successes.

Water was instrumental in the Green Revolution. In the states of Punjab and Haryana, and parts of Uttar Pradesh, production rose and prosperity increased. However, the revolution was based on irrigating more land, applying more fertilizer, and using more pesticides. The chemicals washed from fields into rivers. Irrigated cropland in South Asia increased from 18 percent (132 million hectares; 326 million acres) in 1960 to 32 percent (174 million hectares; 430 million acres) in 1980. The sinking of deeper wells lowered water tables, affecting smaller farmers who could not afford deeper wells.

The environmental problems associated with irrigation and hydropower projects occur in both the source and using areas. Upstream, the storage of water requires building dams and flooding deep valleys. In 1993, local activists, supported by foreign environmentalists, influenced the World Bank to withdraw its funding for a dam on the Narmada River in the power-hungry states of Madhya Pradesh and Gujarat. Since 1947, some 10 million people have been displaced for such projects, but the tolerance is now less. In 2000, work resumed on some of the major sections of this project as the National Hydro Power Corporation became involved. Ninety thousand people being displaced received improved compensation. Even where dams are constructed in sparsely inhabited mountains, they eventually fill with sediment or may be destroyed by earthquakes.

Downstream, the extraction of water for irrigation alters river flow patterns; whereas the Ganges River used to flood regularly, those floods are reduced, and dry season flow is much lower. In areas using irrigation, poor management of water feeds it to fields where the terrain or subsurface conditions are unsuitable. Soils become waterlogged or saline.

Wells already provide nearly half the irrigation water used and remain the main source of irrigation growth. Tube wells drilled by modern rigs increased in numbers from 200,000 in 1960 to over 4 million in 1990. Water is lifted from the wells by diesel or electric pumps, in contrast to the human or animal power used in older wells. Wells have environmental advantages over canal-fed irrigation in that they use local water supplies, keep the water table low, and encourage downward water movement in the soil to avoid salinization. They are increasingly combined with canal irrigation so that local groundwater sources can be replenished.

Figure 7.13 **South Asia: rural pollution.** Water buffalo gathered beside a badly polluted river in the dry season of northern India. Photo: © Robert Weight/Ecoscene/Corbis.

Test Your Understanding 7A

Summary South Asia has a huge and growing population and large numbers of extremely poor people. Its cultural diversity provides an identity but also fuels internal conflicts. Technical and political sophistication and existing infrastructure, however, suggest prospects of better things.

South Asia occupies the Indian subcontinent, hemmed in by the Himalayan Mountains. Passes through the mountains provided routes for invaders. Incoming Muslims confronted the Hindu and Buddhist religions that originated in South Asia. The British Indian Empire was the world's most intense expression of modern colonialism. The region's economy was oriented to the needs of Britain until independence was achieved in 1947.

The physical environment of South Asia is marked by monsoon and arid climatic environments, the world's highest mountain system in the Himalayas, the wide plains of the Indus, Ganges, and Brahmaputra River basins, and the hills and plateaus of the peninsula. The dynamic physical environment produces earthquakes, summer river floods, and tropical cyclones. Human economic development makes use of the mineral and water resources and results in increasing modifications of the natural environment, including irrigation systems and industrial pollution.

Questions to Think About

7A.1 Which major cultural elements contributed to the human diversity of South Asia?

7A.2 Assess the impacts of the British Raj on the region.

7A.3 How would you illustrate the statement that South Asia's natural environments are dynamic and full of contrasts?

Key Terms

Hinduism	Indian National Congress Party
caste order	Muslim League
Buddhism	alluvium
Mughal (Mogul) dynasty	alluvial fan
British East India Company	badlands topography
British Indian Empire	

A Global Role

At independence in 1947, India and Pakistan, later followed by the other new countries after their independence, moved away from the mercantile colonial involvement in the global economy that they believed had held them back economically. They preferred to attempt self-sufficiency and isolated themselves from world markets by protective tariffs, export taxes, and an emphasis on internal investment. For 40 years, economic growth was modest and not enough to establish greater material wealth. The 3.5 percent GDP growth per year on average from the 1950s to the 1990s that Indian leaders saw as a success did not match East Asia's achievement of twice that rate over the same period.

Complex bureaucratic rules and a lack of basic education, health care facilities, transportation, or power supplies left these countries on the margins of the world economic and political systems. In the bipolar Cold War world, even the largest country, India, did not hold a major strategic position by virtue of its military power or even its policies in comparison with the countries of East Asia, Southeast Asia, or Southwest Asia (see Chapter 8). The countries of South Asia developed close relations with the Soviet Union and China but did not adopt their political systems and so did not represent a threat to the West.

Since 1990, the loss of Soviet support and trading forced South Asian countries to enter the global economy. India doubled its rate of economic growth in the 1990s, although it took time to reorient its institutions and bureaucratic outlook. These countries became more open to foreign investments and internal private enterprise but attracted less than China. This approach to global economic involvement also leaves the region without a global city-region in the top groups. Mumbai and Delhi have some global corporate services such as accounting, advertising, banking, and law, while Bangalore has global connections through its high-tech industries. The countries of South Asia remain among the world's poorest (Figure 7.14), as is shown by their low possession of consumer goods (Figure 7.15).

Subregions

South Asia's largest countries only came into being in the last half century. Those in the Mountain and Island Rim have a longer history but were also an outcome of the British Raj that treated virtually the whole region as a colony or as vassals dependent on the goodwill of British administrators. After the 1947 breakup of the British Indian Empire and the formation of the republics of India and Pakistan, there were three subregions (Figure 7.16):

- *The Republic of India,* the world's second-largest country in total population

- *Bangladesh and Pakistan,* two Muslim countries, each with over 100 million people, resulting from the breakup of the original Pakistan in 1971

- *The Mountain and Island Rim,* comprising Afghanistan, Nepal, and Bhutan in the mountainous north and Sri Lanka and the Maldives in the island south

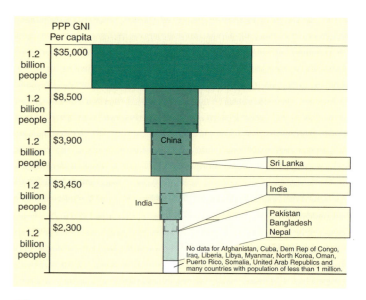

Figure 7.14 **South Asia: country incomes compared.** The countries are listed in the order of their PPP GNI per capita. Source: Data (for 2000) *World Development Indicators,* World Bank, 2002.

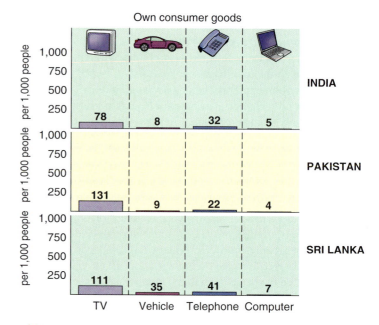

Figure 7.15 **Southern Asia: ownership of consumer goods.** How do these figures compare with other Asian countries? Source: Data (for 2000) *World Development Indicators,* World Bank, 2002.

The population distribution in the region is of high densities except where the natural environment is very difficult (Figure 7.17). The deserts of Pakistan and the Himalayan Mountains form a low-density contrast with the rest of the region.

India

In 2001, the Republic of India's population of 1.033 billion was second only to China's, but its total GDP placed it twelfth in the

Figure 7.16 **South Asia: major geographic features, countries, and cities.** The Indian subregion dominates the region by its size and centrality; Bangladesh and Pakistan are two Muslim countries, created as one at independence but since broken apart; Afghanistan, Nepal, and Bhutan form the Northern Mountain Rim, while the Maldives and Sri Lanka are the Southern Island Rim. This is a region of huge, growing cities.

world. The combination of size and productive potential continues to make India—firmly among the world's poorer countries—a locally dominant presence for other countries in the region. When the British Indian Empire was partitioned at independence in 1947, India gained much of the best agricultural land, controlled most water resources, had most of the manufacturing industry, and included the largest cities in South Asia.

The Republic of India

India became an independent country in 1947 as a republic with a federal union government. The former princely states and British Indian provinces were admitted as states within India's federation (Figure 7.18).

From the start, every adult had a vote. In both federal and state governments, proportions of seats and roles were reserved for the scheduled castes ("untouchables") and the scheduled tribes. The scheduled tribes are also former outcasts. They had no place in society before independence and were greatly exploited but now have protected status. Today, both scheduled groups are significant political forces, although many social taboos still operate against them, especially in rural areas. Women also have a stipulated minimum number of seats in local government.

The upper house of the union government is formed of state-nominated representatives; the lower house, or people's assembly, is directly elected from India-wide constituencies. The head of the federal government is a largely ceremonial

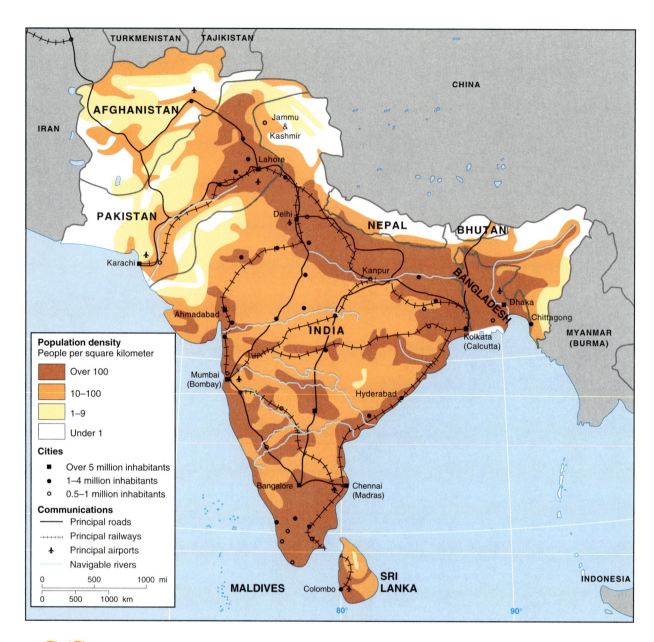

Figure 7.17 **South Asia: distribution of population.** Note the contrasts within South Asia, linking the areas of high population density to river valleys and of low population density to arid areas and mountains. Source: Data from *New Oxford School Atlas*, p. 98, Oxford University Press, U.K., 1990.

president. There are 25 states and 7 union territories, each with its own elected assemblies; the state governors are appointed centrally to represent the union. State boundaries relate to the distribution of languages, reflecting the huge range of languages spoken in the country. English often plays an important part as a common means of communication. The states' roles in administration, law and order, education and welfare, and economic development are supported by some local taxes and grants from central government, but there are never sufficient funds to carry out all their responsibilities. Each state is divided into village-based administrative units.

As in other federal countries, relations between the central government and the states are complex. The union government can declare a state of emergency and depose a state

government. It has responsibilities for defense, foreign affairs, and currency. It raises most of the taxes and makes grants to the states. In the 1990s, states increased their functions, particularly in the delivery of social programs including health and education. Some states took initiatives in foreign relations and defense and strove to attract foreign capital, but others did not, intensifying the uneven nature of Indian economic development.

The number of states increased after independence, mainly as the result of the original states not coping with the interests of ethnic groups (Figure 7.18). States vary in their approaches to government. For example, Bihar, the poorest state, retains remnants of its feudal legacy with land ownership by elites, political and administrative corruption, and inefficiency; private armies control local areas. It finds it difficult to attract for-

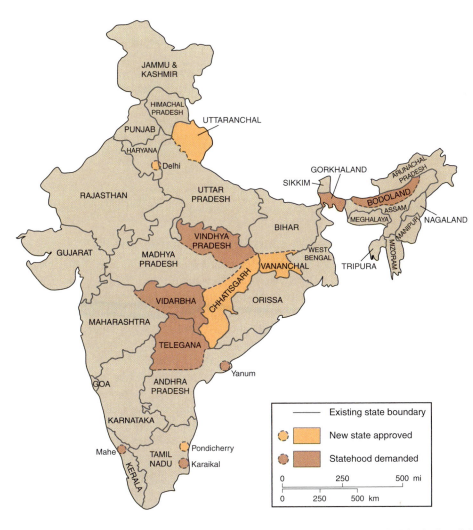

Figure 7.18 **India: new states.** In 1956, Madras Province was divided into Tamil Nadu and Andhra Pradesh. In 1960, Bombay Province became the states of Gujarat and Maharashtra. In 1998, the union government approved the formation of three new states— Uttaranchal (Uttarakhand), Vananchal (Jharkhand), and Chhatisgarh—to be carved out of Uttar Pradesh, Bihar, and Madhya Pradesh, respectively. The status of a state of the union was also conferred on Delhi and Pondicherry. Source: Based on information published in *Frontline,* 31 July 1998.

eign investment. On the other hand, nearby West Bengal instituted land reforms that led to considerable economic and social gains based on smaller farm units and higher levels of popular participation. Some states are governed by the Hindu nationalists, others by secular socialists.

Self-Sufficiency or Global Involvement?

In the 1990s, an economic revolution transformed Indian politics, social relationships, and culture. Basic assumptions and expectations changed radically.

Ending Self-Sufficiency and "Slow" Development

At independence, India's leaders wanted to establish a democracy among a multireligious and multilingual people and return to precolonial "glory." First, the prime minister, Jawaharlal Nehru, transferred land from the aristocracy to the peasants in a move to rid India of the rural feudalism that was encouraged in the colonial era, when princes and rulers took 20 percent of the national income while millions of poor people without rights starved. Second, the government planned to make India an almost self-sufficient industrial power, but its measures resulted in overcapacity, uneconomic units, corruption, and a fall in industrial productivity. The low priority given to food production led to shortages and famines.

Although the private economic sector remained important, contributing 75 percent of total output, foreign competition was excluded by high import duties. The need to control external inputs led to what is variously termed the "license raj," "permit raj," or "quota raj"—requiring many bureaucratic permissions before any economic activity could be undertaken. A huge civil service grew to administer and regulate all aspects of the economy. This approach had some successes, and mutually beneficial trade was established with Soviet bloc countries.

This policy, however, led to dependence on limited internal funds for investment in manufacturing enterprises and infrastructure support. Power shortages led to regular outages of two to three hours per day. The government encouraged firms to employ as many people as possible to the point of overemployment, while the technology applied in textile, vehicle, and other manufacturing became outdated. Education and health facilities were not geared to supporting economic growth in agriculture, industry, or services. The system dampened entrepreneurial enthusiasm and was open to bribes and other forms of corruption. Government investments in large, often unprofitable, and virtually unmanageable manufacturing and infrastructure projects (such as large dams) absorbed much public money.

New Approaches

Nehru's daughter, Indira Gandhi, succeeded him as prime minister in 1966 and held office until 1977. Although her policies continued the central planning, she reversed the neglect of agriculture by encouraging Green Revolution techniques supported by better distribution facilities in some areas. The famines of the 1960s ended.

Later, in the 1980s, after she was reelected prime minister, Indira Gandhi and her son and successor, Rajiv, began to open India to world economic forces, but the internal subsidies and the huge costs of repeated elections in a country with so many people took India into deep indebtedness. A major failure of centralized government was that it did not redistribute growing income—from rich to poor and from urban to rural areas. Furthermore, the 1980s changes allowed in only foreign assembly industries and contributed little to government or personal income.

Crisis and Reform

In 1990 and 1991, the Gulf War brought increases in the price of oil and the end of remittances from Indian workers in the Persian Gulf countries. India faced a balance of payments crisis with no source of borrowing or help from the Soviet Union. India was forced to accept the structural adjustment measures recommended by the World Bank that would open it to the global economy. In 1991, reformist parties took over the government as a reluctant response to external insistence on dismantling the still-popular government food and health care subsidies and public-sector involvement. Change came more slowly than the reformists had hoped, but by the late 1990s, more foreign capital was being invested in revitalizing industry and infrastructure, and new entrepreneurs emerged from all kinds of business sizes and social classes.

Structural adjustment after 1991 and the revolution it brought to India's economy were welcomed by India's middle classes, multinational corporations, overseas Indian investors, and others who gained immediately from it. The package to ease the immediate balance of payments crisis involved liberalizing foreign trade and internally deregulating markets. Measures to achieve this included devaluing India's currency, the rupee (making exports cheaper and more competitive but raising the prices of imports such as food), reducing loans to small businesses and grants to scheduled classes and scheduled tribes, and increasing casual employment. As the Indian government lost its command of the internal economy and politics, the poor lost the support they had received from the government.

The élite Brahmin groups who controlled the government ministries and made money from that now saw privatization and market links as the way to become wealthy. They had the links abroad, often confirmed by higher education in the United States or United Kingdom, had taken on global cultural values, and spoke English. Many who had developed the Indian high-tech and computer software skills, building on previous links with defense research and development in the Bangalore area, moved to the United States and set up companies there, leaving the Indian firms to compete with even cheaper high-tech developers in such centers as St. Petersburg, Russia.

The new politics became increasingly caste-based and regional. The Brahmin-dominated Congress Party gave way to local groups in the states, and some countrywide political groupings began to have greater roles.

In 1998, the new government of the Bharatiya Janata Party (BJP), with a strong Hindu nationalist mandate, detonated five nuclear weapon devices, causing horror around the world and drawing sanctions from major funding nations. This proved popular with many Indian voters as a gesture of identity and confidence in a changing world.

People

Religion, Language, and Ethnicity

India's population is predominantly Hindu: it is the religion of over 80 percent of the population (see Figure 2.19). India is also a country of sizable minorities, with over 11 percent (120 million) Muslims (making it the fourth-largest Muslim population in the world) and just over 2 percent (approximately 20 million) each Christians and Sikhs. The proportions of Buddhists and Jains are under 1 percent.

The religious diversity continues to be a source of contention. In 1980, extremist Sikhs occupied their Golden Temple in Amritsar and used it as a base for killing rival Sikhs until they were ousted in 1984 by a major military offensive. In revenge, they murdered Prime Minister Indira Gandhi, after which Sikh areas of Delhi were burned by mobs. In 1992, militant Hindus pulled down a Muslim mosque at Ayodhya, a place that is thought by some to be the major historic center of Hinduism and the birthplace of the god Rama. In 2002, Hindu-Muslim tensions erupted in violence in Gujarat as Hindu mobs burned Muslim areas after 58 Hindu activists were killed in a train fire.

The diversity in languages is even greater. **Hindi** is spoken by only 30 percent of the population, mainly in the north. It is merely one of 15 official languages recognized by Indian states and one of three most used in movies and TV (followed by Tamil and Teluga). Of the official languages, the Indo-Aryan languages are prevalent in the north and those of Dravidian origin in the south. In all, there are over 1,600 languages and dialects in India. English remains the common language used by the legal system, is spoken by the educated elite—often being used in modern Indian literature and university courses—and is a common form of communication in commerce and national politics. Speakers of minority languages are at an increasing disadvantage as globalization affects more of India.

Caste remains a complex and unequal set of practices, relationships, and prejudices based on traditional and hereditary membership of corporate "communities." The caste order remains of great significance to Indian people in a modernizing era. Rapid urbanization blurs some of the caste divides and suggests a "forward movement of the backwards," but the English-language press still mocks the lower castes and their "vulgar, country ways." In mid-2001 in Alipur, Uttar Pradesh, a young couple was hanged by their parents because they insisted on going ahead with an intercaste relationship that their families tried to ban; several of the family members faced murder charges.

The question of "untouchability" strikes at the heart of the Indian social system. The word itself was avoided as insensitive but is used again in anger or merely to make others recognize the existence of the extremely poor and rejected. Mahatma Gandhi preferred *harijan* (meaning "children of God"), but this was seen as paternalistic and impractical and has been used in abuse. Today, *dalit* (meaning "ground down" or "oppressed") is more commonly used. Officially, people in this group are the scheduled castes.

Dalits have long questioned their social position, with rebellions since the A.D. 600s. In the late 1800s, activism among dalits in northern India produced organizations based on anti-Brahmin ideology. After independence, positive discrimination for the newly designated scheduled castes allowed their first access to government employment and rural land ownership. Although the Indian Congress Party trumpeted its efforts to assist them, special favors, such as reserving college and civil service places for them, tended to emphasize the caste contrasts. The few that obtained reserved places or jobs often used their new status to further political ambitions for the group, such as Jagjiven Ram, a dalit who became India's president in the 1990s.

Today, untouchability practices are most obvious in rural areas, where one's caste status is known to all. Dalits are still barred from using the same wells as Hindu castes, dalit homes are still burned, and caste-based political control remains in many areas. Caste-based violence gets increasing publicity. In Bihar, war broke out when the former landowning families, confronted by dalit-based guerrilla groups, set up their own private armies. The Indian government remains in a difficult position. Having had little success in reducing the ingrained distinctions within Indian society, it now depends on the votes of those who wish to revive *Hindutva*—the view that Hindu culture equates with the Indian way of life. In this view, foreigners include Muslims, Christians, and others.

Impact of Birth Control Policies

India's population of 1.033 billion in 2001 is the world's second largest after China's, and its rate of increase is greater. A sharp population increase began after 1920, when the Indian Empire population numbered around 250 million people, but the British government did nothing to slow the rise. By 1961, India's population was 440 million—at a time of frequent famines. In the 1960s, Indian government estimates of future population growth caused it to institute policies to slow population increase.

Attempts to spread the use of birth control methods began in the 1950s but gained acceptance slowly among a largely illiterate people. Male sterilization in the 1970s was widely rejected because it meant economic suicide for families: the threat of its forced imposition was a major political issue in the 1977 elections that ousted Indira Gandhi's government until 1980. The greater effectiveness of family planning in the 1980s and 1990s built on rising levels of literacy, increased urbanization, and improvements in the status of women. Hindu women

of higher castes use family planning more than those among the poorer castes and less-educated Muslims. More women adopt family planning, including sterilization, once they have given birth to a boy, and this is the most successful aspect of the program. Across the whole population there is an increased will to lower the number of children.

Many families in rural areas, however, continue to want large families to provide extra manual labor and care of parents in later life. Moreover, the 120 million Muslims disapprove of birth control. The new urban middle class is the main group in which there are signs of the Indian population stabilizing around fertility rates of two to three children, but this group forms a small minority of the population.

Estimates for 2025 raise India's population to 1.363 billion. Growth continues because of the huge number of people in the child-producing ages (Figure 7.19) and the increased life expectancy. In the demographic transition process (Figure 7.20), India remains in the phase of population increase.

The Indian age-sex diagram records fewer females than males. It has been suggested that there are millions of missing females in India because of female infanticide, excused by some as saving the girl from a lifetime of suffering. In 1997, the *Times of India* estimated that mothers or village midwives kill 16 million girl babies a year. In a 1994 attempt to stop this practice, the Indian government outlawed abortions after discovering that most aborted fetuses were female. Laws now prohibit doctors from revealing the sex of an unborn baby or carrying out sex-determination tests, although the law is flouted among wealthier families. The poor cannot afford the operations and carry out infanticides. In part of Rajasthan state, there are only 550 women to every 1,000 men, and newborn girls are generally buried without question.

The 2001 Indian census of population showed that India has a sex ratio of 107 males per 100 females, slightly lower than in

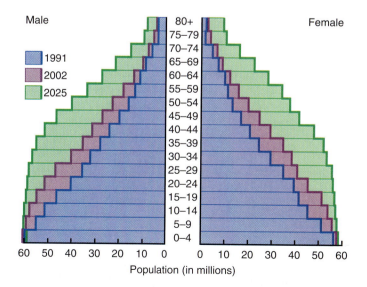

Figure 7.19 **India: age-sex diagram.** The large numbers of young people will produce continuing rapid population growth when they reach childbearing age for many years to come—even if fertility declines. Source: U.S. Census Bureau, International Data Bank.

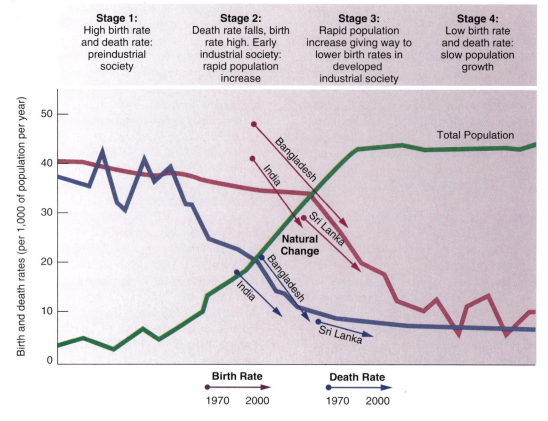

Stage 1: High birth rate and death rate: preindustrial society

Stage 2: Death rate falls, birth rate high. Early industrial society: rapid population increase

Stage 3: Rapid population increase giving way to lower birth rates in developed industrial society

Stage 4: Low birth rate and death rate: slow population growth

Figure 7.20 **South Asia: birth rates and death rates related to demographic transition.** All countries have a wide gap between falling birth rates and death rates. Sri Lanka is farthest ahead in reducing this gap.

1991 but higher than that of its neighbors. Within India, sharp differences exist across the states, from 95 males per 100 females in Kerala to 116 in Haryana. Furthermore, in the under-7 age population, the sex ratio increased from 105.8 in 1991 to 107.8 in 2001, with the sharpest rises in the more prosperous states.

Divided Urban Population

Indian cities continue to hold under 30 percent of the country's population, but levels of urbanization are increasing faster than total population growth. Mumbai, Delhi, and Kolkata (Calcutta) are set to become three of the world's 10 largest cities by 2015, with 26 million, 17 million, and 17 million people, respectively.

Urban areas provide most of India's economic opportunities. New jobs in factories and offices, better access to education and health services, and most opportunities in the informal sector of the economy occur in urban—especially large urban—areas. Indian (and other cities in the region) have different patterns of land use and buildings depending on the influence of colonial builders (Figure 7.21a, b). A typical colonial city center (Figure 7.21c) contrasts with modern high-rise and bustee settlements (Figure 7.21d). The total population of urban areas is projected to rise from 250 million people in the mid-1990s to over 600 million by 2025. From 1980 to 1996, the number of cities with over a million people jumped from 12 to 30. The largest cities are becoming immense. In United Nations estimates for 2000, Mumbai had 18 million people,

Kolkata 13 million, and Delhi 12 million. Other major cities include Chennai (Madras) (7 million), Hyderabad (7 million), Bangalore (6 million), and Ahmadabad (4 million).

Indian Urban Landscapes

The urban landscapes of South Asia reflect the waves of cultural influences that washed across the region. Many ordinary buildings were constructed of materials such as wood that do not have a long life, but most towns have more permanent religious, royal, or military buildings that have lasted for centuries (Figure 7.22). Buddhist buildings range from stupa temples in central India to the fortified monasteries of Bhutan. Hindu temples are often very ornate (see Figure 7.3a). Muslim mosques and tomb gardens (such as the Taj Mahal) also provide distinctive landscape elements.

Older precolonial sections of towns in South Asia have high densities of housing with poor transportation access along winding alleys. Shops and artisan workshops encroach on the walkway. Few buildings are more than two stories, except in Kolkata and Mumbai, and there is no clear distinction of residential, commercial, and industrial functions. People of similar caste or trade live and work together in localities. Up to the mid-1990s, the wealthy lived in the centers of such towns and the poorer people toward the outskirts, but many of the wealthy have moved to new suburbs.

During the British rule from the 1700s to the 1900s, new aspects added to Indian towns included a central market, administrative offices, and frequently a clock tower. British-built

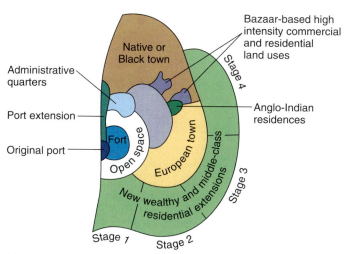

(a) Colonial-based South Asian City

Physical Space

◼ Bazaar-based traditional city from the pre-colonial times with rich in zone 1 and poor in zone 3

◼ New post-independence extensions with extensive squatter settlements

◼ Squatters/slums

+ Chowk or crossroads

✚ High-intensity commercial and residential land uses

■ Wholesale market

Cultural Space

◼ Religious and linguistic clusters and Untouchables

(b) Bazaar-based (traditional) South Asian City

(c)

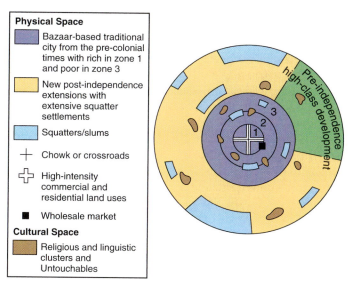

(d)

Figure 7.21 **South Asian cities.** (a) General layout of colonial port city. (b) Traditional inland city. (c) Former British colonial buildings in the center of Bangalore, southern India. People wait for buses as some enter an auto-rickshaw. (d) Slum shantytown conditions in Mumbai with dwellings made of sacking, plastic and metal sheeting, and boards of various types. Contrast the foreground dwellings with those in the distance. Source: (a), (b) From *Human Geography* by Fellman, © The McGraw-Hill Companies; (c) © Michael Bradshaw; (d) © B. P. Wolff/Magnum Photos, Inc.

churches and civic buildings were European in style. New residential areas built on European plans with wide streets and open places separated business and residential functions. Such areas are now occupied by professional and businesspeople. New government-dominated planned cities such as New Delhi adopted wide avenues and massive public buildings. Britain built and developed major ports such as Kolkata, Mumbai, and Chennai, initially as a means of establishing trading security. The ports were then joined to inland centers by railroads. Where the railroad passed through a precolonial town, the town grew in administrative and trading significance; where one was bypassed, it often declined.

During the British Raj, hill station towns provided escapes for British families from the worst of the summer heat and humidity and the diseases that spread in that season. Simla, due north of Delhi, is typical of these hill stations and is still accessed by a mountain railroad line built at great initial expense. The hill stations often resemble small British towns in their architecture. They are now tourist centers and have English-speaking schools.

After independence and partition, cross-border migrants moved into large cities and were accommodated in refugee camps that themselves became cities—Uhlanagar near Mumbai now has 600,000 people. The contrasts increased between the neglected districts of the poor and the well-kept suburbs and new cities. Some smaller cities, such as Bhopal, had no real planned infrastructure, and even in major cities, funds for maintaining infrastructure were scarce. The extensive shantytowns in South Asian cities (see Figure 7.21d) are landscapes that reflect dire poverty and lack of planning.

Modern cities have major industrial areas, including steelworks and aluminum plants, chemical factories, textile mills and garment-assembling factories, vehicle makers, and electrical goods production. Concentrations of factories are most common around Mumbai in the Indian states of Gujarat and Maharashtra.

Urban Contrasts of Wealth and Poverty

The cities bring together the contrasts of wealth, poverty, and status in Indian society. The clogged street traffic, high-rise

Figure 7.22 **South Asia: urban landscapes in Delhi.**
The main entrance to the Red Fort, constructed in the A.D. 1500s by the Mughal emperors. The massive fort walls enclose 6 hectares (15 acres) of land, including several palaces and water gardens. Photo: © Michael Bradshaw.

office blocks and prestige apartments, international hotels, and TV ads for consumer goods portray a society moving into affluence. There is not, however, a large affluent class consuming at Western levels. Estimates of middle-class numbers suggest as many as 300 million Indians are included, but this is the widest definition. Western ideas of social class related to employment in an urban-industrial society scarcely apply in India. Three broad groups of Indian urban dwellers can be identified.

The *very wealthy* include around 1 million people (a large number, but only 1% of the total urban population) who consume at the level of Western countries, buying imported cars, appliances, clothing, cosmetics, and food. Although insignificant in numbers, they are visible politically. The group includes those with "old money" acquired from land ownership or established industries. After independence, wealth taxes, ceilings on land holdings, and other policies aimed at creating a more egalitarian society reduced the prominence of this group. They are joined by the newly wealthy, who include those who profited from the "black" economy, to which influential public office, unlicensed business, trade in drugs, smuggling, or extortion contributed. Newly wealthy people are also emerging from the middle classes through the high-tech industries such as the Bangalore software industry. Many Brahmins, the traditional priests who became university professors under British rule, apply their knowledge and personal links in this group. Wealthy traders also emerge from *jati* clans, such as jewelers and money lenders. Liberalization in the 1990s and a 1997 amnesty on undeclared wealth made it possible for such families to invest in entrepreneurial projects or multinational corporations, or to spend their money on luxury holidays, interior design, and imported goods.

The very wealthy live in guarded colonies, bus their children to private schools and send them abroad for higher education, use air-conditioned cars, and have backup water and power supplies. They pay for the best hospital care (including surgery in foreign facilities) and travel first class by rail or air. They are cosmopolitan in education, livelihood, and social activities, increasingly breaking with traditional Indian culture. In particular, they are shifting away from the social conscience seen at independence and in the Congress Party policies. Today's wealthy young go to Western-style nightclubs and drive sports cars. They take leading roles in the media and literature, including the high-quality free press. Few are involved in right-wing politics but profess allegiance to Gandhian principles and concerns over poverty, the environment, human rights, and charity work. Such attitudes are not always practiced.

The "patriarchal joint family" consisting of the family head, several sons and their wives, unmarried daughters, and elderly relations was typical of landowning families and family businesses where the sons worked with the father and the whole was taxed as a unit. Smaller nuclear families are becoming the norm, with, at most, the addition of an elderly parent or unmarried daughter, an arrangement that is more closely tied to professional and managerial roles.

The importance of marriage is reflected in huge wedding celebrations, for which wealthy families save and invest. The wife takes a dowry, often in the form of jewelry, with her into the marriage, but in the wealthier families, she commonly retains the right to it. Even where a dowry is not involved, presents to the bride are generous, allowing her to acquire some security in a male-dominated society.

Around 10 million households with up to 100 million people constitute the *urban middle classes,* having jobs in factories or as typists, social workers, elementary school teachers, police, or telephone operators. Those in government posts saw their incomes erode as a result of the 1990s reforms, but job security outside government is poor and competition fierce.

Girls in middle-class households marry in their teens and men in their twenties, nearly all by arrangement. Dowries transfer money to the man's family, often for use as his sisters' dowries. For example, a man with a safe job in Tamil Nadu state may command a dowry of US $1,000, 100 grams of gold, household goods, and a car. Families with daughters start saving at birth but still require loans. Wives may be badly treated if their dowries are not up to expectations, and they may be murdered by being burned alive so the husband can remarry for a better dowry. It is also a wifely duty to have sons; families with more daughters than sons are often implicated in dowry murders.

Households dependent on government jobs, such as in the railways or police forces, occupy subsidized housing in colonies of two- to three-bedroom houses or apartments. At the other end of the middle-class housing spectrum, many live in overcrowded slum tenements. These families buy some consumer durables (TV, radio, tape recorder), but refrigerators are luxuries. They buy few imported goods and obtain food from local markets. Convenience foods (in the Indian context including, for example, packaged flour, instant coffee, or baked loaves) are uncommon, and women expect to spend most of their time buying and preparing meals based mainly on grains and vegetables. If there are spare funds, they are used to educate the sons in (often poorly run) private English-language schools.

Life is a struggle against impoverishment. While the gulf between this group and the more affluent widens, they make sure that poorer groups do not intrude. Many of them belong to the "forward" (opposite of "backward") castes, including Brahmins, administrators, and shopkeepers. They resent the reservations for the "backward" castes in civil service jobs. The middle classes join political movements—particularly the Hindutva—that reaffirm the worth of national culture. This group is fertile ground for right-wing Hindu chauvinism, voting for the BJP party, and becoming involved in anti-Muslim and anti-Christian demonstrations. Major consumers of media products in film, TV, and popular press, they are part of an emerging mass culture drawing on both global and pan-Indian values. While younger members of the group try to emulate those who benefit from globalization and liberalization in the economy, others look back to a "Golden Age" of Indian culture before the Mughal emperors or the British Raj and try to restore Indian "purity" in the face of Kentucky Fried Chicken and Miss World competitions. They celebrated the testing of the nuclear "Hindu bomb," regarding it as a national achievement and statement of growing economic and political power.

An even larger group forms the *marginalized urban poor,* struggling for existence and living in shantytown slums (bustees) or on the sidewalks. They work in the unorganized and unprotected (informal) economy for whatever income they can obtain. Children and elderly women are an integral part of the work force. Children have to work for their living—in tea stalls, selling newspapers, scavenging, begging, or in industrial workshops. Few can afford to obtain a free education, and there are few schools in the slums. Crime is the only way most young people can exist.

Slums vary in type and organization. Many slum dwellers are rural migrants with no title to housing or rights to establish businesses. Some slums have brick or block dwellings with windows, doors, and corrugated roofs; others are shacks made of any available material, often at risk from fire. Some are recognized by municipalities and have piped water, legal electricity connections, and shared sanitation. Many do not. There is no street lighting, cleaning, or repairs. Municipal policy is to control and raid, rather than protect, the slums. From time to time, governments clear slum areas but do not provide alternative accommodations for the people living there.

Water is supplied by tanker, and there are few toilets apart from open spaces and the gutters. Yet homes are kept clean. Disease is endemic (diarrhea, tuberculosis) or occurs in rampant outbreaks of cholera, hepatitis, and typhoid. The poorest ragpicker settlements adjoin and merge into refuse heaps.

Caste remains the most significant social division among the very poor, and bustees are often dominated by one group. Most slum dwellers are dalits, but some are Brahmins—remaining aloof in their poverty. Bustee inhabitants of all castes have voting rights and are organized by politicians as vote banks in exchange for small favors and gifts. During the 1990s, however, government grants to help poor people were reduced or ended, and many in-town slums were cleared for up-market developments.

Women in bustees face frequent attacks from the high proportion of men, most of whom leave their wives behind in villages. Some women work as domestic laborers in middle-class homes, where they are housed comfortably and protected but have little personal freedom. Many others enter prostitution, often bonded and with little chance of escape. HIV/AIDS is a major scourge, as are drug dealing and drinking illicit alcohol.

Another group of destitute poor is those with physical and mental disabilities—who were previously looked after in their villages but are marginalized beggars in the cities, rejected by relatives. Some are hidden in huge mental institutions where they are controlled but not treated. When the very poor are suspected of even minor offenses, they are imprisoned for long periods up to years, unable to afford bail.

Mainly Rural Population

India's population remains largely rural (Figure 7.23). Between 1960 and 2001, the population in rural India decreased from 82 to 72 percent of the country's total, and their share of India's total GDP went from 50 to under 30 percent. Rural areas lag behind urban areas in education and health-care provision and still have large proportions of the semisubsistence and low-paid farming lifestyles that are linked to long-term poverty. By 2025, it is projected that India's rural population will be down to 55 percent of the total, but this will comprise a greater number of people than at present (762 million instead of 630 million) because of the overall population increase.

The ways of life in rural India altered substantially during the late 1900s. Even the remotest village is part of the global economy, often selling cotton crops to European mills to pay for polyester clothing. There is still a huge diversity from state to state, across areas of different climate and topography, and under different social structures, indigenous cultures, local politics, and market conditions (see Figure 7.24). Most villages remain extremely poor. Although Mahatma Gandhi envisaged

Figure 7.23 **South Asia: rural landscape.** Lands around the market town of Pushkar, Rajasthan, western India: farmed plains contrast with dry Aravalli Hill pastures. Photo: © Brian Vikander/Corbis.

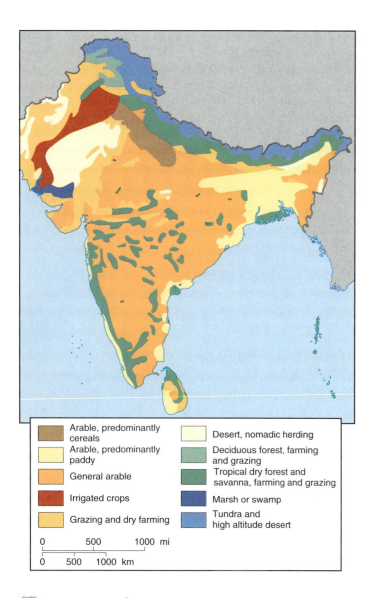

Figure 7.24 **South Asia: distribution of agricultural land uses.** Relate the main types and intensities of land use to climate and relief conditions. Source: Data from *New Oxford School Atlas*, p. 98, Oxford University Press, U.K., 1990.

Legend:
- Arable, predominantly cereals
- Arable, predominantly paddy
- General arable
- Irrigated crops
- Grazing and dry farming
- Desert, nomadic herding
- Deciduous forest, farming and grazing
- Tropical dry forest and savanna, farming and grazing
- Marsh or swamp
- Tundra and high altitude desert

are frequent. A manual laborer pays up to four times his annual wages for a daughter's dowry. Money lenders help with loans, but land is mortgaged and often lost, while members of the family may be taken illegally into bondage.

Countermovements exist to combat poverty. The most encouraging are the women-run cooperatives and small rotating loans projects linked to literacy campaigns (see the "Grameen Bank" section, p. 304). From the 1970s, the union (federal) government provided nonagricultural income opportunities for landless peasants in its five-year plans. Many projects were successful, but scandals affected others. In a Bihar fodder scam, a group of landowners and bureaucrats sold cattle several times among themselves, taking advantage of grants for veterinary care, subsidized fodder, and long-distance transport.

Rural education is poor in amount and quality, apart from the notable exception of Kerala State, where all children are in school. The lowest educational involvements are in the "backward" castes. Without literacy people cannot pursue their rights or take advantage of helpful projects.

Not all Indian farmers are poor. The Green Revolution, especially in Punjab, Haryana, and parts of Uttar Pradesh, led to the emergence of wealthy farmers who enlarged their holdings and invested in commercial production. They employ labor from the poorer states and campaign for even larger government subsidies for seeds, pesticides, fertilizers, and irrigation water. Agribusiness replaces semisubsistence farming, bringing higher living standards, mainly for farmer-owners and merchants. The wealthiest farmers provide the largest number of members in the union government's people's assembly.

Indians Abroad

Around 15 million Indians live abroad, with the largest groups in the late 1990s in Nepal, South Africa, Malaysia, Sri Lanka, the Persian Gulf states, the United Kingdom, and the United States. Under the British Empire, Indians provided a source of cheap labor and were moved to work on plantations and railroad construction in the Caribbean, Eastern and Southern Africa, Burma, Malaysia, and Pacific islands such as Fiji. Many stayed as shopkeepers and businesspeople.

More recently, flows of workers to the oil-producing countries of the Persian Gulf and of professionals to the United Kingdom and United States for better-paid opportunities have created a significant "brain drain." Many of the Indian "diaspora" come from Gujarat and increasingly work in high-tech centers such as Bangalore before moving abroad.

Economic Development

In 2001, India's total GDP was twelfth in the world, ahead of that of the Russian Federation, Australia, South Korea, and the Netherlands, and almost equivalent to Mexico's. All those countries, however, had much smaller populations than India. Many Indians remain very poor, tied to traditional low-productivity farming and work in urban factories.

Parts of India's economy are expanding, becoming more competitive commercially, and diversifying. In the early 2000s, India's broadening outlook toward global trade included

an India of "village republics" (*panchayati ra,* "local government") based on cooperative peasant holdings, it happened only in a few places such as western Bengal.

Few old landowning élites remain. The "zamindar" system, under which the British Raj gave "aristocracy" land grants to élites who practiced a feudal way of life tying cultivators and craftsmen to their village, ended after independence. Land reforms put a ceiling on the amount of land one family could own, although they were not universally applied. Today, communities of landowning peasants are the norm, although many scarcely rise above subsistence.

Rural poverty almost inevitably involves debt. Laborers have little chance of escaping a lifetime of poverty, and debts are inherited by children. Cash incomes are small, and often unexpected obligations, such as sickness, the death of plowing oxen, or a demand for a dowry on the marriage of a daughter, increased private investments to and from the United States, increased involvements in Europe, and trading and investment agreements with Japan and China. In 2002, the Indian government scrapped export restrictions, expecting to double its sales abroad by 2007 and favoring agricultural products and computer hardware. It also abolished price controls on the telecommunications and oil and refining industries, apart from the subsidies on cooking gas and oil.

The economic profile is moving toward that of a more prosperous country. Between 1965 and 2000, farming decreased its contribution to GDP from 44 to 28 percent; industry rose from 22 to 29 percent but then declined a little as services rose from 34 to 46 percent. While India's economy grows and diversifies, however, many of its people remain poor, with a 2001 human poverty index value of 34.3 percent.

Economic Restructuring in the 1990s

In the 1990s, India restructured its economy and adopted policies to raise production and exports. While businesspeople and those with private capital profited from the new economic environment, few of the poor did. Many civil servants became disgruntled as they lost their jobs that had enabled them to add to their basic income from the allocation of privileges. The union government was torn between the need to attract outside capital for private enterprises and the need to retain votes through guarantees of job security in state-owned enterprises. It shifted back and forth in its policies through the 1990s. In 2001, the emphasis on the lowering of import duties, tax rates, and interest rates was renewed.

Further economic modernization depends on India's ability to supply the power, telecommunications, and road transportation that will attract and keep productive businesses. In the late 1990s, plans to expand power generation were behind schedule and power shortages still affected many areas. Plans to privatize telecommunications were held up by bureaucratic and legal squabbles.

Farming and the Green Revolution

On average, India now produces enough food to support the basic needs of its population without famines—although prices and methods of distribution do not guarantee that everyone is fed adequately. The vagaries of the monsoon, however, still determine the harvest: good years enable India to export food; bad years require imports. Important regional variations exist in land use (see Figure 7.24).

The Green Revolution began in 1966 (see Chapter 6), when high-yielding varieties of wheat and rice, developed at research institutes in Mexico and the Philippines, were introduced as part of Indian agricultural policy. Indian production of food grains almost quadrupled, from 50 million tons in the 1950s to over 190 million tons in the late 1990s. India became the world's third-largest exporter of rice in good monsoon years (after the United States and Thailand) and fourth-largest producer of wheat.

Benefits from the Green Revolution were geographically and socially uneven. In Punjab and Haryana, wheat and irrigated rice yields increased rapidly. The larger land holdings and

wealth of the Sikh and Jat farmers enabled them to purchase the expensive seeds, fertilizers, pesticides, and tractors. The local infrastructure of available irrigation water, 90 percent rural electrification, 70 percent all-weather roads, good bus and truck transportation, many medium-sized towns, and regulated grain markets provided a beneficial basis for the farmers' investments. Per capita incomes increased, but so did family size, reducing the economic impact of the increased grain harvest.

Elsewhere in India, the outcomes of the Green Revolution were less positive. Results were negligible in areas where irrigation water was not available, farm units were small, and infrastructure and access to markets at home and abroad were poor. Where the yields did not increase enough to match the increased costs, debt often put farmers out of business. Many landless laborers lost their jobs. Furthermore, the switch to intensive grain production and the elimination of peas and beans increased levels of malnutrition among local populations.

Even in Punjab and Haryana, diminishing returns from grain because of land degradation and global market competition caused farmers to consider other ways of increasing their income. In the 1990s, some farmers diversified into crops for export, causing India's agricultural exports to quadruple in value. The new crops included durum wheat (used in pasta), vegetables, fruit, and flowers. Agribusiness linked these crops to local processing facilities for making pasta, french fries, tomato puree, and wines. Nearly three-fourths of the exports go to the arid, oil-rich countries around the Persian Gulf.

India is the world's third-largest cotton producer, after the United States and China, but has the largest acreage under the crop and low yields. In early 2002, the Indian government approved the use of genetically modified "BT cotton" seed that is resistant to the boll worms that destroy 15 percent of annual yield. It is expected that the new seeds will double India's cotton production and reduce pesticide usage. The decision was taken against environmentalist objections based on "transgenic pollution." Further government measures encourage the wider role of modern agribusiness in grains, oils, cashew nuts, and butter.

Traditional Farming Areas of India

India has some long-established specialist areas for commercial crops:

- Tea continues to be an important plantation crop in the Assam hills of northeastern India. Multinational corporations control marketing, processing, packaging, and distribution of the product in the world's wealthier countries. The domination by external firms results in low prices for producers and low local wages.

- The coastal deltas around the peninsula were drained for rice growing and cotton production. The most successful farming areas in southern India rival output from Punjab and Haryana in the northwest. They have electrification, good roads, and local markets.

- Jute is a commercial fiber crop on the western margins of the Ganges-Brahmaputra Delta. Its growth there was developed when independence separated the Kolkata factories from the Bangladesh jute-growing areas.

Farming in other parts of India is less intensive and commercial. The Ganges River plains east of Delhi retain traditional farming methods for semisubsistence on mainly small farms. Large areas of the hilly peninsula have inadequate irrigation and grow coarse grains such as millet and sorghum that withstand dry conditions.

There are many landless laborers who find it increasingly difficult to make a living, especially since land reforms strengthened the middle peasantry and their ownership of land. Millions of people live in a poverty trap with few prospects of improvement.

Large numbers of livestock are a major feature of India. The 200 million cattle are the largest number for any country in the world. Many are used as working animals, since most religious and high-caste Hindus do not eat any meat, and all Hindus and Sikhs avoid beef. India, however, is a meat exporter. Many water buffaloes work in the wetland environments of rice-growing areas (see Figure 7.25). The laws against killing cattle because of their sacred status allow many to wander freely through villages and suburban areas. Cows are kept for milk and making ghee (liquid butter), both of which are an increasingly important part of the "white revolution" of Indian diets. Dung from cattle is an important fuel and a flooring plaster.

Mining

The most mineral-rich part of India (see Figure 5.10) is the Chota Nagpur Plateau in the poorer northeastern part of the peninsula, where plentiful iron ore and coal deposits formed the basis of local steelmaking. India's rocks also contain other metal ores such as copper, gold, and manganese. In the 1990s, India raised its coal mining output to around 200 million tons each year, making it one of the world's leading producers.

Figure 7.25 **South Asia: rice paddy.** Intense summer rainfall and high river flows make it possible to grow rice in paddies. Water buffaloes and humans are the main forms of labor and energy used in cultivation, especially in the poorer areas. Photo: USDA.

India also has the world's fifth-largest reserves of bauxite (the ore of aluminum) and has the advantage that the metal can be extracted from this ore at lower temperatures than the ore in many other major deposits requires. The thick sediments of the Ganges lowlands contain deposits of natural gas and oil, and by the late 1990s, India produced 75 percent of its oil needs.

Fishing and Forestry

India is one of the top 10 fishing countries in the world, and fish is important in the Indian diet. Many coastal areas depend on income and food from the fishing industry. Most of the industry still uses traditional methods of catching, with mainly unskilled labor and small boats.

Most of India's forest resources were cut centuries ago, and many villages now find it difficult to produce sufficient firewood for their needs. Forests still grow on the southern slopes of the Himalayas, from which softwood is cut, and on some of the hilly areas of the peninsula where hardwoods such as teak are exploited.

Growing Manufacturing

India has a large and growing manufacturing sector ranging from small-scale craft industries (Figure 7.26) to large-scale steelworks and modern high-tech facilities. The distribution of manufacturing in India reflects historic craft specialties, raw material locations (mines, farms), the British legacy of transportation hubs, ports, and administrative centers, varied encouragements offered by the union and state governments since independence, and entrepreneurial skills.

Small-Scale Manufacturing

Small-scale industries in India employ 140 million people in units with only a few employees and produce 35 percent of all manufactured goods. While foreign firms want to view India as a major emerging market for their own benefit, the union government and some state governments wish to retain control over the type and rate of economic growth and so favor small local businesses over large investments by multinational corporations. Two specialties in Uttar Pradesh illustrate the range and conditions of such industries:

- The *Kanpur area* specializes in tanned leather goods. Its small tanneries invest little in technology, training, health and safety, or environmental treatment of the toxic effluent. They pay low wages, employ children, and rely on unskilled, destitute labor for their "quality handmade" goods. Workers take home one-tenth of the wages similar workers earn in the United States or Europe, working barefoot without protective clothing in an unfiltered atmosphere of toxic fumes and liquids to make safety boots for global markets.

- *Mirzapur* is the center of carpet-making that began in the A.D. 1500s, when the Mughal emperor brought in weavers from Persia. Most modern factories set up by British owners closed following labor union militancy over the application of labor laws. Only basic processes such as dyeing wool and packing carpets are still carried on in factories.

Figure 7.26 **South Asia: craft-based industries.** Silk weaving for sari cloth, Kanchipuram. Photo: © Robert Holmes/Corbis.

Around 10,000 weavers work on 3,500 looms in surrounding villages and rely totally on this income but face increasing competition in world markets.

Large-Scale Manufacturing

As part of its self-sufficiency and import-substitution policies after independence, the Indian government developed a large-scale, heavy industrial sector of steel, chemical, and aluminum works. The main steelworks were built near the large iron and coal deposits in Vananchal, Chhatigarh, and Orissa (see Figure 7.18).

Aluminum works are distributed more widely across India close to bauxite deposits and hydroelectricity sources. Since the Indian economy was opened to foreign investment in 1991, the aluminum industry received major investments from multinational corporations.

Car and vehicle assembly became major parts of the self-sufficient Indian manufacturing diversity. Up to the late 1980s, the state-owned factories producing Ambassador cars (based on a 1960s British model in the foreground of Figure 7.3b) and Tata trucks (at Pune) dominated production. After the 1990s reforms, the output and range of manufacturers increased with new auto and truck factories linked to major multinationals such as General Motors, Ford, Chrysler, Peugeot, Volvo, Hyundai, Mitsubishi, Daewoo, and Volkswagen. The Indian market leader, Murati, has links to Japanese Suzuki and produces the most popular car.

Textile Manufacturing

Textiles and garments comprise over one-fourth of India's exports, and sales doubled in the 1990s. Most of these exports go to the United States and European Union countries. Collaboration between Indian manufacturers and European and U.S. customers brought maker and market closer together. Although cotton remains the dominant textile fiber grown and manufactured in India, coarser textile fibers including coir (from coconuts) and jute are cheaper than cotton and are blended with cotton and synthetic fibers for use in denim cloth.

Engineering textiles used in road construction and agricultural textiles that help to prevent soil erosion on slopes are other jute and coir products. The traditional process of manufacturing coir from coconuts polluted large areas of coastal lagoons during the retting (soaking) phase. It lasted 10 months and left quantities of waste products that could not be disposed of. New concrete tanks and machinery for retting cut the time and improve the quality of both the fiber and the local environment.

High-Tech Industries

Modern high-tech industry is a major development. Its total output increased from U.S. $2 billion in 1995 to U.S. $12 billion in 2001. In the Bangalore area in Karnataka state, southern India, the state government attracted multinational corporations such as 3M, AT&T, Digital, Ericsson, Hewlett-Packard, IBM, Motorola, and Texas Instruments. The area was the location for many large public companies established after independence. The main new industries are electronics, computer engineering, software and services, telecommunications, aeronautics, and machine tools. In the 1980s, Bangalore became India's first "electronic city," and parallel developments are taking place at Mysore and Dharwad.

Although information technology industries in the United States suffered a recession in 2001, India's low-cost industry was less affected, switching interest from the United States to

Europe and developing data and call centers and other business-based processing software. At this time, India entered agreements with China to share software development. Fears that India's lead in this area would be lost were countered by the opportunities for building expertise in the growing Chinese market.

India's English-speaking, college-educated, low-cost work force is the basis of a new service industry in "back office" work such as call centers and other information-technology-enabled services. For example, GE Capital Services opened its Indian call center in the mid-1990s and employs more than 5,000 people collecting from delinquent credit card users. British Airways runs frequent flier programs and handles computer message errors. American Express has a large office in Delhi. Savings of up to 50 percent on costs may grow as India's high telecommunications charges fall. Independent call centers have had more fluctuating fortunes, as in the market for U.S. medical transcriptions, although this area is growing fast. Companies are now developing a range of activities, including data entry and conversion, rule-set processing (e.g., can an airline passenger upgrade to business class?), problem-solving (e.g. should an insurance claim be paid?), direct customer interaction, and expert knowledge services using specialists linked to databases.

Mumbai: Center of Manufacturing

The main geographic concentration of manufacturing growth in India is around Mumbai, formerly Bombay, in both its own state of Maharashtra and adjacent Gujarat. Maharashtra state accounts for over one-fourth of India's manufacturing output and is home to over half of India's top 100 companies. Mumbai handles one-fourth of India's trade in its port and 70 percent of its stock exchange transactions. It is the headquarters of nearly all of India's commercial banks. As well as having factories making a wide range of products from textiles to pharmaceuticals. Mumbai is the center of India's very active movie industry (often called "Bollywood"). Over half of the 500 to 600 titles produced annually are filmed in this area. Alongside the growth of new industries in the Mumbai area, however, are many old textile mills with aging and inefficient machinery.

Gujarat, to the north of Mumbai, is the second most industrialized state. The coastal zone between Ahmadabad and Mumbai is marketed as the "Golden Corridor," and over 70 percent of its recent investment is in the southern sector close to Mumbai. Ahmadabad was once the main textile manufacturing center in Asia, but textiles now compose only 10 percent of its manufacturing output, equal to engineering. New investments in the 1990s almost equalled those in Maharashtra, with many in petroleum, petrochemicals, and pharmaceuticals. The people of Gujarat have a reputation for business acumen, and many Indian businesspeople in other countries come from this state.

Northeastern India: The Kolkata Focus

The other main manufacturing region in India is in the northeast—in West Bengal around Kolkata (Calcutta) and around Jamshedpur, where the steel industry is important. Kolkata remains the Indian headquarters of jute manufacturing, which is entering a revival after years of decline and labor union strangulation. At partition in 1947, Kolkata's mills were separated from the main production areas of jute in the Bangladesh section of the Ganges Delta. Although Indian farmers in West Bengal were encouraged to grow more jute, India still buys large quantities, especially of higher qualities, from Bangladesh.

The Jamshedpur iron and steel area, 250 km (150 mi.) west of Kolkata, is an enclave within a rural and mining region of great poverty. India's first iron and steelworks opened at Jamshedpur in 1914, built by the textile entrepreneurial Tata family, whose interests later expanded into engineering, truck manufacture, hotels, software, and an airline. After independence, output increased from 10 million tons in 1980 to over 20 million tons in the late 1990s. The Jamshedpur factory operates well-maintained old blast furnaces together with the most modern steelmaking and steel-rolling mills and produces steel at a cost undercut only by the best South Korean mills.

Service-Sector Growth

The Indian service sector is expanding rapidly. Government employment remains a major source of jobs for the well-educated middle class, although the economic reforms will reduce this source of jobs. India is well supplied with some groups of professionals, from lawyers to accountants and economists.

The education and health sectors both need further development to deliver a good basic education to the whole population and to improve health care and public health levels. The scope clearly exists for the expansion of employment in these areas, but it needs the political will to shift investment into both physical infrastructure and human resources.

The facts that 300,000 children die of diarrhea each year and infant mortality remains high highlight the problem of health in India. Public health services do not work well. For example, Delhi generates nearly 4,000 tons of trash each day but clears only 2,500 tons, leaving the rest on the streets. Of 1,800 tons of sewage, only two-thirds are collected.

Education is good in parts of India. Kerala claims 90 percent literacy in contrast to the overall Indian figure of half that level. Education policy in India needs to achieve a better balance between the basic levels of education that are still lacking and the oversupply of university graduates. The elementary, secondary, and higher education sectors divide the education budget into three almost equal parts, and higher education is subsidized for all who qualify.

India's tourist industry attracted 2.6 million visitors in 2000, up from 1 million in 1980. With its riches of historical buildings, mountains, and beaches, together with its exciting culture, India is a magnet to many from wealthier countries. Visitor numbers could increase with improved infrastructure and public health provisions, as shown by the Chinese.

Reducing Poverty

When India became independent in 1947, its leaders blamed colonialism for the poverty of its people. Over the next 50 years, famines ceased, literacy rates doubled, life expectancy

Test Your Understanding 7B

Summary India has the world's second-largest population and dominates South Asia in land area, population, resources, and economic activity. India remains a largely rural country, although the major cities such as Mumbai, Kolkata, and Delhi are growing rapidly. Most of the people are Hindus and still affected by the caste system.

India's economy is characterized by low productivity in agriculture and manufacturing that produces goods mainly for local markets, but the northwestern agricultural area and the west coast industrial area around Mumbai are developing rapidly.

Some Questions to Think About

7B.1 What are the advantages and disadvantages of the Green Revolution in India?

7B.2 How do the poorer and richer areas of India contrast with each other? What factors are responsible for the geographic differences?

7B.3 What were India's reasons for conducting nuclear tests in May 1998?

Key Term

Hindi

rose from 33 to 59 years, infant mortality fell (from 165 per 1,000 live births in 1960 to 70 in 2001), income poverty was reduced, and overall income growth was substantial. Yet in the late 1990s, India's human poverty index was 36.7 percent, 60 million children under age 4 were undernourished, 50 percent of the population was illiterate, women remained at a disadvantage in society, and rural poverty affected about 40 percent of the population.

Economic growth over this period explains half of the reduction in poverty based on income. Other factors affecting income poverty are less clear. For example, the state of Kerala is a paradox with one of the lowest incomes per capita and slow economic growth but some of the best education and health-care statistics and a human poverty index of 15 percent, compared to one of 50 percent in another poor state, Bihar. Haryana has the fastest economic growth but discrimination against women is such that the ratio of females to males in the state is 86 to 100.

🌐 Bangladesh and Pakistan

Bangladesh and Pakistan share much of the same cultural heritage as the Republic of India and were part of the British Indian Empire until 1947. They were partitioned at independence on the basis of the dominance of Islam. Their central areas occupy lowlands along the Indus River valley and at the mouth of the Ganges and Brahmaputra Rivers (see Figure 7.10a).

After independence in 1947 until 1971, West and East Pakistan were a single country. When the people of East

Pakistan became convinced that the Punjabi military leaders controlling Pakistan were taking East Pakistan's resources to benefit West Pakistan, they revolted. With India's military help, they established Bangladesh as a separate country.

Both countries have internal tensions between rich and poor and between those who migrated in at the time of independence and those who already lived there. The wars with India, particularly over Kashmir, and the civil war prior to Bangladesh's independence from the rest of Pakistan sapped the economies of both countries and led to the need for extensive reconstruction.

Countries

Both Bangladesh and Pakistan are among the world's poorest countries. The differences between them, however, outweigh the influence of their common Islamic faith. Bangladesh is a secular republic, while the Islamic Republic of Pakistan is influenced by Islam. Pakistan is a country of arid lowlands and high mountains (Figure 7.27), whereas Bangladesh is mostly low-lying apart from the small hilly eastern region inland of Chittagong and is well watered by rain and rivers. In Pakistan, the management of scarce water resources is a major problem, while in Bangladesh, flooding is a major problem.

Bangladesh had to build a new country after the 1971 civil war, integrating Muslims expelled from Myanmar and Bengalis returning from India. Huge quantities of aid came from the

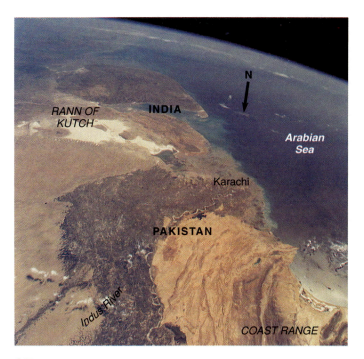

Figure 7.27 **South Asia: lower Indus River valley, Pakistan.** A space shuttle view of the Indus floodplain (dark area of irrigated fields) and the Makran Mountain Range in the foreground. Beyond the Indus is the Thar Desert and the (white) salt marsh of the Great Rann of Kutch in India. Pakistan's largest city and port, Karachi, lies in the western (near) end of the Indus floodplain. Photo: NASA.

world's wealthier countries, and its internal politics involved military takeovers, coups, and assassinations. It was not until the 1990s that major changes to this pattern appeared possible under a settled democracy.

Although Pakistan had an established base of agricultural and manufacturing products at independence, its people are little better off than the Bangladeshis (see Figure 7.2). After its defeat and loss of East Pakistan in 1971, Pakistan's rulers worked out an Islamic religious base for its governmental, industrial, financial, and educational institutions. The 1973 constitution included a basis of Islamic law.

Pakistan maintains a rivalry with its more powerful neighbor, India. When India tested its nuclear devices in 1998, Pakistan followed suit and also received international condemnation. Pakistan continued to dispute the Kashmir territory with India, training terrorists to disrupt Indian control (see the "Point-Counterpoint: Afghanistan-Pakistan-Kashmir-India" box, p. 274).

Pakistan's views of another neighbor, Afghanistan, shift from time to time. During the 1979 occupation of Afghanistan by the Soviet Union, around 3 million Afghan refugees fled to Pakistan, living mainly in camps close to the Afghanistan border. During the 1980s, the Pakistani Jemiat-e-Ulemi-i-Islam religious party taught and trained Afghan refugee children, who later flocked to support the Taliban. Pakistan encouraged the Taliban group in Afghanistan throughout the 1990s by recruiting supporters and supplying armaments. This was partly from religious sympathies, partly to ensure a friendly country to the west so that it could challenge India over Kashmir, and partly hoping that an oil pipeline could be built from Central Asia across Afghanistan and Pakistan. The association with the hard-line Islamic Taliban group, however, fed back into Pakistani politics with demands for increased Islamic influence, often backed by terrorist activity.

In October 1999, General Pervez Musharraf seized power and dismissed the Pakistan parliament. Although the Pakistan Supreme Court backed this move, it insisted that an election be held within three years: the general was reelected in 2002. The events of September 11, 2001, placed Pakistan in the forefront of international actions against Afghanistan's Taliban rulers and Bin Laden's al-Qaeda organization. Despite internal demonstrations in support of the Taliban, the Pakistan government led by President Musharraf joined the international antiterrorist collaboration, a step that led to a rise in its international status, removing sanctions and gaining foreign aid. However, much of the remote Pakistan territory in the mountainous Northwest Frontier province bordering Afghanistan remained a haven for Afghan-Pakistani terrorists. Pakistan military control there is incomplete and remains a security problem.

People

Ethnic Character

Bangladesh is the country of the Bengalis. **Bengali** is the language of 98 percent of the people, who originated from an Indo-Aryan stock that mixed with local ethnic groups. The **Biharis** form a smaller group of non-Bengali Muslims who speak Urdu and migrated to East Pakistan after independence in 1947.

As the site of the main historic entry routes of different groups of people to the Indian subcontinent, Pakistan has greater ethnic diversity than Bangladesh. **Urdu** is the official language of the country, a form of Hindi with Arabic script used by Muslim people before independence. The 1947 **Muhajir** immigrants from Hindu India to Pakistan speak Urdu. **Punjabi** and **Sindhi** are also widely spoken in Pakistan. Many Pashtun groups live in the hills bordering Afghanistan. Rivalries among groups and the elitism of Urdu-speaking feudal landowners and military officers continue to dominate social and political life in Pakistan.

Urban and Rural Population

The two main concentrations of people in Pakistan are in the upper Indus River valley, where irrigation-based farming is labor-intensive, and around Karachi near the coast (see Figure 7.17). The areas of unirrigated desert and mountains are sparsely peopled. The whole of Bangladesh, apart from hilly areas in the extreme northeast and northwest, has high population densities with almost as many people as Pakistan on one-sixth of the land area.

Like India, both Bangladesh and Pakistan remain largely rural countries, with 33 percent of the population living in the towns of Pakistan in 2001 and only 21 percent in Bangladesh. However, both figures understate the true extent of urbanization. The major cities struggle to cope with the arrivals of rural people and the shantytowns and social tensions that the influx creates. Although separated from the political power centers farther north, Karachi, with 12 million people in 2000 and growing rapidly toward a 2015 projection of 19 million, is the largest city in Pakistan and its major port and commercial center (see the "Personal View: Pakistan" box, p. 302). Violence in the 1990s among Muslim groups was made worse by the glaring contrasts between rich and poor, the large numbers of unemployed teenagers, and the availability of guns following years of warfare in Afghanistan.

Hyderabad is an industrial center near Karachi. Other major Pakistani cities are inland, where Lahore (6 million people) is the major center of Muslim culture and Faisalabad (over 2 million) is the cotton industry center. The new capital of Islamabad (1 million) is close to the Kashmir border, while Peshawar (2 million) is the commercial center of the far north at the eastern end of the Khyber Pass into Afghanistan.

Bangladesh, with a smaller urban population, has fewer large cities. Dhaka, the capital, had 12 million people in 2000. Chittagong (3.5 million) is the main port, and Khulna (1.7 million) is a growing industrial center.

Population Dynamics

Bangladesh had the slightly higher population until 1980. By 2001, however, Pakistan's population exceeded that of Bangladesh (145 million compared to 132 million), and estimates for 2025 suggest a further increase in the difference, with

Pakistan having 250 million and Bangladesh 180 million people. The greater rate of population increase in Pakistan is the result of a continuing high total fertility rate of nearly 6, compared to one of 3.3 in Bangladesh following a major family planning thrust. Both have life expectancies in the upper fifties (Figure 7.28).

Given its smaller area, Bangladesh has greater problems of feeding its growing population, an issue that stimulated its government to institute effective family planning programs in the 1970s. This led to the significant reductions in fertility. By the late 1990s, half the married women used contraception, up from 8 percent in 1975. Despite a continuing subordinate role for women, new policies encourage their social and economic development. The loss of farming land and lack of farm jobs for adults forced migration into cities, where jobs demand education qualifications that most migrants do not have. In the 1980s and 1990s, the shortage of jobs in Bangladesh resulted in many laborers seeking employment outside the country, such as in the Persian Gulf countries and Malaysia.

Pakistan's main need of an educated work force is undermined by the combination of influences from Islamic religion, feudal landlords, and poor government that kept Pakistan's school population low until the late 1990s. Its enrollment rates in primary and secondary education were only 30 percent, compared to 70 percent in India and over 80 percent in Indonesia and China. In the mid-1990s, aid donors' demands led to new national policies for education, health, and population. Government funding shifted from elite university education toward universal primary education.

Economic Development

Bangladesh and Pakistan are among the world's poorest countries. The human poverty index is just under 50 percent for both countries. Pakistan's GDP total is one-third greater than that of Bangladesh, but both have adverse trade balances and depend on aid donations.

Although British governments built irrigation canals and railroads so that the Indus Valley could be developed as a source of cotton for British industry, both they and the Pakistan government from 1947 to 1971 largely ignored the East Pakistan (Bangladesh) economy. Only after the 1970s could the Bangladesh government develop its highly populated area in the face of few local resources and major environmental hazards.

In Pakistan, the literacy rate doubled from 1980 to the late 1990s, although it was still only 45 percent; electrification of rural households rose from 16 to 61 percent; and terrestrial television now reaches 75 percent of city dwellers and 50 percent of the rural population. In late 2001, as a result of its support for the U.S. antiterrorist coalition, sanctions against Pakistan (imposed after its 1998 nuclear tests) were dropped, and it was given increased access to European markets as part of fighting the drug trade.

Prominence of Agriculture

Both Pakistan and Bangladesh are essentially agricultural countries that are economically dependent on weather and international commodity markets. Large areas are still dominated by low-productivity farming systems.

In the 1990s, almost 75 percent of the Bangladeshi labor force was engaged in farming. Their products made up 31 percent of GDP, down from 50 percent in 1980. Bangladesh grows over half of the jute that enters world trade (Figure 7.29), as well as rice, tea, and sugarcane. As in India, livestock are kept mainly as beasts of burden. Expansion of the farming area in Bangladesh occurred on new delta islands formed by deposition at the mouth of the Ganges River and reclaimed by building costly dikes. These areas are below sea level and vulnerable to ocean surges during tropical cyclones.

In Pakistan, agriculture occupies only the land that is neither too arid nor too steep (Figure 7.30) but employs nearly 60

Figure 7.29 **South Asia: jute harvest near Tangail, Bangladesh.** The fiber is stripped from the plant stalks using the plentiful water of the Ganges River delta area. What is jute used for? Photo: © James P. Blair/National Geographic Image Collection.

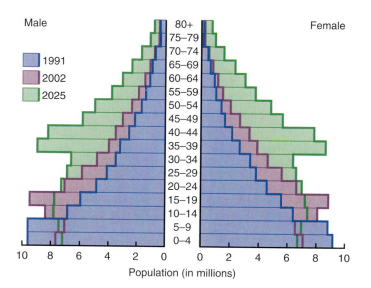

Figure 7.28 **Bangladesh: age-sex diagram.** Is there evidence of falling births? Source: U.S. Census Bureau. International Data Bank.

Personal View

PAKISTAN

Pakistan is a country of contrasts. Karachi, the main port and commercial center and the largest city (by far), is on the southern coast (Box Figure 1). Inland lie the poor agricultural districts and the older northern cities such as Lahore, the center of the historic Mughal Empire. Also in the north are Islamabad, Pakistan's capital city, and Peshawar at the gateway of the Khyber Pass into Afghanistan, Pakistan is a poor country, but many wealthy people live there, with the highest proportion in Karachi. The contrasts between rich and poor affect many aspects of life in the country.

Ali and his family live in Karachi. Following partition in 1947, his Muslim parents moved from what is India today to be in the new country of Pakistan. Ali is a banker whose education was in the United States and United Kingdom. He works for an international bank that has its Pakistan headquarters in Karachi—like most international banks and finance houses. This makes Karachi a very cos-

mopolitan city, one that most resembles Western cities in its metropolitan characteristics. Ali calls Karachi the country's "melting pot" where wealthy and poor and varied ethnic groups meet.

The wealthy in Pakistan have the lifestyle of the wealthy in all countries; the poor are very poor. Around 10 percent of the population possesses 90 percent of the country's wealth. Despite high import duties on cars and other goods, a favorite vehicle among the materially wealthy is the Toyota Landcruiser, which costs $65,000 in the United States but $130,000 in Pakistan. And yet, everything has to be paid for in cash, including goods and houses: there is no system of credit, and being wealthy involves having ready cash.

The poor in Pakistan find it difficult to gain any income. Most still work in the fields of large landowners for tiny wages. They employ their children to work alongside them so that it is difficult for the children to gain an education. Some migrate into towns. In smaller towns, many children also work in the carpet and sportswear factories. In the cities, such as Karachi, there are many poor, attracted by

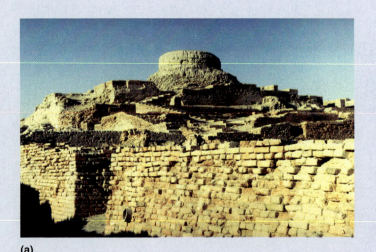

(a)

(b)

Box Figure 1 **Contrasts in Pakistan.** (a) Archaeological remains of the ancient city of Mohendojaro. (b) Muslims offer prayers at a mosque in Karachi. Photos: (a) © Munir Khan/Harappa.com; (b) © Reuters NewMedia Inc./Corbis.

percent of the labor force, and its products accounted for 26 percent of GDP in the 1990s. Cotton remains the chief commercial crop, supported by government credits to farmers so that they will continue to produce the raw material of the main manufacturing sector in the country. Heavy rains in the three years from 1992 to 1994 and poor crops threatened the country's major industry and caused farmers to question whether they should grow different crops. The battle between new virus-resistant strains and new viruses continues. Other land-intensive crops, such as wheat, rice, and sugarcane, are grown as alternatives to cotton. Some farmers are beginning to look to more profitable labor-intensive crops, such as vegetables and fruit.

Much of Pakistan's best farmland is owned by a small group of wealthy people who treat it as a source of political power by controlling local voting rather than as a basis for

greater productivity. By supporting politicians, they gain the patronage of many jobs in the government bureaucracy as well as the ability to veto any measures that might change their dominant place in Pakistani society. Although Pakistan governments began to implement land reform by appropriating land and breaking it into smaller holdings, the feudal landlords continue to own more land than has been reallocated to peasant farmers.

Mining and Forestry

Neither Bangladesh nor Pakistan developed other primary products to a great extent. Both have small reserves of oil and natural gas, although neither exploited them fully until the 1990s because they did not wish multinational corporations to carry out discovery and production. Pakistan's important chemical industry is built on deposits of gypsum, rock salt, and the perception of job opportunities and better access to education and health care. But housing and other facilities are insufficient to maintain them. Many live in shantytowns, which spring up wherever there is a space and without legal land title. Many are built of cement and blocks with asbestos sheeting roofs. A family of eight or more live in the space of a one-car garage. Such shantytowns have no sanitation, water may come from a communal tube well with a tap that is opened for a short time each day, and electricity is stolen by plugging into the local power line. From time to time, a government edict leads to the clearing of a shantytown area of several hundred homes.

For the poor, there are government schools and hospitals, but neither provide a thorough service and both are short of funds. The government of Pakistan spends over half its income on servicing its debts and another one-third on its military. That leaves 10 percent to be spent on government, health, education, and other needs. Health and education each receive just over 1 percent of the national budget, which is small since most businessowners avoid paying taxes.

All sections of society are united in their desire to consume. From the poorest to the richest, people save to purchase radios, TV sets, washing machines, refrigerators, stereos, cars, jewelry, and houses. There is an ascending order of purchases. The differences between wealth and poverty, however, lead to high levels of crime. The materially wealthy live in houses enclosed by concrete walls and employ guards.

Ali and his family live in a house that his parents and brother's family also once occupied. He regrets his separation from the wider family and thinks it important for older family members to live with and pass on traditional ways to the young.

The rich-poor divide is perpetuated by the access to better education and health care that the middle and wealthy classes can afford. Private schools teach all lessons in English (apart from those teaching Urdu, the main local language), and students continue to take British examinations at 16 and 18 years old, facilitated by the British Council. Pakistani universities have a strong reputation, and students often find it as difficult to secure a place in them as in the U.S. or U.K. institutions that many opt for.

Good health care is available but expensive. For example, the Aga Khan Foundation hospital in Karachi has modern facilities and a specialist medical staff that rivals the best hospitals in the United States. Aftercare and nursing education, however, are not so well regarded.

Many aspects of living in Pakistan reflect the continuing influence of the British colonial occupation, although it is over 50 years since the Raj ended. Karachi itself was a fishing village that the British made into a major port and city. Its purpose was to export the cotton grown on the inland fields irrigated by British engineers in the late 1800s. Along with the taking of British school examinations and the wider continued use of English, especially in the commercial community, other relics of the Raj include the legal calendar and the popularity of clubs. Ali is a member of one club where he swims and plays squash. It is very exclusive. Founded in the 1850s as a British officers' club, new members have to wait up to 10 years for entry following a series of taxing interviews with committee members. The British also left behind what has become the national game, cricket. It is not over-stating the case to say that the mood of the country depends on whether the national team wins.

Pakistan is also becoming increasingly cosmopolitan in terms of the global economy. Some 90 percent of the goods available in Western supermarkets and stores can be bought in Pakistan shops—although there are no supermarkets, and most shops are family-owned. They remain open from 9 A.M. to 9 P.M. (sometimes later). An increasing range of food is available in restaurants. The McDonalds, Wendy's, and Pizza Huts are now joined by a range of Italian, Chinese, Japanese, and Middle Eastern specialities. Going out to eat and the accompanying socialization have become a major pastime for the growing middle and upper classes in Karachi. When visiting Pakistan, however, it is wise to remember that informal parties—even wedding receptions—do not begin until well after the stated invitation time. Dinner is often served at 1 A.M. Business meals, however, are prompt.

Ali and other Pakistanis link many of the recent problems of their country to a combination of huge debts that cannot be repaid and the U.S. influences stemming from the wars in Afghanistan. The United States resisted the Russian invasion of Afghanistan in 1979 to 1989 by getting Pakistan to arm and train groups to fight in Afghanistan. Many attracted to this role came out of Muslim fundamentalist schools that themselves were supported by U.S. funds. These funds also provided massive loans to the Pakistan government, mostly spent on the military. Since the Russian withdrawal, Pakistan has been plagued by an abundance of arms, with each home having two or three guns, and by an increase in political unrest stimulated by the fundamentalist groups.

---

soda ash. Fishing in the coastal areas of both countries is poorly organized.

Increasing Manufacturing and Service Industries

In Bangladesh, the slow expansion of agricultural production at 2 to 3 percent per year cannot support the faster population growth, and manufacturing growth needs to rise above the current 7 percent increase per year. The jute industry requires restructuring, but the Bangladesh government faces the opposition of trade unions in trying to privatize the industry. However, Bangladesh's garment industry expanded in the 1990s, becoming a major exporter. Bangladeshi apparel manufacturers work closely with their retailing customers in Europe, the United States, and East Asia, although some countries object to the use of child labor and the low wages: Bangladeshi apparel workers are paid 9 to 20 U.S. cents per hour compared to payments of 20 to 30 cents in Pakistan and India and $8.42 in the United States. Enterprise zones at Chittagong and Dhaka (Dacca) are further attempts to expand industrial income and employment.

Pakistan has a longer experience of manufacturing, which grew from 15 percent of GDP in 1965 to 25 percent in the 1990s. In exports, primary products decreased from 48 to 11 percent of the total between 1975 and 1995, while manufactured goods rose from 52 to 89 percent. Textile manufacture, especially of cotton goods, dominates other industries that include food processing, chemicals, and car assembly. The vested interests of tractor manufacturers within Pakistan prevented the introduction of cheaper tractors made in Poland and Belarus.

Both countries lack many basic provisions that might encourage more manufacturing industry to locate there. Power supplies are subject to shortages, although Pakistan hopes to

Figure 7.30 **South Asia: Pakistan rural area.** The Tupdopdan Mountains in the Karakoram Range, northern Pakistan.
Photo: © John Corbett/Ecoscene/Corbis.

double its power provision with the aid of private investment. In the 1990s, it completed major hydroelectricity projects in the Himalayas, while a new thermal power station is being built with foreign capital and expertise at Hub near Karachi.

Service industries employ a growing proportion of the labor force in Pakistan and are expanding in Bangladesh: both countries produce around half of their GDP from such industries. Health, education, and financial services are growth areas. Both countries attract few tourists.

Grameen Bank

Bangladesh developed the concept of microcredit—making small loans to individuals and groups to enable them to start businesses. The idea began in 1974, when Professor Muhammad Younis, an economist at Chittagong University, lent US $25 of his own money to 42 basket weavers and found they could then pull themselves out of extreme poverty. In 1983, he found the Grameen Bank. In 2001, the bank's 1,170 branches in Bangladesh covered 40,000 villages and had nearly 2.5 million members.

Loans are available for income-generating activities such as simple processing (rice husking or lime-making), manufacturing (pots, weaving, garment sewing), storage, marketing, and transportation. Average household incomes of Grameen Bank members are 25 to 50 percent higher than their neighbors'. Landless people gain the greatest benefits, and the shift from agricultural wage labor to self-employment in petty trading brought a feeling of social improvement. A related result has been better employment conditions and wages for agricultural laborers as their numbers dwindle.

Controversially, women were allowed equal access to the loans, and by 2001, 94 percent of borrowers were women. Also confounding early critics, the loans are repaid at better rates than other types of loans. Similar methods are now applied in 58 countries, and the World Bank acknowledges that millions of individuals worked their way out of deep poverty using the loans, although the total effect on each country's economy is small.

Economic Restructuring

Bangladesh adopted structural adjustment measures imposed from outside. In the early 1990s, it reduced trade restrictions and encouraged external financing of garment factories, doubling its export of goods. During the 1996 elections, political unrest included strikes affecting half of the 1 million poorly paid garment workers, but the subsequent economic reforms and new government soon returned people to work and the economy to growth. Privatization continues, but severe infrastructure bottlenecks (roads, power stations, communications) slow the economy's expansion. Losses by state-owned enterprises continue to be a heavy burden for the limited national budget to bear.

In Pakistan, the growth in output and productivity in manufacturing increased jobs, even in rural areas, at a rate that equaled the rate of new entrants to the labor force, but such gains did not benefit the poorest groups of people. The new government elected in 1997 gave signs that it would help to make its producers more competitive internationally. It also wished to tackle problems that kept so many of its people poor and socially deprived. A military government replaced the elected government before these aims were met. The detonation of nuclear weapon devices in 1998 was a response to the Indian demonstration of military power, but, although Pakistan was supported by Chinese technology in this field, it was a costly event, financially and politically.

Poverty and Social Services

In Bangladesh, the problems are mainly those of overpopulation (very high densities with a low resource and infrastructure base). Forty percent of Bangladeshi children are not at school; only 24 percent of women and 48 percent of men are literate. Fewer than half the population has access to sanitation or legitimate electricity supplies and only 14 percent to garbage disposal.

In Pakistan, the inequitable distribution of wealth still leaves large numbers of very poor, since the average per capita income is higher than that of Bangladesh. Pakistan lagged behind other low-income countries in social conditions in the 1990s, with only 50 percent of primary-age children at school (31% of girls), high infant mortality (85 per 1,000 live births, compared to the average of 60 in all low-income countries), continuing high total fertility, and low literacy (23% for women, 49% for men). Following the failure of earlier efforts, the 1994–1998 five-year plan included a social action program that claimed major advances in the first three years, but the program suffered from slow bureaucratic responses and a lack of community participation.

Mountain and Island Rim

The countries of the northern mountains—Afghanistan, Nepal, and Bhutan—have environments that isolate them from global connections and foster internal strife through tribal rivalries. Events in these mountain countries are influenced not only by what happens in the rest of South Asia but also by interactions with neighboring countries in Iran, Russia, Central Asia, and China. The island countries of Sri Lanka and the Maldives have easier ocean connections with the global economic system but experience other problems that restrict their fuller development.

Countries

In 2001, Afghanistan, Nepal, and Sri Lanka each had 20 million to 27 million people. Bhutan and the Maldives had tiny populations of fewer than 1 million people between them. The mountain countries are among the world's poorest, while the Maldives and Sri Lanka have better living conditions. Following their independence, the Maldives changed from a Muslim sultanate under British protection to a republic. The government called in Indian troops in 1988 to avert a coup by a group using Tamil mercenaries.

Afghanistan

From its independence in 1920, internal feuding and external influences held back attempts to modernize Afghanistan. Tensions with Pakistan emerged over the northwest frontier lands that Pakistan swiftly annexed after its own independence in 1947. Afghanistan then linked more closely to the Soviet Union until 1979, when the latter invaded and occupied Afghanistan. During the period of Soviet occupation, local warlords, or mujahideen (mujahedin), supplied with weapons by the United States through Pakistan, fought the Soviet armies and established their own military units. After the Soviet Union withdrew in 1992, the local mujahideen and their armies fought each other for control of the country. Most government systems broke down. Education became irregular, while public health and health care were weakened: aid agencies provided most of what treatment Afghans obtained.

In the late 1990s, the Taliban group conquered virtually all Afghanistan, although opposed fiercely by northern tribes who resisted the imposition of hard-line Islamic tenets. From the west, Iran supported Shia groups, including Hizbollah, but failed to overthrow the Taliban. The rest of the country lived under the repressive Taliban order, which expelled several international aid agencies. Iran and Uzbekistan closed their borders with Afghanistan. However, Pakistan supported the regime and continued its contacts for some weeks after September 11, 2001.

The association of Afghanistan's Taliban rulers with Osama bin Laden's terrorist network began during the Soviet occupation in the 1980s, forming a group of "Arab Afghan" exiles from all over the world who shared the militant Wahhabist Islamic beliefs (the literal Sunni Muslim group that dominates Saudi Arabia; see Chapter 8) and planned a universal Islamic country. The 1992 withdrawal of the Soviet forces added credibility to the group, which claimed it was responsible for the victory and set up al-Qaeda and its cluster of military-ideologic camps. Most Afghans disliked the Taliban's Wahhabist approach and were quick to revolt after September 11, 2001. U.S. attacks on Taliban strongholds and al-Qaeda positions in late 2001 built on the internal opposition to the Taliban, deposing the Taliban government and establishing a new government of national unity.

Afghanistan faces a difficult and uncertain future in the light of its history of factional strife, occasional oppressive rulers, devastated economy, and difficult relationships with surrounding countries. In 2001–2002, the U.S. reliance on local warlords that saved U.S. casualties gave the impression of an uprising against the Taliban but also provided the warlords with new weapons and opportunities to settle blood feuds. Moreover, Iran and Pakistan continue to support groups around Herat in the west and the former Taliban groups around Kandahar in the center. Tajiks and Uzbeks expelled Pashtuns from the north in a phase of ethnic cleansing. The Afghan International Office for Migration encourages the return of qualified Afghans who fled the Russian- or Taliban-dominated country but has difficulties in attracting professionals who settled comfortably in the United States, Europe, or Australia, where they have higher salaries and prospects of asylum or citizenship.

One significant post-Taliban change that tells us much about Afghans is the return of the ancient aggressive game of *buzkashi* that the Taliban banned. It centers on a headless sheep or goat carcass that horsemen attempt to seize and carry to the far end of a field and place in a chalk circle. The game results in many injuries, although whips are forbidden and horses are trained not to stamp on people. Warlords own their own teams and gain popularity from victories.

Nepal and Bhutan

Nepal emerged from isolation as a democracy in the 1980s, but alternative political parties were not allowed until 1990. In 2001, Nepal's monarchy was almost all murdered by the heir to the throne in a family argument, while left-wing guerrillas caused major disruption in occupying rural areas and attacking remote army and police barracks. The mountainous country is susceptible to earthquakes, landslides, and floods. Its low-lying Terai Plain, where over half the population lives, gives way northward to hilly lands and the highest peaks of the Himalayan Mountains.

Bhutan remains a tiny buffer state between India and China. It has few cross-border contacts, preferring to maintain its own culture and to keep out Westerners.

Sri Lanka

After independence, Sri Lanka's government by the dominant **Sinhalese** people instituted reforms that were disliked by the Tamil minority, leading to a still unresolved civil war. From a position where Sinhalese and Tamils lived side-by-side in Colombo, the capital, vicious civil war resulted in increased separations of people, over 60,000 deaths (Figure 7.31), and

the emigration of many Tamils to India. The Tamil Tigers (Liberation Tigers of Tamil Eelam, where *eelam* means "homeland") became one of the most aggressive terrorist groups fighting for what they saw as their minority rights and built links with other terrorist organizations. Although 50 million Tamils live in India, which at first gave training support to the Tamil Tigers, India banned the organization from 1992 after it assassinated the former Indian Prime Minister Rajiv Gandhi in 1991. In early 2002, one outcome of the global war on terrorism was a cease-fire leading to the first talks in seven years between the Tamil Tigers and the Sri Lankan government.

People

Ethnic-Based Conflicts

The mountain and island countries of South Asia have local fault lines between dominant ethnic groups and minorities. Muslims dominate in the west (Afghanistan and Maldives), Hindus in Nepal, and Buddhists in the east (Bhutan, Sri Lanka).

- Afghanistan is 99 percent Muslim, 84 percent Sunni, and 15 percent Shiite. Its diversity of ethnic groups based on language includes 50 percent speaking the main language, an Afghan form of Persian. Pashtuns dominate the south, while groups of Central Asian peoples, such as the Tajiks and Uzbeks, inhabit the north (see Figure 7.8).

- The Maldives is another Muslim country. Arab traders established that faith from the 1100s. The islands have a diversity of ethnic groups from southern India, Sri Lanka, Arabia, and Africa.

- Nepal is the only country in the subregion where nearly 90 percent of the people are Hindus. It also includes small proportions of Buddhists who have had a major impact on the local character of Hinduism. Like Afghanistan, however, Nepal has ethnic diversity arising from peoples moving into the mountainous country from all directions—Tibetans, Indians, and many local hill tribes.

- Bhutan is a Lamaistic Buddhist country like its neighbor, Tibet (see Chapter 5), but 35 percent of its population are Hindus from Nepal and India. The main group of people, the Bhote, drove out Indian peoples in the A.D. 800s and built impressive fortified monasteries in the 1500s.

- Sri Lanka has a Buddhist majority, which is linked to the dominant Sinhalese-speaking group. Nearly 70 percent of Sri Lankans are Buddhists, 15 percent Hindus (mainly Tamil), and 8 percent each Muslims and Christians. The Muslims and Tamils are the main business groups, and since independence, both have been subject to repressions by the dominant Sinhalese.

Population Distribution and Dynamics

In the mountainous countries, population's main concentrations are in lower areas that are accessible to major routeways. In Afghanistan (see Figure 7.17), the valleys around the capital, Kabul, the northern plains bordering on Turkmenistan, and southwestern valleys connected to Iran have the highest con-

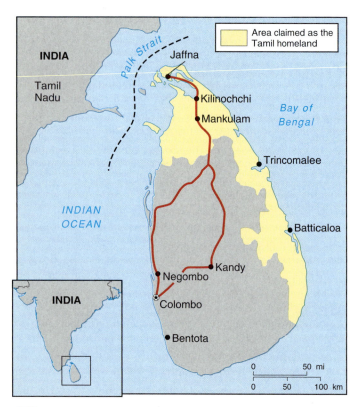

Figure 7.31 **Sri Lanka: basis of civil war.** The Tamil population lives mainly in the north and northeast of the island. Source: *The Economist*, 5 December 1998.

centrations. The Himalayan states have most people near the Indian border. Sri Lanka's population is focused on the rainier southwestern coastlands around Colombo.

Only the Maldives and Sri Lanka have urbanization rates as high as 20 percent. Kabul (Afghanistan; 2.5 million people in 2000), Colombo (Sri Lanka), and Kathmandu, capital of Nepal, are the largest cities of the subregion and contain extensive shantytowns.

Bhutan, the Maldives, and Afghanistan populations grow most rapidly at around 3 percent per year. Sri Lanka's population increases at only 1.2 percent. In most countries, the total fertility rates are declining but remain over 6 in Afghanistan and over 5 in the Maldives and Bhutan. All the countries remain at a stage when the population increases rapidly and there are many young people. For example, Nepal has a strong family planning program, widely disseminated, but only 29 percent of women and 2 percent of men use birth control methods; fertility rates decline slowly. This is linked to Nepalese female adult literacy rates, which were 3.7 percent in 1971 but rose to only 15 percent by 2000.

Economic Development

The physical and political isolation of the mountain countries, together with continuing internal conflicts, limit economic development in the subregion. Agriculture remains important in all the countries, and most people still depend on subsistence farming and occasional wage labor. The highland environments provide massive hydroelectric potential, but construction of the facilities is expensive, and the cost and environmental impacts are high.

The rebuilding of Afghanistan will be a major project in the foreseeable future. Afghanistan is destitute and a threat to its neighbors as long as it remains in chaos. Although the Afghans must determine the strategy, an international coalition, rather than previous attempts by single countries to influence them, needs to support the development of a self-sustaining government. This will involve a major United Nations role and the provision of resources over five to 10 years costing billions of dollars. Repaired roads, power lines, schools, hospitals, functioning police and judicial systems, an acceptable currency, and a central bank are needed alongside housing, factories, and offices to replace the piles of rubble. Millions of landmines need clearing. In 25 years of fighting, most doctors and teachers left the country.

Although the United Nations program averted postwar famine in Afghanistan by trucking in food, deliveries went mainly to the towns, while inaccessible mountain areas continue to be underfed after years of war. As many of those who lived in the Pakistan refugee camps returned to Afghanistan in early 2002, others replaced them, fleeing from northern areas.

Agriculture Base

In Nepal, 82 percent of workers are in agriculture, mainly on the Terai Plain along the Indian border. Noncommercial agriculture is based around the nomadic keeping of sheep in Afghanistan and the growing of rice and other cereals elsewhere. None of these countries is self-sufficient in food.

Sri Lanka (tea, rubber, and coconuts) and the Maldives (coconuts) have commercial plantations that sell cash crops to world markets. These provide 35 percent of Sri Lanka's exports, down from over 90 percent in 1950 as a result of diversification in the economy since independence. Sri Lanka's adoption of new rice varieties in the late 1960s converted the major food deficit—which arose out of the colonial focus on export crops—into a virtual self-sufficiency in rice today.

In a politically motivated act in 1998, the Taliban destroyed the fertile central farming region of Afghanistan, the Shardi Plain near Kabul. They cut fruit and nut trees, burned villages and wheatfields, blew up irrigation channels, and left crops to wither. Around 200,000 people from the region took refuge in Kabul or fled northward.

Despite international pressures, opium poppies remain an important crop in Afghanistan, which in the 1990s supplied 70 percent of the world's heroin as the country's other farm products dwindled through war and Turkey, Iran, and Pakistan enforced strict drug-control laws. Wheat was replaced by poppy fields, leaving city dwellers hungry. Afghanistan's isolation, disrupted government, lack of alternative commercial products, and warlord control of the drug trade make it difficult to stop the production of opium and heroin and their entry to illegal drug traffic routeways. Although Afghanistan was threatened with the withdrawal of aid because of its increasing production of these drugs, payments continued for fear of angering the warlords and as part of projects to persuade peasants to stop growing poppies. In 2000, the Taliban government stopped the growing of poppies as ungodly, and by the following year, few poppy fields remained. After September 11, 2001, however, the Taliban stopped cooperating with the United Nations, and planting took place on a large scale. When the Taliban government was deposed in late 2001, efforts began to prevent the 2002 crops from flooding the European market after the March harvest and as the remnant of the Taliban releases its stockpile, worth billions of dollars (around US $4 billion in Pakistan and US $40–80 billion in Europe). Such sales could result in social disruption and enable the Taliban remnant to buy more weapons.

Mines, Forests, and Fisheries

Apart from natural gas in northern Afghanistan (which was formerly exported by pipeline to Russia), these countries are poor in mineral resources. Several produce gemstones. Sri Lanka has a major source of graphite.

The forests in Nepal, where massive deforestation has occurred, and Sri Lanka are cut and used locally, mostly for firewood. Fishing is important to the economy of the Maldives, where tuna provide 60 percent of exports.

The loss of woodland was another outcome of the crisis in Afghanistan. From 1980 to 2001, its woodland area was reduced by five-sixths, being cut for firewood or by Pakistan-based loggers. The resultant barren hillsides succumbed to rapid soil erosion and gave rise to rapid runoff and flooding. Only a few projects aim to support new plantings or the use of alternative fuels.

Small-Scale Manufacturing

Only Sri Lanka produces more than 10 percent of its gross domestic product by manufacturing. Much of the manufacturing that exists in the subregion is in small-scale factories or **cottage industries** based on the dispersal of manufacturing processes in homes. Local agricultural processing together with the making of carpets and footwear are most common industries.

Nepal (Mahakali project) and Bhutan (Chukha) encouraged external investment in dams for hydroelectricity and other uses. This resulted in additional local industry—making garments in Nepal and steel alloys in Bhutan.

In Afghanistan, Kabul's industrial district had 200 factories in 1992, but they were reduced to 40 under the Taliban, of which only 6 still operated in late 2001. The 1992 fighting destroyed most of the Jangalok steelworks outside Kabul, and the machinery was taken to Pakistan. Even the Army Factory that made cannons, guns, and ammunition is largely abandoned. One factory sends its employees to gather discarded plastic shoes and melts them to produce new cheap shoes.

Sri Lanka has a more diversified manufacturing base than the other countries as the result of government policies since independence, including the production of steel, textiles, tires, and electrical equipment. By the 1990s, however, expansion placed pressure on existing transportation and telecommunications facilities, while power shortages reduced output and exports of garments and other textiles.

Tourism

The mountain and coastal environments of these countries suggest tourist potential, but only the Maldives, Nepal, and Sri Lanka have developed tourist industries since the 1970s (Figure 7.32). Before the recent internal strife, Nepal attracted half a million tourists in 2000. Sri Lanka attracted as many but has also been affected by civil war.

In the Maldives, tourism attracted numbers that increased from a few thousand in 1980 to nearly half a million visitors in 2000. Provision spread to new islands, while older facilities were refurbished. By the late 1990s, the limits of expansion were being reached, signaled by coastal pollution, groundwater contamination, and the need to bring in labor to cope with the increasing demand. The tourist industry employed over 25 percent in the mid-1990s, up from only 4 percent of the total labor force in 1985.

Other countries are too isolated or unsafe for tourists. Bhutan keeps down the numbers of tourists entering to reduce the impact of foreign cultures on its people.

Figure 7.32 **South Asia: tourism in Nepal.** The peak of Annapurna in the Himalayan range rises behind the hotel in Kaski village. How would a trekking holiday in Nepal differ from one in the Rockies or Appalachians? Photo: © John Burbank/The Image Works.

Test Your Understanding 7C

Summary Pakistan and Bangladesh were a single country from partition in 1947 until 1971. They remain largely agricultural, although Pakistan has a more diversified economy, with a long-established cotton textile industry.

Afghanistan, Nepal, and Bhutan are mountainous countries on the northern margins of South Asia. They have limited connections to the world economy. Wars and political tensions have destroyed Afghanistan.

Sri Lanka and the Maldives produce plantation crops and fish for sale on world markets and have growing tourist industries. Sri Lanka's attempts to diversify its economy have been hampered by civil war.

Questions to Think About

7C.1   To what extent were Pakistan and Bangladesh incompatible as parts of a single country?

7C.2   Write a short Point-Counterpoint that focuses on the Sri Lankan civil war.

7C.3   Assess the political and economic needs of Afghanistan.

Key Terms

Bengali

Biharis

Urdu

Muhajirs

Punjabi

Sindhi

Sinhalese

cottage industry

Online Learning Center   www.mhhe.com/bradshaw

Making Connections

The Online Learning Center accompanying this textbook provides access to a vast range of further information about each chapter and region covered in this text. Go to www.mhhe.com/bradshaw to discover these useful study aids:

- Self-test questions
- Interactive, map-based exercises to identify key places within each region
- PowerWeb readings for further study
- Links to websites relating to topics in this chapter

Chapter 8

Northern Africa and Southwestern Asia

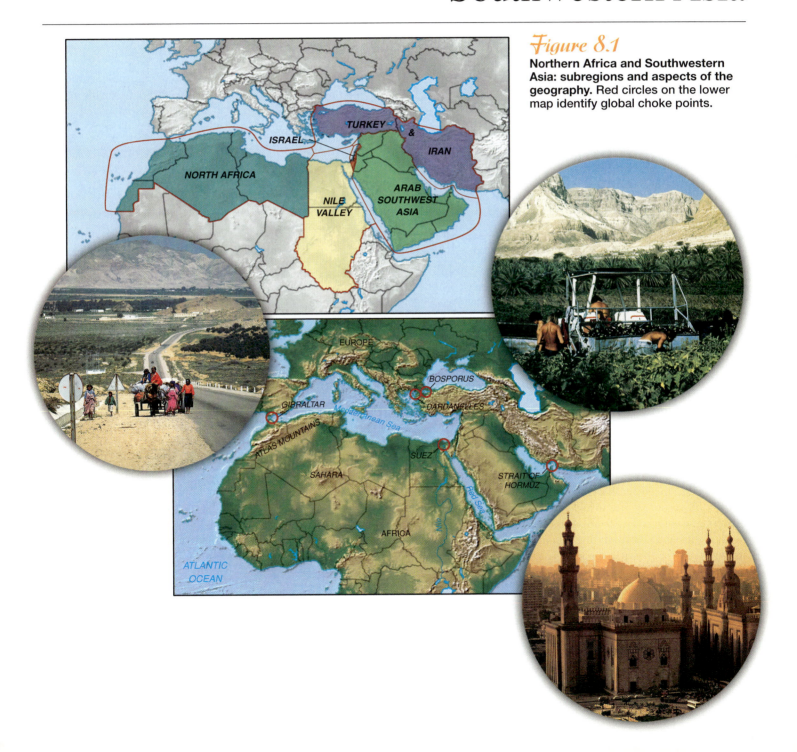

Figure 8.1

Northern Africa and Southwestern Asia: subregions and aspects of the geography. Red circles on the lower map identify global choke points.

TURKEY & IRAN

ISRAEL

NORTH AFRICA

NILE VALLEY

ARAB SOUTHWEST ASIA

EUROPE

BOSPORUS

GIBRALTAR

DARDANELLES

Mediterranean Sea

ATLAS MOUNTAINS

SUEZ

STRAIT OF HORMUZ

SAHARA

Red Sea

Nile

AFRICA

ATLANTIC OCEAN

In the News

This region incorporates Africa north of the Sahara and what Westerners commonly called the "Middle East" (Figure 8.1). The region is sometimes referred to by the acronym MENA (Middle East and North Africa). Its boundaries are coastlines, a line drawn through the Sahara in the south, and the mountain ranges marking the northern and eastern limits of Turkey and Iran. Sudan is included in the region because of its strong links with Egypt along the Nile River valley.

The countries of northern Africa fronting the Mediterranean Sea and those of southwestern Asia occupy nearly 12 percent of the world's land and have 6 percent of the total population (Figure 8.2). However, the region is one of the most influential parts of the world, often at the center of historic events. Today's strategic economic interests make it one of the most politically unstable regions.

The region is often in the news: issues include the Israeli-Arab conflict, the geopolitical implications of two-thirds of the world's oil resources occurring in the Persian Gulf area, the rivalry between Iran and Saudi Arabia, the sanctions applied to Saddam Hussein's Iraq and the Islamic Republic of Iran, the struggles of the Kurdish people, and the politics of water resources. Each issue ignites local wars and has the potential for wider disputes. Many of the issues are closely linked and often reflect the dominance of the Islamic religion in the region and attitudes its adherents take in facing the challenge of Western world dominance.

Each issue has a geographical significance and background. Northern Africa and Southwestern Asia is a region of wide extent from east to west; it is geographically at one of the world's great human and physical junctions, providing the connecting tissues between three continents. The location of this region in the midst of others, influenced internally by such strong cultural developments, made it an initial starting point and later crossroads for many directions in human history. From 1950, oil wealth and then resurgent Islamism brought back global significance after centuries of being overshadowed by Western economic development that left most people in this region materially poor. The region's largely arid natural environment raises water resources to the same level of political significance as oil.

In the Middle: Exchanges, Unities, and Diversities

Past, present, and future interactions among the countries of Northern Africa and Southwestern Asia and between this region and the surrounding regions in the rest of Africa, Europe, and Asia place this region "in the middle" in many senses:

- The *Middle East* is a term that originated from an early 1900s French view of the "Near East" incorporating modern Turkey, Syria, Lebanon, and Palestine, and the British concept of the "Middle East" that included Egypt, Arabia, and the Persian Gulf area. The "Far East" (Pakistan and beyond) indicated greater distance from Europe. The designation "Southwestern Asia" for the eastern part of this region is more appropriate today in the context of a global rather than a Eurocentric view.

- In the *geologic past,* the land areas of the region formed as the continental masses that are now Africa and Arabia rammed into the southern edges of Europe and Asia, raising mountains along the line of collision. Deep seas accumulated oil in sediments worn from the mountains.

- The region was the *center of the earliest urban civilizations and several world religions.* The urban civilizations of Mesopotamia (approximately modern Iraq) and Egypt developed in major river valleys. Judaism, Christianity, and Islam all originated in this region before spreading worldwide.

- Countries in the region contain relics that reflect different *colonial interventions* from the Ottoman Empire and European countries. In the 1920s, the boundaries of the countries were imposed, as were many of the original constitutions as republics (after French influences) or monarchies (British). For many inhabitants, these political boundaries prevent long-established interactions. Until the early 1900s, nomadic pastoralists migrated over huge distances. Such movements are now thwarted by the boundaries.

- The region is increasingly connected more widely to the *global economy* and its political impacts. During the 1900s, the possession of over two-thirds of the world's known oil reserves made the lands bordering the Persian Gulf of immense strategic interest to major oil users. The Persian Gulf became the focus of a huge worldwide ocean network of oil distribution. The employment of expatriate workers overcame the labor shortages in a sparsely popu-

Figure 8.2 Northern Africa and Southwestern Asia: data for regions and subregions.

Subregion	Land Area (km²) Total	Population (millions) Mid-2001 Total	2025 Est.	GNI 1999 (US $ million) Total	GNI PPP 1999 Per Capita	Percent Urban 2001	Human Development Index Rank of 175 Countries	Human Poverty Index: Percent of Total Population
North Africa	4,751,440	75.1	104.5	100,020	4,620	63.0	100.5	26.9
Nile River Valley	3,507,260	101.6	145.8	95,979	3,460	35.0	131.0	35.0
Arab Southwest Asia	3,726,330	98.6	178.1	184,138	4,620	75.4	79.1	20.2
Israel, West Bank, Gaza	21,060	6.4	8.9	99,574	18,070	65.0	23.0	no data
Iran, Turkey	2,427,450	132.4	173.6	300,219	5,980	65.0	90.5	18.6
Total or Average	14,433,540	414.1	610.9	779,930	7,350	60.7	84.8	25.2

Source: Data from *Population Reference Bureau 2001 Data Sheet*; *World Development Indicators,* World Bank, 2001; *Human Development Report,* United Nations, 2001.

lated area and introduced new Western and Eastern cultures in many Persian Gulf countries. The affluent lifestyles of oil-rich elites fueled reactions by local poorer people against them and materially wealthy countries.

- The region's strategic importance within the global economy and its central position on significant sea routes focus attention on the many *global choke points* linking suppliers and markets through Northern Africa and Southwestern Asia (Figure 8.1). The most important are the Suez Canal (Egypt) that controls routes from Europe to Asia and the Strait of Hormuz (Iran and Oman) in the east at the entrance to the oil-rich Persian Gulf area.

- The region is also "in the middle" in terms of material *wealth and poverty.* Apart from Israel, which approaches the income of wealthier countries, countries in the region are dependent on the fluctuating value of their oil sales or on economies that produce countries with a lower-middle income (World Bank classification; see Figure 2.6). Even at the height of the oil-price boom in 1980, many people remained poor as the wealth was not widely distributed; lower oil prices since allowed little improvement for the poor.

- The region remains the *center of the Islamic religion,* which extends into countries of South and Southeast Asia, Africa South of the Sahara, and to significant Muslim populations throughout the world. As the world economic system increasingly involved this region in the 1900s, nationalistic leaders tried to achieve Western modernization, while Islamic religious leaders tried to resist or provide alternatives to the external influences of the global economic system and Western culture. Unsatisfied by such moves, militant groups caused increasing agitation.

- The Muslim reactions against Western influences often center on the 1948 establishment of the *country of Israel* as a homeland for Jews in the midst of Muslim countries that opposed its presence from the outset.

The unifying elements of the human geography of Northern Africa and Southwestern Asia include culture (Islamic religion and Arabic language), economy (oil), and environment (arid). A major common concern is that the water resources, which

scarcely meet the needs of the present population and economy, will become more crucial as the projected total population rises from 420 million people in 2000 to over 600 million by 2025.

Within these unifying elements, there is much variety that gives different characters to places in the region, including the natural environments from mountains to river plains, the distribution of oil and water resources, and the political and economic conditions. Neither the Arab nor the Islamic worlds act as united wholes. Internal attempts to unify this region have not succeeded to date, and some Muslims feel that it is in the interests of the United States and Europe to foster divisions instead of helping to unify a region that might be a major rival based on the economic potential of oil and the religious motivation of Islam.

Cultural and Political History Within a Wider World

This region provided a center of diffusion for early technical developments, the base for three monotheistic religions (Judaism, Christianity, and Islam, related to each other through worshipping the God of Abraham), the source of further technical and artistic developments in the medieval Muslim empires, and a focus of geopolitical strategies by major world powers in the 1800s and 1900s. Being "in the middle" caused it to act as a source of diffusion outward and as an inward-directed focus of external influences. The cultural factors of religion and language are as fundamental as politics and economics in this world region.

First Civilizations

The Tigris-Euphrates River valley of Mesopotamia (modern Iraq) and the Nile River valley formed two of the world's early cultural hearths (see Chapter 1). Early human achievements diffused to the surrounding continents. Even in ancient times of slow and limited transportation, this region acted as a hub with constant movements of people to and from Northern Africa, Europe, and China. The internal empires of the Assyrians, Babylonians, and Egyptians gave way to external control from

Greece and Rome, and then the Byzantines before the A.D. 600s expansion of the Muslim Arabs.

Religions

Religion and related traditions give identity to the countries of Northern Africa and Southwestern Asia. Islam is dominant today, but other world religions that preceded Islam trace their origins to the region and still exist here. Judaism, Christianity, and Islam share a common monotheism, religious figures, and religious texts.

The early animist religions had many gods linked to natural phenomena. Some human emperors were treated as gods. After approximately 1000 B.C., dominance by religions with many gods gave way to religions based on a single god. Beginning with Judaism, those monotheistic religions diffused from their hearths in southwestern Asia to Europe, Africa, and Asia (Figure 8.3).

Judaism

Judaism is a religion whose adherents worship Yahweh, the only God, creator, and lawgiver. It began in the area now known as Israel and the Palestinian territories some 2000 years B.C., where Abraham and his descendants settled after moving from Mesopotamia. Jewish beliefs focus on the historic role of family based on the line from Abraham, persecution beginning with slavery in Egypt, redemption as Moses led the people out of Egypt, and their occupation of the promised land. Yahweh intervened to support and punish through times of trouble and deportation, promising a messiah to save Jews from domination by others. At the time of Jesus (Christ means "messiah"), many Jews expected a military-leader messiah to end the Roman occupation. In A.D. 70, the Roman army destroyed Jerusalem and dispersed Jews through southwestern Asia, northern Africa, and Europe to form a major diaspora.

Christianity

Christianity stemmed from new interpretations of the beliefs and teachings of Judaism in the early years A.D., beginning with the life, teachings, death, and claimed resurrection of Jesus of Nazareth. It gave a fresh impetus to monotheism against the many gods of the Greeks, Romans, and Arabs. Christian churches spread across southwestern Asia and into Africa and Europe. In the late A.D. 300s, the Roman Emperor Constantine declared it to be the empire's official religion. Controversies over how much Jesus was man and how much God resulted in divisions among the churches of this region. A later division occurred between the eastern (Orthodox) and western (Catholic) groups of churches in Europe. From A.D. 395 to 1453, the eastern church was centered in Constantinople (modern Istanbul, Turkey). The largest Christian sect in this region today is the Coptic Church, with a pope who resides in Alexandria, Egypt.

Islam

In the early centuries A.D., the region came under Roman (Byzantine) and Persian control, but from the 600s, the expansion of Muslim Arabs made it once again a major world center.

Islam ("submission to the will of Allah") arose in Arabia during the early A.D. 600s. **Muslims** are "those who submit to Allah," or followers of Islam. In writing the **Qu'ran** (holy book), Muhammad, a trader in Mecca, claimed direct revelations from Allah, the only God. At first, his message was too revolutionary for his fellow Arabs, who preferred the many traditional gods, and he was forced to move to Medina. Muhammad's message, however, soon gained converts in an area as a local monotheistic religion that could stand with Judaism and Christianity through historic links to Abraham.

After Muhammad's death in 632, the area ruled by Muslim Arabs spread rapidly westward to northern Africa and Spain as well as eastward into central Asia, incorporating Berber (North Africa), European (Spain), Turkic (Central Asia), and Indian people. Muhammad's teachings diffused widely as conversions of local peoples to Islam followed the Arab invasions. The eastward extension of Islam into Central Asia and India spread the faith beyond Arabs to Persians, Turks, and Indians. It led to the Muslim "Golden Age" of artistic and scientific achievements alongside further military expansions from the 800s to the 1100s. New forms of art and architecture developed, and Muslim mosques remain dominant features of town landscapes, often doubling as centers of religious and secular activities (Figure 8.4). One of the oldest is the Al-Aqsa Mosque in Jerusalem, begun in 692. Islamic visual art, often characterized by abstract patterns of religious significance that avoid depicting human forms, ranges from pottery and metalwork to paintings and textile furnishings. Calligraphy became the most important form of art because of the religious significance of the Qu'ran. Poetry and music were generally more important than the visual arts but less significant in the landscape.

Early divisions within Islam continue to be significant today. The **Sunni Muslims,** or Sunnites, are the majority, or orthodox, Muslim group and base their way of life on the Qu'ran, supplemented by traditions. Political power was at first given to a succession of leaders, or caliphs, descended from the historic Muslim leaders. Many Sunni Muslims are moderates in maintaining traditional Islamic practices. Exceptions include the Wahhabi sect that dominates Saudi Arabia and the Taliban faction that ruled Afghanistan from 1996 to 2001.

A minority of Muslims also follow the Qu'ran but dispute the caliph leadership succession accepted by Sunnites. They look to descendants of Ali, the fourth caliph of Islam and a cousin and son-in-law of Muhammad. His son, Hussein, was defeated and killed by the Sunni caliph of Damascus in 680. Supporters of Ali as the only *imam*—authoritative interpreter of the Qu'ran—are known as "partisans of Ali," "Shi'at 'Ali," **Shia Muslims,** or Shiites, and look to the return of the twelfth *imam,* Ali's descendant, Muhammad al-Mahdi, who disappeared in 874. For most of their history, Shia Muslims were known for commemorating and lamenting Hussein's death by flagellation and withdrawal from the world as they awaited al-Mahdi's return. Politically, they were quiet, regarding political control as evil and looking to clerics, supported by alms, for leadership. Today, Shiites comprise large proportions of the population in a few countries, including 90 percent of the

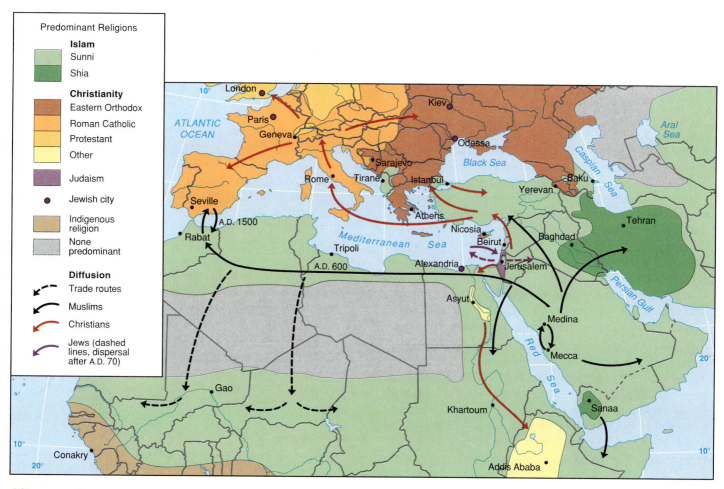

Figure 8.3 **Major religions and movements of diffusion within and from Northern Africa and Southwestern Asia.** Muslim Arabs conquered the area shown in green in the first century after Muhammad emerged as a new religious leader around A.D. 600. Islam also spread over trade routes. Christianity became the religion of the Roman Empire, spreading through Europe. Jews were dispersed at the fall of Jerusalem in A.D. 70, and the modern country of Israel was not formed until 1948.

Iranian population. From 1979, Shiites governed Iran, tried to establish Iranian isolationism from Western ways, and supported extremist groups such as Hamas (Palestine) and Hezbollah (Iran, Syria).

Gender inequalities are a continuing feature of Muslim countries. Male attitudes and legal processes act against equal opportunities in the interests, it is said, of protecting women and in opposition to Western values that allow women greater freedom. As with other aspects of Muslim culture, local interpretations and outcomes vary. Considerable debate exists in Muslim countries over what the Qu'ran says on these matters. Saudi Arabia boycotted the 1995 fourth World Conference on Women in Beijing on the grounds it was anti-Islamic. In Iran, it is a criminal offense for a woman not to wear a headscarf. And yet, Muslim-dominated countries such as Turkey, Pakistan, Bangladesh, and Indonesia have elected women leaders in recent years. While Saudi Arabia excludes women from all public activity, Tunisia promoted women's rights after its 1957 independence. Between 1990 and 1999, illiterate adult females (over 15 years) fell from 66 to 57 percent in Egypt (men from 40 to 34%), from 49 to 34 percent in Saudi Arabia (men from 22 to 17%), and from 62 to 40 percent in Oman (men from 33 to 21%).

Figure 8.4 **Islam and architecture.** Mosques are a common feature of urban landscapes in Muslim countries. Older (1300s) and more recent (1800s) mosques tower over Cairo, Egypt.

Languages

In Northern Africa and Southwestern Asia, language is an extremely important cultural feature, being often linked to religious faith.

- **Arabic** is the main language of most countries in the region, spoken by just under 50 percent of the people. It originates from an Afro-Asiatic language and spread out of the Arabian Peninsula with the Muslim conquests in the A.D. 600s. Arabic has a special place in all Islamic countries because it is the preferred language in the Qu'ran and Muslim prayers. The spread of Islam aided the unification of Arab peoples and involved the imposition of Arabic on others. "Arabs," who were once defined as living in the Arabian Peninsula, now include all those using Arabic as their first language—over 180 million people in this region. The major dialect differences from Morocco in the far west to Egypt, Syria, Iraq, and Arabia in the east are lessening through radio and television. The Egyptian Arabic dialect is most commonly used in the news, media, and movies.

- **Berber** languages are Afro-Asiatic languages spoken by 40 percent of the peoples of Morocco, 30 percent of Algerians, and smaller groups in surrounding countries. Berber languages survived the Arabic onslaught in places such as the hilly areas of the Atlas Mountains. Modern discrimination by governments usually defines the Berber language as illegal. Most Berbers now speak or write Arabic as well.

- **Hebrew** is the official language of Israel. It is related to Arabic, also coming from an Afro-Asiatic (or Semitic) language. The language of the Hebrew Bible and religious services is an archaic form, superceded in everyday usage by modern Israeli Hebrew. The movement devoted to creating a Jewish state from the 1800s, known as Zionism, revived the use of Hebrew, modernizing it for secular usage to provide a common religious and secular element among Jews from many countries.

- **Persian** is an Indo-European language, introduced by Eurasian invaders around 3000 B.C.—their Aryan origin being reflected in the name "Iran." These invaders included the Medes, Parthians, and Scythians, who formed the basis of successive Persian empires. After the Arab invasions in the A.D. 600s, Arab words were introduced and the Persian language was written in Arabic script to form modern Persian, or Farsi. Only 58 percent of Iranians speak Persian today, with 26 percent speaking Turkic dialects and 9 percent Kurdish, but it is the country's official language and is linked to the Shiite Islamic beliefs of the leaders.

- The **Kurdish** language is akin to Persian and kept alive by peoples in the hilly lands of southeastern Turkey, northern Iraq, western Iran, and parts of Syria. The Kurds in these countries press for their own country but are subject to intense discrimination. The numbers of Kurdish speakers are unknown because the language is at the heart of demands for a separate nation-state: non-Kurds talk in terms of 10 million people, while the Kurds claim their numbers to be over 20 million.

- **Turkish** languages are part of the Altaic family (see Figure 2.18), which spread westward from Central Asia in the A.D. 700s. Turkish speakers, using Roman script rather than Arabic, are the majority in Turkey. Groups of people with Turkish affinities also live in the Central Asian republics that were part of the former Soviet Union (see Chapter 4). Proposals to form a united Turkish-speaking country, or nation-state from these various elements have not received wide support. The Azeri (or Azerbaijanis) are Turkish speakers who use Arabic script. They are a minority group in Iran and the majority in Azerbaijan. Attempts by the former Soviet Union to integrate its Azeris failed because many are Shiites who look to a home in Iran.

Persian Dynasties

Invading Mongols ruled Persia until the 1400s, when the Shia-dominated Safavids gradually reclaimed the Persian area and established a dynasty that lasted until 1722. The Shiite *ulama* (cleric scholars) asserted an early doctrine that only they were true interpreters of the Qu'ran and therefore leaders of Islamic communities. By 1722, they took political control from the Safavids but had no military force to resist the Afghans, who rose against the governor of Kandahar and then took much of Persia. Chaotic government followed.

Ottoman Turks

In the 1500s, the Turkish Ottoman Empire under Sulayman the Magnificent conquered the Mediterranean coastlands as far as Morocco and also took much of Balkan Europe (see Chapter 3), besieging Vienna in 1529. The early vigor of the Ottomans subsided into maintaining their position and even turning on other Muslims as they established control of Arab lands, but the overseas colonial expansion of northern European countries and accumulation of wealth and power eroded their empire. The Ottomans did little after the 1700s to encourage modernization or involvement in the expanding global economic system.

European Colonies and Protectorates

From the 1800s, the North African coastlands became colonies or protectorates of France (Tunisia, Morocco, Algeria), Italy (Libya), and Britain (Egypt, Sudan, and much of the southern Arabian Peninsula and Gulf coasts after the 1869 opening of the Suez Canal). Russia took land in the north. Some of the more isolated tribes in Arabia became largely independent under their own kings or sultans. During World War I, the British and French built on renewed Arab nationalism to expel the Turks from the areas that became Palestine, Syria, Lebanon, Jordan, Iraq, and the Arabian Peninsula.

The Arab Southwest Asian countries gained independence in different ways from colonial overlords. The area remained under the gradually disintegrating Ottoman Empire until World War I, after which the League of Nations made Syria and Lebanon French protectorates, while the rest of the area was placed under British protection. The French encouraged republican governments. British protection guaranteed the survival of the kingdoms they established, including Jordan (still a monarchy today), Iraq (a monarchy until 1958), and the small emirates along the Persian Gulf under traditional local rulers, or emirs. In the 1970s, seven emirs united as the United Arab Emirates (UAE). Animosities against former Turkish and European colonial powers that were generated by colonial decisions such as the establishment of the country of Israel remain significant factors in understanding internal tensions.

Strategic Role of Oil

The discovery of oil and the founding of new countries after World War I brought the region into world prominence again. World War II and the Cold War from 1950 to 1990 placed this region at the center of competing interests. Countries aligned themselves with one side in exchange for military and material support. Wars between Israel and the Arab countries and the continuing search for or resistance to peace became part of this tension. However, the conflicting approaches of German Nazism, Soviet Communism, and Western liberal democracy caused many in this region to look for new ways to unite the area or reassert the power of Islamic ideas.

Oil Resources

The huge oil and gas production from Saudi Arabia and the Gulf states amounted to 22 percent of world oil production from the 1960s to the 1990s. Despite such extraction and the addition of new producers in Latin America, Asia, and Africa, the proportion of total known world reserves located around the Persian Gulf increased (Figures 8.5a and b). Continued discoveries increased reserves from 61 (1965) to 65 (2000) percent of world total reserves despite increased usage (from 215 billion to 560 billion barrels). These figures do not include

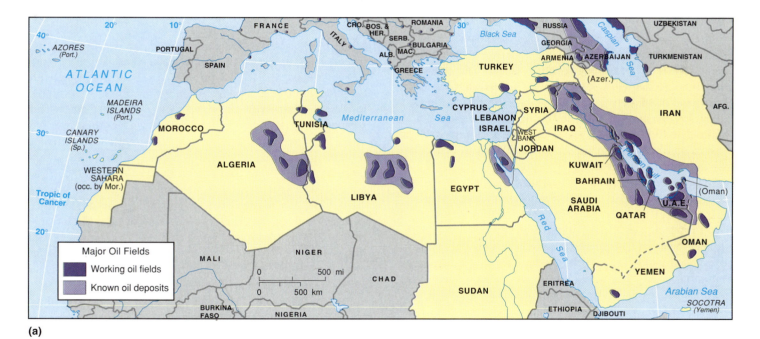

(a)

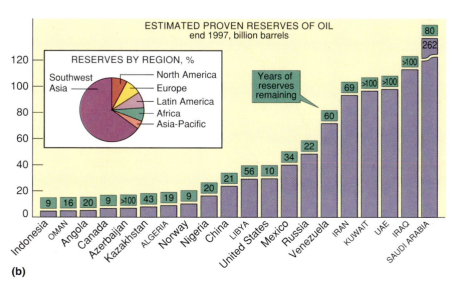

(b)

Figure 8.5 **Oil resources in Northern Africa and Southwestern Asia.** (a) Map of major oil fields. (b) World oil resources at the end of 1997, showing the continuing significance of this region. The share, 65 percent of world reserves, is a little greater than 10 years ago, and reserves in this region are being used more slowly, while many wealthier countries have only 10 to 20 years of reserves. Note which countries in the region have major oil reserves and which have little or no oil. Source: Data from *The Economist*.

Iran, Libya, Egypt, or Algeria, which are also major oil and natural gas producers in this region. New discoveries of oil fields in the Caspian Sea area (see Chapter 4) may rival the Gulf in output, and Iran and Turkey may try to attract oil pipeline traffic across their territories from this new source.

Not all the countries of Northern Africa and Southwestern Asia, however, are major oil producers. Morocco, Turkey, Israel, and Jordan are not. Tunisia, Sudan, Syria, and Yemen produce and export modest amounts. Countries that import oil face the burden of purchasing oil, whatever its price, and many went into debt during the high prices in the 1970s. At the same time, they often received an economic boost from the wages of workers who filled the major oil producers' need for labor. From the 1970s, the oil-rich countries of the Gulf also provided aid to other Arab countries and causes. Both the aid and demand for foreign workers from other Arab countries were much reduced after the 1991 Gulf War, when oil prices were also lower.

Organization of Petroleum Exporting Countries

From the early 1900s, international oil companies kept oil prices low to consumers in the world's wealthiest countries by paying little to the producing countries. In 1960, the producers around the Persian Gulf, together with Venezuela, formed the **Organization of Petroleum Exporting Countries** (OPEC). Today, 75 percent of OPEC membership still comes from Arab countries. OPEC's main purpose is to work as a **cartel**—an organization that coordinates the interests of producing countries by regulating oil prices.

The successful oil embargo during the Arab-Israeli War in 1973 established OPEC for a few years as the controlling factor in world oil distribution. Prices rose fourfold. Revenues of the oil-producing countries boomed in the mid- and late 1970s, and they invested their money locally in new roads, hospitals, government buildings, airports, and military hardware, as well as in projects and banks outside the region. During this period, countries such as Saudi Arabia, Kuwait, United Arab Emirates, and Iraq gained over 90 percent of their export earnings from oil sales.

The wealthier countries that used most of the oil were affected first by higher oil prices and endured economic recession. It was soon clear, however, that the high oil prices and recession in the wealthier countries also affected adversely the world's poorer countries. Western oil companies opened new oil fields outside the OPEC area, including those in the North Sea (Europe) that had previously been too expensive to develop. In the 1990s, Russia exported increasing quantities of oil and natural gas as its main source of currency. A world oil glut brought very low market prices and financial problems to the Arab producers. By the late 1990s, the OPEC oil producers and major industrial countries saw that their interests lay in a moderately high, but stable, oil price.

Pan-Arabism and Islamism

In the mid- and late 1900s, postindependence movements, generally opposed to Western economic colonialism, sought to unite Arab peoples on the basis of (mostly secular) nationalism into a single country with increased world influence. None succeeded in uniting the Arab world.

Arab League and Palestine Liberation Organization

The **Arab League** was created in 1945 to encourage the united opposition of Arab countries to the establishment of Israel. Its seven founding members were the only independent Arab countries at the time, but its membership increased to 21 as more gained independence. The members of the Arab League eventually included the **Palestine Liberation Organization** (PLO), a political organization providing an umbrella for many smaller groups that demand a country for Palestinians.

The Arab League linked Arab countries in the western Mediterranean with those in the eastern Mediterranean and southwestern Asia. After Egypt's 1979 accord with Israel, the Arab League expelled Egypt and transferred the Arab League headquarters from Cairo to Tunis.

One Arab Country

From 1958 to 1961, Egypt and Syria joined as the United Arab Republic under the leadership of Colonel Jamal Nasser with the intent of persuading other countries to commit their futures to a single **Pan-Arab country.** It did not attract others and soon broke up. Disunity, military defeats, and tensions over the Israel-Palestine issue and among countries with different resource bases weakened the Arab League and its ability to foster unity among Arab countries in the 1980s and 1990s. The final blow to the Arab League came in the Gulf War of 1990–1991, when one Arab country (Iraq) invaded another (Kuwait) and was defeated by a coalition of other Arab countries backed by the United States and other Western countries. Many Arabs felt betrayed by both Saddam Hussein's invasion of another Arab country and their own need to rely on outside help to put things right. It was clear that the interests of individual countries remained more significant than an overriding Pan-Arabism, feelings that were repeated in later Arab economic summits.

Following a decade of little Pan-Arab activity, an Arab League conference in 2002 agreed to offer Israel recognition as a country if Israel would hand back the occupied territories, especially the West Bank and Gaza, to form a country for Palestinians. Although this agreement signaled a potential change by the Arab countries, it came at a time when Israel was responding with massive force against the perceived threat of suicide bombings in its cities and was unlikely to agree to give up the lands it won in the 1967 war (see the "Point-Counterpoint: Palestinians Versus Israel" box, p. 320).

The Impact of Islamism

In the 1970s and 1980s, as individual countries preferred their independence to a Pan-Arab identity, pressure increased to base political actions on religious affiliations. The influence of extremist Islamist groups grew in opposition to the global political and economic order.

Political Islamism: Organization of the Islamic Conference

As the moves toward the unification of Arab countries died away, links through the Islamic religion began to replace them. In 1970, foreign ministers of Muslim countries set up the **Organization of the Islamic Conference** (OIC), which now has 45 members, including such countries as Pakistan, Indonesia, and even Nigeria. It is more successful in advancing individual member countries' interests than in defining and pursuing a common agenda to rival Western-dominated globalization. One of OIC's most important affiliates is the Islamic Development Bank, which is dedicated to economic development among OIC members.

Although closer economic integration was encouraged from the late 1970s, most Muslim countries carry out most of their trade with the materially wealthier countries in Europe, the United States, and Japan. For example, the December 1997 meeting of the OIC in Tehran (Iran) was unusually well attended by all members but focused on pragmatic world trading issues rather than on Islamic religious policies. It was chaired by that country's new moderate president, who spoke against plans for Muslims to unite against Israel and the West, urging the conference to avoid rejection of the West and calling for a "dialogue of civilizations."

Attempts at Politicizing Islamism

In the 1970s, when the decay of private religion was predicted around the world, Islamic political groups began to expand, basing their ideology on the Qu'ran, interpretations of *jihad* (applied widely by Muslims to many efforts, including an anti-smoking campaign in Saudi Arabia) as holy war, and references to past Islamic triumphs. The ideas came from 1930s writers such as the Sunnites Sayyid Qutb of the Muslim Brotherhood in Egypt and Mawlana Mawdudi in Pakistan and the Shiite Ruhollah Khomeini in Iran and contradicted the basic peaceful nature of Islamic beliefs and practice. Persecuted under nationalistic Arabism and secular governments, they called for the reestablishment of an Islamic country, rejecting the traditional view that Islam relegated political combat to a secondary concern. They hated the fragmentation of Islamic lands into separate countries, in which the religious establishment was thrust aside by those educated in Westernized ways. They likened the situation to the ignorance and barbarism among Arabs before the early 600s, when Mohammad brought the worship of Allah and new purpose to Arabs.

The main modern flowering of political Islam occurred after the 1967 and 1973 Israeli-Arab wars and the OPEC price rises that brought huge surpluses of dollars to the region. Its main success was the 1979 Islamic revolution in Iran under Khomeini, which managed to gain support from radical young people in the cities, the more conservative, God-fearing middle classes that valued a strict religious and social order, and professionals who had become wealthy from oil income but had no political voice. Islamic law (*sharia*) became official in Iran in 1983, with over 100 offenses carrying the death penalty, but it was not implemented fully because of differences between the ruling elites of clerics and laymen.

Although the success in Iran stimulated Muslims in other countries to consider similar action, most Muslim countries, including Libya, Syria, and many small Gulf countries, stifled such ideas and imprisoned activists. Throughout the 1980s, Iran and Saudi Arabia struggled for dominance. As Iran tried to export revolution, the Saudi princes wished to maintain the system by which they had grown rich. When Saddam Hussein of Iraq declared war on Iran in 1981, he had support from the Saudis and other Gulf countries as well as Western countries. Iran responded with terrorist acts, such as the taking of Western hostages in Lebanon and disrupting pilgrimages to Mecca. In 1989, when an armistice was reached after huge Iranian losses of life, Khomeini declared a *fatwa* (legal ruling) against a British citizen, Salman Rushdie, the author of *Satanic Verses,* a novel that criticized Islam. This action suggested that Islamic law was valid in other countries.

The 1979 invasion of Afghanistan by the Soviet Army generated another jihad financed by the Gulf and Western countries. This unified Islamists around the world, taking over from the Palestinian cause for some years. International brigades came from Egypt, Algeria, the Arabian Peninsula countries, Pakistan, and Southeast Asia and were trained in guerrilla warfare as part of the Islamist armed struggle. In 1989, the Soviet Army withdrew from Afghanistan, causing Islamists to declare a great victory and gain immense self-confidence.

The 1989 fall of the Berlin Wall that signified the beginning of the end for world Communism gave Islamists hope that their way might fill the vacuum left by Communism, but the Gulf War of 1990–1991 and subsequent failures to impose their military solutions in Bosnia, Algeria, and Egypt were major blows for Islamists. In particular, the Gulf War demonstrated how Saddam Hussein of Iraq miscalculated possible Arab support in threats to invade Saudi Arabia and expel the "infidel" Americans who guarded the soil on which so many Islamic holy places were built. That war and the cessation of the Afghanistan jihad ended the Saudi Arabian-led consensus, and a radical Islamic fringe turned against the Arab kingdoms and their international networks, losing the support of the middle classes. Having defeated one major power, radical Islamists turned against the United States and Arab countries with rulers who repressed them.

Islamist failures caused most countries that had supported guerrillas to withdraw from extremist actions. In Iran in 1997, the moderate Muhammad Khatami was elected president with 70 percent of the vote, including strong support from the middle classes and younger elements; he was reelected overwhelmingly in 2001. These votes reflected a distaste for strict Islamist and anti-Western policies, but Khatami had to battle with the country's Guardian Council of clerics and jurists who have power to examine and override legislation they consider to be against Islamic principles. Turkey and Pakistan turned from Islamist leaders to moderate military or secular leaders. In Jordan, radicals lost their parliamentary seats when Jordan began negotiating with Israel after the Gulf War.

Islamists frustrated by political impasse allowed terrorist actions to continue. Attention reverted to Israel and Arab resentments at the lack of movement by the U.S.-supported

ISRAELIS VERSUS PALESTINIANS

[The following is an attempt to stimulate informed discussion of a sensitive issue. The views expressed in the table at the end are those of imagined disputants, not the authors of this text. The most extreme views are held by hard-liners on either side, but that includes only about one in five of the people in Israel and the Palestinian Territories.]

Perhaps the most significant world conflict today is between the Jews living in Israel and the Arab Palestinians living in the occupied Palestinian Territories of the West Bank and Gaza. Although the United States is at the forefront of trying to reconcile the two sides, the Arab countries see it as the main supporter of Israel. The conflict's origins go back in history, before Israel was established as a new country in 1948. The continuing resistance of Palestinians to the heavily armed Israelis suggests an irreconcilable conflict, at least one that will not be resolved until an independent homeland is created for the Palestinian Arabs.

The land in question is a small part of what is known as the "Fertile Crescent" that cradled and linked the early civilizations of Mesopotamia and Lower (northern) Egypt. The hilly coastal lands provided a home for the ancestors of the Israeli nation. When they escaped from slavery in Egypt, they took the land from the Canaanites to establish their own country. Conflicts with neighbors continued, including with the coastal Philistines and the major inland empires of Babylon, Assyria, and Persia. Always fiercely independent, many Jews were taken into slavery each time their country was occupied. They returned and fought to establish their independence, but Roman armies later subdued the country. Eventually, the Romans tired of Israeli plots and rebellions, sacked Jerusalem, and dispersed most of the people in A.D. 70 to create the Jewish diaspora.

For the next 2,000 years, Jewish communities established themselves in most European, North African, and even Asian countries.

Small in number and exclusive in outlook, they often attracted suspicion, jealousy, and opposition, becoming subject to oppression and segregation in ghettos. The vacated lands in Southwestern Asia became known as Palestine and were occupied by people from surrounding areas, mainly Arabs, together with the few Jewish families who had not been evicted. In the A.D. 600s, Arabs quickly converted to Islamic beliefs and Muslim armies spread Islam. Jerusalem became the third most holy site for Muslim pilgrimages after Mecca and Medina. Challenges by European armies that were called "crusades" had only temporary effects but resulted in deep hatreds between Muslims and Christians.

From medieval times, the Turkish Ottoman Empire governed these lands without any great emphasis on economic development. Siding with Germany in World War I, Turkey lost most of southwestern Asia to groups of Arab tribes who had fought alongside the British and French to free the lands. After the war, protectorates were given to France and the United Kingdom, which established the present Arab countries. These countries gained independence after World War II.

Palestine was one of the territories under British mandate. Although the authorities resisted too many Jews coming to Palestine, numbers had been building since the late 1800s, when the Zionist movement emphasized the need for a Jewish homeland after oppression in many countries of eastern Europe and consequent migrations to the United States. Jewish voices became more powerful. The Holocaust of World War II, when Germany and its allies murdered millions of Jews, led to further pressure for establishing a new country of Israel on Palestinian land. Arabs rejected the 1947 United Nations Partition Plan. In 1948, when Jews took matters into their own hands and declared Israel's independence, Egypt, Syria, Jordan, Palestine, and Iraq declared war, but Israel succeeded in resisting their attacks. Millions of Arab Palestinians were displaced in temporary camps in Jordan and southern Lebanon.

Since 1948, in further wars between Israel and Arab countries, the Israeli military power usually triumphed and took over more Arab lands. "Terrorist" activities by Palestinians—seen by

(a)

(b)

Box Figure 1 **Jerusalem: aspects of conflict.** (a) Jerusalem from the Mount of Olives: The Dome of the Rock and the buildings of modern Jerusalem. (b) Israeli soldiers relaxing in the Old City. Photos: © Michael Bradshaw.

them as attempts by an oppressed people to establish their own country—were ruthlessly countered by Israeli military action (see Box Figure 1). In the 1970s and 1990s, there were attempts, led by U.S. presidents, to reconcile differences. After the 1967 war, Israel returned the Sinai Peninsula to Egypt but retained control of the Israeli-occupied Palestinian territories of the West Bank and Gaza. The Jordanian and Egyptian governments gave up their rights to the West Bank and Gaza lands (which were previously parts of their territories) as a basis for creating a new Palestinian country by negotiation.

The following views are those that might be expressed by an Israeli Jew and Arab. Is either totally right? Should the Arabs recognize Israel and live peaceably with it, or should the Jews give back the land they govern in Israel? Should the Arabs stop attacking Jews, or should the Israeli army stop entering the Palestinian territories? Does a knowledge of geography and the distributions of people and resources help in searching for answers to these questions? You may wish to add other opposing views to this list. Is there any prospect of a way ahead that does not involve bloodshed?

A HARD-LINE JEWISH VIEW

We were here first and have a longer history of occupying this land: it is ours, as claimed by Abraham and Joshua.

Our religion started first in this area. Our holy sites include Hebron (where Abraham is buried) and Jerusalem, both in, or partly in, the West Bank.

The United Nations agreed to the partition of Palestine and recognizes Israel with its Jewish majority.

We have returned some territories we took in the 1967 war but hold on to the rest as a matter of national security and survival.

Arabs deny our rights to exist and want to wipe us off the map; we are exercising our right to defend ourselves. When joined, the surrounding Arab countries outnumber us, so we have to ensure strong security.

We regard them all as possible terrorists who blow up our restaurants and nightclubs, kill our athletes, assassinate our leaders, and drive suicide bombs into our neighborhoods.

Jerusalem is our real capital city and is central to the Jewish faith. Some Jews want to remove the mosque on the holy mount.

No strong Palestinian nationality was expressed here before 1948. This area of land was merely a British protectorate carved out of the former Ottoman Empire, and most people knew of the intention to create a land for Jews. The present Palestinians are Arabs who should have been taken in by existing Arab countries. They have invented Palestinian nationalism as part of a plot to eliminate Israel.

They are poor workers and earn only low wages. We go out of our way to employ them, but it would be better to employ only Jews.

The Palestinian Arabs have not repaid all the help we have given to them, raising their well-being above that of other Arabs in this region.

A HARD-LINE ARAB (PALESTINIAN) VIEW

Our ancestors were here from the time of Abraham, whom we also recognize as a father of our people.

Jerusalem and Hebron are sacred to Muslims.

The decision resulted from a combination of weak Arab support and strong U.S. and British pressure. We were ignored, although we made up most of the population here in 1948. Israel has taken large areas of land from us that were not part of the UN plan.

After the 1967 war, the United Nations ordered Israel to hand back the occupied territories, but it has not done so more than 35 years later.

We were forced out of our land, we lived in crowded camps without amenities, and we now live in poverty. The Jews close checkpoints with no notice and interrupt our lives. We cannot argue with them because they are supported by the United States, and we now hate that country as well.

They refuse to recognize our presence and nationality, suppressing our language, religion, and culture. Most of us want to live peaceful lives, but they treat us all as spies and criminals, abusing our human rights.

Jerusalem is sacred to us, and any moves to destroy the mosque would be a declaration of all-out war that would unite Muslims.

We want our own lands and independence from Israel.

They are well fed and materially wealthy. If we want to study for the qualifications that would earn us better jobs, we cannot do so in our country and have to go elsewhere. A doctor friend of mine, who works in a Jerusalem hospital, had to go to Greece to qualify.

We can do nothing that is legal to improve our lot, so it is not surprising that some of us take to the gun and bomb.

Israeli government to accommodate the wishes of Palestinians. The media-highlighted poverty in Iraq as Saddam Hussein remained in power 10 years after the Gulf War was also blamed on the United States. But terrorism, despite resorting to the events of September 11, 2001, and suicide bombings in Israel, gained nothing. Many Muslim countries held memorial services for the victims of 9/11, including Iran in its capital, Tehran.

🌐 Natural Environments

Alongside the cultural and political conditions, the natural environments of Northern Africa and Southwestern Asia are basic to the region's geographic distinctiveness and internal diversity. Hot, dry plains contrast with snow-capped mountain ranges and well-watered river valleys (Figure 8.6).

Dry Climates and Desert Vegetation

Arid climatic environments, in which evaporation rates are greater than precipitation, dominate virtually the whole region (Figure 8.7). The region boasts the world's highest recorded shade temperature (58°C, or 136°F, at Al'Aziziyah, Libya), but the lack of cloud cover makes nights cool or cold, and freezes are possible in winter.

Most places have some rain, but it falls irregularly and with increasing uncertainty as aridity increases. The rainy winters of the coasts of North Africa and the eastern Mediterranean lands are caused by the seasonal southward shift of midlatitude frontal weather systems. The mountains of Turkey and Iran force air to rise, often adding lift to the frontal zones, and precipitating rain and snow. The winter snowfall provides a source of meltwater that feeds the Tigris-Euphrates River system in spring. In southern Sudan, summer rains come from the northward movement of the equatorial rainy belt.

The arid climatic environments support only drought-resisting desert plants that form a partial ground cover. Large areas of desert have little vegetation and are gravel-strewn, rocky, or sand-covered: sand seas cover one-fourth of the Sahara, and most of it has a rocky or gravel-covered surface. The vegetation cover thickens and becomes denser on uplands, where there are higher precipitation totals and lower rates of evaporation. In the uplands of North Africa, Lebanon, western Syria, Turkey, and Iran, grassland and woodland vegetation increase with altitude as temperatures and evaporation levels fall and make the low-to-moderate precipitation more effective.

Soils are poor and undeveloped through most of the region: the best occur in the rainy coastal areas and along the valley floors of rivers where annual floods deposit fertile allu-

Figure 8.6 **Arid lands: parts of Egypt, Israel, Jordan, and Saudi Arabia.** A space shuttle view taken over the northern Red Sea across desert lands to the Mediterranean Sea. On the left, the delta mouth of the Nile River and the Suez Canal; in the center, the Sinai Peninsula; on the right, the Gulf of Aqaba leading northward to the Dead Sea. Photo: NASA.

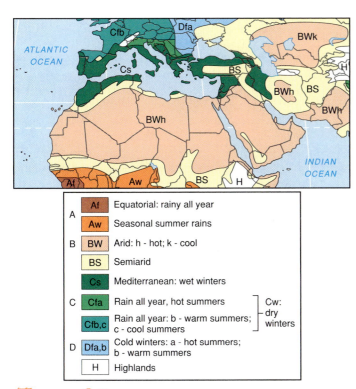

A	Af	Equatorial: rainy all year
	Aw	Seasonal summer rains
B	BW	Arid: h - hot; k - cool
	BS	Semiarid
	Cs	Mediterranean: wet winters
C	Cfa	Rain all year, hot summers
	Cfb,c	Rain all year: b - warm summers; c - cool summers
D	Dfa,b	Cold winters: a - hot summers; b - warm summers
	H	Highlands

Cw: dry winters

Figure 8.7 **Northern Africa and Southwestern Asia: climate regions.**

vium. Soil erosion following plowing for grain and cultivation on steep slopes is a major problem in most of the region.

Climate change left its mark on the region and affected the history of human settlement. The present aridity is one stage in a sequence of fluctuations between more and less arid conditions. Pollen analysis, animal distributions, dried lakes, cave paintings, and written records suggest that the climate was more humid and cattle herders more widespread across northern African and Arabia a few thousand years ago. The stocks of underground water that accumulated in humid phases provide much of the water being used today, but current precipitation levels do not replenish them. The present aridity began some 5,000 years ago, forcing many people into the watered valleys of the Nile and Tigris-Euphrates Rivers, thus concentrating populations. The desert margins continue to fluctuate. They retreat in series of wetter years and advance in drier years—or as human actions remove vegetation and lower groundwater levels.

Clashing Plates

Contrasting landscapes from high mountain ranges and extensive plateaus to wide river valleys (Figure 8.8a) stand out where an arid climate supports little vegetation. It is more common to see bare rock than grassed or wooded slopes. Such landform contrasts are related to geological activity along tectonic plate boundaries (Figure 8.8b) as well as to climatic conditions.

The African and Eurasian continents and the Arabian Peninsula are the tectonic plates of this region. They comprise ancient rocks that were worn down over many millions of years. Such rigid sections of Earth's crust underlie the exten-

sive plateaus of northern Africa and Arabia. The ancient rocks are often covered by flat layers of sedimentary rocks, such as the limestones and sandstones forming the plateaus on either side of the Nile River valley.

Collisions of the African and Arabian plates with the Eurasian plate formed the highest mountains: the Atlas in North Africa and the Taurus, Zagros, and Elburz ranges in Turkey and Iran. The Atlas Mountains rise to nearly 4,000 m (12,500 ft.) and the Elburz to over 5,500 m (17,000 ft.). Both are parts of the young folded mountain ranges that extend from North Africa and Southern and Alpine Europe, through southwestern Asia, to the Himalayas in South Asia. Rocks forming in the seas that filled the basins between the plates were caught up as the collision closed the seas and raised them as parts of the mountains. Volcanic eruptions along the geologically active belt injected mineralized fluids that solidified as veins in the rocks. The compression folded and lifted up these rocks, and large quantities of eroded rock were deposited in downwarped sections of the crust, where oil and natural gas could be trapped. The Persian Gulf area became the environment for the world's largest concentration of these fuels. Whereas the Mediterranean Sea and the Tigris-Euphrates River valley occupy the site of a closing ocean, the Red Sea is considered to be an opening ocean (see Chapter 9).

Steep slopes and mountain ranges also occur along the lands bordering the Red Sea. The tectonic plates here are pulling apart to form a new ocean. Rising molten rock beneath the surface first raised the crustal rocks into a dome that cracked to form a rift valley. Volcanic activity occurred along its margins as the molten rock erupted at the surface. Then the sides pulled away from the center and the Red Sea opened. At its northern end, the rift splits in two around the Sinai Peninsula (see Figure 8.6). The eastern branch forms the valley into which the Jordan River flows from the uplands around the Sea of Galilee toward the Dead Sea some 400 m (1,312 ft.) below sea level.

Major River Valleys

Rivers are unusual in the arid climatic environment of Northern Africa and Southwestern Asia. Most flow from water sources in the surrounding mountains or from rainy areas outside the region.

Tigris-Euphrates Rivers

The highest flows in the Tigris-Euphrates River system occur when the snows melt on the high mountain ranges of Turkey and Iran. In spring, the Tigris and Euphrates flood, supplying water to arid parts of Syria and Iraq. The floods gouge sediment from the upland areas, depositing the coarser materials in giant fans where they leave the mountains and the finer sand and mud across the Syrian and Iraqi lowlands.

Nile River

The Nile River system has two major headwater branches that begin in the rainy equatorial area around Lake Victoria (White Nile) and the seasonal rainy area of the Ethiopian Highlands (Blue Nile). Two major Nile River branches join near Khartoum (Figure 8.9). The White Nile River, flowing from

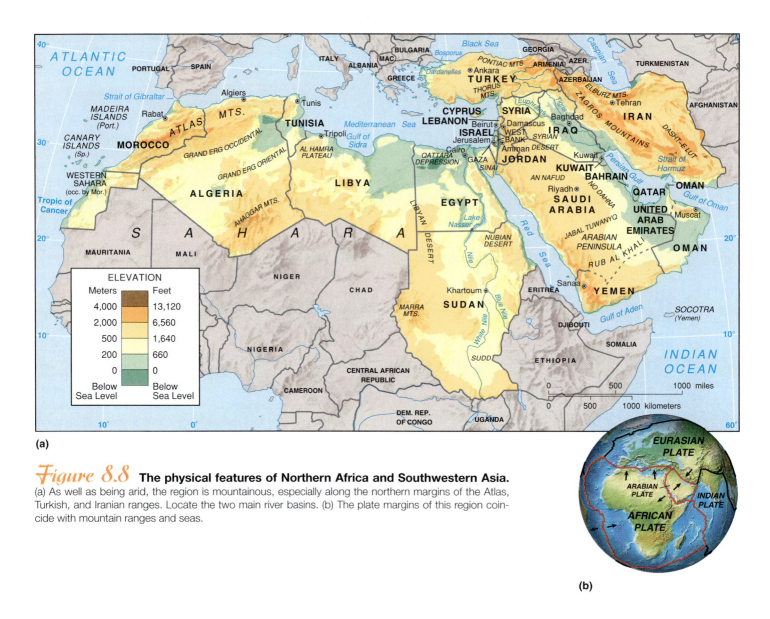

(a)

Figure 8.8 The physical features of Northern Africa and Southwestern Asia.

(a) As well as being arid, the region is mountainous, especially along the northern margins of the Atlas, Turkish, and Iranian ranges. Locate the two main river basins. (b) The plate margins of this region coincide with mountain ranges and seas.

(b)

Lake Victoria on the equator, has tributaries that are fed by rains throughout the year and supply a fairly constant flow of water. On reaching southern Sudan, the White Nile River flows through the Sudd swamp in which it loses nearly half its water by evaporation. The Blue Nile River flow is more important because it is less subject to evaporation and produces the annual September Nile River flood in Sudan and Egypt.

A narrow zone of a few kilometers wide in a valley carved by the river into the surrounding plateaus allows cultivation and settlement as the river flows northward across Sudan and Egypt. Away from the Nile River on either side and northward of central Sudan is arid land that has little prospect of development apart from the river waters.

This flow pattern varies. In 1962, unusually heavy rains led to a massive rise in the level of Lake Victoria, supporting continually high flows in the White Nile River during years when Ethiopian droughts caused the Blue Nile River to contribute much less than normal. By the 1980s, however, the flows from Lake Victoria returned to normal, but the Blue Nile

River continued to be low. The lower flows from both major sources coincided with increasing demands on the water in the source countries of Uganda, Tanzania, Kenya, and Ethiopia. The combination began to threaten the long-term availability of water to Egypt and Sudan.

Natural Resources

Water resources have been highly significant throughout this largely arid region's history. The major civilizations of the Nile and Tigris-Euphrates River valleys were based on irrigation water, and there were many local developments bringing water from its source to areas of use (see the "Personal View: Oman" box, p. 328).

The renewable supplies that come from rain and snowfall feeding rivers are meager over most of this region. Other sources include underground reservoirs that accumulated over millennia (virtually nonrenewable) and desalination plants that make seawater usable (expensive). The demands are mainly from irrigation farming, the oil industry, and growing urban centers.

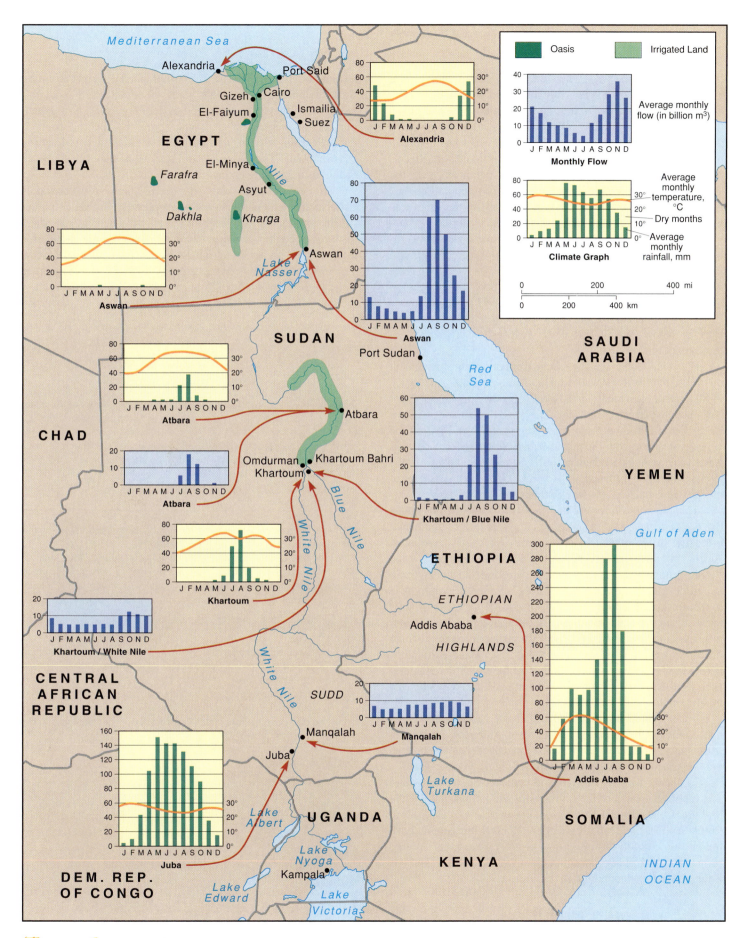

Figure 8.9 **The Nile River valley: rainfall and river flow compared.** The relatively small areas of irrigated land are shown. How do the climate graphs relate to the river flow at the places shown? The rainfall at Addis Ababa links to the Blue Nile flow at Khartoum. The White Nile flow is low but consistent. How do the flows in the White and Blue Nile affect flow through Egypt?

Political issues arise when the precipitation falls in one country and feeds rivers flowing through another. The Euphrates River rises in the Turkish mountains and flows through Syria and Iraq, but Turkey's planned water resource projects assume that it will use most of the water and that only modest amounts will be released downstream. The Nile River's two main branches both rise outside the region but form the sole source of water for Egypt and its large population. The historic development in lowland areas of irrigation using these water resources brought increased economic wealth.

Although Egypt once divided these waters with its neighbor Sudan, it is now having to take account of increasing use in Ethiopia and the other headwater countries, including Uganda, Kenya, and Tanzania (see Chapter 9). By 1988, the expectation that the Aswan High Dam (completed in 1970) would ensure Egypt's water needs for 50 years was proved wrong, and discharges below Aswan had to be reduced to maintain flow throughout the year.

Egypt and Sudan seek to enlarge the scope of their agreement on the use of Nile River waters by involving Ethiopia—the source of 80 percent of the long-term Nile River flow—and other upper basin countries, such as Tanzania, that wish to extract water from Lake Victoria. The outcome will cause tensions and reduce the water available to both Egypt and Sudan. Sudan seeks to develop water projects in its southern region, where easing the flow of the White Nile through the swamps would reduce evaporation, but this project is held up by a lack of funds and the civil war.

The competition for use of Nile River waters demonstrates that political and social factors are as important as environmental provision in the economic development of natural resources. Action on such projects is delayed by unforeseen events. Political upheavals in Ethiopia, Sudan, and the upper basin halted development in those countries, allowing Egypt to have maximum use of Nile River water in the 1970s and 1980s. At present, international funding is not available for major water projects in the upper basin to meet the rising needs there. Such needs, however, must eventually increase water extractions before the Nile River reaches Sudan. At the same time, population growth in the lower basin countries adds pressures for making more water available there.

Turkey, Syria, and Iraq compete for the waters of the Tigris and Euphrates. In Israel, the Jordan River system supplies many of the highly productive farms, while groundwater stored beneath the West Bank hills is a further source of political controversy.

In Saudi Arabia, most water comes from underground sources. The growth of the oil industry and urbanization in the arid areas placed greater demands on water resources, leading to increased groundwater extraction. Coastal settlements have built costly **desalination plants** to provide fresh water. A large proportion of the land is without fresh water, however, and has little potential for economic development.

Environmental Problems

The rapid population growth and attempts to raise economic output focused attention on whether the region is becoming "overdeveloped." Certainly, compared to underdeveloped

Africa South of the Sahara with its untapped mineral resources, water availability, and large areas of uncultivated land, this region has little space to expand agriculture, relies on fluctuating oil revenues with few other potential income sources, requires foreign labor, and builds new cities in sensitive arid environments. The limits of land and water may bring greater problems than using up the oil.

Arid environments not only pose problems of water supply but also are particularly fragile in response to human activities. In this region, the availability and management of water resources become growing political issues as the population increases and higher living standards raise the demand for water.

Irrigation farming in arid areas requires good management if the high rates of evaporation are not to draw so much salt to

Test Your Understanding 8A

Summary The countries of Northern Africa and Southwestern Asia have a strategic geographic position between Europe and the rest of Africa and Asia. The region has been subject to clashes of tectonic plates and of people and religious allegiances throughout human history.

The region has a strategic significance greater than its numbers of people and arid environments might suggest because it contains the world center of expanding Islamic influences, the major repository of world oil resources, and control points on world sea routes. The Arab language and Islamic religion are important to most countries, although Turkey and Iran have their own languages. Israel stands apart from the rest in both language and religion.

Oil production provides a basis for economic development in Algeria, Libya, Saudi Arabia, Iraq, Kuwait, and the Persian Gulf states, but the geographic outcomes vary from country to country. An arid environment is common in these countries, influencing where people live.

Questions to Think About

8A.1 How do language and religion contribute to the culture of peoples and countries? Use examples from Northern Africa and Southwestern Asia.

8A.2 What are the advantages and disadvantages of dependence on oil production by countries in this region?

8A.3 To what extent does water limit economic development in the region?

Key Terms

Judaism	Turkish
Christianity	Organization of Petroleum Exporting
Islam	Countries (OPEC)
Muslims	cartel
Qu'ran	Arab League
Sunni Muslims	Palestine Liberation Organization (PLO)
Shia Muslims	Pan-Arab country
Arabic	Organization of the Islamic Conference
Berber	(OIC)
Hebrew	desalination plant
Persian	salinization
Kurdish	

the surface soil that crop productivity is reduced or ended. This process is known as **salinization.** Under careful management, maintaining good drainage allows water to flush the salts downward. Adding too much water waterlogs the soil and concentrates salts at the surface. In ancient irrigation schemes, 60 percent of the land in the Tigris-Euphrates River lowlands became unusable because of poor management. Modern usage led to loss of farmland for similar reasons.

The oil industry is a major polluter throughout the world, and the concentrations of production and distribution centers in the Persian Gulf area pollute the atmosphere and waters of areas around oil wells, surface seepages, and where unwanted gases are flared off. Leakages pollute the waters near ocean terminals. Particles in the atmosphere cause fogs that worsen respiratory complaints. Even before the Gulf War, the Persian Gulf had lost most of its plant and animal life as a result of pollution since the first oil production there earlier in the 1900s. At the end of the Gulf War, the retreating Iraqis set oil wells on fire, adding carbon gases and particles to the atmosphere, but the particles fell on desert sands. They also released a huge oil slick, damaging the plant and animal life, but it had less impact than it would have if the Persian Gulf had not already been heavily polluted.

Global Connections

Following a medieval history of trading connections across Asia, eastern Africa, and the Indian Ocean, the region's rulers stood back while Europeans monopolized the extension of trade and political dominance in the 1800s and early 1900s. After World War II, the huge increase in demand for oil and the newly independent Arab countries gave the region a new strategic significance. Wars between Arab countries and Israel, and Cold War rivalry between the United States and the Soviet Union added to

that significance. The 1973 war with Israel triggered massive increases in the price of oil, and the oil-producing countries in this region built huge financial surpluses in the 1970s. Later falls in oil prices brought problems of indebtedness, but many of the region's countries remain wealthier than other formerly poor countries in the world (Figure 8.10). Ownership of consumer goods (Figure 8.11) reflects a range of material wealth across the region. Istanbul, Turkey, has the greatest claim to being a global city-region, but Tel Aviv-Jaffa (Israel), Cairo (Egypt), Riyadh (Saudi Arabia), and some of the Gulf country cities such as Abu Dhabi have growing business services with global connections.

Subregions

The countries of Northern Africa and Southwestern Asia have differences in national income and the economic well-being

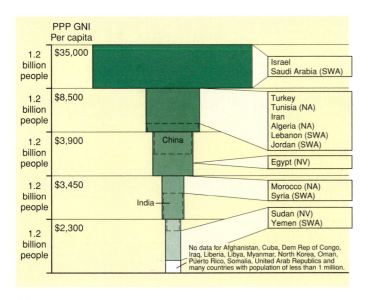

Figure 8.10 **Northern Africa and Southwestern Asia: country incomes compared.** The countries are listed in the order of their PPP GNI per capita. Source: Data (for 2000) from *World Development Indicators,* World Bank, 2002.

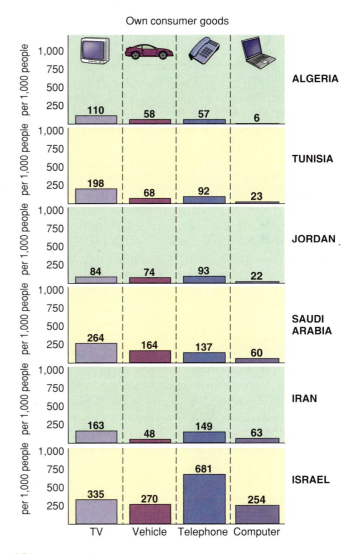

Figure 8.11 **Northern Africa and Southwestern Asia: ownership of consumer goods.** Compare these levels with those of other countries in Africa. Source: Data (for 2000) from *World Development Indicators,* World Bank, 2002.

OMAN

Oman is a country at the entrance to the Persian Gulf with a long Indian Ocean coast (see Box Figure 1). It is an oil producer and has a history of trading across the Indian Ocean, still retaining links in eastern Africa, where Zanzibar Island was owned by the sultans of Muscat, and in the island of Gwadar off the Pakistan coast. Although the 2001 population of Oman was only 2.4 million people, it is a country of contrasts, and its peoples live with both increasing global connections and continuing local voices. Traditional ways meet the modern in Oman.

Oman is a mainly hot and arid country, most of which hardly sees a cloud, let alone rain, from one year to another. After a rare rain, the normally dry valleys fill with water that may spread out over the flatter coastal areas to depths of 30 cm (1 ft.) for a short while. Lower land separates mountains in the north facing the Gulf of Oman and in the Dhofar region in the south. It is 10 hours' drive from Muscat to Salalah across desert landscapes or around 90 minutes by air. Dhofar is the exception to the unrelenting arid climate, with its August monsoon (*khareef*) that brings rains, cool air, and mists to some of Earth's hottest places.

Traditional Oman

The traditional past of Oman is reflected in the agriculture of areas near the coast, where narrow aqueducts (*falajs*), constructed centuries ago by Persians, bring water from springs in the hills, often by underground routes. The water is used to grow dates of many varieties and colors, and fruit from limes to papaya and coconuts in the far south (Dhofar). Fishing for tuna and kingfish, among others, is also part of the traditional economy. Omani food combines rice, meat, and fish in spicy dishes. The brown *halwa* dessert is made from spices, sugar, and nuts.

Until modern changes arrived, settlements were small villages, and port towns were homes to merchants and traders as well as farmers and fishers. The towns have souks (traditional markets) where foods, spices, and craft goods are sold. Standing above some villages and ports are forts, often constructed by Portuguese traders (see the banknote in Box Figure 2). Although Arabic is the national language today, many Omanis speak Swahili (former eastern African Creole language of trade) or Baluchi (language of western Pakistan). The landlord of the expatriate university teacher mentioned later has Swahili as his first language (and mother and brothers who live in Zanzibar), English as his second, and Arabic as his third.

In the traditional clothing, men wear the white "dishdash" robe, sandals, and an embroidered cap (*kuma*) or turban (*masa*), or sometimes both (see Box Figure 3b). The women wear long dresses and headscarves. They love bright colors but wear a black outer robe (*abaya*) in public. It is not customary for all women to cover their faces: this is not demanded by Islam but is a traditional extra feature for some Omanis and in countries such as Saudi Arabia. The Ibadi Muslim sect that most Omanis follow allows more freedom for women.

Modern Oman

Modernized Oman emerged after 1970, when the present sultan, Qaboos bin Said, deposed his father with the help of the British. He used the oil and gas wealth developed by Shell, British Petroleum,

(a)

(b)

Box Figure 2 **Views of Oman provided by banknotes.** (a) Omani banknote with the front information in Arabic and the head of Sultan Qaboos bin Said, ruler of Oman since 1970, date palms, and a *falaj* water channel. (b) Omani banknote, reverse side in English. View of Nizwa with its souk, mosque, and fort.

Box Figure 1 **Map of Oman.**

(a)　　　　　　　　　　　　　　　　(b)　　　　　　　　　　　　　　　　(c)

Box Figure 3　　**Oman and its people.** (a) Coastal mountains southeast of Muscat. (b) Omani young men in traditional clothing in the Eid souk at Nizwa. (c) Thuwara spring above Nakhl, a vegetated valley where the stream flows from time to time (wadi). The valley is popular with Omanis and expatriates. Photos: © Gerald Selous.

and Petroleum Development of Oman, the local oil company, in a paternalistic development of the economy that is referred to as a "renaissance" leap forward. In particular, the oil income funded new hospitals, including the Sultan Qaboos University teaching hospital, which is rated highly by the World Health Organization. The university dates from 1986 and has schools of agriculture, arts, commerce, education, engineering, medicine, and science. The growth of higher education in Oman also attracted private institutions such as the Caledonian College of Engineering linked to a Glasgow, Scotland, university. New elementary and secondary schools continue to be built in which male and female enrollments are equal and boys and girls are taught separately. The high levels of achievement by many girls mean that they have to obtain better grades than boys to enter the most prestigious higher medical education—which also has strictly equal male and female enrollments and where the sexes are taught together.

Oman is unusual among the Gulf countries in remaining outside the Organization of Petroleum Exporting Countries in order to preserve what were at first thought to be limited oil reserves. For a time, it also supported Arab agreements to work with Israel, for example in the field of water resources.

Modern technology is used wherever possible. Cell phones are common with a system of pay as you go called *hayyak* with no deposit or monthly bills. Computers are commonly available, and e-mail is much used. Banks include the multinational Hong Kong Shanghai Banking Corporation (HSBC) and Bank Muscat. Modern shopping includes French-owned supermarket chains (Carrefours, Prisunic) and American fast-food chains (McDonald's, Pizza Hut, Starbucks, Subway, and others).

Cultural tourism is a recent trend aimed at wealthy visitors from the United States, Japan, and Europe that takes advantage of good airline linkages, the sunny climate, the quaint souks, and the historic forts.

Oman has 75 percent of its population claiming a local origin, higher than most Gulf countries. Only 25 percent are foreigners, including low-wage workers from South Asia and the Philippines employed in factories, shops, gasoline stations, hotels, and gardens. They may live in outside rooms attached to larger homes or in group accommodations. One villa with four apartments housed 141 young Sri Lankan women for a year who were bused each day to work in a clothing factory. Many overseas workers come from Kerala state in southern India and are known by their language as "Malayalees." There is also some abject poverty, including female beggars, although begging is prohibited by the government because in a Muslim country, women in need should ask for *zakat* from the mosque. Some poor people live in suburbs isolated from the rest of the towns.

Newer suburbs have large white villas with walled gardens and ornate gates. A strip of blue-and-white tiles along the tops of walls is common; roofs are flat; and some windows may have stained glass. These are the homes of the wealthier Omanis, many of whom take bank loans and rent the property to expatriate engineers, teachers, geologists, diplomats, military advisers, or airline pilots to help pay off their loans.

The Student and the Expatriate University Teacher

Ahmed is a medical student, studying in what is regarded as the "cream" of Omani professional courses. He is one of seven children, and going to university in Muscat was his first time away from home. He began the medical training directly out of high school. It took some adjustment and several weekend trips home—a three-hour drive taken in relatives' cars or taxis. Public minibuses called *biaza* are cheap and stop anywhere along the main routes.

In his first year, Ahmed lived in a hostel for men. He enjoyed sports and a variety of clubs in mixed groups on campus. In Oman, men and women are taught together in universities. At Sultan Qaboos University, the men enter lecture rooms from the front and the women from the back. The men have walkways at ground level, the women on the level above. All students wear national dress, except when engaged in sports. The women at university may not wear face coverings for security and identity reasons.

The facilities for study are as good as anywhere in the world, with modern laboratories and huge libraries. Everything is free. There are no tuition fees, room and board costs, or charges for books. Ahmed now lives off the university campus and receives 120 rials (US $300) a month toward his expenses.

Fred is a United Kingdom national who teaches English as part of the medicine faculty team of eight with colleagues from India, Pakistan, the United States, and Canada. His team is part of a group of 150 teachers of English who provide courses for almost all the students and use 70 percent of all classroom space.

Fred lives in a rent-free villa. It is about 10 km (6.2 mi.) outside Muscat, giving him a daily 10-to-15-minute drive. He leaves his auto about a mile from his office and walks for exercise—although the early morning effort means he has to dry out in the air conditioning. He has to wear light clothing and a hat to protect himself from the strong sun. He works from 8 A.M. to 4 P.M., with an hour for lunch, from Saturday to Wednesday and has the local Thursday and Friday Muslim "weekend."

Living in Oman has many pleasant features for Fred. The traditional Omani natural and cultural environments are overlaid with the trappings of the West and with welcome local features such as cheap fuel and an absence of taxes. There are reminders of the United Kingdom in the BP gasoline stations, British Airways flights, the embassy, and the British Council (which funds cultural links to the U.K.). There is the thriving Protestant church. The supermarkets contain all—and more than—the selections of those in the U.K. When in the U.K., the HSBC bank gives Fred an immediate readout of his Omani account in rials (1 rial = UK £1.80).

He also has many opportunities that living in the U.K. would not offer. Cinemas show not only the latest Western films but also Indian and some Arabic films. There are camping opportunities ("wadi bashing") and other outdoor activities in the cooler months (between November and March), visits to the old forts, diving, snorkeling, rock climbing, and eating out in a wide range of cheap cosmopolitan restaurants.

Omani Women: The Traditional and Modern

From the 1970s, the sultan of Oman encouraged women to be more involved in society. Today, Omani women both maintain the traditional culture and, particularly the younger ones, look for jobs appropriate to highly educated people. In 1975, there were hardly any schools, and girls did not attend them. Since then, all girls and boys go to school.

Although many hard-working and intelligent women emerge with college educations, they find it difficult to get jobs in what remains a man's world. Many women become teachers or doctors, and others are now allowed to be waitresses in restaurants or nurses in a hospital, overcoming previous taboos against such work (outside the home, at least). Omani women are particularly successful in businesses and banking. There is a specific place for them on government advisory councils, where they are the ones concerned with social and family affairs. As with other aspects of this thrust to involve women, including allowing them to drive cars, the Omani approach is different from that found in Saudi Arabia and many of the Gulf countries.

Women continue traditions through their roles of raising children and providing hospitality, and in dressing and adorning themselves for special occasions. In the past, most Omanis were brought up in large families of nine or more children that kept the mothers in the home for much of their lives. Although some women became physically weak and ill, and were advised by their physician not to become pregnant, they continued to have children if their husband and extended family expected it. The extended family and wider kinship group forms the Omani "tribe," and most Omanis continue to marry within that circle.

Newly married couples will often live in the same house as the man's parents, and this cements family relationships. Since young people marry within a known group, there is no arrangement of marriages. This system is considered to work well, but some educated families now question the wisdom of closely related marriages if specific physical or mental defects continue through interbreeding. Moreover, many modern families, encouraged by the government, have a maximum of three children, spaced out over a woman's childbearing years.

Women's clothing is traditional, colorful, and elegant with different emphases in specific parts of Oman. Traditional jewelry is based on older, chunky silver craftwork linked to Oman's silver-mine products, and each of many rings worn has a meaning. Today, gold is more common. Henna, a copper-colored extract from leaves, is applied to feet and hands in intricate patterns for religious feast and wedding days during a preceding "henna day" time of preparation. Perfumes based on oils, as well as Western commercial products, are most important: many shops in the souks sell the oils in glass bottles. Their application plays an important part in social life, as does the burning of frankincense (bakhour) from trees grown around Salalah. All houses have a burner that is lit in the late afternoon before visitors come.

New visitors are welcomed with a special ceremony that brings together many aspects of Omani traditions. On entry, women and men go to separate rooms. The lady of the house brings into the women's room a large round tray on which are placed oranges, dates, and other fruits, a coffee jug and cups, and a water bowl (augmented today with a box of paper tissues), and all the guests and family gather round. They are offered fruit and newly baked bread, rinsing their hands. Coffee is poured into very small cups, that are refilled until a person shakes the cup to indicate he or she does not want any more. Then the incense burner is brought in, and each lady stands and places it under her outstretched skirt until the smoke rises upward and out through her neck and into her hair. All the company take part in this one by one. Finally, a tray of perfume bottles is passed, and guests apply a mixture before they are "free to leave" (i.e., the party is over). This complex welcome makes newcomers feel like royalty and may lead on to a continuing series of visits with others in the neighborhood. Later visits to the same household will be less formal.

The sultan of Oman's aim is to modernize his country without losing the values inherent in the traditional ways of life. Although the country remains exotic in so many ways, its people can take advantage of educational opportunities and health care. Omanis remain conscious of their antecedents and see modern life as an extension of the old. Each holiday and religious festival involves returning to their village and meeting the extended family.

---

demonstrated by the ownership of consumer goods. The arid environments of much of the region are a major cause of population concentrating (Figure 8.12) along the coasts and major river valleys, as well as in the uplands of Turkey and Iran and southern Sudan, where there is more rain. The coastal cities contain nearly half the total population, combining port, government, manufacturing, and tourist functions. The region is divided into five subregions (see Figure 8.1):

- The countries of *North Africa* face the Mediterranean Sea west of Egypt and have desert interiors. This subregion is sometimes referred to as the Maghreb.

- The *Nile River valley* links Egypt and Sudan.

- The region's heartland of *Arab Southwest Asia* extends from Syria in the northwest to Oman in the east. The western section consisting of Syria, Lebanon, Jordan, Palestine,

and Israel, sometimes including Turkey, is referred to as *Al-Mashrig* ("Levant" in English).

- *Israel and the Palestinian Territories* form a continuing focus of tensions within the region.

- *Iran and Turkey,* two of the largest countries in population and area, form the northern and eastern borders and have mostly mountainous terrains. They differ from the rest and each other in language, with Arabic being replaced by Farsi and Turkish, respectively.

🌐 North Africa

The four countries of North Africa are Algeria, Libya, Morocco (with Western Sahara), and Tunisia (Figure 8.13). In 2001,

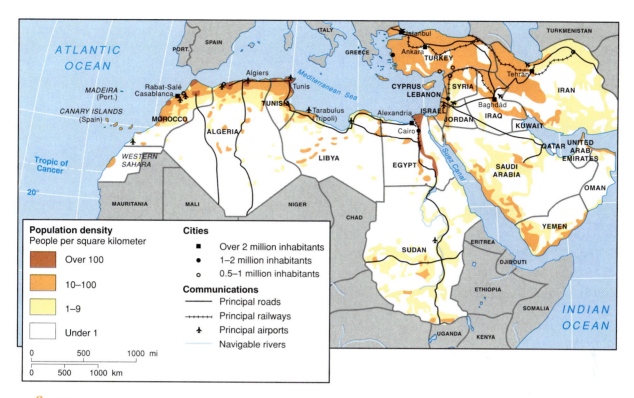

Figure 8.12 **Northern Africa and Southwestern Asia: population distribution.** Note where the highest and lowest densities of population occur in each subregion. Source: Data from *New Oxford School Atlas,* p. 98, Oxford University Press, U.K., 1990.

Algeria and Morocco had around 30 million people each, Libya had 5 million, and Tunisia nearly 10 million (Figure 8.14). Over 80 percent of Algeria's and Libya's territories are desert, but Morocco and Tunisia do not extend so far into the arid Saharan environment. The northern parts of Morocco, Algeria, and Tunisia are dominated by the Atlas Mountains (Figure 8.15), and that area is known as the Maghreb. It includes high ranges (Mount Toubkal in Morocco is 4,165 m, or 13,665 ft., in altitude), broken by internal plateaus and river valleys. The harsh, largely arid, and often mountainous natural environments of the North African countries restrict agriculture and most human settlement to a small percentage of the territory along the northern coasts and in the immediate mountainous hinterland. Problems of water supply affect all these countries.

Although located at a considerable distance from the heart of the Muslim/Arabic world in Southwest Asia, this region shares an adherence to Islam as the almost exclusive religion, and Arabic is even more dominant as the official language. In 1979, the link with other Muslim Arab countries was strengthened by the move of the Arab League headquarters from Cairo (Egypt) to Tunis.

North Africa faces Europe across the Mediterranean Sea. Many of its past political ties and present economic links are northward to Spain, France, and Italy. Algeria, Morocco, and Tunisia, in particular, retain close ties with France and have strong links to markets in Europe for selling products, buying goods, and sending emigrant labor. Libya sells much of its oil and gas to Italy. The European Union's 1986 extension to include Spain, Portugal, and Greece—which compete with North Africa for similar agricultural produce markets—placed stresses on these relationships.

Countries

Political History

The countries of North Africa have similar political histories. In the 800s B.C., the coast was settled by traders from the eastern Mediterranean who established the kingdom of Carthage. This was conquered by Rome, then overrun by Vandals in the A.D. 400s, but retaken by the Christian Byzantines. Arabs invaded in the A.D. 600s, converted the local Berber tribes to Islam, and imposed new ways of life. The Muslim invasion of Spain followed (see Chapter 3), producing the architectural riches of Córdoba, Seville, and Grenada as reflections of the wealth and civilization they brought, including toleration of local Christians and Jews. By 1500, internal divisions and weaker control allowed Spanish Christians to expel the Muslims, forcing over a million to move back to North Africa.

From the 1200s, pirate groups controlled most of the subregion's port cities. The coast (apart from Morocco) became known as the Barbary Coast, making the western Mediterranean dangerous for merchant shipping and contributing to the decline of Mediterranean ports in Southern Europe. The pirates continued their activities even after the Ottoman Empire took control of this area in the 1500s. During the

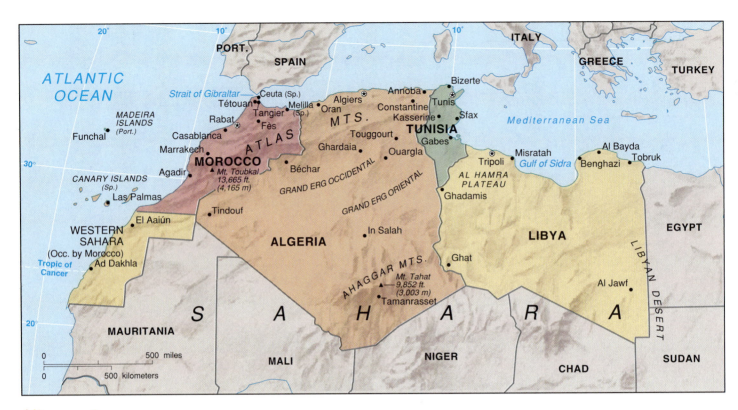

Figure 8.13   **Northern Africa: countries, physical features, and main cities.**

1800s, European countries and the United States stopped the pirate attacks, and France annexed territory in Algeria and later in Tunisia and Morocco.

In Algeria, Tunisia, and Morocco, French settlers and local people established commercial farms that grew citrus fruits and produced wine for European markets. The settlers took political and administrative control of the colonies. After World War II, nationalist groups fought for and obtained independence—in 1956 in Morocco and Tunisia, and in 1962 in Algeria. Libya remained poor, a mainly desert area of little economic or political outside interest until Italy occupied it in 1911. After World War II, Italy was replaced by a British-French protectorate until Libyan independence was achieved in 1952.

Recent Shifts

Independent Algeria has been ruled by democratically elected governments with socialist policies based on central planning. The first-round election success of the fundamentalist Islamic Salvation Front (FIS) in the 1992 elections led to an army takeover of the government and civil war with the dispossessed Islamic militants. Terrorist activities and army repression through the 1990s devastated Algeria's economy and people with over 100,000 deaths and the army's destruction of the FIS

Figure 8.14   **Data for North Africa and Nile River Valley subregions.**

Country	Capital City	Land Area (km²) Total	Population (millions) Mid-2001 Total	2025 Est.	GNI 1999 (US $ million) Total	GNI PPP 1999 Per Capita	Percent Urban 2001	Human Development Index Rank of 175 Countries	Human Poverty Index: Percent of Total Population
Algeria, Democratic and Popular Republic of	Algiers	2,381,740	31.0	43.2	46,548	4,840	49.0	109	28.8
Socialist People's Libyan Arab Jamahirya	Tripoli	1,759,540	5.2	8.3	No data	No data	86.0	65	16.4
Morocco, Kingdom of	Rabat	446,550	29.2	40.5	33,715	3,320	55.0	126	39.2
Tunisia, Republic of	Tunis	163,610	9.7	12.5	19,757	5,700	62.0	102	23.1
Egypt, Arab Republic of	Cairo	1,001,450	69.8	96.2	86,544	3,460	43.0	120	33.0
Sudan, Republic of	Khartoum	2,505,810	31.8	49.6	9,435		27.0	142	36.8

Source: Data from *Population Reference Bureau 2001 Data Sheet; World Development Indicators,* World Bank, 2001; *Human Development Report,* United Nations, 2001; Microsoft Encarta (ethnic group, language, religion).

Figure 8.15 **Northern Africa: the Atlas Mountain environment.** A Berber village in the Atlas Mountains, with desert vegetation (cacti, thorn bushes) in the foreground, olive trees and arable land around the village, and trees planted on the hillside to prevent flash floods by slowing the passage of rainwater into the gullies. Photo: © Wolfgang Kaehler/Corbis.

as a political party. In early 2000, an amnesty for armed terrorists was mainly successful, reducing monthly deaths from 1,000 to 200. After September 11, 2001, Algeria regained some international credibility for its own war on terrorism, although its terrorists never expressed enmity to the United States and fed on the disregard of social needs and human rights by the military government. In 2001, anti-Islamic Berber youths in eastern Algeria demonstrated another source of disquiet in the country.

Morocco and Tunisia, which border Algeria on either side, view the self-destruction in Algeria with fear and take steps to avoid involvement. Morocco has political stability under its moderate but strongly involved king, Muhammed VI. His predecessor, King Hassan II, introduced gradual political reforms in the 1990s and gained international Muslim credibility through his mediating role in Arab issues and the construction of a massive new mosque. The former Spanish Sahara, an arid land known as Western Sahara, was annexed by Morocco in 1976, but the Saharawi people and their independence movement, the Polisario Front, seek control over their territory. By 2002, the United Nations had proposed a plan that seemed

Country	Ethnic Groups (percent)	Languages O=Official	Religions (percent)
Algeria, Democratic and Popular Republic of	Arab 83%, Berber 16%	Arab (O), Berger dialects	Muslim 99%
Socialist People's Libyan Arab Jamahirya	Arab-Berber 97%	Arabic (O), Italian, English	Muslim 97%
Morocco, Kingdom of	Arab-Berber 99%	Arabic, local dialects	Muslim 98%
Tunisia, Republic of	Arab-Berber 98%	Arabic (O), French, Berber	Muslim 98%
Egypt, Arab Republic of	Egyptian-Bedouin-Berber 99%	Arabic (O), English, French	Christian 6%, Muslim 94%
Sudan, Republic of	African 49%, Arab 39%, Nubian 8%	Arabic (O), Nubian, Dinka, others	Christian 5%, local 25%, Muslim 70%

to accept Moroccan control with limited local autonomy and a referendum in Western Sahara that would include all residents such as the recent settlers from Morocco. While UN resolve weakens, so does that of Polisario's supporter, the Algerian government, which was bought off by French aid. A further step was taken when Morocco awarded licenses to two international oil corporations for exploring areas offshore of Western Sahara.

Tunisia had 30 years of one-party rule when Islamic extremists were repressed but women's rights were established. Women in Tunisia have a better position than those in other Arab countries, resulting in lower adult illiteracy (40% in 1999) and a prohibition on polygamy.

Libya remains under the strong direction of Colonel Muammar al Qadhafi, who seized power in 1969 and runs the country as a mixed military, socialist, and Islamic republic. In the 1970s and 1980s, he supported anti-West terrorism, including the destruction of a PanAm airliner over Scotland. The U.S. bombing of Tripoli and UN sanctions from 1992 reduced such activities. More recently, Qadhafi moved toward a role as mediator and leader of new political linkages in the rest of Africa (see Chapter 9). After September 11, 2001, he signaled an intention to be part of the allied antiterrorism campaign and offered Libyan assistance to the United States, but he also took the opportunity to post on website lists of so-called anti-Libyan agitators who have little or no involvement in terrorism.

People

Ethnic Variety

Arab and Berber people with Muslim faiths constitute over 95 percent of the populations of these countries. Speakers of local dialects and traditional groups of nomadic pastoralists now make up a small proportion of the population. Colonial legacies include many French speakers among educated middle classes in the Mahgreb countries, while Italian has a similar role in Libya.

Urban Population Growth

The urban areas of North Africa contain nearly half the total population (over three-fourths in Libya). However, the decline in employment opportunities in the rural areas has been only partly compensated by the growth of urban employment in factories and offices. In 2000, the largest cities included Algiers (Algeria, 2 million people), Casablanca (Morocco, 3.5 million), Tripoli (Libya, 1.8 million), and Tunis (Tunisia, 2 million). None of these had global-city status. Rabat (Morocco) had 1.5 million people, and cities with just under 1 million people included Oran (Algeria), Benghazi (Libya), and Marrakech and Fez (Morocco).

Before the late 1900s urbanization, the region had a number of ancient but relatively small towns. The older towns that were central to agricultural and trading economies survive as enclaves within newly expanded cities, some much changed by clearing crowded buildings to make way for new highways. High densities of homes, commercial premises, and public

buildings inside city walls marked the traditional small towns of the region and their central medinas (Figure 8.16). **Medinas,** named after the sacred Muslim city in Saudi Arabia, are historic sectors of cities, valued for their distinctive structure and social fabric. Their labyrinthine alleys, souks (marketplaces), and artisan shops relate them to the past and attract tourists. Within these sectors existed a rigid pattern of land use; prestigious craft workers, such as religious artisans, had premises close to a central mosque, castle, and square; those in lowlier occupations, such as leather workers, existed on the edges. The World Heritage list of historic cities contains many medinas across the whole region, from Fez and Marrakech in the west to the tall, brown mudhouses of Yemen in Southwest Asia. Today, city walls remain only in cities where tourist interests are economically important, such as Fez and Marrakech (Morocco). Elsewhere, they were replaced by wide boulevards or other buildings.

As the populations grew and crowded into cities, the expanding demand for housing by poorer people was often more than governments could meet. Shantytowns are features of cities across the subregion, known as *bidonvilles* in Casablanca, Morocco, and as *gourbivilles* in Tunis.

Population Dynamics

The populations of all the North African countries continue to grow rapidly. The total population of the subregion rose from 50 million in 1980 to 75 million in 2000 and could be over 100 million by 2025. Total fertility rates of around 7 in 1965 fell by half by 2001—except for Libya (3.9) and Western Sahara (6.8). Tunisia reduced population growth through policies instigated soon after independence, including forbidding polygamy, setting a minimum age for marriage, and instituting a successful family planning program. Morocco set up a program to empower women that included family planning, and maternal and child services. Algeria began its program later.

The demographic transition diagram (Figure 8.17) shows how the birth rates fell sharply from 1970 to 2000; the Algerian age-sex diagram shows the preponderance of young people in the population (Figure 8.18). One advantage of the current population growth is that there are few older dependents, but these will increase in coming years, as life expectancies are now around 70 years.

Employment and Migration

The current rapid population growth creates problems for the region's education systems and employment prospects. Only in Libya is there a shortage of labor and need for immigrants. The other countries have more entrants to the labor force than jobs. Algeria and Morocco have more than 20 percent unemployed. And yet, despite intensive education programs to provide primary school literacy and to redress the male-female differences in opportunity, shortages of skilled labor continue. Female literacy, although improving, lags behind that of men: for example, in Morocco in 1990, 53 percent of adult men and 25 percent of women could read; by 1999, 61 percent of men and 35 percent of women could read. Levels are somewhat higher in the other countries.

Figure 8.16 **Northern African landscapes: old town souk.** Small shops selling craft goods along narrow streets close to the market-place in Fez, Morocco. Photo: © Dave Bartruff/Corbis.

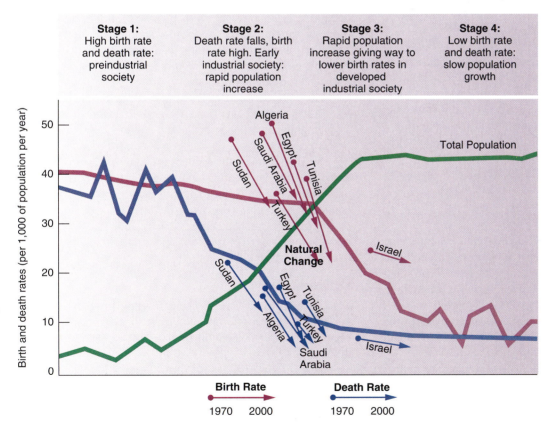

Figure 8.17

Northern Africa and Southwestern Asia: demographic transition. Birth and death rates in selected countries. Compare the experiences of these countries with those in other parts of Africa. Source: U.S. Census Bureau. International Data Base.

In the 1970s, many North Africans migrated to France and other European countries, where low-paying, mainly manual, jobs were available, but more recently, there have been fewer vacancies, and resistance to immigration from North Africa has increased. Today, there are around 1.5 million North African workers who send home wages from European countries, espe-cially France and Spain. When workers return from Europe, they often set up a store or business and buy a good house. In the 1980s, migrant workers from Morocco and Tunisia also worked in Libya and in the Persian Gulf. Each country has a growing university-educated group, but these people find few well-paid professional employment outlets; that, combined

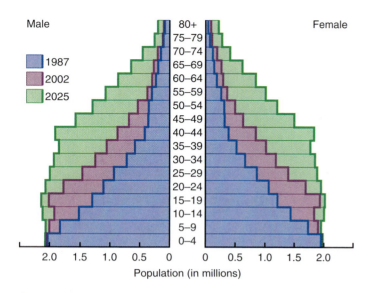

Male

1987
2002
2025

Female

80+
75–79
70–74
65–69
60–64
55–59
50–54
45–49
40–44
35–39
30–34
25–29
20–24
15–19
10–14
5–9
0–4

2.0   1.5   1.0   0.5   0   0   0.5   1.0   1.5   2.0
Population (in millions)

Figure 8.18 **Algeria: age-sex diagram.** Source: U.S. Census Bureau. International Data Bank.

with the oppressive political regimes, spurs many to emigrate to Europe and the United States, creating a significant brain drain.

Economic Development

Algeria, Morocco, and Tunisia are rising middle-income countries, but their economic growth faltered in the 1990s. Algeria's civil strife halved its per capita income, while Morocco's and Tunisia's rates of increase slowed. Libya does not report income, but its oil reserves and a relatively small total population put it into the upper-middle income group of countries. In spite of UN sanctions, Libya's oil and gas are sold to Europe, and the income is used to subsidize housing, basic foods, oil products, education, and health care. The subsidies make up for low wages.

Colonial Influences and Farming

Many of the problems facing North Africa stem from the type of economy established by colonial countries. The colonial system involved land appropriation for European settlers and local elites, who used irrigation water for intensive commercial farming tied to markets in Europe. Manufacturing was not encouraged. Later, oil exploration became important once the world market was established, and the French discovered and developed the major oil and gas resources.

The modern economies and overseas markets reflect colonial systems. Export crops, such as citrus fruits and olive oil, are grown on larger holdings of over 50 hectares (124 acres). Although disrupted and often pushed off the best land, the traditional farming sector remains a major employer, growing cereals on holdings of 5 to 10 hectares (12 to 25 acres) (Figure 8.19). Yet North African countries still need to import up to 50 percent of their food needs, with Libya needing up to 75 percent. The growth of private, as opposed to collectivized, agriculture in the 1980s led to increases in the

commercial production of vegetables, fruit, and dairy and poultry products.

Although a few Italian settlers took up land in Libya before 1945, the main development of farming there was after 1969, when rotary-sprinkler irrigation was introduced along the coast and at a few inland sites to grow cereals. Oil revenues funded this investment, and the production costs are high while yields are not. Libyan irrigated farming is thus uneconomic, and farmers are upset when priority for water use is given to urban inhabitants.

Morocco's Primary Products

Only in Morocco is as much as half the population still dependent on agriculture, although it constitutes only 15 percent of GDP. This reflects the arid environments of the subregion that restrict cultivable land areas (Figure 8.20), as well as the growing returns from oil production, manufacturing, and service industries in the other countries.

Morocco's farmers use irrigation water to produce citrus fruits, vegetables such as tomatoes and potatoes, and cut flowers for European markets. At present, farmers take 85 percent of the water used in Morocco, but this is expected to fall as domestic and industrial demands increase. At a time when Moroccan farmers need to invest more capital to improve productivity and quality, the shift toward urban living and water usage is bringing a crisis to farms, especially in drier southern Morocco. It may be necessary for farmers to switch to more valuable crops or those that require less water, but the markets for such crops are not obvious.

Morocco's economy also rests on other primary products. Its exports of phosphate (for fertilizer) are the most important. Its fishing industry has grown to the point where it provides 15 percent of Moroccan exports. At present, Morocco is contesting the management of fishing grounds off western Africa with the European Union and particularly Spain. The main fish caught are squid (for export to Japan) and tuna. Morocco continues to export cork from the bark of oak trees in the northern area of the country. Algeria and Tunis also export phosphate rock, but it is a less important component of their overall exports.

Oil and Gas

Oil and natural gas are the chief mining products of North Africa. It was only after independence that Algeria, Libya, and Tunisia discovered their oil reserves or were able to exploit them.

In Algeria, which has the fifth-largest natural gas reserves in the world and is the second-largest natural gas exporter, revenues from oil and gas make up over 50 percent of the annual budget and 95 percent of export earnings. Algeria is so massively indebted as a result of loans raised on the strength of its oil income that it now has difficulty attracting foreign investment. Pipelines bring the oil and natural gas from interior locations to coastal ports and refineries (Figure 8.21).

Libya is similarly dependent on its oil income. With its small population, Libya suddenly gained great riches that were

Figure 8.19 **Northern Africa and Southwestern Asia: rural landscapes.** Bedouin women taking goods to market near Foudouk al Aouerab, Tunisia. The cultivated valley behind them and the bare hillsides above are typical of North Africa and much of the wider region. Photo: © Kess van der Berg/Photo Researchers, Inc.

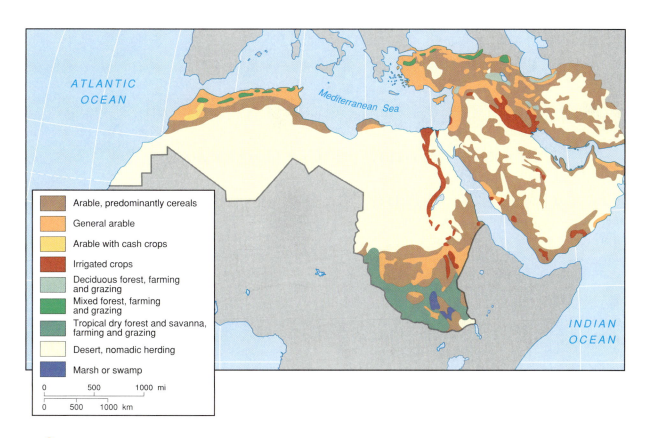

Figure 8.20 **Northern Africa and Southwestern Asia: agricultural land use.** Note the extent of desert and other dry-farming uses. Data from *New Oxford School Atlas,* p. 98, Oxford University Press, U.K., 1990.

nationalized after an initial phase of development by multinational oil companies. The country's smaller debts and fewer people enable the Libyan economy to adjust to the ups and downs of oil income.

Expanding Manufacturing and Services

Manufacturing expands in all countries but is overshadowed by oil income in Algeria and Libya. Morocco's manufacturing sector is the least developed but includes substantial craft industries. All countries have basic food processing, constructional materials, and textile industries located in and around major cities. Except in Morocco, oil-refining and petrochemical industries are important. Algeria also has light industries, including the manufacture of electrical components, Libya makes steel and aluminum, and Tunisia has a small steel industry.

North African countries have growing government bureaucracies and service occupations in education and health care. In Morocco and Tunisia, tourism is a major source of income, based on their sunshine, coastal locations, historic and cultural features, and stable political environments. Most tourists come from northern Europe, with its cool and rainy climate and high personal incomes. Tunisia had over 5 million visitors in 2000, up from 1.6 million in 1980 and 3 million in 1990, while Morocco had over 4 million. Terrorist activity in Algeria after 1992 ended the envisaged tourist expansion there, and there were fewer than 750,000 visitors in 1999. Libya does not encourage tourism.

Trade with Europe

The trade of the North African countries is tied to Europe, both the source of most imports and the destination of most exports. Algeria and Libya supply one-fourth of the European Union's oil and natural gas, while Morocco sells over half its phosphate exports to the EU. Morocco, Algeria, and Tunisia sell agricultural products and textiles to EU markets, but 1992 EU rules made it more difficult for these exports to penetrate the EU countries. This is partly because the addition of Portugal and Spain to the EU led to surpluses of products that are also the main outputs of North Africa. Such setbacks came at an unfortunate time when remittances from workers in Europe were unlikely to increase and compensate for the lost export earnings.

Economic Policy Changes

Governments trying to mastermind economic development in the North African countries placed economic growth high on their agendas. The setting of five-year plans, however, became wastefully bureaucratic and inflexible at a time of rapid change. Only the oil-exporting countries receive adequate foreign currency to pay for major developmental projects. Algeria, however, borrowed so much on the strength of its oil sales that its debts rose to one-fifth of its GDP. Despite the smaller role of Tunisia's oil income, drought and locust attacks caused a financial crisis when its revenues decreased as a result of lower world oil prices in the 1980s. The collapse of world phosphate prices and the costs of conflict over Western Sahara upset Morocco's overambitious planning in the late 1970s, leading to a massive increase of its debts.

In the 1980s, the International Monetary Fund required Tunisia, Algeria, and Morocco to carry out measures that would cut internal budget deficits by reducing the public sector of the economy and encouraging trade in export products. The resultant unpopular measures of structural adjustment included cutting civil service employment and removing subsidies on basic foods (flour, oil, and sugar), and led to poverty increases and riots. In many places, literacy and access to health care fell.

In the 1990s, North African countries began to privatize large sections of their economies. Tunisia is farthest ahead and even has its own stock exchange. Morocco followed with unparalleled sales of state holdings that reversed the previous policy of "Moroccanization" up to 1989.

Figure 8.21   **Northern Africa: oil refinery at Skikda, Algeria.** Oil refineries take up huge areas of land, here necessitating reclamation from the sea. They include ocean terminal facilities, storage tanks, and the complex pipework of the processing plant. Photo: © SuperStock, Inc.

🌐 Nile River Valley

The Nile River provides a life-giving water supply to Egypt and Sudan that has sustained a human presence in the dry eastern Sahara since the early days of farming and civilization (see Figure 8.9). The histories of the two modern countries were linked over many centuries by the Nile flows. The 1959 Nile Waters Agreement called for sharing the use of water between them, with Sudan receiving only 13 percent of the annual flow. These two lower-valley countries now contest the use of the water with those in the upper Nile River watershed, such as Ethiopia, Uganda, and Tanzania.

Although the Nile River waters provide a basis for living in an arid climatic environment, the resource is finite. Pressures imposed by increasing populations make it difficult for Egypt and Sudan to continue to modernize and diversify their economies. Both are poor countries (see Figure 8.14) with major balance of payments problems. Neither produces sufficient export income to pay for the imported needs of its people. Both rely on aid from the world's wealthier countries and the major Arab oil producers, but the contributions from these sources were reduced in the 1990s.

Countries

The countries of the Nile River Valley—Egypt and Sudan (Figure 8.22)—share the river waters and problems of developing their economies. Their geographic positions, political environments, and products, however, provide contrasts. Egypt is the largest Arab country in population (70 million in 2001) and one of the foremost in international relations. It retains control over the major global choke point of the Suez Canal, which has been widened to accommodate increasingly large shipping vessels. It also controls the Sinai Peninsula that is the land route from Africa into Southwest Asia. By contrast, Sudan has twice the area of Egypt, but only half the population, and has a more isolated position relative to global trading.

Political History: From Empires to Colonies

The earliest Egyptian empire dominated the lower Nile River valley from around 3200 B.C. The centers at Memphis in the north and Thebes in the south of modern Egypt expanded influ-

Figure 8.22 **Nile River Valley: countries, physical features, and main cities.**

ence over the Nubian kingdom in the area of modern Sudan. The complex imperial history ended when Alexander the Great conquered Egypt in 332 B.C., and the Romans and Byzantines imposed control up to the A.D. 500s. During the later part of this period, the Christian Coptic church spread through the area and southward into Ethiopia.

In the A.D. 600s, Muslim Arabs conquered Egypt but did not fully control the area of modern Sudan until the 1500s. Later, Mongols and Turks invaded Egypt, and it became part of the Ottoman Empire in the 1500s. Egypt's position at a crossroads of international trade brought prosperity to its markets. In the 1800s, the United Kingdom and France vied for control of the country. The British encouraged Nile River valley farmers to grow cotton for British textile mills. The opening of the Suez Canal in 1869 enhanced Egypt's strategic position and French involvements, but the British then occupied Egypt until the latter gained full independence in 1952.

After becoming a Muslim country in the 1500s, Sudan extended its territory southward, although many of the tribes in the southern area retained their traditional beliefs and some became Christians. In the 1800s, Egyptian forces attacked Sudan, but it held out until the British joined in an 1899 invasion. After this, Sudan came under the joint control of the United Kingdom and Egypt. British engineers constructed irrigation works in the Gezira area south of the junction between the White Nile and Blue Nile Rivers.

Sudan regained its independence in 1956, despite Egypt's claims to keep control. The tensions resulting from the period of Egyptian occupation of Sudan and fears of future Egyptian expansion remain in the minds of many Sudanese and affect negotiations over the use of Nile River water.

Recent Events

The year 1956 marked a turning point in Egypt's modern history. After centuries of dependence on the annual Nile River flood—for food and, more recently, for growing cotton to be exported to the British textile industry—Egypt became a socialist state with new priorities focused on its own internal needs. The civil service expanded to implement the centralized direction of civil affairs.

The first major goal was to make the water and power supplies secure to support farming and manufacturing expansion. The 1959 **Nile Waters Agreement** to share the river's waters with Sudan led to the building of the Aswan High Dam to store sufficient water and generate electricity. On completion of the dam in 1970, Lake Nasser behind the dam stored three times Egypt's annual water usage (Figure 8.23).

President Gamel Abdul Nasser hoped Egypt would lead the Arab world and for a few years combined Egypt and Syria in a single country. By getting the Soviet Union to help build the Aswan High Dam, he also exploited Cold War politics. Under Nasser's leadership, Egypt developed more rapidly than the rest of Africa in the 1960s and 1970s. Nasser died in 1970, and by the late 1970s, Egypt's foreign policy shifted by recognizing Israel. Then generous aid came from the United States, but Egypt was isolated from the rest of the Arab world.

Sudan is the largest country by area in Northern Africa and Southwestern Asia and the poorest. In 1999, Sudan had less than one-eighth of Egypt's annual product. When Sudan achieved independence from Egypt and the United Kingdom in 1956, the seasonal rains and fertile soils of the southern region gave hope for economic potential. The new leaders began to develop areas that could sustain rain-fed agriculture along the Nile River valley and on sandy areas in the west of the country. The full potential of such developments was not realized—partly because of droughts and partly because of political conflict. The Muslim leaders of the national government tried to impose their language, religion, and sharia law on Christian and traditional religious communities in the south.

Sudan is part of the southern Saharan rim and has conditions somewhat similar to those of its neighbors Chad and Ethiopia. With its low income, narrow economic base, continuing high rates of population increase, low life expectancy, and civil war, Sudan resembles other African countries more than those of northern Africa. Over a million refugees who fled to Sudan from Ethiopia's famines compounded its problems. In the 1980s and 1990s, civil war between the Arab Muslims of the north and the black African Christians and people of traditional faiths in the south cost as many as 2 million lives through fighting and related famine, plus led to the displacement of many other people. Arab Muslim soldiers take southern Christians as slaves, some of whom are bought and freed by U.S. Christian organizations—although so doing may raise the price of slaves and increase the trade.

In the early 1990s, the (rebel) Sudan People's Liberation Army lost much of its influence by brutally attacking its own people to extract funds. After developing a friendlier political wing in 1994, it attracted foreign military support and besieged Juba, the main city of southern Sudan. The war prevented crop planting, and many people died of hunger and disease. In 1999, Sudan's government moved toward multiparty politics internally and better international relations, but the discovery and extraction of oil from the disputed lands resulted in further conflicts in the early 2000s.

Figure 8.23 **Nile River Valley: Aswan High Dam, Egypt.** An earlier dam was built on the site in 1902 and extended in 1943. The massive modern dam was completed in 1970, causing the relocation of 90,000 people and the submersion of archaeological sites. Photo: © Adam Woolfit/Corbis.

People

Population Distribution and Dynamics

The populations of Egypt and Sudan are concentrated along the Nile River valley (see Figure 8.12), as both countries depend on the Nile River waters in the desert environment. Population distribution follows the water sources and transport lines.

The 2001 total of 100 million people in the two countries was up from 37 million in 1960 and 60 million in 1980, and may be close to 150 million by 2025. The large groups of young people will produce continuing high rates of population increase as they reach child-bearing age. The rising numbers of people will place further pressures on only 5 percent of the land that is cultivated in each country.

Egypt is making progress in reducing rates of population growth (Figure 8.24). In the late 1970s, worries over its ability to support the projected population increases led to mass media and government support for family planning, backed by U.S. aid. Total fertility rates dropped from over 5 to 3.5 and to under 3 in cities in 2001. Further progress depends on raising the status of women in Egyptian society. In 2001, only 43 percent of women, compared to 66 percent of men over 15 years old, could read and write. In Sudan, overall population growth remains over 2 percent per year, a reduction from the 1990s; literacy rates are similar to those in Egypt.

Egypt is an increasingly urbanized country, with almost half its population living in towns and cities. Cairo, which has some of the features of a global city (Figure 8.25), has expanded to the foot of the pyramids, and its population in 2000, including Giza, was nearly 11 million. Some of its new suburbs ease pressures in the central area (Figure 8.26). Alexandria, the main Egyptian port, had over 4 million people in 2000. Sudan had only 27 percent of its population in towns, of which Khartoum, the capital, was by far the largest with 2.7 million people.

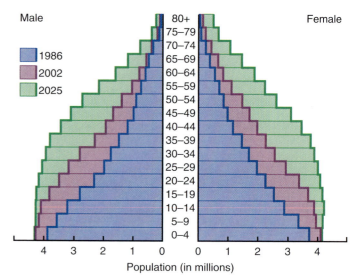

Figure 8.24 **Egypt: age-sex diagram.** Source: U.S. Census Bureau. International Data Bank.

Figure 8.25 **Nile River Valley: Cairo, Egypt.** Cairo, on the Nile River, is a modernizing city—the political and economic center of Egypt.

Economic Development

Egypt was the one country in northern Africa that doubled its GDP in the 1990s. Sudan is poor and still has a tenuous hold on external economic links because of its narrow economic base, internal conflict, and isolation from major transportation routes.

Economic Diversification in Egypt

Egypt's agriculture moved from concentrating on cotton exports at the beginning of the 1900s to increasing food production during the 1950s, although it is still not self-sufficient in food production. After independence, the landholdings were controlled in size to allow more people to share Egypt's most crucial resource. The Agricultural Reform of 1952 limited holdings to 80 hectares (200 acres), and in 1969, the largest were reduced to 20 hectares (50 acres). The irrigated area rose steadily until the 1970s, and further expansion took place outside the Nile River valley in the early 1980s at a time when Egypt was able to use more than its agreed-upon share of water. Productivity increased as the application of fertilizers and double cropping allowed some areas to harvest two crops within a year. Egypt remains a major world exporter of cotton and produces sugar, rice, vegetables, and fruit. After 1987, however, some land was withdrawn from cultivation because of subsidence near the Nile River delta coast along with upstream salinization, waterlogging, and shortages of water. Such withdrawals highlighted the limits of Egypt's land resources.

The main hopes for Egyptian development based on the generation of electricity at Aswan and its potential to supply power for manufacturing were not fully realized. The manufac-turing base of iron, steel, and chemicals production, together with the assembly of cars, food processing, and bus, textile, tire, and television manufacturing, is more diverse than that of most other African countries, but the factories are aging. The largely nationalized industries need modernization and considerable investment. Egypt's armament production received a boost in the Iraq-Iran War (1980–1988) but was less busy in the 1990s.

However, Egypt received new income from these major fluctuating sources. First, Egypt's agreement to a cease-fire and closer relations with Israel earned it massive aid from the United States, amounting to $10 billion per year, or 10 percent of Egyptian GDP, at its height. This made it second to Israel in U.S. aid. The Gulf countries paid for Egyptian military

Figure 8.26 **Nile River Valley urban landscapes: new suburbs.** High-rise apartment blocks in the Maadi district of Cairo, Egypt. Photo: © Derek Berwin/The Image Bank.

involvement in the Gulf War. When President Anwar al-Sadat switched the focus of his country's external relationships from the former Soviet Union to the United States, Egypt became the forerunner of other Arab countries—a pragmatic rather than an ideologic policy. Through the 1980s and 1990s, this switch brought economic benefits to Egypt, but future continuation depends on the world and regional political situations.

Second, the wages sent home from 2 million Egyptian workers in the Persian Gulf equaled agriculture and manufacturing as a source of income for a few years in the 1980s. They declined, however, when over a million workers with their families returned to Egypt during the Gulf War. Few went back to the Gulf in the 1990s. Within Egypt, the Nile River valley upstream (to the south) of Cairo is designated "Upper Egypt." Incomes in Upper Egypt are half those of the rest of the country, and this area has the poorest measures of infant mortality, adult literacy, health services, schools, and unemployment. Many Egyptians who worked abroad came from Upper Egypt, and those who returned at the time of the Gulf War came back to a penniless area. Much of the civil unrest in Egypt and nearly all the terrorist killings occur there.

Third, tourism based around Egypt's historic treasures, such as the pyramids and temples along the Nile River, formed another growth area until Islamic fundamentalists began to attack foreigners in the 1990s, causing a major slump in 1993. By 2000, over 5 million tourists visited Egypt, up from 1.25 million in 1980, but the September 11, 2001, events halved numbers in the 2001–2002 season. The Suez Canal revenues, static for a long time at just below $2 billion per year, also fell after September 11 as world trade slowed for a while, and the low level of Egypt's oil exports does not make up for these losses of income.

Sudan's Economic Plight

In contrast to Egypt, Sudan never had much support from the United States or the Soviet Union, and the little it received from other Arab sponsors fell away after the Gulf War. Any wages sent home by its small number of workers overseas also declined in the 1990s. Economic disaster was enhanced by the attempts to integrate the southern part of Sudan into an Islamic state by force. Fifteen years of civil war destroyed production in the south and diverted northern resources to military expenditure. The internal policies, civil war, and failure to implement required reforms interrupted external funding from the World Bank.

Until the 1990s, Sudan's main products were from farming. Cotton provided 50 percent of exports in better times, and the southern part of the country produced livestock products for export. Although Sudan is normally able to feed its population, civil war, drought, and the pressures of refugees from Ethiopia combined to cause famine from the 1990s. Sudan's manufacturing sector is at an early stage of development. Textile factories, paper mills, and an oil refinery, all based around Khartoum, contrast with small-scale production for local needs in the rest of the country and together make up a small part of the country's total product.

Discoveries of oil in Sudan in the late 1990s enabled the country to supply its home market and export enough to offset

Test Your Understanding 8B

Summary North Africa has close ties with the former colonial powers in Europe, particularly France. It continues to supply it with energy but faces economic difficulties as the European Union expands to include countries that market similar agricultural products.

The Nile River Valley countries of Egypt and Sudan still rely heavily on the river's water, but the supply cannot be increased to match the population growth. Economic diversification in Egypt scarcely keeps income above population growth rates. Civil war slows development in Sudan.

Questions to Think About

8B.1 Which five geographic characteristics do you think make each of the North African countries distinctive?

8B.2 Outline the causes of the annual Nile River flood. Why should Egyptians be pessimistic about their future water supply?

Key Terms

medina   Nile Waters Agreement

one-fourth of its import costs after a pipeline to the coast was opened in 1999. The oil deposits occur in the middle White Nile River basin in southern Sudan. Chinese, Malaysian, Canadian, and Swedish oil companies share the development. Although oil may be part of the solution to Sudan's problems, it also brings new problems. In 2001, there was evidence that the Sudanese government used the oil income to fund armaments purchases, killed or removed inhabitants from the oil-producing area and oil pipeline zone, and used the cleared area as a base for attacking other parts of the south.

Arab Southwest Asia

Arab Southwest Asia is the heart of the Arab and Islamic worlds. It comprises the Arabian Peninsula and the Fertile Crescent that includes the Tigris-Euphrates River basin and the Lebanon coast (Figure 8.27). The world centers of Islam that are focuses of Muslim pilgrimage are located in the Arabian Peninsula at Mecca, where Muhammad was born in A.D. 570, and at Medina, which became his power base after he was expelled from Mecca. Despite having small numbers of people, the countries of this subregion play a major part in world affairs because of their oil wealth and involvement in the Arab-Israeli peace process.

Oil Wealth

Arab Southwest Asia contains two-thirds of the world's known oil reserves (see Figure 8.5), giving it a huge strategic significance within the world economic and political systems. Sales of oil and oil products brought great wealth to some countries from the late 1970s, when world oil prices quadrupled.

Figure 8.27 **Arab Southwest Asia and Israel: countries, physical features, and cities.** The Fertile Crescent stretches from the Mediterranean coastlands through Syria and Iraq to the Persian Gulf.

Although new discoveries of oil reserves more than compensate for current extraction, most of the 1990s were not easy for even the wealthiest oil countries. After building huge financial reserves when oil prices rose during the 1970s, the oil-rich countries of Arab Southwest Asia took on major commitments to increase military strength, build internal infrastructure, and provide welfare services. They made donations to the oil-poor Arab countries. Paying for the Gulf War (1990–1991) and subsequent continuing military expenditures at a time of low oil prices drained their financial reserves, forcing cutbacks in expenditures by the mid-1990s. Such cutbacks caused their inhabitants to question the quality of leadership, particularly where hereditary princes continued to maintain extravagant lifestyles and remain unresponsive to their people. The poorer countries received fewer donations from the wealthier and missed the money sent home from their citizens who worked in the oil-rich countries but returned home during the Gulf War. The late 1990s and early 2000s, however, brought relief to the oil-producing countries in the form of higher oil prices.

By the late 1990s, faced with continuing U.S. punitive actions against Iraq, many of the Gulf countries began to regret and reduce their support for such Western intrusions in an Arab country. They wished to demonstrate their independence from the United States in the face of public sympathy toward the impoverished Iraqis. By late 2002, however, the other countries encouraged Iraq to accept U.N. weapons' inspectors once again.

Countries

The Arab Southwest Asia countries of Bahrain, Iraq, Jordan, Kuwait, Lebanon, Oman, Qatar, Saudi Arabia, Syria, United Arab Emirates (UAE), and Yemen are differentiated by natural resources of oil and water, emphases within Islam, and forms of government. The differences in natural resources and economic management produce a wide range of economic status—from countries that remain poor to those that rival the world's wealthiest (Figure 8.28).

Oil and Water

Some countries, including the Gulf countries of Bahrain, Kuwait, Oman, Qatar, Saudi Arabia, United Arab Emirates, and Yemen, produce oil, but have little water. Lebanon has some water but little or no oil; Iraq and Syria have both oil and water; Jordan has little of either. These differences in crucial natural resources cause tensions among countries that share an Islamic faith and Arabic language.

Although major oil producers redistributed some of their oil wealth in the 1970s and early 1980s, the donor countries found that low oil prices and indebtedness in the late 1980s and 1990s forced them to reduce such support. Tensions over delivery and control of the aid grew between the donor Arab countries (mainly Gulf countries) and those that receive aid (Jordan, Lebanon, Syria, and Yemen). Other tensions arose from the lack of both skilled and unskilled labor in the oil-rich Gulf

Figure 8.28   **Data for the Arab Southwest Asia subregion, Iran, and Turkey.**

Country	Capital City	Land Area (km²) Total	Population (millions) Mid-2001 Total	2025 Est.	GNI 1999 (US $ million) Total	GNI PPP 1999 Per Capita	Percent Urban 2001	Human Development Index: Rank of 175 Countries	Human Poverty Index: Percent of Total Population
Bahrain, State of	Manama	680	0.7	1.7	No data	No data	88.0	37	9.8
Iraq, Republic of	Baghdad	438,320	23.6	40.3	No data	No data	68.0	125	No data
Jordan, Hashemite Kingdom of	Amman	89,210	5.2	8.7	7,717	3,880	79.0	94	9.8
Kuwait, State of	Kuwait City	17,820	2.3	4.2	No data	No data	100.0	35	No data
Lebanese Republic	Beirut	10,400	4.3	5.4	15,796	No data	88.0	69	11.3
Oman, Sultanate of	Masqat	212,460	2.4	4.9	No data	No data	72.0	89	23.7
Qatar, State of	Doha	11,000	0.6	0.8	No data	No data	91.0	41	No data
Saudi Arabia, Kingdom of	Riyadh	2,149,690	21.1	40.9	139,365	11,050	83.0	78	No data
Syrian Arab Republic	Damascus	185,180	17.1	27.1	15,172	3,450	50.0	111	20.1
United Arab Emirates	Abu Dhabi	83,600	3.3	4.5	No data	No data	84.0	43	17.7
Yemen, Republic of	Sanaa	527,970	18.0	39.6	6,088	730	26.0	148	49.2
Israel	Tel Aviv	21,060	6.4	8.9	99,574	18,070	91.0	23	No data
Iran	Tehran	1,648,000	66.1	88.4	113,729	5,520	64.0	95	20.4
Turkey	Ankara	779,450	66.3	85.2	186,490	6,440	66.0	86	16.7

Source: Data from *Population Reference Bureau 2001 Data Sheet*; *World Development Indicators,* World Bank, 2001; *Human Development Report,* United Nations, 2001; Microsoft Encarta (ethnic group, language, religion).

countries and from the other countries' reliance on wages sent home by their migrant workers in the Gulf countries.

In the future, as populations continue to increase rapidly, the differences in water distribution may become more vital than the possession of oil reserves. Many countries' water use already exceeds renewable levels. The Tigris-Euphrates River system is the major surface-water source in the subregion but is increasingly the basis of tension between Turkey, where the headwaters rise, and the downstream countries of Syria and Iraq.

Governments

The governments of the countries in this subregion mostly move slowly, if at all, toward democracy. Many of the oil-rich countries remain dominated by ruling families or dictators: Saudi Arabia is the only country named after a family, the Sauds, who still rule with the title of king and support a huge range of related princes. Other Gulf countries have sheikhs, emirs, and sultans with similar roles, and Jordan has a king. Syria and Iraq have military dictatorships. In Lebanon, there is a greater degree of democracy after a destructive civil war in the 1980s and 1990s between the Muslim and Christian groups spurred on by Islamic extremists.

Syria's dictatorship under Hafez Assad lasted for 30 years. The minority Shiite Alawin people took control of the army and the socialist Baath Party. Supported by the Soviet Union during the Cold War, Syria fostered extremist groups that took part in the Lebanese civil war and actions against Israel and European countries. After the Soviet link ended in 1991, Syria's inward-looking economy suddenly sought to enter the global economic system. Privatized businesses expanded their activities. However, the poor banking system and lack of political freedom attracted little foreign investment, and economic

growth stood still as the population increased. Following Hafez Assad's death in 2000, his son Bashar slowly relaxed the government's tight grip on society and the economy.

The tensions in the subregion led to conflicts among the countries. The Gulf War in 1990–1991 resulted from Iraq's invasion of Kuwait. Internal conflicts sprang up in Oman from 1970 to 1975 and in Yemen in the 1980s. The latter resulted in the merger of North Yemen with the Yemen People's Democratic Republic in 1990. The 1994 renewed outbreak of violence between the two parts of Yemen highlighted remaining differences; victory went to the northern Islamists over the southern socialists. In 2002, the Yemeni government avowed support for the U.S. war against terrorism. After siding with Saddam Hussein of Iraq in the Gulf War, it lost foreign aid and saw its workers deported from Saudi Arabia. When the USS *Cole* was bombed in the Yemeni harbor of Aden in late 2000, Yemen was suspected of housing al-Qaeda terrorists, especially as the government scarcely controls a number of rural hideouts.

People

Ethnicity

The people of this subregion are mainly Arabs and Muslims, but there are important variations. Iraq and Syria have Kurdish groups in the north along the Turkish and Iranian borders. Lebanon contains Christian groups as well as Muslims. Yemen contains groups of African origin. The Gulf countries have important groups of Asian people, mainly foreign workers with temporary residence but some who are allowed to stay longer, while Oman has many links from its former trade with eastern Africa (see Chapter 9).

Country	Ethnic Groups (percent)	Languages O=Official	Religions (percent)
Bahrain, State of	Bahraini 63%, Asian 13%, other Arab 10%, Iranian 8%	Arabic, English, Persian, Urdu	Christian 9%, Muslim 85%
Iraq, Republic of	Arab 75%, Kurd 15%, Assyrian, Turkana	Arabic, Kurdish, Assyrian, Armenian	Christian 3%, Muslim 97%
Jordan, Hashemite Kingdom of	Arab 98%	Arabic (O), English	Christian 8%, Muslim 92%
Kuwait, State of	Kuwaiti 48%, other Arab 35%, Asian 9%, Iranian 4%	Arabic (O), English	Muslim 85%
Lebanese Republic	Arab 95%, Armenian 4%	Arabic (O), English	Christian 30%, Muslim 70%
Oman, Sultanate of	Omani Arab 75%, Pakistani 21%	Arabic (O), English, Urdu	Muslim 98%
Qatar, State of	Arab 40%, Pakistani 18%, Indian 18%, Iranian 10%	Arabic (O), English	Muslim 95%
Saudi Arabia, Kingdom of	Saudi 73% (90% Arab), foreign workers 27%	Arabic (O), English	Muslim 100%
Syrian Arab Republic	Arab 90%, Kurd 7%, Armenian	Arabic (O), Kurdish, French	Christian 10%, Muslim 89%
United Arab Emirates	Emiri 19%, other Arab 23%, Asian 50%	Arabic (O), Persian, English	Muslim 96%
Yemen, Republic of	Mainly Arab, with African-Arab and Indian	Arabic (O)	Muslim dominant
Israel	Jewish 82%, non-Jewish (mainly Arab) 18%	Hebrew (O), Arabic, English	Jewish 82%, Muslim 14%
Iran	Persian 51%, Azerbaijani 24%, Kurd 7%	Persian (Farsi), Turkic, Kurdish	Muslim 99%
Turkey	Turkish 80%, Kurd 17%	Turkish (O), Kurdish, Arabic	Muslim 99%

Highly Urbanized

The population of Arab Southwest Asia clusters along the coasts and in the Tigris-Euphrates floodplain (see Figure 8.12). The 11 countries have relatively small populations. In 2001, Iraq, Saudi Arabia, Syria, and Yemen had 17 million to 24 million each. The rest each had populations of 5 million or fewer. Many coastal cities grew historically as trading centers. The main exceptions to this pattern of coastal cities occur in Saudi Arabia around the capital city, Riyadh, and the religious centers of Mecca and Medina. The desert environment causes agricultural settlement to be linked to water availability, leaving large areas with few people, especially in the center of the Arabian Peninsula.

Arab Southwest Asia is highly urbanized. Apart from Syria and Yemen, all countries have over 70 percent of their populations living in towns. This high level reflects the direct shift from nomadism to urbanism in countries that lack farmland, changes brought by the oil industry, and the development of urbanized manufacturing and government employment. Between 1950 and 2001, Kuwait and Qatar went from 50 percent to over 90 percent urban, while Saudi Arabia went from 10 to 83 percent urban. The largest cities are all capitals, including Baghdad (Iraq, with 5 million people in 2000; Figure 8.29), Amman (Jordan, 1.5 million), Beirut (Lebanon, 2 million), Riyadh (Saudi Arabia, 3.8 million), Damascus (Syria, 2.3 million; Figure 8.30), Abu Dhabi (UAE, 1 million), and Sanaa (Yemen, 1.3 million). Other large cities include the port of Basra and the oil center of Mosul (1 million) in Iraq; Irbid in Jordan; the cities of Jeddah (1.8 million), Mecca (1 million), and Medina in Saudi Arabia; Aleppo (Halab, 2 million) in Syria; Dubai in UAE (Figure 8.31); and Aden, the port and commercial center of Yemen. Of these, Abu Dhabi, Dubai

(both in UAE), and Riyadh have some characteristics of global cities.

The rate of urban expansion means that many cities are dominated by new buildings. Amman, Jordan, lost most of its historic small-town character as it grew to be a capital city with over a million people; land speculation combined with lack of planning replaced the old town with wide avenues and featureless blocks. In Kuwait City after the destruction of the 1990–1991 Gulf War, only the ruler's palace remained of the old town.

As traditional trading and craft industries gave way to manufacturing and service industries, built-up area landscapes changed

Figure 8.29 **Arab Southwest Asia: Baghdad, Iraq.**
Modern bridge over the Tigris River, with view looking from city center to new buildings on the west bank of the river. Photo: © Guy Thouvenin/Photo Researchers, Inc.

Figure 8.30 **Arab Southwest Asia: Damascus, Syria.** View across suburbs to the multistoried buildings of downtown at the foot of a hill. Photo: © Charles and Josette Lenars/Corbis.

with the construction of prestigious government offices in capital cities, high-rise apartments, hotels, and offices just outside the old centers. Beyond are extensive suburban spreads of housing, factories, and shopping facilities. Many older town areas were abandoned to poorer people when merchants and businesspeople moved to more spacious homes and commercial premises in the new suburbs. The factories and supermarkets changed patterns of working and social movements (see the "Personal View: Oman" box, p. 328). New industrial towns were built near the Gulf oil installations: Al-Jubayl and Yandu in Saudi Arabia, for example, are becoming as large as longer-established cities. Segregation by wealth and social group is common within the cities, since the Arabs often live apart from immigrant workers.

Population Dynamics

Population growth in Arab Southwest Asia became rapid from a low base earlier in the 1900s. In 1970, the total population was around 30 million, growing to nearly 50 million by 1980 and 100 million by 2000. The estimated total for 2025 is 180 million people. Such an increase will place massive pressures on local resources, especially water.

Population growth is maintained by some of the world's highest total fertility rates of over 6.5 children per woman (Oman, Saudi Arabia, Yemen) and immigration. In most of the smaller and richer Gulf countries—Bahrain, Kuwait, Qatar, and UAE—2001 fertility was under 4. Lebanon's fertility rate

Figure 8.31 **Arab Southwest Asia urban landscapes: Gulf waterfront.** Dubai, United Arab Emirates, where modern office and apartment blocks line Dubai Creek. Photo: © Christine Osborne/Corbis.

is 2.5—recovering after a long period of civil war. High birth rates combined with falling death rates result in rapid population growth (see Figure 8.17), large proportions of young people, and life expectancies of around 70 years in most countries. In Yemen, life expectancy remains lower, at 60 years. In Iraq, it fell below 60 after years of war and deprivation.

In many Arab Southwest Asian countries, the high population growth reflects high immigration rates. The outlay of rapidly growing oil income on construction works required more labor than the many small oil-producing countries possessed. Instead of developing skills internally, they imported labor from other parts of the region and from South and East Asia. Foreign labor made up nearly 30 percent of the total labor force there in the mid-1980s. Jordan, Lebanon, and Yemen exported many workers to the Gulf countries and depended on the wages they sent home.

Immigrant labor formed large proportions of the population in Kuwait, Qatar, United Arab Emirates, and Saudi Arabia. Figure 8.32 shows how the presence of male foreign workers affected the age and sex structure of Kuwait. The reconstruction of Kuwait after the 1991 Gulf War also depended on immigrant labor. Although the origins of foreign labor shifted from other Arab countries to South Asia in the 1990s, the total remains high. In 1995, estimates put the number of migrant workers in the Gulf countries at around 1.6 million, over 90 percent coming from India, Pakistan, Bangladesh, and Egypt.

Economic Development

Arab Southwest Asia has major economic contrasts between countries with high oil revenues and those that produce little or no oil. The different levels of income are reflected in the ownership of consumer goods (see Figure 8.11) and total income. The Saudi Arabian economy produces about three times the total income of any other country in the subregion ($139 billion in 1999), while Jordan, Lebanon, Syria, and Yemen have GDPs of $15 billion and under per year.

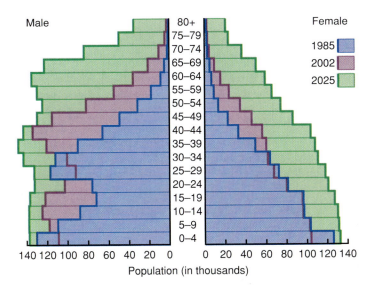

Figure 8.32 **Kuwait: age-sex diagram.** Source: U.S. Census Bureau, International Data Bank.

The richer countries bordering the Persian Gulf (Bahrain, Kuwait, Oman, Qatar, Saudi Arabia, and UAE) sit on huge oil reserves, and exploiting them brought high incomes to the population in the 1970s and early 1980s. Then as oil prices fell until the late 1990s, incomes fell. They have small total populations and rely on immigrant labor. Their oil wealth was not distributed widely or evenly among their populations, either through increased wages or social services such as education and health care. Women have restricted roles in the economies in many of these countries. The growing numbers of educated young people find declining local job opportunities as government income from oil falls and more people are competing for those jobs. The menial, low-wage jobs and much of the commerce, especially in retailing, remain the province of the Indian and wider Asian diasporas.

The countries of Lebanon, Jordan, and especially Yemen are much less developed. They have more available labor and some capacity for agriculture. Iraq and Syria are the exceptions to this rich-poor duality, since until the mid-1980s, they both had oil revenues, significant water resources and agriculture, and large labor forces. The Iran-Iraq War, Gulf War, and subsequent UN sanctions, however, almost destroyed Iraq's economy.

Impacts of Fluctuating Income on Oil-Producing Countries

The exploitation of oil brought huge changes to many countries in the subregion. Previously, the lands of Arab Southwest Asia had economies based on low-intensity farming wherever water was available, pastoralism where water and vegetation were scarce, and extensive empty desert areas. A few crops, such as dates and citrus fruits, were exported from the best-watered places in southern Iraq and along the Mediterranean coasts.

The income of oil-producing countries changed their economies by adding wealth but creating dependence on oil. OPEC's actions in the 1970s that resulted in huge rises in the price of oil brought the newly rich oil-producing countries of the Gulf to a strategic place in the global political and economic systems. In 1965, Saudi Arabia's GDP was $2 billion at today's price equivalents, but it rose to $156 billion in 1980, when oil prices reached their highest point, fell to $105 billion in 1990, and then rose to $139 billion in 1999 as oil prices recovered again partly because of OPEC pressures. Kuwait's GDP rose from $2.1 billion in 1965 to $28 billion before the Gulf War in 1990.

Internally, the income from oil enriched the ruling elites and was used partly to generate industrialization, intensify agricultural output, provide more and better roads, airports, health services, and education, and increase living standards. Up to one-third in some countries went to purchasing military hardware. The oil income made it possible to shift economies toward sustainable development based on diversification. A **diversified economy** is one in which manufactured goods are more important than primary products and a variety of manufactured products is joined by a growing service sector (see Chapter 2).

As world oil prices fell in the 1980s, many of the oil-producing countries borrowed heavily to maintain the levels of internal investment that had been possible in the 1970s. Falling oil prices in the 1980s and 1990s, for example, cost Saudi Arabia

$2.5 billion per year for every $1 drop in the price per barrel; its GDP in the early 2000s was half that at the peak in the early 1980s. As crude oil and oil products still make up between 80 and 95 percent of exports from many oil-producing countries that have not greatly diversified, the price declines devastated internal programs of change, although military expenditures fell less.

Kuwait, a country with a very small population, had less need or opportunity to invest its huge oil wealth internally, and by the late 1980s, its overseas investments generated more income than oil sales. Such income was the major source of Kuwaiti rebuilding funds for several years after the destruction in the Gulf War of 1991. Iraq's oil production was also reduced by military action during the Gulf War. Following the war, with continuing international suspicions over the Iraqi production of nuclear, biological, and chemical weapons, heightened by the refusal of Saddam Hussein's regime to allow UN inspections, Iraq was subjected to sanctions that prevented exports and imports apart from food and medical supplies. After 1994, Iraq could have sold some oil on world markets but did not choose to do so until 1998, and in 2002 it threatened to withhold its quota in protest against Israeli oppression of the Palestinians. Named as one of the "Axis of Evil" rogue countries by U.S. President George W. Bush because of its refusal to allow UN inspections, Iraq expected further attacks from the United States and its allies. However, Iraq would likely prove more difficult to overthrow than Afghanistan. Its army was stronger and its internal opposition (the Shiites in the south and Kurds in the north) weaker. Turkey remained opposed to arming the Kurds, who also form a strong and aggressive community in Turkey and seek an independent country of their own.

Meanwhile, by the late 1990s, Kuwait's government largely restored the oil production that was wiped out in the 1991 Iraq invasion. The 742 oil wells torched by the retreating Iraqi armies were repaired and output restored to pre-1990 levels. The costs of reconstruction and funding the Gulf War coalition, however, reduced the country's overseas assets from $100 billion to $35 billion. As the oil output increased, the employment and social welfare conditions of its people improved. The dominance of the ruling Al-Sabah family in both political and economic areas was reduced by an outspoken legislature and demands for the privatization of Kuwait Airways and the government oil company. Other Gulf countries also faced major changes. During the 1980s and 1990s, some smaller states, such as Bahrain, Qatar, and Oman, began to experience falling oil production, while the UAE had a dramatic rise in output. Saudi Arabia took over the position of leading oil producer from Iraq and Kuwait.

The future may bring more income to those countries as OPEC tries to raise world oil prices by restricting production. In addition, the massive Gulf oil reserves contrast with the relatively small reserves being used up rapidly elsewhere in the world. Unless alternative energy sources are developed economically, the Gulf countries could again have the power to set world prices.

Diversification in Oil-Producing Countries

The oil-exporting countries spend a large proportion of their income on food imports, a need that grew as the population increased. Although major investments were made in home-based irrigated agriculture, costs are very high and production is uneconomic compared to world prices.

Manufacturing development began with the building of oil-refining capacity to replace the direct export of crude oil to European refineries. Petrochemicals and other "downstream" activities based on the materials produced from oil took advantage of the value added by processing the raw materials: the value of the processed materials is greater than the combined cost of the materials and the manufacturing. Bahrain and UAE used the local energy source to operate aluminum smelters; Oman built a copper smelter, while Qatar and Saudi Arabia produce steel. Several countries also manufacture construction materials, including cement, to provide the materials for upgrading roads and building new housing and factories. Further diversification into food and consumer goods industries increased the proportion of GDP from manufacturing. Services also became important in many of these countries where banking and government employment are increasing rapidly.

From the 1980s, the Gulf countries invested some of their oil revenues in higher education facilities, and the many graduates are now employed in banking, education, health care, government, and new industries, including the media and information technology. In 2002, Dubai's Internet City had 200 firms employing 4,000 workers and linked similar centers from Europe to Bangalore, India. The adjacent Media City, Festival City, conference center, expanding major airport and booming Emirates airline, luxury housing, and high-quality tourist facilities all signal Dubai's investment in the future and make it an example for others, although it threatens to take up most of the Gulf's business opportunities.

In Saudi Arabia, new industrial towns are being developed at Al-Jubayl on the Gulf coast and at Yanbu on the Red Sea. Al-Jubayl is halfway toward its planned population of around 300,000 and has major industrial zones with airport, port, and highway linkages. It will cover an area the size of a large U.S. city such as Atlanta. From the late 1990s, Saudi Arabia began to allow foreign investment, including the construction of a giant desalination plant by Japan's Sumitomo Corporation, the building of 3,000 schools by a U.S. consortium, and natural gas exploration linked to power, desalination, and petrochemical developments by international oil corporations. Yet the Saudi Arabian government still controls two-thirds of the internal economy, and its refusal to open internal markets denies it a place in the World Trade Organization.

In 1981, the Gulf oil countries, except for excluded Iraq, formed the **Gulf Cooperation Council** in the context of the Iran-Iraqi War. It focused on common economic and political interests. Saudi Arabia dominates the organization, which was effective in bringing together other countries to resist the 1990 Iraqi invasion of Kuwait.

Tourist and Pilgrimage Sites

The hot, sunny climate and status as a major stopover for international passenger airlines from Europe to Asia, as well as the cultural attractions of the subregion, formed the basis for a growing tourist industry in some countries that contain religious centers or have a welcoming political stability. Bahrain and UAE each had over 2 million tourists in 2000 following international marketing of their sunny climate and shopping opportunities. Tourist numbers tripled in the 1990s.

Although it refused entry to tourists, Saudi Arabia had 4 million foreign visitors in 2000. Some were businesspeople, but most were pilgrims destined for the Islamic sites in Mecca and Medina that every Muslim wishes to visit at least once in their lifetime.

Countries with Little or No Oil

Syria has small oil deposits in its eastern area that are its major export at present. Failure to find new sources will lead to the end of oil income in the early 2000s. Syria also has lands watered by the Euphrates River on which its farmers grow cotton, cereal grains, and fruit. It is not self-sufficient in food, however. After centuries of soil erosion on its northern hills, the slopes are being reseeded to expand local sheep farming. In the 1990s, Syrian markets were opened to foreign products, and private enterprise and exports were encouraged. Syria's tourist industry grew in the 1990s with the greater openness to foreign visitors, rising from 0.5 million in 1990 to 1 million in 2000. The interest is particularly in religious and historic sites, and most tourists come from Europe.

Jordan, Lebanon, and Yemen continue to have small economies with limited prospects of expansion. When Israel appropriated the West Bank area in 1967, Jordan lost half its agricultural land and much of its water and labor. Jordan's high debts led to austerity measures and social unrest. Exports of phosphate, its main source of income, are affected by fluctuating world prices. Attempts to reestablish manufacturing industries around Amman when Israel took the West Bank were not successful. Aid income from oil countries supported Jordan as a "frontline state" adjoining Israel. After the Gulf War, Arab countries withdrew their aid, overseas workers returned home, and a flood of refugees arrived from Iraq. As in Syria, Jordan's tourist industry expanded in the 1990s, from 0.5 million to 1.4 million visitors, with special interest in historic sites such as the rock-hewn city of Petra.

In 1986, Yemen experienced a political crisis that led to uniting the two parts of the country in 1990. The discovery of oil in the north gave hopes of an improving future, but renewed internal conflict in 1994 showed that major difficulties still stand in the way of concerted efforts to improve the economy. Yemen remains the poorest country in the subregion.

Lebanon was an agricultural country before the civil war that began in 1975 tore it apart. Its mountainous landscapes and banking system had earned it the title, the "Switzerland of the Middle East." Most of its income was generated from banking and financial corporations based in its capital, Beirut, the commercial center for much of the region and an early global city. This status was lost when many Beirut financiers fled abroad during the civil war, although some are beginning to return. By the early 2000s, Beirut was being rebuilt (Figure 8.33), and much of the financial capital

(a)

(b)

Figure 8.33 Arab Southwest Asia: Beirut, Lebanon. (a) Devastation of the city during the civil war in the 1980s. (b) The world's largest rebuilding project: work being carried out by the Solidère Corporation in Central Beirut, late 1997. Photos: (a) © SuperStock, Inc.; (b) © Will Yurman/The Image Works, Inc.

that went abroad during the civil war was returning. Farming production was being restored in the coastal plain (fruits, vegetables, tobacco) and the Bekaa Valley (grains).

Prospects for Regional Interchanges

The economies of all Arab Southwest Asian countries are linked closely to the world's wealthier countries and their demands for oil and other products. Although an Arab Common Market was established in 1965, only Iraq, Syria, Jordan, and Egypt are full members. Thus, the countries of Arab Southwest Asia have little trade with each other. They remain vulnerable to price instability in world oil markets, and some face European Union protectionism against agricultural imports. In the 1990s, increases in trade with East Asia offset oil exports and stimulated hopes that markets there might develop further, reducing dependency on Europe and the United States. Egypt, Israel, Jordan, and Syria are at present considering new forms of economic cooperation with each other. Iraq is now isolated by United Nations sanctions but could make a major mark on the world economy when sanctions end.

Israel and the Palestinian Territories

Israel is a major anomaly in southwestern Asia. It is a unique example of a country created by the United Nations for a particular ethnic group, despite opposition from those living in and around it (Figure 8.34). Subsequent territorial disputes, terrorism, and the creation of refugee groups have major geographic impacts locally and implications throughout the world. The Palestinian territories of the West Bank and Gaza are Israeli-occupied territories following conquests in the 1967 war, although the United Nations ruled that they should be returned to Syria, Egypt, and Jordan. This subregion is marked by a dual society of different opportunities for Jew and non-Jew.

In 1998, Israel celebrated 50 years of very hard work, growing prosperity, democratic government, and the settling of more than one-third of the world's 15 million Jews in a separate homeland. Still surrounded by antagonistic countries and subsidized heavily by the United States, Israel has a strong economy and a powerful military, including a range of nuclear weapons. None of the Arab countries has the same firepower, although Iraq has attempted to build it.

Countries

Origins of Israel as a Modern Country

Israel is a symbol to many Jews of a safe haven from persecution. For some, it reestablishes a religious national way of life after centuries of dispersal; to other Jews, it is a distinctive secular ethnic community. Many Jews place the making of a religious or ideological statement above achieving economic growth for its own sake. To Palestinians, Israeli Arabs, and increasing numbers of Muslims worldwide, the establishment and defense of Israel and the need to free Palestinians focuses their need to assert an Islamic presence.

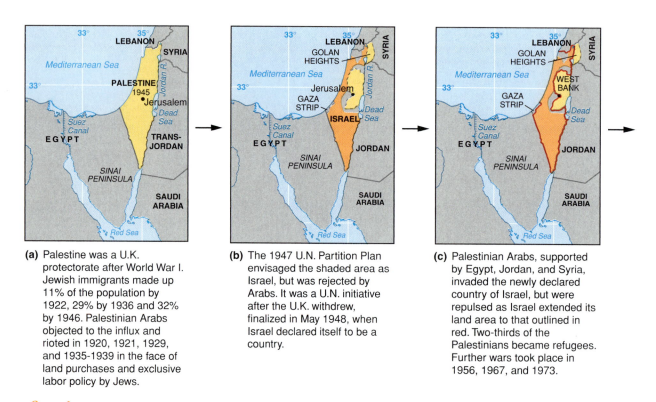

(a) Palestine was a U.K. protectorate after World War I. Jewish immigrants made up 11% of the population by 1922, 29% by 1936 and 32% by 1946. Palestinian Arabs objected to the influx and rioted in 1920, 1921, 1929, and 1935-1939 in the face of land purchases and exclusive labor policy by Jews.

(b) The 1947 U.N. Partition Plan envisaged the shaded area as Israel, but was rejected by Arabs. It was a U.N. initiative after the U.K. withdrew, finalized in May 1948, when Israel declared itself to be a country.

(c) Palestinian Arabs, supported by Egypt, Jordan, and Syria, invaded the newly declared country of Israel, but were repulsed as Israel extended its land area to that outlined in red. Two-thirds of the Palestinians became refugees. Further wars took place in 1956, 1967, and 1973.

Figure 8.34   **History of modern Israel: a timeline of changing maps.** Source: © The Economist Newspaper Group, Inc. Reprinted with permission. Further reproduction prohibited. www.economist.com

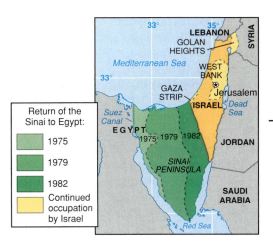

(d) After the 1967 war, Israel extended its frontiers to include the West Bank (from Jordan), the Sinai Peninsula (from Egypt), and the Golan Heights (from Syria). In 1973 a Syrian and Egyptian attack failed, leading to a desire for a diplomatic settlement, the 1978 Egypt-Israeli Camp David Peace Accord in which Israel returned Sinai. In 1978, Israel attacked refugee camps in southern Lebanon.

(e) After the Accord, Israel returned Sinai to Egypt in a phased withdrawal. It held on to the West Bank and Golan Heights.

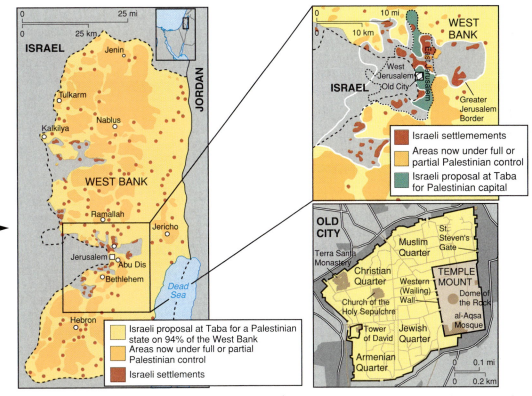

(f) Prolonged Palestinian and Arab resistance in the late 1980s gave way to diplomacy after the Gulf War (1990–91). From 1993, the United States fostered talks that led to partial Palestinian control of West Bank and Gaza. Israel expected to return over 90% of these lands to form a Palestinian country.

(g) The Jerusalem area with proposals that were moving forward in 2000, before the later hardline position of Ariel Sharon's Israeli government ended diplomatic means.

The idea of a separate country for Jews arose out of Zionism, a movement of the 1800s that sought to link Jews in a common secular country based on political self-help. Increasing anti-Semitism, including the imprisonment and murder of Jews, led wealthy Jews to finance initial waves of settlers, numbering 60,000 from 1880 to 1914, in Palestine. The settlers taught the modern Hebrew language in schools and established socialist institutions such as labor organizations and farming in communal **kibbutzim.** In the kibbutzim, land is communally owned and decisions are made collectively. They aimed to prepare for mass migration into a new, independent country as a safe haven from a persecuting world.

During World War I, the United Kingdom, France, Italy, and Russia made secret agreements to divide the Ottoman Empire after the defeat of Germany and its allies. Of these, the Sykes-Picot Agreement recommended internationalizing Palestine and giving much of Turkey to Russia, the Persian Gulf area to the United Kingdom, and Syria-Lebanon to France. The United Kingdom negotiated separately with Arab leaders, promising them a separate country on lands won from the Turkish Ottoman Empire.

The 1923 Paris Conference settlement incorporated the 1917 British Balfour Declaration that favored "the establishment in Palestine of a National Home for the Jewish people." It proposed that the British carve an area for this purpose from within its protectorate of Palestine. Arabs agreed to the principle on the basis of their obtaining a new Arab country out of the old Ottoman Empire lands that Turkey lost in World War I. However, many Arabs expressed disappointment when the future Saudi Arabia was the sole politically independent Arab country and the rest of the former Ottoman Empire was divided into separate protectorates under European ruling countries.

As the political futures of parts of Southwest Asia were being decided, many Jews fled Nazi Germany and Europe. Before 1940, 350,000 Jews moved to Palestine despite prohibition under the British mandate (Figure 8.34a). The largely impoverished Jewish immigrants were supported from Germany and donations by Jews in the United States. After World War II and the genocide of the Nazi holocaust, a million more joined them. Pressure mounted for an independent Israel. Violence among Jews, Palestinians, and British forces increased. The United Kingdom could no longer control the protectorate and handed its jurisdiction to the United Nations, leaving in 1948 even though Arabs had rejected a 1947 United Nations Partition Plan (Figure 8.34b) that was a prelude to the creation of an independent state of Israel.

After the partition plan was rejected, Arab protests led to war when the Jewish settlers declared the independence of Israel in 1948. The Arab forces of Palestine and surrounding countries attacked the Jewish settlers. Israeli forces not only repulsed the attackers but also expanded the country's original territory by the end of the war in 1949 (Figure 8.34c). During the conflict, some 600,000 Palestinian refugees were forced out or fled for their lives and moved mainly into Jordan and Lebanon. By 2000, some 3.6 million Palestinians were living in Jordan (1.5 million), West Bank and Gaza (1.4 million), Lebanon and Syria (600,000 together).

Further Wars and Negotiations

Arab countries refused to accept the existence of independent Israel and fought it unsuccessfully again in 1956, 1967, and 1973. In 1956, Israel assisted France and the United Kingdom in attacking Egypt, and in 1967 the Israelis attacked first amid rising tensions. During the 1967 war, Israel extended its territory southward into the Gaza Strip and Egypt's Sinai Peninsula, eastward across the West Bank area of Jordan, and northward to the Golan Heights of Syria (Figure 8.34d). Despite UN resolutions, Israel refused to give up the occupied lands, apart from Sinai (Figure 8.34e), for security reasons—at least until surrounding countries accepted the existence of Israel and renounced military activity against it. Syria would not acknowledge Israel's existence, so Israel held on to the Golan Heights. Unsuccessful in open warfare, Arab groups, both secular and religious, turned to terrorist methods in the 1970s and 1980s.

In 1994, following the Oslo Accord between Israel and the Palestine Liberation Organization, a Palestinian Authority was established and given limited jurisdiction and autonomy in Gaza and the West Bank. Israeli settlements in these areas (Figure 8.34f) were maintained, however, and their populations expanded. After 1996 changes in the Israeli government, progress toward the development of a Palestinian country halted, as Israel resisted further devolvement of power and transfers of land to Palestinians. Terrorist activity by Hamas increased. The Israeli military entered Palestinian territories to punish people for atrocities, such as suicide bombings, committed against Israelis.

After more than 100 years of conflict over the position and status of Israel, many outside the situation see an almost insoluble problem (see the "Point-Counterpoint: Israelis Versus Palestinians" box, p. 320). Palestinians and other Arabs continue to wish for Israel's destruction, causing Israel to focus on defending itself. Revenge for one set of killings sets off more revenge, leading to economic devastation and cultural warfare that could become widespread.

Continuing Stresses

Israel established its political presence in the region, making the unpromising environment more productive and enhancing the quality of life of its people until it contrasted greatly with that of surrounding Arab countries. Yet, many of its citizens found it difficult to celebrate wholeheartedly their country's 50th anniversary in 1998 as they were still the target of terrorist acts that many Palestinians regard as steps toward freedom. Strong disagreements exist among Israeli Jews over issues that intertwine in practice to generate a pluralistic but polarized society:

- Whether the territories gained in 1967 should be traded for peaceful relations or kept as part of Israel

- How relationships among nationalism, government, land, and religion should be resolved in conflicts between Jewish religious and secular interests

- How to manage the move from a rural focus that made the desert bloom to a high-tech, urbanized country

- How the original melting pot required by national cohesiveness in the face of early enemies can accommodate the divisive ethnic interests of Russian, Moroccan, Western, and Asian Jews as well as Israeli Arabs

The democratically elected Israeli government reflects the complex grouping of interests, conservative and liberal, religious and secular, as well as those of varied ethnic origins. Many parties receive small percentages of the election votes, forcing alliances in the formation of governments and giving small parties scope to assume greater power when voting is close.

Jerusalem presents a major problem for Israeli-Palestinian negotiations. After Israel took East Jerusalem in the 1967 war, it confiscated land in the occupied area and made Palestinians second-class citizens by denying them property rights. Although Israel proclaimed Jerusalem to be its capital in 1950 and built the Gnesset (parliament) in Jerusalem suburbs, the United States and most other countries do not recognize Jerusalem as the Israeli capital and maintain embassies in Tel Aviv.

Limited Resources

Israel, Gaza, and the West Bank occupy a small area of land between the Mediterranean coast and the Jordan River valley (see Figure 8.27). The terrain includes a coastal plain and mostly hilly land with lower areas around Lake Tiberias (Sea of Galilee) and along the Jordan River to the Dead Sea (see Figures 8.6 and 8.8a). Although Israel receives winter rain, the total rainfall is low, and summer drought brings water shortages. Great efforts made it possible to supply water to dry areas and to manage the environment efficiently. Despite such careful management, internal groundwater sources are now fully used, and Israel relies on external sources for 25 percent of its water. Brackish (slightly saline) water is used for some farming but can damage the soils if too much is used over many seasons. Urban water supply relies increasingly on coastal desalination plants.

The water shortage enters politics. Israel retains jurisdiction over the West Bank and the Golan Heights both for defense and access to water. Peace negotiations with Jordan included agreement on the use of the Jordan River basin waters, which have become central in the negotiations over the future of the West Bank.

People

Ethnic Differences

Israel's population is composed of Jews and Israeli Arabs, all having the same rights. The Arabs and some of the Jews trace their family existence in the country to before Israel's independence. The Semitic Sephardic Jews form the majority, but the Ashkenazi Jews of European origin play important roles in politics. Other Jews are of Asiatic origin or arrived from Russia in the 1990s.

Arabs compose 90 percent of the population in the occupied Palestinian territories of the West Bank and Gaza and resent not having full governmental powers. As soon as a terrorist or external threat occurs, Palestinians are confined to their living areas. As Israeli government policy, Jewish settlements were estab-

lished with Israeli government funding in both the West Bank and Gaza (see Figure 8.34f) and populated by hard-line Jews committed to retaining these areas. Lines of Israeli settlements now separate the Palestinian areas within the two territories.

Rural Versus Urban Emphases

The Israeli population density is higher than that of other countries in the region because of the country's small size, special ethnic characteristics, and urban-industrial economy. Most people live along the coastal plain, but political considerations support settlements close to frontiers. The West Bank and Gaza Strip are also densely populated, mainly by Palestinians but with some Jewish settlements. A major source of the tensions between Arabs and Jews is the continued building of Jewish settlements in the Palestinian areas. The intrusive Jewish settlements split the Palestinian territories and cost the Israeli government heavily to build them and maintain their security. Palestinians are mainly restricted to low-wage jobs and denied access to higher-earning professions: Palestinians who wish to become doctors, for example, have to train abroad.

Much of the initial thrust of Israeli settlement was in rural areas on the kibbutzim, motivated by the need to occupy its new territory and by Zionist socialist principles. Kibbutzim provided a spiritual, agricultural, and social basis for the economy and in the formative days of Israel as a country tied families to the land (Figure 8.35). More recently, development of the rural areas has been through cooperative (*moshav*) farms and smallholdings (*moshawa*). Many Israelis, however, moved into the growing towns, and less than 5 percent of the Israeli labor force now works on kibbutzim or *moshav* farms.

Israel changed its goals as increasing numbers of immigrants came from urban areas in other parts of the world and sophisticated manufacturing and commercial functions developed. By 2001, 91 percent of the Israeli population lived in towns. Tel Aviv–Jaffa, with over 2 million people, is the largest urban area, with some features of global cities, while Jerusalem and Haifa are other large cities. Jerusalem has immense significance for Jews, Muslims, and Christians. Relations in and around the city remain tense over such issues as the holy sites and Israeli prevention of Arab movements at times of emergency.

Population Dynamics

Variations in immigration make Israel's population growth irregular. The annual growth rate fell from 2.8 percent in the period 1965 to 1980 to 1.8 percent in the 1980s but rose again in the mid-1990s to 3.3 percent as a result of immigration. In the early 1990s, over 1 million Russian Jews moved to Israel. In 2001, Israeli natural increase was 1.6 percent. Life expectancy was almost 80 years. The 2001 resident Israeli population of 6.4 million included 18 percent non-Jews, most of whom were Arabs. Israel's population structure differs from that of the Arab countries of this region (Figure 8.36) because of the fluctuating arrival of immigrants and its higher proportion of older people.

The Palestinians living in Gaza and the West Bank have a more typical Arab-country population structure and lower life

Figure 8.35 **Israel: communal farming.** Harvesting eggplant on Kibbutz Ein Gedi, located between the desert hills and the Dead Sea. Bananas are in the background. Photo: © George Holz/The Image Works.

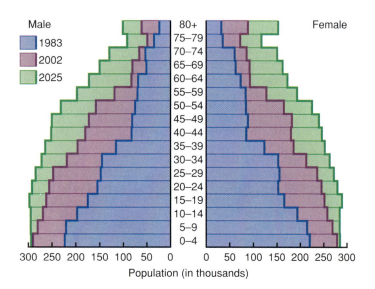

Figure 8.36 **Israel: age-sex diagram.** Source: U.S. Census Bureau, International Data Bank.

expectancy. Gaza has one of the highest total fertility rates in the world, around 6, compared to 3 in Israel as a whole in 2001, and annual natural increase of 3.7 percent.

Economic Development

Israel's economy places its per capita income in the same league as countries of southern Europe. Ownership of consumer goods is high (see Figure 8.11). In the mid-1980s, the steady growth and diversification of the economy stalled in crisis. Israel's large 1985 budget deficit and rapidly depleting currency reserves required U.S. assistance to reduce foreign debts to manageable proportions. Israel remains a highly centralized and heavily taxed country—an outlook supported by workers and managers fearing job losses, an influential group of politicians, and a large bureaucracy.

Diversified, High-Tech Economy

Israel's economy is diversified. The agricultural sector, using intensive reclamation and irrigation farming methods, produces fruits, vegetables (Figure 8.37), and flowers for export to Europe but now constitutes only 5 percent of total GDP.

Israel possesses a well-educated work force and access to foreign aid and investments. Providing a sophisticated defense capability prepared a generation of engineers for work in Israel's high-technology industries. Although the influx of Russian Jews in the early 1990s generated worries about rising unemployment, they brought professional skills and caused a new boom in house building.

Manufactures make up 44 percent of exports and include diamonds, machinery, military equipment, and chemicals. In the 1990s, Israel became a major center and leader of high-technology development in manufacturing areas such as telecommunications, electronic printing, diagnostic imaging systems for medicine, and data communications. Such products accounted for 50 percent of industrial output in the later 1990s, compared to 15 percent in 1990. Most of the manufacturing industries are based around the coastal cities such as Tel

Figure 8.37 **Israel: agriculture.** Sprinkler irrigation on a farm in the Negev Desert of southern Israel. Assess the importance of the presence or absence of water for landscapes in this region. Photo: © Paul Souders/Corbis.

Aviv and Haifa. Industrial estates along the borders with Gaza and the West Bank employ cheap Arab labor.

Services

Israel gains over half of its GDP from the services sector. The quality of education and health-care provision places it on a par with Western Europe and ahead of most Arab countries. The financial sector grows increasingly as Israel's economy opens to privatization and foreign investment.

Tourism is a major industry that attracted over 2.4 million visitors in 2000 but was interrupted by conflict and war. Many visitors combined visits to, for example, Jerusalem, Petra (Jordan), and the pyramids (Egypt), but the 2002 invasions of Palestinian areas caused both Jordan and Egypt to close their borders with Israel and suspend diplomatic relations, effectively ending such combined itineraries.

Diversifying Trade Links

Israel is now actively pursuing free-trade policies and has opened its borders to products from neighboring Jordan and countries farther away in Asia and Latin America. In a peaceful time, Israel could have a leading economic role in the eastern Mediterranean. Conversely, conflict necessitates more arms spending, interrupts trade, and generates long-term hatreds.

At present, half of Israel's trade is with the European Union, and the proportion is rising. Its attempts to join the EU as a trading partner—with access to the free-trade area, rather than as a political partner—move forward more slowly than Israel would like. Israel is particularly keen for its agricultural and telecommunications products to gain equal entry alongside those of European countries. It argues that joining the European free-trade area group of countries would compensate for opening its home markets to Jordanian produce and the future return of the Golan Heights to Syria.

Poverty in Gaza and the West Bank

While Israel has a growing economy that places it ahead of its neighbors in development and lifestyles, the Palestinian areas of Gaza and the West Bank continue to have poorer conditions for human development. Palestinians accuse the Israelis of paying unequal attention to the needs of Palestinians in these territories compared to Israelis—a form of apartheid. However, many Israelis accuse the Palestinians of a lack of enterprise. The contrasted positions and qualities of life form a basis for intergroup violence.

After Israeli military action in the Palestinian lands in the early 2000s, the Palestinian economy disintegrated, leaving Palestinians impoverished. Destruction of infrastructure that had been built with contributions of billions of dollars of international aid in the later 1990s, cessation of economic activity, growing numbers of dependent orphan children and older people without family support, and the loss of almost all the ambulances resulted in a political, economic, and social disaster.

Although Gaza is home to many professional people, its infrastructure is crumbling, 40 percent of its population lives in refugee camps, and at least 50 percent of the work force is unemployed. Access to jobs in Israel and remittances from workers overseas are now reduced and heavily controlled. Many families depend on emergency food supplies. Israeli settlements in the Palestinian territories have a separate existence, occupying one-fourth of the land.

In the West Bank, Palestinian farmers increasingly find themselves close to Israeli settlements that have been built among their villages since 1967. Israeli settlers are attracted by generous financial incentives from the government, which also supports their high quality of life. Tensions are high, but the Palestinian groups opposed to peace, such as Hamas, take no part in consultations and planning. Large sums of money pledged by the United Nations to support the Israeli-Palestinian agreements and provide facilities for Palestinians arrive slowly or await fuller plans for their use to minimize corruption. Some projects—such as the provision of water, transportation facilities, and school buildings—easily attract investments, but funds that can be used for paying teachers, health programs, power provision, and waste disposal are more difficult to obtain, thus threatening the successful management of the projects.

In 2002, in reaction to the "Passover Massacre"—suicide bombing in which about 20 Israelis were killed celebrating the most important Jewish festival—and after months of partial and intermittent destruction of Palestinian facilities, Israeli army units entered West Bank towns including Ramallah and the Jenin refugee camp—by now built up with permanent housing—searching for Palestinian terrorists. The alternation of suicide bombings by Palestinian extremists and repressive Israeli military responses continued with little indication of a peaceful resolution.

⊕ Turkey and Iran

Turkey and Iran occupy the northern and eastern margins of this region (Figure 8.38). The two countries are influential in Southwestern Asia and the wider world, sharing economic leadership of the region with Saudi Arabia, Egypt, and Israel. Both Iran

Figure 8.38   **Turkey and Iran: countries, physical features, and cities.**

and Turkey had 2001 populations of 66 million people—together comprising nearly one-third of the total population of the Northern Africa and Southwestern Asia region (see Figure 8.28).

Turkey and Iran have crucial strategic positions between the southern boundary of Russia and Neighboring Countries (see Chapter 4) and the Persian Gulf oil fields. Their control of three major choke points on sea routes—the Bosporus and Dardanelles in Turkey linking the Black and Aegean Seas and the Hormuz Strait at the entrance to the Persian Gulf—adds to their strategic significance. The increasingly democratic nature of their governments, after years of military and religious dictatorships, contrasts with those of other countries of Southwestern Asia and gives them greater potential for success within the world economic system, although Iran is still the subject of U.S. sanctions.

Iran and Turkey share other characteristics. They are largely mountainous countries along the plate collision margin between Arabia and Asia. This makes them subject to frequent and devastating earthquakes. The mountains also cause the uplift and cooling of humid air, attracting precipitation, much of which falls as winter snow. Meltwater in the spring and early summer feeds rivers flowing southward to the Persian Gulf and northward into the Black and Caspian Seas.

Both countries have Kurdish minorities and must deal with the fight for Kurdish independence, a problem they share

with Iraq and Syria. All these countries resist demands for a separate country of Kurdistan.

Countries

Although Iran and Turkey have some similarities, they are long-term rivals with very different religious emphases, types of government, basic resources, and approaches to the world economy. Iran's main language is Persian (Farsi), and its Muslims belong to the Shiite group, while Turkey's people speak Turkish and are mainly Sunnite Muslims. Memories of historic rivalries color relations between the two countries.

In the late 1990s, Turkey diversified its economy in the context of its secular and democratic government and slowing population growth. In 1979, Iran moved from a military dictatorship to an Islamic constitution, but its suspected support for terrorist activities led to U.S. sanctions. Iran spent major resources on the 1980–1988 war with Iraq and has to contend with a faster rise in population than Turkey.

Historic Rivalries

Although both Iran and Turkey are Muslim countries, both are proud of their historic and pre-Islamic traditions. Modern Iran occupies part of a wider Persian Empire, conquered by the Greeks under Alexander the Great but successful in resisting expansion by the Roman Empire. In the Muslim conquests around A.D. 650, the new Islamic rulers took on Persian court manners. In the 1000s, Turks, followed by Mongols from Central Asia, overran the country—which became a zone of conflict between Turks and Afghans.

In the later 1800s, rivalry between Russia and the United Kingdom raised Iran's significance in world geopolitics. British discoveries of oil in Iran and its extraction from 1901 began the southwestern Asia oil era. During World War II, Britain and the Soviet Union occupied parts of Iran to guard its oil fields against German attack.

Interior Turkey had its own civilization, developed under the Hittites around 1600 B.C. Later, Persians, Greeks, and Romans took it over. It was Christianized as part of the eastern Roman Empire of Byzantium. The Muslim eruptions during the A.D. 600s reduced Byzantine influence to the peninsula that is modern Turkey, but Turks from Central Asia overran most of it after 1000, superimposing Islam on Christian traditions.

The Ottomans led the Turks besieging Constantinople, but it did not fall until 1453. In the meantime, they bypassed this city and, with help from Arab Muslims, conquered much of southeastern Europe (see Chapter 3). In the 1300s, they reversed this direction and attacked the Arab world, bringing it under subjection by the late 1500s. By the late 1800s, this system began to break down, and the alliance of the Ottoman Turks with Germany in World War I led to defeat, destruction, famine, and the death of almost one-fourth of the Turkish population. This defeat freed other areas in Northern Africa and Southwestern Asia and resulted in the formation of the nationalist, secular state of modern Turkey.

Modern Shifts

In the 1900s, Iran was ruled by a shah who had elevated himself from a military career to royal status. He and his successor son kept the country largely under military control but with a semblance of freedom to adopt Western education and develop a wider range of economic activities. Iran's oil wealth and control of the Strait of Hormuz (Figure 8.39) resulted in involvements with the global economy but in less economic diversification at home. In 1979, led by the Ayatollah Khomeini, nationalist religious leaders seized political power, the shah fled, and Iran shifted to Islamic religious leadership and isolationism.

The new constitution of the Islamic republic in Iran depends on a Shiite interpretation of Islamic government. Clergy

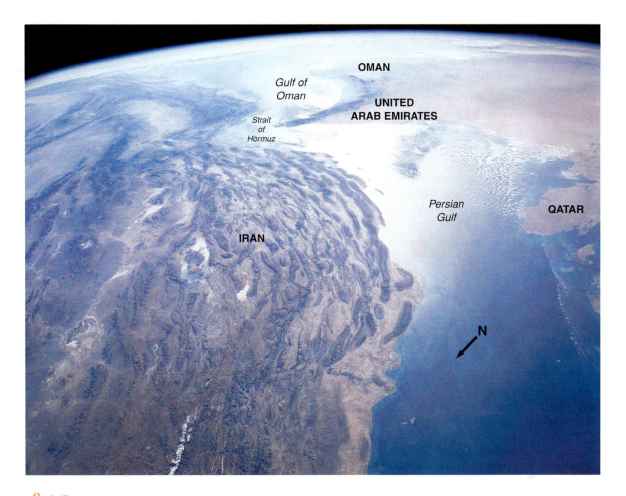

Figure 8.39 **Iran and the Persian Gulf.** Space shuttle view of the Strait of Hormuz with the Zagros Mountains of Iran to the left (north) and the United Arab Emirates and Oman on the south side of the strait. All marine traffic into and out of the Gulf must pass through the strait, which has great strategic significance. Photo: NASA.

are expected to establish a just social system and implement Islamic laws. Four institutions resulted:

- The *Council of Leadership* formed of three to five leaders elected by expert theologians.

- A legislature that includes the *Council of Guardians,* formed of 12 theologians and jurists serving six-year terms to examine legislation and its compliance with Islam and the constitution. It also includes the *Islamic Assembly (Majlis),* which has 270 members elected for four years, with five seats reserved for non-Muslims and two for Armenians.

- The *Executive,* consisting of the president, who is elected for four years and approved by the Council of Leadership, and the Council of Ministers, which the president coordinates. The president appoints a prime minister with *Majlis* approval, and the prime minister appoints the Council of Ministers with *Majlis* and presidential approval.

- The *Supreme Judicial Council* is Iran's highest judicial authority and is composed of a chief justice and state prosecutor appointed by the Council of Leadership plus three judges appointed for five-year terms by their colleagues.

In practice, this system raises tensions between the president and the members of the Islamic Assembly, all of whom depend on a popular vote, on the one hand and the religious priorities of the controlling Council of Leadership on the other. Many Iranians see the latter as obstructionist over internal reforms and external relations, preventing the elected president and *Majlis* from enabling Iran to take a full place in the modern world.

Turkey's political system is very different from that of Iran. After the disastrous defeat of the Ottoman Empire in World War I, Turkey became a nationalist and secular republic, putting the country before religion in questions of government. From the 1920s until 1936, the new leader, Mustafa Kemal Ataturk, ruled with a single-party government. Turkey gradually modernized and became increasingly involved in the global economy, although most economic developments came after World War II, in which it remained neutral. After the war, pressures from the Soviet Union caused Turkey to become a member of NATO (see Chapter 3). The country received U.S. support as it provided locations for U.S. Air Force facilities. The control of the Bosporus made it possible to monitor Soviet ship movements from Turkey's southern flank of the Black Sea. U.S. bases in Turkey were close to the Soviet heartland. Iran avoided close links with the West, however, and developed agreements with the Soviet Union, especially after 1979.

In the 1960s and 1970s, Turkey's actions in support of Turkish people against the majority Greek population on the island of Cyprus resulted in an arms embargo and sanctions against it, as well as sparking resistance to Turkey's membership of the European Union in the 1990s.

Following the breakup of the Soviet Union in 1991, rivalry between Turkey and Iran continued as both sought links with Muslim countries in the former Soviet republics of the Caucasus Mountains and Central Asia. Turkey claims historic and language links with these new countries, but Iran is developing more practical links with the landlocked countries in trade and transportation (see Chapter 4).

In the late 1990s, both Iran and Turkey faced tensions between liberalizing Western influences and pressures from Islamic-oriented political groups. Iranians in particular live with cultural confusions. In Iran, the strictest Islamic rules are imposed by a fundamentalist few on a moderate majority. The election of a more liberal cleric as president in 1997 and again in 2000 resulted in conflicts between his edicts and those of the conservative Council of Guardians. The complexity of cultural differences within Iran is magnified by access to Western media through CDs and DVDs bought in Turkey or Dubai and local websites. Teenagers forbidden to gather with the opposite sex do so on Friday mornings, when the morals police are at the mosque. Many women suffer in marriages where Islamic interpretations of laws by male clerics sanction male brutality, but others do not and are able to manage businesses. In many parts of the country, the heavy Islamic law punishments are frequently avoided or applied lightly.

Turkey faces continuing challenges from Islamic political groups to its long-term secular state principles. When a militant Islamic group was elected to government in the late 1990s, the military took over and installed a secular group in power.

People

Ethnic Differences

Both Iran and Turkey have varied ethnic compositions. Iran's population is just over half Persian, one-fourth Azeri (Shiite Turks who make up the majority in Azerbaijan), and about 7 percent Kurds. Turkey has around 80 percent Turks and 17 percent Kurds. Although Turkish policy toward the Kurds in southeastern Turkey is harsh, many Kurds live fully integrated in communities in other parts of Turkey. In 1999, pro-Kurdish Hadep Party members were elected as mayors in the major cities of Izmir, Adana, and Mersin, being valued for looking after the interests of all constituents, whether Turks, Arabs, Kurds, Jews, or Orthodox Christians.

Urban Growth

The upland plateaus of Turkey and western Iran support a moderate population density. The melting snows of the mountain ranges (see Figure 8.8) water this area in spring. Eastern and southern Iran is mainly desert with low population densities. Government policy in both countries supports the extension of settlements into sparsely inhabited areas.

Both countries have rapidly expanding urban areas, with people migrating to towns from the countryside. In Iran, such migration resulted from greater government investments in urban jobs in the manufacturing and services sectors instead of agriculture. The Turkish urban population increased from 25 percent of the total in 1945 to 66 percent in 2001. Urban jobs increased in Turkey, but agriculture remains important, and many migrate to the largest cities to escape the internal war against the Kurds in southeastern Turkey.

In Iran, where 64 percent of the population was urban in 2001, Tehran (7 million people in 2000), the capital, is the largest city. The isolation of Iran from much of the global economy leaves Tehran with few global city features. Other large Iranian cities include Meshed (Mashdad, 1 million), Isfahan (2.5 million), Tabriz (1.5 million), and the port of Abadan. In Turkey, Istanbul is the largest city and has a population that grew to around 10 million in 2000. Of all the cities in this region, it ranks highest in global city-region characteristics. Ankara (3 million), Turkey's capital, was built after 1923 in the formerly "empty" interior. Izmir (2.4 million), Adana, and Bursa are other Turkish cities with over 1 million people each.

Population Dynamics

Iran's population total caught up with Turkey's from 1980 (Iran 39 million, Turkey 45 million) to 2001 (both 66 million). By 2025, projections suggest 88 million people for Iran and 85 million for Turkey. When the Iranian religious leaders gained power in 1979, they encouraged more births but later reversed the policy. They now find that the preponderance of younger adults is a major source of opposition, both politically and culturally.

The legacy of past increases is large populations that need rapid economic growth to raise material prospects and reduce political pressures. Although birth rates are falling in both countries, death rates have fallen faster (see Figure 8.17). The age-sex diagram for Turkey (Figure 8.40) shows the results of the high birth rates, followed by slowing rates of increase.

Economic Development

The two countries have different types of economic development based largely on oil income (Iran) or water resources (Turkey). While Iranian national income fluctuated after 1980 with world oil prices, Turkey's real income rose steadily to exceed that of Iran in the 1990s—partly as a consequence of lower oil prices and the sanctions affecting Iran. Ownership of consumer goods in Turkey is twice the rate of that in Iran (see Figure 8.11), and Turkey's total 2000 GNI was the highest in the region (see Figure 8.28).

While Iran blocked greater economic involvement with the global system in an attempt to be self-sufficient as an Islamic power, Turkey depended on IMF encouragement and support that demanded reforms of public ownership and trading arrangements. In 2001, Turkey faced economic crisis after internal political disagreements caused a failure to implement the economic reforms that were conditional for the payment of IMF loans. After September 11, 2001, however, Turkey gained renewed Western support as the first Muslim country to offer troops for the United States-led Afghanistan campaign.

Oil or Water

Iran has 10 percent of world oil reserves and experienced rapid oil income growth from the 1960s, reaching 40 percent of the value of national production in the oil boom of the 1970s. Much of the additional wealth was at first spent on encouraging urban-industrial, rather than agricultural, expansion and importing foreign goods. The takeover of the country by religious leaders in 1979, U.S. sanctions, and the war with Iraq in the 1980s led to isolation from the world. The

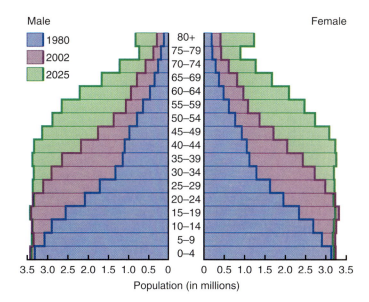

Figure 8.40 **Turkey: age-sex diagram.** Source: U.S. Census Bureau. International Data Bank.

low oil prices of the 1980s and 1990s worsened Iran's financial position. In the 1990s, however, Iran began to look outward and restructure its economy and world links. The country sought links with major multinational oil corporations for marketing and technical updating, but Iran's constitution forbids foreigners from owning mineral concessions. Although Iran possesses other mineral resources, few have been explored or extracted.

Turkey has little oil but invested heavily in the development of its water resources for more agricultural output and hydroelectricity. Industrial expansion increased from 1950 and especially in the 1970s and 1980s. The 1990s were a period of political and economic confusion, partly due to the internal war against the Kurds that absorbed large amounts of money and manpower.

Agricultural Contrasts

In agriculture, Turkey used mechanization and fertilizers to increase yields. Half the working population is engaged in farming, and the opening of new areas to irrigation in southeastern Turkey (Figure 8.41) enabled the country to increase its production of cotton, soybeans, grains, fruits, and vegetables. The agricultural sector produces nearly 20 percent of Turkey's exports by value.

Although a third of Iran's workers are farmers, agricultural production is much less sophisticated than in Turkey, and Iran needs to import much of its food. Water projects in Iran are concerned mainly with urban-industrial requirements. Land reforms in Iran designed to please the peasant farming community often conflict with hopes of modernizing the sector and increasing productivity.

Manufacturing Differences

Iran was the first major oil exporter in the Persian Gulf, but internal economic development was slow until the 1960s. Craft

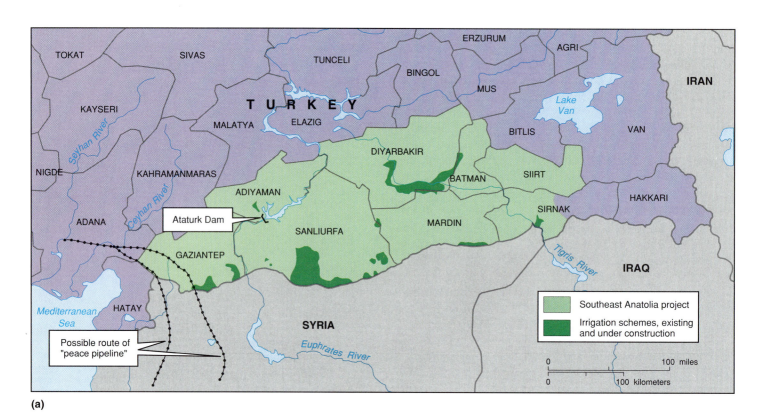

(a)

Figure 8.41 **Turkey: Southeast Anatolian water project.** (a) Storage of water behind huge dams, such as the Ataturk Dam, makes it possible to irrigate large new areas of land and provide hydroelectricity for local industry. Such uses of water, however, deprive Syria and Iraq of water they have been using from the Tigris and Euphrates Rivers. The "Peace Pipeline" was part of a project to supply water to Israel and its neighbors but was not implemented. (b) The Ataturk Dam on the Upper Euphrates River as it neared completion in 1992. Source: (a) © The Economist Newspaper Group, Inc. Reprinted with permission. Further reproduction prohibited. www.economist.com (b) Photo: © Ed Kashi.

(b)

industries, such as carpet making and weaving, remain significant. Oil revenues, which now provide 90 percent of exports, were used to fund public-sector manufacturing ventures such as oil refineries and petrochemical plants on the Gulf coast, the iron and steel works at Isfahan, and machine building at Arak. In the 1960s and 1970s, new roads were built to link the capital, Tehran, with the Gulf coast ports.

The private sector, often supported by public investment, developed car assembly, textiles, leather goods, plastics (Figure 8.42), and other light industries around Tehran and other major cities. Small industrial estates were established in most medium-sized and larger towns. In the early 2000s, Iran launched its own car, to be produced mainly by Iran Khodro, one of the world's top 20 carmakers. The motor industry makes up 20 percent of Iran's manufacturing output but stalled behind protectionist trade barriers, relying on old designs.

The first Turkish manufacturing developed out of mining for chromite and copper in the mountain ranges. In the 1990s, gold was added to mineral exports. Import substitution industries, such as car assembly, textiles, and steel, were also established.

As an oil importer, Turkey's manufacturing economy fared better than that of Iran in the later 1900s and is more diversified. The problems of high oil prices in the 1970s caused Turkey to reform its economy. Economic growth was not smooth, but industrial output increased in the 1980s and 1990s at a time of low oil prices. Between 1980 and 2000, real GDP per capita rose by 50 percent while Iran's stayed the same. Turkey diversified into light industrial products, which now make up some 80 percent of exports. In the 1980–1988 Iraq-Iran War, Turkey supplied both nations with food and manufactures in exchange for oil.

Services and Tourism

Turkey has a more developed services sector than Iran. Government employment is very important in the economy. International tourism grew and annually brings in over a billion dollars. In 2000, Turkey attracted 9.6 million visitors (up from 1 million in 1980). The development of tourist resorts along its sunny coasts and the availability of historic, often religious, sites made Turkey a major venue for Europeans. The rising income from tourism partly compensated for the falling level of wages sent home by 1.5 million Turkish workers in European and Arab countries. In the late 1980s, they contributed nearly half of Turkey's foreign exchange income. Turkey is the only country in Northern Africa and Southwestern Asia (outside of Israel) with such a diversified economy.

Test Your Understanding 8C

Summary Arab Southwest Asia is the heartland of the Arab world but has internal divisions based on whether oil is produced; few countries have sufficient water for agriculture. Israel faces a future of continuing conflict with its Palestinian population and its neighbors and, like them, has problems of finding resources to handle an increasing population. Israel has a Western economy with strong human resources and technologic skills.

Turkey and Iran are large but very different countries from historic times to the present cultural and political regimes. Turkey's water resources are the basis for much of the country's development, and it has a more diversified economy than most countries in the region. Iran's oil income and reserves provide continuing potential.

Some Questions to Think About

8C.1 How do the economies of the oil producers and non-oil-producing countries in this region compare with each other in terms of income fluctuations and levels of diversification?

8C.2 What has been the impact of wars and civil strife on economic development in the countries of this region?

8C.3 How do Israel's geography and development contrast with those of the other countries of this region?

Key Terms

diversified economy   kibbutzim
Gulf Cooperation Council

Figure 8.42 **Iran: manufacturing.** The Irana plastics factory, in which the machinery is basic and aging. Photo: © Don Smetzer/Stone/Getty.

Online Learning Center

www.mhhe.com/bradshaw

Making Connections

The Online Learning Center accompanying this textbook provides access to a vast range of further information about each chapter and region covered in this text. Go to **www.mhhe.com/bradshaw** to discover these useful study aids:

- Self-test questions
- Interactive, map-based exercises to identify key places within each region
- PowerWeb readings for further study
- Links to websites relating to topics in this chapter

Chapter 9

Africa
South of the Sahara

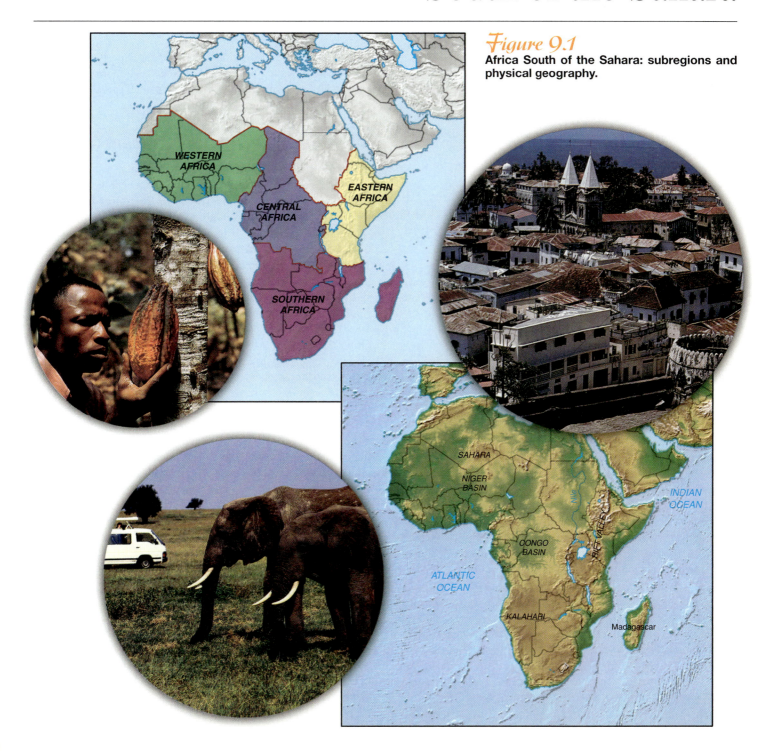

Figure 9.1
Africa South of the Sahara: subregions and physical geography.

WESTERN AFRICA

EASTERN AFRICA

CENTRAL AFRICA

SOUTHERN AFRICA

SAHARA

NIGER BASIN

CONGO BASIN

RIFT VALLEY

INDIAN OCEAN

ATLANTIC OCEAN

KALAHARI

Madagascar

A New Dawn?

A Heritage of Resources and History

Africa South of the Sahara covers most of the world's second-largest continent (Figure 9.1). It has a great diversity and beauty of natural landscapes from plateaus to rift valleys and mountains; its climates range from tropical arid to equatorial rainy; its ecosystems include tropical rain forest, savanna grasslands, and desert—a stock of biodiversity. The rocks underlying these landscapes are mostly ancient and contain mineral resources including the world's largest deposits of bauxite (for aluminum), cobalt, copper, gold, and diamonds as well as newly valuable minerals such as platinum and coltan (tantalum, used in mobile phones). Increasing amounts of oil are being found around the coasts.

Africa was the cradle of the human species. Skeletons of the oldest known *Homo sapiens* are found only in Africa. People in other continents are probably all descended from one group that moved out of Africa after the initial phase of human genetic diversification. Modern Africans number over 600 million people.

The history of African peoples before the coming of Arabs and Europeans was rich and sophisticated. It included organized kingdoms and empires with trading connections across the continent (Figure 9.2). However, the extension of global links with Arab, Asian, and European people in the last 1,500 years was often to the disadvantage of Africans. The slave trade and the arrogance of the external powers brought undeserved low expectations of African abilities. African people, whose ancestors were forcibly moved as slaves, now form significant populations in other parts of the world. In the Americas, particularly Brazil and the Caribbean but also the United States, they contributed lasting cultural features despite local discrimination.

Folklore and folk tales persisted from the rich African storytelling traditions. Musical influences from spirituals and blues to reggae affected modern Western music, while African art underlies many modern abstract paintings and sculptures.

The region itself has clear boundaries consisting mostly of coasts. The northern boundary through the Sahara reflects the impacts of the later 1800s and early 1900s colonial period. Trade routes and connections moved away from the land crossing of the Sahara that connected northernmost Africa with the rest of the continent and toward the ocean routes linked to the expanding European global economy. The Mediterranean countries became increasingly linked to the Arab world, while Africa South of the Sahara experienced selective exploitation by European colonial powers. The Sahara remains a significant boundary and transition zone, although some countries could be placed in different regions. Thus, Mauritania is included in Africa South of the Sahara, but not Sudan. The former has long links with Western Africa, while the latter is part of the lower Nile River valley and linked closely to Egypt.

The Challenges of the Present

Today, Africa South of the Sahara faces many problems. Materially it is the world's poorest region, where half the population lives on 65 cents (U.S.) a day or less. Most of its countries are not gaining in wealth as many are in the rest of the poorer world. Some African countries lost ground in the later 1900s. In 1955, Africa South of the Sahara accounted for 3.1 percent of world trade; by 2000, its proportion had fallen to 1.2 percent. In general, despite continuing support by former colonial countries such as France and the United Kingdom, this region's increasing problems were mostly ignored by the West. In 2000, the U.S. Congress passed the Africa Growth and Opportunity Act, which reduced import duties on African imports, especially in textiles and clothing, but this was a small step toward more open economic relations.

Major internal problems include the civil strife that holds back many countries politically, economically, and socially. There were reports in 2001 of a slave trade operating to force children to work on cocoa plantations in West Africa and to export them to Europe and the Arab countries. Always a problem, tropical diseases, such as malaria and river blindness, together with the new scourge of HIV/AIDS, infect many people across the continent. Environmental problems include the expansion of the Sahara as the climate becomes more arid.

 # African Cultures

Africa was the original stage for the generation and interaction of human cultures and has since mixed Africans, Arabs, Asians, and Europeans, all of whom call Africa South of the Sahara home. The great variety of traditional indigenous cultures, the Arab Muslim cultures entering from medieval times, the European colonial cultures, and the Westernizing cultures of the late 1900s all left their marks on the region's geography.

(a)

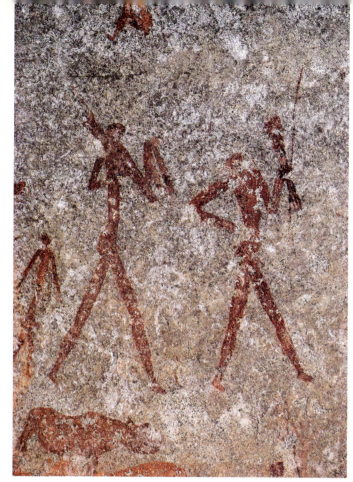

(b)

Figure 9.2 **Ancient African achievements.** (a) The ruins of Great Zimbabwe record the existence of a major trading center from around A.D. 1100. It was based on a sophisticated political organization and economy before European exploration and colonization. Each layer of rocks in the walls is approximately 7 to 10 cm (3–4 in.) high. Great Zimbabwe became a symbol to both sides in the 1960s struggle for majority rule in Southern Rhodesia (now Zimbabwe). For Africans, it was a symbol of African historic achievement; for white Rhodesians, it symbolized the unthinkable triumph of black Africans. (b) San (Bushmen) rock paintings that are found in caves in Southern Africa and are up to 6,000 years old. Photo: (a) © Michael Bradshaw.

Ethnic Diversity and Shared Cultures

The members of ethnic groups that form the basic social and political units of the indigenous people share kinship and territorial links (Figure 9.3a) and frequently a language, a culture, and political and economic institutions. European colonists called these groups "tribes," but it is difficult to define *tribe* with precision. Sometimes distinctive appearance marks out a tribe, such as the tall Masai herders of the grassy plains or the small pygmy groups of the forest. However, tribal identities often override physical and cultural differences; separate identities may occur within groups having similar physical characteristics; and in some places, individuals may move from one group to another.

The diversity of groups is reflected in language. More than 1,000 languages are spoken in Africa South of the Sahara (Figure 9.3b). The largest group of closely related languages is the Niger-Congo group that includes the languages of Western Africa and those of the Bantu groups in the south. The Afro-Asiatic group is important along the Saharan margins across the continent. Smaller groupings include the Nilo-Saharan languages in the north and the Khoisan languages in the far south.

Many shared features of indigenous African groups extend more widely than tribal loyalties.

- In traditional African beliefs, there is a close relationship between humans, nature, and spiritual forces. Gods and spirits influencing human success or failure inhabit rivers, rock outcrops, and tree groves. Such beliefs are collectively known as animism.

- All humans are seen as part of a continuing chain of life, with reverence for ancestral spirits and large families. The elderly are highly respected. Childlessness is a tragedy and large families a blessing. The wider family supports each member through difficult times.

- Artistic expressions in sculpture, music, dance, and story-telling are central to African cultures, linked to reverence for elders and educating the young in traditions.

- Wisdom and strong leadership are respected. Traditionally, the chief is often a feared, trusted, and spiritual leader, acting for the common welfare.

- Most Africans still get their living from the land, through cultivating, herding, or hunting (Figure 9.4), although urban populations are expanding rapidly. Land is traditionally in communal, not individual, ownership—an inheritance from the past and responsibility to the future.

External influences added to many of the indigenous cultures, especially in language and religion. With so many indigenous languages, trading languages became significant. Older creole languages, which combine several languages, often beginning with people of mixed descent or through trade,

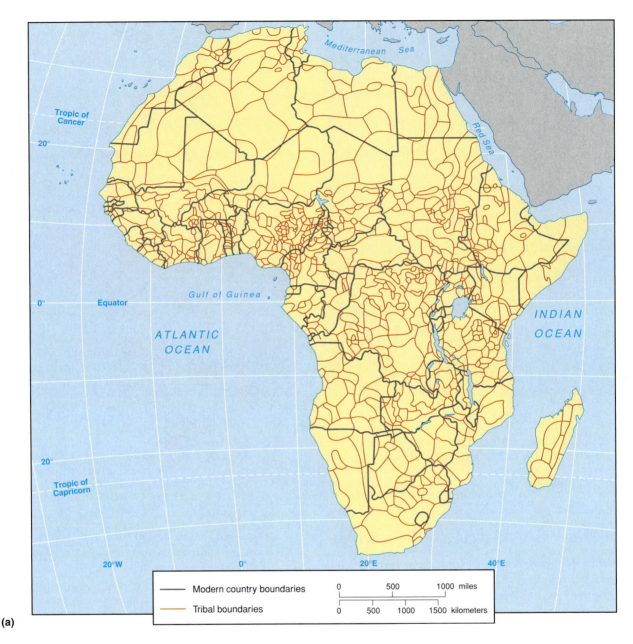

(a)

Figure 9.3   **Africa South of the Sahara: cultural features.** (a) The relationship of colonial-imposed boundaries of the modern countries to ethnic group areas. (b, page 367, top) Ethnic groups, languages, and precolonial kingdoms. In the zone between the area dominated by the Afro-Asiatic groups of northern Africa and the area dominated by the Niger-Congo peoples to the south, a diversity of languages exists. The advance of the Afro-Asiatic groups into regions south of the Sahara in medieval times helped to encourage southward movements of Bantu groups. By the late 1800s, they were replacing longer-established groups in the far south. Source: (a) From *World Regional Geography: A Question of Place,* by Paul W. English and James A. Miller, 3d ed., 1989, © Paul W. English and James A. Miller. (b) From *The Changing Geography of Africa,* by A. T. Grove, 2d ed. Oxford University Press, 1993.

include Swahili (combining African, Indian, and Arabic elements) in Eastern Africa, while English, French, and Portuguese are still used in former colonies. English, French, and Swahili became common in the TV and newspaper media.

The great diversity of religious allegiance is a further characteristic of this region. There are many versions of traditional animistic religions, together with strong influences from Islam, especially in the north and east of Africa South of the Sahara, and Christianity—Catholic and Protestant—elsewhere. In fact, many faiths coexist without strife between their members. Some syncretism (mixing of religious forms) occurs among Muslim,

Catholic, and traditional expressions of faith. Conflicts often result where political interests take up religious differences.

African Empires

Indigenous African groups, particularly in Western Africa (Figure 9.3b), established empires based on the wealth created by trade in salt, gold, and slaves. The Western African empires of Ghana (A.D. 700–1240), Mali (1050–1500), and Songhai (1350–1600) had widespread influence. When the Malian Muslim emperor, Mansa Musa, visited Cairo, Egypt, in 1324,

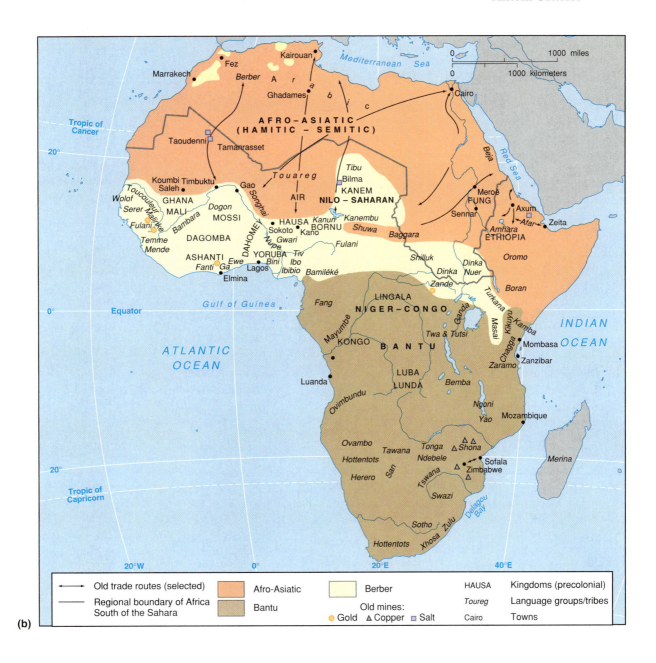

Legend (map):

→ Old trade routes (selected)	Afro-Asiatic
— Regional boundary of Africa South of the Sahara	Berber
	Bantu
	Old mines: ● Gold △ Copper ■ Salt

HAUSA	Kingdoms (precolonial)
Toureg	Language groups/tribes
Cairo	Towns

Figure 9.4 **African villages.** (a) A village in Chad with the Guera Mountains behind. (b) A feeding station at a time of drought and local famine, Chad. These pictures emphasize the closeness of people to the land, communal facilities, and the local extent of most interactions. Photos: © MAF.

(a)

(b)

on his pilgrimage from western Africa to Mecca, he had 500 porters, each bearing a golden staff. In southern Africa, Great Zimbabwe (see Figure 9.2) was the center of a widespread trading empire.

As an example of the influence of these empires, Timbuktu in modern Mali lies near the northernmost bend of the Niger River on the southern margin of the Sahara (Figure 9.5a). It grew as a market for local crops and cattle. By the A.D. 700s, its salt trade linked mines in the Sahara with markets down the Niger River. Local gold fields made its merchants very wealthy. Universities with libraries containing large numbers of imported books were established at Timbuktu and Djenne before any existed in northern Europe. Scholars from Greece, Egypt, and Arabia were employed as teachers.

Muslims in the North and East

The Arab Muslim expansion from the A.D. 600s (see Chapter 8) had wide geographic impacts. Culturally, Muslims brought their Islamic religion but also accommodated local practices, including polygamy and the use of the African drum. The introduction of camels enabled Arab and African traders to exchange goods along a few well-marked routes across the Sahara, reaching a major level of activity from 900 until the 1800s. In the west, the routes went south from Fez and Marrakesh (modern Morocco) to the middle Niger River valley; in the center, routes connected Tripoli (modern Libya) and the Lake Chad area; in the east, the Nile River routeway connected Egypt, Sudan, and Ethiopia. Later, Islam diffused along the same routes, setting off holy wars as Fulani warriors and zealots searched for grazing lands.

Trade across the Red Sea from Arabia to the Horn of Africa entrenched Islam and Arab culture in eastern Africa by the 1300s. Arab traders with connections to India and farther east spread from the Red Sea down the east African coast and its islands, establishing trading ports such as Zanzibar. The Arabic language came to coastal parts and led to Swahili developing as a creole trading language. Overall, Muslims had significant impacts on trade and added their influences to African culture. Arabs developed the slave trade, selling around 10 million Africans through their networks.

Colonial Regimes

European influence in Africa grew from the mid-1400s (Figure 9.5b). With improved ship technology, merchants began trading along the African coasts in order to reach India without going through the Muslim countries of Southwest Asia. Coastal trading posts in Western Africa built forts to protect the Europeans. They were points where ships could be loaded with the local gold, ivory, and palm products in exchange for alcohol, guns, and sugar. Sectors of the West African coast were called the Ivory Coast, Gold Coast, and Slave Coast for their chief products. Portuguese ships took gold from eastern Africa to pay for the silks and spices of Asia (see Chapters 5 through 7).

(a)

(b)

Figure 9.5 **Africa South of the Sahara.** (a) A Moroccan sign that tells camel caravans it will take them 52 days to reach Timbuktu across the Sahara. (b) Early European map, c. 1570, showing a good knowledge of the African coast.

Slave Trade

The Portuguese were the first European colonists, reaching northern Angola in 1483 and establishing relations with the local Kongo ruler. They visited the Mozambique coast in 1498 and ousted the Arab traders who had built coastal cities as centers for their Indian Ocean trade. The slave trade devastated both Angola and Mozambique as over 3 million slaves were taken to Brazil from Angola alone.

From the 1600s, the European trading routes included the Americas, where colonies produced sugar and tobacco and later cotton. When these colonies needed labor, European ships transported slaves from Africa to the Americas (Figure 9.6), often using the sources of supply pioneered by African empires and Arab traders. Slaves entering the Atlantic trade came mainly from the coastal areas of Western Africa, Central Africa, and what is now Angola.

Europeans justified enslaving peoples of different skin color and culture through assumptions of racial superiority. Eventually, somewhere between 6 million and 30 million (with 10 million as a widely accepted figure) African slaves were transported across the Atlantic Ocean to the United States, the Caribbean, and Latin

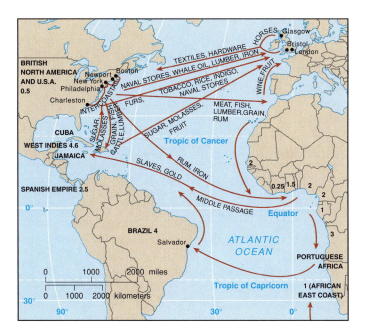

Figure 9.6 **Atlantic Ocean: the slave trade.** The 1700s Atlantic economy, trading colonial commodities with home countries and bringing slaves from Africa to the Americas. Numbers in millions from Africa are estimates of slave movements to sectors of the Americas. The British ended most of the Atlantic slave trade in the early 1800s. The country boundaries are those of today. Source: Data from Hugh Thomas, *The Slave Trade,* Simon & Schuster, 1997.

America. The slave trade slowed after Britain abolished slave shipments in 1808, but some countries did not abolish slavery until 1880 (and it continues today in parts of Africa and Asia). Humanitarian efforts in Europe and the United States during the 1800s resulted in the return of African families freed from slavery to people the new country of Liberia and the port of Freetown in Western Africa. However, the returnees found it difficult to integrate with the Africans who had not been enslaved.

Explorers and Colonies

From 1884 to the 1960s, a relatively short period, colonial occupation and government took over from traders and explorers, leaving major impacts on the region. European countries professed an intent to "civilize the dark continent" through missionaries and administrators, but they were also motivated by commercial interests and competition for overseas possessions.

As the slave trade declined, all of Africa South of the Sahara, except Ethiopia and Liberia, became colonies of European countries. Explorers such as Henry Stanley (United States) and David Livingstone (United Kingdom) penetrated the inland regions along major rivers, meeting many of the African empire rulers and providing knowledge from their travels. While they did so, the Industrial Revolution developed in Western Europe from the early 1800s, resulting in demands for raw materials such as tropical tree crops and minerals. In 1884–1885, European countries competing for world power met in Berlin and divided the continent into French, British, German, Portuguese, Belgian, Italian, and Spanish spheres of influence (see Figure 1.14)—omitting any consultation with Africans.

Colonists Who Settled

In a few places, Europeans stayed for the longer term and made lasting impacts on the local political and economic life. In the far south, the Dutch settled from the mid-1600s around their port of Cape Town to support ships trading to and from the East Indies (now Indonesia). Farmers came to produce crops and meat to supply the port and town. When Britain bought the Cape Colony from the Netherlands in 1814, the Dutch settlers sought interior isolation to preserve their culture and undertook their "Great Trek" to the Orange and Vaal River valleys.

In the late 1800s, the United Kingdom developed the swath of land between Angola and Mozambique. The British colonization primarily aimed to develop mineral wealth, but sometimes it coincided with cries of help from indigenous peoples who perceived the threat of Portuguese colonial repression to be greater than that of British administration and loss of land to farming settlers. For example, tribes in Malawi (1891) and Zambia (1889) requested British protection, while the Ndebele in western Zimbabwe signed contracts with Cecil Rhodes to allow his British mining company to exploit the minerals on their land (1888). Settlers came to take up the good farming land, particularly in what was Southern Rhodesia (modern Zimbabwe).

Portugal discouraged settlement by Europeans in its colonies until the early 1900s, when there was a push toward greater economic exploitation of mineral resources and plantation crops led by Portuguese technology, finances, and administrators. Little attempt was made to provide schooling for the African population, and any dissension was brutally repressed. As European settlement increased, African opposition gained strength, with rival groups fighting each other in both Mozambique and Angola. In 1975, a revolution in Portugal led to the end of a dictatorship that had held onto the colonies and a new government that brought independence to Angola and Mozambique. As most people of Portuguese origin left in the late 1970s, their scorched-earth policy destroyed cropland and machinery, leaving the two countries poorly prepared for independence.

Colonial Government

Some colonial powers, such as France and Britain, provided basic education and health care. They also gave limited experience for Africans in local government, especially where traditional political institutions were in place and the rulers amenable. Elsewhere, military strength imposed control. The new university colleges, hospitals, and other social infrastructure built by the colonizers in the centers they established provided a basis of professionals, educated in European ways, for the new countries after independence and maintained a European presence. The French policy of assimilating Africans into the French way of life linked their colonies more closely to French institutions. However, Portugal, Belgium, and Germany (before 1918) provided little infrastructure outside the mining and plantation areas and limited the extent of education prospects.

Colonial powers legitimized their rule by claiming to bring peace to ethnic rivalries but often took sides that deepened the rivalries. Local ethnic groups were labeled as "dependable" or "lazy," encouraging regional differences of investment and forceful recruitment of some groups as laborers. Moreover, all colonial powers frequently generated so much hostility that guerrilla groups led the push toward independence.

Colonial governments built ports to link with the home country. Railroads connected the ports to interior mining and commercial crop regions. Water projects supplied commercial farming areas occupied by European settlers. Most of the wealth derived from the export products supported the growth of major industrial enterprises in the colonizing countries that grew to multinational corporations such as Unilever, Nestlé, Firestone, Dunlop, Michelin, and Hershey.

Overall, African colonies remained marginalized in the world economy by the emphasis on exporting raw materials to manufacturers in colonizing countries. African labor was used on farms and in mines, often in brutal conditions. Where local labor was insufficient, migrants were used—often under force in new forms of semislavery. The emphasis on commercial crops for exports and the attractions of the towns caused a decline in subsistence farming of traditional food crops and increasing food shortages. The mercantile colonial system ruled out manufacturing, apart from local needs, in case it should compete with the products of the colonizing countries. The discouragement of manufacturing enterprises in African colonies made them dependent on European products. Preferences for European products and ways of living brought cultural as well as economic dependency.

Independence and After

Agitation for independence began in the 1920s and was enhanced by the Italian invasion of Ethiopia in 1936, World War II isolation, and the example of independence gained by India and Pakistan in 1947 (see Chapter 7). Nationalist movements, often led by those who had studied in Europe or the United States, achieved independence, beginning with Ghana in 1957 and ending with Namibia in 1990.

On gaining political independence, most countries of Africa South of the Sahara had limited experience of government and a restricted source of income. In many African countries, the methods of the authoritarian colonial governments and bureaucracies were followed, producing sequences of dictatorship, single-party, and military rule. External pressures and internal difficulties further slowed economic, political, and cultural development.

Colonial Country Borders

At independence, the borders of colonies and the colonial administrative districts became the borders of new countries and often of their internal subdivisions. The imposed boundaries paid little attention to ethnic (tribal) territories (see Figure 9.3a). One result was the creation of many tiny countries without the resources or population numbers to become major players in the global economy. Some countries, mostly small (e.g., Somalia, Swaziland, Botswana), are dominated by a single tribal group and have few minorities. But most countries have several different ethnic groups, and some have hundreds. Nigeria has three dominant groups (Hausa with 35 million

people in the north, Yoruba with 25 million in the southwest, and Igbo with 20 million in the southeast), but there are also 300 smaller groups—some of which have more members than the entire population of smaller African countries. Such divisions often became the basis of conflict with government and opposition representing two major ethnic groups, as with the Shona and Ndebele in Zimbabwe.

Multinationals and Raw Materials

At the time of independence for most African countries in the 1950s and 1960s, Western economists claimed that countries would grow economically by following the world's wealthier countries in moving from agriculture and mining into manufacturing and services. Few African countries followed this progression.

Multinational mining companies and makers of coffee, tea, and chocolate products continued to buy African raw materials cheaply. For example, multinational aluminum manufacturers helped arrange funding for the Volta River project in Ghana. It generated hydroelectricity to refine bauxite, the ore of aluminum, as cheaply as possible to provide an exportable product that could be finished in the wealthier countries. However, such reliance on producing raw materials keeps local incomes low. In 2002, Nigerian and Ghanaian coffee producers received around 50 U.S. cents per pound. Each cappuccino served in U.S. coffee chains takes about 1 ounce of coffee, bringing 4.5 U.S. cents to producers, but customers pay around $2. Most of the markup goes to the coffee traders, blenders, grinders, and retailers in the United States (and other wealthier countries). Purchasers of African raw materials often maintained low world prices by opening up new areas of production as growing markets absorbed what established areas produced. Soaring raw material prices in the 1970s (minerals) and in the early 1980s (beverages) were short-lived but often enabled the producer countries to take out loans; when the prices fell, such countries faced debts that they could not repay.

Cold War and Subsequent Pressures

During the Cold War (1950–1990), the United States and Soviet Union encouraged conflict by propping up bad governments or supporting rebel groups. They sold weapons to help the groups maintain or achieve power. For example, Ethiopia became a battlefield as Soviet forces supported its Communist government against rebels from Somalia or Eritreans fighting for their independence.

In Southern Africa, the external activities of South Africa destabilized neighboring countries. South Africa gained external support by labeling those antagonistic to its apartheid system as Communists. When Angola drew in Soviet and Cuban forces, South Africa supported the opposition. Although the Cuban forces withdrew from Angola in 1990, the civil war continued until the 2002 death of the opposition leader. In 1980, the other countries in Southern Africa formed the Southern Africa Development Coordination Conference (SADCC) to oppose South African apartheid. In 1992, the name was changed to **Southern Africa Development Conference** (SADC), and the focus moved toward reducing economic dependence on South Africa. After the abandonment of apartheid, SADC in 1994 welcomed the membership of South Africa to strengthen mutual linkages in trade and to help control traffic in illegal drugs and arms.

When the Cold War ceased in the early 1990s, the powers that treated African countries as being for or against them relegated them to having little or no strategic interest. Their heavy debts made it difficult for African countries to attract new investment or aid. By the late 1990s, African countries attracted only 3 percent of the foreign investment flowing to developing countries, while countries in Latin America attracted 20 percent and those in East Asia about 60 percent.

Population and Environmental Pressures

From 1980 to 2001, the total population of Africa South of the Sahara rose from around 380 million to 638 million at rates of increase that equaled or exceeded rates of economic growth. By 2000, the annual rate of population increase in Africa South of the Sahara remained around 3 percent, while that of agricultural production growth was only 2 percent. Death rates in most countries plummeted as modern medicine was applied, but birth rates in many countries fell slowly if at all (Figure 9.7). Per capita food production dropped by 12 percent between 1960 and 2000 in contrast to the Green Revolution increases in Asia.

Population control is one major key to Africa's future. Efforts to encourage smaller families had only moderate suc-

cess because of cultural, economic, and often political pressures that regard large families as a high priority. Larger families bring more support to older family members and have more children to work on farms. They are seen as a sign of male virility, and women have little say in the numbers of births. Sadly, the most likely curb on population growth is the HIV/AIDS pandemic, which is increasing death rates and reducing life expectancies in many African countries.

As if such economic and political problems were not enough, a majority of scientists now agree that environmental changes are occurring through global warming. The consequences for Africa are likely to be worst in the southern part of the continent, the subregion with greatest prospects of a better future, and along the **Sahel** (the southern margin of the Sahara). Droughts got worse and more frequent after 1980 and affect both farming and water supplies for industry.

Governments and Global Institutions

The combination of low prices in world markets, external interference in their politics and economics, and internal ethnic rivalries resulted in many unstable governments and a trend to oppressive and corrupt dictatorships in many countries. In the 1990s, partly in response to external pressures as well as internal democracy, many countries moved rapidly from single-party to multiparty constitutions and open elections. It is now conceivable that entrenched leaders may be removed by ballots instead

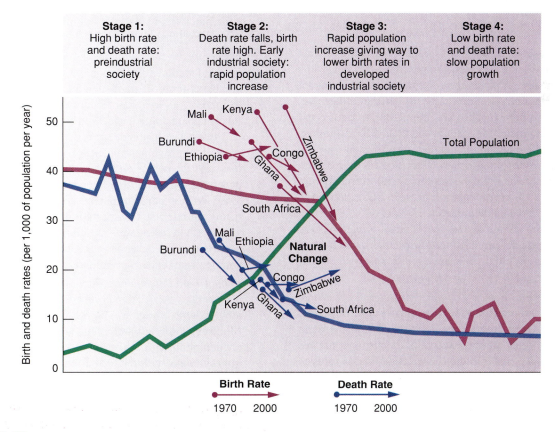

Figure 9.7 **Africa South of the Sahara: population change linked to demographic transition between 1970 and 1992.**
Although the death rates fell comparably to the experience of core countries in the 1800s, many birth rates remained very high, despite some dropping sharply. Some death rates increased, especially in countries subject to HIV/AIDS.

of bullets, but established leaders often found it difficult to hand over power. The full implementation of the new constitutions is often hesitant, people are discouraged if new governments do not produce economic growth, and tribal rivalries surface in open conflict.

In the 1980s, the World Bank and the International Monetary Fund—the two major lending institutions for developing countries—established new guidelines for grants and loans. This was a response to the low rates of success achieved by previous loans and the large proportions that were absorbed by high exchange rates and the cost of internal government bureaucracies. To date, these structural adjustment policies have had little success in the countries of Africa South of the Sahara, although Ghana, Tanzania, Burkino Faso, Nigeria, and Zimbabwe came closest to following their precepts. The policies also had negative impacts by bringing about an overreliance on commercial crops for export. The reduction in government employees reduced health care and education provision. For most countries in the region, structural adjustment is not yet appropriate and each country needs specific measures. At present, countries in Africa South of the Sahara seem to lose out whether they decide to follow structural adjustment policies or not. If they do not adopt them, they lose access to funds from the World Bank, International Monetary Fund, and aid agencies. If they do adopt the stringent policies, they often alienate their people.

Can Africa Claim the Twenty-First Century?

Assuming we wish to end poverty around the world, Africa South of the Sahara presents a challenge. Many of the world's biggest development problems (see discussion of development in Chapter 2, especially Figure 2.13) are concentrated there. Lagging primary school enrollments, high child mortality, and endemic diseases—including malaria and HIV/AIDS—impose barriers to growth. Population numbers grow faster than the economic output. Many countries in Africa South of the Sahara are short of the basic human resources and infrastructure needed to achieve slower population growth, faster economic development, and political democracy.

Some external (France and the United Kingdom, the World Trade Organization, and the World Bank) critiques of Africa's problems highlight four areas of action, but external support is insufficient to meet the needs expressed.

1. *Improved governance and conflict resolution* is the most basic need. Civil conflicts, such as those in Sierra Leone, Rwanda, the Democratic Republic of Congo, and Angola impose huge costs at home and in neighboring countries through deaths, maimings, destruction, and refugee migrations.
2. *Investment in people.* The vicious circle of high fertility and mortality, low enrollments in education (especially of girls), high numbers of young people who are dependent on the working-age group, little action against HIV/AIDS, and low savings lies behind much of Africa's slow or static development.

After independence, all countries increased educational achievements. By the 1990s, countries such as Botswana, Cameroon, Kenya, South Africa, Zambia, and Zimbabwe had virtually total enrollment, male and female, in elementary schools, up from 50 percent in 1965. Burundi, Chad, and Mauritania made major strides by increasing primary education from 10 to 20 percent in 1965 to 50 to 70 percent in the 1990s. However, Burkina Faso, Ethiopia, Guinea, Mali, and Niger still have under 30 percent of children in elementary school, and education of females lags behind that of males.

Increasing numbers of people in the region also have opportunities to become fully literate in secondary school and to earn higher academic qualifications (see the "Personal View: Ghana" box, p. 373). Many who gain higher qualifications are disappointed to find few jobs available in their home countries. African doctors, lawyers, and airline pilots are increasing in numbers, but many become part of the brain drain by finding employment abroad in the world's wealthier countries. Emigration from Africa to the United States more than doubled in the 1990s, from 15,000 to around 40,000 per year, disproportionately in the professional, managerial, and technical occupations, and continues to grow. Although they may benefit home African countries if they send money to their families or eventually return with skills and experience, many decide to live permanently in the wealthier countries. Meanwhile, African countries pay expatriates from wealthier countries high salaries to carry out professional jobs.

3. *Economic diversification* makes countries more competitive in world markets. The region's countries need new products and better terms of trade, new incentives, and wider access to markets in wealthier countries. The perceived risks of investing and doing business in Africa slow job creation.

Internal reforms needed include the reduction of corruption, improvement of infrastructure and financial services, and the provision of better access to the information economy. All countries need improved infrastructure. They do not have sufficient all-weather roads or other forms of internal transportation. They lack clean water supplies and adequate sanitation. There are shortages of electricity and telecommunications. Ports are poorly equipped and expensive to use.

4. *Reduced aid dependence, debt, and stronger intraregional partnerships.* Africa remains the world's most aid-dependent and indebted region. World Bank and International Monetary Fund aid and loans, together with development cooperation with the European Union, continue to improve links with the former colonial countries but bring few advantages to African countries. Programs of debt relief became more significant in the 1990s as aid donors channeled investments to countries but still insisted on approved development policies to avoid corruption. The World Trade Organization tries to help African and other developing countries improve their access to markets for agricultural products and gain a reduction of farm subsidies

Personal View

GHANA

When meeting Yaa Boadi, one is readily drawn in by her natural, high-wattage smile—an appealing West African trait. But it is not merely because she is Ghanaian that this young woman smiles so freely: it is also because of the way she feels about her life—one in which she has made remarkable strides from her rural, impoverished upbringing.

At the age of 26, Boadi's most vivid childhood memories are of beginning each day by carrying pails of water on her head. Sometimes she made three grueling trips to the village well before school.

Baodi's village, Nkawkaw, was on a red-earthed mountain in the eastern region of Ghana. It is inaccessible to motor vehicles, so getting there involves a walk of 30 minutes or more along a footpath from the main highway linking the regional capital of Kumasi with the national capital, Accra.

During the dry season, it could be so cold in the morning that Boadi warmed herself beside an outdoor fire before setting off for water. Then it could become so hot and dusty during the day that her grandmother would rub cocoa butter into her black hair as a moisturizer.

Water collection also became more arduous during the dry season. As the level in the well dropped, villagers would jostle in line. When it was her turn at the well, Boadi tried to lower her pail gently to avoid stirring up the muddy bottom.

Inevitably, the well dried up before the spring rains returned, forcing villagers to travel far to a deeper source or to buy water from a government truck in a distant town.

On a Saturday morning recently near the central market in Accra, Boadi wore jeans and a royal blue, tie-dyed shirt with white lace embroidering the collar and short sleeves. She inherited this striking shirt from her father, a man she hardly knew.

While she was still a toddler, her parents left home: her father took a teaching position in Cameroon. Boadi and a brother were raised by their grandparents. Lots of cousins lived in the compound, as well as Boadi's uncles and aunts. Everyone slept on floor mats in two rooms, except her grandfather, who had his own room and a mattress. Boadi says it was a typical, traditional, village upbringing.

In a region where many girls never attend elementary school, very few attend college, and fewer still ever achieve professional status, a doting uncle encouraged her to achieve. She could become an engineer one day, he told her. Then she could return home to build wells, so that other girls would no longer have to haul water.

Baodi made it to the university in Kumasi, the regional capital, one of only four women to enroll for civil engineering alongside 36 men and the only one in her class to stick with it as others switched to less rigorous majors. In time, Baodi found she could hold her own with male classmates from more privileged families who had been groomed in prestigious secondary schools.

On graduating four years ago, she became the first professional woman hired by an Accra engineering company, and she has worked on design teams for World Bank-funded national highway projects. She enjoys her work but also detects some bias when the men get all the field assignments while she remains office-bound with design work.

She has also become active in a support group for women engineers in Ghana and has traveled to South Africa to speak to an association of women engineers. Boadi was among seven Ford Foundation fellows selected from almost 600 applicants in Ghana. She plans to enroll in a master's degree program at the University of Southampton (United Kingdom), focusing on engineering for rural development. Within a few years, she hopes to form an NGO dedicated to attracting funding for infrastructure projects in neglected regions of Ghana. Perhaps, she says, one day she'll return home to build wells—just as her uncle foretold.

[Excerpted from "Getting the best out of Africa," by Todd Shapera, *Financial Times* (London) *Weekend,* 27–28 April 2002.]

---

in wealthier countries, but countries such as the United States make small concessions and increase their own farm subsidies to maintain their farming communities.

Partnerships among African countries could be more significant. In July 2001, the heads of government meeting in Lusaka, Zambia, changed the "Organization of African Unity" to the "African Union." At the same conference, the Millennium Action Plan, proposed by President Thabo Mbeki of South Africa, had a twofold thrust: on the one hand, it restated the policies previously urged by Western countries and institutions: better government, more democracy, respect for human rights, market reforms, and recognition of the advantages of globalization. On the other hand, the plan highlighted the need to reduce poverty by improving education and public health. It asked for continuing aid together with the removal of trade barriers and agricultural subsidies in richer countries, but subsequent actions suggest that the world's wealthier countries demand the first part but contribute little to the second. When the wealthiest (G8) countries met in 2002 and discussed African needs, they offered $1 billion of the $64 billion requested at a time when the United States increased its own farm subsidies by $190 billion.

Natural Environments and Resources

The tropical conditions in Africa South of the Sahara include the range of climatic contrasts from arid to all-year rain, ecosystems from desert to rain forest, some of the world's longest rivers, and some of its grandest scenery. The range of big animals, from herbivores such as elephants and gazelles to carnivores such as lions, attracts tourists from around the world. Water and mineral resources are among the world's most impressive.

Tropical Climates

The climates of Africa South of the Sahara—apart from the extreme south—range from the tropical arid environments of the Sahara, Kalahari, and Namib Deserts, through tropical seasonal to the equatorial climatic environments (Figure 9.8a). The southern tip of the continent has warm midlatitude climates. These contrasts affect human livelihoods through the diseases of the wet tropics, the lack of water in many areas, and their effects on soils. A satellite view (Figure 9.9) shows the contrasts as expressed by the dense thundercloud clusters over

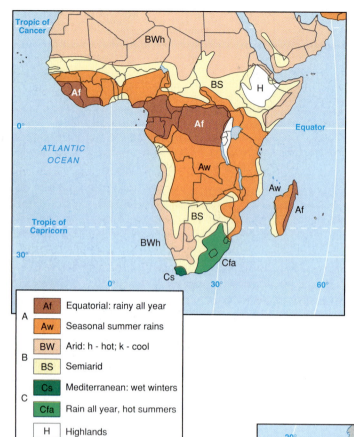

(a)

A
Af	Equatorial: rainy all year
Aw	Seasonal summer rains

B
BW	Arid: h - hot; k - cool
BS	Semiarid

C
Cs	Mediterranean: wet winters
Cfa	Rain all year, hot summers
H	Highlands

Figure 9.8 **Africa South of the Sahara: climate and vegetation regions.** (a) Tropical climates dominate the region. (b) Major natural vegetation types. Compare the two maps.

the equator, the cloudless arid areas, and the midlatitude weather systems with their frontal cloud patterns.

The equatorial climatic environment dominates the basin of the Congo River in Central Africa. Temperatures remain high all year, and rain comes in all months. The large rainfall totals and huge basin area within this environment cause the Congo River to carry to the oceans the second greatest volume of water in the world (after the Amazon River in South America; see Chapter 10). The high plateaus in Eastern Africa, however, restrict the African extent of the equatorial rains and may suffer droughts outside of rainy periods in April and November.

In Western Africa, the coastal areas have a monsoonlike climatic environment with contrasting wet and dry seasons caused by alternating airflow directions. Moist southerly air from the Atlantic Ocean brings heavy summer rains. In winter, dry northerly winds, known as *harmattan*, blow from the Sahara. Annual rainfall totals fall with distance from the ocean.

Most of Africa South of the Sahara has a tropical seasonal climatic environment in which temperatures remain high throughout the year and rains come in the season when the sun is high. Rainfall totals range from low to moderate, and all areas are subject to variations from year to year. The dry seasons increase in length and the rains become more variable closer to the arid areas.

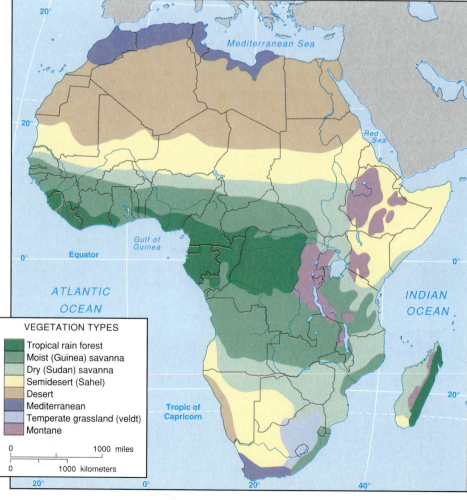

VEGETATION TYPES
- Tropical rain forest
- Moist (Guinea) savanna
- Dry (Sudan) savanna
- Semidesert (Sahel)
- Desert
- Mediterranean
- Temperate grassland (veldt)
- Montane

(b)

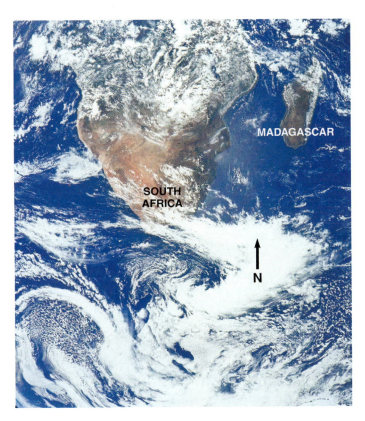

Figure 9.9 **Southern Africa: weather patterns.** A space shuttle photo shows a clear distinction between the cloudy (rainy) zones along the equator in the north. Moist air rises near the equator, condensing into clouds and spreading northward and southward at higher levels in the atmosphere. The air then descends at around 30 degrees of latitude, drying out and preventing cloud formation over the deserts. To the south of the continent, the swirls of midlatitude weather systems and frontal clouds over the ocean move from west to east. Photo: NASA.

The tropical arid climatic environments of the Sahara, Kalahari, and Namib Deserts have little rain in any season. The little that falls is rapidly returned to the atmosphere by high rates of evaporation.

Winters become cooler toward the southern tip of Africa. Winter cold in southwestern Africa comes from a combination of heat loss through clear skies in the dry climatic environment and the cooling of air above the cold Benguela ocean current along the Atlantic coast.

Changing Climates

The present arrangement of climatic environments in Africa is part of a sequence of changing climates. The best-known example of change is that over periods of a few decades, the southern boundary of the Sahara shifts north and south by up to 160 km (100 mi.). This affects the Sahel, a zone of very low rainfall (see Figure 9.8a), where annual amounts of rain are crucial to plant life and human activities. Drier periods cause the grasses covering old sand dunes to die and expose the sands to movement by winds. In wetter periods, the grasses extend their coverage.

Over longer periods, the Sahara boundary shifted greater distances. Five thousand years ago, the Sahara was more humid,

as shown by the dry lake and riverbeds around Lake Chad that were filled with water and the remains of settled human communities in the now arid areas. Around 3000 B.C., increasing drought forced people southward toward the Western African coasts and eastward into the Nile River valley. The margins of the Namib and Kalahari Deserts in Southern Africa experienced similar changes, but the numbers of people living there were fewer than along the southern edges of the Sahara.

As the deserts retreated or expanded, the areas affected by seasonal or equatorial climates occupied more or less land respectively. Such shifts affected patterns of human occupation and migration, but little evidence exists concerning specific changes.

Uncertainty over the climatic future fuels debates about whether the drying of the continent is a human-induced or a natural phenomenon. While people can do little about the natural changes, it might be possible to modify their own activities and so reduce the impacts of change. For example, where overgrazing or the removal of vegetation for cultivation caused the expansion of desert conditions, some African countries are now implementing tree-planting programs.

This region is likely to face the worst effects of global warming, especially in the expanding drought-stricken areas. The region that contributes least to additional greenhouse gases may thus be the greatest sufferer of the consequences of an enhanced greenhouse effect—in contrast to materially wealthy regions that create most of the added greenhouse gases.

Ancient Rocks, Plateaus, Rifts, and Volcanoes

Extensive plateau surfaces on ancient rocks dominate African landscapes (Figure 9.10a). **Plateaus** are elevated areas with relatively flat tops. The main reason for Africa's extensive plateau landscapes is that most of the continent is not crossed or touched by tectonic plate margins (Figure 9.10b) that would disrupt the land surface.

Some 200 million years ago, Africa was at the center of a huge worldwide continent (see Figure 6.9c). The world continent then broke apart, and its fragments moved to form the present continents, separating from Africa by divergent plate margins. The rift valleys that cross Eastern Africa from north to south are linked to the divergent plate margin activity along the line of the Red Sea. **Rift valleys** lie along zones where Earth's crust arched, pulled apart, and broke, collapsing the central part. The rift valleys have deep, elongated lakes and volcanic activity along their margins. Geologists identify the links from the East African rift valleys across Ethiopia, the Red Sea, and the Dead Sea–Jordan River valley line (see Chapter 8) as being part of a new ocean that is widening.

The highest mountains in Africa include Mount Kilimanjaro (5,895 m; 19,340 ft.) and Mount Kenya in Eastern Africa, which are extinct volcanoes close to the rift valley. The eruption of Mount Niriganga that engulfed the town of Goma, Democratic Republic of Congo, in January 2002 was part of this pattern. The Ethiopian Highlands are built partly of lava flows, also linked to rift valley volcanic activity. The Cameroon Highlands on the Nigeria-Cameroon border include active volcanoes that are

(a)

Figure 9.10 **Africa South of the Sahara: major geologic features.** (a) Map of relief (elevation) and rivers. Identify the highest points, which are mainly of volcanic origin. (b) Africa and tectonic plates: what is the significance of divergent margins (arrows moving apart) and convergent margins (arrows moving together)?

(b)

distant from plate boundaries and rift valleys, and result from the local melting and eruption of deeply buried rocks.

Ancient Landscapes

The long-term molding action of rivers and wind on stable rocks created the plateau surfaces of Africa South of the

Sahara. The surfaces are formed by erosion that lowers whole landscapes (planation) in seasonally wet tropical climates and contrast with the deep valleys incised by powerful rivers or glaciers in midlatitudes. Consistent high temperatures and the availability of water in deep soils cause rapid rock breakdown by subsurface chemical action. During the dry season, the grasses and small shrubs die and the soil surface dries, detach-

(a)

ing particles. At the start of the wet season, the combination of bare, loose soil and water carries fine sand and clay across the landscape and into the rivers. This process lowers the landscape as a whole and leaves individual hills with bare rocky sides, or inselbergs, as prominent landforms. Rocky slopes characterize the steps from one plateau level to another.

The major rivers such as the Niger, Nile, Congo, Zambezi, and Orange flow across the plateaus, enabling boat transportation. Rivers such as the Congo and Niger are much used for internal transportation. River navigation is often made difficult, however, by seasonal variations of flow related to rainfall or by waterfalls and rapids where the rivers descend from one plateau level to the next or from a plateau level to the coast.

Long-term climatic changes did not greatly alter the set of processes producing the extensive plateaus. The most recent glacial phases produced shifts in the climate zones in Africa but left few glacial landforms, except on the highest mountains. The ice caps on top of Mounts Kilimanjaro and Kenya are now melting rapidly. The desert landscapes probably formed under seasonal rainfall conditions during periods of greater humidity. Increasing aridity led to the desiccation of soil and drying of the rivers. The wind then blew away the finest soil particles and concentrated the sandy fractions into large dune seas.

Forests, Savannas, and Deserts

The natural vegetation of Africa South of the Sahara follows closely the patterns established by the climatic regime (see Figure 9.8b). Equatorial climatic areas are covered by dense tropical rain forest containing a huge variety of tree and other plant species and of birds and insect species (Figure 9.11a). The seasonal climate areas between the deserts and forests are characterized by savanna grasslands in which there is varying amounts of tree cover (Figure 9.11b). The savanna grasslands are noted for their large herbivore animals such as elephants, giraffes, zebras, and a variety of deerlike forms, such as antelopes and springboks, together with their predators, the lions, leopards, and wild dogs. Arid areas are deserts with sparse vegetation growth and a few small animals (Figure 9.11c).

The majority of soils in Africa, especially in the tropics, are poor in nutrients because of rapid chemical weathering and removal of the nutrients by water flowing through the soils. Some are workable for agriculture, however, if care is taken to cope with the high clay and iron contents that are liable to becoming cemented as they dry out. More fertile areas occur around some of the volcanoes, where soils form from the breakdown of rocks rich in plant-supporting nutrients. Overall, however, physiological population density (number of people per area of arable land) is high in equatorial Africa.

Resources

Africa has a wealth of natural resources, but, by definition, their use is a matter of economic and cultural demand (see Chapter 2). Much resource development in this region is stimulated by foreign investment for external demands. Many

(b)

(c)

Figure 9.11 **Africa South of the Sahara: contrasting ecosystems.** (a) Congo. Small village by a river in the heart of tropical rain forest—an ecosystem based on plentiful rainfall. (b) Savanna woodland of the tropical summer rains zone with Mount Kilimanjaro in the background. (c) Rocky desert with vegetation supported by subsurface water. Photo: (a) © Torleif Svensson/The Stock Market/Corbis.

African countries depend for much of their foreign income on mining and exporting the minerals contained in the ancient rocks. Overall, Africa's potential resources are underused, and the economic benefits from their exploitation seldom return to Africans. The extraction of minerals, the production of commercial crops, timber, and fish, and the development of tourism are seldom carried out by African-owned companies, and most profits go to companies in the materially wealthier countries. Some large deposits, such as the iron ores of Equatorial Guinea, remain unused at present because of civil wars or the lack of internal transportation.

Until the 1990s, many African countries possessed few sources of fossil fuels. Known deposits of oil and natural gas were limited to a few areas of coastal subsidence and sediment accumulation in such areas as the Niger River Delta. In the 1990s, evolving offshore oil exploration technology identified major oil fields along the Atlantic coast of Africa. Gabon and Angola became important producers based on multinational oil corporation investments in areas judged to be safe from internal conflicts. In the early 2000s, the United States and European countries encouraged these African producers to leave OPEC so that they could increase their output—and provide an alternative and less regulated source, along with Russia. In the 1980s, South Africa developed coal mining when international sanctions prevented imports, but coal deposits in other countries remain remote from transportation.

The tropical climates make it possible to grow a variety of crops, including yams, rice, cassava, corn, millet, and sorghum that are the basis of life for many Africans, who also sell them in the growing urban markets. Commercial crops for export to midlatitude countries are grown separately, including bananas, cocoa, coffee, tea, palm oil, rubber, cotton, tropical fruits, and peanuts. More research and farmer education is focused on the export crops than on the subsistence crops.

Forest resources are extensive in the equatorial countries, and plentiful fish resources occur in the major rivers and in the areas of cold ocean currents off northwestern and southwestern Africa. Both forests and fisheries are experiencing increased rates of exploitation and depletion through demand from the world's wealthier countries.

Africa also possesses the natural resources such as sunshine, coastal beaches, and large numbers of animals now protected in the national parks of countries such as Kenya, Tanzania, Zimbabwe, Botswana, Namibia, and the Republic of South Africa that form the basis for tourism. Tourist interests also center on scenic wonders such as Victoria Falls on the Zambia-Zimbabwe border and historic sites such as the former slave markets of eastern and western Africa. Many opportunities for tourist development, however, await political stability and the development of transportation facilities. Moreover, tourism brings mainly low-wage employment to local Africans.

Environmental Problems

The environmental problems of tropical Africa are associated with drought, poor soils, loss of wildlife, and diseases affecting humans and livestock.

Drought and Desertification

Water resources range from plenty in the Congo River basin to scarcity in many areas where there is seasonal rainfall, frequent droughts, or continuous aridity. Shortages of water through longer dry seasons or periods of dry years appear to be increasing in the seasonal rainy areas of Africa. Such shortages lower crop productivity and hydroelectric project efficiency. The Sahel zone suffered increasing droughts from the 1970s that indicated a change in climate combined with overgrazing and removal of woody plants for firewood—a case of desertification. It caused many livestock deaths in Western Africa and Chad and local famine conditions.

Soil Quality Losses

Many tropical soils lack nutrients or are difficult to work with simple tools. The decline in soil quality and workability following agriculture is one of the most widespread problems in Africa. For example, removal of the forest cover in the Ethiopian Highlands led to rapid erosion of the soils, forcing people to move elsewhere and impose further pressures on land resources in already crowded areas. Soil exposure in the seasonal rainfall areas may lead to a cementing of the soil into hard laterite that resists plowing. In semiarid areas, the overgrazing of slow-growing vegetation and the removal of woody plants for firewood expose soils to erosion by water and wind.

Wildlife

The tropical African flora and fauna are threatened by expanding logging, farming, and poaching. The level of the problems varies from place to place. For example, the killing of elephants for their ivory tusks in Kenya caused a fall in the elephant population and international action to ban the ivory trade in the late 1900s. Meanwhile, in Southern Africa, better conservation policies resulted in an expansion of the elephant population until it exceeded the carrying capacity of the land and forced governments to institute annual culls. The ivory from the culls fills many warehouses but cannot be sold because of the world ban.

Tropical Diseases

Despite massive efforts and advances in health care, tropical diseases remain among the main environmental problems of Africa for people and food production. Malaria, river blindness, and other diseases favored by the tropical climates remain endemic, while cholera flares up from time to time. Some areas of savanna grassland have very low populations because of the prevalence of sleeping sickness affecting humans and heavy tsetse fly infestations infecting cattle; people escaping from that threat move into river valleys where river blindness attacks. Diseases such as measles and tuberculosis are returning in the wake of HIV/AIDS, which weakens immune systems.

Great strides in immunology and antibiotics allow recovery today from many previous killer diseases such as cholera, various fevers, and even malaria. New deadly viruses, such as Ebola that affected the Democratic Republic of Congo in 1995, however, keep appearing and cause immediate panics. Other major diseases are still killers on a larger scale than the Ebola virus. Sleeping sickness, for example, kills 200,000 people a year in the Democratic Republic of Congo. Although a cure for sleeping sickness is known, few people are treated for it. The high cost of continuing treatments for diseases such as sleeping sickness and malaria makes them too expensive for most Africans. The continual risk of infection from waterborne diseases where water supply is untreated causes death rates to remain higher than those in other parts of the world. In many parts of Africa, a lack of sufficient food causes malnutrition and makes people less resistant to disease.

HIV/AIDS Pandemic

By 1998, the World Health Organization listed AIDS as the fifth main cause of global deaths with an expected rise to third by 2005. By destroying the immune system, HIV/AIDS makes its victims more liable to diseases such as tuberculosis, pneumonia, toxoplasmosis, fungus infections, and cancers. It is transmitted in blood and other bodily fluids, including through semen during sexual activity, unhygienic injecting needles, and transfusions of infected blood. It is passed from infected mothers to babies at birth or in breast milk. The causes of HIV/AIDS spread include poverty, the breakdown of traditional family support systems, the apartheid policy that brought miners into male-only camps serviced by prostitutes, continuing promiscuity at a time when traditional polygamy gives way to the taking of sexual partners outside monogamous marriages, and mistaken government policies.

Although reduced in Europe and North America in the 1990s by expensive triple-drug therapy, the disease diffused rapidly through Africa from the 1960s. Africa South of the Sahara has the world's highest and most rapidly increasing concentration of this disease (Figure 9.12). At present, Southern Africa has the greatest number of cases, affecting up to 40 percent of the population aged 15 to 49 years. It spreads quickly in cultures that value male sexual prowess.

In 2000, nearly 4 million new infections and 2.5 million deaths from HIV/AIDS occurred in this region—over two-thirds of the world's declared totals, although some countries inside and outside Africa do not report the full extent. Life expectancies in the worst affected countries of Botswana and Zimbabwe rose to 60 years in 1990 but fell to 40 years in the following decade as HIV/AIDS took hold. Other African countries experienced smaller but growing impacts unless special measures were adopted (see the "Point-Counterpoint: HIV/AIDS" box, p. 30).

As well as the demographic impacts, HIV/AIDS has many social implications. Half the miners in South Africa are HIV carriers, and thousands of orphans, often carriers themselves, constitute a growing need for help in the region. Throughout Africa South of the Sahara, the lives of young, often skilled workers are being shattered. Military personnel, migrant miners, and their wives and prostitutes have the highest proportions of infection. The lack of medical understanding and panic at not being able to do anything about it generate false taboos and myths, such as the one that men can cure themselves by having sex with a virgin girl, a basis for many child rapes.

Possible measures for controlling the spread of AIDS are complex and focus on treatments that recognize human frailty

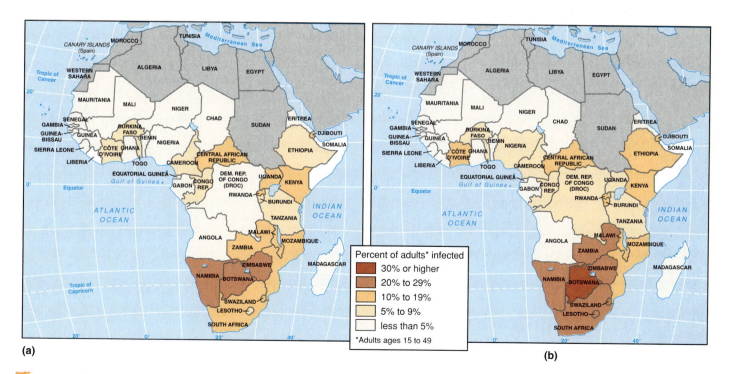

(a)   (b)

Figure 9.12 **Africa South of the Sahara: HIV/AIDS.** The percent of adults (age 15 to 49) infected with HIV/AIDS. Some countries are unlikely to report the full extent. What factors might account for the higher prevalence in Southern Africa? A high incidence is linked to falling life expectancy and slower population growth. (a) Data for 1997. (b) Data for 1999. Note: Congo is the former Zaire. Source: Data from United Nations and Population Reference Bureau.

by providing cheap or free condoms and needles because HIV is spread by sexual activity and drug addicts reusing needles. These treatments gain over moralistic "better behavior" policies. In Senegal, religious groups supported a program that included sex education in schools, the "social marketing" (at low prices) of condoms and needles, and a focus on at-risk groups (prostitutes and young men in the army). HIV infections remained below 2 percent, compared to nearly 11 percent in nearby Côte d'Ivoire. Successful actions resulted from prompt responses, open discussion of sensitive issues, and repeated targeting of vulnerable groups. Even where the cost of the drugs is reduced to affordable levels, African countries still spend more on armaments than on the clinics and laboratories needed to administer and monitor the use of the drugs. HIV/AIDS requires global action—as important as controlling terrorism, the armaments and drugs trades, and slavery.

Test Your Understanding 9A

Summary Africa South of the Sahara has a rich history and a wealth of natural resources, but it is one of the world's poorest regions and faces major problems of engaging with the global economic system. The legacy of slavery, colonialism, and ethnic divisions is a negative influence on modern political, economic, and social conditions.

The African natural environment is dominantly tropical in climate and vegetation, from rainy equatorial climate with tropical rain forest, through areas of seasonal rain and savanna grassland, to arid areas of desert vegetation. The ancient rocks contain massive mineral resources, while the scenery and animals attract many tourists. Droughts, poor soil quality, and diseases are negative factors, and this region is the world's worst for the spread of HIV/AIDS.

Questions to Think About

9A.1 How can a region with such a rich history and wealth of resources end up as the poorest in the world?

9A.2 How do the histories of Asian and African people compare in their relationships with European colonizers?

9A.3 How are climate, natural vegetation, and landforms linked in Africa South of the Sahara?

9A.4 What would be the Western countries' response if they experienced the African levels of HIV/AIDS?

Key Terms

Southern Africa Development   plateau
   Conference (SADC)   rift valley
Sahel

🌐 Africa, Globalization, and Localization

To date, Africa South of the Sahara has been a loser in the global economy. The countries of this region produce only 1 percent of global gross national income (Figure 9.13). Most people have little access to the consumer goods that are the

signs of Western-oriented well-being and a financial ability to engage with global connections (Figure 9.14).

Much of Africa South of the Sahara remains a plantation or quarry providing cheap raw materials to the materially wealthy Western world—often a hangover in underdevelopment from trade established in colonial times. Some African countries and cities are more integrated parts of the global economy. Ports and airports link commercial farms and mines to markets in wealthier countries. Cities such as Nairobi (Kenya), Johannesburg and Cape Town (South Africa), and Lagos (Nigeria) exhibit some signs of becoming global city-regions as they provide headquarters for the subregional centers of commerce, multinational corporations, and nongovernmental organizations. For many from outside, the region contains intriguing tourist opportunities because it has warm climates, historic relics, and especially, big animals.

Many African countries are part of the global economic system as dependent debtors and recipients of aid. Their postindependence focus on self-sufficiency to develop the well-being of their people did not enable them to emulate the growth realized in the countries in East and Southeast Asia or Latin America. They often spent their small resources on what they perceived to be important in raising their country's identity through lavish new capital cities and military purchases.

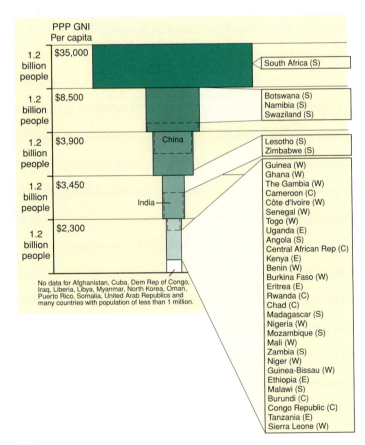

Figure 9.13 **Africa South of the Sahara: country incomes compared.** The countries are listed in the order of their PPP GNI per capita. (C) = Central Africa; (W) = Western Africa; (E) = Eastern Africa; (S) = Southern Africa. Sources: Data (for 2000) from *World Development Indicators*, World Bank, 2002.

However, poorly managed projects, together with the corrupt siphoning of funds into personal bank accounts abroad, in some countries led international lenders and bankers to impose stricter conditions. Moreover, in countries with shortages of educated potential leaders, African professionals left in a continuing brain drain for the higher salaries they could earn in wealthier countries.

Within the countries, most people have local rather than global orientations. Many rural Africans live their lives with little reference to the global economy, although it now reaches into the remotest villages through the use of motor vehicles or clothes made of synthetic fibers. Small towns have increasing involvement in the global economic system as they experience a growing mobility of people, increased levels of commercial exchange, and rising demands for the consumer goods seen advertised on global TV channels. Some villages mushroom into small service centers with rapidly built shops and market stalls where food and small consumer goods are sold and buses bring people from surrounding rural areas.

🌐 Subregions

Consideration of the cultural and natural environments of Africa South of the Sahara leads to the study of four subregions with distinctive geographic patterns of human geography (Figure 9.15):

- *Central Africa:* the countries of the Congo River drainage basin
- *Western Africa:* the countries in the western bulge of Africa
- *Eastern Africa:* the countries of the eastern Horn of Africa
- *Southern Africa:* the countries south of Congo and Tanzania, influenced strongly by the Republic of South Africa

The population distribution within the region (Figure 9.16) provides an introduction to the subregions. Central Africa has the lowest density of population in Africa South of the Sahara, on average 18.5 per km^2 in 2001 (47 per mi.^{2}). Although low, it

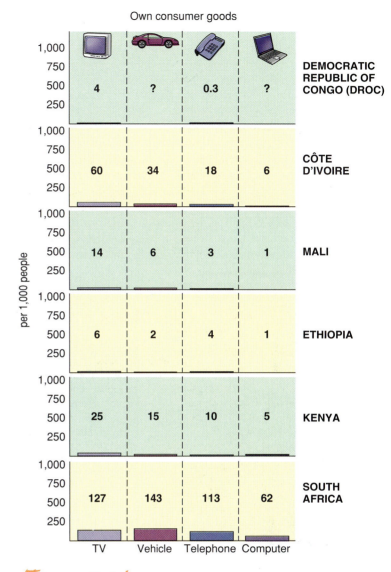

Figure 9.14 **Africa South of the Sahara: consumer goods ownership.** It is very low in African countries, apart from South Africa: compare levels with those in other world regions. How does this diagram reflect Western values? Are these items the sole criteria for prosperity? Source: Data (for 2000) from *World Development Indicators,* World Bank, 2002.

Figure 9.15 **Africa South of the Sahara: subregional and total data.** Further details at country level are included in each subregion section.

Subregion	Land Area (km²) Total	Population (millions) Mid-2001 Total	2025 Est.	GNI 1999 (US $ million) Total	GNI PPP 1999 Per Capita	Percent Urban 2001	Human Development Index Rank of 175 Countries	Human Poverty Index: Percent of Total Population
Central Africa	5,419,340	100.0	180.9	20,326	1,832.5	33.4	147.2	44.4
Western Africa	6,120,380	239.6	393.2	70,813	1,167.1	34.1	157.8	49.7
Eastern Africa	3,644,040	167.8	283.0	33,819	866.0	28.4	157.7	39.0
Southern Africa	6,571,950	130.1	155.9	166,733	3112.7	33.3	135.8	31.3
Total or average	21,755,710	637.5	1,013.0	291,691	1,744.6	32.33	149.6	41.1

Source: Data from *Population Reference Bureau 2001 Data Sheet*; *World Development Indicators,* World Bank, 2001; *Human Development Report,* United Nations, 2001.

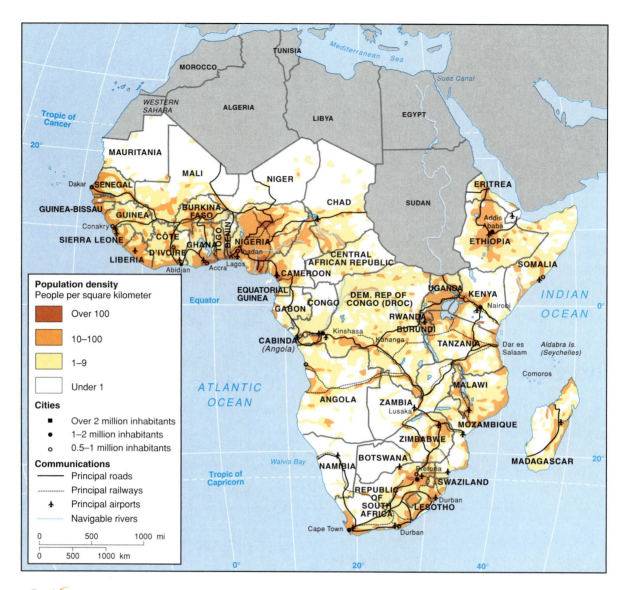

Figure 9.16   **Africa South of the Sahara: population distribution.** What is the relationship between the population densities and the natural environments (see Figures 9.8 and 9.10). Source: Data from *New Oxford School Atlas,* p. 98, Oxford University Press, U.K., 1990.

will almost double by 2025 to 33.5 per km² (85 per mi.²). At present, the small countries of Burundi and Rwanda have some of the highest population densities among African countries. The Sahara of northern Chad, Mauritania, Mali, and Niger and the tropical forests of other countries have large areas with less than one person per km². Only coastal Cameroon, the mining areas of southern Democratic Republic of Congo (DROC), and the immediate vicinity of towns and cities have more than 10 people per km².

Western Africa is the most populous subregion and densities exceed 100 people per km² along most of the coast from Nigeria to Côte d'Ivoire. Between the coasts and interior desert, areas of moderate population, including the highest rural densities in interior Nigeria, have better transportation links and more commercial farming.

Most people in Eastern Africa—in areas with population densities of 10 and more per km²—live in the better-watered upland areas of Ethiopia, Kenya, and Uganda, including the

lands bordering Lake Victoria, as well as along the major routes linking inland areas of commercial activity to ports. Few inhabit the semiarid areas between the Ethiopian uplands and Indian Ocean coast.

In Southern Africa, the population is very sparse over a large area inland from the desert-bordered southwestern coasts. The main populated area is in South Africa between the mining and industrial area around Johannesburg and the southern coasts from Cape Town to Durban. Other well-populated areas are along the coast of Mozambique and the railroad lines to inland mining centers in Zimbabwe and Zambia.

In addition to the variations in population distribution, some of the subregional linkages are expressed in efforts to cooperate through trade. The regional trading groups (Figure 9.17) are attempts by countries to work with each other, often along subregional lines, and to compete with other world regional groupings of countries. Unlike the EU (see Chapter 3) or NAFTA (see Chapters 10 and 11), however, the African groups are

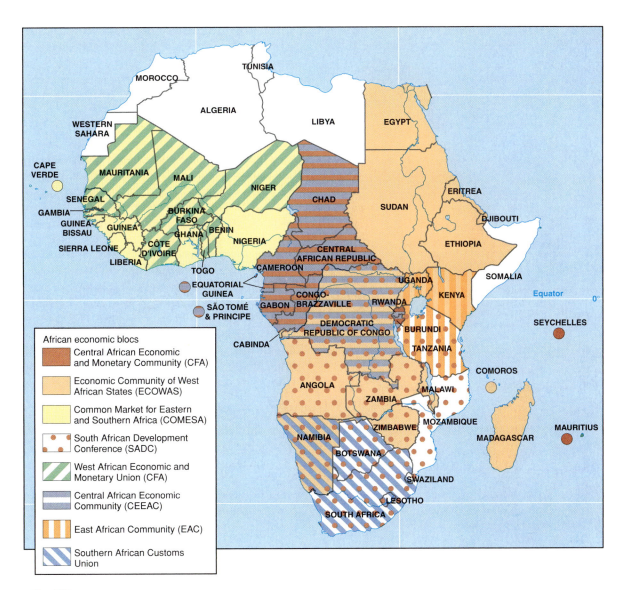

Figure 9.17 **Africa: economic blocs.** Many have little influence on world markets. There are also overlaps between regional groupings that cause problems: for example, the ECOWAS considers monetary union, but the former French colonies in the French franc (CFA) zone are expecting to change to euros. Source: © *The Economist Newspaper Group, Inc.* Reprinted by permission. Further reproduction prohibited. www.economist.com.

loosely organized and include complex overlaps. They may start up with enthusiasm but then remain dormant or achieve little for want of political support from members or credibility among the wealthier countries, or difficulties in administrating their activities.

 # Central Africa

Central Africa (Figure 9.18) is the least-developed subregion of the least-developed world region. Most countries in this subregion remain isolated from world trade networks. The natural environment, marked by equatorial rains, dense forest, and diseases, was sparsely populated before European intrusions and proved difficult for European colonial powers to develop. Belgium and France were the main colonizers, but they encouraged limited development except for selected extractive

industries. Since independence, most governments struggled to cope with the legacy of problems, and widespread civil warfare in this subregion destroyed the prospects for raising economic output and improving people's livelihoods.

Only two small coastal countries, Gabon and Equatorial Guinea, with a combined population of under 2 million, escape the worst poverty (Figure 9.19), although the distribution of income from mineral exports is uneven. Cameroon and the Congo Republic also have coastal belts with some direct links to international markets. The Democratic Republic of Congo, Central African Republic, Chad, Burundi, and Rwanda all rely for transportation links to those markets on the Congo River waterway, which is interrupted by rapids from its mouth to its source. In the 1990s, when DROC disintegrated through civil wars, countries such as Chad, Central African Republic, Burundi, and Rwanda had to find other routes—through Cameroon to the west or Kenya to the east.

Figure 9.18 **Central Africa: main features.** The map shows the countries included in the subregion, the major cities, and rivers.

Countries

After independence, the Central African countries mostly became dictatorships through the Cold War period, aligned to the Soviet Union or the United States, whatever the internal methods of governing. This provided some stability, but the autocracy was often harshly administered and rebel groups

were encouraged by the opposition. In the early 1990s, political order broke down with the collapse of economies and the disruption of social order.

The combination of large populations, centrality in the Congo River basin, and many mineral resources might be expected to make the Democratic Republic of Congo a focal country in this subregion. It is almost twice as large as the next country in area, Chad, and with 54 million people (2001), had more than three times the population of Cameroon, the only other country with over 10 million people. However, instead of forming a central driving force for development in the subregion in the late 1990s and early 2000s, DROC was largely taken over by private armies and military from surrounding countries. Government of the whole territory is virtually impossible at present.

Soon after independence in 1960, the Democratic Republic of Congo army under Mobuto Sese Seko seized control from the Communist Patrice Lumumba. Supported by the United States, he remained in power, building up his family positions and private bank accounts abroad until a 1997 rebellion deposed him. Mobuto's robbery of the national wealth resulted in a description of his government as "kleptocratic." Laurent-Désiré Kabila led the 1997 rebellion from his base in eastern DROC, using exiled **Tutsi** and local related groups with the backing of Rwanda. The demoralized DROC armies offered little opposition, Mobutu went into exile, and Kabila became president.

Kabila could not control DROC from the capital, Kinshasa, and other countries became involved, widening the impact of the internal conflict. Zimbabwe and Namibia sent troops and equipment to support Kabila because of old Marxist ties. Uganda and Rwanda backed the eastern (Tutsi) groups to attack the Kabila regime, which used **Hutu** men as the basis of the Congolese army. Angolan troops entered from the west,

Figure 9.19 **Central Africa: data for countries.** The religious groups listed here are not always exclusive. For example, in Africa many devout Catholics also follow traditional religious practices.

Country	Capital City	Land Area (km²) Total	Population (millions) Mid-2001 Total	2025 Est.	GNI 1999 (US $ million) Total	GNI PPP 1999 Per Capita	Percent Urban 2001	Human Development Index Rank of 175 Countries	Human Poverty Index: Percent of Total Population
Burundi, Republic of	Bujumbura	28,000	6.2	10.5	823	570	8	170	46.1
Cameroon, Republic of	Yaoundé	475,440	15.8	24.7	8,798	1,490	48	134	38.1
Central African Republic	Bangui	622,980	3.6	4.9	1,035	1,150	39	165	53.6
Chad, Republic of	N'Djamena	1,284,000	8.7	18.2	1,555	840	21	162	52.1
Congo, Republic of the	Brazzaville	342,000	3.1	6.3	1,571	540	41	135	32.3
Congo, Democratic Republic of	Kinshasa	2,344,860	53.6	106.0	No data	No data	29	140	No data
Equatorial Guinea, Republic of	Malabo	28,050	0.5	0.9	516	3,910	37	131	No data
Gabonese Republic	Libreville	267,670	1.2	1.4	3,987	5,280	73	124	No data
Rwanda, Republic of	Kigali	26,340	7.3	8.0	2,041	880	5	164	No data

Source: Data from *Population Reference Bureau 2001 Data Sheet*; *World Development Indicators,* World Bank, 2001; *Human Development Report,* United Nations, 2001; Microsoft Encarta (ethnic group, language, religion).

outwardly to support Kabila but mainly cutting off the supply routes to Angolan rebels. Once there, all the countries extracted mineral wealth, from diamonds to gold and copper.

After Kabila was assassinated in 2001, Joseph Kabila, his son, replaced him, although he still did not have countrywide political support. Ethnic rivalries again erupted into local wars, such as the Lunda fighting the Luba in the south (Katanga) and attracting different external country support. By 2002, as many as 2 million Congolese people had been killed, together with many thousands of refugees, military, and aid workers from other countries.

Of the other Central African countries, the landlocked countries (Chad, Central African Republic, Burundi, Rwanda) are among the world's poorest countries. Vital roads, railways, and river facilities linking coastal ports to inland centers deteriorated as a result of conflict and lack of maintenance, restricting movements of goods and people.

Chad, Central African Republic, and Congo had 30 years of upheaval after their independence from France in 1960. Chad experienced wars between the Libyan-backed Muslim northerners and the southerners; Central African Republic had misrule and misspending by a flamboyant dictator, and although it opted for an elected civilian government with a new constitution in 1995, military rebellions from 1996 dislocated the economy; Congo had an authoritarian Marxist government from independence but also opted for more democratic status in the 1990s until civil war returned the former president in 1997. Gabon and Equatorial Guinea retain long-term autocratic government.

Attempts to establish groupings of cooperating countries in this subregion, such as the Central African Economic Community and the Central African Economic and Monetary Community (see Figure 9.17), made little progress because of the civil strife.

People

Ethnicity

The peoples of Central Africa are almost all of black African ethnic groups who hold to their long-term allegiances. The positive roles of extended family and kinship connections work together with normally harmonious interreligious group relations in societies subject to major stresses. Most people belong to groups pushed southward into the forests when Muslims took over the northern parts of the region. Small numbers of the pygmy groups displaced or ethnically cleansed by the incursions continue to decline in numbers and now make up less than 2 percent of the population.

The ethnic mosaic of Central Africa, as of all Africa South of the Sahara, sits uneasily with country borders determined by colonial powers in the late 1800s and confirmed at independence. Politics in these countries is increasingly about ethnic rivalries. Each ethnic group recognizes a territory, but a country may have from tens to hundreds of such groups. The borders cut some of the territories in half (see Figure 9.3a), and there are many cross-border links. For example, the Fang people dominate Equatorial Guinea, a small country, but Fang people also live in neighboring Cameroon and Gabon. In Gabon, a pact among smaller groups keeps the Fang—the largest group—out of government. Some of the worst ethnic warfare occurred in Rwanda and Burundi in the 1990s (see the "Personal View: Rwanda" box, p. 386).

The proportion of Muslim population declines southward in the subregion (Figure 9.20). Countries with more balanced influences of traditional, Muslim, and Christian groups in this region often saw them working together and levels of syncretism in which religions borrowed from each other. Traditional practices are often adopted or allowed by Muslim and Roman Catholic groups.

Country	Ethnic Groups (percent)	Languages O=Official	Religions (percent)
Burundi, Republic of	Hutu 85%, Tutsi 14%	Kirundi (O), French (O)	Christian 67%, local 32%, Muslim 1%
Cameroon, Republic of	200 groups; Fang, Bamileke, Fulani	English (O), French (O), 24 African	Christian 53%, local 25%, Muslim 22%
Central African Republic	Baya 34%, Banada 27%, Mandjia 21%	French (O), Sango (N)	Christian 50%, local 24%, Muslim 15%
Chad, Republic of	Muslim groups in N, non-Muslim in S	French (O), Arabic (O), Sara in S	Christian 25%, local 25%, Muslim 50%
Congo, Republic of the	Kongo 48%, Sangha 20%, Teke 17%	French (O), Kikongo, Lingala, Teke	Christian 50%, local 48%, Muslim 2%
Congo, Democratic Republic of	Over 200 African groups; Mongo, Luba, Kongo	French (O), Lingala (N), others	Christian 70%, local 20%, Muslim 10%
Equatorial Guinea, Republic of	Fang 80%, Bubi 20%	Spanish (O), Fang	Officially Roman Catholic; local practices
Gabonese Republic	Fang 30%, Eshira 20%, M'bele 15%, Kota 13%	French (O), Fang, others	Christian 70%, local 10%, Muslim 20%
Rwanda, Republic of	Hutu 90%, Tutsi 9%	Kinyarwanda (O), French (O), Kishwahili	Christian 74%, local 25%, Muslim 1%

RWANDA

Rwanda is one of the smallest and poorest countries in Africa, but it has often been in the world news over the last 30 years because of civil wars and horrific tales of hatred expressed in violence. Ntwari is a Rwandan who grew up in the country during this period. His experience of living in Rwanda reflects some of the events that affected people's lives and changed the human geography of the country in the later 1900s.

Rwanda is a hilly-to-mountainous country close to the equator (known as "the land of a thousand hills"), but it is landlocked and surrounded by other countries—Congo, Uganda, Tanzania, and Burundi (see Box Figure 1). The highest volcanic peak, Karisimbi (4,507 m, 14,787 ft.), is on the edge of the rift valley that contains Lake Kivu, and most of the country is over 1,220 m (4,000 ft.). This elevation modifies the equatorial climate, bringing temperatures down to just over 20°C (70°F) and resulting in a distinctive type of tropical rain forest vegetation.

Ntwari grew up in southern Rwanda, one of nine children on a typical family plot of land with the house surrounded by banana trees and segments of the land devoted to crops such as corn, beans, potatoes, tropical root crops, a little pasture for a few head of livestock, and land for growing the cash crop, coffee or cotton. From time to time, cousins might come and stay for a year or so, but any grandparents still alive maintained their own plots. Some friends lost their parents in the civil war, but the children continued to live on the family plot, with teenagers raising their younger brothers and sisters.

With a rapidly growing population, the family plots typical of Rwanda cut into almost all the preexisting forest. Even the Akagera National Park on the eastern border was largely taken over by farmed plots of land. Although the plots can provide both subsistence and cash-crop income, many on steeper slopes suffered soil erosion, reducing the harvest as soils lost their nutrients or flooding destroyed crops. Low world market prices for coffee reduced incomes and made families indebted. Rwanda became dependent on international aid to supply food and to fund rural development programs fighting soil erosion. Although the fertility rate remains high, total population growth slowed in the 1990s because of the civil war, a high incidence of HIV/AIDS, and the continuing impact of poor nutrition and tropical diseases such as malaria.

Ntwari went to the local first-level school up to age 15. There was no public transport, and so he walked the 5 km (3 mi.) to and from school each day with his friends. He was one of three from his class of 45 to go on to the second-level school up to age 21. Again, one of a small proportion who progressed further in their education, he then studied for a bachelor's degree in sociology and anthropology at the National University in Butare. Jobs were available for his fellow students in government and aid agencies, and for Ntwari in teaching. There was even a shortage of educated local personnel, requiring expatriates to be brought in.

Following a struggle for independence that involved a civil war and increasing tensions between Hutu and Tutsi peoples, Rwanda gained its independence from Belgium in 1959. Many of the former ruling Tutsi groups, including the king, went into exile in Europe and America. At first, people from the south, the Nduga, controlled the country's government through their political party, the Democratic Republican Movement. A mixture of Hutu and Tutsi peoples living together, they provided most university students and so gained most government jobs. In 1973, a military coup d'état was led by northerners, the Rukiga, mainly of Hutu peoples, who governed for 20 years through their political party, the Republican Movement for National Development. During this period, the president took control of the military and trained a militia force of northerners. At the same time,

southerners, backed by the exiled Tutsis who had formed the Rwandan Patriotic Front (RPF), trained their own militias. While the northerners were in government, the mixture of southern peoples were increasingly labeled as "Tutsis," "friends of Tutsis," or simply "enemies of the state."

Changes in the 1980s and 1990s resulted in the greater social and political polarization of groups of people within Rwanda. Rural conditions still predominate over most of the country, with traditional ways of life continuing and a social life based around neighborhood parties. As Kigali, the capital, expanded its role after independence, the increasing numbers of educated young people in government jobs there established new social groups based on professional and business interests or on college links. Increasingly, too, social gatherings brought together peoples in groups from either the north or south of the country. This division reflected increasing political rivalries.

In 1990, matters came to a head as the RPF invaded Rwanda through Uganda in the north. They advanced southward, pushing back the Rwandan forces into the southern part of the country. This led to a massacre of southerners, who were suspected of working closely with the RPF. Ntwari, working for a church helping increasing numbers of fleeing refugees in Gitarama, was beaten up because he had taught in a school with a Tutsi head teacher (who was murdered) and had a Tutsi wife (many of whose family were killed).

Peace-keeping efforts prevailed to bring this part of the conflict to an end, but strife erupted again in 1994, when hopes for a democratic government were destroyed by the assassination of the newly elected Rwandan president. The country once more descended into civil war. Ntwari and his wife moved westward to a town near the Congo border, where they stayed for a month. Then the former (northerner-based) official Rwandan army and militias disintegrated

Box Figure 1 **Rwanda.** This tiny country is located where Central and Eastern Africa meet, but has a mighty impact on its neighbors.

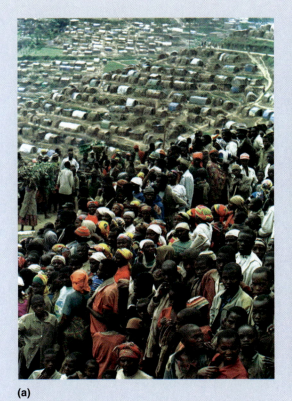

(a)

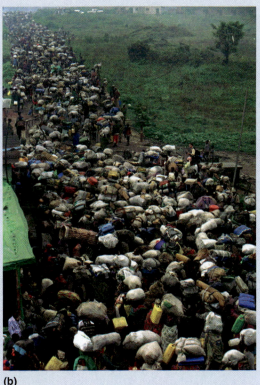

(b)

under the RPF attacks and fled into Congo. Ntwari and his wife moved with them and entered a refugee camp at Bukavu. Almost all the people in the camp were Hutus who hated Tutsis. When Ntwari's wife was threatened with death, she fled and was cared for in hiding by a local family.

Many Tutsis were killed at this time. The remainder fled into the hills or into Tanzania. The intervention of the largely French "Operation Turquoise," backed by the United Nations, saved many Tutsis from the slaughter. They were brought together in camps within Rwanda.

Ntwari decided that there was no immediate future in Rwanda or a refugee camp outside the country. Although most of the nonmilitia Hutus from the camp returned to Rwanda (Box Figure 2), he found his way via a boat across Lake Tanganyika into Tanzania and then took a train and bus to Nairobi, Kenya. Having found that he could gain admission to Kenya through this route, he returned to bring out his wife. They persuaded the United Nations Refugee Commission that they would be in danger if they returned to Rwanda and were airlifted from Bukavu camp to Nairobi. Once there, they looked for an opportunity for Ntwari to pursue his studies in another part of the world.

The Hutu-Tutsi ethnic differences are blamed for the civil conflict, but it is clearly not just a tribal war. Colonial and postindependence events heightened previous rivalries. Before the colonization of this area by Europeans, the Tutsi tribe had the status of nobility, but people from other tribes could be transferred into the "Tutsi" group following marriage or effective military service, for example. The leadership group became broadly based in tribal terms and was identified as *imfura*—civilized people able to take leadership and speak in public.

The European influence was partly political, with a short-lived German occupation (1897–1918) and a longer Belgian protectorate (1918–1959), and partly religious. The Roman Catholic Church and Protestant missionaries had major influences, probably the greatest in any African country. In 1900, the white-robed and white-faced European Roman Catholic priests ("white fathers") entered the country. In 1943, the Rwandan king was baptized and declared his lands to be a Christian country. New social status was gained by those who

became like Europeans in income or education: *umunzunga* indicates a person who is both educated and wealthy; *umusilimu* is one who is educated and not wealthy but still dresses like white people (e.g., a school teacher).

The various strands of social and cultural development led to friction between the better-educated and ethnically more varied southerners and those in the north who felt underprivileged. The military coup in 1973 and the civil war in the 1990s arose from such frictions and destroyed the fledgling economy of this small, poor country.

Ironically, despite attempts to throw off foreign influences, the country remains even more dependent on outsiders and particularly on the European countries and aid agencies that bring funding and expertise for water projects, house building, road and bridge construction, and forestry and advice for farmers. The Roman Catholic Church still runs most of the hospitals and clinics, although other churches are also involved. Any use of modern technology, from computers to telecommunications, occurs in the offices of United Nations and European agencies. All these organizations and projects are designed to work in partnership with local Rwandans and provide jobs for many of those educated to college level. But the civil war destroyed or set back many projects that had begun to improve people's lives.

Rwandans identify a need for improved education as a basis for overcoming the problems they face. This is in the widest sense. There is a need for people to understand the origins of frictions among groups of people. Who arrived first in the country—the Twa (pygmies), Hutu, or Tutsi? What were the grounds of claimed superiority? What did each group contribute? How can tolerance be reestablished? What does *democracy* mean and how can it be restored? What is the nature of development, and how can Rwandans take advantage of offered aid? What are the advantages of family planning in a country where population growth exceeds growth in economic provision? What is the truth about HIV/AIDS in a country with a high incidence and growing mortality rate? Unless the Rwandan people can come to terms with such issues openly, knowledgeably, and democratically, the future of the country will remain bleak.

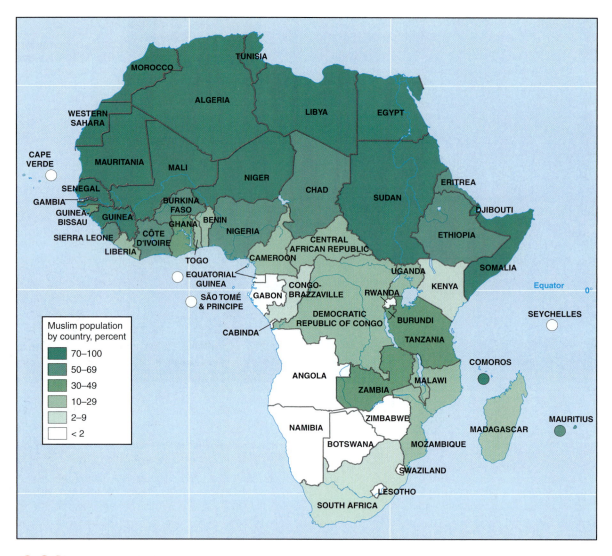

Figure 9.20 **Africa South of the Sahara: proportion of Muslims in population.** The remainder are Christian or of local traditional faiths. Countries in Western Africa are often Muslim in the north and Christian/traditional in the south. Source: © *The Economist Newspaper Group, Inc.* Reprinted by permission. Further reproduction prohibited. www.economist.com.

Urban and Rural Distribution

Most people live in rural areas: over 90 percent in Burundi and Rwanda and 60 to 80 percent in Central African Republic, Chad, Congo, DROC, and Equatorial Guinea. The high proportion of rural living is linked to the dominance of subsistence farming in the economy and the lack of urban employment apart from service industries and government jobs. Urban populations are growing, however, as people migrate into towns for perceived safety in a time of civil war, making official figures of urbanization lower than the fact. The informal economy becomes the only urban means of livelihood for these people who are cut off from their home food production.

The largest towns are often country capital cities, but none can claim the status of a connected global city. Kinshasa (DROC, 5 million in 2000) is the largest but has lost its former role as a trading center for much of the subregion because of the country's political chaos. Other capitals include Yaoundé (Cameroon, 1.4 million), N'Djamena (Chad, 1 million), and Brazzaville (Congo, 1.2 million). Libreville (Gabon) was established for freed slaves

from the Americas in the early 1800s. Douala (Cameroon, 1.7 million), Pointe Noire (Congo), and Matadi (DROC) are ports, while Lubumbashi (DROC, 1 million) is an inland mining town.

Brazzaville (Congo) and Kinshasa (DROC), on opposite banks of the Congo River, illustrate different forms of urbanization and their causes. Brazzaville became the chief city of French Equatorial Africa, where French diplomats and educated African elites built large houses on quiet avenues. Its population grew after 1960 independence, from around 100,000 to over 1 million today. Kinshasa reflects grandiose Belgian purposes—continued in the Mobutu era—with its wide boulevards and postindependence construction of prestigious tall office blocks. It grew from around 100,000 people at independence to over 5 million people, with huge sprawling shantytowns.

Population Dynamics

Despite the deaths from warfare, ethnic cleansing, and HIV/AIDS, and the disruption of so many lives, the populations of Central African countries increase rapidly. From 54 million people in 1980, the total rose to 100 million in 2001 and will likely almost double again by 2025. Such population increases place massive pressures on available resources.

The high rates of population increase, between 2.5 and 3.0 percent per year, result from continuing high fertility rates at a time of falling death rates (see Figure 9.7). The age-sex diagram for Cameroon (Figure 9.21) shows how the new births and young age groups dominate the populations. Throughout Central Africa, fertility remains high and life expectancies are low, between 40 and 55 years. Although up from 35 to 42 years in 1960, they are down from the early 1990s. Flows of refugees within DROC and from Rwanda and Burundi into DROC must affect these figures, but no statistics are available. Refugees often get sucked into military activities, including guerrilla groups, adding to disruption in countries without means of resettlement apart from special camps.

Economic Development

Although many countries have mineral deposits and tropical timber resources in demand by wealthier countries, the combination of physical isolation, repressive colonial history, continuing civil strife, and lack of foreign investment hinders the economic development of Central Africa. The Central African countries now produce a smaller proportion of total African output than they did in the mid-1960s. Only a tiny group of people in these countries can afford consumer goods, as shown by the Congo (DROC) figure (see Figure 9.14).

In real terms, incomes in almost all the countries except the oil-producing Equatorial Guinea and Gabon fell since 1980. Despite improvements in education and health since the 1960s, levels of all measures of life in Central African countries hardly rose. Human poverty indices are all over 30 percent, with those of the Central African Republic and Chad over 50 percent. Several countries depend almost totally on external

aid, since their own products earn little income from world markets and do not supply internal needs.

In the absence of the wealth common in Western countries, many Africans create ways of living that enable them to survive and even enjoy life. Items that might be regarded as waste become children's toys; bicycles or walking are common modes of travel, while cheap rides on crowded vans and buses enable wider circulation; and loads from water to many goods and possessions are often carried on people's (especially women's) heads.

Dominant Agriculture

Agriculture remains the main economic occupation of over 70 percent of the populations of these countries. Most farming is for subsistence, using traditional methods to grow tropical root crops, cereals, fruits, and vegetables. Some people continue practices of gathering forest products or of shifting agriculture (Figure 9.22). Only a small part of the surface is cultivated (Figure 9.23) because of a combination of soils that are difficult to work with, plant and animal diseases, and poor transportation facilities. For example, although much of DROC has potential for cultivation, only 3 percent is used. Cattle farming in Central Africa is severely restricted by diseases, particularly those borne by the tsetse fly. Modern veterinary treatments for animals are expensive and seldom used.

Commercial farming for export crops is poorly developed in Central Africa apart from tropical tree crops such as cacao (for chocolate), coffee, and rubber that are grown on plantations. Plantation agriculture was introduced by colonial powers to produce tropical foods and raw materials cheaply for European markets. Tree-based plantation crops replicate the form of the natural equatorial forest vegetation but not its variety of species and may fall victim to disease. However, they have the advantage that they reduce the soil fertility over 30 to 40 years instead of the 5 to 10 years under field crops. Farmers in the drier northern regions of Chad and the Central African Republic produce cotton for export, but poor transportation

Figure 9.22 Central Africa: traditional farming in tropical rain forest. Some 85 km (55 mi.) south of Kinshasa (DROC), the rain forest is being cleared by crude "slash and burn" methods. Some of the smaller wood is taken home for fuel. Crops will be grown for a few years until the soil nutrients are exhausted, then the land will be left to return to forest, often in a degraded form. Photo: © James P. Blair/National Geographic Image Collection.

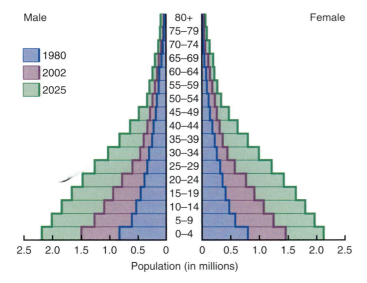
Figure 9.21 Cameroon: age-sex diagram.
Source: U.S. Census Bureau. International Data Bank.

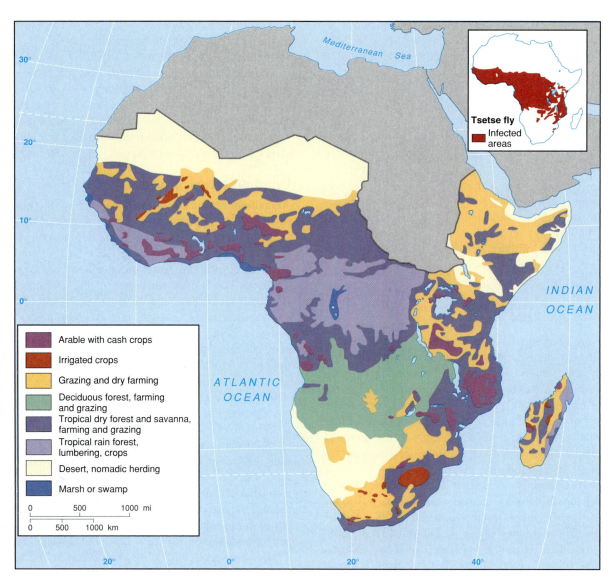

Figure 9.23   **Africa South of the Sahara: land use.** Note the distribution of intensive, commercial uses and of unusable land or low-intensity uses. The insert shows the area affected by tsetse flies, an area where cattle rearing is difficult. Source: Data from *New Oxford School Atlas*, p. 98, Oxford University Press, U.K., 1990.

from these landlocked countries makes competing in world markets difficult.

Most commercial farming was established in the European colonial era, and its continuation is often under threat. Although some countries maintained and extended such production after independence, DROC nationalization in the 1970s resulted in a decline of commercial farming within its borders. Even where plantation culture continues, growing a single commercial crop (monoculture) may bring in a good income at a time of high world market prices, but farmers suffer when low prices combine with lower yields because of declining soil quality.

Despite the dominance of farming occupations, most Central African countries find it difficult to feed their populations. Government policies maintained low prices for food in urban areas but lowered farmer incomes. The food imports made necessary by the consequent fall in local produce add to overseas debts.

Forestry, Fishing, and Mining

Timber products from the tropical rain forest, including mahogany and ebony wood, are of increasing importance to the exports of Cameroon and Congo, while Gabon produces a softwood that is used in plywood. Deep-sea fishing increases in the coastal countries such as Cameroon and Equatorial Guinea. Chad obtains fish from its lake.

Mining brings in most foreign exchange to the countries of Central Africa. It could bring more if transport infrastructure were available or world markets paid higher prices to justify investments. Before the civil disorder, southern DROC was one of the world's major copper-mining regions, and the country mined diamonds and cobalt and produced some oil. Gabon's relatively high per capita income comes from its oil and manganese production, while oil makes up 90 percent of Congo's export revenues. Cameroon, Chad, and Equatorial Guinea also have oil deposits awaiting investment before exploitation. Gabon possesses the world's largest unexploited iron ore deposit.

Emerging Manufacturing

Manufacturing industries are little developed in Central African countries. Internal markets are tiny because the countries (apart from DROC) have small total populations and few people who are wealthy enough to purchase consumer products. Most factories are small and engaged in part-processing of local mine, forest, and farm products or in making bulky products that do not stand transport (e.g., cement, bottled drinks). Cotton textiles are made in Chad from the local crop.

The relative size of economies in this subregion is reflected by the fact that, of the Central African countries, only Gabon generates more than 250 kilowatt hours (kWh) of electricity per capita each year, compared to 12,000 kWh in the United States; Burundi, the Central African Republic, and Chad generate less than 30 kWh per capita. The huge water resources of the Congo River basin, estimated to contain one-sixth of the world's potential hydroelectricity, could be a source of power for manufacturing and other forms of economic development. The Inga Dam downstream from Kinshasa harnesses some of the Congo River power, but the electricity from Inga is carried overland southward to the mines in southern DROC without intermediate allocations of electricity to local uses.

Transportation Is Vital

Prospects for economic development rest on improving internal transportation links, but so far, Gabon is the only country to invest some of its oil revenues in constructing all-weather roads and a railway into the interior. Elsewhere, the extensive navigable waterways of the Congo River system are the main means of inland transport, but the railways on either bank linking Kinshasa and Brazzaville with ocean ports are narrow gauge, limited in capacity, and unconnected to each other. The road from Kinshasa to the port of Matadi below the rapids, on which trucks once took five hours to cover the 350 km (220 mi.), was in such poor repair by the mid-1990s that trucks needed up to five days for the trip.

French Links

One of the major external economic factors affecting many countries of Central Africa and Western Africa is their history as French colonies and continuing financial links with France (Figure 9.24). Instituted in 1948, the **Communauté Financière Africaine** (CFA) franc shared by 14 African countries, including Chad, the Central African Republic, Congo, Equatorial Guinea, Gabon, and Cameroon, continued after independence. Its value was tied to the French franc and guaranteed by the Bank of France. This link held currencies at artificially high exchange rates but allowed French companies to retain dominant positions in local contracting. The high exchange rates, however, made it difficult for French-speaking African countries to export goods, used up their foreign exchange reserves on imports, and made them increasingly dependent on France. In January 1994, the French government, in line with other economic moves encouraged by the World Bank and IMF,

Figure 9.24 **French franc (CFA) area of Africa.** These countries, formerly French colonies, were supported by French loans and their currency was linked to the French franc. The French government devalued the CFA franc by 50 percent in 1994, causing major changes in economic policies in the countries. The 2002 French adoption of the euro currency resulted in some confusion among CFA countries over their regional or European links.

devalued the CFA franc by 50 percent. Initially, this brought hardship to people in the countries that depended on France, especially by raising the cost of imports. For its part, France wrote off the debts of the poorer countries, and the World Bank made increased grants available.

By 2002, most CFA countries saw some benefits in lower inflation and better foreign currency reserves, although economic growth was not as high as in other African countries. The introduction of the euro currency in the European Union resulted in confusion over the future of the CFA franc and the moves of the Economic Community of West African States (ECOWAS) toward monetary union in Western Africa (see Figure 9.17).

Western Africa

The external orientation and environmental setting of Western Africa contrast with those of Central Africa. Its long coastline makes Western Africa (Figure 9.25) physically more open to the world than Central Africa and was the first part of Africa to receive European maritime contacts. From the equatorial all-year rains of the south-facing coasts to the margins of the Sahara in the north, the natural vegetation reflects the increasing aridity northward. A belt of tropical rain forest along the south-facing coasts merges northward through savanna grasslands, where seasonal rains support frequent trees, to dry savanna with little rain and semidesert (see Figure 9.11b). Human groups responded differently to these natural zones, with the African empires based in the moister parts of the savanna grasslands.

This subregion experienced more indigenous historic development than other parts of Africa South of the Sahara. It became the center of wealthy, mainly Muslim, kingdoms along the southern margin of the Sahara (Ghana, A.D. 700–1240; Mali,

1050–1500; Songhai, 1350–1600). The trans-Sahara trade exchanged gold, ivory, and slaves from Western Africa for armaments, books, and textiles from Southwest Asia and Europe.

When Europeans began trading along the coasts of Western Africa in the 1400s, they were also at first interested in the local gold and ivory. The exploration of the Americas and the devel-

Figure **9.25** **Western Africa: main features.** The map shows the countries included in the subregion, the major cities, and rivers.

Figure 9.26 Western Africa: country data.

Country	Capital City	Land Area (km²) Total	Population (millions) Mid-2001 Total	2025 Est.	GNI 1999 (US $ million) Total	GNI PPP 1999 Per Capita	Percent Urban 2001	Human Development Index Rank of 175 Countries	Human Poverty Index: Percent of Total Population
Benin, Republic of	Porto-Novo, Cotonou	112,620	6.6	11.7	2,320	920	39	155	50.9
Burkina Faso, People's Dem Rep of	Ouagadougou	274,000	12.3	21.6	2,602	960	15	171	59.3
Côte D'Ivoire, Republic of	Abidjan	322,460	16.4	25.6	10,387	1,540	46	154	46.8
Gambia, Republic of The	Banjul	11,300	1.4	2.7	415	1,550	37	163	49.9
Ghana, Republic of	Accra	238,540	19.9	26.5	7,451	1,850	37	133	36.2
Guinea, Republic of	Conakry	245,860	7.6	12.6	3,556	1,870	26	161	50.5
Guinea-Bassau, Republic of	Bisau	36,120	1.2	2.2	194	630	22	168	51.8
Liberia, Republic of	Monrovia	97,750	3.2	6.0	No data	No data	45	No data	No data
Mali, Republic of	Bamako	1,240,190	11.0	21.6	2,577	740	26	166	52.8
Mauritania, Islamic Republic of	Ouackchott	1,025,520	2.7	5.4	1,001	1,550	54	149	47.5
Niger, Republic of	Niamey	1,267,000	10.4	18.8	1,974	740	17	173	65.5
Nigeria, Federal Republic of	Lagos	923,770	126.6	204.5	31,600	770	36	146	38.2
Senegal, Republic of	Dakar	196,720	9.7	16.5	4,685	1,400	43	153	49.6
Sierra Leone, Republic of	Freetown	71,740	5.4	9.9	653	440	37	174	57.7
Togo, Republic of	Lomé	56,790	5.2	7.6	1,398	1,380	31	143	38.4

Source: Data from *Population Reference Bureau 2001 Data Sheet*; *World Development Indicators,* World Bank, 2001; *Human Development Report,* United Nations, 2001; Microsoft Encarta (ethnic group, language, religion).

opment there of sugar, tobacco, and cotton plantations led to demands for labor that could not be filled locally or by indentured Europeans. From the 1600s to the early 1800s, the trans-Atlantic slave trade dominated the exploitation of Western Africa. It enriched European shipowners and merchants, who financed the growth of ports in Europe to dominate world trade. It influenced the underdevelopment of Western Africa by taking around 10 million people away and focusing on the extraction of high-value minerals, ivory, and slaves. Local payments for slaves enabled the internal African slave-trading aristocracies to buy guns and luxuries from Europe, dominate their African neighbors, and discourage European incursions. When the Atlantic slave trade ended after 1808, some slaves were repatriated from the United States to Liberia and from British colonies in the West Indies to Freetown (in modern Sierra Leone).

From a phase of trading at the margins of Africa, internal interventions increased. During the later 1800s and early 1900s, French and British companies established plantations to produce cocoa and palm oil in the coastal forests, while mining companies explored the interior. Roads and railroads connected interior mines with ports to facilitate exports.

In 1884, the Berlin Congress allocated African lands to the European colonial powers. Occupying and controlling the new colonies in Western Africa, however, was often difficult and met strong resistance into the 1900s. Colonies became inland extensions of coastal bases, and boundaries between the colonial territories divided tribes and African territories. For example, the Kanem-Bornu sultanate, founded around A.D. 1100, was divided among Nigeria, Cameroon, and Niger/Chad. Tensions also continued among the colonizers, and after World War I, Germany lost its colonies (Togo and Cameroon) to the British and French.

Countries

The 15 countries of Western Africa range from Nigeria, with Africa's largest population (127 million in 2001), to Guinea-Bissau and The Gambia, which had just over 1 million people each. Besides Nigeria, only Ghana (19.9 million), Côte d'Ivoire, Burkina Faso, Mali, and Niger had over 10 million people (Figure 9.26).

The largest country, Nigeria, ranges across all three natural climate/vegetation zones, and others with a southern coast—Benin, Togo, Ghana, Côte d'Ivoire—also have a strip of rain forest backed by moist savanna. The west-facing coastal countries—Liberia, Sierra Leone, Guinea, Guinea-Bissau, Senegal, The Gambia, and Mauritania—become drier northward. The large land-locked countries of Mali, Burkina Faso, and Niger have only dry savanna and semidesert zones. Attempts to bring together the countries of Western Africa for their mutual economic benefit centered in the Economic Community of Western African States (ECOWAS) and the West African Economic and Monetary Union, but these links grew slowly and intermittently.

Fourteen countries achieved independence around 1960, with Guinea-Bissau becoming independent of Portugal in 1974. Although many countries are small in area with small

Country	Ethnic Groups (percent)	Languages O=Official	Religions (percent)
Benin, Republic of	42 African groups: Fon, Adju, Yoruba	French (O), Fon, Yoruba	Christian 15%, local 70%, Muslim 15%
Burkina Faso, People's Dem Rep of	Mossi 25%, Gurumsi, Senufo	French (O), Sudanic tribal languages	Christian 10%, local 40%, Muslim 50%
Côte D'Ivoire, Republic of	60 African groups, inc. foreign: Akan, Kru, Mande	French (O), Akan, Dioula	Christian 12%, local 65%, Muslim 23%
Gambia, Republic of The	Mandinke 42%, Fulani, Woluf	English (O), Mandinke, Woluf, Fula	Christian 9%, Muslim 90%
Ghana, Republic of	Fanti, Ashanti, Gadangbe, Ewe	English (O), various African	Christian 24%, local 38%, Muslim 30%
Guinea, Republic of	Fulani 35%, Malinde 30%, Sousson 20%	French (O), African languages	Christian 8%, local 7%, Muslim 85%
Guinea-Bassau, Republic of	Balanta 27%, Fula 23%, many others	Portuguese (O), Kriolu, French, local	Christian 5%, local 65%, Muslim
Liberia, Republic of	African groups 95%, U.S. Liberians 5%	English (O), local Niger-Congo groups	Christian 10%, local 70%, Muslim 20%
Mali, Republic of	Mande 50%, Peul, Voltaic, Tuareg/Moor, Songhai	French (O), Bambara, others	Local 9%, Muslim 90%
Mauritania, Islamic Republic of	Maure (Arab-Berber) 80%, Fulani, Wolof	Arabic (O), Wolof, French	Muslim 100%
Niger, Republic of	Hausa 56%, Djena 22%	French (O), 10 other official	Christian/local 20%, Muslim 80%
Nigeria, Federal Republic of	Hausa 65%, Fulani, Yoruba, Ibo	English (O), Hausa, others	Christian 40%, local 10%, Muslim 50%
Senegal, Republic of	Wolof 36%, Fulani 17%, Serer 17%	French (O), Wolof, Serer, others	Christian 2%, local 6%, Muslim 92%
Sierra Leone, Republic of	Mended, Temeden, Krioles, others	English (O), Krio, others	Christian 10%, local 30%, Muslim 60%
Togo, Republic of	37 African groups: Ewe, Kabre	French (O), Ewe, Kabre, others	Christian 35%, local 50%, Muslim 15%

populations, attempts to join Senegal and Mali, and later Senegal and The Gambia, collapsed as politicians failed to work together across the established borders. Except for Côte d'Ivoire, all countries until the early 1990s had long periods of control by dictators of military or socialist background. After that, most countries moved to multiparty democratic constitutions. These worked most successfully in Ghana, Mali, Senegal, and The Gambia. In Guinea, Guinea-Bissau, Mauritania, Niger, and Togo, the military retained real control, the move to democracy ended after military coups, or the elections were contested as having been unfairly conducted by the majority party. Even in Côte d'Ivoire, with its long history of democracy, internal north-south ethnic and immigrant worker quarrels damaged its stable, tolerant, and prosperous image by a military coup in 1999, continued ethnic and antiforeign violence in 2000, and reports of child slavery on cocoa and coffee farms in 2001.

In the 1990s, civil war in Liberia, Sierra Leone, and Guinea-Bissau disrupted their economies and social systems and threatened to spread into Guinea. In Liberia, the differences between local people and the slaves resettled from the United States in the 1800s were reflected in military rank. Officers came from the minority group descended from returning slaves and nonofficers from the majority indigenous groups. Sergeant Samuel Doe of the indigenous group led a coup in 1980 and became the new president. Further coups occurred into the 1990s. In 2000, President Charles Taylor supplied arms to Sierra Leone rebel groups in exchange for diamonds that he banked. The civil strife in Sierra Leone continued and spread into Guinea, where the resettling of refugees sparked external attacks and encouraged local dissidents. The combination of traditional intergroup rivalries, colonial favoritism, Cold War antagonisms, and personal ambitions lies behind current conflicts in Nigeria, Liberia, and Sierra Leone.

The largest and most populous country in the region exemplifies many of the problems facing the countries of Western Africa. Nigeria experienced so much bad government in the 1980s and 1990s that it moved from an oil-rich, middle-income country to become one of the world's poorest. A lot of the problems began in the continuing north-south ethnic and religious conflict. The northern Muslim Hausa and Fulani people took control of the military and much of the rest of the country, with the southern Yoruba being outvoted and the Igbos being widely disregarded after losing the 1960s civil war in which they had tried to assert their independence. While the 1966–1979 military rule brought life to the economy and united peoples with very different interests inside the country, the military rule from 1984 to 1999 was a disaster. It deepened cultural divisions by favoring the north and mismanaged the economy through neglect and corruption. In 1998, it was estimated that three-fourths of official GDP was generated by the informal economy. Some US $12.4 billion of government funds was paid out without proper accounting, while the military dictator in the mid-1990s stole $3 billion. Northern Muslims dominated the military governments, imposing repressive and manipulative regimes. They made Nigeria a member of the Organization of the Islamic Conference, annulled the 1993 presidential election, and repressed the peoples of the oil-producing Niger River delta region.

After 1999, Nigeria made a hasty transition from military to democratic government without an opportunity to develop countrywide political parties. It had to deal with the fragile federal system, the military undermining of the judicial system, and the reduction of the police force (in case it competed with the military).

The democratic government of Nigeria elected in 1999 faced huge problems over the legitimacy of the 1999 constitution that was hastily put together by the military before they ceded power.

The main political challenges included:

- Several states in the Muslim north declared in 1999 and 2000 that Islamic religious (sharia) law should take precedence over the colonially inherited common-law order, raising questions about the relationships between public institutions and religious traditions throughout Nigeria.

- Most of Nigeria's oil comes from the delta of the Niger River, inhabited by several small, poor ethnic groups. The oil revenues, however, go to the federal government, which takes the largest share and divides the rest among the 36 states on the basis of population and area. The oil-producing areas demand that more funds be returned to them, and local people cause pipeline disruptions. The federal government scarcely listens and replies with repressive measures such as the murder of the activist Ken Saro-Wiwa, who protested oil company and government attitudes.

- The 17 southern states complain that the central government makes most of the important decisions without listening to the states. They want more local power over the police force, education, revenues from resource exploitation, and infrastructure provision. But the 19 northern states resist this, since they are poorer and more dependent on central government funding.

People

Ethnicity

Ethnic differences in Western Africa have important political and social consequences. Many countries established in the former colonial order contained mostly unhappy amalgamations of inland Muslim groups of former pastoral people in the savannas and the sedentary coastal animist and Christian groups. However, many countries, such as Ghana and the landlocked countries, maintain good intergroup relations.

Urbanization

Urbanization is rapid, although no country in Western Africa outside the deserts of Mauritania had half of its population living in towns in 2002, and Burkina Faso and Niger were less than 20 percent urban. Compared with Central Africa, urbanization in Western Africa still partly reflects perceptions of safety but is more related to job opportunities and health and education provision in towns. Today's major cities nearly all began as ports or port-related colonial developments, although there are few natural harbors. Lagos (Nigeria, 17 million people in 2000), Abidjan (Côte d'Ivoire, 3 million), Dakar

(Senegal, 2 million), Accra (Ghana, 2 million), Freetown (Sierra Leone), Monrovia (Liberia), and Conakry (Guinea, 2 million) are of this type. Inland centers such as Ouagadougou (Burkina Faso, 1 million) and Bamako (Mali, 1 million) began as ancient trade centers at the southern end of trans-Sahara routes and became modern capital cities of the interior countries. All these are primate cities, much larger than any others in their countries.

Many of the oldest African urban landscapes occur in Western Africa. They include the trading centers of Timbuktu (Figure 9.27), Sokoto, and Kano at the southern end of trade routes across the Sahara. These towns are still dominated by craft workshops rather than modern industry. Ibadan, the center of the Yoruba culture in southwestern Nigeria, preserves a pre-European urban landscape linked to a series of surrounding towns. Ibadan has a central palace and nearby market at the focus of streets radiating toward other towns. Fortifications were added later to resist Muslim attacks. The modern populations of these older towns remain much more ethnically unified than those in many of the newer towns. Islamic cities along the Sahel zone (e.g., Timbuktu, Kano, Sokoto, N'Djamena) have central markets, mosques, citadels, and public baths.

Cities built on the modern trade routes or as ports may retain older sections but show evidence of colonial government buildings and postindependence, often grandiose, public buildings and housing. Accra, Ghana (Figure 9.28), and Kano, Nigeria (Figure 9.29), illustrate many of the features typical of West African cities—the precolonial walled city, religious segregation, colonial expansion, and features of independence.

Population Dynamics

The population of Western Africa is increasing rapidly. The 2000 total of over 230 million compared to just over 150 million in 1980 and could reach 400 million by 2025. Although death rates continue to fall, birth rates scarcely change (see Figure 9.7). The age-sex diagram for Nigeria (Figure 9.30) shows the results of high birth rates in very large numbers of children and young people. As death rates fell across this subregion, life expectancies increased from 35 to 50 years in 1960 to between 40 (Niger) and 55 (Ghana) years in 2000 but remain low by world standards.

Economic Development

The newly independent Western African countries began with much greater hopes of an improving future than those in Central Africa or many other poorer countries outside Africa. The combination of restricted economic bases, rapid population growth, drought hazards, political conflict and mismanagement, and fluctuating world market conditions got in the way. Almost all countries in Western Africa experienced declining real income from 1980, and all were in the World Bank's lowest group in 2000. Benin, Burkino Faso, Guinea, Guinea-Bissau, Mali, Niger, and Sierra Leone have some of the highest human poverty indices in Africa, between 50 and 66 (Niger) percent. The low levels of consumer goods ownership reflect this outcome (see Figure 9.14).

Western Africa supplies world markets with a wide range of products from farms, forests, mines, and ocean waters. Although some countries diversified their output and are not as dependent on one product as they were, all primary products remain sensitive to world markets and exchange rates. After the 1970s, when the prices of oil, metallic minerals, and agricultural exports remained high, the 1980s were disastrous for many countries in Western Africa as world prices fell. Crops are affected by weather as well as market fluctuations. From the 1970s, droughts affected peanut and cattle production in the northern parts of the subregion. Furthermore, former Western European buyers of palm oil and peanuts as sources of vegetable oil began to produce their own oil crops, causing a decline in demand for palm oil and peanuts.

Primacy of Agriculture

Agriculture remains the main source of employment (50–90%) and income (25–50% of GDP) for most countries. For the majority of people, this means subsistence—based on growing crops in the south but increasingly on herding livestock toward

Figure 9.27 **African urban landscapes: precolonial.** The old buildings of Timbuktu, Mali, founded in the A.D. 1000s at the center of a former African kingdom and on trans-Saharan trade routes. Compare this urban landscape with others known to you. Photo: © Charles and Josette Lenars/Corbis.

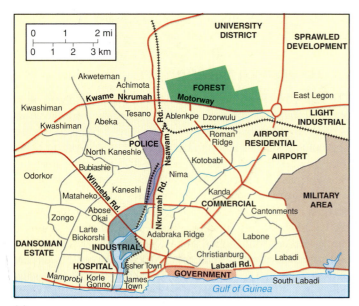

Figure 9.28 **African urban landscapes: Accra, Ghana.** (a) The port city of Accra, Ghana, expanded from a village for colonial trade, with port facilities along the lower river, flanked by government offices to the east and poor housing to the west. Account for locations of military, airport, and university districts. (b) A generalized pattern based on Accra that is repeated (with variations) in other Western African port cities. Source: (a) Data from S. Aryeetey-Attoh, *Geography of Sub-Saharan Africa*, Fig. 7.4, pp. 192–93, Prentice-Hall, 1997.

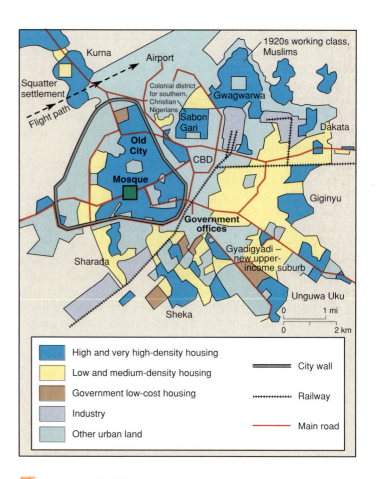

Figure 9.29 **African urban landscapes: Kano, Nigeria.** The inland center of Kano, northern Nigeria, reflecting over 1,000 years of change. The old city was built in the early A.D. 1100s and enlarged several times, with a central market linking trade routes northward across the Sahara and southward to the coast. The colonial period disrupted former trade patterns, but the railroad made Kano the major center in northern Nigeria with exports of cotton and peanuts and led to the establishment of government offices, industrial and residential areas outside the walls. Since independence, manufacturing, education, and civil service jobs have increased, giving rise to the new residential areas ranging from high-income to poor squatter shantytowns. Source: Data from R. Stock, *Africa South of the Sahara*, Vignette 13.1, p. 199, and Fig. 13.5, p. 201, Guilford Press, 1995.

the drier lands of the north. Of the main tropical commercial agriculture producers, Nigeria, Ghana, and Côte d'Ivoire are world leaders in palm oil, cacao (Figure 9.31), rubber, tropical fruits, rice, and coffee. Benin and Togo produce smaller amounts. Senegal, The Gambia, Nigeria, and Mali in the drier zone export peanuts, and Mali also exports cotton, but the interior countries have few commercial farm products apart from their livestock, which provided meat for the coastal countries until the major droughts of the 1970s decimated the herds.

Through the 1980s, Côte d'Ivoire diversified its established products such as cocoa, coffee, and palm oil (of which it is Africa's leading producer) into bananas, pineapples, cotton, and rubber. In the 1990s, the government privatized the largest rubber plantations. For a few years, growers benefited from the

CFA devaluation, with exported crops earning up to twice the previous price.

Liberia's civil strife disrupted its output and trade in rubber produced on plantations established by the Firestone Rubber Company in 1926. Plantation owners moved their operations to countries such as Nigeria and Côte d'Ivoire.

Until the 1990s, Ghana's agriculture sector grew slowly. The crop marketing boards established in colonial times set quotas and took a percentage of income for administration. By contrast, Nigerian farmers enjoyed greater incomes after the government abolished its crop boards and their charges. Nigeria invested more in crop research and farm management education and has a better rural infrastructure, enabling farmers to get their produce to market more easily.

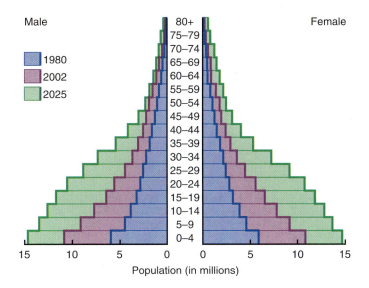

Figure 9.30 **Nigeria: age-sex diagram.** Source: U.S. Census Bureau. International Data Bank.

Agricultural development policies shifted from large externally funded projects to smaller-scale projects that involved the local farmers more centrally. In the 1970s, countries that neglected food production found themselves paying for expensive food imports. Attempting to produce more wheat, corn, and rice for local and national markets, they invested in large-scale water projects in the drier parts of northern Nigeria, Senegal, and The Gambia. However, poor management of human and natural resources resulted in lower-than-expected crop yields, higher costs, and disaffected farmers who were given little latitude for their own decision-making. The sudden shift to commercial crop production using high-priced new strains of crops and heavy applications of fertilizer was unsustainable environmentally and financially.

Subsequent development projects in the 1980s focused on involving local communities in sustainable development based on small-scale and appropriate technology that fed people as well as producing crops for sale in world markets. Funds were used to sink water boreholes worked by hand pumps and to provide rural infrastructure, advice, and credit. Better projects combined commercial and subsistence crops, as well as row and tree crops with some pastoral farming.

Forestry, Fishing, and Mining

Forest products remain significant in Côte d'Ivoire and Liberia. In many countries of Western Africa, however, the former coastal forests were largely cut and replaced by tree-crop plantations. In Nigeria, attempts to maintain the remnant of woodland in national forest reserves face continual challenges by farmers and timber companies.

Fishing is of growing importance in countries such as Mauritania, Senegal, and Guinea-Bissau, which are close to the Cold Canaries Current and the nutrients it brings to the surface. However, their fishermen face increasing competition from Spanish trawlers. The fishing licenses countries negotiated with

Figure 9.31 **Western Africa: plantation agriculture.** Cacao trees in a small plantation near Oyo, Nigeria. The pods contain beans from which chocolate and cocoa are made. They grow directly from the trunk of the tree. Photo: © W. Robert Moore/National Geographic Society.

the European Union bring in revenue that makes up one-third to one-half of their foreign exchange earnings. But enforcing the agreements is difficult, and EU boats—along with those of the Japanese, Russian, and Chinese—take more than their allocations, depleting the stocks. Lake and river fish are important foods in the interior countries such as Niger and Mali.

The main mineral developments are the oil and natural gas fields of southern Nigeria (Figure 9.32) and smaller quantities

Figure 9.32 **Western Africa: Nigerian oil refinery.** Africa's main oil field is located in the tropical rain forest of the Niger River delta. Notice the storage tanks, the refinery area on the left, and the stack flaring off natural gas. Imagine a discussion between two local people—one who works at the facility and one suffering from pollution and low income. Photo: © Mike Wells/Stone/Getty.

in Ghana and Côte d'Ivoire. Oil was discovered in Nigeria in 1956, and production expanded to a maximum of 114.2 million tons at the height of world demand and high prices in 1979, making it a major world producer, before falling back to 73 million tons in the mid-1980s. Oil provides over 90 percent of Nigerian exports, but the oil income was used for grandiose infrastructure projects funded by loans that turned into debts. Dependence on oil income reduced enthusiasm for economic diversification and selling other products in world markets. Despite the great hopes Nigeria placed on using its oil income to boost its economy, its living standards are now lower than before the oil boom of the 1970s.

Most countries have mineral resources that provide income. In Ghana (formerly the colony of Gold Coast), production in the gold-mining industry increased in the 1990s following major new investments. Bauxite mining linked to local smelting of aluminum powered by the Volta Dam remains a major output. Guinea possesses one-third of the world's reserves of high-grade bauxite and is one of the world's top producers. Guinea also has major gold and diamond reserves and iron mines near the Liberian border, where production was halted by civil war. Iron ore is mined in Mauritania and Liberia. Nigeria produces tin, and Niger has important uranium deposits. Phosphate deposits are mined in Senegal and Togo, and others exist in Mali. The Sierra Leone diamonds are in the control of rebel groups that trade them for arms with Liberia's rulers.

Ghana and Côte d'Ivoire wish to use the natural gas that was previously flared off from Nigerian oil wells to fuel thermal power stations. Plans to build a gas pipeline along the coast await agreement on whether it should be on land (Nigeria's preference for increased control) or offshore along the seabed (Ghana's preference to diminish terrorist actions).

Manufacturing

The manufacturing sector produces only 5 to 15 percent of Western Africa's GDP. Most countries have import-substitution factories for food and drink products, construction materials, and other low-price or high-bulk goods for local markets. Export crops and minerals are sometimes processed locally in oil-refining, textile, fertilizer, aluminum, rubber, or fishmeal factories. Nigeria and Côte d'Ivoire have car assembly factories. The lowest proportions of manufacturing industry occur in the landlocked countries.

As a result of the devaluation of the CFA franc, some local manufacturing industries developed in Western Africa. One of the most obvious is the manufacture in Côte d'Ivoire of the highly colored printed cloth used in women's dresses across Western Africa (Figure 9.33)—after years of importing it from Europe.

Services

The service sector of the economy in Western African countries grows slowly, largely through the government employment that is a major factor in many countries. However, structural adjustment demanding the reduction of civil service bureaucracies causes pain to many countries because of the importance of family and political links into such employment.

Tourism, a service industry that is part of the development strategy of many poor countries, was little developed in Western

Figure 9.33 **Ghana.** Girls wearing kente cloth, the signature fabric of the Ashanti people (Kumasi area). The cloth is handwoven in strips that are then sewn together. Photo: © Lisa Kahn Schnell.

Test Your Understanding 9B

Summary Central Africa is isolated from the world economy by its physical environment and by the warfare that consumes the Democratic Republic of Congo, the largest country. Both physical and human resources are poorly developed. Most people depend on subsistence agriculture. Historic slavery and colonial experiences did not prepare the countries for an independence that would bring people a full life.

At independence, Western Africa was involved in the world economy through its tropical tree crops and mineral exports, but civil wars disrupted development. Connections with European markets since the A.D. 1400s have been marked by unequal exchange. Commodity trade established in colonial times continues with tropical crops and minerals but is increasingly disrupted by civil wars in the subregion.

Questions to Think About

9B.1 Summarize the differences between Central and Western Africa through five geographic characteristics (e.g., distribution of people, ethnic groups, climates and landforms, resources, economic activities, governments).

9B.2 Why is there a lack of engagement with the rest of the world in Central Africa?

9B.3 Why do many Western African countries have poorly developed manufacturing sectors?

9B.4 Why are tropical tree crops an uncertain source of income for Central and Western Africa?

Key Terms

Tutsi
Hutu

Communauté Financière Africaine

Africa, apart from servicing small groups of travelers crossing the Sahara. When African Americans began seeking their family roots increased numbers of groups visited this subregion.

In 2001, Nigeria began to talk about encouraging tourism, despite comments such as those from Chinua Achebe, the country's greatest living novelist, saying that, "It is a measure of our self-delusion that we can talk about developing tourism. Nigeria is one of the most corrupt, insensitive, inefficient places under the sun. It is dirty, callous, noisy, dishonest, and vulgar. Only a masochist with an exuberant taste for self-violence will pick Nigeria for a holiday." Other countries, such as Ghana, Côte d'Ivoire, and The Gambia developed tourist industries based on sun and historical features.

Some Western Africans are financing small-scale enterprises by microlending. Small sums of money are lent to poor people to help them set up or expand a business. This program began in the 1970s in Bangladesh (see the section on the Grameen Bank in Chapter 7) and Latin America. In Burkina Faso, a small restaurant in a marketplace near Ouagadougou is built simply of log and thatch and serves simple meals based on rice. The owner took out a small loan to buy the rice wholesale and now employs seven people, pays her children's school fees, and owns a motorcycle. As elsewhere, microlending funds mainly businesses run by women, who are likely to use the additional income to feed and clothe children. Such lending offers independence and rewards enterprise.

🌐 Eastern Africa

Eastern Africa's first orientation to the outside world was toward the Arab countries bordering the Red Sea and the Indian Ocean. Arab traders established a series of centers along the eastern coast of Africa, often on islands such as Zanzibar and Pemba. From the A.D. 700s, trading increased, with an emphasis on ivory and gold until the 1700s and 1800s. As many as 5 million slaves were exported to Arabia, Persia, and even China. An African-Arabian Islamic culture developed with Indian and Persian influences and leaders constructing impressive palaces. Communication through their distinctive Swahili creole language became common throughout Eastern Africa. European colonists brought in laborers from India and redirected trade toward European countries, links that were often maintained after independence.

Countries

The countries of Eastern Africa form two groups (Figure 9.34). In the north, Ethiopia is partly surrounded by Eritrea, Djibouti, and Somalia. To the south are the former British colonies of Uganda, Kenya, and Tanzania occupying a plateau area that is broken in the west by major rift valleys and has volcanic peaks rising above (Figure 9.35). These countries have a variety of political arrangements and varied economic and cultural conditions (Figure 9.36), and most are affected by continuing internal conflicts.

Figure 9.34 **Eastern Africa: main features.** The map shows the countries included in the subregion, the major cities, and rivers.

Figure 9.35 **Eastern Africa: Mount Kenya.** A space shuttle view shows the remnant of former glaciers near the peak (white), the open forest of the upper slopes, the rain forest on the lower slopes, and the surrounding savanna that has been largely converted to farmland. Photo: NASA.

Figure 9.36   Eastern Africa: country data.

Country	Capital City	Land Area (km²) Total	Population (millions) Mid-2001 Total	2025 Est.	GNI 1999 (US $ million) Total	GNI PPP 1999 Per Capita	Percent Urban 2001	Human Development Index Rank of 175 Countries	Human Poverty Index: Percent of Total Population
Djibouti, Republic of	Djibouti	23,200	0.6	1.1	511	No data	83	157	40.8
Eritrea, Republic of	Asmara	121,140	4.3	8.3	779	1,040	16	167	No data
Ethiopia, Federal Democratic Rep of	Addis Ababa	1,100,760	65.4	117.6	6,524	620	15	172	55.8
Kenya, Republic of	Nairobi	580,370	29.8	33.3	10,696	1,010	20	136	28.2
Somali Democratic Republic	Mogadishu	637,600	7.5	14.9	No data	No data	28	No data	No data
Tanzania, United Republic of	Dar es Salaam	945,090	36.2	59.8	8,515	500	22	156	29.8
Uganda, Republic of	Kampala	235,880	24.0	48.0	6,794	1,160	15	158	40.6

Source: Data from *Population Reference Bureau 2001 Data Sheet*; *World Development Indicators,* World Bank, 2001; *Human Development Report,* United Nations, 2001; Microsoft Encarta (ethnic group, language, religion).

Ethiopia and Neighboring Countries

The uplands of Ethiopia (see Figure 9.10a) force moisture to rise and cause rains that feed into the Blue Nile River (see Chapter 8) and support agriculture. The surrounding lands are mainly arid.

Ethiopia has the distinction of being the only African country apart from Liberia in Western Africa that was not colonized by a European power. Although occupied by the Italian military from 1936 to 1941, it was soon returned to the Emperor Haille Selassie's rule by British forces at the outset of World War II. Selassie reigned from 1930 to 1974, when a military coup brought a Communist government to power. After further coups, drought, and huge movements of refugees in and out of the country, rebel groups replaced that regime in 1991. A new constitution in 1994 led to multiparty elections.

The small countries to the north and east reflect attempts to expand or displace the Ethiopian monarchy, who, with their people, converted to the Coptic Church in the A.D. 300s. Muslims failed to dislodge the monarchy on many occasions, although they devastated much of the area in 1523 and established coastal settlements that are now the countries of Eritrea, Djibouti, and Somalia. Ethiopia remained an isolated feudal country.

In the late 1800s, Britain expanded the port of Aden on the Arabian Peninsula coast to guard the approaches to the Suez Canal and refuel its ships on the way to and from India. It established British Somaliland on the African coast. Italy colonized Eritrea and the Indian Ocean coast of present Somalia in the late 1800s.

After World War II, the United Nations federated Eritrea with Ethiopia and joined British and Italian Somaliland as Somalia. The former British Somaliland declared its independence from Somalia in 1991 but is not recognized by other countries. In the early 2000s, Somalia was in chaos after a United Nations force, largely made up of U.S. military, had left in 1995 without restoring order. The southern part of Somalia is ungovernable and in a state of anarchy marked by random banditry and fighting among clans. The official government controls only part of Mogadishu. A loose alliance of warlords based in Baidou and supported by Ethiopia controls most of southern Somalia. Puntland, the northeastern corner of Somalia, remained peaceful throughout the 1990s, exporting livestock and frankincense. After September 11, 2001, however, activities by Islamist extremists gave the local political struggles an international emphasis, involving Ethiopian support against the extremists.

Disappointed at being reduced to a province within Ethiopia, Eritreans fought Ethiopian and Soviet Union forces. After the latter departed and the Communist government in Ethiopia was deposed, Eritrea gained independence in 1993, making Ethiopia landlocked. Eritrea is Africa's newest country. Tensions between Ethiopia and Eritrea continued, and costly border wars from 1998 to 2001 sapped both countries' resources and made it difficult to implement the new Ethiopian constitution.

Former British East Africa

The countries of Kenya, Tanzania, and Uganda present contrasts in physical and human geography to those grouped around Ethiopia. On the west, streams flowing into Lake Victoria and out into the rift valley are sources of the White Nile River. Although occupying an equatorial position, the plateau receives only moderate rainfall totals, which also fall off toward Ethiopia and Somalia in the north.

Before the European colonial era, the region was a major crossroads of African peoples. Cattle herders from the Nile River valley, such as the Masai, lived in tension with farming Bantu peoples, such as the Kikuyu of Kenya, migrating from Western and Central Africa. On the east, ports established by Arabs as part of their Indian Ocean trading system were dominated by the sultans of Oman and Muscat until the late 1800s.

When European colonists arrived in the late 1800s, they met less organized resistance to settlement than in Western Africa. Britain got involved initially to end the slave trade based on Zanzibar Island under the sultans of Oman and Muscat (Figure 9.37), who also controlled the Indian Ocean trade in cloves and palm oil. In 1886, Britain annexed Kenya

Country	Ethnic Groups (percent)	Languages O=Official	Religions (percent)
Djibouti, Republic of	Somali 60%, Ethiopian 35%	French (O), Arabic (O)	Christian 6%, Muslim 94%
Eritrea, Republic of	Tigray 50%, Tigre-Kunama 40%	Tigre, Afar, others	Christian, Muslim
Ethiopia, Federal Democratic Rep of	Oromo 40%, Amharan 32%, Tigray	Amharic (O)	Christian 45%, local 12%, Muslim 40%
Kenya, Republic of	Kikuyu 21%, Lui 15%, Luhya 14%	English (O), Swahili (O)	Christian 70%, local 10%, Muslim 6%
Somali Democratic Republic	Somali 85%	Somali (O)	Muslim 99%
Tanzania, United Republic of	120 African culture groups	Swahili (O), English (O)	Christian 43%, Muslim 35%
Uganda, Republic of	Ganda, Kamora, Teso, etc.	English (O), Luganda, Swahili, others	Christian 66%, local 18%, Muslim 16%

and Uganda and in 1902 built a railroad from Mombasa to Lake Victoria in Uganda. This encouraged British settlers to farm the fertile Kenya highland area and lands around Lake Victoria. In Uganda, local cotton was processed for export.

A smaller number of German colonists settled German East Africa for coffee and tea production, but after World War I, it became a British protectorate as Tanganyika. The British did little to develop Tanganyika and did not encourage white set-tlement there. Tanganyika, Kenya, and Zanzibar gained their independence in the early 1960s following guerrilla warfare in Kenya. Tanganyika and Zanzibar united to form Tanzania in 1964, although a Zanzibar independence movement still exists. Zanzibar was one of the richest countries in Africa before it became part of Tanzania.

After independence from the United Kingdom in 1963, Kenyans elected their governments. President David Toriotich

Figure 9.37 **Eastern Africa: Zanzibar.** Zanzibar was a center of Arab trade along the coasts of Eastern Africa and across the Indian Ocean before Europeans arrived. The view from the top of the Arab fort takes in part of the old city. Now that Zanzibar is part of Tanzania, many in the city campaign for its independence to be renewed. Photo: © Volkmar Kurt Wentzel/National Geographic Image Collection.

arap Moi has been in power since 1978, but his reelection in 1997 was marred by violence and fraud. The country is beset with violence, often related to tribal rivalries and corruption. The potential successors to arap Moi have strong support from their own tribal groups, but each faces opposition from the others. Political uncertainty over the succession is enhanced by a growing gulf between urban and rural societies because of uneven effects of past policies. A country that has enjoyed more economic progress than many in Africa is now declining in prosperity.

In Tanzania, one-party, socialist rule continued until the first multiparty elections in 1995. Zanzibar has its own president and legislature for internal matters. In Uganda, dictators Idi Amin (1971–1979) and Milton Obote (1980–1985) both governed by oppression, killing hundreds of thousands of people. The first postindependence multiparty elections were held in 1996. The Tanzanian and Ugandan legislatures are elected by the whole adult population, but some seats are reserved for women.

In early 2001, Kenya, Tanzania, and Uganda reestablished the East African Community that they abandoned in 1977. It is intended as a means of long-term economic integration, starting with a customs union. Fears have already arisen of Kenya products replacing local ones in Tanzania and Uganda. Further movement toward political union, including an East African court of justice and legislative assembly, is a distant dream.

People

Ethnicity

The coastal people in the north are mainly Muslims, while those in Ethiopia and the southern countries are mainly Christian or traditional animist groups. Arabs, Arab-African mixes, and South Asians are common on the eastern coasts of Kenya and Tanzania, with Asians taking many of the business opportunities. Many Ugandan Asians were expelled by Idi Amin and moved to the United Kingdom or to other countries. Languages include many of local significance, but those with wider use range from Arabic in the north and Amharic in Ethiopia, to Somali, Swahili, and English.

Population Dynamics

The population of Eastern Africa is growing rapidly. The total rose from 91 million in 1980 to 168 million in 2001 and could reach over 280 million by 2025. In 2001, total fertility rates varied from 4.4 (Kenya) to 7.3 (Somalia). With death rates falling (see Figure 9.7), population increase is rapid. Like most other African countries, the youthful element in Ethiopia remains high (Figure 9.38). Life expectancy rose from 35 to 40 years in 1960 to 40 to 55 years in 2001. In Uganda, it declined to 42 years because of HIV/AIDS.

Apart from Djibouti, where almost all the people are urban, the other countries remain largely rural with fewer than 20 percent living in towns. The largest cities are Addis Ababa (Ethiopia, 2.6 million in 2000) and Nairobi (Kenya, 2.3 million), Dar es Salaam (the current capital of Tanzania), Kampala (Uganda, 1.2 million), Djibouti, and Mombasa (the main port

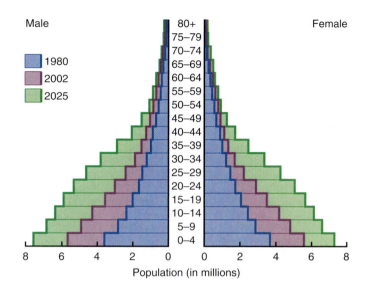

Figure 9.38 **Ethiopia: age-sex diagram.** Source: U.S. Census Bureau. International Data Bank.

of Kenya). Their fast growth combines migration from rural areas and natural population increase. It produces extensive shantytown areas, contrasting locally with the concentration of impressive office towers in Nairobi.

Economic Development

In the period 1960 to 1992, all the countries in Eastern Africa made limited progress in human development basics such as education and health care. The three northern countries had human poverty indices of over 50 percent. Only Kenya and Tanzania had indices of less than 30 percent. All countries face the disadvantages of inadequate infrastructure and the relicts of conflict (Figure 9.39).

Agriculture Is Central

In the general absence of mineral resources and the slow development of manufacturing and service industries, most countries in Eastern Africa depend on agriculture as an economic base (see Figure 9.23). They earn foreign exchange by exports of farm products.

Coffee makes up 90 percent of Ethiopian exports; coffee, tea, and tobacco are 97 percent of Uganda's exports; coffee, tea, sisal, cotton, cashews, and cloves are 85 percent of Tanzania's exports. Even in Kenya, which has greater economic diversity and output, coffee, tea, sisal, fruit, and vegetables provide around 50 percent of exports. In the 1990s, higher-priced tea, fruit, and vegetable exports replaced much of the former coffee and sisal exports from Kenya. The early 1980s boom in beverage crop prices followed a shortfall in world production, but production rose again and prices fell, remaining low in the 1990s.

The arid areas in the north of the subregion are not able to grow crops and rely on exports of livestock. Before the current anarchy, Somalia's livestock made up 65 percent of its total exports, augmented by bananas grown in the south. Few of the

(a)

(b)

Figure 9.39 **Eastern Africa.** (a) A road in Tanzania—straight but potholed and dusty in the dry season and a river in the wet season. (b) Ugandan farmer tends his crop as children play on an abandoned tank, a relic of recent oppression. Photos: © MAF.

arid countries balance their imports by exports, and most rely on external aid.

Most people in Eastern Africa gain a livelihood by subsistence farming or nomadic herding, but both are subject to the vagaries of weather. Long droughts in the region since the 1970s led to major famines in Ethiopia in the 1970s and 1980s. The central government not only fanned civil war but for long prevented aid getting to famine-affected opposition areas. The famine of 1983 to 1985, when over half a million people died of starvation or diseases contracted at the feeding sites, attracted worldwide publicity. This famine arose after soil erosion in the formerly fertile highlands area of Ethiopia led to forced migrations of people northward into the zone of drier conditions and civil war (see the "Point-Counterpoint: Famine in Africa" box, p. 404). Ethiopian farming continues to be poorly developed because of droughts, bureaucratic obstruction, little land-tenure security, and poor infrastructure (such as roads and telephones). Many parts of the country remain susceptible to famine.

Kenya has the most diversified farming community. Following independence, the government redistributed as smallholdings the land from the European-farmed highlands around Nairobi. They encouraged Kenyans to grow commercial export crops, such as coffee and tea, and food crops for the home market. Today, Kenya provides some comfort for those who see more hope than alarm in Africa. For example, in the Machakos district east of Nairobi, a region of periodic drought that suffered famine and soil erosion in the colonial period, land-reform smallholders invested in high-value crops, earning money to support this by off-farm jobs. Former badlands were transformed into terraced hills and fenced fields. Good road links to Nairobi markets made it possible to sell fruit and vegetables there. By the 1990s, the area had 1.4 million people, and output had risen 15 times with yields per hectare (2.47 acres) increasing up to tenfold.

The Tanzanian government applied socialist principles instead of the Kenyan emphasis on small, commercial family farms. It regrouped the scattered farmers into communal farms and centralized villages, and guaranteed low prices in food markets across the country. Many farmers lost interest in increasing their productivity, and the Tanzanian economy failed to grow. Although it also exports some of its crops, its total exports are less than half those of Kenya. From the late 1980s, however, efforts to improve farm management included more use of fertilizers. Experimental corn crops raised yields from 2 to 3 kilograms per hectare in 1988 to 18 to 20 kilograms in 1994. Lack of storage, the high cost of fertilizer imports, and poor roads are problems in the way of further advances.

Variable Manufacturing Development

Manufacturing makes up between 5 and 13 percent of GDP in Eastern African countries. Although several countries gave emphasis to manufacturing in their early development plans, only Kenya achieved much. In addition to processing its crops, Kenya has oil refineries and medium-sized consumer goods factories. It is a growing center of production and distribution for the surrounding countries. The other countries do not have the home demand for manufactured products or the foreign exchange needed to buy machinery and replacement parts. Most countries have small-scale enterprises based on processing local crops, manufacturing foods and drinks, and producing goods for local markets. Ethiopia has metal and chemical industries that were built in the Communist era, but such large-scale industries contrast with the more common small-scale craft manufactures.

Services and Tourism

For most countries, the service economy is poorly developed apart from extensive government bureaucracies. Once again, Kenya is ahead of the others. Its relatively stable political environment

Point COUNTER Point

FAMINE IN AFRICA

Famine is a severe shortage of food occurring over a wide area and causing deaths by starvation and by diseases that take advantage of the body's reduced ability to fight them following malnutrition. Famines occurred in various parts of the world throughout human history, but in the last 20 years, most have occurred in Africa South of the Sahara. Despite huge injections of aid from other parts of the world, many countries remain in danger of famine from year to year. It is likely that Africa will continue to be the hungry continent into the new century as its population growth outstrips economic growth. Famine and chronic malnutrition will continue to be major problems.

Famine results from shocks to natural and human systems that disrupt food production or distribution. Natural shocks include drought, river flooding, insect plagues, and plant diseases. Human shocks

include wars, civil conflict, widespread poverty, inefficient food distribution, and population growth that exceeds the ability of a country or region to feed the extra mouths. It is often not so much the immediate result of such shocks but the failure of human systems to cope with them that causes famines. For example, recent African famines have occurred on the semiarid margins of deserts where droughts that are part of climate changes have had greater famine impacts than before. The reasons for these greater impacts include pressures from increasing populations, civil wars that disrupt transportation systems, and the abandonment of long-established systems of cultivation and livestock keeping that made allowances for coping with drought.

The much publicized famines in Ethiopia and Somalia occurred when stresses resulting from drought were made worse by the difficulties of getting food to populations cut off from supplies by civil war (Box Figure 1). Warring sides stopped the delivery of aid as a

(a)

(b)

Box Figure 1 **Images of famine.** (a) A refugee camp in Kenya near the border with Somalia during a drought. (b) Starving people in the Wamma camp, Kismayo, Somalia, in 1992. Photo: (a) © MAF; (b) © Norbert Schiller/The Image Works.

attracted offices of the United Nations and the regional and continental headquarters of multinational corporations and nongovernmental organizations to Nairobi.

Tourism—the world's largest industry—provides a major potential for earning foreign currency through attracting visitors to view the unrivaled scenic grandeur and wildlife. The commitment of African governments to conservation in designated game and national parks resulted in Africa having a higher proportion of such land uses than any other continent. Unfortunately, the governments have little money to spend on maintaining the parks and their wildlife, and realistic manage-

ment policies. For example, elephant numbers in Africa are around 700,000—down from several million in the 1800s and reduced by poaching, particularly in the countries of Eastern Africa.

Although many of the staff of the parks and hotels are poorly paid by the standards of materially wealthy countries, tourism brings an injection of foreign currency. Like other African products, however, tourism is subject to fluctuations according to cycles of prosperity and recession in the wealthier countries. Kenya made the most progress in its tourist industry, attracting almost a million visitors in 2000. In years weapon. By mid-1992, up to 4.5 million people in Somalia faced starvation. Baidoa and Bardera, two small towns west of the capital Mogadishu, lie in some of the country's best farmland. After refugees from other areas moved in, famine struck both. At Baidoa, where the population was swollen to 60,000, over 100 died each day; in Bardera, 30 to 40 died each day. Neither town had community kitchens, and the distribution of dry food was irregular. Farther inland, Belet Uan had to rely on daily flights sponsored by the Red Cross to bring in food. Distribution was slowed by the stipulation that all food had to go through Mogadishu instead of being landed directly at points along the coast. Half the food distributed by truck was lost to looters.

Famines in Zimbabwe and southern Sudan occurred in 1991 and 1992, when the drought impacts were increased by the arrival of hundreds of thousands of refugees from war in Mozambique or famine in Ethiopia, respectively. Famine in Mali and Niger occurred in the 1970s when livestock herds had to be slaughtered because the grass had dried up rapidly when overgrazing was followed by drought.

Famines get emergency attention from governments and from international aid agencies. Governments can usually cope with the onset of famine if they have distribution systems that deal equally with urban and rural areas. Famines are more prevalent in rural areas where provision is uneven, transportation is poor, and it is difficult for the people to migrate rapidly into the better provisioned urban areas. Rural areas are also at a disadvantage because many people there are undernourished compared to those in towns. Poor nourishment gives famine conditions a start as the young and old succumb rapidly to starvation and killing diseases. One of the best means of using limited resources to control famine in southern Zimbabwe in 1991–1992 was the provision of lunches for elementary school children by aid agencies. This policy kept most children attending school and ensured a good level of nutrition. In previous famines, children were neglected at home and many died. Agencies also provided seed for planting corn in the following year, leading to a more rapid return to economic health in rural areas.

International agencies bring food and medical aid to emergency situations. They are limited in what they can do by response time and access to affected areas. It takes time to assemble staff and purchase, ship, and deliver emergency food to the needy country. Once there, internal transportation is often so poor that only limited quantities can be shipped rapidly. Such emergency food often brings wheat flour and milk powder not normally consumed by local people, who may need a cultural shift to continue consuming such food.

It is clear that better systems are required to cope with the threat of famine. Early-warning systems are being implemented in some areas by monitoring nutrition levels. Improvements in education, population control, political stability, and infrastructure in the poor countries of Africa are basic to reducing the threat of famine in the long term. For some years to come, it will be necessary to improve short-term aid delivery.

Mass starvation through famine is, however, not the main problem of hungry Africa. Inadequate food supplies affect all parts of the region long term. Malnutrition is at crisis levels, linked to urban poverty and poor harvests in rural areas. Malnutrition forms a basis for short-term dramatic famine episodes. The causes are complex and wide-ranging, from the decline in environmental quality, the greater value placed on inexpensive food for urban areas and commercial crop production, and the lack of rural-urban transportation facilities, to the growing population creating demands that exceed supplies.

Complete the following table:

FAMINE DUE TO NATURAL CAUSES	FAMINE INDUCED BY HUMAN ACTIONS
Widespread famine occurs when there is drought and is a matter of climate and climate change. It is a major problem in the Sahel region of Africa.	
Local famine also arises because of pests such as locusts, plant diseases, river flooding, and destructive weather systems—also natural environmental factors.	

when coffee and tea suffer poor prices, tourism is Kenya's main source of foreign exchange—making up 20 percent of export earnings in 1990 but only 10 percent in 1999. People from the wealthier countries of Europe, the United States, and Japan visit Kenya mainly to see the many animals in the national parks (Figure 9.40). The hotels in Nairobi have a twofold clientele of tourists and businesspeople. Other countries, especially Tanzania, also have tourist industries, but the egalitarian Tanzanians resisted building large luxury hotels, and their tourist industry remained small until the government encouraged a major expansion of park resorts in the mid-1990s. It had half a million visitors in 2000, up from 80,000 in 1980, and tourism receipts made up over 60 percent of export earnings.

 # Southern Africa

Southern Africa (Figure 9.41) stands at the southern margin of the tropical climatic environments and includes warm midlatitude conditions. Seasonal tropical climates in the northern parts

Figure 9.40 **Eastern Africa: Masai Mara National Park, Kenya.** Tourists view elephants from the safety of their bus. Tourism based on such viewings is a major source of foreign income in Kenya and other countries in Africa, but its methods are often criticized by those who wish to preserve the animals.

have summer rains that decrease southwestward toward the aridity of Namibia and western Botswana. The Republic of South Africa has midlatitude climates with winter rains at the Cape of Good Hope and summer rains along its southeastern coasts. This made it particularly attractive to European colonial settlement. A series of plateaus is drained by major rivers such as the Zambezi (Figure 9.42), Limpopo, and Orange-Vaal system and cut by the southernmost extension of the East African rift valleys.

Southern Africa has the greatest potential for leading the rest of Africa into a better future. It includes countries that have some of the best records of economic progress in the continent based on mineral wealth, diversified agriculture, and manufacturing. Over a third of Africa's rail mileage is in this subregion.

Such potential, however, rests on a fragile immediate past and present. For most of the second half of the 1900s, civil wars devastated Angola and Mozambique. When the Republic of South Africa was isolated from the rest of the subregion and the world by its apartheid policy and resulting sanctions, the separated white and black communities diverged further in

well-being. Lesotho, Madagascar, Mozambique, Angola, Malawi, and Zambia remain among Africa's poorest countries (Figure 9.43). In the early 2000s, Zimbabwe's leader, Robert Mugabe, made decisions that destroyed his country's improvements in economic and human development and led to bloodshed, famine, and bankruptcy.

The future of this region depends very much on the Republic of South Africa after its rejection of apartheid policies, its democratic elections of a truly national government, and its restoration to the world economic system following decades of isolation and sanctions. South Africa has one-third of Southern Africa's population but produces three-fourths of its GDP. Many families in the rest of this subregion derive considerable income from the wages sent home by their members who work in South Africa.

The Republic of South Africa's presence dominated the subregion even through the apartheid years, largely as a perceived adversary. Countries such as Botswana, Malawi, Zambia, and Zimbabwe made public pronouncements against apartheid but maintained economic relations with South Africa. Lesotho and Swaziland, virtually encircled by South Africa, maintained political relations. To destabilize its potentially antagonistic neighbors, the South African military supported guerrilla groups in the civil wars in Angola and Mozambique. Namibia was occupied by South African troops until 1990, when the United Nations finally made Namibia independent.

Countries

The countries of Southern Africa include war-wrecked Angola and Mozambique and countries that have had more stable governments such as Botswana, South Africa, and Zimbabwe.

Republic of South Africa

The evolution of the Republic of South Africa is particularly significant in the light of the country's role in the whole subre-

Figure 9.41 **Southern Africa: main features.** The map shows the countries included in the subregion, the major cities, and rivers.

(a)

Figure 9.42 **Space shuttle photo: the coast of Mozambique, Southern Africa.** (a) The false-color reds are from the reflection of infrared radiation from growing vegetation, picking out wetlands near the coast. Red colors are absent from areas of seasonal burning. The smoke from burning areas shows the constancy of the trade winds blowing from the Indian Ocean. The map (b) provides details and scale. Photo: (a) NASA, Michael Helfert.

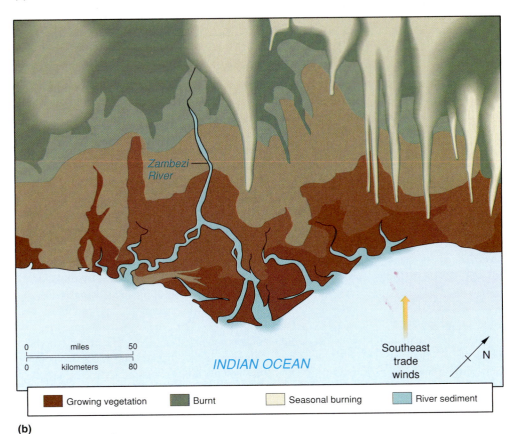

Zambezi River

	miles	50
0		
0	kilometers	80

INDIAN OCEAN

Southeast trade winds

N

■ Growing vegetation   ■ Burnt   □ Seasonal burning   ■ River sediment

(b)

gion. The Dutch built Cape Town as a port on the route to the East Indies and established settlers on cattle farms, forcing the indigenous Khoikhoi northward. Following pressure on land at the Cape Colony in the 1700s, some Boers (Dutch for farmer) pushed inland. They retained their Dutch-based language (Afrikaans) and were later called Afrikaners. In places, they enslaved the Khoikhoi, whom they called Hottentots. In 1814, the British purchased the Cape Colony from the Dutch, bringing in new colonists who demanded the use of English, the end of slavery, and the protection of the Khoikhoi. Many Boers undertook their "Great Trek" northward to join

others in the Orange and Vaal River valleys, where they established the Orange Free State and Transvaal. The Boers displaced the Ndebele people, who moved north of the Limpopo River, and the Zulus, who moved southward into Natal, creating tensions with the local tribes.

The Boers declared a South African (Afrikaner) Republic in their new lands, but the discovery of diamonds and gold there from the 1860s and the occupation of South West Africa (modern Namibia) by the Germans in 1884 caused Britain to annex Bechuanaland (modern Botswana) to block German-Boer links. It also extended protectorates to Basutoland (modern

Figure 9.43   Southern Africa: country data.

Country	Capital City	Land Area (km²) Total	Population (millions) Mid-2001 Total	2025 Est.	GNI 1999 (US $ million) Total	GNI PPP 1999 Per Capita	Percent Urban 2001	Human Development Index Rank of 175 Countries	Human Poverty Index: Percent of Total Population
Angola, Republic of	Luanda	1,246,700	12.3	20.5	3,276	1,100	32	160	No data
Botswana, Republic of	Gabarone	581,730	1.6	1.2	5,139	6,540	49	122	27.5
Lesotho, Kingdom of	Maseru	30,350	2.2	2.4	1,158	2,350	16	127	23.0
Madagascar, Republic of	Antananarivo	587,040	16.4	30.8	3,712	790	22	147	No data
Malawi, Republic of	Lilongwe	118,480	10.5	17.1	1,961	570	43	159	42.2
Mozambique, Republic of	Maputo	801,590	19.4	21.6	3,804	810	28	169	49.5
Namibia, Republic of	Windhoek	824,290	1.8	2.0	3,211	5,580	27	115	25.0
Republic of South Africa	Pretoria	1,221,040	43.6	35.1	133,569	8,710	54	101	19.1
Swaziland, Kingdom of	Mbabene	17,360	1.1	1.4	1,379	4,380	25	113	27.6
Zambia, Republic of	Lusaka	752,610	9.8	14.3	3,222	720	38	151	38.4
Zimbabwe, Republic of	Harare	390,760	11.4	9.5	6,302	2,690	32	130	29.2

Source: Data from *Population Reference Bureau 2001 Data Sheet*; *World Development Indicators,* World Bank, 2001; *Human Development Report,* United Nations, 2001; Microsoft Encarta (ethnic group, language, religion).

Lesotho) and Swaziland after Boer attacks. In 1899, the Boers declared war on the British, who had instigated local conflicts. Although the British soon took the major centers, a costly and inconclusive guerrilla war followed.

After the South African (Boer) War, the four colonies—Cape, Natal, Orange Free State, and Transvaal—joined in 1910 to form the Union of South Africa as a self-governing dominion within the British Empire. However, although South Africa became a self-governing country in 1910, the rights of black Africans were never part of the arrangement, and black struggles for self-determination took longer than in other African countries. White Afrikaners led the South African governments and imposed the apartheid policy in 1948. Apartheid laws and linked informal measures increased the separation of white, black, and mixed races. Black Africans and other "colored" (mostly Asian) people lost many human rights. Each had to carry an identity pass. At one level, "petit apartheid" legalized segregation equivalent to that in the American South up to the mid-1960s; at another, "grand apartheid" relocated black people to "independent" homelands. Their economic role was as labor for white people. Each group had a designated housing area, and each black person was assigned to a homeland area. Germany lost Namibia after World War I, and it became a protectorate of South Africa in 1920. South Africa refused to give up the territory when the United Nations tried to cut this tie after apartheid became South African policy.

In 1961, South Africa declared itself to be a republic without ties to the United Kingdom. From the 1960s, while policies of racial separation became less significant in other countries, including the United States, the Republic of South Africa continued and extended the separation. For a time, the West tolerated South African apartheid, but then it largely isolated South Africa from the world political, economic, and sports competition systems.

The combination of internal resistance to apartheid by leaders of the African National Congress (ANC), such as Nelson Mandela, and external isolation and sanctions forced the relaxation of the apartheid policy in the 1990s, signaled by the release of Mandela from long-term imprisonment. Despite many forecasts that South Africa would descend into anarchy and a civil war bloodbath, the transition orchestrated by Nelson Mandela and F. W. de Klerk (the South African president at the time) worked, and both received Nobel Peace prizes. Democratic elections in 1994 resulted in a "Government of National Unity"—the first by the majority racial group in South Africa—with Mandela as its first president. South Africa continues to struggle with the geographic expressions raised by apartheid, including the townships, homeland policies, and separate racial schooling. The opposing communities of wealthy whites and poor blacks, of ANC Xhosas and Inkhata Freedom Party Zulus, of local Africans and those from other countries, and the parties that represent Afrikaners, other whites, and Coloureds barely tolerate each other.

The Truth and Reconciliation Commission, however, dealt with many sensitive issues that whites and blacks held against each other in a positive manner, although black Africans gave it greater credibility than many whites. White-owned businesses now employ black Africans at all levels, and restaurants mostly place profits over prejudices. When Mandela retired in 1998, other ANC leaders continued the reconciliation policies—although demands increased from black Africans for greater equality.

South Africa has extremes of wealth and poverty that bring economic development difficulties (Figure 9.44). Apartheid created a huge black underclass that now has a vote but will take decades to enjoy improved conditions. Income growth continues, and there has been no need to resort to external aid. Some benefits, such as domestic water and electricity

Country	Ethnic Groups (percent)	Languages O=Official	Religions (percent)
Angola, Republic of	Ovimbundu 37%, Mbundu 27%	Portuguese (O), Bantu languages	Christian 85%, local 10%
Botswana, Republic of	Tswana 75%	English (O), Setswana	Christian 50%, local 50%
Lesotho, Kingdom of	Basutho 79%, Ntguni 20%	Sesotho (O), English (O)	Christian 79%, local 20%
Madagascar, Republic of	Malaysian, Indonesian, coastal groups	French (O), Malagasy (O)	Christian 41%, local 52%, Muslim 7%
Malawi, Republic of	Chewa, Nyanja, others	English (O), Chichewa (O)	Christian 70%, Muslim 20%
Mozambique, Republic of	Makua-Lomwe, Yao, others	Portuguese (O)	Christian 30%, local 60%, Muslim 10%
Namibia, Republic of	Orambo 50%, white 6%, mixed 7%	English (O), local, Afrikaans	Christian 90%
Republic of South Africa	African 75%, European 14%, mixed 8%	Afrikaans (O), English (O), Zulu, Xhosa	Christian, Hindu, Muslim, Jewish
Swaziland, Kingdom of	African 97%, European 3%	English (O), Siswati (O)	Christian 60%, local 40%
Zambia, Republic of	African 96.7%, European 1.1%	English (O), Ichbula, others	Christian 60%, Muslim-Hindu 40%
Zimbabwe, Republic of	Shona 71%, Ndebele 16%, others, European 2%	English (O), Chishona, Sindebele	Christian 20%, syncretic Christian-local 50%

supplies, are now available to areas of poorer housing; a growing black middle class enjoys the fruits of affluence; and the African National Congress-led government is not opposed disruptively. However, jobs are not expanding at a time when 20 to 30 percent of adults have no formal employment. In this context, many white managers and professionals moved out of the country for apparently better opportunities, living quality, and less crime in Australia, Canada, the United Kingdom, and the United States. This "white flight" reduced other employment opportunities in South Africa, although some potential leavers stayed for the sunny, mild climate and lower costs of living.

Other Former British Colonies

In 1953, the white settlers in Zimbabwe first tried to dominate the former colonies of Nyasaland (now Malawi) and Northern and Southern Rhodesia (now Zambia and Zimbabwe) by establishing a united Federation of Rhodesia and Nyasaland, but it was resisted by black Africans and dissolved in 1963. Zambia and Malawi achieved independence in 1964 on the breakup of the Rhodesian federation, Botswana and Lesotho in 1966, and Swaziland in 1968. In 1965, the whites in Southern Rhodesia unilaterally declared independence, but Britain and other countries did not accept this move and imposed sanctions. Years of internal strife and international sanctions delayed Zimbabwe's full independence until 1980.

Former Portuguese Colonies

In Angola, the recognized government since independence in 1975 controlled only part of its country until 2002. The National Union for the Total Independence of Angola (UNITA) opposed the socialist government that was backed by the Soviet Union and Cuban military. Until 1990, South Africa backed UNITA, after which it funded itself by illegal diamond sales. Following a 1994 peace accord and elections, fighting resumed in 1998, although the death of UNITA'S leader in 2002 may lead to peace if UNITA shifts from being a military to a political group. For most Angolans, their economy is in disarray, and millions of landmines make farmers reluctant to return to their fields. In the late 1990s, the promise of more secure offshore oil income attracted external investment to Angola, but enthusiasm waned in the early 2000s as few productive wells emerged.

Current Governments

In the early 2000s, governments in Southern Africa fall into three main types with implications for human development and human rights.

- In the monarchies of Lesotho and Swaziland, hereditary kings are divesting their powers to democratic systems. Lesotho's first multiparty constitution was agreed upon in 1993, retaining the king as figurehead. Internal disruption in 1998 destroyed much of Maseru, the capital. In Swaziland, student and labor unrest in the 1990s added pressures for democratic reform.

- The Republic of South Africa (since 1996) and Botswana (since 1966) elect democratic governments from multiparty systems. In South Africa, the change from a minority white-led government to an elected black government resulted in a renewed acceptance and encouragement of the government by other countries. The new government, however, faced the task of refocusing the country's education, health care, housing, and economic strategies that had depended on and supported the white population.

- From independence in the 1960s and 1970s, the other countries had long-term single-party governments, in some cases moving slowly toward multiparty systems. Angola has a long-term socialist government, but it is opposed violently by the UNITA party. Mozambique, another ex-Portuguese

(a)

(b)

Figure 9.44 **Southern Africa: contrasting environments.** (a) Desert near Usakos, Namibia; an area largely empty of people. Describe the barrenness. (b) An early morning view of Cape Town and Table Mountain, South Africa. The site provided the foundation for one of Africa's major ports. The development of the area is constrained by the landforms. It is the midlatitude southernmost tip of Africa, with a climate marked by dry summers and cool, wet winters. Photos: (a) © Don L. Boroughs/The Image Works; (b) © James L. Stanfield/National Geographic Image Collection.

colony, made a successful transition from civil war to democracy in the 1990s, while Madagascar, Malawi, and Zambia moved from restrictive single-party governments toward multiparty democracies with varied degrees of openness. Madagascar added to its problems of underfunded health and education facilities by the erratic imposition of reform measures that are opposed by antigovernment strikes and demonstrations. Namibia was released from links to apartheid-ridden South Africa in 1990 and moved toward democracy, although still dominated by a single party. Zimbabwe has been dominated by the government of Robert Mugabe since its 1980 independence, although internal opposition to repressive measures is growing.

People

Ethnicity

The people of Southern Africa include a wide variety of ethnic groups. The cultural variations within Southern Africa reflect the last four centuries of history that witnessed major movements of people of all colors.

The low densities of the San and Khoikhoi hunting and collecting peoples that occupied the area were gradually displaced by the advance of agricultural and cattle-herding Bantu tribes from the north, reaching the present South Africa in the 1700s. By the early 1800s, the Bantu peoples in southernmost Africa—the Swazis, Zulus, Xhosas, and Sothos—fought each other for territory and built strong kingdoms.

The level of European presence varied. The Portuguese were the earliest colonists (modern Angola and Mozambique), but few settled and most of those left after independence in 1975. In southernmost Africa, the Dutch settlers stayed as the Boers and British settlers also built up numbers. The British introduced people from Asia, particularly India, to carry out some laboring and administrative roles. Today, South Africa is the only country in Africa South of the Sahara with a sizable non-black population, having 75 percent black, 14 percent white, 3 percent Asian, and 8 percent of mixed race.

The island country of Madagascar has a mixture of African and Asian people and cultural influences. Peoples from the East Indies (Indonesia today) and Africa occupied Madagascar, but the Portuguese, French, and English all attempted to colonize the island. They found it difficult to overcome the forces of the powerful rulers in hilly and forested terrain. By the end of the 1800s, the French established a colony, but internal dissent gradually rose until independence was granted in 1960.

Urbanization

The Republic of South Africa is 50 percent urban. The South African proportion living in towns might be higher without the apartheid policies that forcibly sent black Africans to their tribal homelands. Large parts remain rural. It is likely that urban growth will be more rapid now that freedom of movement is allowed. Botswana, Zambia, and Zimbabwe are 30 to 40 percent urban because of their mining towns and urban industries. Other countries are less than 25 percent urbanized, since the lack of manufacturing gives their towns less drawing power.

South Africa has most of the large cities in Southern Africa, including Cape Town (3 million people in 2000), Johannesburg (2.3 million, with another 3 million in the nearby Rand areas), Durban, Pretoria, and Port Elizabeth (each with over 1 million people). Johannesburg is the only African city of global significance, classed on the levels of global business services with Boston, Amsterdam, and Melbourne (see Figure 2.17). Maputo (Mozambique, 3 million people in 2000) and Luanda (Angola, 2.5 million) are also large cities that grew mainly as a result of migration from rural areas to national capitals during the recent civil wars. Harare (1.7 million) and Bulawayo are the largest cities in Zimbabwe, while Lusaka (Zambia, 1.6 million) and Antananarivo (Madagascar, 1.5 million) are capital cities with large government bureaucracies.

Shantytowns are a feature of towns—as they are of growing cities in many developing countries—and their inhabitants are often involved in the informal economy. Shantytowns are unplanned, constructed of any materials that come to hand—from packing cases to cement blocks and corrugated iron—and basic in their services. Some condemn families to a hopeless future of poverty. In many cases, however, people move into them on arriving from a rural area but eventually find better accommodations. Governments may supply utilities and build schools, hospitals, and roads to integrate shantytowns with the rest of a large urban area, usually after a considerable period in which they become established. Shantytowns were encouraged

under apartheid in South Africa (Figure 9.45), but the long process of improving housing conditions there is under way.

Population Dynamics

The population of Southern Africa increased from 81 million in 1980 to 130 million in 2001 and may rise to over 150 million by 2025—slower than expected in the mid-1990s because of HIV/AIDS. Birth rates fell in all countries after 1970 (see Figure 9.7), but high proportions of younger age groups dominate the population structures even in the more developed South Africa (Figure 9.46). Differences of lifestyle affect life expectancy (higher for whites) and total fertility (lower for white women).

Although the Zimbabwe population grew faster than the African average of over 3 percent in the mid-1980s, its 2001 increase was under 1 percent, a rate that could soon be outstripped by economic progress if the political environment improves. Zimbabweans were reluctant to take up birth control because they saw it as a white colonialist plot to reduce black population growth. In the 1990s, a rural-based family planning

Figure 9.45 **South Africa: Soweto Township, near Johannesburg, in 2002.** The contrasts within Soweto can be seen between the squatter shacks in the foreground and the more permanent housing in the rear. The sprawling township became a symbol of the struggle by South African blacks against the racist apartheid rule. Now it is an important tourist destination. Little change appears to have occurred in the 10 years since the end of apartheid. However, postapartheid restrictions on labor migration and the high AIDS/HIV infection rates among their workers have prompted mining companies to replace hostels with family homes in some areas. Photo: © AP/Wide World Photos.

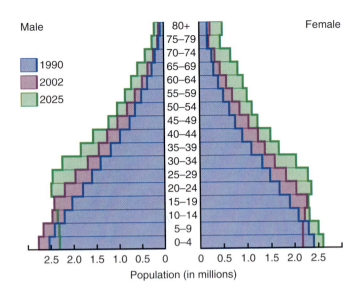

Figure 9.46 **South Africa: age-sex diagram.** Source: U.S. Census Bureau. International Data Bank.

network was successfully introduced, and a motivation campaign targeted at men helped to reduce the beating of wives discovered to be using contraceptive methods. This also made men aware of the need to control HIV infection by using condoms. In the late 1990s, HIV/AIDS was at its worst in Zimbabwe and Botswana (see Figure 9.12), increasing death rates and lowering life expectancy to around 40 years. Life expectancy is almost 55 years in South Africa but dropping because of HIV/AIDS (see the "Point-Counterpoint: HIV/AIDS" box, p. 30). By 2002, 38 percent of Botswana's population aged 15 to 49 years were HIV-positive, and tuberculosis deaths tripled in a decade as a result of weakened immunity. The disease affects what was an economically improving country, tripling health expenditure to provide universal availability of retroviral drugs and build four dispensing centers.

Economic Development

Southern Africa has 20 percent of the population of Africa South of the Sahara and produces over 50 percent of the total GDP. The Republic of South Africa has a major impact on these subregional totals. In the early 2000s, however, the subregional economic environment deteriorated with the political one through the Zimbabwean collapse, Zambian reliance on stalled copper mining redevelopment, and spreading HIV/AIDS.

Agricultural Diversity

Agriculture is the economic mainstay in the very poor countries of Lesotho, Madagascar, Malawi, Mozambique, and Swaziland, where over 85 percent of the population gains a living from subsistence farming. Even in countries with mining or tourism as important sectors, subsistence farming occupies over 50 percent of the work force. Corn is a common staple food, grown on thousands of small farms in Southern Africa.

Farming remains significant in the more diversified economies of South Africa and Zimbabwe, where commercial farming related to world markets is expanding. Landscapes of commercial farming exist alongside those of semisubsistence (Figure 9.47). South Africa made itself self-sufficient in food during the years of international sanctions against its apartheid policies. It produces a variety of temperate grains, vegetables, and fruits, together with sugar, cotton, and livestock products. Commercial crops provided nearly 40 percent of Zimbabwe's exports by value until the early 2000s destruction of commercial farms and the ousting of their white owners. Although Zimbabwe's tobacco production fluctuated, commercial farmers diversified into vegetables and flowers for European markets. Livestock products are also important in Zimbabwe and Botswana.

In Malawi, tobacco, tea, and coffee make up 70 percent of the country's exports: its tobacco output is just over half that of Zimbabwe, the subregion's leader. The two countries produced over 80 percent of African tobacco exports in the 1990s. The political turmoil in Zimbabwe in the early 2000s spread to Malawi, where financial reserves and corn stockpiles became exhausted and left a famine-prone situation. In Madagascar, coffee, sugar, vanilla, cloves, and cacao make up 70 percent of the exports. In Zambia, the potential for commercial farming was neglected in the years of high copper prices, and only one-fifth of its good arable land is used. The opening of Zambian frontiers to trade with Zimbabwe in the 1990s flooded Zambian markets with cheaper Zimbabwean grain and meat. In Mozambique, agricultural production declined to less than three-fourths of the 1980 level during the civil war and food imports were necessary—although by the late 1990s, there were signs of a return to self-sufficiency in food.

Land ownership reform is a major issue in the parts of Southern Africa where white settlers took land. Newly independent governments have to weigh the political advantages of returning that land to black Africans against the loss of income from commercial farms. In Zimbabwe, the commercial farms that remain are efficient, employ local labor, and provide valuable export crops. When the government took over some commercial farmland for redistribution in the early 1980s, many independent family farmers on smaller plots were more productive than large commercial farms because they used the land more intensively—albeit for subsistence rather than commercial export crops. In other cases, soil erosion caused the abandonment of the smaller landholdings.

In the early 1990s, droughts affected the southern part of Zimbabwe where most of the communal areas and small farms are situated but had less impact on the commercial farms in the center and north. The commercial farms gained further advantages when marketing controls were dropped as part of structural adjustment and small farmers lost various forms of government assistance. However, these events emphasized the differences between the two types of landholding and created pressure from a growing population for political action to convert more large commercial farms to smallholdings. In the late 1990s and early 2000s, the Zimbabwean government allowed the appropriation of the white-run commercial farms, and groups of former rebel fighters ("war veterans") took over a number of them, expelling or killing the occupants and black farm workers.

(a)

(b)

Figure 9.47 **African rural landscapes: commercial and subsistence farming.** (a) The Hex River valley, Cape Province, South Africa: commercial vineyards with large cultivated areas and a wine product that is sold overseas. (b) Terraced farmlands in the Shona area of the Eastern Highlands of Zimbabwe. Contrast the two landscapes. Photos: © Don L. Boroughs/The Image Works.

Mining Wealth

Mining dominates the economies of Angola, Botswana, Namibia, South Africa, Zambia, and Zimbabwe, involving them deeply in the global economic system (Figure 9.48). The possession of such resources, however, is not always a recipe for prosperity, since cycles of economic boom and recession in wealthier countries cause fluctuating global markets to affect local employment and income. Mining also has negative environmental impacts, including air and water pollution and the desertification of local areas poisoned by fumes. Without stable political conditions, adequate transportation facilities, and good management of the national economy, attracting multinational corporation investors is difficult. Multinational corporations, however, do not always return wealth they gain to local communities, while their license payments and taxes go to central governments that may direct spending away from the mining areas.

The exploitation of minerals and their export emphasizes the position of Southern Africa in fulfilling the demands of the world's wealthier countries. South Africa is the world's top producer of platinum and one of the major producers of gold, diamonds, iron ore, and several strategic metal ores. Platinum is of growing importance because of its use in aerospace construction and pollution control (catalytic converters) in modern vehicles. South Africa has 40 percent of the world's gold reserves and 90 percent of the platinum reserves. It is the world's largest exporter of gold and platinum. Other minerals include chrome, manganese, and vanadium, which are important in special steels. Mining products, often refined in South Africa, make up two-thirds of the country's exports.

The diamond industry depends on the fashions that keep diamond jewelry in demand at high prices around the world. The South African firm of De Beers manages to maintain a cartel of producers who restrict production to this end, although challenges are increasing from new producers in other parts of Africa.

South Africa's mining industry, however, is subject to growing challenges. In 1970, it mined 70 percent of the world's

Figure 9.48 **Southern Africa: mining.** The Okiep Copper Mining Company at Nababiep, Northern Cape Province, South Africa. This is typical of many mines in the subregion: the piles of waste after the ore is separated from its containing rock, the pithead buildings, the smelter with its tall stack; location in a rural area with the surrounding vegetation killed by gases from the smelter. Photo: © Hubertus Kanus/Photo Researchers, Inc.

gold but only 27 percent in 1995. Over that period, Australia, Canada, Russia, and the United States increased their shares, and new mines opened in other African countries such as Mali and Ghana. As the world gold price fell, South Africa's costs of deep mining rose. The major South African corporations now invest in gold mines in other parts of Africa, rather than in expanding their own mines. Employment in South Africa's mines is still a major source of income in the homes of migrant workers, mainly from Lesotho, Swaziland, Botswana, and Malawi, but these workers are also at the greatest risk from HIV/AIDS, which affects around one-third of the mineworkers, adding US $10 to the cost of producing one ounce of gold.

In some cases, mineral wealth is now returned to the indigenous peoples. In the 1990s, the Bafokeng people (just west of Pretoria) changed from watching multinational mining companies extracting wealth from platinum mines on their land to obtaining an agreement that ceded to them 22 percent of mining profits and a million shares in one company. The mas-

sive income was invested in community facilities and basic necessities (water, sanitation, electricity, housing, education).

Namibia is the world's fourth-largest exporter of nonfuel minerals and one of the major producers of uranium, together with diamonds, zinc, lead, and tungsten. In Botswana, diamonds make up 80 percent of exports (Figure 9.49), and the country became wealthy by exploiting these resources from the late 1970s.

Zambia was formerly a major world producer of copper, sharing a rich ore field with southern Democratic Republic of Congo (DROC). During the political chaos in DROC, Zambia became the region's main cobalt exporter. The need to restructure its copper industry lies at the heart of the Zambian future. Nationalization in Zambia led to a halved mineral output, and in the 1980s, income was reduced by falling global copper prices. The resulting lack of investment to replace exhausted mines, combined with heavy debts, made the Zambian mines uneconomic in competition against Chilean mines that produce more copper with half the workers (see Chapter 10). By the early

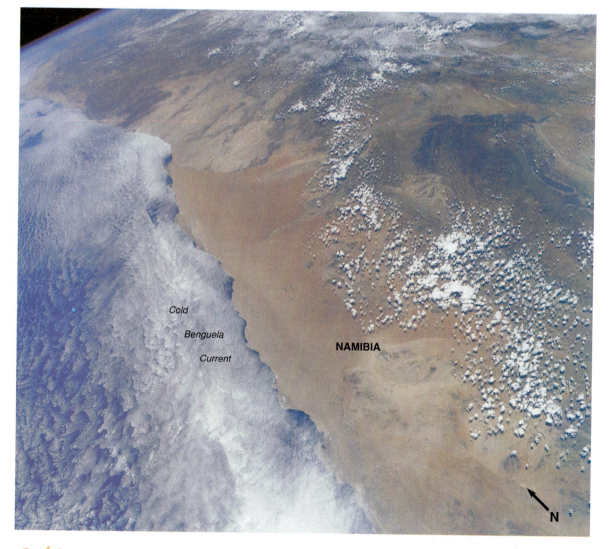

Figure 9.49 **Southern Africa: Namibian Desert.** The red sands of the desert dunes are a distinctive feature of this arid area, made drier by the offshore cold Benguela current that flows northward toward Angola in the distance. The clouds provide some rain to the higher inland areas, the cold ocean current is a rich fishing ground, and some of the world's richest diamond mines are found in the foreground of this space shuttle view. Photo: NASA.

2000s, foreign corporations bought Zambian copper mines, expecting to redevelop them but with much smaller work forces.

Zimbabwe's mineral output, making up 40 percent of its exports, includes coal, gold, chrome, and nickel in a wide belt along a major igneous intrusion, the Great Dyke, between Bulawayo and Harare. After some years of standstill, output increased in the 1990s, when a new platinum mine was the largest foreign investment in Zimbabwe in 25 years. It was expected to make Zimbabwe the world's second-largest producer of platinum after South Africa, but political problems inside Zimbabwe caused it to close in 2000.

Angola has oil fields along its northern coast, together with diamond mines inland. After the civil war destroyed its northern land-based oil installations, oil companies recognized the deepwater (over 1000 m, or 3,281 ft., deep) offshore resources as being of major world significance. Oil exports rose from 6,000 metric tons (6,612 U.S. tons) in 1980 to 35,000 (38,570 U.S. tons) in 1998 (20% of Africa's total oil exports) and made up 90 percent of Angola's export revenues as nine new oil fields came on stream. Angola was once the world's fourth-largest diamond producer and could be higher than that when its huge deposits are exploited after UNITA occupation of the interior diamond lands ends.

Manufacturing Contrasts

South Africa dominates manufacturing in Southern Africa, together with the linked power utility and transportation systems. As soon as mining began a century ago, there was a need for engineering services and chemical supplies, since the ores had to be refined near the mines to reduce the bulk for transportation. During World War II, sources of machinery and other manufactured goods in Europe were cut off, and South Africa diversified its manufacturing base into food products, textiles, clothing, armaments, and motor vehicle assembly in a process of import substitution. International sanctions resulting from its apartheid policy caused South Africa to increase its manufacturing base further.

South Africa raised its coal output from 50 million tons in 1970 to 200 million tons by the 1990s and developed technologies for using coal as a source of chemicals. The main coal-mining area is on the high veld east of Pretoria, where coal-burning plants generate 80 percent of South Africa's electricity. They cause acid rain downwind.

The great majority of South Africa's manufacturing is concentrated in the urban-industrial area around Johannesburg and around the ports of Cape Town and Durban. During the period of apartheid policy and the establishment of tribal homelands, factories were built just outside the homeland areas for access to cheap labor.

Six large South African corporations run almost totally by white South Africans continue to dominate mining and manufacturing. The current South African government has a commitment to enable black people to enter the white-controlled business world throughout the country. At present, most black businesses, apart from a major brewery and an insurance company, are small and engaged in trade rather than manufacturing.

The restoration of South Africa's access to world markets following the end of apartheid and the free elections of 1994 exposed its manufacturing industries to competition from cheaper overseas products. It is hoping to export some of its own products, especially armaments, which have been in demand from other countries and previously were exported in a clandestine manner. Unemployment remains a great threat to political stability in the country as the business sector expands cautiously and the government spends money on training but not on creating many new jobs.

As South African markets open to foreign manufactures, the large local business conglomerates and continuing political uncertainty deter some foreign investments. Multinational corporations, however, bought back companies they pulled out from in the 1980s. New investment was slow to start, but by 2000, companies such as Levi Strauss, Nestlé, Coca Cola, Toyota, other auto manufacturers, and some South Korean electronics manufacturers established factories. There is considerable capital within South Africa for investing in local and wider industrialization and infrastructure, but much goes abroad for investment, including to other African countries, China, and Vietnam.

Zimbabwe has the only other development of diversified manufacturing in Southern Africa. Industrialization began before the fight for independence, when cotton textile plants (Kadoma) and an iron and steel mill (Kwe Kwe) were established. During the period of independence declared unilaterally by the white minority, factories were built to produce goods that were kept out by sanctions. Manufacturing grew from 5 percent of GDP in 1965 to 26 percent in the 1990s. With the reopening of its South African trade and easier access to world trade, Zimbabwe now faces the need to move from import-substitution industries to compete with other world manufacturers. The country possesses good power sources from its own coalfields at Hwange and hydroelectricity at the Kariba Dam. Ten years of drought in the 1990s lowered the Kariba lake level to the point where only a small fraction of the potential hydroelectricity could be generated, but better rains filled it in 2000. Having one of the highest HIV/AIDS infection rates in the world (25% of the population aged 15 to 45) also weakens the economy, and its earlier progress is now changed to a static or declining position.

Among the other countries, agricultural products are processed in Malawi and Swaziland, and minerals are refined in Namibia and Zambia. Portugal did not encourage manufacturing in Angola or Mozambique until the 1960s, but by independence, Angola had vehicle assembly and chemical factories, while Mozambique produced steel and textiles using hydroelectricity from Cabora Bassa Dam on the Zambezi River. After independence, skilled managers and technicians departed, some sabotaging the factories, and further decline occurred during the civil wars. In the mid-1990s, it was estimated that Mozambique industries operated at less than half their capacity, although restoration of some with South African assistance began later in the decade.

Services

Service industries, from shops and banks to government jobs and tourist facilities, are increasing in importance throughout Southern Africa, but especially in South Africa. Tourism is a growing feature of the economies in Botswana, South Africa, and Zimbabwe. In 2000, South Africa received over 6 million tourists, Zimbabwe 1.8 million, and Botswana 800,000. In Zimbabwe, tourism was the fastest-growing sector of the economy before the political troubles of the late 1990s: the country attracted more tourists to its side of the Victoria Falls than Zambia did to the north. As with Kenya, there is a special interest in the national parks, where large animals are protected and able to live in something like natural conditions. Offshore island countries, such as the Seychelles and Mauritius in the Indian Ocean, are also important tourist venues.

Outlook

In the early 2000s, grounds for optimism in the economic future of this subregion and the significance of South Africa included the fact that infrastructure destroyed during civil wars in Angola and Mozambique was being reconstructed. In the latter, transmission lines take power from the Cabora Bassa Dam on the Zambezi River across 1,400 km (868 mi.) of land containing few people to reconnect South African users—and bring income to Mozambique. Port facilities at Maputo, Beira, and Nacala in Mozambique were expanded, together with the repair of railroad links into Zimbabwe and Malawi. The line from Lobito (Angola) into the southern mining area of Democratic Republic of Congo is being rebuilt. The development of major industrial zones along new roads between Johannesburg (Republic of South Africa) and Maputo (Mozambique) and between Windhoek (Namibia) and southern Angola testifies to increasing confidence by external investors

and to the creation of regional marketing and development strategies.

Test Your Understanding 9C

Summary Eastern Africa received cultural influences from Arabia and was later colonized mainly by the U.K. and Italy after the opening of the Suez Canal. Plateau landscapes are carved by rift valleys and topped by volcanic peaks. Arid conditions and dangers of drought are common. Kenya has the most diversified economy based on agriculture, tourism, and service industries. Civil war and single-party governments hold back other countries.

Southern Africa contains poor countries such as Angola, Malawi, Madagascar, and Mozambique but also the wealthiest African country of South Africa. Botswana, Namibia, Zambia, and Zimbabwe have mineral wealth. This subregion has subtropical and midlatitude climates but also suffers from aridity and drought. It is where the future of Africa South of the Sahara may be worked out.

Questions to Think About

9C.1 Suggest five geographic characteristics that help to distinguish Eastern and Southern Africa (e.g., population, cultural history, economic products, government, environment).

9C.2 Assess the impact of the precolonial and colonial histories on the present human geographies of the countries in Southern and Eastern Africa.

9C.3 Discuss whether or not increasing democracy is making a better life for the peoples of these two subregions.

Online Learning Center

www.mhhe.com/bradshaw

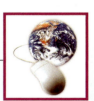

Making Connections

The Online Learning Center accompanying this textbook provides access to a vast range of further information about each chapter and region covered in this text. Go to **www.mhhe.com/bradshaw** to discover these useful study aids:

- Self-test questions
- Interactive, map-based exercises to identify key places within each region
- PowerWeb readings for further study
- Links to websites relating to topics in this chapter

Chapter 10

Latin America

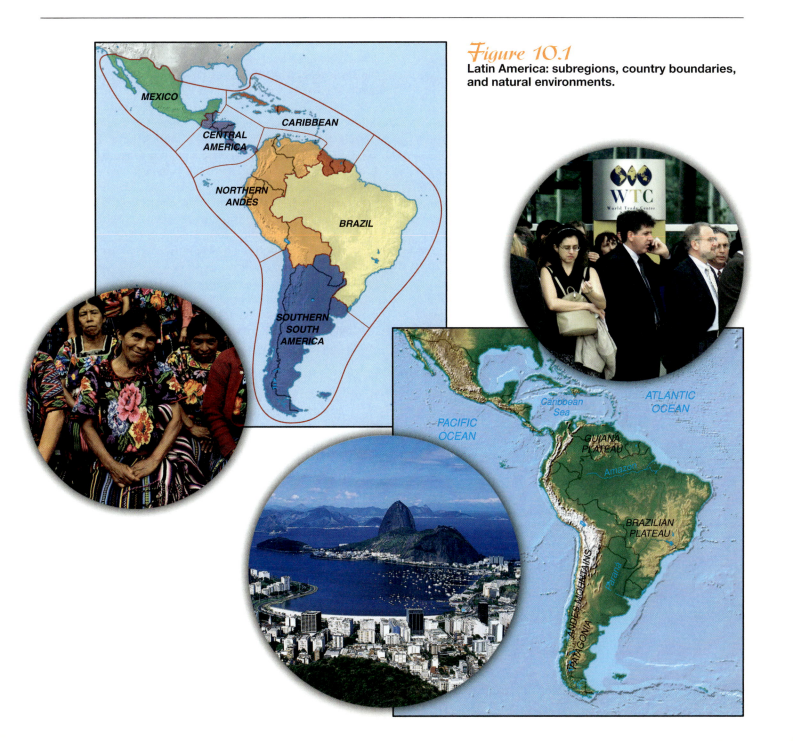

Figure 10.1
Latin America: subregions, country boundaries, and natural environments.

This Chapter Is About . . .

Spanish and Portuguese colonial impacts on human geography

Tropical, Southern Hemisphere, and mountain natural environments

Six subregions:

- Mexico: Regional power, U.S. neighbor, people on the move, increasing global connections and local resistance
- Central America: Diversity amid a common past
- The Caribbean: Island countries, contrasts, and isolation
- Northern Andes: Mountain environments and the international drug trade
- Brazil: The sleeping economic giant, national pride, and controversial resource use
- Southern South America: Temperate lands and a European culture

Point-Counterpoint: Tropical Forests and Deforestation

Personal View: Bolivia

Contrasts Amid a Common History

Bustling cities, ancient ruins, tropical beaches, and volcanic eruptions are but a few of the dramatic aspects of the human and physical geography of Latin America. The region includes the country of Mexico, Central America from Guatemala in the north to Panama in the south, the islands of the Caribbean, and the countries of continental South America (Figure 10.1). Latin-based Romance languages (especially Spanish but also some Portuguese and French) dominate throughout the region and provide a degree of unity among the varied populations.

Patterns established during the colonial era created a legacy of sharp contrasts in the social health and material wealth of the burgeoning population of the region. Spanish and Portuguese colonizers forced a relationship between the region and Europe in which natural resources were extracted and sent to Spain and Portugal. Based on the European-created export structure, a feudal system of land ownership was imposed that lasted into the 1900s. Europeans dispossessed indigenous people of wealth and land, forcing them to the bottom of a rigid social stratification and causing fierce social tensions.

In the late 1900s, several countries and subregions in Latin America were rising out of extreme poverty and into greater levels of global economic activity (Figure 10.2), while others continued a long cycle of economic challenge. The total GDP of the region rose very slowly during the 1980s and 1990s. In 1998, only Brazil, Argentina, and Mexico produced 1 percent or more of world GDP. Brazil had the world's eighth-largest economy during the later 1990s. In 2000, Brazil, Argentina, Mexico, Chile, Uruguay, Venezuela, and some Caribbean islands including Jamaica, and Trinidad and Tobago were in the World Bank's upper-middle group of countries (see Figure 2.6).

Geographic Contrasts

Latin America is a region of exciting contrasts in its human and physical geography. Countries range in size from population giants such as Brazil (172 million people in 2001) and Mexico (100 million) to a sizable group of much smaller countries whose populations number less than 10 million, to the Caribbean Basin, where many countries have fewer than 100,000 people. Latin American countries range from high to low incomes, from dependence on a single economic product to a diverse and integrated economic base, and from involvement in to isolation from the global economic system. The second-highest mountain range in the world contrasts with the huge, low-lying basins of the world's largest river system. The Earth's largest tropical rain forest is situated across the mountains from one of the world's driest deserts. The great range in latitude present in the region produces a variety of climate regimes, which are further altered dramatically by changes in elevation and proximity to mountain ranges. Some countries in the region, such as Mexico, have both tropical beaches and snow-capped mountain peaks.

Contrasts continue within the countries—inside cities and between the growing cities and hinterland rural areas. Mexico, Brazil, Argentina, Peru, and Chile contain large urban-industrial areas around their major cities, such as Mexico City, São Paulo, Rio de Janeiro, Buenos Aires, Lima-Callao, and Santiago-Valparaíso. The São Paulo and Mexico City metropolitan areas each contain more than 20 million people and are

Figure 10.2 Latin America subregions: key indicators.

Subregion	Land Area (km^2) Total	Population (millions) Mid-2001 Total	2025 est.	GNI 1999 (US $ million) Total	GNI PPP 1999 Per Capita	Percent Urban 2001	Human Development Index Rank of 175 Countries	Human Poverty Index Percent of Total Population
Mexico, Central America	2,477,700	137.8	189.9	488,307	4,796	53.0	86	17.9
Caribbean Basin	689,910	33.3	39.8	37,440	6,872	53.6	67	14.1
Northern South America	4,718,100	115.2	161.9	255,958	4,120	71.0	74	15.5
Brazil	8,511,970	171.8	219.0	730,424	6,840	81.0	79	15.8
Southern South America	4,108,000	62.0	79.5	374,677	8,370	80.0	49	8.4
Totals or Average	**20,505,680**	**520.1**	**690.1**	**1,886,806**	**6,200**	**67.7**	**71**	**14.3**

Source: Data from *Population Reference Bureau 2001 Data Sheet*; *World Development Indicators,* World Bank, 2001; *Human Development Report,* United Nations, 2001.

two of the world's largest urban centers. Extreme contrasts between the materially wealthy and materially poor of the region are most dramatic within Latin America's cities.

An international demand for selected agricultural products and mined resources from the region connects some rural areas to the global economic system while leaving others isolated from international activities. Brazil's coffee, soybean, and citrus fruit crops remain an essential component of its export economy. Argentina and Uruguay produce and export high-quality meat and wool from efficient commercial farms. Chilean farms exploit seasonal markets for fruit and vegetable products in Northern Hemisphere countries. Much of the exported Brazilian iron and gold, Chilean copper, and Venezuelan oil are extracted from rural areas. Although selected rural areas are enjoying global attention, widespread static or declining rural economic activity remains interspersed throughout Latin America. Heated debate, both within the region and in the outside world, surrounds the best future use of undeveloped areas.

Geographic Conflicts

Many of Latin America's historically disputed boundaries continue to cause friction among countries in the region today. The boundary dispute between Ecuador and Peru in the western Amazon River basin included an outbreak of fighting between the two countries in early 1995. Disputes continue between Venezuela and Guyana over the Essequibo River valley. Guatemala took many years to recognize Belize, and a segment of the Guatemalan population continues to assert their country's ownership of Belize. Memories and impacts of past conflicts still sour international relations among Chile, Bolivia, and Peru (the Pacific War), and between Paraguay and its neighbors.

Some of the larger Latin American countries had geopolitical ambitions to extend their territories in the later half of the 1900s. Argentina's expansionist stance threatened Paraguay and Uruguay and later extended to a dispute with Chile over the ownership of the Beagle Channel in the far south. In 1981, Argentina occupied the Falkland (Malvinas) Islands in an attempt to displace the United Kingdom. Other regional countries appear to maintain expansionist attitudes today. Brazilian leaders talk about "living frontiers," meaning that they do not see them as fixed. Since the 1950s, Brazil's military expanded the road network across the Amazon River basin for internal security reasons. Surrounding countries, however, see the network as a threatening gesture. Mexico and Venezuela used their oil wealth at times of high oil prices to influence other countries. Such actions threatened their smaller neighbors, who saw hidden agendas in proposals for increased economic cooperation.

🌐 Regional Cultural History

The cultural history of Latin America provides a vital foundation for building an understanding of the contemporary regional geography. The colonial occupations that dramatically reduced the numbers of Native American peoples and con-signed the survivors to an underclass left a legacy of ethnic strife and hindered economic development. Habits and attitudes born under colonialism affected the region's countries for decades after independence.

Pre-European Peoples

The indigenous peoples present in Latin America when the Europeans arrived migrated to the region several thousand years earlier through western North America. While many indigenous groups had a village-based subsistence economy that supported only modest numbers of people, several urban-based civilizations or empires emerged. Urban-based civilizations in Latin America prospered through tightly structured societies and large-scale agricultural production (Figure 10.3). Estimates of the numbers of people living in Latin America before European entry range widely, with 50 million serving as a conservative mid-ground figure.

The Lasting Influence of the Maya, Aztec, and the Inca

The **Maya** civilization rose to regional prominence through agricultural surplus and a rigid social structure. The Classic Period, in which the Maya flourished, occurred from the A.D. 200s through the 900s. The Classic Maya based their regional structure on **city-states,** individually ruled and independently functioning urban centers, primarily located in the Yucatán Peninsula lowlands of southeastern Mexico, the Petén of Guatemala, parts of Belize and El Salvador, and eventually in parts of western Honduras. During the pinnacle of the Classic Maya civilization, cities flourished with abundant food supplies, extensive construction of urban structures, religious, royal, and athletic ceremonies, and detailed attention to the arts and sciences. City-states of the Maya region conducted trade, competed with each other in athletics, and arranged marital unions among their royal children. The city-states also competed fiercely at times for regional resources and often waged bloody warfare with one another. The growing populations of the Maya cities, coupled with environmental challenges such as drought, placed increasing burdens on the surrounding environment for food and other natural resources. The decline of the civilization occurred over several decades as agricultural harvests diminished and nutritional deficiencies began to take a toll on the crowded urban centers. The Maya imprint remains etched in the cultural landscape today in the form of pyramids and other ceremonial structures (see Figure 1.13). There are people throughout the Yucatán and northern Central America today who tie their ancestral heritage to the Maya of the Classic Period.

In the 1100s, a group of tribes from central and northern Mexico settled in the area of modern Mexico City. By 1325, the **Aztec** dominated the region with a hierarchical society centered on the massive urban capital of **Tenochtitlán.** Unlike the preceding Maya civilization, where numerous individually ruled city-states existed simultaneously, the Aztec Empire was governed by one king from one capital. The capital of the empire was built on marshland and lakeshore in the Central Valley of Mexico. This city had many civic, commercial, ceremonial, and

Figure 10.3 **Latin America: pre-European economies and empires.** A variety of Native American groups occupied the region before A.D. 1500. Some lived at subsistence level; others established far-reaching empires.

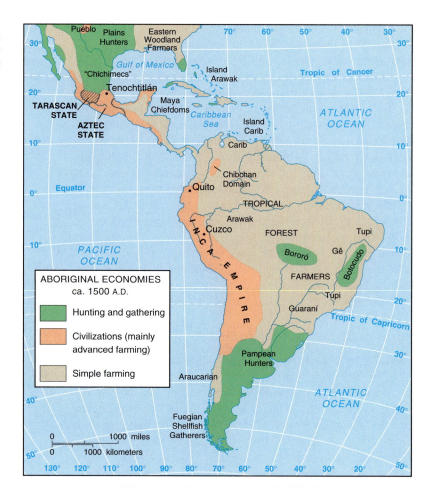

residential structures built on islands connected throughout the marshes of the valley by causeways. By the 1400s, the Aztec ruled most of Middle America and gathered great tribute riches in their capital. The elite of the empire converted acquired lands and peoples to their systems and used human and natural resources to further their goals. The Aztec forced conquered peoples into slavery, which fueled widespread animosity toward the empire's leadership. In the 1500s, when the Spanish conquistadors reached what is now Mexico, their goal was to topple the Aztec leaders and transform the Aztec cultural landscape to one suited for the glory of Spain. The Spanish conquistadors used the animosity held by many peoples of the region to assist in overthrowing the Aztec Empire.

Far removed from the Aztec Empire, the warlike **Incas** formed a strong society based in the southern part of modern Peru around the mountain city of Cuzco during the early centuries of the second millennium A.D. During the 1400s and early 1500s, the Inca Empire expanded dramatically and covered a region some 4,300 km (2,800 mi.) in length from southern Colombia to central Chile. Inca rulers established a well-connected system of roads running the length of their territory, which facilitated the empire's resource use, military control, and social order. The agriculturally based society also organized gold and silver mining and a system of messengers and transportation that maintained administrative cohesion. The tightly ruled empire functioned from a hierarchical system of noble leadership believed to be the offspring of the Sun God.

All resources of the controlled territory were considered to be the property of the Inca leaders. The official language of the Inca Empire, **Quechua,** remains widely spoken today among the indigenous people of Bolivia, Ecuador, and Peru.

Subsistence Cultures

The Arawak and Carib groups of the islands and South American mainland areas of the Caribbean gained subsistence by forest hunting and fishing. The Araucarians of modern central Chile and the Pampean tribes of modern Argentina had relatively strong economies. Sharing a similar fate with the larger indigenous empires of Latin America, most of these subsistence groups were decimated by or largely absorbed into the colonial cultures.

Spanish Colonization

Christopher Columbus embarked from Spain in 1492 on the first of four trans-Atlantic voyages. His ships called upon several Caribbean islands as well as a large stretch of the east coast of Central America and the coastal waters of northeastern South America. Spanish military and missionary groups soon followed. Within 50 years of the initial voyage of Columbus, much of the region was conquered and occupied. The only exceptions were some of the smaller Caribbean islands, the far southern extremity of South America, and the territory that became Brazil. In 1494, Spain and Portugal signed the **Treaty** of **Tordesillas**, which established a demarcation line (approximately 46°W longitude) between their global spheres of interest, giving the eastern quarter of South America to Portugal.

The initial Spanish occupation around 1510 centered on the large Caribbean islands of Cuba and Hispaniola. Inspired by tales of gold and other vast riches, Hernán Cortés led a group of conquistadors in 1519 on an expedition to the heart of the Aztec Empire (Figure 10.4). The Aztec thought the conquistadors were gods, which permitted the Spaniards to infil-

trate the hierarchy with relative ease. The conquistadors combined the multiple advantages of being revered as deities, having relatively more advanced weaponry, and the accumulated animosity of the numerous enemies of the Aztec leaders to overthrow the empire by 1521. The Spaniards captured Aztec lands and established the former Aztec capital as the core of Spanish colonial America. A Spanish viceroy began to govern the colony of New Spain in 1535. Spain claimed the remainder of Mexico and Central America, northern Colombia, northern

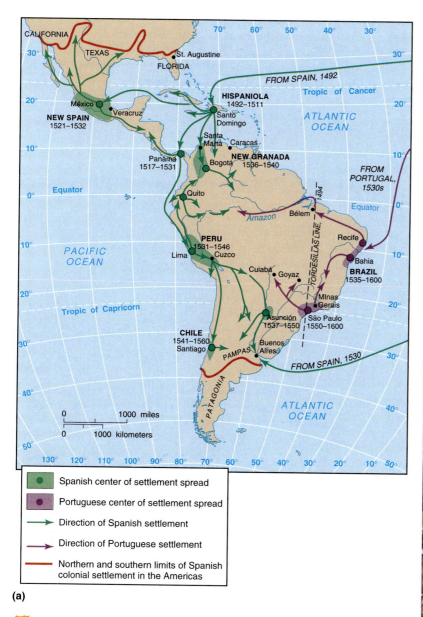

(a)

(b)

(c)

Figure 10.4 **Latin America: colonial conquest and settlement.** (a) During the 1500s, Spanish and Portuguese adventurers established bases and conquered huge areas rapidly. The areas of initial conquest became centers of administration and development. Scenes of the Aztec capital of Tenochtitlán (b) and the interaction of the Conquistadors with indigenous Mexicans (c) depicted in the artwork of celebrated Mexican painter Diego Rivera. Photos: © 2002 Banco de México Diego Rivera & Frida Kahlo Museums Trust. Av. Cinco de Mayo No. 2, Col. Centro, Del. Cuauhtémoc 06059, México, D. F. Schalwijk/Art Resource, NY.

Venezuela, and Caribbean islands such as Jamaica and Trinidad. Expeditions from New Spain pushed northward into the southern parts of the modern United States from the 1530s but found no gold.

In 1532, a small Spanish force invaded and captured Cuzco, the capital of the Inca Empire. It founded new cities to administer the gold and silver mining industries and facilitate the transportation of resources across the Panamanian isthmus and onward to Spain. Spanish expeditions in southern South America in the 1530s did not lead to material occupation or integration into Spain's colonial system until much later (Figure 10.5). Lima and its nearby port of Callao, which linked the resource wealth of Peru to Spain, became the capital of the Viceroyalty of Peru that included most of northern and western South America. The Viceroyalty of New Granada, centered on Bogotá, was established in 1717 and that of La Plata, centered on Buenos Aires, in 1776.

The Spaniards conquered most of the region in the 1500s and imposed a high degree of control on their colony of New Spain by establishing an oppressive system of agricultural production and tightly connected urban settlements and ports. Several forces hindered the Native American inhabitants in forming strong resistance to Spanish intentions in the Americas. First, after the conquest, the lack of immunity to European diseases drastically reduced indigenous populations. Then, surviving communities were converted by the Roman Catholic church. They were forced to give up their subsistence agricultural practices to provide crops and livestock for the food needs of the foreign city dwellers. Mining settlements in the Andes Mountains used various forms of slavery to coerce the local people.

The colonial period in Latin America left a legacy of underdevelopment because it primarily focused on mining for export and the imposition of large landed estates with a strict feudal system. The native peoples and many of mixed ancestry were relegated to being landless laborers. By the end of the Spanish colonial period, antagonisms between privileged and underprivileged groups were ingrained. Subsequent history is largely a record of such antagonisms at work.

Spanish Control

Spain granted large land areas to nobles, soldiers, and church dignitaries, in conjunction with the responsibility for political control, establishment of settlements, and exploitation of natural resource wealth. They were given jurisdiction over the native peoples to use them as laborers. This feudal **encomienda,** or tribute system, was applied throughout the Spanish colonies for more than 200 years. Although some efforts were made in Spain to pass legislation calling for the fair treatment of native peoples, such laws were not applied in a meaningful manner in the colonies. In addition to the subjugation of native peoples, the Spanish eventually imported more than 1.5 million people from the African continent to labor as slaves in mining and agriculture.

Spanish colonial agriculture developed large production estates, or **haciendas,** to cultivate crops and livestock products for local or domestic markets. Landholders often advanced credit to laborers, causing them to be indebted to the hacienda. Coastal areas around the Caribbean were home to plantations, which produced agricultural products for export to Europe and elsewhere outside of the region.

Inefficient and corrupt colonial administration of the region established a foundation for pervasive and lasting tension between various socioeconomic groups. The native peoples were dispossessed of land and rarely afforded educational opportunities. At the other end of the socioeconomic scale, the **peninsulares**—Spaniards born in Spain who were living or working in the New World—took the highest offices and largest land grants. **Criollos**—Spaniards born in the colonies— and the increasing numbers of **mestizos** (mixtures of European and Native American ancestry) had fewer privileges and became resentful as their links to Spain weakened.

Portuguese Colonization

Portuguese colonial efforts in Brazil moved at a slower pace than Spanish efforts in the region until the French, in the 1530s, attempted to colonize in areas previously claimed by Portugal.

Figure 10.5   **South America: colonial divisions.** The jurisdictions established by Spain and Portugal formed the basis of country boundaries at independence but were poorly defined. The red line divides areas of Spanish (to the west) and Portuguese (to the east) influence. The pink shading highlights early attempts to extend Brazilian territory westward. Buenos Aires became the center of the Viceroyalty of Río de la Plata.

Then the Portuguese stepped up their settlement and political organization of the region. A Portuguese governor general was installed in 1549. The city of Salvador became the first capital and northeastern hub of Portuguese Brazil, while São Paulo was established much farther south to solidify the claims and settle southern lands. The Portuguese established a settlement along Rio de Janeiro Bay after expelling French settlers from that portion of the region. The bayside settlement of Rio de Janeiro became the Portuguese colonial capital in the late 1700s.

In the 1600s, the discovery of gold inland of Rio de Janeiro led to increased Portuguese immigration. This in turn set off expeditions to explore and claim the interior. From 1680, various attempts were made to push the Portuguese boundaries southward to the Plata estuary, where Spanish settlement was slow. In 1750, Spain agreed to permit Brazilian interior expansion westward of the Tordesillas line—which was really an expression of Spain's inability to prevent what was already occurring.

The Portuguese process of occupying Brazil resembled that of Spain in its colonies, although it had some distinctive features. Colonial Brazil was economically and socially controlled by an elite class who purchased or captured millions of African slaves and transported them across the Atlantic to work on large sugar plantations along the northeastern coast. The Portuguese were responsible for the forced migration of more people from the African continent than any other single colonial power (see Figure 9.6). Estimates suggest that more than 4 million Africans were shipped to Brazil to fill the labor needs of economic development. Large numbers of present-day Brazilians trace their ancestral heritage to African slaves. Farther south in the São Paulo and Rio de Janeiro areas, Europeans who were not part of elite colonial circles had more opportunity to establish their rights to new lands, leading to a greater degree of personal autonomy.

Other European Colonies

French, Dutch, and British attempts to colonize Latin America came later and were largely resisted by the Spanish and Portuguese. There were limited incursions on lands controlled by the dominant colonizers primarily in and on the periphery of the Caribbean Basin.

The French took the western one-third of Hispaniola in 1697. An independence movement in 1804 expelled the French and led to the creation of the new country of Haiti. The French also took control of the remainder of Hispaniola (Santo Domingo) from Spain in 1797 and maintained a presence until 1809. When Santo Domingo became independent in 1821, Haiti invaded and occupied it until 1844, after which the independent country of the Dominican Republic was declared. Cuba and Puerto Rico remained Spanish colonies until the United States backed a Cuban uprising in 1898 and won the subsequent Spanish-American War. Cuba became independent a few years later. Puerto Rico became a "commonwealth territory" of the United States in 1898, and its people became U.S. citizens in 1917.

Some of the smaller Caribbean islands that were ignored by Spain had their first European settlement by the French, Dutch, or British. The material wealth of Europeans tied to the region prospered in the 1600s due to dramatic increases in the land area used for sugar cultivation and the amount of African slave labor imported to the region. Many of the islands changed hands, as when Britain took Jamaica (1670) and Trinidad (1797) from Spain.

The Guianas were within the Spanish realm on mainland South America, but little interest was shown in the swampy coasts and forests until the Dutch, British, and French settled them. The Dutch settled first in 1581, eventually establishing what is today the country of Suriname. The British took the western area in 1814 and established a colony (today the independent country of Guyana). The French settled the eastern part in the 1600s, where the infamous penal colony of Devil's Island deterred further immigration into French Guiana until the prison was closed in 1938.

The British settled along parts of the east coast of the Central American mainland and on islands just offshore. British influence proved to be lasting in the western Caribbean in Belize and parts of Honduras. British control in Belize lasted until well into the 1900s, and English-speaking residents of the Bay Islands of Honduras are today a lasting geographic remnant of British influence in the region.

Independence

The initial wealth that attracted Europeans came from exports of high-value minerals such as gold, silver, and gemstones from Latin American colonies to Europe. When the flows of such wealth slowed in the 1700s, combined with the devotion of resources to fight the Napoléonic conquest of Spain in the 1790s, the colonial power of both Spain and Portugal weakened considerably. The eroding strength of those European colonial empires enabled many Latin American countries to gain independence in the early 1800s. The revolutions often began with uprisings that were initially suppressed by Spanish soldiers but led to concerted drives centered on Middle America and Andean sections of South America. The new countries emerged from administrative divisions within the Spanish viceroyalties. The colonial boundaries between these divisions were poorly surveyed, if at all, forming a basis for subsequent disputes.

Independence from Spain

Independence from Spanish rule was often followed by a period of successive attempts at different types of political arrangement. The establishment of the modern countries in the region unfolded slowly. When Mexico separated from Spanish rule in 1821, it covered an area from the present-day southwestern United States through the modern Costa Rica–Panama border. The Central American region separated from Mexico in 1823 as the United Provinces of Central America, which then further divided in 1838 into the Central American republics existing today. Northwestern South America functioned apart from Spanish rule as Gran Colombia from 1822 to 1830. In 1830, the independent countries of Colombia (including Panama), Ecuador, and Venezuela emerged. Panama did not gain independence from Colombia until 1903, when U.S. interest in building a canal across the isthmus facilitated that separation. The remainder of Spanish South America splintered into several countries independent of the Spanish crown during the 1810s and 1820s.

The newly independent countries were poorly prepared to capitalize on their sovereignty. The next 150 years after independence were marked by political instability punctuated by short periods of economic growth. Spain left behind a geographic system based on mineral exploitation, transportation to a limited number of ports, and large estates engaged in raising livestock. The postindependence period created rivalries and boundary disputes between countries. Rigid social class stratification fueled deep and lasting internal resentments. In most former Spanish colonies in Latin America, government control alternated between the elite conservative landowning group, often allied to the military, and the more liberal criollos who wanted a wider spread of democratic involvement. Many countries had long periods of dictatorship after conservative-liberal tensions made democratic government almost impossible.

Brazilian Independence

Independence came in the 1820s to Brazil, which had a stronger foundation in place for modernization than other postcolonial territories. In 1807, the Portuguese royal family evacuated from Europe to Brazil when Napoléon threatened to take their country. Rio de Janeiro became the temporary capital of Portuguese government. Although the royal family returned to Portugal after Napoléon's defeat, the prince regent returned to Brazil in 1816 at a time of flourishing revolutions in the Spanish colonies. In 1822, he proclaimed Brazil's independence from Portugal and himself king as Pedro I. Under King Pedro II (1840–1889), the Brazilian economy grew rapidly with the construction of railroads and ports and the expansion of mining in the east-central parts, commercial farming of coffee in the south, and rubber collecting in the Amazon River basin. In 1889, a military coup made Brazil a republic, but falling prices of coffee and rubber led to widespread unrest and a period of dictatorship. Brazil encouraged immigration and received many people from Germany, Italy, and Japan in the early 1900s. They moved mainly into the southern parts of the country, where environmental conditions were somewhat similar to those of their homelands.

Economic Colonialism

Although Spanish and Portuguese rule was broken, Europe remained the primary market for Latin American exports. Britain in particular established a relationship of economic colonialism with several Latin American countries that lasted until the early 1900s. Areas targeted for economic development and control by the British were set up for the export of raw materials and products to supply Britain's industries. The British built railroad networks and port installations aimed at developing the production of minerals, cotton, beef, grain, and coffee. Elite families in Latin American countries who worked within the system often sent their children to schools in Europe and the United States and banked their wealth in those countries.

Continuing External Influences

Import Substitution

The economic depression of the 1930s in the United States and Europe significantly disrupted trade and economic activity for Latin America. Following on the heels of the global economic depression was World War II in the 1940s. The combination of these two dramatic events caused many decision-makers in Latin American countries to strive to be more internally self-sufficient. The countries of the region established the goal of becoming less dependent on selling unprocessed or unrefined raw materials in exchange for high-priced manufactured goods from industrial

---

Test Your Understanding 10A

Summary   Latin America includes Mexico, Central America, the Caribbean Basin, and South America. Although many parts of the region are a legacy of influence outside of Europe's Iberian Peninsula, a degree of unity exists in contemporary Latin America centered on the Spanish and Portuguese languages and an adherence to Catholicism. While the region is tied by common colonial legacy, it contains dramatic variations in both its human and physical geography. Latin America is plagued by a widening gap of human wealth and economic development in some of the world's largest cities and resource-rich farming regions. Despite considerable growth in many areas, Latin America contains expansive impoverished rural areas throughout the region.

Many Latin American countries are small, especially the Caribbean islands, but historic animosities and the production of competing goods often prevent them from working together for economic growth. The economic and cultural history of Latin America led to slow modernization, internal and international antagonisms, and reliance on one or a few export products.

Questions to Think About

10A.1 Compare and contrast the location, time period, social structure, and regional influence of the Maya, Aztec, and Inca. What role (if any) did the Spanish conquistadors play in the evolution of each civilization?

10A.2 What were the primary historic events that imposed a Hispanic culture on Latin America? How did that culture affect postindependence changes and attitudes? Which countries did not experience Hispanic influence during the colonial era, and how did this affect their contemporary human geography?

10A.3 Compare and contrast the roles of peninsulares, criollos, mestizos and African slaves in the social structure and economic development of colonial and postcolonial Latin America.

Key Terms

Maya	encomienda system
city-state	hacienda
Aztec	peninsulare
Tenochtitlán	criollos
Incas	mestizo
Quechua	nationalize
Treaty of Tordesillas	

countries. Under import substitution, Latin American countries attempted to use their raw materials in their own internal production of various manufactures for domestic markets. Governments established high tariffs, quotas, and bureaucratic barriers on goods arriving from countries outside of the region. Many of the new industries were owned by the various country governments, and others, established earlier by foreign interests, were **nationalized,** or taken over by the governments of the Latin American country in which they operated. By the 1970s, this policy resulted in the rapid growth of a few major urban centers in each country. Because of their larger home markets, the most populous countries (Argentina, Brazil, Colombia, Mexico) produced the most under this system. Economic growth continued to be uneven.

 # Natural Environment

Latin America is a region of dramatic physical geography, including the world's second-highest mountain system, the Andes, the world's largest river, the Amazon, and the world's largest remaining tropical rain forest region, the Amazon. Although the region extends through almost 90 degrees of latitude—the greatest north-south distance of any major world region by some 25 degrees—the majority of Latin America lies within the tropical latitudes. When coupled with high mountain ranges, the latitudinal expanse of the region results in a wide variety of climates, natural vegetation types, and soils.

Tropical and Southern Hemisphere Climates

Middle America and the Caribbean Basin

Nearly all of **Middle America** (Mexico, Central America, and the Caribbean Basin) lies within the tropics (Figure 10.6). The dominant east-to-northeast trade winds that blow across the Gulf of Mexico and especially the Caribbean Sea bring these areas consistent warmth and humidity. Temperatures remain around 30°C (86°F) through most of the year. The northern Caribbean islands, Mexico, and even parts of northern Central America are occasionally affected by modified cold air masses, locally referred to as "*nortes,*" that move down from continental Canada and the United States during the Northern Hemisphere's winters. In Middle America, the rainfall generally increases southward, from the arid region that straddles the Mexico–U.S. border toward the coasts of Nicaragua, Costa Rica, and Panama, which receive over 2,600 mm (100 in.) of rain each year.

Most rain falls in the summer and fall, when heating of the lower atmosphere causes humid air to rise, cool, and condense into clouds. When condensed moisture in the clouds becomes heavy, gravity pulls the moisture to the surface as rain or some other form of precipitation. On Caribbean islands with mountains, windward slopes facing the east or northeast trade winds add to the uplift of the humid air and increase precipitation totals. Where the air descends, warms, and becomes less humid, less rain is likely. For example, in Jamaica, the wind-

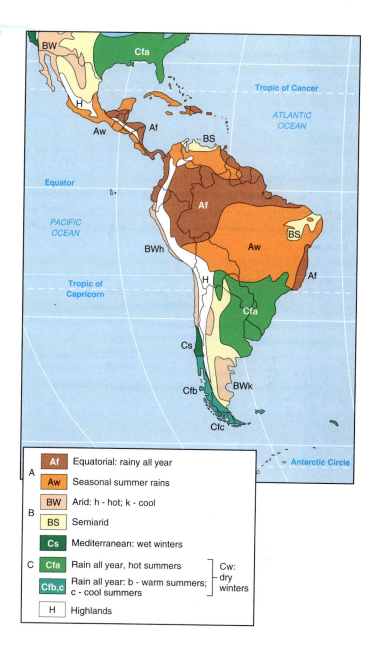

Figure 10.6 **Latin America: climate regions.** Note the contrasts.

A	**Af**	Equatorial: rainy all year
	Aw	Seasonal summer rains
B	**BW**	Arid: h - hot; k - cool
	BS	Semiarid
C	**Cs**	Mediterranean: wet winters
	Cfa	Rain all year, hot summers
	Cfb,c	Rain all year: b - warm summers; c - cool summers
	H	Highlands

Cw: dry winters

ward northeast coast receives over 3,300 mm (130 in.) a year, while the leeward southern coast requires irrigation for farming in areas where the annual rainfall is less than 750 mm (31 in.).

Middle America is a region of annual hurricane activity. The North Atlantic hurricane season runs from June through November. Early in the season, hurricanes form in the Western Caribbean and Gulf of Mexico. During the seasonal peak in August and September, hurricanes form in the eastern Atlantic near the Cape Verde Islands west of the African continent. The storms move westward into the Caribbean Basin and curve toward either the northwest or north (affecting the U.S. mainland or shipping channels) or continue westward, striking Central America or the Mexican Yucatán Peninsula. Late in the annual hurricane season, storms once again form mostly in the western

Caribbean and Gulf of Mexico. Only the southernmost countries of Middle America lie outside the hurricane zone. Hurricanes cause both a dramatic loss of life and extensive damage to crops, livestock, personal property, and the vital regional tourism industry. Recovery is often slow in small countries with few resources. For example, in 1998, Hurricane Mitch caused extensive damage and loss of life that devastated the already poverty-stricken countries of Honduras and Guatemala (Figure 10.7).

Northern South America

All the northern Andean countries lie within the tropics. Temperatures fluctuate little throughout the year, but drier periods and seasons increase in length and severity away from the equator. The amount of rainfall at a location depends on whether the land faces the direction from which moist air arrives. Humid airflow from the Atlantic Ocean into the Amazon River basin carries heavy rains far westward across the continent to the east-facing slopes of the Andes. Slopes that face away from the winds, sheltered valleys and basins, and land at high altitude may require irrigation for effective agriculture. On the Pacific coast, rainfall is high on the slopes facing the ocean north of the equator, but southward the offshore winds and cold Peruvian ocean current create an arid climate on the adjacent land.

Northern Brazil experiences an equatorial climate in which the temperatures hover in the low 30s°C (80s°F), humidity is high, and rain falls in all seasons. On the Brazilian and Guiana Plateaus, the rains are more variable and more seasonally concentrated in the high-sun period when evaporation and rising air currents are most intense. The variability is particularly marked in the northeastern corner of Brazil, where severe and prolonged drought occurs cyclically, roughly every 10 years. In southernmost Brazil, the climate becomes midlatitude in type, with cool winters and a shorter growing season.

El Niño

The **El Niño phenomenon** is regarded as a significant feature in explaining connections among worldwide climatic environments (Figure 10.8). The major 1997–1998 El Niño event brought drought to Middle America and northern South America, and exceptionally heavy rains to the deserts of Peru and Chile. The same El Niño event was blamed for unusual weather around the world. Droughts in Indonesia and Australia coupled with drought-based fires in Florida and unusually hot weather in southern Europe were indirectly linked to the event.

The basic features of El Niño are understood. The warm current is linked to atmospheric and oceanic movements across the Pacific, the world's largest ocean, and is monitored by satellite. Every two to five years, easterly winds that usually push the cold Peruvian current westward die down in the tropical eastern Pacific. Warmer water from the western Pacific flows eastward, eventually reaching the coasts of western North, Central, and South America. Warmer and more humid air masses off the west coasts of the Americas alter the regional climate. Warm water also dramatically changes the marine ecosystem, and dur-

Figure 10.7 **Hurricane Mitch damage.** Aerial view of a Honduran coastal village damaged by Hurricane Mitch in 1998. Photo: © Yann Arthus-Betrand/Corbis.

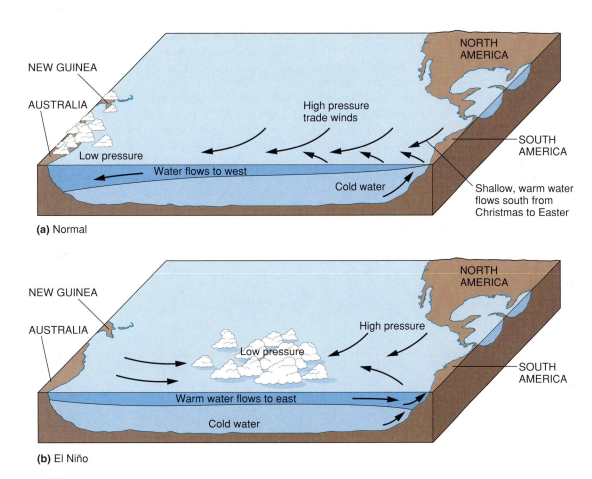

(a) Normal

(b) El Niño

Figure 10.8 **The El Niño effect.** How do the changes in wind direction from a normal circulation pattern (a) to the El Niño conditions (b) affect the surface waters off the coast of western South America? Source: From *Foundations of Physical Geology* by Bradshaw & Weaver. Copyright © The McGraw-Hill Companies.

ing strong El Niño events, it may wreak havoc on the fishing industries of countries in western South America.

The 1997–1998 El Niño event was the largest since 1982–1983, and its impacts on people's lives in Latin America were diverse. Although some areas enjoyed advantages, such as freshwater reservoir replenishment and a prolonged winter ski season due to increased snow depth on the Chilean Andes, many outcomes were negative. Droughts caused lower farm yields in Central America and northern South America, while floods destroyed crops in Argentina and Uruguay and washed out roads in Ecuador and Peru. Destructive fires burned forest and savanna vegetation on the Brazil-Venezuela border and in Central America. Water shortages disrupted transportation in the Panama Canal as well as the output of some hydroelectric plants, resulting in temporary power cuts.

Southern South America

The climates of Southern South America range from deserts in northern Chile and southern Argentina to one of the world's stormiest and wettest regions in southern Chile. The Andes ranges are not as important for human settlement as they are farther north but affect the climates of the lands on either side.

In northern Chile, the Atacama Desert continues the Peruvian Desert southward between the high Andes and the cold Peruvian current. Winds blow almost parallel to the coast

or offshore, pushing the cold current northward along the coast. The cooling effect of the ocean on atmospheric temperatures makes it difficult for air to rise and cause precipitation.

In southern Chile, by contrast, the midlatitude westerly winds bring precipitation and high winds at all seasons of the year. Between the desert and the stormy southern region of Chile is a transition zone of dry summers when the arid climate moves south and wet winters when the storm tracks move north. This zone is similar in climatic regime to the countries around the Mediterranean Sea in Europe (see Chapter 3).

On the eastern side of the Andes is another contrast between northern and southern lands. In the north, the warm Brazilian current in the Atlantic Ocean allows high rates of evaporation offshore, and trade winds blow humid air into the continent. The highest rainfall totals are near the coast and decline inland, so that the Andean foothills are often arid. In the far south of Argentina, Patagonia has a climate that is driest near the coast and becomes somewhat more humid toward the mountains. The westerly airflow descends after crossing the Andes, warms up, and becomes drier. Patagonia is often cited as an example of rain shadow conditions in the lee of a mountain range, in contrast to the very wet Pacific slopes of the Andes facing the prevailing westerly winds in southern Chile. The strong winds blowing across Patagonia, after crossing the Andes, often produce very dry conditions at the surface.

High Mountains and Island Chains

The major relief features of Latin America (Figure 10.9a) were formed by a combination of clashing tectonic plates along the western and northern margins and of precipitation runoff in major rivers and mountain glaciers. Along the west coast, the South American plate overrides the Nazca plate (Figure 10.9b). The convergent plate margin marks the line of the Andes Mountains and causes earthquakes and volcanic eruptions. The tectonic pat-tern is more complex in Middle America, where the North American, South American, Caribbean, and Cocos plates meet.

Insular and Mainland Middle America

Mainland Middle America is formed of two relief provinces. High-altitude plateau lands between the eastern and western Sierra Madres dominate northern Mexico. The collision of the North American and Cocos plates created the plateau. The plateau rises over 2,000 m (6,000 ft.) and contains shallow

Figure 10.9 **Latin America: physical features.** (a) A relief map, showing the mountain ranges, plateaus, and river basins. (b) The tectonic plates and their boundaries.

basins that become northward more arid than the surrounding plateau but are often connected by rivers, forming fertile areas for farming. The western slopes facing the Pacific Ocean are steep, but those on the east are less steep with wide coastal plains. This province terminates at the Tehuántepec isthmus, where the land narrows in southern Mexico.

South of the Tehuántepec isthmus, the Caribbean plate collides with the Cocos plate, forming a single spine of mountains along and parallel to the Pacific coast. There are very narrow coastal plains and some areas without any flat land between mountain and ocean. Eastward of the mountain spine, west-east ranges are separated by deep basins, often occupied by rivers flowing into the Caribbean Sea. A large limestone platform emerged from the seafloor to form the Yucatán Peninsula. The ridges and basins can be traced eastward across the floor of the Caribbean Sea and beneath the island arcs on its eastern edge.

Middle America is subject to earthquakes and volcanic eruptions where tectonic plates collide. A major earthquake centered off the west coast of Mexico devastated Mexico City in 1985 (Figure 10.10), and earthquakes twice leveled the Nicaraguan capital city of Managua in the 1900s. There are 25 active volcanoes between northern Mexico and Colombia that periodically spew lava and ash on surrounding areas.

Insular Middle America is also affected by the clashes of tectonic plates. The North American plate drove into the Caribbean plate, producing intense volcanic activity along the plate margin and forming the Lesser Antilles arc. The eruption of lava and ash continued into the 1900s at Martinique (1902), Mount Soufrière on St. Vincent (1979), and the Montserrat eruption of 1995 that eventually forced the evacuation of the island. In the Greater Antilles, the plate movements caused a mass of continental rock to founder, leaving only the highest points above sea level. The Bahamas and an outer group of islands including Anguilla, Barbuda, and Barbados are flat limestone islands constructed of coral reefs on top of subsiding former volcanic peaks (Figure 10.11).

Andes Mountains

The Andes Mountains affect all aspects of the physical environment of western South America. The collision of the South American and Nazca plates produced a volcanic and earthquake-prone western mountain range and a folded and faulted eastern range. The Andes rise to over 6,500 m (20,000 ft.) in Argentina, Chile, Peru, and Ecuador. In Bolivia, Peru, and Ecuador, the central Andes have two main ranges, the Cordillera Occidental (west) and Cordillera Oriental (east) (Figure 10.12). Between the two ranges, a high plateau, the **Altiplano,** is widest in Bolivia and narrows northward into Peru. In Peru, rivers cut deep gorges as they flow northward and eastward to join the Amazon River tributaries.

In Colombia, the Andean ranges splay out northward into three cordilleras—the Occidental, Central, and Oriental. The Atrato River separates these from a coast range, while the Cauca and Magdalena rivers separate the three main ranges. The Pacific coast has only narrow coastal plains. In northern Venezuela, the Cordillera Oriental branches into a further series of lower ranges. Along the Caribbean coast, the mouths

Figure 10.10 **Mexico City: earthquake damage.** Rescue workers search for victims in a building completely demolished during an earthquake that devastated parts of Mexico City in 1985. Photo: © Owen Franken/Corbis.

of the Colombian rivers, Lago de Maracaibo, and the Orinoco River delta provide limited areas of lower coastal land between the ranges. Islands such as Trinidad and the Dutch Antilles are extensions of mainland geologic structures.

The southern Andes Mountains dominate the landscapes of Chile and the western parts of Argentina, with their highest points constituting the border between the two countries for most of its length. The highest point in the whole range, Cerro Aconcagua (6,959 m, 22,831 ft.), occurs where the central Andean ranges narrow in width southward from about 500 km (300 mi.) to about 150 km (100 mi.). The lowest point for crossing the Andes between the main centers of population in Chile and Argentina is the Uspallata Pass at 3,841 m (12,600 ft.). To the south of this pass, the Andean peaks get lower, continuing for a further 2,700 km (1,800 mi.) to Tierra del Fuego at the southern tip of the continent. The snow line on the Andes gets lower toward the southern tip of the continent, where glaciers descend to sea level (Figure 10.13). The former more widespread glaciations left behind moraine-dammed lakes and fjords.

Figure 10.11 **Caribbean Basin: coral islands.** The Exuma Cays, Bahama Islands, are formed of coral reef limestone and rim the Great Bahama Bank for over 150 km (100 mi.). Photo: © Bruce Dale/ National Geographic Image Collection.

Figure 10.12 **South America: Andes Mountains.** The major features of the central Andes Mountains related to their formation along a destructive plate margin. The Nazca plate plunges beneath the South American plate, causing volcanic activity in the western cordillera along the coast and uplift of the eastern cordillera and the high plateau (Altiplano) between the two ranges.

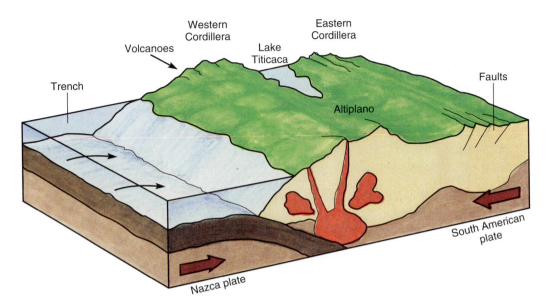

On the Chilean side, the Andes come close to the Pacific Ocean in the north. Southward, a coastal range is separated from the main Andes ranges by a series of basins and then a wide continuous valley south of the Chilean capital, Santiago. The coastal range and the valley get lower in height alongside the main Andes range and are drowned by the ocean south of Puerto Montt. On the Argentinean side of the Andes, the front ranges in the north are broken by deep, river-carved valleys. Their eastern margins have large alluvial fans formed by the deposition of rock material eroded from the mountains. These fans mark both sides of the Andes throughout Chile and Argentina.

Broad Plateaus

Broad plateaus and wide river valleys dominate Brazil's physical environment. Locally, relief is sharp near physical transition zones, as when one travels from the coast to the first plateau level or from one plateau level to the next. More generally, the traveler in Brazil is struck by small differences in the elevation over great distances.

Figure 10.13 **Southern South America: Andes Mountains.** The Moreno glacier, Glacier National Park, Santa Cruz province, Argentina. The glacier is 5 km (3 mi.) wide, and its front melts on entering Lago Argentino. Such glacial features increase in significance in the southernmost part of the Andes Mountains. Photo: © James P. Blair/National Geographic Image Collection.

The main relief features of Brazil consist of the ancient rocks of the Brazilian Highlands, which are topped in the southeast by layers of lava flows, and the similar ancient rocks of the Guiana Highlands on the Venezuelan border to the north. The Guianas have low coastal plains that were formed by the deposition of sediment brought to the Atlantic Ocean by the Amazon River and then moved westward along the coast by offshore currents. Inland, these countries rise to the Guiana Highlands plateau. In southern Argentina, the Patagonia Plateau is cut deeply by rivers draining eastward from the Andes.

Major River Basins

Three major river basins between the high mountains and lower plateaus dominate South America. Tributaries of the Orinoco River primarily drain the largest areas of lower land in Colombia and Venezuela. The Amazon River tributaries flowing from the Andes are muddy "white water" rivers in contrast to the black rivers, which contain little sediment as they flow from the plateaus. The contrast extent is visible where the "black" Rio Negro joins the muddy Solimões just below Manaus in the center of the Amazon basin (Figure 10.14). The Amazon River is navigable well into Peru. In Manaus, Brazil, the Amazon River is 2,500 km (1,500 mi.) from the ocean, 15 km (10 mi.) wide, and over 50 m (160 ft.) deep.

In southern Brazil, rivers drain south to the Paraná-Paraguay River system. The southward flowing rivers present great hydroelectricity potential in the waterfalls at lava plateau breaks. In Argentina and Uruguay, the pampas plains were formed by deposits of the Paraná-Paraguay River system and overlaid by fine windblown loess. These river and wind deposits blanket ancient rocks that occasionally pierce them, forming hilly areas and low ridges in the northern part of Uruguay and on the southern margins of the pampas.

Natural Vegetation

Rain forest is the natural vegetation where tropical rainfall is plentiful and distributed through the year, as in the Amazon

Figure 10.14 **Brazil: Amazon rain forest.** A LAND-SAT satellite view of the area around Manaus, Brazil, in July 1987 (approximately 150 km, or 100 mi., across). Unbroken tropical rain forest is shown as red. The wide, black river is the Río Negro that contains little silt. The blue river is the Solimões branch of the upper Amazon, which brings large quantities of silt from its upper reaches in the Andes Mountains. Manaus is the light-colored area just west of the Negro-Solimões confluence, with radiating roads leading to and from it north and south of the river. Small white areas with shadows to the west and east are clouds. Photo: Image courtesy of Space Imaging, Thornton, CO, USA.

River basin, along the northern Pacific and Central American coasts, and on some Caribbean islands (see the "Point-Counterpoint: Tropical Forests and Deforestation" box, p. 434). Such vegetation spreads several thousand meters up the east-facing slopes of the Andes. The world's largest expanse of tropical rain forest covers most of the Amazon River basin and extends along the eastern coastal lowlands of Brazil to the Tropic of Capricorn. Soils beneath the forest vary, but the areas of good soils are small, apart from the flooded areas close to sediment-carrying rivers.

Tropical grasslands or shrub vegetation communities are present where tropical rainfall is seasonal or significantly lower on average during the year than in rain forest areas. On the Brazilian Highlands, dense deciduous woodland gives way to more open woodland with increasing proportions of shrubs and grasses. Soils are generally poor beneath the natural vegetation and need treatment for agriculture, but some of the lava flows capping the plateau in southeastern Brazil produce easily worked soils.

Cold ocean currents, arid air, and the rain shadow from easterly winds produce deserts along the central west coast of South America. The Atacama Desert here is one of the driest places on Earth. Vegetation ranges from tropical plant species in the north to midlatitude types in the south.

In the southern part of South America, cooler midlatitude conditions coupled with varying levels of annual precipitation create a range from bare desert, through semiarid bunch grasses and drought-resisting plants, to tall grasses and forest in humid areas. Before European settlement, central Chile had natural vegetation of trees and shrubs that could grow in a regime of wet winters and dry summers. The plentiful precipitation of southern Chile supports natural vegetation of beech and pine forests. The pampas region of central Argentina and Uruguay is named after the tall, lush grasses that grew there before the region was plowed.

The majority of the high mountain ranges of Latin America are located within tropical latitudes. The latitudinal position of the Andes Mountains coupled with their very high altitude results in a series of vertical zones with distinctive climate and vegetation regimes. The greatest number of distinctive vegetative zones is at the equator. The **altitudinal zonation** in the region is directly linked to the variety of crops that may be cultivated (Figure 10.15). The lowest 1,000 m (3,000 ft.) have warm to hot conditions and tropical forest in the *tierra caliente*. The next 1,000 m (3,000 to 6,500 ft.) have mild to warm temperate conditions and deciduous forest in the *tierra templada*. Between 2,000 m and 3,000 m (6,500 to 10,000 ft.), the *tierra fría* has cold to mild temperature conditions and pine forests. Above this zone, there is grassland, and at the greatest heights, in the *tierra helada*, where snow lies all year, even on the equator. Climbers of the Andes near the equator pass through vegetation zones in a few kilometers that would require a sea-level trek from the equator to the poles. The similarities to sea-level climatic environments and vegetation zones are not complete, however, since temperatures in mountains near the equator vary little from month to month at all altitudes, while winds increase and the air gets thinner at higher altitudes.

Natural Resources

The natural resources of Latin America include minerals, soils, forests, water, and marine life. The Andes Mountains and the ancient plateau rocks contain considerable resources of metal ores that were among the early attractions for European settlers. They are still not fully exploited where surface transportation is poor. Rivers flowing from the uplands deposited alluvial concentrations of heavy metal-ore minerals in the adjacent lowlands. Sedimentary rock basins between the mountains and in offshore areas became petroleum and natural gas reservoirs, especially in eastern Mexico, northern Venezuela, Colombia, the offshore areas of northeastern Brazil, and along the eastern slopes of the Andes in Ecuador, Peru, and Argentina.

The alluvial soils of the Amazon River basin, the weathered lava surfaces in southern Brazil, and the pampas of Uruguay and central Argentina form large areas with predominantly good soils for farming. The soils on the eastern plateaus are generally low in plant nutrients, while the steep slopes of the Andes confine potential farming areas to small valleys, high basins, and plateaus.

The water resources of Latin America are huge, including the world's largest river (by volume), the Amazon, which carries more than twice the amount of water of the next largest river, the Congo in Africa. Many Latin American countries currently generate a large proportion of their power as hydroelectricity, and they continue to explore the expansion opportunities of this power source. Although some areas, such as northern Mexico, northeastern Brazil, the Peru–north Chile coastlands, and southern Argentina, are arid, much of the region has plentiful precipitation to support agriculture. For example, rivers flowing from melting snows on the Andes water the arid coast of Peru and the oasis settlements of northern Argentina.

The combination of soil and water conditions in much of the region fostered the early development of domesticated crops.

TROPICAL FORESTS AND DEFORESTATION

Deforestation, or the permanent clearing of forest vegetation, is a centuries-old land modification practice. Societies in all world regions and in both temperate and tropical latitudes used forest resources for fuel, shelter, and transportation. Dramatic increases in the permanent clearing of tropical forests, and especially tropical rain forests, in recent decades is causing significant alarm among a growing global body of scientists, medical researchers, government officials, and environmentalists. Global communities concerned with the potential ecological and human health detriment from tropical forest clearing call on the governments practicing or permitting tropical deforestation to cease. Locally, where deforestation is taking place, governments, business communities, and farmers angrily retreat from the global community and assert their sovereign rights to resource use.

The controversy surrounding the permanent clearing of tropical forests is a prime example of the antagonistic relations that may form when globalization forces and local traditions or preferences work against one another. Complicating the struggle between the global and local in tropical deforestation is that globalization forces are specifically one of the major causes of deforestation. Why is tropical deforestation such an emotionally charged issue that seems to pit environmental groups against international development organizations, developing country governments against developed country governments, and global corporations against grass-roots human rights advocates? The answer lies in attempting to understand the scientific significance tropical forest biomes play in terms of global ecology, global climate, and human biology in a politically charged and economically challenged environment.

Tropical forests may exist in many forms depending on wind flow, elevation, proximity to large ocean systems, and global air circulation patterns. Many tropical forests, commonly called tropical seasonal forests, have a distinct dry season during which much of the vegetation may loose its greenery. High elevation and warm moist airflow result in cloud forests, or rain forest ecosystems produced through orographic lift. True tropical rain forests exist in tropical latitude regions where high temperatures and high levels of humidity year-round produce a fairly consistent precipitation and vegetation cover each month. The three most significant locations of tropical rain forest in the world are in Southeast Asia (see Chapter 6), Central Africa (see Chapter 9), and the largest, the Amazon River basin in South America (Box Figure 1).

Tropical rain forest is the dominant vegetation and climate regime in the Amazon River basin of Brazil and adjacent parts of its neighboring countries, including eastern Colombia, Ecuador, Peru, Bolivia, southern Venezuela, and the Guianas. The tropical rain forest of the Amazon basin is the largest in the world. Although other locations in Latin America contain similar vegetation and climatic conditions, including the Pacific coast of Colombia, along the eastern coasts of the Central American countries, and down the northeast coast of Brazil, the Amazon dominates in size and global impact.

Tropical forests contain the highest plant and animal species diversity per unit of land of any ecosystem or biome in the world. Seventy percent of the world's known plant and animal species reside in tropical forests, with hundreds of distinct species existing in relatively concentrated areas. Contemporary global health care depends on the species diversity of tropical rain forests. Existing treatments and promising cures for various forms of cancer come from tropical forest species. Numerous human health issues and aging conditions may be treated, cured, or slowed through medicines derived from the plants there. Medical research communities assert the need to explore the unknown potential hidden in the diversity of tropical rain forest vegetation. International pharmaceutical research, develop-

ment, and sales generate hundreds of millions in revenue from products related to tropical rain forests. The potential benefit of hundreds to thousands of species present in the tropics has yet to be identified. Permanent clearance could eliminate countless medicinal cures and treatments yet undiscovered.

Trees absorb carbon dioxide from the atmosphere and return oxygen to the atmosphere. In vegetative respiration, trees and other forms of vegetation "breathe in" carbon dioxide, converting it to energy and "exhale," or release, oxygen. Tree respiration takes place in the troposphere, the lowest level of the Earth's atmosphere, where humans live and breathe. Although the scientific community is far from a complete understanding of the relationship between carbon dioxide and the extent of global warming, there is a general consensus that warming is taking place and carbon dioxide is a major culprit. Removing huge tracts of tropical forest eliminates a primary source of oxygen and a mechanism for removing the carbon dioxide at the same time. Burning forests, or drowning them under dammed waters, both common practices in Latin America, contributes to the release of carbon dioxide and methane, both considered to be culprits in global warming.

Tropical forests provide a number of local environmental stabilizers. Tropical forests hold moisture and release it into the local and downwind environment. They regulate stream flows and prevent downstream flooding. Tropical rain forests provide a source of water when global cycles bring drier periods to a region. Through the process of evapotranspiration, tropical rain forests can provide humidity and rainfall to the region they inhabit as well as far downwind. Tropical soils are not nutrient rich like many temperate soil regions. The vegetation in the tropical forests stores the life-sustaining nutrients within its green cover, so that clearing the forest removes the local nutrient base and diminishes the potential for future species diversity. The vegetation keeps the local soil intact. In its absence, soil will wash through the watershed and out through the drainage system. The trees on poorer soils maintain their existence by circulating the nutrients without letting them enter the soil: when leaves fall, insects and fungi soon break them down so that roots near and above the surface can capture the chemicals again.

Government debt, crowded population densities in the east, millions of materially impoverished and jobless people, farmers with no land to farm, and a wealth of unexploited natural resources prompted the Brazilian government to promote the development of the Amazon Basin. The Brazilian government is hoping huge mining projects located in forested areas will bring in foreign capital and ease Brazil's massive foreign debt. The government claims sovereign rights to exploit its resources however it deems appropriate. Brazilian political leaders assert the need to develop resource potential fully in order to bring down an indebtedness structure that is so severe they are barely able to make the interest payments. Brazilian officials point to countries such as the United States where unencumbered resource exploitation led to material wealth and prosperity. Many local and international environmental groups are calling for international reduction of Brazilian debt to help the country shift its development efforts away from the Amazon.

The Brazilian government claims it learned lessons from prior projects such as the Fordlândia experiment to produce rubber in the 1930s and the Jari development for wood pulp in the 1970s, which failed because of plant disease or inadequate soil surveys. However, critics believe that the government will repeat past mistakes and jeopardize the future of the Amazon Basin. Development in the vicinity of mining operations leads to an influx of people to provide services. As supporting industries grow, more people move to the region. As roads are cut into the forest, farmers and ranchers settle along both sides. Poor farming practices coupled with the nature of the soils and vegetation often lead to rapid nutrient depletion, and the farms are either sold to large-scale enterprises or

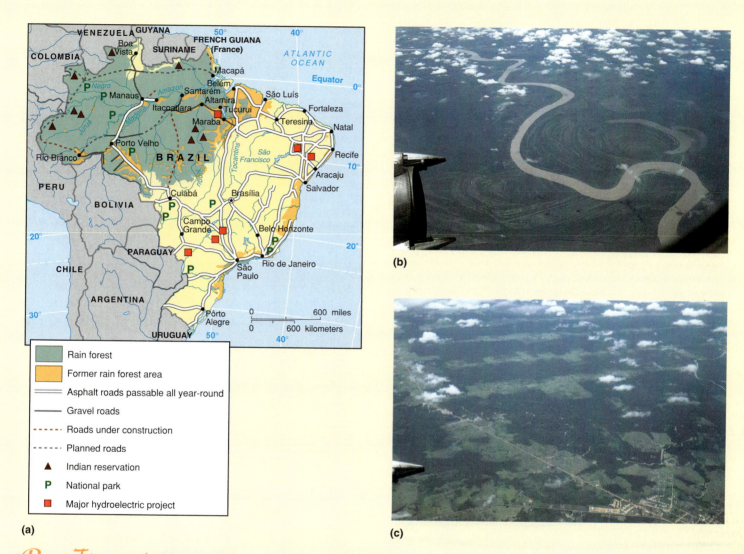

Box Figure 1 (a) Map of the Amazon River basin in Brazil and the extent of the tropical rain forest. (b), (c) Aerial views of the tropical rain forest—(b) where uncut forest is crossed by the Madeira River and (c) in Rondonia state, where the forest is cut into as farms are established along the roads. Photos: (b) and (c) © Michael Bradshaw.

Legend:
- Rain forest
- Former rain forest area
- Asphalt roads passable all year-round
- Gravel roads
- ----- Roads under construction
- ----- Planned roads
- ▲ Indian reservation
- P National park
- ■ Major hydroelectric project

abandoned for new land. As new roads are built, new settlers move in, existing farmers move to more productive land, and the rain forest vegetation slowly disappears.

The government claims such projects produce jobs that employ people who would otherwise have no work or income. People working the mines and providing services to the developing regions are grateful for the opportunities presented. Some farmers claiming land in the region assert their pride in land ownership and their ability to be self-employed. In places where better soils coincide with a good local knowledge and where farmers are not too indebted, there has been some success in growing commercial crops of beans and vegetables. In most parts, the cut areas gave way to cattle ranching, but few cattle can be supported per hectare and the carrying capacity soon declines to uneconomic levels.

Settlement in the Amazon River basin has often been uncontrolled, without proper surveys or understanding of the rain forest ecosystem, and it often results in armed combat over land rights. The Brazilian government believes development of the region will lead to national stability and will diminish any border discrepancies neighboring countries may have over Amazonian lands. The Brazilian army remains a major influence on Amazonian development, but one that is

seldom considered. Arising from the border wars that drained Brazilian manpower earlier in the 1900s, the military encouraged cutting of the rain forest for strategic reasons. When in power, military dictators instigated projects such as the Trans-Amazonica Highway and encouraged families to settle the area as part of the "great march westward." A network of military posts is still being built along the western border of Brazil, linked by ground and air transportation. Neighboring countries fear a loss of their remote lands in the Amazon if Brazilian development pushes too far to the west.

The future of the tropical rain forest in Latin America will be determined by the ability of local and global governments to understand the ecosystem better and cooperatively look for egalitarian uses that may serve the needs of countries like Brazil while sustaining the ecological and human health value of the rain forest system. The following matrix provides some of the positions asserted by those urging tropical rain forest conservation and even preservation, and those who believe resource exploitation in the rain forest is in the best interest of the local and national communities. Read the conflicting views and consider how you might help to resolve this controversial, emotionally charged, and complex issue. Are there other factors for or against tropical rain forest resource use and deforestation you might include?

CONSERVE TROPICAL RAIN FOREST RESOURCES	USE TROPICAL RAIN FOREST RESOURCES
Tropical rain forest (TRF) resources provide a sink for carbon dioxide. Burning TRF vegetation adds carbon dioxide to the Earth's atmosphere. TRF areas are a source of oxygen in the lowest level of the Earth's atmosphere, where humans live and breathe.	There is incomplete carbon dioxide data for the Earth's atmosphere. Large portions of the Earth's surface are unreported. Ocean exchanges with the lowest levels of the Earth's atmosphere are more significant than TRF exchanges.
There is tremendous biodiversity in the plant life present in TRF ecosystems.	There is no conclusive evidence that TRF clearing will permanently change the biodiversity of the Earth as a whole.
Many medical treatments are derived from TRF products, and many disease cures come from TRF products, including current treatments and potential cures for cancer patients. Destruction may eliminate many undiscovered cures and treatments. TRF-derived pharmaceuticals earn billions internationally each year.	Medical treatments come from many sources. Many treatments and cures may be synthetically generated in laboratories and do not need naturally growing species from TRF.
Governments permit the rapid clearing of TRF resources and sell them internationally, claiming rights to destroy domestic resources that impact the entire Earth. Yet, the same governments may be corrupt and waste other resources and spend their cash foolishly.	Debt-ridden and impoverished countries need to and have the right to use their natural resources for their own best interest. The wealthier countries of the world obtained high material living standards by depleting much of their own and others' resources as they grew. Now those countries want to hold countries with TRF resource wealth back.
Indigenous tribes and local people are displaced by TRF clearing. In some cases, bloody conflicts ensue while government officials turn a blind eye.	Growing countries need to push their frontiers and develop their resources. "Productive" members of society have a right to use land in a manner that will benefit them and their country.
TRF resources provide an increasing tourism revenue potential. During the 1990s, travel to natural areas in the tropics was one of the fastest-growing components of the global travel industry (which is the world's largest industry). Although some governments assert their ability to balance resource clearing for export sales and development goals with conservation for tourism growth, few have shown a true commitment to achieving such a balance.	Governments have the right to determine how they will earn revenue from their resources. Governments of TRF resource-wealthy countries assert their ability to balance resource depletion and extraction with conservation and replenishment.
TRFs provide a natural habitat for species found only in this biome. Removing the TRF would eliminate habitat and cause permanent loss to global species diversity. Loss of species could alter the ecological balance of the Earth.	A good source of income in a debt-ridden country with a large materially poor segment in its population is far more important than the conservation of a bird or a tree.

Pre-Columbian cultures in Latin America cultivated corn, potatoes, sweet potatoes, manioc, and beans, among numerous others. Many of the crops cultivated in Latin America were not grown in other parts of the world at the time. The early globalization connections created during the colonial era facilitated the diffusion of crops from Latin America to other parts of the world, where they remain major components of agriculture today.

Marine life provides some of the world's richest fisheries along the western coast of Ecuador, Peru, and northern Chile. Other fishing grounds in the Caribbean and off southern Argentina are being developed. Fish production for North American markets is a growing source of income to many countries in Central America and the Caribbean. Such resources are vulnerable, however, to human overuse and environmental conditions such as the El Niño effect.

Environmental Problems

Natural hazards such as earthquakes, volcanic eruptions, and hurricanes bring destruction and death to the Andean and Middle American mountain ranges and to the Caribbean islands. Many environmental problems, however, are related to European colonial patterns and subsequent decades of political instability and corruption. Two of the more serious environmental issues in Latin America are soils and air quality. Perhaps the most serious and globally controversial environmental issue centers on deforestation in the Amazon tropical rain forest (see the "Point-Counterpoint: Tropical Forests and Deforestation" box, p. 434).

Soil Erosion

Soil erosion is a major problem in many countries of Latin America, especially where increasing populations place additional pressure on subsistence lands. The primary subsistence crop is corn, which is commonly grown in rows up and down a slope, making it easy for intense rainstorms to wash soil away. As economic and population pressures force more subsistence farmers onto the hillsides, the elimination of the natural vegetation removes one of nature's mechanisms for holding soil in place on steep slopes.

Small Caribbean islands colonized for intensive commercial agriculture became notoriously liable to soil erosion and other forms of environmental degradation. The introduction of sugarcane and banana plantations dramatically altered the local environments. After more than 300 years of commercial agriculture, soil fertility has decreased to the point where large applications of fertilizer are necessary. The development of resort hotels and the construction or expansion of supporting

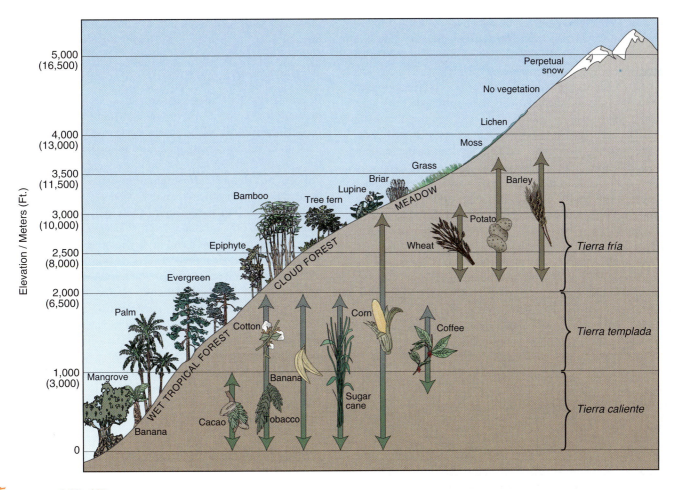

Figure 10.15   **Northern Andes: altitudinal zonation of vegetation and crops.** As they cross the equator, the Andes have maximum height and the greatest number of climatic environments from mountain foot to peak, from coastal mangroves and tropical rain forest to perpetual ice and snow cover. How do the crop zones relate to the vegetation zones?

Test Your Understanding 10B

Summary Latin America's climatic environments are mainly tropical, ranging from the equatorial climates of the Amazon basin to seasonal rainy areas and the deserts of northern Mexico and the Peru-Chile coasts. Midlatitude climatic environments affect the southernmost countries.

Tectonic plate movements gave rise to the Andes Mountains and the complex pattern of mountain ranges and islands in Middle America and the Caribbean Basin. Brazil is dominated by the Guiana and Brazilian Highlands plateaus, which are separated from each other and the Andes by the Orinoco, Amazon, and Paraná-Paraguay River systems.

Latin America has resources of metal ores, petroleum, water, good soils, and major forests. It is subject to earthquakes, volcanic eruptions, hurricanes in the north, and storms in southern Chile. Human activities have degraded water, air, soil, and forest resources.

Some Questions to Think About

10B.1 What are the primary natural hazards that occur in Latin America? Which areas are prone to these hazards and why?

10B.2 What are the major mineral products of Latin America, and how are they distributed geographically?

Key Terms

Middle America	*tierra templada*
El Niño phenomenon	*tierra fría*
Altiplano	*tierra helada*
altitudinal zonation	temperature inversion
tierra caliente	

infrastructure on many islands caused vegetation loss and accelerated soil erosion. Clearing natural vegetation on the more hilly islands of the Caribbean causes soil to wash rapidly down the hills and into the surrounding sea, resulting in the accumulation of sediment in the coral reef areas surrounding many of the islands. The accumulation of sediment in a reef system may ultimately kill the coral. Fertilizers used to compensate for lost soil nutrients from rapid runoff also enter the sensitive coral reef system, further degrading the health of the reefs.

Air and Water Pollution

Air and water pollution results from mineral extraction and refining, and from the concentration of human activities in urban areas. Mexico City suffers more than any other metropolitan area in Latin America from air pollution. Mexico City is a very densely populated urban center situated in a bowl-shaped depression known as the Central Valley of Mexico. The city sits on the valley floor at approximately 2,200 m (7,000 ft.) above sea level. Mexico City and the valley floor are surrounded by much higher mountains, which virtually form walls around the urban expanse. The air in and just above the city is often trapped in the valley and becomes extremely stagnant. Vehicular exhaust, coupled with industrial and domestic pollutants, becomes trapped in the valley's stagnant air mass and settles near the valley floor where millions of people breathe. Pollutant levels exceed maximum tolerances by several times and do not decrease despite stringent government measures. Naturally occurring **temperature inversions,** or periods during the winter months when cold, dense air remains "trapped" at the surface under warmer air for several days or even weeks, further complicate the air pollution phenomenon of Mexico City. Although migration and natural increase rates of population are now slower than they were a few decades ago in the Central Valley of Mexico, population numbers continue to challenge pollution-reduction efforts.

🌐 World Roles

The United States and Latin America

In 1823, the United States staked its claim on Latin America by formulating the **Monroe Doctrine,** in which it vowed to resist intervention by other countries in Latin American affairs. Its own influence was greatest in Middle America until the mid-1900s. The United States intervened in the affairs of Cuba, Haiti, and the Dominican Republic on several occasions. From the early 1900s, the decline in the availability of British capital coupled with growing production in the United States led to a regional shift of economic dominance toward the United States. During the later 1900s, the impact of two world wars, the depression of the 1930s, and political and corporate intervention from the United States caused most Latin American countries to look northward for trade, political support, and elements of popular culture, including fashions in clothing, motion pictures, music, and fast foods. Miami, Florida, with its large Hispanic population, regional air and sea transportation networks, and growing banking and financial services sector, became a strong cultural link between the United States and Latin America in the second half of the 1900s.

Latin American connections with the United States strengthened in the 1990s. The acceptance by Mexico, the United States, and Canada of the North American Free Trade Agreement (see the "Point-Counterpoint: NAFTA North American Free Trade Agreement" box, p. 504) enhanced trade and business connections among them. By the end of the 1990s, a unique countertrend changed the dominant unidirectional flow of influence between the United States and Latin America. In the 1980s and 1990s, Latin American popular culture began to rival U.S. homegrown popular culture as the Latin-based segment of the U.S. immigrant population grew rapidly. Latin American

influences in the restaurant industry thrived, and Latin American popular music topped the U.S. charts.

Financial Dependence

In the 1970s, the rising cost of oil on world markets caused dramatic price increases for fuel in non-oil-producing countries. As world oil prices continued to rise, large quantities of petrodollars were urged on Latin American countries in the form of low-interest loans. Such loans were used to pay for oil imports and major infrastructure projects—and to improve some government officials' overseas bank accounts.

The recession in the more materially wealthy countries caused by the high oil prices led to weakened markets for products from Latin America and higher interest rates on the loans. The combination of debts and falling export income resulted in many Latin American countries defaulting on debt repayments by the mid-1980s. The Brazilian government had to devote its large overseas trade balance, generated by import-substitution industries, plus sales of mineral and farm products, to servicing its extensive international debts. Brazil's debt servicing reduced its ability to invest in domestic production, thereby dramatically diminishing economic progress. The high interest rates caused foreign investment and aid to dry up in the late 1980s, and debt liabilities forced countries to emerge from reliance on their internal markets.

Political Change and the Global Economy

The PPP GNI (Figure 10.16) and consumer goods ownership data (Figure 10.17) for Latin American countries reflect a wide range from economic giants such as Brazil to materially impoverished countries such as Honduras. The current global connections of many economies in the region are partially a legacy of past economic institutions and international relations and partially a response to 1990s changes in governmental approaches to participation in the world economy. Elected governments were in power everywhere (except Cuba), and emphasis shifted from inward-looking policies to the need for cooperation. The policies that focused on government-run industry and the protection of domestic products through tariffs, quotas, and red tape, or **protectionism,** involved high levels of government intervention and highlighted the lack of capital available for internal investments. These policies gave way to those of structural adjustment that encouraged less government intervention, more foreign investment, and more exporting industries. The new policies also opened markets to foreign products and privatized government corporations. Such policies were partly forced on Latin American countries by world bankers as a means of reducing debt burdens or attracting new foreign investment. As with other simplistic approaches to development, this economic restructuring that was superimposed on the Latin American regional culture was subject to potential disasters, as Mexico found in 1994–1995, when its economy opened too rapidly, sucking in imports and capital investments and creating a huge trade imbalance.

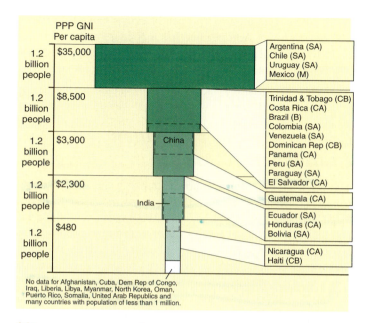

PPP GNI Per capita

- 1.2 billion people — $35,000 — Argentina (SA), Chile (SA), Uruguay (SA), Mexico (M)
- 1.2 billion people — $8,500 — Trinidad & Tobago (CB), Costa Rica (CA), Brazil (B), Colombia (SA), Venezuela (SA), Dominican Rep (CB), Panama (CA), Peru (SA), Paraguay (SA), El Salvador (CA)
- 1.2 billion people — $3,900 — China — Guatemala (CA)
- 1.2 billion people — $2,300 — India — Ecuador (SA), Honduras (CA), Bolivia (SA)
- 1.2 billion people — $480 — Nicaragua (CA), Haiti (CB)

No data for Afghanistan, Cuba, Dem Rep of Congo, Iraq, Liberia, Libya, Myanmar, North Korea, Oman, Puerto Rico, Somalia, United Arab Republics and many countries with population of less than 1 million.

Figure **10.16** **Latin America: country national incomes compared.** The countries are listed in the order of their PPP GNI per capita. Subregions are indicated parenthetically. Source: Data (for 2000) *World Development Indicators,* World Bank, 2002.

Global Cities

Although many cities in Latin America interact and trade on a global level, the region's two giants, Mexico City and São Paulo, are the dominant forces responsible for inserting the Latin American region into globally connected world systems (Figure 10.18). Both Mexico City and São Paulo have populations in the range of 18 million to 20 million people. Extensive rural-to-urban migration, migration within the metropolitan areas, and high rates of natural increase created very dynamic urban populations for both cities, which challenge demographers in determining exactly how many people reside in each. Rapid economic development in both Brazil and Mexico led to dramatic growth in the largest cities of both countries. The decrease in nationally controlled industries with a corresponding increase in private investment opened a floodgate for the larger cities to establish connections with the industries, governments, and investors of cities, countries, and regions throughout the world.

Both Mexico City and São Paulo have diverse populations, with each housing people from all regions of their respective countries. The cities are representative of the different dialects, social customs, religious practices, and regional pride existing throughout their respective countries. The expanse of trade and interaction with multinational corporations from all world regions, as well as relationships with government officials from various countries of the world, furthers the diverse composition of people and customs present in each urban center. The size of each, the diversity of activity within, the international trade between each and other major world centers, and the attraction of the cities to business and leisure visitors from across Latin American and around the globe create environments unparalleled in the other metropolitan centers of Latin America.

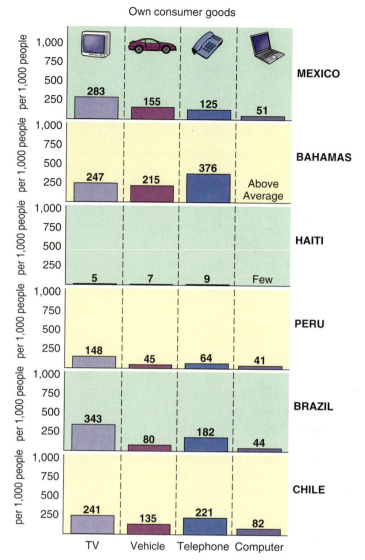

Own consumer goods

MEXICO — TV 283, Vehicle 155, Telephone 125, Computer 51

BAHAMAS — TV 247, Vehicle 215, Telephone 376, Computer Above Average

HAITI — TV 5, Vehicle 7, Telephone 9, Computer Few

PERU — TV 148, Vehicle 45, Telephone 64, Computer 41

BRAZIL — TV 343, Vehicle 80, Telephone 182, Computer 44

CHILE — TV 241, Vehicle 135, Telephone 221, Computer 82

(per 1,000 people)

Figure **10.17** **Latin America: ownership of consumer goods.** Contrast conditions in materially wealthy, moderately wealthy, and very poor countries. Source: Data (for 2000) from *World Development Indicators,* World Bank, 2002.

Mexico City is the political, industrial, financial, and cultural capital of the country. São Paulo is the financial, industrial, and cultural capital of Brazil. Although Brasília is the political capital, São Paulo wields strong political power within the Brazilian system. Both Mexico City and São Paulo are media centers for their countries, and both serve as their countries' media connections to the world. Although each global city-region houses representatives of all subsets of each country, the residents of each take pride in being part of the most influential metropolitan center of their country.

Regional Links

The 1990s saw the rise and regeneration of regional trade groupings. In 1997, trade between Argentina, Brazil, Paraguay, and Uruguay (MERCOSUR) grew by 25 percent, while that in Central America and Andean South America grew by 10

Figure 10.18 **Global connections and local voices in São Paulo, Brazil.** Brazilian business personnel wait outside Brazil's World Trade Center in São Paulo after a bomb threat forced the evacuation of an adjacent shopping and office complex on September 25, 2001. The bomb threat was one of several that followed the devastating attacks on New York City and Washington, D.C., on September 11, 2001. Photo: © Reuters New Media/Corbis.

percent. The Caribbean area awaited its request for parity with NAFTA, a proposal that did not progress through the U.S. Congress. The countries of Southern South America opened talks with the European Union, their biggest trading partner and primary source of investment since 1996. An early 1990s proposal by U.S. President George H. W. Bush to establish a Free Trade of the Americas trade bloc by 2005 remained largely disorganized by 2002 and unlikely to be established by the original target date.

🌐 The Subregions

Latin America may be divided into several subregions. The most geographically obvious division first is between Middle America and South America. The Middle American subregions consist of Mexico, the seven Central American countries from Guatemala and Belize south through the isthmus to Panama, and most of the islands in and surrounding the Caribbean Sea. The South America component consists of traditional continental South America and some of the islands just offshore in the southern Caribbean.

- Mexico dominates Middle America in population, land area, and economic might and will be examined as an individual subregion.
- The second subregion consists of the seven Central American countries that share a common history, related cultural development patterns, and similar physical geographies.
- Although Central American countries and some of the Caribbean Island countries share some similarities, the differences are significant enough to merit examining the Caribbean area as a distinct subregion.
- The countries of the northern Andes subregion include Colombia, Bolivia, Ecuador, Peru, and Venezuela.

- Brazil covers an area that nearly equals half of the continent of South America and has a population that is nearly half the total population for the continent, supporting the need to examine Brazil as an individual subregion.
- Paraguay, Uruguay, Chile, and Argentina comprise the countries of the final subregion, Southern South America.

Latin American Population Distribution Patterns in the Subregions

The Mexican population is concentrated in the central region from Guadalajara in the west through centrally located Mexico City to Veracruz on the east (Figure 10.19). This central plateau and the valleys cutting into it formed both an indigenous and a colonial hearth, and more recently became the center of industrial developments and government functions focused on Mexico City. Northern cities, such as Monterrey, and cities along the U.S. border stand out in areas with sparse rural populations. Moderate densities of largely rural population occur on either side of the central urban belt. The areas with fewest people include the arid northwest and much of the Yucatán Peninsula.

In Central America, the main concentrations of people are in and around the largest cities—often in the highlands and closer to the west coast, where temperatures and soils are better for cultivation. Eastern coastal areas and parts of the narrow isthmus of Panama have fewer people in areas of dense forest cover and humid wetlands or mangrove marsh.

In countries of the northern Andes Mountains, population distribution reflects the Spanish pattern of colonial settlement. Some concentrations of people are in mountain valleys or on plateaus where midlatitude crops could be grown close to the equator (see Figure 10.15). Other population clusters are on the coasts, having developed as port cities to link the home country of Spain with the interior sources of minerals. The interior lowlands have few people, reflecting the formidable barrier presented by the Andes chain.

The distribution of population in Brazil is a combination of both historical and contemporary regional development goals. The highest densities are in the southeast, around and inland of São Paulo and Rio de Janeiro. Moderate densities occur in a band parallel to the coast from the southeast around Pôrto Alegre to west of Fortaleza in the north. Farther inland, the very low densities of the Amazon rain forest area create a major geographic contrast within the country.

The main population centers in Southern South America are around the Río de la Plata estuary (Argentina, Uruguay) and in central Chile. Smaller centers occur in the irrigated farming oases of northern Argentina and around Asunción, capital of Paraguay.

— 🌐 Mexico

The burgeoning population of Mexico exceeded 100 million in the year 2001, making Mexico the most populous country in Middle America and second largest after Brazil in all of Latin

Figure 10.19 **Latin America: distribution of population.** Explain the distribution of heavily and lightly populated parts of this region. The red lines identify the subregions. Source: Data from *New Oxford School Atlas*, p. 98, Oxford University Press, UK, 1990.

America (Figure 10.20). Mexico's economy is the largest in Middle America due in part to its physical size, industrial development history, and vast oil reserves. Although Mexico serves as a regional leader in Middle America, the government continues to look northward for economic growth prospects. The population growth and economic size of Mexico result in an increasing amount of global attention focused on its role in the larger Latin American sphere.

Mexican independence from Spain came in 1821 through a series of rebellions. Political instability resulted in numerous power struggles and dozens of government turnovers in the

decades after the end of Spanish rule. Stability came in 1871 with the harshly repressive dictatorship of Porfirio Díaz at the cost of many social freedoms. Porfirian oppression planted the seeds for revolution that would shape Mexico's political culture for most of the 1900s. Revolutionaries who overthrew the Porfirian government in 1911 created a new land tenure system with the purpose of easing rural poverty. The plots of land created in 1917 became known as *ejidos,* which were state-owned rural cooperatives developed to provide landless peasants with a degree of control over the land they farmed. The government granted the right to use land to individual *ejido* farmers, or

Figure 10.20   Mexico and Central America: population, economic, and cultural data.

Country	Capital City	Land Area (km²) Total	Population (millions) Mid-2001 Total	2025 Est.	GNI 1999 (US $ million) Total	GNI PPP 1999 Per Capita	Percent Urban 2001	Human Development Index Rank of 175 Countries	Human Poverty Index: Percent of Total Population
United Mexican States	Mexico City	1,956,200	99.6	130.9	428,877	8,070	74	50	10.6
Belize	Belmopan	22,960	0.3	0.4	673	4,750	49	83	
Costa Rica, Republic of	San Jose	51,000	3.7	5.0	12,828	7,880	45	45	4.1
El Salvador, Republic of	San Salvador	21,040	6.4	9.3	11,806	4,260	58	107	20.6
Guatemala, Republic of	Guatemala City	108,890	13.0	22.1	18,625	3,630	39	117	28.3
Honduras, Republic of	Tegucigalpa	112,090	6.7	9.8	4,829	2,270	46	114	24.8
Nicaragua, Republic of	Managua	130,000	5.2	8.6	2,012	2,060	57	121	28.1
Panama, Republic of	Panama City	75,520	2.9	3.8	8,657	5,450	56	49	9.0

Source: Data from *Population Reference Bureau 2001 Data Sheet*; *World Development Indicators,* World Bank, 2001; *Human Development Report,* United Nations, 2001; Microsoft Encarta (ethnic group, language, religion).

ejiditarios, as well as the theoretical right to pass land use to their offspring. However, ownership of the land remained with the government. The *ejido* system resulted in the fragmentation of landholdings as the government took control away from some landowners and gave it to *ejiditarios*. In many cases, the government placed already impoverished farmers on marginal land, exacerbating their situation. Although rights to individual property ownership were defined more clearly in the 1990s, many rural or materially poor farmers continued to struggle while large-scale agribusiness took control of more land.

Another long-lived institution arising from the Mexican revolutionary period from 1910 to 1920 is the Institutional Revolutionary Party (*Partido Revolucionario Institucional*, **PRI**). The PRI grew out of the revolutionary movement into the most powerful and lasting political party in the history of Mexico, dominating Mexican politics for more than 70 years. The party controlled the government, nationalized industries, and allegedly fixed many political votes to remain in power. In July 2000, a historic election overturned the PRI's dominance of Mexican politics with the election of Vicente Fox, who represented the National Action Party (Partido de Acción Nacional, **PAN**). Fox and the PAN inherited a vastly corrupt system supported in part by graft and money from illegal drug smuggling to the United States.

Serious debt problems in the 1980s contributed to a reversal of extreme protectionist government policies concerning Mexican industries, leading to policies permitting foreign competitors into Mexican markets. The Mexican government significantly reduced tariffs and trade restrictions, cut inflation, attempted to reduce deficit spending, privatized telecommunications, banking, and agriculture, and reduced a wide range of central government controls. Mexico signed the North American Free Trade Agreement (see the "Point-Counterpoint" box, p. 504), with Canada and the United States in 1992, and after passage in the respective legislatures, the pact was formally implemented in 1994.

NAFTA increased the domestic demand for imports, making them much less expensive due to significant tariff reduc-

tions, and caused a rapid rise in Mexico's overseas trade deficit. By insisting on keeping an exchange rate with U.S. dollar parity, Mexico could not adjust to the demands placed on it. Foreign investors lost confidence and began to withdraw their capital. In 1993, Mexico attracted US $75 billion of foreign investment, mainly from the United States and Japan. In 1994, it attracted $60 billion, but in 1995, the net flow was just over $1 billion. At the end of 1994, Mexico had to devalue its currency. From 1999 to 2000, the confidence that had suffered a sharp loss appeared to be recovering through further U.S.-supported reform measures and increased demand for inexpensive Mexican goods in overseas markets.

Other factors of Mexican social, political, and economic conditions, however, may also discourage foreign investments. While a few people increased their wealth, the materially poor gained little from the changes. Larger companies were privatized to the advantage of their new owners but shed labor to become more productive. By the new century, many smaller local companies in Mexico were not nearly prepared to compete in the world economy.

It was also difficult for many Mexicans to adjust to the new norms of world market competition after decades of state control, protection, and provision. The three areas in which Mexico needed to make rapid changes to compete in world markets—increase productivity throughout the economy, decrease the budget deficit, and maintain civil peace—all confronted contrary forces that slowed progress and pointed to the long-term nature of Mexico's problems.

The 1994 challenge mounted by the Zapatista National Liberation Army in the southern province of Chiapas drew attention to the social costs of economic reform. The **Zapatistas** made the point that the Native Americans, who comprise one-third of the Chiapas population and were already very materially poor, were further disadvantaged by the economic restructuring and NAFTA. They called NAFTA a "death sentence." The Native Americans in Chiapas are only a part of the 13.5 million Mexicans who live in "extreme poverty" and a

Country	Ethnic Groups (percent)	Languages O=Official	Religions (percent)
United Mexican States	Mestizo 60%, Native American 30%, European 9%	Spanish (O), local, English	Roman Catholic (nominal) 89%
Belize	Mestizo 44%, Creole 30%, Maya 11%	English (O), Spanish, Mayan	Protestant 30%, Roman Catholic 62%
Costa Rica, Republic of	Mestizo and European descent 95%	Spanish (O), English	Roman Catholic 95%
El Salvador, Republic of	Mestizo 94%, Native American 5%	Spanish, native tongues, English	Protestant 25%, Roman Catholic 75%
Guatemala, Republic of	Mestizo 56%, Native American 44%	Spanish (O), 20 Native American dialects	Roman Catholic, some Protestant, local
Honduras, Republic of	Mestizo 90%, Native American 7%	Spanish, local dialects	Roman Catholic 97%
Nicaragua, Republic of	Mestizo 69%, European 17%, African 9%	Spanish (O), local	Protestant 5%, Roman Catholic 95%
Panama, Republic of	Mestizo 70%, Caribbean 14%, European 10%	Spanish (O), English, Creole, local	Protestant 15%, Roman Catholic 85%

further 23.6 million who are "poor." In the beginning of the 2000s, the Chiapas situation remained unresolved. Reforms promised in 1996 bought time for the government but were not implemented. In response, Zapatistas set up their own autonomous municipalities in defiance of government orders. The Mexican government maintains a large military presence in the region but does little to stimulate dialogue. **Revolutionary tourists,** supporters from other countries who travel to the region and offer their fighting services to the cause, add to the perpetuation of the Zapatista revolution.

Regions of Mexico

Mexico extends from the subtropical latitudes along its border with the United States to the tropics in the southern reaches of the country (Figure 10.21). Mexico is a country of mountain ranges, plateaus, basin-shaped lowlands, and coastal plains. Most of southern Mexico receives adequate rainfall; much of northern Mexico along and near the U.S. border, however, is quite arid.

Mexico's dynamic human geography continues to evolve new spatial patterns. From the 1500s to the early 1800s, ranchers and miners settled northern areas as part of the colonial-period expansion of empire that took Spanish soldiers and Roman Catholic priests from the Mexico City area into California, New Mexico, and Texas. Today, foreign-owned factories are widespread. The western coastlands have fishing, commercial agriculture, and a burgeoning tourist industry on the Baja Peninsula. Central Mexico is the home of the bulk of Mexico's population and is the seat of Mexican political, cultural, and economic activity, centered on Mexico City. Urbanization and industrialization extend westward to Guadalajara and eastward to Veracruz. The east coast plains around the Gulf of Mexico and the Yucatán Peninsula became a region of oil and natural gas production and tourist opportunity in the second half of the 1900s. The indigenous of the Yucatán region and the Mexican state of Chiapas do not feel as

though they are included in the Mexican mainstream. The southern mountains and the border region of Chiapas are the remotest parts of the country. Slow economic development and feelings of political and cultural isolation contributed to the Zapatista uprising in the region.

People

Mexico's population grew rapidly to around 70 million in 1980 and over 100 million in 2000. Although its fertility rates and birth rates are declining (Figure 10.22), Mexico's population is projected to rise to nearly 131 million by 2025. The age-sex diagram (Figure 10.23) shows a decline of births that was achieved through a well-developed family planning program, halving total fertility rates from about 6 in 1976 to under 3 in 2000. The very low death rate and increasing life expectancy, however, maintain a gap between births and deaths that keeps the population totals increasing.

Mexico's population is highly urban. From 1970 to 2001, estimates of the proportion of the Mexican population in towns increased from 59 to 75 percent. Mexico City grew from a population of 500,000 in 1900 (2.5 percent of Mexico's population) to more than 18.5 million in 2000 (slightly more than one-fifth of the country's total population, making it a primate city). Other "million cities" included Guadalajara (4 million), Cuidad Juarez, León, Monterrey (3.5 million), Puebla (2 million), and Tijuana (Figure 10.24). In Mexico, the stimulus for movements of people to cities is the increasing growth of urban-based manufacturing and service jobs. The official rates of unemployment, however, remain high, and many jobs are in the informal sector.

Migration from rural to urban areas is not the only significant movement of Mexican peoples. Many Mexicans migrate to the United States seeking employment opportunities and a better quality of life. Some migrants to the United States are legal, and others are not. Many Mexicans living in the United States support family members who remain in Mexico. The

Figure 10.21   **Mexico: a diversity of regions.** Northward, the climatic environments are drier, but economic opportunities increase close to the U.S. border. Southward, the land narrows and localities become increasingly remote from the influence of the Central Valley of Mexico.

remittance of cash from Mexicans working in the United States to family members back home is a significant source of foreign exchange for Mexico.

Mexico's population is dominated by three groups: Native Americans (around 30 percent), Europeans (9 percent), and people of mixed Native American and European ancestry known as mestizos (60 percent) (see Figure 10.20). The Native Americans are particularly numerous in the southern province of Chiapas, where language dialects of Mayan origin are commonly spoken in preference to Spanish, the official language. Almost 90 percent of Mexicans are nominally Roman Catholic, although some personal links to the Catholic church are eroding as established lifestyles are disrupted by moves to urban centers.

The combination of Mayan or Aztec ancestry, Spanish colonial influences, and, more recently, interaction with the United States has led to a distinctive Mexican culture that diffused into surrounding areas of the southwestern United States and Central America. Mexican cities are built in architectural styles that reflect Spanish colonial influence on the region (Figures 10.25a and b). Many public and government buildings are adorned with beautiful murals painted on the exterior walls facing garden courtyards. Other nationalistic trends are expressed in music, motion pictures, and literature. Many Spanish-speaking countries in the Caribbean and Central America look to Mexico for elements of popular culture. Mexican soap operas, or *telenovelas,* are commonly viewed in places like Honduras and Puerto Rico. Mexican food, based on corn tortillas, local vegetables, and chilies, is one of the most distinctive national cuisines.

Mexico City

Mexico City is by far the largest urban conglomeration in Middle America. Its site was a major population center before the arrival of the Spaniards and even before it became the Aztec capital. It continued to be the focus of road and rail networks within New Spain and independent Mexico. Mexico City is the political capital and media center of the country, and has three-fourths of Mexico's manufacturing industry and nearly all of its commercial and financial establishments.

During the 1950s, 1960s, and 1970s, up to 1 million people per year migrated from rural areas in Mexico to metropoli-

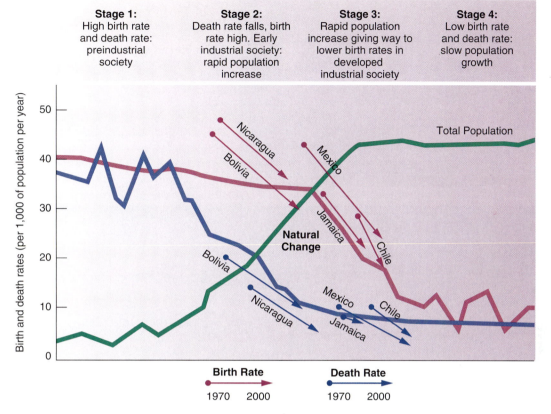

| Stage 1:
High birth rate
and death rate:
preindustrial
society | Stage 2:
Death rate falls, birth
rate high. Early
industrial society:
rapid population
increase | Stage 3:
Rapid population
increase giving way to
lower birth rates in
developed
industrial society | Stage 4:
Low birth rate
and death rate:
slow population
growth |

Figure 10.22 **Latin America: demographic transition.** How does the process vary among Latin American countries?

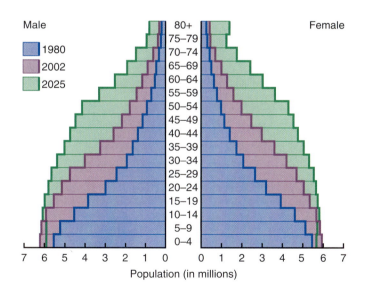

Figure 10.23 **Mexico: age-sex diagram.** Source: U.S. Census Bureau. International Data Bank.

tan Mexico City. Housing supplied by government efforts combined with that from private construction, yet development projects could not nearly provide adequate accommodation for almost one-third of the rapidly growing city's population. The city attracted rural migrants with the expectation of greater economic opportunities, better education, more diverse recreation and cultural choices, and more substantial health care

services. Squatter settlements exploded in many parts of the Central Valley of Mexico in and around the city. The massive shantytown satellite city of Nezahulacóyotl, now with more than 1.5 million people living on flood-prone lands, is the most notorious squatter settlement resulting from Mexico's migration decades. Squatter neighborhoods often lack amenities such as electricity, water, sewage, and even paved streets. In some cases, once such shantytowns were established, the authorities began to pave the streets, put in utilities, and provide access to schooling and health care, but amenities are often slow to arrive for many, and millions live well below poverty standards.

The overall result of such rapid growth is a combination of overcrowding, congestion, and air and water pollution. The physical geography of the Central Valley of Mexico coupled with the intensely crowded living conditions make the urban center one of the most polluted, in terms of air quality, in the world. Although residents of metropolitan Mexico City have greater access to health care than rural Mexicans, infant mortality rates and other social health indicators are among some of the worst in the country due in large part to poor air and water quality. Depletion of underground water resources during the past several decades presents another problem for the city. Many areas within the urban system are subsiding, some more than 5 m. Engineers struggle to stabilize historic structures that have been slowly sinking for decades. As job markets also failed to cope with the population explosion of the city, the informal sector of the economy grew from around 4 percent to 26 percent of people of working age from 1980 to the 1990s.

Figure 10.24 **Mexico: urban landscape.** The Gran Plaza in the city center of Monterrey, northern Mexico. The older government and cathedral buildings on the left contrast with the newer ones on the right and the multi-level development in the foreground. Photo: © Sergio Dorantes/Corbis.

(a)

(b)

Figure 10.25 **Colonial architecture.** The Spanish colonial imprint is visible throughout the cultural landscape of Mexico. Colonial plazas and churches, as here in Guanajuato (a) and (b), dominate contemporary urban settings.

Economic Development and the Human Landscape

Mexico is by far the most economically developed of the countries of Middle America. In 2000, Mexico had a total GDP that was over 82 percent of Middle America's total GDP. Mexican GDP per capita is one of the highest in Latin America and near the top of the World Bank's upper-middle income group. Ownership of consumer goods was well above that in the countries of Central America apart from Costa Rica (see Figure 10.17). Mexico's HDI and GDI are much higher in rank than Brazil's but not as high as Argentina's.

Mexico has a diversified economy. Peasant farmers cultivated corn before the Mexican Revolution of the early 1900s, while most of the country was covered by large wheat or livestock-producing haciendas (Figure 10.26). In the middle decades of the 1900s, rapid urbanization occurred, as well as some land redistribution and agricultural irrigation. Farming became commercial with the growing urban market at hand, while foreign companies built packing plants for canned and frozen vegetables. Farming remains important today, employing one-fourth of Mexico's labor force.

Manufacturing became a major source of employment together with increasing numbers of jobs in services and government. Tourism is a major and growing source of income for Mexico, with 20 million visitors in 2000 (up from 17 million in 1990), which is millions more than anywhere else in Latin America. Mexico City, Pacific coastal resorts such as Cabo San Lucas, Mazatlán, and Acapulco, well-preserved ruins of Mayan city-states, and the Caribbean coastal resorts of the Yucatán Peninsula are major tourist attractions for Mexicans, Latin Americans, and visitors from North America, Europe, and Asia.

Variations in the development of manufacturing define economic regions within Mexico. Nearly two-thirds of the Mexican population lives in the urban region surrounding Mexico City that includes Guadalajara, Veracruz, and León. The majority of Mexico's economic power and activity is centered in this megalopolis-type urban expanse similar to the northeastern United States and to parts of Japan and Western Europe. Northern Mexico became the main growth area in Mexico's economy for new industrial development in the 1970s and 1980s, building on already established industries such as steel and iron production in Monterrey.

More recently, the towns along the U.S. border, from Tijuana in the west to Matamoros in the east, experienced very rapid growth. After 1965, the **maquiladora** program made it possible for foreign-owned factories sited in Mexico to import components for assembly without customs duties. Such factories used cheap local labor to assemble goods that could be exported to the United States. U.S. corporations involved in textiles, apparel, electronics, and wood products built their factories in these towns, where there was rapid growth through the 1990s. Asian and European corporations joined U.S. facilities in producing under *maquila* laws in the northern zone.

The concentration of so much economic activity along the border, however, brought increased air and water pollution.

Medical research continues to indicate a correlation between the industrial expansion of the maquiladora region and increased rates of cancer and other diseases among the people on both sides of the Mexico-U.S. border. Both Mexico and the United States are implementing programs to reduce industrial pollution from both sides of the border.

The *maquila* industrial zone enhanced global connections, especially those established with the United States. A human landscape now exists in which U.S. popular culture is incrementally prevailing upon the Mexican people of the zone. Local Mexican voices are talking about Halloween instead of *Día de Los Muertos* (Day of the Dead) and Madonna instead of Mexican pop star Thalia, and they are increasingly using English to do so.

The combined forces of the maquiladora program and NAFTA opened a floodgate of economic globalization in Mexico. Ford, GM, Nissan, IBM, Whirlpool, Kodak, and Caterpillar are just a few of the hundreds of major foreign companies operating within Mexico's borders. Furthermore, the irrigated northern coastal areas became a primary center of modern agriculture, producing winter vegetables for U.S. markets, in addition to cotton, sugarcane, and rice.

While diversified urban-industrial economies develop in Mexico's center and north, the east coasts facing the Gulf of Mexico supply much of Mexico's wealth from large oil and natural gas fields. Although discovered early in the 1900s, this wealth was not effectively tapped until the 1970s, when oil went from 2 to 80 percent of Mexico's foreign exchange earnings. During the high oil-price period of the late 1970s, Mexico borrowed heavily, and falling prices and demand in the 1980s left it deeply in debt. Oil production had a major impact on the economy of the coastal area between Tampico and Campeche, and recent finds extended this affected area inland. During the 1980s, new refineries and petrochemical industries were built. In the late 1990s, oil formed less than 40 percent of exports as more oil was used by growing industries within Mexico, and the products of those industries made up a higher proportion of exports.

Southern Mexico is less developed than northern areas of the country and more reminiscent of the countries of Central America. Most of the region is remote and populated by Native American subsistence farmers who grow corn on the hillsides and some wheat on the valley floors. Overgrazing by sheep is common. Poor transport facilities slow development in this region. Contrasting with this poverty, the government-planned tourist resorts such as Acapulco bring American tourists to the coast and new highway links connect the resorts to airports. Unfortunately, sewage disposal from the hastily built hotels and surrounding shantytowns has polluted the bay at Acapulco, and a 1997 hurricane devastated the city's materially impoverished neighborhoods.

Traditional Native American culture is least disturbed in Chiapas, Yucatán, Oaxaca, and the southernmost Mexican states. Over a million Native Americans in Chiapas do not speak Spanish. Rapid tourism development in parts of the Yucatán Peninsula is dramatically altering the human geography for some while furthering a feeling of isolation for

Legend:

- Arable, predominantly cereals
- Arable, predominantly paddy
- General arable
- Arable with cash crops
- Irrigated crops
- Grazing and dry farming
- Desert, nomadic herding
- Mixed forest, farming and grazing
- Tropical dry forest and savanna, farming and grazing
- Tropical rain forest, lumbering, crops
- Coniferous forest, lumbering
- Marsh or swamp
- Tundra and high altitude desert

0   500   1000 mi
0   500   1000 km

Figure 10.26 **Latin America: major land uses.** Compare this map with that of population distribution in Figure 10.19. Are there any links? What factors are missing to explain the distribution of population? Compare it also with the climate map (Figure 10.6). Source: Data from *New Oxford School Atlas,* p. 98, Oxford University Press, UK, 1990.

others (Figure 10.27). The far south remains very remote, with few people or commercial activities. In recent years, government investment in Chiapas resulted in the irrigation of farmland from the Angostura Dam waters that occupy 90 km (60 mi.) of a once-farmed valley. Chiapas also houses thousands of refugees who fled earlier civil disturbances in Guatemala and El Salvador. The combination of dispossessed refugees and local disillusionment with the Mexican treatment of these states contributed to the Zapatista rebellion and regional unrest.

Central America

Central America is a land bridge, or **isthmus,** that narrows in width from the Mexican border in the northwest to the Colombian boundary in the southeast. The Central American isthmus provides a narrow physical boundary separating marine environments of the Caribbean Sea/Atlantic basin to the east and the Pacific Ocean to the west. Seven independent countries occupy the land of Central America (Figure 10.28). These states have a common Hispanic culture and geographic proximity. Although Belize was a British colony, linked to Jamaica from the late 1800s until 1981, it has a relatively high percentage of people who trace their ancestry to the African continent. Its mainland location, increasing use of the Spanish language, and Roman Catholic populace link Belize to Central America rather than to the Caribbean Basin.

The seven Central American countries had a total population of 38 million people in 2000 and are among the most materially poor countries in the Americas. Many of the countries of this subregion depend on the foreign capital generated from a limited production of export crops. They are sometimes termed **"banana republics"** because of their economic dependence on one or two export crops, among which bananas figure strongly. Dependence on one primary export crop and slow economic growth were linked to decades of regional political instability. Although Central America forms a land

Figure 10.27 **Cancún resort development.** Tourism growth supports the diffusion of resort hotel development along the Caribbean coast of Mexico's Yucatán Peninsula, while economic conditions worsen in some interior villages.

Figure 10.28 **Central America: major geographic features, countries, and cities.**

bridge between North and South America, the physical and human geography of the subregion presents a barrier to the development of strong travel and communication connections between the core areas of the two continents. Periods of instability, dire rural poverty, drug trafficking, and the nearly impenetrable environmental conditions in the Darien of eastern Panama complicate efforts to form substantial links to South America. In the late 1990s, further setbacks to these countries included widespread forest fires linked to an extended period of drought that destroyed crops and the 1998 devastation following Hurricane Mitch.

Countries

Regional independence from Spain came with a brief union between contemporary Mexico, El Salvador, Guatemala, Honduras, Nicaragua, and Costa Rica as the Mexican Empire from 1821 to 1823. In 1823, the incipient Central American countries separated from Mexico as the United Provinces of Central America, which lasted until 1838, when they each became independent. Panama existed as a province of Colombia until a U.S.-brokered canal deal created a separate country in 1903. Belize remained within the British sphere until 1981. Since 1838, unity has not figured prominently in the interactions among the countries of the subregion.

The four largest countries to the south of Mexico—Guatemala, El Salvador, Honduras, and Nicaragua—had 2001 populations ranging from 5 million (Nicaragua) to 13 million (Guatemala) (see Figure 10.20). Coffee plays an important role in the cash crop export-oriented economies of each. Civil unrest and natural disasters have wreaked havoc on the social fabric, physical infrastructure, and economies of all four countries. Deadly conflict, stemming in part from dramatic inequalities in land ownership in Guatemala, El Salvador, and Nicaragua, focused attention on disparity issues but did little to bring about an egalitarian solution.

In Guatemala, Native Americans comprise approximately half the population yet have little economic or political clout. The subregion's largest indigenous population percentage exists in Guatemala, where attempts by native peoples to improve their living conditions sparked fierce government repression. In 1980 and 1981, 150,000 of Guatemala's indigenous peoples fled into Mexico as government forces burned down hundreds of their villages.

Although El Salvador experiences frequent earthquakes, land reform conflict in the 1980s proved far more costly to the country than any natural disaster. The El Salvadoran civil war erupted when landowners of vast coffee estates challenged changes imposed by a reformist government. Thousands of people migrated in an attempt to escape the bloodshed. Such large numbers of war refugees moved to the capital, San Salvador, that the economy and infrastructure of the city virtually collapsed. Recovery continues today.

Decades of oppression under the Somoza family dictatorship in Nicaragua instigated revolutionaries to overthrow Somoza control. The Marxist-oriented Sandinista revolutionaries took control in 1979 and inflicted another form of oppression and corruption on the people of Nicaragua for the next 11 years. The United States imposed a trade embargo on the Sandinista government, which added to the country's economic disruption and already impoverished social conditions. The United States also backed armies *(Contras),* partially comprised of former Somoza military personnel, who were fighting against the Sandinistas. The Sandinista party was eventually voted out of office in Nicaragua in 1990.

Civil unrest in El Salvador, Guatemala, and Nicaragua eased considerably in the 1990s. The three countries began the slow process of repairing the physical and social damage inflicted by years of conflict. It remains difficult for the governments of these three countries to attract external investment because of their recent violent histories, and tourists are slow to return and spend their money.

Belize, Costa Rica, and Panama have the smallest populations of the Central American countries. Belize is by far the smallest country in the region, having only 300,000 citizens. The country became independent in 1981 after a long period of British colonial rule. Although years of British tenure established an English language base and some cultural elements different from those of neighboring countries, the Hispanic influence of the region exerts strong forces on Belize and is causing the development of a more regionally uniform human geography. Many Guatemalans assert a historic right to the lands and waters of Belize. Guatemala's formal claims eased in 1991, and the issue seems to be fading with time, although oil deposits off the Belize coast could renew claims for some Guatemalans.

Panama became a separate country when its territory was carved out of Colombia in 1903, in preparation for the construction of the Panama Canal. The newly independent country had an immediate economic and military dependence on the United States, which resulted in a relationship in which Panama functioned more like a U.S. colony than as an independent country. The construction of the canal brought in people from the Caribbean islands, Asia, and southern Europe. The Canal Zone created a dramatic dichotomy in the human landscape for Panama. The corridor along the canal is relatively wealthy and developed in contrast to areas of Panama to the east and west. Panama City, the country's capital, retains a cosmopolitan population and an international role in trade and finance. The United States virtually ran the country until 1979 and then invaded in 1989 to protect strategic and economic interests.

People

The population of Central America is projected to grow by 60 percent in the next 30 years. By 2025, the 2001 total of 38 million may become 60 million people. Such growth will place increasing stress on the already pressured natural environment and the political, economic, and social systems of these countries.

Very high fertility rates in the 1960s and 1970s—over 6 children per woman, apart from Costa Rica (fewer than 5)—continue to impact the population growth within the subregion. By the year 2001, fertility rates significantly decreased for the majority of the countries of Central America, but they remain high relative to those in the United States, Canada, and most European countries. In 2001, rates of natural increase remained between 2 and 3 percent, except in Costa Rica and Belize, where they fell below 2 percent (see Figure 10.22). The age-sex diagrams for Central American countries are similar to those of other materially poor countries throughout the world, with large numbers of young people and small percentages of late middle-aged and elderly groups.

Urbanization is lower in Central America than in Mexico, remaining under 40 percent in Guatemala, under 50 percent in Belize, Costa Rica, and Honduras, and under 60 percent in El Salvador, Nicaragua, and Panama. Most countries have a single major, or primate, city with populations that are several times that of the second-largest city. The largest cities are the capitals of San Salvador (1.4 million in 2000, El Salvador), Guatemala City (over 3 million, Guatemala), Tegucigalpa (almost 1 million, Honduras), Managua (almost 1 million, Nicaragua), Panama City (1.1 million, Panama), and San José (350,000, Costa Rica).

The populations of Central American countries are comprised of racial and ethnic groups that became social classes. Native American, or indigenous peoples, remain in large numbers in western Guatemala (Figure 10.29) and comprise significant minorities in the other countries of the subregion. Many indigenous groups in Guatemala and other Central American locations do not speak Spanish as their first language. Differences in language, social customs, and livelihood separate such communities from those with political power and economic wealth. Indigenous communities of Central America often feel they are regarded by other inhabitants as groups that should remain outside the evolution of the mainstream. Costa Rica has the smallest indigenous population of the subregion.

Along the east coast of Nicaragua and the northern and eastern coast of Honduras are many communities of English-speaking Afro-Caribbean peoples, mixtures of Miskito Indians and blacks from Jamaica and other Caribbean islands, and even pockets of English-speaking whites. The Spanish largely ignored coastal activity in this region during the colonial era, which opened the coasts to decades of influence and control by the British. The British forced the migration of blacks from the

Caribbean to work primarily in agricultural production. The historical combination of African and Caribbean cultural influence coupled with the imprint of British rule created a very distinctive cultural landscape region that remains strong today.

European descendants and mestizos hold the political power and economic wealth of Central America. Land tenure was the basis of power and wealth for many years, but during the later 1900s, power passed to the owners of manufacturing industries and financial and legal services in the modern economy—and sometimes to the military. Smaller groups of people within Central American countries include some of African origin who came after escaping slavery in the Caribbean, central and southern Europeans who migrated to Central America between or after the two world wars, and Asians, Jamaicans, and Barbadians who came to work on construction projects such as the Panama Canal or, more recently, to establish factories in Costa Rica or Panama.

Economic Development

In contrast to Mexico's diversifying economy, the Central American countries have a much narrower economic base, resulting in slower economic and social infrastructure development. In Central America, the major difference in economic geography is between the more populous centers of Native American and Hispanic culture in the uplands and the less populated Caribbean coastal areas with their American-owned plantations growing bananas, other fruits, and sugarcane.

Guatemala is a country of three parts. First, in the uplands northwest of Guatemala City, the largest Central American concentration of Native American peoples lives in subsistence conditions, often landless and in extreme poverty. An uprising in the early 1980s resulted in government repression and destruction of what little infrastructure existed. Second, there are the people in the southern part of the country who adopted a more European lifestyle. This area's commercial agriculture is based on coffee on the Pacific slopes and bananas and sugarcane near the Caribbean (Figure 10.30). Guatemala City has manufacturing industries, but recent influxes of people swamped the job markets and half of the working-age population is unemployed. Third, the lowland Petén area of the north is forested and less populated than parts of the country to the south. Dramatic archaeological relics of the Maya civilization attract numerous tourists (nearly 1 million in 1999) and scholars from all global regions. Although tourism brings some revenue to the Petén, much of the forest resources are being cleared, which could ultimately hurt the tourism industry.

El Salvador has the densest population in the subregion. Most of the people are concentrated in the western highlands, where the dominant crop, coffee, is grown. Light manufacturing industries were established from the 1960s. Civil war over land redistribution since the 1980s disrupted both rural and urban economies. Thousands of Salvadorans fled the country during the civil war and took up residence in the United States. These foreign-based communities send regular payments to relatives remaining in El Salvador. Such remittances are currently El Salvador's largest source of foreign revenue.

Figure 10.29 **The peoples of Central America.** Indigenous people comprise a significant portion of the population of Guatemala. Patterns and colors present in clothing signify the home village or region for many native Guatemalans. Photo: © Bill Gentile/Corbis.

Figure 10.30   **Central America: Pacific coast.** An eastward-looking space shuttle view of Guatemala and southern Mexico. The cities are in the highlands, where temperatures are modified by elevation. Photo: NASA.

Honduras is one of the most materially poor countries in the Western Hemisphere. Barely 25 percent of its land can be used for farming. The ranching economy established in colonial times persists in the parts of the west, and large banana and pineapple plantations dominate the Ulua River valley and the northern coastal plain. Coffee became a primary export in the 1960s, and after Hurricane Mitch devastated the banana industry, coffee is the leading source of foreign income today. Shrimp farming on the Pacific Ocean coast is a new export specialty that is finding markets in the United States. Tourism is a growing industry, attracting almost 400,000 visitors in 1999 (a 100 percent increase from 1990 figures), centered on Maya ruins in Copán, nature tourism-related mountain hikes and river rafting, and a relatively thriving scuba diving and beach attraction along the north coast and in the Bay Islands. Much of the former forest cover was logged, and even the secondary growth of pines is being removed rapidly. Manufacturing was encouraged after 1950, and a growing maquiladora industrial belt is forming around the city of San Pedro Sula.

Nicaragua developed a mixture of agriculture and modern manufacturing industries. The U.S.-imposed trade embargo that followed the Sandinista revolution in 1979 devastated the economy, giving it the lowest GDP in the subregion in the late 1990s and undercutting Honduras as the poorest country in Central America. Coffee, cotton, bananas, and sugarcane are important export crops. The population is concentrated around the lakes of Nicaragua and Managua in the structural depression that is subject to earthquakes. Other parts of the country have poor transportation links to this region. Tourists, once frightened away by conflicts and natural disasters, are slowly returning to the country. The potential for tourism growth is significant, and the government is encouraging investment and exploring development opportunities. It may be some years until the stigma of warfare is removed enough for large numbers of visitors to enjoy the diversity of the country.

The United States relinquished control of the Panama Canal Zone in 1999 to the government of Panama. The country has economic contrasts between the urban belt including Panama City and the corridor along the Canal Zone, and the rural areas on either side of this developed belt. Panama City and the corresponding Canal Zone is a cosmopolitan center of global trade, where banking and financial services create twice the GDP of agriculture. In 1991, Taiwanese investors established a free trade zone in Panama City in which some 8,000 people are employed in clothing, furniture, toy, and bike factories. The products are sold in the U.S. market. In the rural districts to the west, coffee is grown on uplands, and rice and palm oil are produced along the coasts. To the east of the Canal Zone, the Central American Highway tapers off in the impenetrable Darién, an area of extremely dense vegetation between the canal and the Colombian border that prevents completion of effective north-south highway links. Many believe completion of the highway would dramatically increase the flow of illegal drugs northward from the Northern Andes region of South America.

Panama is dealing with the problem of an aging canal that is now too small for many ships (Figure 10.31). Work began in 1991 to widen and lengthen the main locks, and a pipeline now conveys oil across the isthmus to the west of the Canal Zone, virtually eliminating the need for oil tankers to pass through the canal. Although the United States formally closed its military bases in Panama in 1999, it maintains close ties with Panama in an effort to continue the U.S. war on drugs in the region.

Costa Rica, with 3.7 million people in 2001, is the only country in the region that has had a long-term democratic government. This may be related to the legacy of its initial colonial settlement, when there were few Native Americans to be dominated and the people of European origin took up medium-sized farms. As a stable democracy, Costa Rica attracted manufacturing and free trade zones financed by U.S. and Taiwanese corporations. Costa Rica has a more prosperous and diversified farming industry than the other countries, based mainly on coffee production along the Pacific slopes, bananas and cacao on

Test Your Understanding 10C

Summary Mexico is one of the wealthiest countries in Latin America, having a diverse economy, but it has problems adjusting to increased involvement in the world markets and NAFTA. Mexico City is one of the world's largest metropolitan areas. The contrasts of internal development among regions and groups of people within Mexico cause social tensions and fluctuating patterns of economic growth that threaten to slow the process of human development.

Central America comprises several small countries—Belize, Costa Rica, El Salvador, Guatemala, Honduras, Nicaragua, and Panama. Their peoples, with the exceptions of Costa Rica and Panama, are among the poorest in Latin America. Central America's environment consists of mountain ranges and swampy tropical coasts. Its countries produce coffee, bananas, and a few other crops for fluctuating world markets, but many economies have been affected badly by civil wars.

Some Questions to Think About

10C.1 How do the main features of the population and economy of Mexico compare with those of the Central American countries?

10C.2 What impact did civil war have on the societies, population distributions, and economies of the countries of Central America? What is the situation today with respect to the regional conflicts?

10C.3 Do Mexico and the countries of Central America count as the "near abroad" of the United States? (See Chapter 4 for the usage of this term with reference to Russia.)

Key Terms

Monroe Doctrine	Zapatista
protectionism	revolutionary tourist
ejido	maquiladora
ejiditario	isthmus
PRI	banana republic
PAN	

Figure 10.31 **Central America: Panama Canal.** The Miraflores locks with a cruise ship entering to be raised to the level for traveling through the canal. Note the land uses on either side of the canal. These locks are now too narrow for the world's largest ships. Photo: © Morton Beebe-S.F./Corbis.

the eastern coast, and a range of new crops such as tropical fruits, ornamental plants, cut flowers, and tropical nuts. Areas of former rain forest are used for livestock production. Costa Rica has little civil strife. It does not have an army, but the militarized police and relatively good quality of life help to maintain internal peace. Costa Rica has many successful industries, including textiles and pharmaceuticals for export and cement, tires, and car assembly for the domestic market. Costa Rica is one of the top tourist destinations in Latin America. Costa Rican beaches, volcanoes, and culture help the country to attract more than a million tourists per year. An extensive system of national parks incorporates a variety of ecological habitats and microclimate zones (through altitudinal zonation) and protects a vast array of flora and fauna.

🌐 The Caribbean Basin and Environs

The contemporary human and physical geography of the Caribbean Basin reflects dramatic transformations imposed on the region by European countries during the colonial era. The loss of indigenous people to disease, the forced migration of Africans to the basin, and the establishment of extensive sugar cultivation on the islands laid the foundation for the social, environmental, and economic conditions present in the subregion today. The Caribbean Basin consists of a few large islands, several small islands, and three political units on the South American mainland (Figure 10.32). The political geography of the Caribbean is one of the most diverse of any subregion. Caribbean countries vary in size and population from Cuba, with 11 million people in 2001, to hundreds of tiny island countries with few inhabitants (Figure 10.33). The ownership of consumer goods (see Figure 10.17) presents a picture of the basin's economic diversity from poverty-stricken Haiti to the more affluent Bahamas. In 2001, 33 million people lived in the Caribbean Basin. The majority of the Caribbean's residents live on one of the four largest islands, known as the Greater Antilles: Cuba, Hispaniola (consisting of two countries: Haiti and the Dominican Republic), Puerto Rico, and Jamaica.

The smaller Caribbean islands are commonly referred to as the Lesser Antilles. The Lesser Antilles chain is an arc of small islands around the eastern edge of the Caribbean Sea, with the Leeward group in the north (Virgin Islands to Guadeloupe) and the Windward group in the south (Dominica to Grenada). Many of the small island countries of the Caribbean have only a few thousand residents.

North of the Caribbean are the Bahamas, which include some 700 islands, situated to the east of Florida. Another group of islands, including Trinidad and Tobago, lies along the northern shore of South America. Also included in this subregion are the three Guianas—Guyana, Suriname, and French Guiana. Although situated on the South American mainland, their histories, people, and present economies are more connected to those of the Caribbean Basin than to those of other mainland South American countries.

The islands of the Caribbean Basin are situated between continental North and South America. Increasing global connections have made the islands of the Caribbean strategically significant from the colonial era to the present. In the 1500s, they were on the route for Spanish treasure ships between Europe and the mainland of the Americas, and from the later 1600s, they became a major source of European colonial wealth through sugarcane cultivation. The opening of the Panama Canal in 1914 placed these countries on a major world route, further elevating the strategic importance of regional waterways. The involvement of the United States in the Cuban Missile Crisis of 1962, plus U.S. interventions in the Dominican Republic in 1965, Grenada in 1983, and Haiti in 1994 highlighted the sensitivity of Americans to neighboring events. The continued presence of U.S. military bases, including Guantánamo Bay in Cuba, and concerns over drug traffic through the region emphasize the geopolitical significance of the Caribbean Basin.

Countries

The people of the Caribbean have a very strong sense of belonging to the individual island or country of their birth, as opposed to strong feelings of regional unity or identity. Although the majority of the Caribbean residents are of African ancestral heritage, most people identify with their fellow countrymen or women and perceive readily apparent differences between their nationality and those of neighboring countries. A Jamaican would identify with other Jamaicans, and see their national people as different from those of Barbados, rather than perceive any unity from ancestral heritage. National pride is often expressed through rivalries with neighboring island countries. For example, sporting events, such as cricket and soccer, among the people of Jamaica, Barbados, Trinidad, and Guyana are very competitive, prestigious, and emotional affairs.

A major source of variety within the human geography of the region arises out of the different colonial histories of the countries. While Middle America has an almost exclusively Hispanic history, the Spanish, French, British, Dutch, and Danes all influenced the contemporary geography of the Caribbean. Although the Spanish were the first to discover and claim the largest islands, their main areas of interest shifted to mainland Middle America and to South America after their invasion of Mexico in 1519.

Spain retained colonies in a few of the Caribbean islands to supply meat to the mainland empire of New Spain and sugar and tobacco to the home country in Europe. Ports were fortified to protect trade routes. In the 1600s, the Dutch, British, and French established colonies on Bermuda, the Bahamas, the Lesser Antilles, the Guianas, and the offshore South American islands. At first, some were merely bases for pirates preying on Spanish treasure ships. After 1650, the development of the sugarcane industry dramatically elevated the value placed on the region by the European colonizers. Islands of the Caribbean were linked to the Atlantic economy through a pattern of product trade from the Caribbean Basin and Americas eastward to Europe, then sailing southward to collect slaves from western Africa, and westward back across the Atlantic (see Figure 9.6).

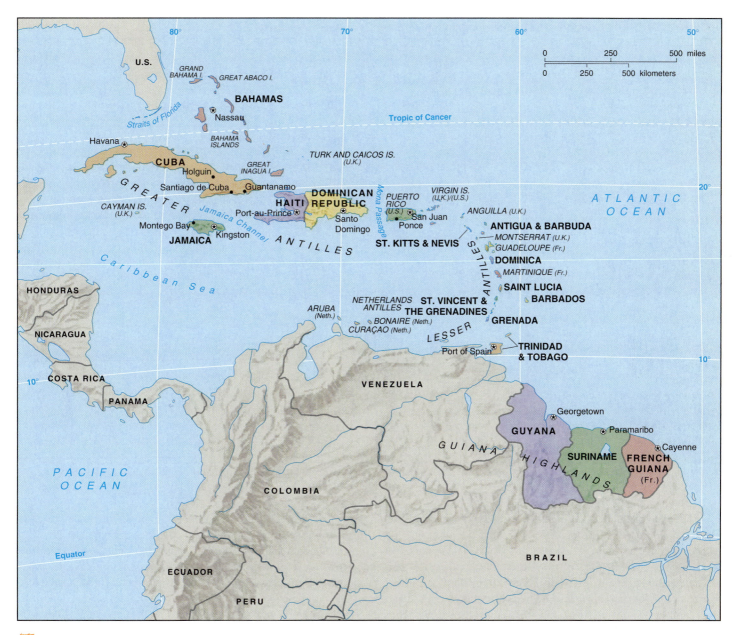

Figure 10.32 **The Caribbean Basin.** Numerous island countries and some overseas colonies.

Throughout the 1800s, as the rest of the Latin American countries gained independence and the slave trade was banned, most of the sugar islands remained colonies. Colonial connections declined in significance as European countries began to cultivate sugar beets. The United States took a growing interest and played an increasing role in the area. The U.S. war with Spain in 1898 led to the freeing of Cuba and Puerto Rico from Spanish colonial rule under U.S. oversight. Cuba took full political independence, while Puerto Rico moved politically closer to the United States. Nearly all the British and Dutch colonies became independent in the mid-1900s, but the remaining French islands—Martinique and Guadeloupe—together with French Guiana remain to this day within the French economic arena, sending representatives to the French parliament and being officially regarded as part of the European Union.

People

Hardly any of the indigenous population of the Caribbean Basin survived the early days of European settlement, succumbing especially to the diseases brought across the Atlantic. Although a few Caribs live on Dominica and St. Vincent and some Arawaks on Aruba and in French Guiana, the vast majority of the regional population traces their origin to the colonial occupation. Residents of the Caribbean speak a variety of languages, from French and English to regional hybrids of Patois and Papiamento. Religious practices are equally diverse. Some religious ties to former colonial rule are strong and reflect in a somewhat pure form the religious practices of various European peoples. Many religious beliefs in the region today, however, evolved from mixtures of practices brought from the African continent with religious ideology imposed by the Europeans.

Figure 10.33   Caribbean Countries: population, economic, and cultural data.

Country	Capital City	Land Area (km²) Total	Population (millions) Mid-2001 Total	2025 Est.	GNI 1999 (US $ million) Total	GNI PPP 1999 Per Capita	Percent Urban 2001	Human Development Index Rank of 175 Countries	Human Poverty Index: Percent of Total Population
Antigua and Barbuda	St. John's	440	0.1	0.1	606	9,870	37	38	2.6
Bahamas, Commonwealth of	Nassau	13,880	0.3	0.4	No data	15,500	84	31	4.7
Barbados	Bridgetown	430	0.3	0.3	2,294	14,010	38	29	No data
Cuba, Republic of	Havana	110,860	11.3	11.9	No data	No data	75	58	17.7
Dominica, Commonwealth of	Roseau	750	0.1	0.1	238	5,040	71	53	No data
Dominican Republic	Santo Domingo	48,730	8.6	12.1	16,130	5,210	61	88	No data
French Guiana	Cayenne	91,000	0.2	0.3	No data	No data	79	No data	No data
Grenada	St. George's	340	0.1	0.1	334	6,330	34	52	10.2
Guyana, Cooperative Republic of	Georgetown	214,970	0.7	0.6	651	3,330	36	99	46.1
Haiti, Republic of	Port-au-Prince	27,750	7.0	9.6	3,584	1,470	35	152	13.6
Jamaica	Kingston	10,990	2.6	3.3	6,331	3,390	50	82	No data
St. Kitts and Nevis, Federation of	Basseterre	360	0.04	0.05	259	10,400	43	51	No data
St. Lucia	Castries	620	0.2	0.2	590	5,200	30	81	No data
St. Vincent and The Grenadines	Kingstown	390	0.1	0.1	301	4,990	44	75	No data
Suriname, Republic of	Paramaribo	163,270	0.4	0.5	No data	3,780	69	64	No data
Trinidad and Tobago, Republic of	Port-of-Spain	5,130	1.3	1.4	6,142	7,690	72	46	3.5

Source: Data from *Population Reference Bureau 2001 Data Sheet*; *World Development Indicators,* World Bank, 2001; *Human Development Report,* United Nations, 2001; Microsoft Encarta (ethnic group, language, religion).

The former Spanish colonies have relatively high proportions of people with European origins. Eighty percent of the population in Puerto Rico counts itself white, as does over 30 percent in Cuba. In the former British, French, and Dutch colonies, this figure is less than 5 percent, and the dominant majority is of African origin. Jamaica, Trinidad, Guyana, and Surinam have substantial Asian populations resulting from British and Dutch shipments of workers from South Asia to replace former slaves after slavery was abolished. Many racial mixtures enhance the cosmopolitan nature of West Indian populations.

Population Dynamics

The total population of the Caribbean Basin almost doubled from 17 million to 33 million people from 1965 to 2001 and is expected to rise to around 40 million by 2025. Population pressure is already severe in many islands, with densities of over 600 people per km² (1,500 per mi.²) in Barbados and over 200 in many of the other islands (see Figure 10.19). Only in the Guianas does population density fall below 5 per km². Jobs available are fewer than the labor force, and there is high unemployment and continuing out-migration.

Population growth rates vary, although the regional averages fell in the late 1900s. In 2000, most Caribbean countries had growth rates of under 1 percent. Although total fertility ranges between 1.5 and 3.0 (Haiti's rate is 4.7), out-migration lowers the overall rates of increase. Jamaica's population trends are compared with other Latin American countries in Figure 10.22.

Other factors influencing population changes include high growth rates in islands such as St. Lucia and St. Vincent, which give rise to large proportions of young people (44% under 15 years in these islands), declining fertility rates partly balanced by reduced infant mortality, and increasing life expectancy to over 70 years. Only Haitians had a 2001 life expectancy under 50 years; the rest were over 65 years. Jamaica's age-sex diagram (Figure 10.34) demonstrates the effects of losing population as migrants moved to jobs in the United Kingdom and United States.

Even within the context of small total populations in many countries, the size of some of the major cities is impressive, with primacy existing on several islands. Havana, the capital of Cuba, had 20 percent of the country's population (2.2 million in 2000), despite the government policy of focusing economic development in rural areas. In the Dominican Republic, Santo Domingo (3.5 million in 2000) and Santiago do los Caballeros (1.5 million) had over a million people each. San Juan (Puerto Rico), Port-au-Prince (1.7 million, Haiti), and Kingston (600,000, Jamaica) all are primate cities.

Migrations in the 1900s

The population of the Caribbean Basin is very mobile. In-migration dominated until 1900, when intraregional movement and out-migration became significant. In the early 1900s, many

Country	Ethnic Groups (percent)	Languages O=Official	Religions (percent)
Antigua and Barbuda	African 96%, European 3%	English (O)	Protestant (main), Roman Catholic, Muslim
Bahamas, Commonwealth of	African 75%, European 15%	English, Creole	Protestant 75%, Roman Catholic 19%
Barbados	African 80%, mixed 16%, European 4%	English	Protestant 57%, Roman Catholic 4%
Cuba, Republic of	African 11%, European 37%, mixed 51%	Spanish	Roman Catholic 40%, (none 55%)
Dominica, Commonwealth of	African, Carib natives	English (O), French patois	Protestant 15%, Roman Catholic 77%
Dominican Republic	African 11%, European 16%, mixed 73%	Spanish	Roman Catholic 95%
French Guiana	African, European, Arawak natives	French (O), Creole	Roman Catholic, Hindu
Grenada	African	English (O), French patois	Protestant 35%, Roman Catholic 60%
Guyana, Cooperative Republic of	African/mixed 43%, Asian Indian 51%	English, Hindi, Urdu	Christian 57%, Hindu 33%, Muslim 9%
Haiti, Republic of	African 95%	French (O), Creole (O)	Protestant 16%, Roman Catholic/voodoo 80%
Jamaica	African 76%, mixed 15%, Asian Indian 3%	English (O), Creole	Protestant 56%, Hindu, Muslim, Jewish 39%
St. Kitts and Nevis, Federation of	African	English	Protestant, Roman Catholic
St. Lucia	African 90%, mixed 6%	English (O), French patois	Protestant 7%, Roman Catholic 90%
St. Vincent and The Grenadines	African, European, Asian Indian, Carib native	English, French patois	Protestant, Roman Catholic
Suriname, Republic of	Creole 31%, Asian Indian 37%, Indonesian 15%	Dutch (O), English, others	Christian 58%, Hindu 27%, Muslim 20%
Trinidad and Tobago, Republic of	African 43%, mixed 14%, Asian Indian 40%	English (O), Hindi, French, Spanish	Christian 60%, Hindu 24%, Muslim 6%

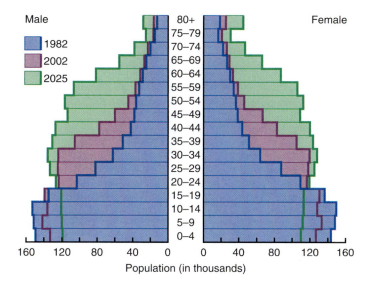

Figure 10.34 **Jamaica: age-sex diagram.** Source: U.S. Census Bureau. International Data Bank.

inhabitants of the older British West Indian island colonies moved to new plantations opening in Trinidad and British Guiana, followed by Jamaicans and Haitians moving to Cuba and the Dominican Republic. The cutting of the Panama Canal attracted workers from Caribbean islands such as Barbados.

Further movements were toward banana plantations established in eastern Central America, U.S. military bases, and new oil-producing centers in Trinidad and Venezuela.

After World War II, Caribbean residents from British, French, and Dutch colonies were attracted to job openings in Europe. The United States and Canada also became recipients of people on the move. Cubans and Puerto Ricans comprise a significant majority of the more than 5 million people of Caribbean origin living in the United States today. The pattern of out-migration eases some of the extensive population pressure on the Caribbean islands and provides a considerable source of foreign cash through remittances, but the exodus often takes away the most active and able young people, leaving behind dependent populations of the very young and the elderly. Although migration to Europe slowed because of immigration restrictions imposed in the 1980s, people from the Caribbean continue to migrate to the United States.

The decline of agriculture as an employer on many Caribbean islands contributed to migration to the largest cities of each country. The push factor of poor rural job prospects is complemented by the pull factors of perceived better employment prospects, social service delivery, and amenities in the towns. Yet most cities were unable to offer the burgeoning population adequate employment opportunities and housing. Congestion, shantytowns, pollution, and crime resulted in many urban centers of the Caribbean.

Economic Legacy of the Colonial Past

Colonial Farming Heritage

The Caribbean colonies of Spain, France, Britain, and the Netherlands assumed a huge significance in the world order of the day when the area took over the leading sugarcane production role from Mediterranean countries in the early 1600s. The heritage from that period still influences island economies today. Sugarcane cultivation began in the large islands of the Spanish colonies, but it was the small islands of the Lesser Antilles and the Dutch colonies that made it a major commercial phenomenon. In Barbados—one of the closest islands to Britain—the forest was cleared, holdings expanded, and African slaves brought in. Other islands followed, with Jamaica becoming the most financially significant sugar producer. After cutting and milling, raw sugar was sent to be refined near the markets in Europe and the United States.

From 1500 to the 1800s, some 10 million African slaves were shipped to the Americas, of which half went to the Caribbean Basin, 39 percent to Brazil, and under 5 percent to British American colonies (and later to the United States). The Africans brought their own social customs, religious beliefs, handicraft skills, and various forms of artistic expression. At the height of sugar cultivation prosperity in the 1700s, the average plantation unit was 200 acres, used 200 slaves, and produced 200 tons of sugar each year.

The British navy stopped the slave trade across the Atlantic after 1807, and slaves were emancipated in British colonies by the 1830s, in French colonies by 1848, and in Spanish colonies by the 1880s. The African American populations increased naturally, but many families left the plantations for their own plots of land. The plantations declined in numbers and output and faced increasing competition from sugar production in other parts of the world. When new plantations opened in Trinidad and the Guianas, labor was imported from Asia. Cuba increased its sugar production in the 1800s, becoming the largest producer of all, and attracted labor from other Caribbean islands. In the 1900s, the family-owned plantations on the sugar islands sold out to major corporations, such as Tate & Lyle (U.K.) in Jamaica and Gulf & Western (United States) in the Dominican Republic. Since the mid-1900s, many island governments took ownership of the sugar industry.

The colonial occupations and economies often took major decisions away from the region and placed them in countries overseas. The interests of external countries and company shareholders often assumed greater importance than local considerations. The colonies were restricted to producing single commercial crops for the needs of overseas markets, and a white elite controlled the imported African- and Asian-origin working populations.

Economic Strategies Following Independence

The majority of the independent countries emerging from colonial rule were small in size and had limited natural resources. Their economies depended on exporting a limited range of crops and minerals and importing manufactured goods, food, and fuel. Multinational corporations often marketed the export products. Demand, as measured by world prices, for the main products—bananas, other fruit, sugar, petroleum, and bauxite—fluctuated.

Caribbean countries adopted economic policies similar to those of other developing countries. At first, manufacturing was promoted through import substitution. Caribbean-based factories were protected from external competition by tariffs, which often led to high prices and low local profits. Together with the small size of local markets, this strategy had little impact on the economies and employment levels of individual countries.

In the 1980s, policies switched to making local sites attractive to investment by overseas manufacturers that would take advantage of cheap locations and labor to produce such goods as clothing, sporting goods, and electronics for export, principally to markets in the United States and Canada. This policy proved more effective in economic growth but did not reduce external dependency. Manufacturing remains a small part of most island economies.

The Challenge of Economic Cooperation

The need for economic integration among the small West Indian countries is clear, but progress has not been easy. Countries of the Caribbean cultivate the same crops, produce the same goods, and largely compete for the same markets. The **Caribbean Community and Common Market (CARICOM)** formed in 1973, taking over from previous attempts at free trade and comprising 13 former British colonies with a total of 5.5 million people. The Caribbean Basin Initiative of 1984 was set up partly to promote economic growth in the Caribbean Basin and partly to protect U.S. interests. The implementation of NAFTA in 1994 fueled Caribbean fears that the region would be shut off from investment and trade opportunities from the United States and Canada. Mexico, the Central American states, Colombia, Venezuela, and countries of the Caribbean formed the **Association of Caribbean States** in 1995. It was proposed by the CARICOM group, and although it comprises the fourth-largest trading bloc in the world, progress is stifled by numerous challenges. Mexico's NAFTA relations with the United States and Canada, a lack of unity in Central America, strained relations between Hispanic countries and English-speaking countries within the Caribbean Basin, dramatic variation in country size, population, and resource base, and the United States' refusal to lift its trade embargo on Cuba are some of the many challenges facing the cohesion of the association.

Tourism

Tourism is increasingly important in the Caribbean Basin, attracting around 15 million visitors in 2000. Localities within the Caribbean Basin cater to a range of tourist interests from the elite to the mass markets. The gleaming beaches and turquoise water of the region are its number one attraction (Figure 10.35). In addition to the famous regional beaches, the islands of the Caribbean also provide unique historical experiences; varied language, food, and cultural practices; and a wide range in topography from the flat Cayman Islands to the rugged hills of Jamaica and Dominica. The United States now provides two-thirds of the tourists, while most of the rest come from Canada, Europe, and Japan. Puerto Rico, the Bahamas, and the Dominican Republic attract the greatest number of visitors, while the Virgin Islands, Antigua, Guadeloupe, and Martinique continue to grow in significance.

Initial tourism development consisted largely of foreign-owned and -managed resorts, many of which were all-inclusive. Foreign control of tourist infrastructure resulted in severe rates of **economic leakage** for many areas. Economic leakage results when the revenue generated flows to the foreign owners and foreign employees of tourism infrastructure (includes hotels, restaurants, gift shops, bars, and dive shops) and thus "leaks" from the local economy. Although many foreign-owned all-inclusive resorts remain in the Caribbean, some governments are increasingly taking control of tourist attraction promotion and operation and attempting to involve local communities in tourist infrastructure investment and management.

The number of cruise ship passengers visiting Caribbean ports continues to rise annually. Although a cruise ship may unload thousands of money-spending tourists into a port town, many of the cruise visitors participate in prepackaged events where revenues generated go to the cruise lines and only a couple of local businesses fortunate enough to be working with the ship's activity planners. Island residents view resort and cruise tourism with mixed emotions. Some see the industry as economically beneficial, providing jobs, housing, and schools, while

others do not reap any direct economic benefits and resent the crowds, resource consumption, and pollution that accompany tourism growth in many areas.

Tourism is an economic activity that is extremely susceptible to external forces. Economic recessions in the wealthier countries from which visitors originate, destination-based crime or terrorism, or media reports of natural disasters such as volcanic eruptions or hurricane damage all may produce immediate reactions in the form of dramatic declines in tourist visits. Although tourism brings in foreign exchange, much of it is used to buy the foods and equipment expected by the tourists.

Island and Country Specialties

Puerto Rico is not an independent country nor is it a state within the U.S. political system. An official relationship between Puerto Rico and the United States beginning in the early 1900s became more formally structured in 1952, with the establishment of "commonwealth" status. Puerto Ricans are considered to be U.S. citizens, yet they do not have true representation in Washington, D.C. Puerto Rico's representatives in the U.S. Congress do not have the right to vote on the passage of bills. Puerto Ricans have open access to migration into the United States, and many have taken advantage of this since the 1950s. U.S. statehood remains an issue with distinct political views regarding Puerto Rico's status. The smallest group would like Puerto Rico to become an independent country, free of any U.S. authority, another group supports statehood, and the largest group currently is content to maintain the commonwealth status. Although the Puerto Rican government declared Spanish to be the official language in 1991, English as a second language is mandatory in the Puerto Rican public school system.

Puerto Rico has a much higher per capita GDP than other Caribbean islands. Although its sugar industry expanded, its economy remained narrow until after World War II, when farming shifted to dairying and manufacturing increased rapidly. Machinery, metal products, chemicals, pharmaceuticals, oil refining, rubber, plastics, and garments provide a diverse product mix and have grown under special U.S. tax laws for the island. Tourism became a leading economic sector for Puerto Rico with 3.3 million visitors in 2000.

The Dominican Republic, comprising the eastern two-thirds of the island of Hispaniola, became independent in 1821 and subsequently suffered from economic stagnation and political instability. As in Cuba, U.S. corporations formed links with the plantation owners, and the U.S. government intervened with military occupations in 1916–1924 and 1965. Spurts of economic development occurred in the mid-1900s, when construction began on irrigation and road infrastructure, and U.S. corporations made industrial investments. Agriculture remains the dominant economic activity, based on sugar, while the Standard Fruit Company owns large banana plantations and the Dole Corporation has pineapple plantations. Meat is exported from cattle ranches, and coffee is grown on the uplands. There is some mining, and oil reserves are present. Most manufacturing is the assembly of imported components for export. In the 1990s, tourism was a new growth industry,

Figure 10.35 **St. Thomas, Virgin Islands.** While tourists enjoy the sea, sand, and sun of the Caribbean, much of the revenue generated from their vacations may not reach the local economy or community. Photo: © Neil Rabinowitz/Corbis.

with 3 million visitors in 2000 and an expanding basis of package holidays. Through the gradual diversification of its economy, the Dominican Republic now has one of the highest incomes of the larger Caribbean countries.

Haiti, occupying the smaller western part of the island of Hispaniola, is the poorest country in the Americas. Life expectancy is low, infant mortality high, and 50 percent of its adults are illiterate. Over three-fourths of the people are crowded onto poor quality lands resulting from an imbalanced division of land resource holdings. Independence in 1804 was followed by the political instability that continues today. The sugar plantations deteriorated, and economic stagnation and decline set in. The United States occupied the country from 1915–1934 and again in 1994 to end military takeovers of the government. In between these dates, corrupt and repressive governments produced little economic development, leaving most people dependent on their subsistence plots of land. Some commercial farming for coffee and cacao produced export income, but a 1980 hurricane destroyed many trees and recovery was slow. In the 1970s, tax incentives attracted U.S. corporations to set up factories in Haiti. Exports of clothes, electronics, and sports equipment became more valuable than farm products. Haiti's mineral resources, however, remain undeveloped. Tourism is not developing in Haiti as it is in other Caribbean countries due to the extreme material poverty, political instability, and a high incidence of HIV/AIDS (see Figure 10.33).

Following the military coup against a democratically elected government in 1992, an international embargo on Haitian products caused the economy to collapse, including the loss of vital aid, default on its public debt, and closure of the new assembly industries. Exports and imports halved. A democratic government was reinstated in 1994 after U.S. intervention. Haitians expect rapid improvements from the new government but face a bleak future with an economic gap that is being filled by drug traffic-related corruption.

Beginning as a minor Spanish colony, Jamaica gained significance as the largest British colony in the Caribbean Basin. Sugar plantations expanded until the abolition of slavery; bananas then became important on the plantations and groups of cooperative farms. Agriculture now provides less than 10 percent of Jamaican GDP, although it still employs one-fourth of the labor force. After World War II, the Jamaican economy diversified. Some 60 percent of its export income now derives from bauxite mines and aluminum-refining plants. Although overseas competition from Australia, Brazil, and Guyana led to a fall in this source of income in the 1980s, the 1990s saw increased output. Jamaican manufacturing is becoming more important, based on the duty-free imports of parts to be assembled for export and textile products. An extensively developed tourism industry along Jamaica's northern coast supported over a million visitors in 1999 and contributed income equivalent to half the total export of goods.

Of the Lesser Antilles, Trinidad and Tobago have the largest area and population. Trinidad has an oil-based economy and a unique racial mix. The British took Trinidad from Spain in 1797, and British settlers developed it as a sugar island. Only one-seventh of the land was settled by the time of slave emancipation in 1834–1838, and only one-fifth of that was cultivated. Most plantation development occurred later in the 1800s, when British economic interests shipped indentured labor from Asia. After serving a set number of years on the plantations, these laborers were paid off with land of their own, and today, their small farms produce most of the sugar grown. The increasing quality of its product enables Trinidad to expand sugar output as other countries diminish their production.

Natural gas and oil production, together with the refining of oil from Southwest Asia, is a major sector of Trinidad's industrial base. In the mid-1990s, oil and natural gas output rose by 10 percent. Manufacturing industries include petrochemicals and steel, and employ 15 percent of the labor force, while government and other service jobs provide 50 percent of employment. Tourism is not yet a rapidly developing industry.

The other islands of the Lesser Antilles are small and have restricted economies. Only 5 percent of the work force on Barbados is currently employed by the sugar industry due to a continual decline in cultivation and production, while 80 percent of the workers are in service jobs (dominated by the tourism industry). Bridgetown has some light industries, and its high level of literacy attracts U.S. data processing facilities. Commercial agriculture is on the decline in many of the other former British colonies, such as Antigua and Barbuda, St. Kitts and Nevis, Montserrat, Anguilla, Dominica, St. Lucia, St. Vincent, and Grenada. Most of these islands are attempting to build their tourism industries.

The former French colonies of Martinique, Guadeloupe, and French Guiana are political subdivisions of the country of France, known as oversees departments. Their residents are French citizens and members of the European Union. They once were wealthy sugar islands, but the output declined with a shift to bananas, melons, and pineapples for the French market in line with European Union policies. The French government is the largest employer, and tourism is increasing. External contacts tend to be with France rather than with other Caribbean islands.

The former Netherlands colonies of Aruba, Bonaire, and Curaçao off the northern coast of Venezuela are low-lying and arid, obtaining their water supplies from desalination plants. They are linked administratively to the Netherlands. Although Aruba has had a separate status since 1986, it is moving back toward closer political ties with the Dutch government. The people are exceptionally cosmopolitan, including descendants of Africans, Indians, Dutch, Portuguese, Danes, and Jews, all speaking an old trading language, Papiamento, as well as Dutch (the language of their government), English (the language of tourist visitors), and Spanish (the language of neighboring Venezuela). Oil refineries linked to the Dutch development of Venezuelan oil, along with tourism and offshore banking, are among the main sources of income.

Guyana (former British Guiana), Suriname (former Dutch Guiana), and French Guiana all have small populations on relatively large areas of land. Dutch drainage engineers in the early 1800s made the coastal plains of the Guianas habitable. Indentured laborers were shipped in from South Asia to work on the vast sugar plantations near the coast. In both Guyana and Suriname, living standards declined after independence as the result of civil strife. Guyana and Suriname export bauxite from inland mines, while French Guiana exports timber to the EU. The reopening of gold mines by Canadian companies in Guyana caused exports to quadruple in the 1990s, but in 1995, the Omai gold mine was closed for several months following a cyanide leak from a settling tank into the local river.

United States' corporate ownership of the sugar industry dominated Cuban economic geography in the early 1900s. When Fidel Castro led a Communist takeover in 1959, ties with the United States and U.S. companies were severed and dependency shifted to the former Soviet bloc countries. Large state farms and cooperative farms replaced confiscated private plantations. Although some farms diversified to produce citrus fruit for Eastern Europe, overall productivity remained relatively low. Manufacturing included import-substitution units for locally needed goods such as textiles, wood products, and chemicals. Cuba exported sugar, tobacco (cigars), and some strategic minerals to the Soviet bloc in exchange for oil, wheat, fertilizer, and equipment. During the 1980s, 85 percent of Cuba's trade was with the Soviet bloc, but this close link was broken with the 1991 dissolution of the Soviet Union. Cuba suffered as much as other Communist countries from the breakup of the Soviet bloc. From 1989 to 1994, Cuban GDP fell by 34 percent as it moved cautiously toward market economy policies. The sugar crop fell from 8.4 million tons in 1990 to 3.4 million in 1995 because of poor organization, bad weather, and shortages of fuel, fertilizer, and willing labor. Most farm workers found growing fruit and vegetables for the black market to be more profitable. The United States maintained its trade embargo through the 1990s. Tourism, however, began to grow as a source of foreign currency, with 1.5 million visitors in 1999. Canadian and other non-U.S. mining companies took over Soviet-instigated mining projects, primarily for nickel, cobalt, and gold. In the short term, Cuba faces major economic and social problems.

Northern Andes

The Andes Mountains are a dominant feature in all five countries of the Northern Andes subregion consisting of Bolivia, Colombia, Ecuador, Peru, and Venezuela (Figure 10.36). The world's second-highest mountain range creates a multitude of local environments at different heights throughout the subregion. The geologically active Andes are in the process of uplift, resulting in frequent earthquakes, landslides, and occasional volcanic eruptions. The dramatic topography of the Andes Mountains isolates large interior sections of the Orinoco and Amazon River basins from global connections.

Many people in the Northern Andes subregion live with a pervasive interaction of traditional culture and globalization. Elements of mountain culture groups exhibit centuries-old social customs played out in quiet agricultural towns tucked into the valleys of the steep terrain. Cash from the illegal drug trade creates a unique elite with dramatically different lifestyles from those in rural mountain villages. Political stability is occasionally challenged by powerful criminal elements

and intermittent border disputes such as the disagreement between Ecuador and Peru over Amazonian territories.

Economically, the countries of this subregion range from South America's poorest (Bolivia) to some of South America's more materially wealthy countries (Colombia, Venezuela) (Figure 10.37). In the later 1990s, only Colombia increased its income per capita; consumer goods ownership remained relatively low (see Figure 10.17), while all aspects of human development have modest scores.

Countries

Although rugged terrain and high relief are common features in each of the Northern Andes countries, topographic and environmental variety is present throughout. The high altitude of the Andes in the western part of the subregion contrasts with and isolates the lower lands to the east. Eastern areas are largely covered by tropical rain forest vegetation and experience high rainfall throughout most of the year. The interior reaches of Colombia, Venezuela, and Bolivia (see the "Personal View: Bolivia" box, p. 464) experience more seasonal rains and a tropical grassland vegetative cover. The most dramatic contrast of the subregion exists on the Pacific coast of Peru, where air circulation patterns, cold currents, and the rain shadow of the rugged Andes combine to produce one of the most arid climates in the world.

Venezuela and Colombia have the highest incomes in the subregion (Figure 10.37). Venezuela's income is directly related to its dominant oil and mineral exports that continue to be in demand. Debts accrued during the expansion of oil production in the late 1970s restricted development in the 1980s, when oil prices declined. Colombia's income strength in the subregion relates to its diverse mix of economic products, including mineral extraction, agriculture, manufacturing, and the drug trade. Dramatic economic growth in Colombia is restricted by political instability and civil strife. Ecuador suffers from a small home market. Peru's economy experienced a series of booms and busts as it developed different export products. Bolivia's mining output of tin and silver is less in demand on world markets than it used to be. It lacks an ocean outlet since a war over phosphate deposits in the 1800s resulted in Chile taking over its route to the sea at Arica. Bolivia has natural gas reserves that are second only to Venezuela's in South America. Bolivia and Chile are beginning efforts to construct a natural gas pipeline leading from Bolivian reserves to Chilean port facilities. The joint venture will export Bolivian natural gas to California.

Although the formal economies of the Northern Andes countries are not thriving, the informal and illegal economies based on a globally significant role in the production of coca and cocaine grew rapidly in the 1970s. Increasing demand from the wealthy U.S. market fostered dramatic growth and continues to support coca cultivation and cocaine production. Government efforts thus far have not been successful in eradicating coca production. The number of routes from Colombian processing centers through the Caribbean Basin to the United States is rapidly increasing. Coca-growing areas are diffusing from the eastern slopes of the Andes into Brazil.

Figure 10.36 **Northern Andes: main geographic features.** Note the position of the Andes Mountain ranges in each country and how they isolate interior lowlands from the main centers of population and trade.

Figure 10.37 **Northern South America and Brazil: population, economic, and cultural data.**

Country	Capital City	Land Area (km²) Total	Population (millions) Mid-2001 Total	2025 Est.	GNI 1999 (US $ million) Total	GNI PPP 1999 Per Capita	Percent Urban 2001	Human Development Index Rank of 175 Countries	Human Poverty Index: 1997 Percent of Total Population
Bolivia, Republic of	La Paz	1,098,580	8.5	13.2	8,092	2,300	63	112	21.1
Brazil, Federative Republic of	Brasília	8,511,970	171.8	219.0	730,424	6,840	81	79	15.8
Colombia, Republic of	Bogotá	1,138,910	43.1	59.7	90,007	5,580	71	57	10.5
Ecuador, Republic of	Quito	283,560	12.9	18.7	16,841	2,820	62	72	16.8
Peru, Republic of	Lima	1,285,000	26.1	35.5	53,705	4,480	72	80	16.6
Venezuela, Republic Of	Caracas	912,050	24.6	34.8	87,313	5,420	87	48	12.4

Source: Data from *Population Reference Bureau 2001 Data Sheet*; *World Development Indicators,* World Bank, 2001; *Human Development Report,* United Nations, 2001; Microsoft Encarta (ethnic group, language, religion).

Drug-related instability in Colombia worsened during the 1990s. Allegations of a political power base that is increasingly controlled by drug money are rampant. Clashes between some government officials who openly oppose the drug cartels and guerrilla groups supporting narcotic cultivation regions often result in kidnappings and executions on all sides. The United States injects large amounts of money into fighting the Colombian drug war, yet marginal government relations between the two countries and corruption within the Colombian political system dilute the value of any external assistance efforts. Cease-fire efforts between the Revolutionary Armed Forces of Colombia, a formidable guerrilla group whose base is in the heart of a prime drug cultivation region, and the Colombian government crumbled in 2002, leaving many in the region and around the world wondering if full-scale combat would follow.

People

Dynamic Growth

In the 1960s and 1970s, the Northern Andean countries had annual rates of population increase that were among the highest in the world at 3 to 4 percent. In the 1980s and 1990s, total fertility rates fell from over 5 to around 3 by 2001, except in Bolivia, where they are just under 4.2. This subregion lags behind others in South America (see Figure 10.22) in bringing birth rates and death rates closer together and continues to have annual rates of natural population increase around 2 percent. The total population of 115 million people in 2001 may grow to 160 million by 2025. The large number of young people shown in the age and sex diagram for Peru points to the likelihood of continuing population growth (Figure 10.38).

Urban Growth

The Northern Andean countries vary in their levels of urbanization. Venezuela was 86 percent urban in 2001, while Colombia and Peru were over 70 percent. Ecuador and Bolivia, however, were around 60 percent. All these figures show increases of 10 to 20 percent since 1970. Peru and Bolivia each have a single primate metropolitan center. Peru's primate urban expanse is the Lima-Callao metropolitan complex, where one-third of the country's population resides (7.5 million in Lima in 2000). This urban complex extends along the Rimac River valley from Lima to the port of Callao. In Bolivia, metropolitan La Paz has 20 percent of the country's total population (1.5 million).

Leading urban centers in Venezuela include Caracas (3 million), Maracaibo (2 million), and Valencia (2 million). Recent economic growth in the Pacific coast city of Guayaquil, Ecuador (2.6 million), positions this metropolitan region ahead of the established highland capital of Quito (2 million). Colombia has a number of major cities that grew up in once-isolated areas and developed in different ways. Bogotá, the capital and largest city (6.3 million), has extensive manufacturing industries as well as the national government bureaucracies. Medellín (3 million) is another major manufacturing center for Colombia, and Cali (2.7 million) is linked to one of Colombia's busiest port areas. Both Cali and Barranquilla (1.7 million) are rapidly growing urban centers.

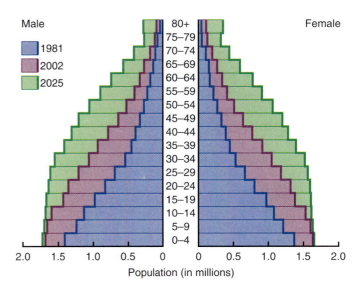

Figure 10.38 **Peru: age-sex diagram.** Source: U.S. Census Bureau. International Data Bank.

Country	Ethnic Groups (percent)	Languages O=Official	Religions (percent)
Bolivia, Republic of	Mestizo 30%, Quechua 30%, Aymura 25%, European 10%	Spanish (O), Quechua (O), Aymura (O)	Roman Catholic 95%
Brazil, Federative Republic of	African 11%, European 55%, mixed 32%	Portuguese (O), Spanish, English	Roman Catholic 73%, other Christian 20%
Colombia, Republic of	Mestizo 58%, European 20%	Spanish (O), Quechua (O), Aymura (O)	Roman Catholic 95%
Ecuador, Republic of	Mestizo 55%, Native American 25%, European 10%	Spanish (O), Native American	Roman Catholic 95%
Peru, Republic of	Mestizo 37%, Native American 45%, European 15%	Spanish (O), Quechua (O)	Roman Catholic 95%
Venezuela, Republic of	Mestizo 67%, European 21%, African 10%	Spanish (O), English, local	Roman Catholic 96%

BOLIVIA

Bolivia is one of the poorest countries in Latin America. Although it is five times larger than Britain, it has only one-tenth of the population. It is a mountainous and landlocked country that long depended on exports of tin and silver for its economy (Box Figure 1). People living in Bolivia experienced important changes after the 1950s that opened new opportunities to them. The following is the personal view of living in Bolivia from Dr. Salinas, who practiced medicine in a mining community for many years.

The 1951 revolution led to huge political changes. Until that time, British mining companies virtually ruled Bolivia, but they were dispossessed when the mines were nationalized in 1952. Although the old mining companies were criticized for the slavelike conditions in which they held their laborers, they left an infrastructure heritage of housing, schools, railroads, and medical facilities that were scarcely improved in the next 50 years.

The new government also instituted land reforms, dividing huge landholdings among those who worked them and setting up cooperative arrangements for purchasing and marketing. This policy was an attempt to diversify the economy away from the reliance on mine products by expanding commercial farming. People of Native American origin, who make up around 50 percent of the Bolivian population, were given the vote for the first time and now wield a major influence in government. For them in particular, the revolution brought access to education for a group that was 90 percent illiterate and had no hope of advancement. Today, many Native Americans are qualifying in the professions, and one has been vice president.

The revolution and its aftermath were not a total success. In the 1980s, the government shut down mines when the costs of producing tin rose to $17 per pound while the world price was only around $4. In the mining town of Catavi, where Dr. Salinas served the mine laborers and the town, the number of mine workers declined dramatically, from 5,000 in 1982 to 500 in 1986. Dr. Salinas's salary fell in real value at a time of rapid inflation, but the mining unions forced him to attend to his patients until his contract expired. In the late 1980s and 1990s, many of the former government-owned mines were privatized, often bought by Japanese companies, but the new owners employed few miners.

Box Figure 1   **Map of Bolivia.**

Bolivia as a whole remains a country of contrasts. There is an affluent class of people that includes businesspeople, professionals such as doctors and lawyers, and military officers (Box Figure 2). Some teachers are in this group if they teach at private schools as well as in the poorly paid, public school jobs. Most of the elite claim Spanish origins, but many are of mixed ancestry. They live in good houses, and many own luxury cars bought mainly from Brazilian suppliers. If not traveling by car, they travel in special-class trains or well-equipped buses.

There is also a working class of factory, mine, and shop workers. They have regular jobs, but wages are low. They travel in the back of trucks or on basic buses. There is also a substantial number of landless and jobless people in Bolivia. Many are rural Native Americans who still do not have access to public education. The lack of facilities in rural areas set off a major migration to the urban areas, especially

---

All the major urban centers of the Northern Andes sub-region grew dramatically when rural-to-urban migration, beginning in earnest in the 1950s, sparked the rapid establishment of shantytowns. In Lima-Callao, rapid population growth outpaced any attempts to develop urban structural amenities, and the prevalence of uncontrolled street vendors led to the closing of many established shops, the loss of local taxes, and the deterioration of much of the city's infrastructure. In the 1970s, it was hoped that the shantytowns might provide a transition to more organized urban living, but such hopes were largely unfulfilled. The impacts of the Marxist Shining Path guerrilla war in the 1980s and early 1990s, which resulted in the deaths of approximately 90,000 Peruvians, a large-scale abandonment of control in the Andean provinces south of Lima, and increased migration into the capital's shantytowns all complicated attempts to organize urbanization better.

Ethnic Contrasts

The Andean countries contain racial and ethnic contrasts that have important economic, social, and political effects. In Colombia and Venezuela, the Native American groups live primarily in highland and interior areas that were not taken over by Hispanic or mestizo groups. In Ecuador, Peru, and Bolivia, indigenous peoples tend to be concentrated in the highlands. Smaller numbers of different Native American tribes occur in the Amazon River lowlands, totally isolated from European intrusions until the 1950s. Current settlement for European descendants and mestizo groups reflects colonial occupations of upland basins and coastal locations that provided overseas links to Spain. Workers from Japan, China, and Africa immigrated to work in the coastal irrigated farmlands of Peru and established neighborhoods near their work sites.

The economic and social differences between European-origin and Native American peoples continue to raise tensions

for access to schooling. When such people reach urban areas, they do not have jobs or housing. They move into the shantytowns, or *barrios parifaricos,* often close to people from their home district, and do whatever they can to earn some money. In the barrios, electricity is available for those who can pay, while water standpipes are shared among some 50 houses. There are stories of Native American families arriving from the country and the parents working in domestic service while the children go to school and become successful professionals. But most Native Americans do not rise so quickly, if at all. Many unemployed peasants obtain a living by begging.

Cutting across these social and economic differences, Bolivia has become a major producer of coca plants for cocaine drugs. When mines closed, many miners moved to the Chaparé area in the eastern lowlands, working on the coca plantations or in the distribution trade. It is said that some school teachers use their vacation periods to earn as much in a month on the coca plantations as they get in the rest of the year by teaching. Drug sales now underpin the Bolivian economy. The very wealthy people who control the production of cocaine sometimes invest in other local facilities in an attempt to improve their image: factories are built, stores sell imported goods bought in exchange for the cocaine, and sometimes schools and medical centers are provided. The cocaine trade has made many of those who were very poor quite rich and has altered the existence of many Bolivians. It has not been uncommon for shabbily dressed men to carry thousands of dollar bills to a vehicle dealer to buy a Mercedes truck.

within the countries of the Northern Andes. Although formal attempts to modernize some Native Americans were made, many native groups prefer to retain their traditional cultural values instead of adopting those of the European descendants.

Economic Development

Land use and trade patterns implemented during the colonial era established lasting trade relations in which most countries of the region relied on cash crop and mineral resource exports to pay for imports that only the most materially wealthy segment of the population could afford to buy. The governments of several Andean countries convened to explore the formation of an economic union to reverse lasting trade patterns and boost internal economic growth. In 1969, five countries (Bolivia, Chile, Colombia, Ecuador, and Peru) formed the Andean Group of trading partners. In the 1970s, Venezuela joined and Chile left.

At first, this group saw its purpose as creating an internal trading area and keeping out foreign goods. It faltered as members disagreed over which country should produce what goods. In 1988, it was relaunched as the **Andean Common Market** with a greater emphasis on integrating with world trade patterns.

Export-Led Underdevelopment

All five of the countries of the Northern Andes subregion suffer from **export-led underdevelopment.** In the late 1800s and early 1900s, the countries of this subregion relied on a series of export-based booms as specific mining and agricultural products came into demand in Europe and North America. Local people and foreigners invested in these products, while the government taxed the exports as the basis of its own income. Any hard currency earned went to pay for imports, on which tariffs were not charged. In the 1930s, attempts at changing this process by encouraging import-substitution manufacturing at home were not very successful because illegal imports undercut local products and only a restricted range of goods was made in local factories. As a result, the Northern Andean countries found that fuller involvement in the world economic system made them vulnerable to external control. Today's dependence on smuggling coca and cocaine arose from the need to find a high-value product with continuing demand in world markets.

Patterns of Economic Diversification

Early colonists in Peru established mines in the cordilleras and cultivated crops along the coast to support the people and pack animals that delivered mined silver and gold to the coast. Later Peruvian export booms from the mid-1800s involved less valuable minerals, such as lead and zinc from the cordilleras, guano fertilizer from offshore islands, and sugarcane and cotton from irrigated coastal lands. Oil from the north, iron from the south, and fishmeal from what became the world's second-largest fishery (Figure 10.39) added new aspects to Peru's income from the 1950s. Although this output appears diverse, each product comes from a particular area and had a limited period of importance.

Peru dramatically opened its economy to external investment in the 1990s, creating a boom in mining exploration. In 1992, Peru opened Latin America's largest gold mine at Yanacocha near Cajamarca. It began production at other mines by the late 1990s and continued to explore opportunities. Peru's rapid economic growth in the mid-1990s was later slowed by the combination of expanding demand for imports, increasing indebtedness, and continuing poverty among a large sector of the population. Peru experienced a recession pattern similar to that which occurred in Mexico. The large sums of money entering Peru, which is the world's largest producer of cocaine, by way of the illegal drug trade may cushion such effects, but wealth distribution related to the drug trade is extremely uneven. Three events in the late 1990s cast doubts on such optimism. First, the 1997–1998 El Niño event brought

disaster to coastal farming and fishing communities. Second, in 1998, the Shell Oil Company canceled development of the Camisea natural gas field in the Amazon River basin, which might have given Peru entry to higher-level manufacturing skills and products. Third, the economic downturn in Asia reduced demand for Peruvian minerals and caused the postponement of mining projects.

Exports of silver and tin dominated the Bolivian economy for decades. High elevation limited extensive agricultural development and impeded economic expansion or diversification. A 1980s fall in world tin prices hurt Bolivia's limited economy. The Bolivian economy today depends on other metallic ores, such as gold, which comprise up to 50 percent of Bolivia's exports. New development is occurring in the eastern interior plains where nutrient-rich soils and improved transportation make it possible to move more soybeans from field to consumer each year. The crops are trucked across the Andes to Peruvian ports for export. Oil from the eastern Andean slopes makes Bolivia self-sufficient in energy.

Bolivia is the world's second-largest producer of cocaine. United States pressure on the Bolivian government to crack down on production in the Chaparé Valley east of Cochabamba has met little success due to internal political disagreement and an economically challenged population. Bolivian economic reform, based on the encouragement of foreign capital investments in state corporations and Bolivian public institutions, has not created a financial environment worthy of challenging the cocaine industry.

Ecuador is a small country with limited upland and interior economic potential. Economic growth occurs primarily on the coast around Guayaquil, where the expansion of fishing for tuna and white fish, together with shrimp farming, is adding to an increasingly diversified base of light manufacturing, commercial farming for sugarcane, bananas, coffee, cacao, and rice, and tourism. The construction of an oil pipeline in the 1970s from the Andes to the port of Esmeraldas increased overall oil production. Petroleum products make up nearly half of Ecuador's exports. New laws opened mining prospects for foreign corporations from Canada, South Africa, France, and Belgium that are testing the viability of gold, silver, lead, zinc, and copper deposits.

Although shrimp farming diversifies the economic base of Ecuador by providing a non-oil export, further expansion is slowed by disputes with environmentalists who claim it is destroying coastal mangroves. The farmers counterclaim that they have to preserve the mangroves in which the shrimp begin life and that they are providing jobs and services in areas that have great poverty, malaria, no electricity, and no health services.

Global attention remains focused on Ecuador's Galápagos Islands, where unique animal and plant species attract scientists and tourists from all parts of the world. The government of Ecuador is caught between the desire to exploit the tourism potential of the islands fully and the need to preserve a world-class natural heritage site as well as to maintain the conditions that would attract tourists in the first place. Visitor quotas, established by the government to protect the ecological balance of the islands, have been increased several times and are often exceeded. The economic potential of the islands could easily

Figure 10.39 **Northern Andes: fishing industry.** Peruvian fishing boats set out to sea at sunset. They fish in the rich grounds of the cold Peruvian current, where nutrients support the food chain from plankton to fish and birds. El Niño events bring warm waters that often kill fish and reduce fishing catches. Photo: © Alain Choisnet/The Image Bank/Getty.

be destroyed if too many visitors degrade the health of the animals and natural resources that make the Galápagos such an attractive destination.

In the early 1900s, Venezuela became a major oil producer, raising its income to the highest in the Northern Andes subregion. Oil production in the Lake Maracaibo region of Venezuela increased during World War II and again when world oil prices rose in the 1970s. The government of Venezuela nationalized the oil industry in 1976 and, under the expectation of increasing oil revenues, simultaneously acquired extensive foreign debt for the purpose of domestic investment. The 1980s brought a period of higher interest rates globally, decreased oil demand, and lower oil prices. Countries such as Venezuela faced the seemingly impossible task of making payments on loans out of significantly reduced oil revenues. Venezuela could barely pay the interest on the loans, thus worsening its debt situation. National debt caused investment in oil exploration to lag in the 1980s. In the 1990s, exploration opened to private companies that are linking to joint production and marketing arrangements with the government as discoveries are proved.

The Caracas area is the focus of manufacturing and service industries. Southern Venezuela is a source of mineral wealth from iron and bauxite mines. The mines provide a basis for a manufacturing zone in Ciudad Guyana on the Orinoco River. The manufacturing zone is powered by hydroelectricity that is generated in the Guiana Highland valleys. This region has potential for future economic development but remains isolated by poor transportation facilities and is affected by Venezuela's shortage of investment capital.

Colombia has a more diversified economy than the other Northern Andean countries. Colombian cities such as Bogotá and Medellín grew from colonial settlements established in elevated basins in the Andes mountains. Local manufacturing developed in the 1800s to supply a growing market. Medellín, and the capital, Bogotá, became the main industrial centers for textiles, steel, agricultural equipment, and domestic goods. In the lower valleys and near the northern coast in contemporary Colombia, large plantations raise tropical cash crops, including cotton, sugarcane, bananas, cacao, and rice for export. Colombia is the world's second-largest producer/supplier of coffee, after Brazil. Coffee, which is Colombia's largest legal export, thrives in the optimum conditions of well-drained mountain slopes, fertile soil, and sufficient rainfall. Colombia began to grow new varieties of corn and rice in the mid-1990s, which were bred for higher yields in the poor soils of the eastern savannas. The new corn and rice, coupled with livestock and locally grown cattle feed, are meeting with success in this sparsely settled area. Marijuana, coca, and poppies are major crops in the more rugged and secluded reaches of the mountains. Colombia is also a mining country with considerable resource stocks of iron, coal, oil, natural gas, gold, and emeralds. Colombia's important overseas trade is regularly hampered by continuous internal political strife. In the 1980s and 1990s, guerrilla warfare, based partly on political repression and partly on strife among the drug syndicates, defeated government attempts to capitalize on the country's strengths.

The Northern Andes and the International Drug Trade

The countries of the Northern Andes comprise one of the world's primary drug-producing regions (Figure 10.40). The initial system in the subregion centered on cultivation areas in Peru and Bolivia, with processing or production centers based in Colombia. The expansion of Bolivian cultivation coverage slowed in the early 1980s, and the Peruvian coverage declined in output following a fungus infestation of the crop. Colombian cultivation increased and expanded to coca, opium poppies, and marijuana, and is now greater than that in Bolivia.

The global demand for cocaine from consumers in the United States, Europe, and Asia has a devastating impact on the local politics, agriculture, and social fabric in the Northern Andes. An excess of production over market demand in the early 1980s caused the price of cocaine to fall by 75 percent. The huge drop in price dramatically opened the international market to those with less disposable income, resulting in a subsequent rapid increase in demand. The United States led efforts

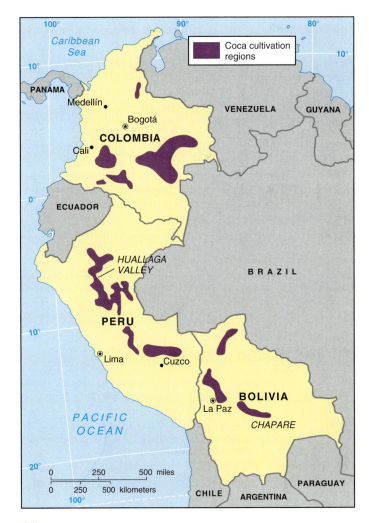

Figure 10.40 **Northern Andes: coca-growing areas.** The dominance of Peru and Bolivia in the 1980s gave way to more dispersed production in the 1990s, including Colombia (previously the main gathering and distribution center) and Brazil.

to prevent drug production in these countries, but its efforts were diluted by limited cooperation from the governments and farmers of Bolivia and Peru. Although Bolivian farmers planted other commercially valuable crops on former coca-growing land as a result of U.S. aid, the acreage under coca remained the same. Peruvian government attempts to halt cocaine production are limited in part by fears of the resurgence of guerrilla groups among disaffected former coca growers. Subsistence and small-scale farmers are attracted to working on large coca farms where average wages are far in excess of what they would otherwise earn. Other farmers are forced by threats of violence to become part of the drug labor force, and in many cases, their farms are forcibly taken over by the drug cartels.

In 1995, Colombia attempted to adopt a tougher stance and announced its intention to eradicate production within two years. Programs of eradication, however, have to compete with the high prices farmers get for their drug crops and with the guerrilla-backed disorder that such farmers may support, including the strategic occupation of isolated oil-pumping stations or airports. Chemical defoliation of coca-growing areas paralleled a doubling of the output of coca. By 1998, the failure of such programs led the United Nations Drugs Control Program to try to establish a new global drugs control convention. It has to contend with the income from coca-growing that lifts farmers out of poverty and the long-established custom of coca-leaf chewing among Native Americans living at high altitudes in the Andes.

 # Brazil

Brazil is the largest country in both area and population in all of Latin America (see Figure 10.37). Brazil has three times the area of Argentina, the second-largest South American country. In 2001, Brazil's population exceeded 170 million, which is much larger than the total population of any other Latin American subregion. The people of Brazil have a strong sense of national pride and a passion for cultural expression. Although the Portuguese language and Roman Catholic church provide some national homogeneity, the diverse contemporary human geography and cultural expression of Latin America's largest country are tangible extensions of the historical mixture of European, African, and indigenous influences. The physical expanse of Brazilian territory incorporates a vast array of topography, climate, and vegetative cover, and holds a diverse and abundant supply of natural resources. Brazil presents an immense economic market, has the greatest economic diversity of any Latin American country, and had the world's eighth-largest economy in 1999. Despite decades of rapid economic growth between 1940 and 1980, and the size of its economy, Brazil faced increasing problems in the 1980s and 1990s that resulted in high inflation, declining economic growth, and growing numbers of materially poor people.

The establishment of sugar plantations along the northeast coast of Brazil during the colonial era began the course of European-induced human and physical geographic change experienced in so many locations in Latin America. What most distinguishes the colonial history of Brazil from that of other Latin American countries is the lasting influence of the Portuguese rather than the Spanish. The Portuguese language and unique blend of Portuguese, African, Caribbean, and indigenous traditions give Brazil a national flavor that is distinctively different from Spanish Latin America. Portuguese economic institutions forced the migration of millions of Africans to Brazil. The contemporary human mosaic includes the largest African and Afro-Caribbean heritage of any Latin American country.

Regions of Brazil

The expansive size and geographic diversity of Brazil dictate further subdividing the country for geographic examination (Figure 10.41). Politically, Brazil is divided into states that are part of a federal government system. Further human geographic patterns relate to natural environment contrasts, the history of European settlement, and political emphases among the different states.

- Nearly 60 percent of Brazil's territory is drained by the Amazon River network. Much of this area is covered by tropical rain forest. Although the Amazon River basin has rela-

Test Your Understanding 10D

Summary The Caribbean Basin includes four large islands (Cuba; Hispaniola—with Haiti and the Dominican Republic; Jamaica; and Puerto Rico), many small islands, and the Guianas on mainland South America. They were mainly settled by Europeans to develop sugar plantations and by imported African slaves.

Most of the former colonies are now independent, and their economies are based on more varied agricultural products, tourism, local mineral products, and some light manufacturing. Some have developed offshore financial services mainly linked to the United States.

The Northern Andean countries of Bolivia, Colombia, Ecuador, Peru, and Venezuela are dominated by the high mountains that isolate internal plains covered by savanna and tropical rain forest from the main centers of population. Their economies are based on mineral production, including increasing oil discoveries, and export crops. These countries include major drug producers, especially of cocaine but increasingly of marijuana and opium.

Some Questions to Think About

10D.1 How are the physical and human geographies of the Caribbean Basin different from those of Mexico and the Central American countries?

10D.2 What is the impact of the Andes Mountains on the human geographies of the subregion's countries? What role does physical geography play in drug cultivation?

10D.3 What local and global relationships are involved in Andean drug cultivation?

Key Terms

Caribbean Community and
   Common Market (CARICOM)
Association of Caribbean States

economic leakage
Andean Common Market
export-led underdevelopment

Figure 10.41 **Brazil: major geographic features.** The states, rivers, and major cities.

tively low population densities, development efforts led by the Brazilian government are increasingly encroaching on tropical forests of the region, and deforestation is rampant in some parts of the Amazon Basin (see the "Point-Counterpoint: Tropical Forests and Deforestation" box, p. 434).

- The northeastern coastlands and plateau were the first settled areas of Brazil. Sugarcane plantations prospered until competition with the Caribbean Basin began in the later 1600s. Periodic droughts devastate the interior of this region (as in the late 1990s), causing many to emigrate to other parts of Brazil.

- The main economic development in Brazil occurred in the southeast around Rio de Janeiro, the colonial and national capital for 200 years, and São Paulo. The early industrial focus of the region centered on mining and commercial agriculture (based on coffee cultivation). Today, this region is a major center of manufacturing and financial service activities for Brazil.

- The three southern states of Paraná, Santa Catarina, and Rio Grande do Sul became centers of growing population and agriculture from the 1930s. After 1950, the coffee crop exhausted the soils of the area west of São Paulo, causing cultivation to spread to the west-southwest along new railroads into northern Paraná until bursts of cold wintry air from the south demarcated its limits. Cattle and a variety

of temperate and subtropical agricultural products, including oranges, are produced in these states, settled largely by immigrants encouraged to move from Germany, Italy, and Japan. Hydroelectricity generated on the Paraná River and its tributaries powers manufacturing in the area.

- Inland of São Paulo, straddling the drainage divide between the Amazon and Paraná-Paraguay River systems, a new area of farming development in Campo Cerrado grew in significance from the 1970s. With improved road communications, this area became one of the world's main soybean producers, with some 6 million hectares (about 15 million acres) growing 12 million tons per year in the 1990s. Farm management, use of machinery, fertilizer inputs, and marketing procedures are highly organized, and yields rival the best in the U.S. Midwest. New towns grew to over a quarter of a million people within 10 years, and Brazil now exports a greater value of soybeans than of coffee.

People

Population Dynamics

Brazil's population is moving out of a period of rapid growth, which produced the relatively young population of today (Figure 10.42) and a population total that rose rapidly to the

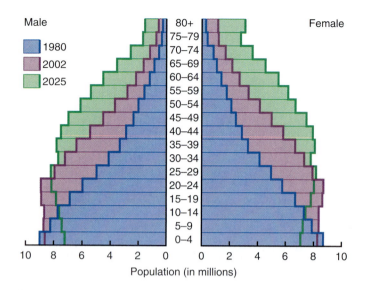

Figure 10.42 **Brazil: age-sex diagram.** Source: U.S. Census Bureau. International Data Bank.

2001 figure of 172 million people. Total fertility dropped from nearly 6 to 2.4 between 1970 and 2001, while infant mortality went from 116 to 35 per 1,000 live births between 1960 and 2001. The rate of natural increase fell from 2.4 to 1.5 percent between 1970 and 2001. The fall in birth and fertility rates (see Figure 10.22) is linked to rising education levels, increased urbanization, and the impact of the media. Although Brazil is the world's largest Roman Catholic country and that church in Brazil has a conservative hierarchy, its local priests are often liberal on birth control matters. Moreover, a high proportion of women choose to give birth by a cesarean operation followed by sterilization, a service available free from the government. Recent studies suggest a significant impact from television on family size choices for many Brazilians. Television programs depicting happy and stable situations for small families seem to influence Brazilian family planning choices. Life expectancy in Brazil remains similar to most other South American countries, at 68 years.

Increasing Urbanization

Brazil has two of the three largest metropolitan areas in Latin America and a highly urban population overall. The proportion of the population living in towns or urban settings increased from 56 percent in 1970 to 81 percent in 2000 (Figure 10.43). São Paulo is one of the largest metropolitan areas in the world with 19 million people in 2000. It started in 1522 as an inland mission station and was a base for exploration of the interior by paramilitary bands in the 1600s. Later, it became the focus of Brazil's world-dominating coffee industry, and railroads were built westward from it to open up new lands. In the 1900s, it developed into the major manufacturing center of Brazil. The state of São Paulo produces nearly half of Brazil's GDP and two-

thirds of its manufacturing output. In the late 1990s, São Paulo's population growth slowed (from 5% to 0.5% per year). New industries tended to be sited elsewhere, often farther out in São Paulo state to avoid the high costs of traffic congestion, pollution, and strong union control (high wages and restrictive working conditions). The provision of housing, schools, and health facilities could not keep up with the metropolitan area's growth.

Brazil's second-largest metropolitan center is the Rio de Janeiro complex (Figure 10.44). Rio de Janeiro, with over 11 million people in 2000, first grew as a port city for early interior gold mine development, becoming Brazil's largest port and capital city in 1763. Today, Rio de Janeiro is a contemporary culture hearth for Brazilian nationals and a major draw for international visitors to the country.

Other major cities include Belo Horizonte (4 million), the early center of mining and now the focus of a series of growing mining and metal-manufacturing towns, Recife (3.5 million) and Salvador (3 million) on the northeast coast, Fortaleza (3 million) on the north coast, and Curitiba (2.5 million) and Pôrto Alegre (3.7 million) in the south. Manaus (1.5 million), in the center of the Amazon River basin, has over half the population of its state, Amazonas. Brasília, the city built in the 1950s and established as the Brazilian capital in 1960, now has 800,000 people living in its planned central city area and 2 million included in the surrounding satellite cities (see Figure 2.21).

Within the Brazilian cities, the shantytowns, known as **favelas** in the southern parts of the country, house millions of materially poor people who cannot be accommodated by the formal patterns of housing construction, infrastructure, and service provision (Figure 10.45). It is estimated that over 7 million people in São Paulo and up to 6 million people in Rio de Janeiro live in favelas. Favelas vary in their character. In and around the city centers, they occupy gaps in the built environment, in which families erect their own minimal accommodations. It is rare for these favelas to have water, electricity, waste disposal, or anything resembling a road. They provide no security for the inhabitants and are often centers of disease and crime. Many such favelas have been in existence for 10 years or more. Old apartment blocks in the city centers, which were abandoned and taken over by impoverished peoples, resemble favelas in many ways,

(a)

Figure 10.44 **Rio de Janeiro.** The dramatic physical geography of Rio de Janeiro contributes to the strong pride Brazilians feel for this city (a), which also attracts millions of visitors from around the world. (b) Contrasting with expensive hotels and condos along Copacabana beach, the Rocinha favela crowds thousands of materially poor Brazilians into marginal structures on a steep hillside. Photo: (b) © Stephanie Maze/Corbis.

(b)

with several families occupying each room. On the outskirts of the cities, different types of favelas are pushing the built-up area outward at a rapid pace. These favelas spring up almost overnight and grow by thousands of people within months. Once established, favelas may be provided with some basic roads and may soon acquire utilities, shops, and schools.

The favelas are also the primary site where thousands of Brazilian street children base their nomadic and often shelterless existence. This serious issue focuses attention on global connections and local voices in a negative manner. Economic, political, and social globalization forces have neither eliminated nor meaningfully diminished the social conditions that combine to place hundreds of thousands of Brazilian youth in jeopardy. Some argue that multinational corporations, international banking and development institutions, and intergovernmental trade relations have increased Brazilian poverty rates rather than reduced them. Brazilian government debt and resource-export economic goals also figure prominently in the social environment that produces so many children of the streets. The socioenvironmental conditions produced from such domestic and foreign relationships include extreme disparities in wealth, high poverty rates, relatively high death rates for women in childbirth, high rates of teenage pregnancy, high illiteracy rates, and high jobless rates. Street children either literally live on the streets with no shelter or work the streets at a young age, trying to earn money for their survival or that of their families. The street children often end up in prostitution, and many contract life-threatening diseases. The most challenging situation is the reported periodic murder of groups of the street children, whose bodies are abandoned in ravines or concealed areas. Although local and global human rights groups are persistent in their attempts to attract significant attention to the plight of the street children, the problem remains very serious.

Brazilian Cities

The built environment of Brazilian cities is distinctive. Large numbers of middle-class people live in inner-city high-rise apartment blocks with external walls, fences, and guards, and internal swimming pools and other services. Others live in single-family homes in the suburbs but guard their possessions with massive steel fences and gates.

One of the most remarkable examples of urban growth is Brazil's planned capital city of Brasília. The central area of Brasília's government complex is flanked by apartment accommodations, local retail and other commercial facilities, and a series of parks and open spaces (see Figure 2.21). Affluent housing for the diplomatic corps and highly paid bureaucrats was built around an artificial lake. Outside the center, often at a distance of a one-hour bus journey, a series of satellite towns support the growing metropolitan population of the planned capital.

Ethnic Variety

Although the large Brazilian population is extremely varied ethnically, the citizens of the country maintain a strong sense of national pride. The original Native American population is much reduced: in the Amazon River basin, it was probably around 3.5 million in A.D. 1500 but is now closer to 200,000. The traditional culture and people's health are increasingly threatened as outsiders invade their territories.

Portuguese and other European peoples make up a major proportion of the present Brazilian population, with highest concentrations in the southern states and elite areas of the cities. The descendants of African-origin slaves form a significant minority proportion of the population along the northeastern coast and have spread into the southern cities. Brazilians of African descent have had a major impact on Brazilian art, food, music, and dance. Immigrants over the last 30 years include large numbers of Japanese, now accepted as a part of the Brazilian people and taking many of the top jobs in business and politics. Japanese Brazilians form the largest Japanese community living outside of Japan. Mixtures of people of African, European, and Native American descent comprise a sizable and growing portion of the Brazilian population (Figure 10.46).

Economic Development

Brazil is a leader in Latin America in terms of targeted economic modernization and development. Brazil's rates of consumer goods ownership (see Figure 10.17) are similar to those of the countries in southern South America, but human development indicators for Brazil are not as high as those for

(a)

(b)

Figure 10.45   **Latin America: urban landscapes.** The San Pedro favela, San Bernadino—part of the urban region centered on São Paulo. (a) The favela becomes established along a grid of unsurfaced roads with houses built of cement blocks and packing cases. There are few services apart from electricity supply. (b) Where the favela joins the main road, the roads are becoming surfaced, and a varied range of services, including shops and schools, are available. Photos: © Michael Bradshaw.

Argentina or Chile. Economic and related social problems stem in part from a long-term dominance of public involvement in the economy and the high rates of population increase in the 1960s and 1970s. Other factors contributing to lower economically related human development indicators include the shorter-term impacts of the 1985 constitutional changes that enshrined provisions for social programs and job security. The constitutional provisions maintained high levels of people in employment but placed a huge burden on the country's finances. In addition to being responsible for social provision administration, the federal government had to distribute a high proportion of its income to Brazilian states. The combination of international debt servicing, the need to cover constitutional demands, financially domestic development projects, and numerous cases of government corruption led to inflation that rose to 50 percent per month in early 1994. In the mid- and late 1990s, the Brazilian government made changes that reduced the inflation rate and altered the constitution to remove many of the 1985 provisions.

Varied Economic History

Brazil's colonial economic development began with coastal sugarcane plantations along the northeast coast, interior cattle ranches, mining for gold, and trading for Amazon River basin forest products. In the 1800s, the area west of São Paulo became the world's chief coffee producer. This required the recruitment of workers from Europe and the building of railroads through British financing. The Amazon River basin became a rubber producer of world importance in the late 1800s, but competition with Asian producers ended that boom at the outset of the 1900s (Figure 10.47).

By the early 1900s, the southeastern coffee-producing states contributed over 70 percent of the Brazilian GDP. São Paulo and its port city, Santos, grew as the hub of this development, but by the mid-1900s, the soils of the older coffee lands became exhausted and new coffee plantations sprang up westward along the railroads into Paraná state, with growth especially around the city of Londrina. Brazil's development of primary products for export paralleled that in many other Latin American countries. The sugarcane, gold, and rubber export

Figure 10.46 **Brazil: the people.** Brazilians watch the 1994 World Cup soccer match between their country and the United States on TV. The crowd includes a variety of peoples with origins in Africa, Europe, the Americas, together with mixtures. Photo: © John Maier, Jr./The Image Works.

surges lasted for a while and then declined; coffee production was maintained only by shifts to new lands.

Economic stagnation in the 1920s and the world economic depression of the 1930s caused a crisis during which the military took over the government of Brazil. From this point, the Brazilian economy diverged from that in many other Latin American countries by basing internal economic growth on a form of state capitalism. The federal government assumed the ownership of major economic activities and established tariffs to protect internal manufacturing. Import substitution reached its greatest level of activity in Brazil aided by the huge internal market. Durable goods such as automobiles and domestic appliances that were made abroad were subject to an 85 percent import duty, while some, such as computers, were excluded altogether.

In 1937, the new government nationalized the iron mines in the state of Minas Gerais. The huge mines around Itabira became 80 percent government-owned, and production was increased with 70 percent now being sold overseas, delivered by mining company ships. A steel industry grew up along the railroad from Itabira to the port of Tubarão, with the largest integrated mills at Volta Redonda being established in 1940 and extended in 1970 to double their output.

Modern Mining

The national iron ore company is developing the iron ore mining Carajás Project in the eastern Amazon River basin. A railroad

Figure 10.47 **Brazil: Amazon River.** The city of Manaus has an opera house that was built early in the 1900s to cater to 2,000 or so rubber barons who lived there and controlled trade in the valuable commodity. Now refurbished, the opera house is in the center of a city of over a million people, with high-tech industries and improved communication by river, road, and air. Photo: © Michael Bradshaw.

...nt in 1985 takes the ore to a port near São Luis on the northern coast for export to Japan, the United States, and Europe, while the Tucurui Dam generates hydroelectricity for mining and industrial needs. The controversial project required the clearance of 3 million hectares (7.4 million acres) of rain forest, and much of that area is now used for ranching and small farms.

Other Brazilian mining developments include the production of manganese (used in hardening steel), tin, and bauxite. One-third of the world's bauxite resources occur east of Manaus in the Amazon River basin. In the upper reaches of the basin, small deposits of gold attract thousands of independent miners, but the use of mercury to separate the gold pollutes the rivers, and this activity forms a major intrusion in the lives of Amazon tribes such as the Yanomami.

Brazil is increasing its production of petroleum. The oil consumers' crisis of the 1970s hit Brazil hard and was responsible for much of its high debt. An immediate response was the development of ethanol fuel from sugarcane. The 1975 policy for ethanol development guaranteed a price at 65 percent of gasoline prices. The aim was for one-third of all cars to use this fuel. Early technical problems were overcome, although ethanol-using cars still have low resale prices. The ethanol policy was hampered by the lack of rural pumps selling the fuel, the cutback in sugarcane production as a result of lower subsidies, and the development of Brazilian oil products. It is now necessary for Brazil to import sugarcane to make ethanol, and the pump prices for ethanol fuels are higher than those for gasoline. Petrobras, the Brazilian state oil company, was established to import, refine, and distribute oil products but now finds itself an oil producer with offshore wells along the eastern and northern coasts and major reserves in the western Amazon. By the mid-1990s, over 60 percent of Brazil's oil needs were produced from domestic sources.

Farming Today

The government continues to encourage agricultural production, but agriculture's proportion of GDP fell as that of manufactured goods rose. Coffee and sugarcane remain important exports, but coffee, once the mainstay of the economy, now makes up less than 10 percent of the agricultural exports. New developments, such as the growing of oranges and other citrus in the south and tropical fruits and nuts farther north, added to the diversity of commercial farming. The major new development in agriculture is the efficient commercial production of soybeans that shot Brazil to the top of international traders in this crop (Figure 10.48). Government-based research, advice to farmers, and marketing facilities contributed to this success. In many localities, however, one commercial crop tends to be dominant, and so the problems of reliance on monoculture in the context of shifting world markets are still felt.

Manufacturing Expansion

The largest manufacturing sector in Brazil is the production of automobiles and trucks, concentrated in the southern São Paulo suburbs. Autolatina, in which Ford and Volkswagen shared ownership with the government and produced cars specifically for the Brazilian market, made about half the annual output of 1 million cars until 1995, when Ford and Volkswagen split to take advantage of new sales opportunities in Brazil and Argentina. General Motors and Fiat (near Belo Horizonte) have similar but separate manufacturing arrangements. In the 1990s, the lowering of import tariffs led to further competition, falling prices, and expansion of Brazilian car sales and production, especially of small cars. Output of 900,000 cars in 1992 doubled by the mid-1990s. By 1995, Renault (France) and Hyundai (South Korea) decided to build new factories in Brazil to take advantage of the growing market and the Mercosul free trade agreements.

The Brazilian government owns many large corporations that employ thousands of citizens. Electrobras is a nationalized corporation that produces and distributes electricity. Small hydroelectricity projects on the plateaus in the east gave way to huge projects in the interior (Figure 10.49). The world's largest hydroelectricity project is at the Itaipu Dam on the Paraná River at the border with Paraguay. It was opened in 1983 and produces one-fifth of Brazil's power (and much of Paraguay's).

Economic globalization connections flourish in the **free trade zone** established in the Amazon River city of Manaus in 1966. In this zone, foreign companies may import materials for assembly without tariffs and export the assembled products to other countries. Manaus is now Brazil's fastest-growing city and the second in manufactured goods value after São Paulo. The local human and physical landscapes are dramatically different from outlying forest areas. There are more than 6,000 factories in the huge industrial parks of Manaus, many of which are the production facilities of foreign-owned companies, including Honda, Sharp, Kodak, Olivetti, Toshiba, Sony, and 3M. Manaus was transformed by the number of visitors from elsewhere in Brazil coming to purchase goods that they were not allowed to import into Brazil. The "electronics bazaar" occupies a maze of streets in the city center. People come to buy the foreign-made computers and domestic electronic goods that Brazilian companies, protected by high tariffs, make poorly and sell at high prices.

Amazon Backwater

The main failure of government efforts is in the Amazon River basin. Efforts to settle the lands along the Trans-Amazonica Highway from the 1950s under the regional development ministry, SUDAM (*Supertendencia para o Desenvolvimento da Amazonia*), saw little progress. Land was not surveyed properly, families with little farming experience were recruited, and bureaucratic convenience caused most of the grants to go to large-scale ranches. When a new road, BR 364, was built into the states of Rondônia and Acre in the western Amazon, thousands of settlers entered the area on their own initiative and cut plots in the forest without government encouragement. Dramatic increases in forest clearing and burning aroused international outrage. Many of these settlers failed to grow anything on their plots. The alternative attraction of the rapid development of the soybean farming in the Campo Cerrado in the early 1990s led to a slowing of such settlement along with the linked forest burning and destruction in the western

Figure 10.48 **Brazil: western Mato Grosse.** A space shuttle photo of new cattle ranches and soybean farms in a 60 km² (40 mi.²) area. The plateau surface has been cleared and the steep, floodable intervening valleys left in forest. Lands along the Río Sangue (bottom) are flooded up to 10 m (40 ft.) deep for three months in the summer. Photo: NASA.

Figure 10.49 **Brazil: hydro-electricity.** The spillway at the Itaipu hydro-electricity station on the Paraná River between Brazil and Paraguay. The 18 massive turbines make this Brazil's largest hydroelectricity investment. Although the electricity is shared between the two countries, Paraguay sells nearly all of its quota to Brazil, where there is greater demand. Photo: © Michael Bradshaw.

Amazon (see the "Point-Counterpoint: Tropical Forests and Deforestation" box, p. 434). In early 1998, devastating fires reversed this downward trend by destroying rain forest in the north of the Amazon River basin along the Venezuelan border.

Another aspect of government activity in the Amazon is the Indian rights agency, **FUNAI** *(Fundação Nacional do Indio),* which seeks to provide education, health care, and support for Native Americans. Unfortunately, it is underfunded and, although it has taken some steps to protect its Indian reserves, ranchers and miners frequently penetrate tribal lands and begin to develop them before FUNAI can act. Numbers of indigenous communities in Brazil are diminishing along with the area of the land on which they are sequestered.

Indebtedness and New Policies

Following its huge infrastructure and state-owned enterprise investments of the 1970s and 1980s, Brazil found itself deeply indebted to world banks in the late 1980s, reaching over $120 billion in the 1990s. The cost of oil imports and the huge public works projects to develop iron mining and hydroelectricity, build Brasília, and construct new roads and airports, together with reported political corruption, contributed to this national debt. Despite Brazil's being by far the most prosperous country in Latin America and having a trade balance in which exports were double that of imports until the early 1990s, most of the surplus income went to the payment of debt interest. Defaults on loan repayments caused wider foreign investments in Brazil to collapse. **Hyperinflation,** or the process in which rising prices (and often wages) feed upon themselves and potentially spiral out of control in a vicious cycle, is another financial problem that is an obstacle to further development, since it stifles savings. Hyperinflation may significantly devalue a currency and wreak havoc on a country's economy. Brazilians became used to living in an environment of high inflation. Large areas of living costs were indexed to this condition. In early 1994, inflation reached 50 percent per month, over 1,000 percent per year, but renewed attempts to bring inflation under control marked government financial policy later in the 1990s.

The 1990s also saw the start of new government policies moving away from the import substitution and state ownership of manufacturing that had been major factors in economic growth within Brazil. Tariffs on goods not produced in Brazil were lowered. For example, the 85 percent levels on foreign autos and appliances fell to 20 percent. Some products, including computers, remained illegal imports unless purchased in Manaus. Foreign investment was encouraged, and a sign of these changes is that Japanese car manufacturers are establishing outlets in the major cities.

Brazilian economic diversity places it on a par with some of the newly industrialized countries of East Asia. It also hides the gap between the materially rich and poor within Brazil that is reflected in the uneven distribution of land, the high levels of unemployment, and the strength of the informal economy. Brazil has the capability to develop its economy further, but current economic and social problems and environmental issues make the future difficult to assess.

 # Southern South America

Argentina, Chile, Paraguay, and Uruguay form the southernmost part of South America and are sometimes called the "Southern Cone" because of their combined shape on a map (Figure 10.50). Physically, this long and narrow subregion extends far south into the mid-latitudes. Average temperatures cool and rainfall amounts increase progressing from the north to the south in the subregion. The Hispanic cultural imprint is a strong common element of the human geography of Southern South America. The majority of the population of each country speaks Spanish as their first language and adheres to Roman Catholicism.

Southern South America has the highest percentage of European descendants of the subregions on the South American continent. Sparse indigenous populations and climatic environments more reminiscent of Europe led to the establishment of pervasive European-based populations in Argentina, Chile, and Uruguay. The levels of consumer goods ownership and human development are among the highest in Latin America (see Figure 10.17), although Paraguay does not reach the same levels as the other three countries. Argentina's strong economy experienced a major economic crisis in 2001 and 2002. Although the Argentine economy showed signs of slow rebounding, recovery will take several years. Argentina's population dominates the four countries of the subregion, with more than half of Southern South America's people living here in 2001 (Figure 10.51).

Countries

Chile, Argentina, and Uruguay have relatively strong export-based economies, but Paraguay is landlocked and has more restricted economic opportunities. While Argentina had 37.5 million people in 2001 and Chile just over 15 million, Paraguay had around 5.7 million, and Uruguay 3.4 million (see Figure 10.51).

The diverse physical and human geographic spatial patterns of Southern South America relate in part to the subregion's extensive latitudinal coverage from north to south, a relatively narrow width, the rugged Andes Mountains and geologic instability of the west, and the influences of the cold Pacific and warm Atlantic Ocean currents. All the physical properties of the region support a wide range of temperature and precipitation regimes resulting in varied settlement and land use patterns.

The physical geography of Southern South America coupled with the historic goals of Spain created a vibrant contemporary human mosaic. The Andes Mountains proved an important dividing factor between types of settlement. Under Spanish rule, Chile was governed from Lima. The lands east of the Andes Mountains became part of the Viceroyalty of La Plata and were first settled with the limited objective of providing food and animals for the Potosí silver mines in Bolivia. The whole subregion lagged behind the Northern Andean countries and Brazil through the 1800s. Despite the small amount of economic development on both sides of the Andes, a series of towns and ports were established early in the 1500s as part of the Spanish control network but grew slowly before independence.

Following independence in the early 1800s, Chile became one country, while Argentina at first incorporated what is now modern Paraguay and Uruguay. Paraguay refused to acknowledge Argentinean control, however, and declared its independence in 1811. With British backing, Uruguay also gained its independence from Argentina by the 1820s. Disagreements over borders continue, although Paraguay was subdued after warfare with neighboring countries in the late 1800s gave the country a diminished population and a smaller land area.

Expanding contacts with Europe led to the development of mineral extraction and commercial farming by the end of the 1800s, although Paraguay's political problems and interior location continued to hold it back. Chile's exports were mainly nitrates and copper, Uruguay's were livestock products, and Argentina's livestock and grain. Since the middle of the 1900s, these countries, again with Paraguay lagging, industrialized with movements of people from rural to urban areas.

People

The population of Southern South America is growing more slowly than in the other subregions of Latin America. In 1930, this subregion had nearly 20 percent of Latin America's total population, but it now has less than 12 percent. The 62 million people who lived there in 2001 could rise to around 80 million by 2025. Rates of population growth and fertility are low (see Figure 10.22), apart from Paraguay, where the 2001 natural growth rate was 2.7 percent per year and the total fertility rate was 4.3. Argentina, Chile, and Uruguay had low annual population increase rates around 1 percent, not very different from those in the 1970s. Total fertility rates in 2001 were under 2.5.

Except for Paraguay (52% urban in 2001), all the countries of Southern South America had over 85 percent of their populations living in towns. As in many countries of Latin

Figure 10.51   Southern South America: population, economic, and cultural data.

Country	Capital City	Land Area (km²) Total	Population (millions) Mid-2001 Total	2025 Est.	GNI 1999 (US $ million) Total	GNI PPP 1999 Per Capita	Percent Urban 2001	Human Development Index Rank of 175 Countries	Human Poverty Index Percent of Total Population
Argentine Republic	Buenos Aires	2,766,890	37.5	47.2	276,097	11,940	90	39	
Chile, Republic of	Santiago	756,950	15.4	18.6	69,602	8.410	86	34	4.8
Paraguay, Republic of	Asunción	406,750	5.7	9.7	8,374	4,380	52	84	16.4
Uruguay, Eastern Republic of	Montevideo	177,410	3.4	4.0	20,604	8,750	92	40	4.0

Source: Data from *Population Reference Bureau 2001 Data Sheet; World Development Indicators,* World Bank, 2001; *Human Development Report,* United Nations, 2001; Microsoft Encarta (ethnic group, language, religion).

America, each country has a primate metropolitan center. Buenos Aires grew from a population of 170,000 in 1870 to around 12.5 million people, one-third of the Argentinean total, in 2000. Montevideo, Uruguay (1.2 million), is home to nearly half the national total. Santiago (5.5 million) has one-third of the Chilean total. Asunción (1.2 million) is the largest city in Paraguay with over 20 percent of the country's total. These cities are the centers of government, manufacturing, and service industries.

Few other cities in any of the four countries approach the size or important role of the national primate cities. In Argentina, the second city is Córdoba, and the third is the inland port of Rosario, both with populations of around 1 million. Other centers include towns in the northwestern irrigated areas such as Tucumán and Mendoza, and small ports in Patagonia. Chile has a series of port towns along its length that have hinterlands in mining, farming, or forested areas.

In all these countries, immigration from Europe since the late 1800s was of great significance to population growth. Chile grew in population from colonial times, but the growth of commercial farming and mining in the late 1880s led to more rapid increases, including some 200,000 German immigrants between 1881 and 1930. Growth was slower at first in Argentina and Uruguay, but immigration became more important after independence. Between 1821 and 1932, 6.4 million people emigrated to Argentina from southern and eastern European countries, and another million to Uruguay. Paraguay received few immigrants but lost around 60 percent of its population in its wars with Brazil, Argentina, and Uruguay in the 1860s and sustained further losses in the Chaco War with Bolivia in the 1930s. After such losses, Paraguay also encouraged migration—with only moderate success.

The proportions of immigrant population give different emphases to the racial and ethnic mixes of the Southern South American countries. In Chile, around 40 percent of the population claims to be of European heritage with the remainder considered to be mestizo. In Paraguay, virtually all the people are mestizos apart from a few people of European or African origin. Argentina and Uruguay have the largest proportions of Europeans and the smallest proportions of mixed populations.

Economic Development

In 1991, Argentina, Brazil, Paraguay, and Uruguay formed the trading group of Mercosur following discussions between Argentina and Brazil since 1988. The countries agreed to cut internal tariffs as intraregional trade increased, despite the problems of the high inflation rates in Brazil. The group produces half of Latin America's GDP, and the European Union—Mercosur's main trading partner—is showing interest in negotiating a closer agreement. Chile and Bolivia are associate members.

Argentina, Chile, and Uruguay followed economic development patterns that are somewhat typical of Latin America. Paraguay always lagged behind the others. Until the mid-1900s, Argentina and Uruguay were the most prosperous. Since the 1950s, the Argentine people have been through a series of economic boom-and-bust periods. By the 1990s, the Argentine economy was very robust and the citizens of the country were enjoying a relatively prosperous standard of living. It appeared to the global economic community that Argentina's troubles were far behind it. A series of both local and global actions combined to devastate the economy in Argentina by the end of 2001 and bring the middle classes into the streets of Buenos Aires in protest. The combination of government policies, global trade relations, and government corruption necessitated devaluation of the Argentine currency. The Argentine economy now may be in an incipient stage of recovery, but any positive change is likely to be slow.

Paraguay

Paraguay remains a materially poor country. The larger portion of the country to the west of the Rio Paraguay, the mainly semiarid Gran Chaco, remains lightly settled apart from military camps and lumber operations. The quebracho ("ax-breaker") tree is its commercial timber product. The tree is very hard and in demand for railroad ties and its tannin extract.

The eastern part of Paraguay is more developed, with forest products being diversified by commercial agriculture. Livestock products and some industrial crops are exported. Paraguay gets its power from hydroelectricity and earns foreign exchange from its share of the Itaipu project on the border with Brazil (see

Country	Ethnic Groups (percent)	Languages O=Official	Religions (percent)
Argentine Republic	Mestizo/Native American 15%, European 85%	Spanish (O), English, Italian	Roman Catholic 90%
Chile, Republic of	Native American 3%, European 95%	Spanish	Protestant 11%, Roman Catholic 89%
Paraguay, Republic of	Mestizo 95%	Spanish (O), Guaraní, Portuguese	Protestant 10%, Roman Catholic 90%
Uruguay, Eastern Republic of	Mestizo 8%, European 88%	Spanish	Protestant 4%, Roman Catholic 66%

Figure 10.49) by selling most of its power to Brazil. Transportation connections with Brazil have improved, and that is now the main direction of trade. Paraguayan border towns such as Ciudad del Este sell cheap consumer goods to Brazilians and are involved in smuggling between Brazil and Argentina.

Colonial Legacy

In the other countries, European Hispanic culture was imposed on diverse groups of Native Americans, including the grassland nomads east of the Andes and the peoples of southern Chile, who resisted the Spanish arrival. During the 1500s, Spanish settlement in Southern South America was hesitant. The initial settlement at Buenos Aires was abandoned in 1641 after five years of occupation. By 1600, however, towns in central Chile and northwestern Argentina were centers of agriculture, and roads linked them to more northerly Spanish settlements. The oasis towns in northern Argentina supplied food and mules to the Bolivian silver mines. These towns gradually extended Spanish control over their hinterlands. Montevideo was established to block Portuguese expansion from Brazil toward the La Plata estuary. Following independence, each new government expressed national aspirations to build their capital cities. Economic growth remained slow, however, and a series of wars, mainly focused on Paraguay in the 1860s, also held back development.

Growing Engagement with World Economy

In the late 1800s, Britain and other European countries recognized the potential of the pampas grasslands as a source of grain and livestock for the expanding demand in the industrial Northern Hemisphere countries. By then, refrigerated ships could carry meat products to these countries. British capital built railroads around Buenos Aires and Montevideo and encouraged the owners of large estates to move from extensive livestock management to intensive grain farming.

To cope with such economic expansion, the governments of Argentina and Uruguay in particular encouraged immigrants from Europe to come and work on the estates. Many of the immigrants soon moved into the towns. The diffusion of commercial farming methods led to the final control of the warlike Pampas Native Americans who had kept settlement at bay until

the 1880s. In similar fashion, German immigrants to the southern part of central Chile in the mid-1800s removed the Araucarian Native Americans who had prevented occupation of the land south of the Río Bío Bío. By the early 1900s, the lands around Buenos Aires and Montevideo vied with some European countries in both the quantity and quality of life.

Economic Isolation

The economic depression of the 1930s and World War II in the 1940s shifted these countries away from dependence on the farm and mine products demanded by Europe. The countries of Southern South America introduced a broader base of manufacturing protected by high tariffs to reduce dependence on Europe and the United States. Import substitution was at first successful, but its impact was reduced after 1950 as the United States and Europe reestablished their manufacturing markets with better and cheaper goods, while prices for primary products fell because of overproduction.

The economic problems were compounded by a series of governments, both civilian and military, that tried to maintain their protected industries in order to reduce dependence on primary product exports. Their reliance on import-substitution products tied to the needs of their own country's limited market, however, made it difficult for them to sell these goods abroad the way Brazil, with its huge internal market and diverse range of products, managed to do in selected sectors. Primary products still constitute a large proportion of the exports of all these countries, although they have also been diversified. Manufacturing constitutes over 25 percent of employment and GDP in Chile, Argentina, and Uruguay but under 20 percent in Paraguay.

Chile

In Chile, the dictator Augusto Pinochet's regime that lasted from 1973 to 1989 established an economy based on open trading, increasing exports, good financial management that withstood the fallout of the Mexican collapse in 1995, and low unemployment. After the 1989 democratization of government, the new government followed similar policies with the addition of a greater emphasis on social programs to reduce the numbers of poor people.

At the same time, Chile's economy diversified. Copper, which made up 80 percent of Chile's exports in the 1970s, was down to 44 percent in the late 1980s, although the amounts increased. Agricultural products such as apples, soft fruits, grapes, and wine, together with fish, lumber, and new minerals, grew in proportion. The agricultural products and lumber come from central Chile, where a large proportion of the population lives. Copper mining is important in the Atacama Desert of northern Chile, with the world's two largest mines at Chuquicamata and Escondido along with a series of new large mines being prospected and developed in the late 1990s. After the highest-grade copper deposits were mined out by the early 1900s, Chile used U.S. technology to mine lower-grade ores. Its companies now lead the world in refining technology.

Uruguay

Uruguay remains a largely agricultural country, with meat and wool exports, but the industrial developments in and around Montevideo, powered by hydroelectricity generated along the Rio Negro, draw workers from the rural areas. Over 80 percent of Uruguayan trade passes through Montevideo, which has over half the country's economic product and population. It developed an offshore banking industry. It attracts tourists to its beaches in an industry that contributes more than either farming or manufacturing to GDP, with over 2 million visitors in 1999 (up from 1.2 million in 1990).

Argentina

In 1995, Argentina negotiated a cooperative venture with the United Kingdom to develop the oil potential of the continental shelf area southwest of the Falkland Islands. Although efforts have not produced oil, the area is believed to hold significant deposits. Drilling began in 1998 north of the islands (outside the Argentinean area), but commercial production is many years away.

Argentina has a larger manufacturing base than Chile. The upturn in world economic activity and economic restructuring in the 1990s resulted in more investment, mainly from foreign sources, in Argentinean manufacturing. Argentina became the fastest-growing economy in Latin America as it established what appeared to be a stable financial system.

Most of the Argentinian factories are based in and around the Buenos Aires metropolitan area. The farm products of the pampas are still important to Argentine trade, but oil and gas resources fuel new industries, mainly on the coast. Argentinean service industries developed as part of this growth. In 1999, over 3 million foreign tourists visited centers on the coast and in the mountains. Some of the older settlements, such as the northern cities of Tucumán, Córdoba, and Mendoza, where the economy is still based on irrigation agriculture, do not grow as rapidly because of the dominance of Buenos Aires in attracting manufacturing investment. Some of the inland centers, such as Jujuy near the Bolivian border, remain poor and overwhelmed by Bolivian migrants. In the late 1990s, the Mexican crisis and fears of open competition with Brazil as Mercosur develops led to a slowing of growth in Argentina, which contributed to the economic crisis of 2001.

Buenos Aires, Argentina

The rapid growth of Buenos Aires followed patterns that are common to other cities in Southern South America. After slow growth in its early history, Buenos Aires expanded in the late 1800s as commercial farming took hold in its pampas hinterland. The built environment developed a trading and industrial waterfront and a central thoroughfare at right angles along the route inland. Rapid growth occurred in both the economy and the immigrant population around 1900. It resulted in the middle class of skilled workers and office workers moving out to new suburbs, while the poor and most affluent remained in the inner city. Amenities came slowly to the suburbs. During the later 1900s, increasing rates of population growth produced squatter settlements and rising inner-city population densities as apartment blocks replaced mansions. Population growth stagnated in the coastal industrial areas. From the 1960s, much of the commercial and industrial activity moved out of central Buenos Aries to the city edges. Attempts were made to divert the overcrowding in the Buenos Aires metropolitan area by placing new projects and development in other centers farther inland from Buenos Aires, but they have not been successful.

Test Your Understanding 10E

Summary Brazil is the largest country in Latin America and contains one-third of the region's population. Its economy is in the world's top 10, but extremes of riches and material poverty exist side by side. Brazil consists of contrasting regions: the sparsely settled tropical rain forest of the Amazon River basin, the industrial cities around São Paulo and Rio de Janeiro, the poverty-stricken northeast, the commercially farmed plateau interior around Brasília and Campo Cerrado, and the southern farming and industrial states. Much of Brazil's economy is controlled by the government through import duties and ownership of corporations that are responsible for utilities, encouraging farming developments, and the high-tech sector.

Southern South America (Argentina, Chile, Paraguay, and Uruguay) has midlatitude climatic environments. The Andes form a distinctive boundary between Chile and Argentina. These countries, apart from Paraguay, have greater proportions of European-origin peoples and more diversified economies than other parts of Latin America outside of Brazil, but they suffered from unstable political conditions and protectionism until the 1990s.

Some Questions to Think About

10E.1 What are the main economic uses of the Amazon River basin? How have attempts to settle the area solved the problems faced?

10E.2 How would you summarize the main features of the distribution of population, ethnicity/ancestral heritage, agriculture, and manufacturing in Brazil?

10E.3 Why did the European settlement of Southern South America occur later than in other parts of Latin America? What effect did this have on those countries' human geography?

Key Terms

favela	FUNAI
free trade zone	hyperinflation

Online Learning Center

Making Connections

The Online Learning Center accompanying this textbook provides access to a vast range of further information about each chapter and region covered in this text. Go to www.mhhe.com/bradshaw to discover these useful study aids:

- Self-test questions
- Interactive, map-based exercises to identify key places within each region
- PowerWeb readings for further study
- Links to websites relating to topics in this chapter

Chapter 11

North America

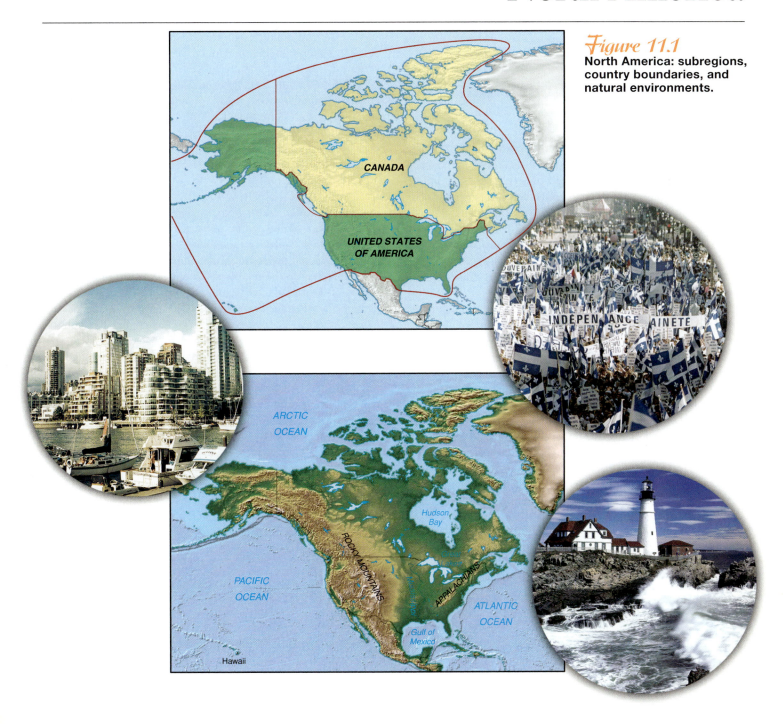

Figure 11.1
North America: subregions, country boundaries, and natural environments.

This Chapter Is About . . .

The intersection of cultures past and present
The wide spectrum of natural environments
The global connections to the region
The material wealth and high living standards of the region
Two subregions:

- United States of America: material wealth, global influence, urbanization, and diversity
- Canada: resource wealth, southern population concentrations, high latitudes, devolutionary pressure and national identity, and living in the shadow of a giant

Point-Counterpoint: North American Free Trade Agreement
Personal View: Canada

Global Prominence and Controversy

North America (Figure 11.1) is part of what Europeans called the "New World" after their explorations in 1492 introduced vast lands west of the Atlantic into the European sphere of influence. The population of the incipient United States of America grew dramatically in the 1700s and 1800s as migrants crossed the Atlantic in search of a new home. By the early 1900s, the region became a new center of globally connected activity, joining the established region of Western Europe. The United States became the world's most important manufacturing country, outpacing both the United Kingdom and Germany. In 1950, the United States produced half of the world's goods, and although the proportion fell as Europe and Japan recovered from wartime destruction, the United States retained its global economic leadership. In 2000, Canada and the United States produced 33 percent of world GNI, Japan produced 14 percent, and France, Germany, Italy, and the United Kingdom together produced 20 percent. These seven countries accounted for two-thirds of the world's total economic output.

The prominent global role thrust on the United States by its political, economic, and military might, and, to a lesser degree, on Canada, often places the region center stage in international controversy. The global influence and connections of the United States are highly sought by many governments, national groups, nonprofit organizations, and corporate leaders. The very same connections and influence are also often met with protest and even violence. Various special interest groups throughout the world disagree with U.S. government policy or intervention in events occurring outside of U.S. borders. Other groups maintain an antagonistic position with respect to U.S. multinational corporations whose external connections exist through foreign operations and investment. Groups who feel threatened by the global prominence of the United States attempt to raise their local voices to the international agenda through protest or demonstration. Canada often finds itself involved in controversy through association with the United States more than through its own policies or actions, which may differ significantly.

Canada and the United States of America are two of the world's largest countries in area. Canada is second after the Russian Federation and the United States is fourth after China (Figure 11.2). In population, however, both the United States with 285 million in 2002 and Canada with 31 million are substantially smaller than global population giants China and India, each with more than 1 billion residents. Thus, the huge output of Canadian and U.S. industries comes from just 5 percent of the world's population. This makes the majority of the people who live in the region extremely affluent relative to global averages.

North America, where English is the language of the majority, is often referred to as Anglo-America, setting the region apart from Latin America, where Spanish and Portuguese (in Brazil) are dominant. Anglo-America is a challenging regional **toponym,** or place name, for several reasons. First, referring to the region as "Anglo-America" erases thousands of years of indigenous history that existed before European colonialism. Additionally, minority languages in North America are increasingly significant, and the cultural influence from various world regions is chipping away at a mainstream "Anglo" society. The increasing percentages of non-English-as-a-first-language residents, the intensity of their linguistic and national pride, and the cultural contribution various groups make to the larger North American composite provide strong arguments against labeling the region "Anglo-America." The French-speaking minority in Canada (25% of the Canadian population) constitutes a formidable force in Canadian government policy, sparking one of Canada's largest domestic controversies. In the United States, the number of Spanish-speaking residents (now the largest

Figure 11.2 **Key indicators for North American subregions.**

Subregion	Capital City	Land Area (km^2) Total	Population (millions) Mid-2001 Total	2025 Est.	GNI 1999 (US $ million) Total	GNI PPP 1999 Per Capita	Percent Urban 2001	Human Development Index Rank of 175 Countries	Human Poverty Index Percent of Total Population
Canada	Ottawa	9,976,140	31.0	36.0	614,003	25,440	78	1	12.0
United States of America	Washington, D.C.	9,809,460	284.5	346.0	8,879,500	31,910	75	3	16.5

Source: Data from *Population Reference Bureau 2001 Data Sheet*; *World Development Indicators,* World Bank, 2001; *Human Development Report,* United Nations, 2001; Microsoft Encarta (ethnic group, language, religion).

minority) continues to increase as immigrant flows from Latin America intensify. Although Mexico has formed alliances with Canada and the United States and continues to influence the human geography of the region significantly, it has greatest cultural affinities with the other countries of Latin America that were colonized by Spain in the 1500s. Mexico will be discussed briefly in this chapter where relevant and was emphasized in greater detail in Chapter 10.

Canada is linked closely to the United States in many ways but asserts its individuality through different global connections and distinctive internal policies. Although the two countries share many commonalities, the Canadian people have a strong sense of national pride, which has thus far prevented their Canadian identity from being overwhelmed by their giant neighbor, the United States of America.

This chapter begins by examining the ways in which European cultures took over the land from indigenous peoples and set the stage for the human and economic development that created the diverse and connected countries of Canada and the United States of America. Next, the natural environmental conditions of both countries are detailed, leading into a discussion of issues stemming from human-environment interaction in the region. Finally, the global connections and economic prominence of the region are outlined prior to a detailed examination of the chapter's two subregions: the United States of America and Canada.

🌐 Regional Culture History

Cultural Intersection and Emergence

The United States and Canada are intricately connected to every other world region. The significance of the economic and political ties they share today with the other countries of the world is partly explained by the ways in which people at different stages in their historical development interacted with each other, the natural environment, and the developing world economy. Although archeologists believe Viking explorer Leif Erickson reached the eastern shores of present-day Canada some 500 years before Columbian expeditions crossed the Atlantic, the Americas, North and South, were virtually unknown to the rest of the world before Columbus's sailings beginning in 1492. The indigenous peoples of the Americas lived in societies based mainly on subsistence and the local communal ownership of resources. Global connections began with the exploration of the region by various European interests. The indigenous Americans throughout the region were overwhelmed by the **cultural hegemony,** or forced conversion of their social customs to the vastly different technologies and social tenets of the Europeans.

In the three centuries after the European "discovery" of the region, North America became a series of colonies and occupied territories governed by the French, Spanish, British, Dutch, Russians, and Swedish. The British became the dominant power by the mid-1700s. Almost immediately, settlers south of the St. Lawrence River valley fought to become the independent United States of America by signing the Declaration of Independence in 1776, and forcing Britain to recognize U.S. sovereignty.

Canada arose mostly out of the colonial territories first established along the St. Lawrence and the Great Lakes and remained a British colony far longer than the United States. It achieved a degree of independence through the British North America Act in 1867, while maintaining legal ties to Britain until 1982, when it gained full control of its constitution. Today, Canada is a fully independent country that enjoys membership in the Commonwealth of Nations.

Native Americans

Indigenous groups of people, who are referred to today as **Native Americans** (**First Nations** in Canada), inhabited North America for centuries before 1500. The first Americans probably migrated from Siberia to Alaska over 20,000 years ago. By 1500, they lived in hierarchically structured ethnic groups, commonly called tribes, adapted in culture and distribution to the physical conditions of terrain and climate. The combination of natural resources, environmental conditions, and a sedentary settlement structure enabled groups such as the Mississippians in present-day southwestern Illinois, to flourish for centuries. The groups in the eastern forests raised corn, beans, and squash, hunted and fished, and lived in medium to large villages. Those in the warmer lower Mississippi Valley used a surplus from farming to support urban development such as Natchez. Farther west, the prairie environment on the plains between the Mississippi River and the Rocky Mountains was inhabited by hunting groups, such as the Dakota, who killed bison to supply their food, clothing, and shelter needs. Their tepee tents could

Ethnic Groups (percent)	Languages O=Official	Religions (percent)
European (British 40%, French 27%, other 20%)	English (O), French (O)	Protestant 26%, Roman Catholic 46%
European 83,%, African 12%, Asian 3%, Native 1%	English, Spanish.	Protestant 56%, Roman Catholic 28%

be moved to follow the bison herds. The introduction of horses by Spanish colonists facilitated hunting, travel, and communications for many western hunting-based societies, thus enabling them to grow in both significance and numbers of people.

Tribal numbers were even sparser farther west in the more arid parts of the mountains and high plateaus and farther north in the colder parts of northern forests and Arctic lands. Some locations in the arid southwest favored the development of irrigation farming and the building of distinctive villages by the pueblo cultures (Figure 11.3). Along the Pacific coast from present-day northern California to southern Alaska, small tribes found plentiful seafood, including salmon and tuna. They carved the tall trees into dugout canoes, plank houses, and totem poles. The Arctic and north Pacific coasts of the continent supported small numbers of Aleuts and Inuits, who hunted and fished the nearby ocean waters.

After the arrival of the Europeans, many Native Americans were killed by the introduction of diseases to which they had no immunity. The survivors were increasingly pushed to marginal lands of the region. This process began in the east and southwest in the 1600s and lasted into the 1800s in the central and northwestern parts of North America. Although many Native American trails evolved into major travel routes for the regional settlers, the indigenous groups had much less of an impact on the evolution of the regional landscapes once the Europeans began their dramatic land use transformations.

Figure 11.3   **North America: Native American settlement.** The Taos pueblo, New Mexico. Such settlements reflected the local climate and materials available. The mud brick houses are typical of arid areas. The beehive ovens (in the foreground) are used for cooking. Winters are often cold, but snowfalls are infrequent. Photo: © Thomas Ives/The Stock Market/Corbis.

European Settlers

Several European countries began to settle the region as early as the 1500s, including Britain, France, the Netherlands, Russia, Spain, and Sweden.

- The French settled the mouth of the St. Lawrence River in present-day Canada from the early 1500s. They established river-related farm settlements along the banks of the St. Lawrence valley. Others explored the interior, helped by Native Americans, to hunt for beaver and other furs. The French traded in and claimed lands through the Great Lakes and along the Mississippi River.

- The Spanish settled parts of present-day Colorado, California, Arizona, New Mexico, Texas, Florida, and the Carolinas from the late 1500s. Spanish territory was known as the colony of New Spain, centered on the area of modern Mexico, during the colonial era. When areas did not deliver the gold that they wanted, the Spanish lost interest in them and left much of the land to be cared for by Roman Catholic missions or to be used for cattle ranches.

- The British first targeted the tidewater lands surrounding the Chesapeake Bay for colonial development. In 1607, they established their first permanent settlement of Jamestown in present-day Virginia, after an earlier failed attempt on Roanoke Island (in present-day North Carolina). The tidewater settlers found tobacco to be the lucrative commercial crop needed to fund the settlement and satisfy investors in England. The local climate and soils were well suited for tobacco cultivation. The settlers also adopted Native American crops for subsistence. Sales of tobacco back in England guaranteed the future of the settlement, attracted more settlers, and led the British government to take over the lands as a crown colony. This early success led to the production on individual plantations of an increasing range of commercial crops, including indigo, rice, sugarcane, and, later, cotton. The plantations and colonies spread southward along the Atlantic coastal plain and in 1619 began importing Africans to do the fieldwork. Slavery became a significant institution by the 1700s, which fueled the development and expansion of agriculture south and westward from the Mid-Atlantic coast.

- Religious freedom-seeking migrants from England on their way to Virginia ignited a second wave of British settlers in 1620. They landed at Cape Cod in present-day Massachusetts and moved across the bay to Plymouth, beginning the settlement of the region that became known as New England. Other groups followed and established a town (township)-based settlement pattern founded on self-governing villages and a subsistence economy. New England's community orientation contrasted with the plantations of the southern settlements.

- In the 1630s, the Dutch were the first to settle in the Middle Atlantic between New England and Virginia around their port town of New Amsterdam (modern New York City). Swedish settlers took small areas along the lower Delaware River but were quickly ousted by the Dutch. Next, the British drove the Dutch from their settlements in the 1660s. The British Duke of York received charge of the Dutch lands and gave his name to the largest city, changing it from New Amsterdam to New York. He paid off his war and land acquisition debts by selling most of his remaining lands to William Penn, the founder of Pennsylvania. This Middle Atlantic area attracted settlers who wished to set up family homes with the least amount of external control in a new country. Many Scots-Irish and Germans came to the region in the early 1700s, establishing a farming system based on growing corn and raising livestock, and spreading it southward along the Appalachian valleys.

After independence in 1783, the three areas of settlements along the Atlantic coast with their different economic and social systems formed springboards for thousands who moved westward within the United States as new lands were acquired. The New England subculture diffused into areas around the Great Lakes. The Pennsylvanian system of individual family farms became the basis of farming in the southern Midwest. The plantation system first established in the Virginia tidewater, along with the accompanying use of slaves, was extended south and then southwestward along the Gulf lowlands to Texas. The North-South tensions that came to a head in the Civil War of the 1860s arose from the spatial diffusion of these early differences.

Wealth of Natural Resources

The early settlers of North America discovered a wealth of natural resources, including seemingly limitless land, good soils, minerals, fur-bearing animals, fish, and timber. These provided a solid foundation for the settlers to establish strong commercial farming and industries.

Primary Products and New Lands

At first, in the 1600s and 1700s, the wealth was unearthed by an agricultural society. The long, warm summers of the widening coastal plain southward from New York encouraged the commercial farming of subtropical crops such as tobacco, sugarcane, rice, and indigo, and, later, cotton. The growing slave trade provided an abundant source of relatively low-cost labor. Slave labor coupled with good climate and soil conditions enabled planters to sell crops to Europe, make payments to investors, and pocket or reinvest the proceeds. New England's timber and pitch provided new ships for the British navy. The coastal harvest of fish in Canada and New England was followed by an inland harvest of animal furs, which further diversified developing patterns of global trade.

After establishing complete independence in 1783, the United States tripled its area by 1850, buying land from France and Spain and acquiring the western third by negotiation and military conquest (Figure 11.4). To ensure the rapid occupation of these new areas, land was surveyed and sold by the U.S. government at lower and lower prices. The Homestead Act of 1862 provided families with very inexpensive or even free farmland, although much land near the railroads was sold by the railroad companies to pay for the track, so homesteading land was often less accessible and poorer in quality. Settlers spread to the vast interior plains. Many found that although the land was free, there were numerous costs associated with the move and establishment of farms. Families able to overcome the expense of starting farms in the interior took advantage of the fertile soils and the warm, moist summer climate, a combination that proved ideal for growing corn and wheat and raising cattle and pigs. By the mid-1800s, the United States exported a range of farm produce to Europe, competing directly with European farmers in bulk grain markets as well as continuing to export crops such as cotton and tobacco that could not be grown in northern Europe. Later, the range of climates and area of land within the North America region made it possible to grow a greater diversity of farm products on a much larger scale than any one country in Europe could manage.

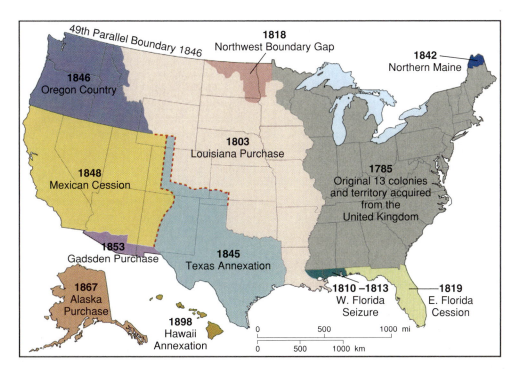

Figure 11.4 **United States: rapid expansion of territory, mainly between 1800 and 1850.** Lands first occupied by Native Americans and later assumed by European colonial Empires, were gradually incorporated into the expanding United States. The eastern lands were acquired from the United Kingdom at independence in 1783. The Louisiana Purchase bought lands that France had just won back from Spain. Texas claimed independence after ousting the Mexicans in the early 1830s but then joined the United States in 1845. The western lands of the conterminous United States were added by 1850, including the southwestern lands acquired through the 1848 Mexican Cession after a rapid U.S. military victory. Alaska and Hawaii came later.

Slower Canadian Changes

The settled region along both sides of the St. Lawrence River and estuary (part of modern Canada) was a materially poor and often environmentally hostile remnant left to British authority after the United States declared its independence. British loyalists who voluntarily emigrated or were forced out of the incipient United States joined small numbers of Native Americans and French-speaking European settlers. Since the French settlers along the lower St. Lawrence resented British rule, the loyalists mostly settled on the east coast or farther inland around the northern and western shores of Lake Ontario. Westward of the Great Lakes there was little settlement in an area still largely administered by the Hudson's Bay Company. The interior remained the realm of Native Americans and isolated groups of French settlers attempting to escape the watchful eye of British rule. When the Hudson's Bay Company fur trappers began to compete with Americans for land along the lower Columbia River near the coast of the Pacific, another British-American war appeared possible. Negotiations in 1846, however, set the boundary between the two countries west of Lake Superior at the 49th parallel.

Until the middle of the 1900s, the natural resources of Canada were primarily comprised of Atlantic and Pacific marine life, the good soils and temperate climate of southern Ontario, the vast prairies opened to farming settlement in the early 1900s, and the timber resources of the St. Lawrence River valley and British Columbia. In the post-World War II era, rich stocks of metal ores in Canada's north; coal, oil, and natural gas in Alberta; and hydroelectric power generation in northern Québec ignited new geographic directions of development and provided the basis for growing Canadian affluence.

Resources for U.S. Manufacturing

The first Industrial Revolution began in Western Europe (see Chapter 3), coinciding with the United States' independence from Great Britain. Independence and British hostility motivated Americans to establish their own manufacturing industries, providing goods that were at first not obtainable from Britain. Textiles, metal goods, and leather goods were among the first industries to develop, often powered by water mills in New England, New Jersey, and eastern Pennsylvania. By 1850, the coal in eastern Pennsylvania and around Pittsburgh replaced charcoal in iron making and, later, steel industries that set off further industrial revolutions. The northeastern United States soon emerged as a dominant region of the country, housing the bulk of the population with a growing material wealth and related power based on manufacturing industries.

The primary symbol of the early steel-based Industrial Revolution—the railroad—spread across the country to the Pacific Coast. The Manufacturing Belt of the northeastern United States initially developed based on the regional resources of coal, iron, and the agricultural produce of the Midwest. The Manufacturing Belt later drew in mineral resources from other parts of the country as railroads linked the mines of the western mountains to the eastern economic and political core. The mining of gold, silver, and then copper and zinc in the new western lands provided capital for further development and widened the scope of the late-1800s metal industries.

In the 1900s, the availability of such natural resources, the taming of rivers for irrigation, navigation, and hydroelectricity, and the discoveries of oil and natural gas drew more people and manufacturing corporations southward and westward from the country's core. Although the emphasis of most Americans with regard to natural resources was on the potential for economic gain, influential groups from the late 1800s persuaded the federal government to set aside large areas of attractive wilderness as national parks, often before much settlement had occurred. Most of these are in the West, but smaller areas of the eastern lands have also been purchased and designated as national parks (Figure 11.5).

Human Resources

The variety and richness of the natural resources—such as minerals, climate, and soils—available within a territory as expansive as North America combined with a diverse stream of immigrants to build a framework of exploration, expansion, and development.

New People, New Skills

Immigrants came initially from northwest Europe. They brought European cultures, capitalist outlooks, and individual skills and initiative. Many soon established businesses that prospered in the political, economic, and social climate fostered by the independence of the United States. Tensions among Protestants, Catholics, and Jews were worked out in the context of economic competition and residential congregations of like-minded people. There was ample space within the developing United States for a variety of ethnically, nationally, and religiously diverse groups to establish themselves. Africans forced to migrate to the region to serve primarily as slaves in the agriculture industry were not among early immigrants who had any real opportunities or future.

Education and Technology

The new United States sought to develop conditions in which its whole population could flourish by making the most of their freedoms under the democratic Constitution. It established compulsory education much earlier than European countries. Its emphasis on technology transfer began with the late 1800s land grants for establishing engineering and farming universities in many states. Such provision ensured that Americans could not only devise innovative technologies but also widely apply them.

Management of Manufacturing

The manufacturing structures and processes in which the developing United States excelled quickly brought it prominence in world economic activity by the late 1800s. The expanding internal market of the United States provided the demands that stimulated many of the developments.

- First, Americans achieved **economies of scale** by building larger factories for increased output, together with access to larger markets for their growing range of products. Such economies cut the cost of individual items by manufacturing large numbers of each item in a single factory.

- Second, **horizontal integration** occurred when financiers bought up several producers of the same product, giving the new owners a large share of the total production of goods and enabling them to set market prices.

- Third, **vertical integration** took this process further by bringing together, in a single corporation, the producers of inputs to a product and the users of the same product once completed. As an example of these processes, Andrew Carnegie built larger steel mills to gain economies of scale; then he bought out other steelmakers in a time of economic recession in the 1870s to integrate the steel industry horizontally. In the next phase of this vertical integration scheme, he purchased both coal and iron mines that provided the raw materials and the heavy engineering corporations that used the steel. By 1900, Carnegie's United States Steel Corporation dominated the industry.

- Fourth, Henry Ford took the concept of the factory **production line** to a new level. He applied it to the assembly of a wide range of components in the production of automobiles. His company produced thousands of a limited number of models for a growing market. His methods, called **Fordism,** were widely adopted by other industries and in other countries.

- Fifth, the making of machine tools to produce machines that manufacture and assemble parts was another feature

that made the United States the world leader in output of manufactured goods by the mid-1900s.

Canada Emerges

Not until the late 1800s did Canada begin the process of becoming a major industrial country. British rule that suppressed internal development, as well as a lack of capital and expertise, hindered the creation and expansion of Canada's global connections. After the British North America Act of 1867, Canada became a dominion within the British Empire with increased responsibility for its own affairs. At that stage, Canadian leaders began to integrate their vast country by building transcontinental railroads to encourage the settlement of the prairie grasslands and the Pacific coast. Manufacturing industries, including wood processing and steelmaking, were established behind tariff walls to protect them from American competition in particular.

The Contemporary Human Mosaic

Patterns of Population Change in the United States

The relatively rapid annual population increase in the United States is unusual among the world's materially wealthy and technologically advanced countries. The U.S. population in 2001 of 285 million could rise to around 340 million by 2025. In 2000, the U.S. fertility rate dropped to 2.1 and the birth rate to 15, while the death rate was 9.

The continuation of relatively high rates of population increase for the United States is primarily a result of immigration, which has been vital to the country's economic growth since the days of European exploration. Immigration currently accounts for a third of U.S. population growth. Following the early domination by people of British origin, Irish and German people immigrated in greater numbers during the middle of the 1800s (Figure 11.6). In the later 1800s, the highest numbers of immigrants came from southern Europe, particularly Italy, and at the turn of the century, large numbers came from the Slavic countries of Eastern Europe, the Balkans, and Russia.

The African American population increased by forced immigration in the 1700s and by natural increase in the 1800s (Figure 11.6). Slaves were freed in the 1860s, but African Americans continued to have higher birth rates and death rates than the European-origin population.

From the mid-1900s, the Hispanic peoples from Middle and South America, and Asians became major sources of immigrants. The Chinese, who came to the United States as laborers in the late 1800s, and the Japanese, who had moved in as farmers in the early 1900s, were later joined by other Asian groups from 1960, particularly from Vietnam, Korea, and India. Most recently, influxes of Africans and the 1990s immigration of Russians following the demise of the Soviet Union further diversified the cosmopolitan American population.

The United States faced a dilemma during the 1900s as to whether it should encourage immigration as a continuing feature of its dynamic society or whether it should discourage it with the perceived objective of preserving the standard of living of its people. Many European countries adopted the latter approach, but they now experience shortages of people in the working age groups (see Chapter 3). Immigrant groups often maintain a younger age for the working population and have higher birth rates than segments of the population established generations earlier. Beginning in the 1920s, U.S. immigration laws gave first preference to Europeans, but from the 1960s, they opened access more widely on the basis of family ties or specific skills. Today, however, several million illegal immigrants, or "undocumented aliens," mainly of Hispanic origin, live in the United States. An examination of immigration trends throughout the development of the United States indicates that immigrants respond more to boom or recession in the United States economy than to legislation.

Increasing numbers of persons living in countries outside of North America apply for immigrant status each year. Democratic freedoms and the attractive quality of life in the world's most affluent region are strong pull factors creating migration streams from all regions of Earth. Total numbers of legal immigrants and refugees living in the United States—the latter being given special status for a limited period because of adverse political conditions in their own country—topped 800,000 by 1997. Illegal immigrant totals were estimated to be well over 1 million. Overall, immigration helped to maintain a U.S. population growth of around 1 percent per year, compared to increases of under 0.5 percent in many countries of Europe and Japan.

Contemporary Immigration Patterns in the United States

The contemporary immigration picture for the United States is extremely diverse. Immigrants today come from all corners of Earth for a variety of reasons. Immigrant residence and ethnic diversity in the United States have established distinctive spatial patterns throughout the county (Figure 11.7). Many highly educated people migrate to the United States to fill demand in computing, high-tech, medical research, and many of the hard

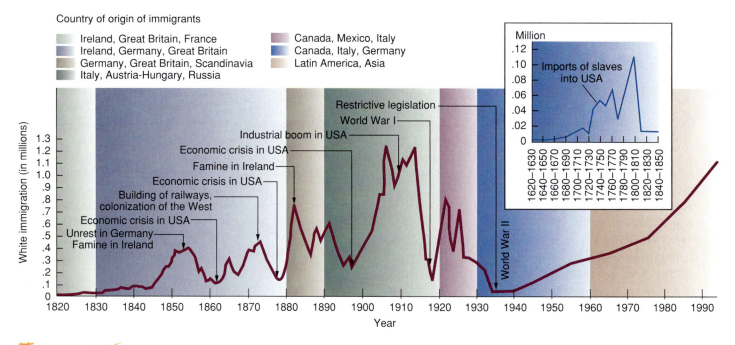

Figure 11.6 **United States: pattern of immigration.** The effect of push and pull factors on immigration fluctuations and the changing geographic sources of immigrants. Compare the size of forced slave migrants to the levels of white immigration up to the mid-1800s. Source: Data from National Academy of Sciences.

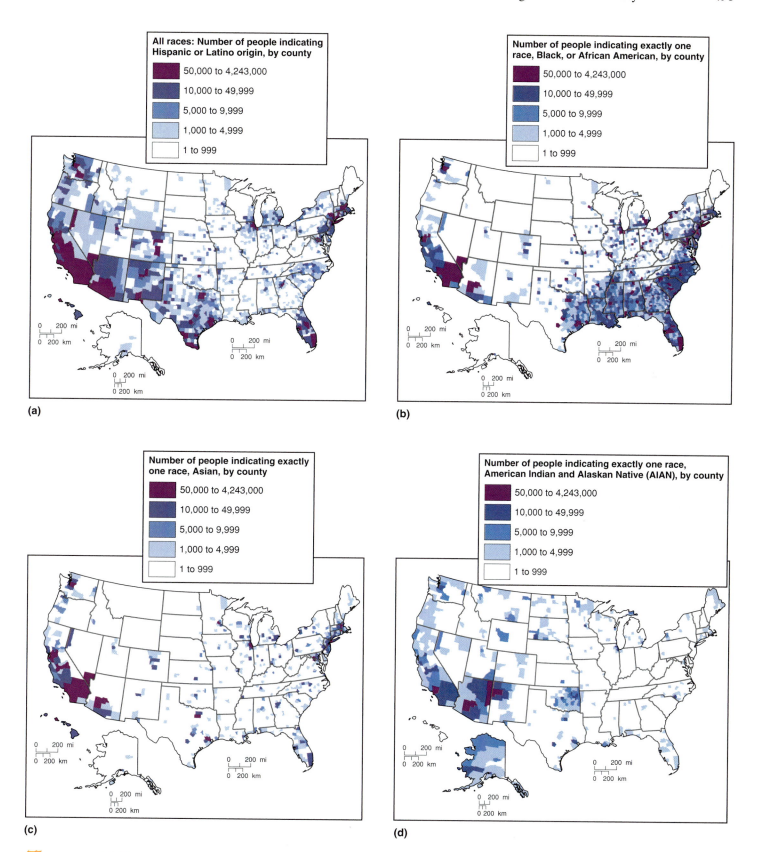

Figure 11.7   Ethnic distribution in the United States. (a) Primary concentrations of Latin Americans include California and the Southwest, Florida, Chicago, and the cities of the Megalopolis. (b) Primary African American concentrations include California, the Southeast from east Texas through North Carolina, Florida, major cities of the Great Lakes, and cities of the Megalopolis. (c) Primary concentrations of Asian Americans include the Pacific Coast, Chicago, and cities of the Megalopolis. (d) Primary concentrations of Native American and Alaskan Native peoples include Alaska, California, Arizona, New Mexico, and Oklahoma. Source: Data from U.S. Census Bureau, Census 2000 Redistricting Data (PL 94-171) Summary File. Cartography: Population Division U.S. Census Bureau. American FactFinder at factfinder.census.gov provides census data and mapping tools.

sciences. Other immigrants come to the United States with no material wealth or education in search of a life different from the environment in their country of origin. Impoverished immigrants often take low-wage positions in agriculture or as housekeeping or maintenance staff in hotels, restaurants, and retail business.

The 2000 federal census counted 281 million people in the United States. Of these, almost 200 million were of European origin, 34.6 million (12.3%) were African Americans, 2.5 million (0.9%) were Native Americans, and just over 10 million (3.6%) were Asians. Some 35 million (12.5%) were Hispanics, an ethnic designation based mainly on speaking Spanish.

Over half of the African Americans in the United States live in the South in both rural and urban areas. The majority of the remaining African American population is clustered in northern and western metropolitan centers. The African American percentage of the total U.S. population is diminishing as the percentages for Hispanic, Asian American and other groups continue to increase.

In the United States, Spanish is used as a home language by 5 percent of the total population of the country. Spanish-speaking peoples are located mainly in the states bordering the Latin American region, including Florida, Texas, New Mexico, Arizona, and California; most large cities of the northeastern part of the United States and in the expansive urban reaches of metropolitan Chicago. The Hispanic population is the fastest growing ethnic group within the United States, and Hispanic Americans are emerging as the largest ethnic minority within its borders. The Hispanic population in the United States grew 58 percent during the 1990s, to 35 million.

Increasing immigration of people from many Asian countries further intensifies the ethnic mosaic of the United States. Asian immigrant settlement patterns are in concentrations along the West Coast of the United States, especially in the dynamic urban centers. There are also large concentrations of Asian Americans in parts of the upper Midwest and in large cities in the eastern United States. The climate and ecosystems of the Gulf Coast region attracted and now house a significant population of people who have emigrated from Vietnam. Vietnamese Americans living along the Gulf coast engage in agricultural and fishing practices similar to those of Vietnam.

The movements of so many different groups of people into and around the United States gave rise to two images of the resultant population mixture. Up to around 1950, immigrants were subject to pressures to conform and be Americanized—the "melting pot" concept. In the later 1900s, greater roles were given to African Americans, Native Americans, and recent immigrants from Latin America and Asia, while the ethnic consciousness of various European groups was revived. The changes led to a new image—the "salad bowl"—in which the variety of contributions is emphasized rather than pressure for conformity or homogeneity.

The First Nation

In the 1970s, the Native American indigenous groups began to increase their proportion of the U.S. population for the first time since a census was taken in 1790. Expected to develop indigenous lifestyles on reservation lands that were often arid and hostile for agricultural productivity, most Native American groups had remained relatively materially poor. Further complicating the reservation lifestyle was an administration structure imposed by and monitored directly from the federal government rather than state or local levels. By the 1970s, some 50 percent of Native Americans were unemployed and 90 percent were on welfare.

Attempts were made in the later 1900s to reduce the poverty imposed on Native Americans who had been previously forced to live on marginal and largely unwanted lands. Helped by the clarification of their rights in Alaska arising from statehood and the selection of state lands from the federal holdings, Native Americans elsewhere in the United States raised questions about the ownership and management of natural resources on their reservations, including minerals and water. They filed lawsuits seeking compensation for being placed on such resource-deficient land. Lucrative deals with governments and water projects, however, did little to change the economic structure of the Native American reservations.

New enterprises brought limited satisfaction. Some Native Americans developed tourist facilities on their reservations, but others chose to forgo such opportunities—often to avoid copying the commercial practices of white Americans. Perhaps the most important source of income to many Native American groups is the series of casino gambling complexes in various states. The main problems are managing the influx of capital and the discouragement of the low-wage jobs that are provided. Casino-related challenges also include an increasing dependence on a single employment and revenue stream, and social ills that often accompany a gaming-based culture. While casino-related development on reservations remains controversial, such enterprises may encourage Native American groups to engage with the wider United States economy and to begin to overcome the isolation and defensiveness they have toward the federal government. To develop their water, minerals, and tourist opportunities, Native Americans need to agree on how to administer and develop the lands that are held in trust for them.

People on the Move: Internal Migration in the United States

The dynamics of population geography in the United States involve internal migrations as well as international immigration (Figure 11.8a). During the 1800s, the main internal migrations were from the Atlantic seaboard westward to the interior and then on to the West Coast. Within the South, the plantation economy was transferred westward from the Atlantic plain to the Mississippi Delta area, taking with it planters and slaves. The northern (mainly white) and southern (mainly black) streams remained distinct from each other during the 1800s. After the Civil War in the 1860s, many African Americans were able to migrate independently, creating a small stream of northward movement that remained relatively insignificant until the twentieth century. The majority of African Americans remained working on southern farmland for several decades following the Civil War.

From the early to mid-1900s, large numbers of African Americans moved from the rural and urban South to northern cities. Those from the Atlantic coastal plain mostly moved to Washington, D.C., Baltimore, Philadelphia, New York, and Boston; those from the Mississippi Delta moved primarily to

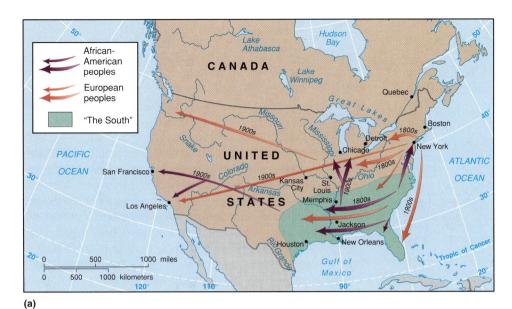

(a)

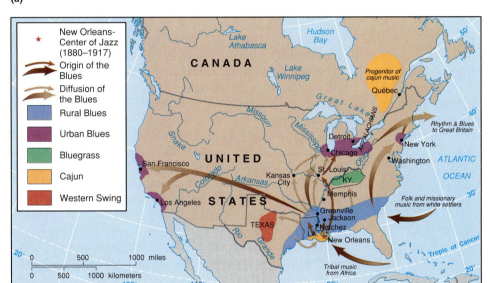

(b)

Figure 11.8 **United States: internal migrations.** (a) The major movements of African and European Americans in the 1800s and 1900s. (b) Compare the developments in blues music and their diffusion to the movement of African Americans.

Cleveland, Cincinnati, Indianapolis, and Chicago. Smaller numbers of African Americans moved to the West Coast. In 1900, nearly 90 percent of African Americans lived in the South, but this proportion declined to just over half by the 1990s. The movements also stimulated the dissemination of cultural features of African American society, such as the varied types of blues music (Figure 11.8b).

Most of the movements of African Americans out of the South were completed by 1970. After that, smaller return movements, often on retirement, balanced or exceeded the northward flow. Movements of white people from the Northeast to both the West Coast and the South continued into the 1980s, but recessions in that decade slowed southern economic growth and reduced such migrations.

Canadian Patterns of Ethnic Integration

Government policy protecting multiculturalism and well-known high living standards combine to make Canada an

extremely attractive destination for peoples emigrating from their homelands. The Canadian federal government instituted policies outlawing ethnic discrimination and protecting the rights of immigrants. For several decades, Canada has been one of the leading immigrant-receiving countries of the world. Contemporary immigration trends for Canada exhibit significant increases in the numbers of people migrating from Asia and the Americas (including the Caribbean) and an overall percentage decrease in the European immigrant population.

The large urban centers of Toronto, Vancouver, and Montréal are experiencing dynamic cultural processes that increase the diversity of each metropolitan area. Canada's largest city, Toronto, attracts more immigrants from all reaches of Earth than any other Canadian city. Asian migration is most notable along the west coast of the country. The city of Vancouver and its hinterlands in the province of British Columbia are the hosts to two types of immigrant from Asia, those who settle year-round and those who use the city and environs as a second home or home for their families while they "commute" to work in Hong Kong.

Centrifugal Forces at Play: The Challenge of Québec and the Rights of Indigenous Canadians

The greatest ethnicity-related challenge facing the Canadian federal government and the unity of the country is the devolutionary pressure from an increasing movement of French-speaking Canadians toward greater autonomy. Approximately 25 percent of Canadians are French-speaking descendants of the French settlers who came to the area during the earliest stages of European development in the region. Most French Canadians are distributed along the St. Lawrence River, primarily in the province of Québec. The **Francophones,** or French-speaking Canadians, have an extremely strong sense of being a separate national group within the larger Canadian frame. They see themselves as having a different ethnic history, different religion, and different culture than the English-speaking Canadian majority (Figure 11.9).

Although the majority of the French descendants adhere to Catholicism, which is in contrast to the primarily Protestant English Canadians, Québec's Francophones see their French language as the strongest symbol they have for discerning their uniqueness from the English-speaking Canadian majority. The perception of national identity loss was enhanced by the choices new immigrants were making in metropolitan Montréal, Québec's largest city and the largest French-speaking city outside of France. Immigrants settling in Montréal, who spoke neither English nor French as their first language, were choosing to learn English rather than French and were choosing to conduct business in English and to send their children to private English-language schools.

The Québec separatist movement gained momentum during the 1980s and became especially strong during the 1990s. Decades of perceived oppression from the federal government as well as the perception of a threat to their cultural identity and language have unified members of the **Parti Québécois,** a political party formed with the sole purpose of achieving Québec's independence from the federal government. The Parti Québécois would like nothing less than an independent country of Québec. A 1995 referendum pushed for by the Parti Québécois resulted in an almost 50 percent vote by the residents of Québec province in favor of separation.

Several groups inside the provincial borders of Québec do not share in the enthusiasm for separation. English-speaking and other business owners fearing revenue losses are adamantly opposed to independence. The more recent diverse immigrant population residing in the metropolitan Montréal area also wishes to remain part of Canada. The most serious internal challenge to the Francophones' movement toward independence comes from the indigenous Canadian peoples known as the Cree, who claim the northern half of Québec as their ancestral land. The Cree do not wish to be part of an independent Québec and threaten to secede from any newly formed Québec state and remain with Canada. The northern half of Québec is the site of significant natural resource potential. Most notably, the generation of hydroelectric power is so successful in northern Québec that the province is able to export energy to the United States and generate profitable revenue streams.

The Québec issue raises questions of whether the varied regions, separated by great overland distances, can continue to provide a complementary unity or will embark on a course that will tear the country apart. Other provinces in Canada have grown weary of the federal government's attempts to appease the provincial government of Québec. The provincial governments of the Prairie Provinces have successfully lobbied the federal government for greater control over their resources and decisions using the perceived "special treatment" given to Québec as a bargaining tool.

Indigenous Canadian issues throughout all Canadian provinces and territories are important for federal government consideration in addition to the situation in Québec. In 1973, the Canadian government opened the possibility of negotiating land claims with organizations representing native peoples. Until that date, little had been done in much of Canada to implement treaties negotiated with native peoples in the 1800s. Now several areas are identified for a

Figure 11.9 **National Pride in Québec, Canada.** French Canadians proudly wave the Québec flag and independence signs during a parade in Montréal. Photo: © Reuters NewMedia, Inc./Corbis.

Test Your Understanding 11A

Summary North America is the world's most affluent region, producing nearly 30 percent of the world's goods and services. The United States and Canada are two of the largest countries on Earth in terms of area, with approximately 15 percent of total land, yet together they house only 5 percent of the global population.

European settlers displaced indigenous groups in North America from the 1500s to the 1900s. Native groups in both Canada and the United States today continue to work toward greater recognition and equality. Although both countries are politically very stable, many French-speaking residents of the province of Québec desire separation from the Canadian federal government.

Questions to Think About

11A.1 What role did Europeans play in shaping the human geography of North America? What impact did European decisions play in shaping current human geography patterns for Native and African Americans?

11A.2 How have immigration patterns changed during the development of the United States and Canada? Which countries were major source areas of immigrants historically, and which are major source areas today?

11A.3 How might future decisions of French-speaking Canadians affect indigenous groups in Québec and the political stability of the country?

Key Terms

toponym	vertical integration
cultural hegemony	production line
Native American	Fordism
First Nations	Francophone
economies of scale	Parti Québécois
horizontal integration	

degree of local government (Figure 11.10). The largest area, Nunavut ("Land of the People" in Inuit language), became a new territory with its own elected government in 1999, although it will remain subject to federal control. Other agreements were reached in northern Québec and with the Inuvialuit people in the northwest Arctic. Further discussions are underway, although many of the smaller claims may take several years to resolve. Some are complex because of overlapping land claims and because bargaining involves the often-opposing interests of native groups, the federal government, and provincial governments.

🌐 Physical Geography and Human-Environment Interaction

North America contains a wide range of natural environments, from rugged high-altitude mountain ranges to broad flat plains and from hot and humid subtropical areas to Arctic cold. The north-south trending mountain systems on the western and eastern sides of the United States and Canada and the intervening broad central lowland are the most distinctive features of relief and have a dramatic impact on much of the environmental conditions in the region. The lowland is drained southward by the Mississippi River and its tributaries or northward to Hudson Bay (Figure 11.11).

While the natural environments of most of the United States are conducive to human settlement, Canada's natural environments present significant challenges because so much of the country lies near or beyond the margins of productive or habitable land. Continuous settlement in Canada is restricted to a narrow zone just north of the USA-Canada border.

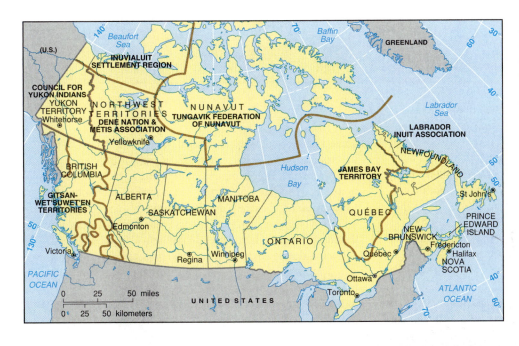

Figure 11.10 **Canada: contemporary First Nation claims.** Major Native American land claims in the Arctic north, Québec, and British Columbia, late 1990s. Some of these areas may obtain a greater degree of self-government than others.

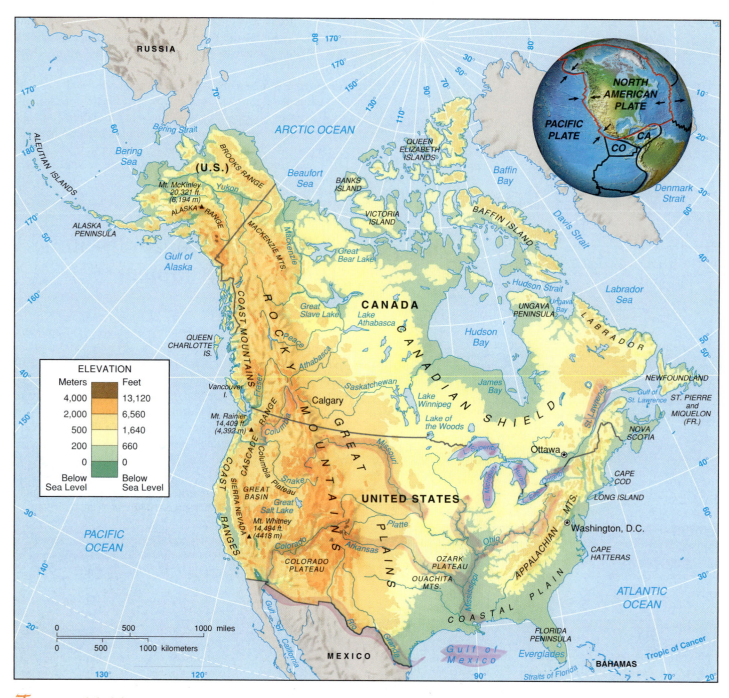

Figure 11.11   **North America: physical features.**

Tropical to Polar Climates

Climatic environments within North America are mainly midlatitude in type (Figure 11.12). The western coast is dominated by the midlatitude west coast and Mediterranean climates that give way eastward to midlatitude interior and midlatitude east coast climates. This is a similar sequence to that found across Europe and northern Asia (see Chapters 3 and 4). The west coast climates of Canada and the United States are narrowed in extent by the north-south mountain ranges.

The southern United States, northern Canada, and most of Alaska lie outside the midlatitude climate regimes. Although the

United States Southwest is situated north of the Tropic of Cancer, the tropical arid climatic environment present in Mexico extends into the region. Southernmost Florida has tropical, humid conditions, and the Gulf Coast from Texas to Florida and the Atlantic Coast as far north as the Virginia tidewater have subtropical conditions that bring hot, humid summers with many thunderstorms and the annual threat of hurricanes. In northern Canada and Alaska, the climate is deeply Arctic in winter and has only short summers.

The north-south central plains enable bursts of Arctic air to move southward in winter, bringing extreme cold especially to the Midwest and northeastern United States but also extending in shorter bursts to the Gulf Coast. Warm, dry southwestern

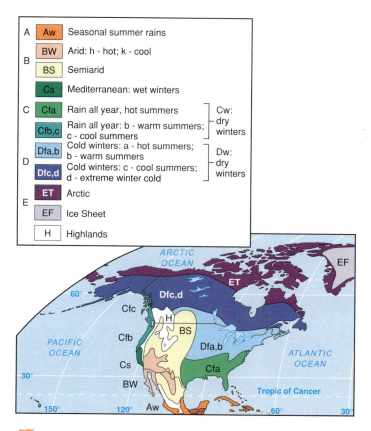

A	Aw	Seasonal summer rains
B	BW	Arid: h - hot; k - cool
	BS	Semiarid
C	Cs	Mediterranean: wet winters
	Cfa	Rain all year, hot summers
	Cfb,c	Rain all year: b - warm summers; c - cool summers
D	Dfa,b	Cold winters: a - hot summers; b - warm summers
	Dfc,d	Cold winters: c - cool summers; d - extreme winter cold
E	ET	Arctic
	EF	Ice Sheet
	H	Highlands

Cw: dry winters

Dw: dry winters

Figure 11.12 **North America: climates.** Source: From *Foundations of Physical Geology* by Bradshaw & Weaver. Copyright © The McGraw-Hill Companies.

air or humid air from the Caribbean and Gulf of Mexico moves northward in spring and summer. The easy confrontation of cold Arctic air and humid subtropical air in the central lowlands leads to the formation of active cold fronts with thunderstorms that may develop into local violent tornadoes mainly in spring and summer. Tornado activity moves northward from Texas in February to the Great Lakes in June. The United States experiences more tornados each year than any other country.

Past climatic changes left their impacts on the present environment. During the Pleistocene Ice Age—the last 2 million years of geologic time—much of Canada and the northern parts of the United States were covered by advancing ice sheets centered on the Hudson Bay area and the western mountains. The greatest effects occurred in the last half million years, but most of the ice melted beginning about 15,000 years ago, leaving glaciers only in the highest and coldest areas. Distinctive ice-eroded landscapes are relic features in high mountains and in northern Canada, while the melting ice sheets left depositional features mainly in the Midwest.

Mountains and Plains

North America is part of the North American plate (see Figure 11.11b). New plate material forms along the divergent margin at the Mid-Atlantic ridge, forcing the plate to move westward and clash with the Pacific plates at the convergent margin along the coast of North America. This pattern explains the differences between the eastern and western coasts of North America. The

mountainous western coasts are affected by frequent earthquakes and periodic volcanic activity, while the less-disturbed eastern coasts are low-lying or hilly with older mountains inland.

Western Mountains

The western third of Canada and the United States is primarily comprised of rugged mountains consisting of several different elements. The highest peak in the region is Mount McKinley in the Alaskan Range (6,194 m, 20,231 ft.). The Rocky Mountain ranges extend from Alaska and northwestern Canada southward to New Mexico. To their west in the United States are extensive high plateaus from the Colorado Plateau in the south to the lava layers of the Columbia Plateau just south of the Canadian border. The Pacific coastal ranges are backed inland by the Californian Central Valley, Puget Sound lowlands, and the drainage area along the British Columbia and Alaskan panhandle coasts. East of these valleys are the Sierra Nevada mountains of California and the Cascades farther north. The Cascades have many volcanoes such as Mount Saint Helens (Figure 11.13), and there are

Figure 11.13 **United States: Mount Saint Helens eruption, 1980.** (a) Before the eruption. (b) After the eruption, with the mountain top blown away and a new small cone forming in the crater. Photos: U.S. Geological Survey.

(a)

(b)

others in southwestern Alaska. Frequent plate movements in Earth's interior make California liable to earthquakes along the San Andreas Fault system as the rock on the Pacific side of the fault moves northward.

Canadian Shield

About half of Canada has the ancient rocks of the Canadian Shield at the surface. These are the oldest rocks in North America and contain major deposits of mineral ores. Covered by ice sheets during the last Ice Age, most of the soil was scraped from their surface. Deposits left by the melting ice sheets over the last 15,000 years are scattered and varied in nature and often separated by large lakes that fill hollows gouged out by the ice. A line of very large lakes follows the approximate boundary between the shield rocks and the overlying rocks, from Great Bear Lake in the northwest to the Great Lakes.

Interior Lowlands

Lowlands with little relief dominate Southern Ontario, the Prairie Provinces, and the central United States. Layers of sedimentary rock cover the shield rocks, forming plateaus and escarpments such as that over which the Niagara Falls plunge. In the United States Midwest and in Ontario, Canada, most of these rocks are covered by deposits from the melting ice sheets. Similar deposits occur over part of the Canadian Prairies, but in areas not covered by the ice sheets, distinctive features such as groups of small mounds and depressions were formed by exposure of the land to an atmosphere of extreme cold.

East of the Rockies, the Mississippi lowlands give the impression of being flat for over a thousand kilometers (620 mi.) but rise gradually eastward and westward to the bordering mountains. In the north, they were covered by rock fragments and particles dropped by the ice sheets that blanketed the area during the Pleistocene Ice Age and by the finer particles blown farther afield by winds and deposited as loess as far south as the states of Louisiana and Mississippi. West of the Mississippi River, the lowlands rise to the Great Plains in the United States and the Prairies in Canada.

Appalachian Mountain System

East of the Mississippi lowlands, the Appalachian Mountains form a continuous chain of rolling hills and mountains extending from northern Georgia into the Adirondacks of New York, the Green and White Mountains of New England, and the Atlantic Provinces of Canada. These mountains are not so high as those on the West Coast—few exceed 2,000 m (6,500 ft.)—and were formed by ancient plate tectonic movements and subsequently eroded over time. In the northeastern United States and eastern Canada, the glaciers scraped away much of the surface rock and soil, carried it southward, and deposited it along the coast in the low ridges (moraines) that form much of Long Island and Cape Cod. West of the Mississippi, the Ozarks and Ouachitas form similar upland areas.

Major Rivers and the Great Lakes

The physical and human geographies of North America are greatly affected by the region's surface hydrology. In particular, the major rivers and the Great Lakes provide sources of water together with water and land transportation routes. The Mississippi River was the basis of early interior transportation and continues that role for bulk materials, though diminished in significance. Its tributary, the Ohio River, was a transportation route at the heart of manufacturing developments in the United States during the late 1800s. The Colorado River provides irrigation water for the arid Southwest. The Columbia River is used to irrigate farmland and to generate hydroelectricity in the northwestern United States. In the 1950s, the Great Lakes–St. Lawrence Seaway brought easier trade and associated industries deep into the interior of Canada and the United States, although this function is much reduced owing to the increased size of ocean ships. In northwest Canada, the Mackenzie River is a major summer transportation route.

The largest area of lowland and fertile farmland occurs in the combined drainage basin of the Mississippi-Missouri-Ohio Rivers and the Great Lakes, covering nearly one-third of Canada and the United States. Although this is a continental interior region, the rivers provide relatively easy outlets to world ocean trade routes to the Gulf of Mexico in the south and along the St. Lawrence and Hudson Rivers to the Atlantic Ocean in the east. Much of the lower relief along the Mississippi Valley was formed by sand and clay deposited from meltwater, as the ice sheets dissipated (Figure 11.14) and the ocean level rose.

The Great Lakes were formed by processes that began as the Pleistocene ice sheets advanced over Canada and the United States. Lobes of ice gouged depressions between festoons of marginal moraines built from rock fragments dropped at the margins of melting ice masses. As the ice sheet retreated northward at the end of the last glacial phase, meltwater accumulated in the depressions, which became the Great Lakes. At first, the meltwater overflowed southward into the Mississippi system; later, the lake water drainage was limited to flowing eastward into the Hudson and St. Lawrence Rivers. Today, the Great Lakes function as inland seas with coastlines, ports, and recreation areas. In 1959, the St. Lawrence Seaway was opened as a cooperative project between the Canadian and United States governments. Through it, ocean-going ships could reach Chicago and Duluth. The original locks are too small to accommodate modern bulk-carrying ships and are still closed by ice for three months of every year. Current efforts to expand the locks should enable the system to accommodate larger ships.

From the early 1900s, the Mississippi-Missouri River system was controlled for flood protection, improved water transportation, and hydroelectricity generation by building a series of dams. The flood protection system prevents most floods from spilling onto surrounding land; however, extreme rain events prove to be too much for the levees and dams to handle (Figure 11.15). The dams prevent silt and clay reaching the mouth and so reduce deposition on the delta south of New Orleans. In recent years, the delta surface subsided along its southern margins because it did not receive sediment to build it to sea level.

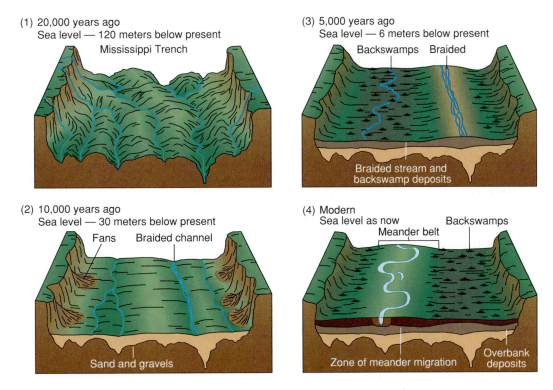

(1) 20,000 years ago
Sea level — 120 meters below present
Mississippi Trench

(2) 10,000 years ago
Sea level — 30 meters below present
Fans   Braided channel
Sand and gravels

(3) 5,000 years ago
Sea level — 6 meters below present
Backswamps   Braided
Braided stream and
backswamp deposits

(4) Modern
Sea level as now   Backswamps
Meander belt
Zone of meander migration   Overbank deposits

Figure 11.14   **North America: impact of ice meltwater on the landscape.** Stages in the evolution of the lower Mississippi River valley. (1) Large flows and low sea level resulted in the Mississippi River carving a deep channel at first. (2) and (3) As the sea level rose and the flows lessened, the channel filled with gravel, sand, and mud. Finally (4), flows from the Great Lakes were diverted eastward, and the Mississippi flow was reduced to its present levels.

Natural Vegetation and Soils

The combinations of terrain and climate produced a range of natural vegetation and soils distributed in a north-south pattern similar to that of Russia and its neighbors (Figure 11.16), although the east-west extent is less in North America. The hot deserts of the Southwest support mainly drought-resistant varieties such as cactus and low shrubs and little else. The subhumid Great Plains and Prairies of the western Mississippi River basin and south-central Canada

Figure 11.15   **United States: flood hazard.** The 1993 floods on the Mississippi River near St. Louis. (a) Before the floods. (b) During the floods. The reddish areas are highways and buildings. The darkened areas on (b) show the extent of flood water cover. Photo: Courtesy of Space Imaging, Thornton CO, USA.

(a)   **(b)**

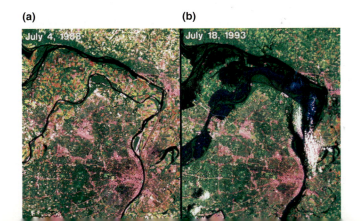

had a natural vegetation of prairie grasslands that are now largely plowed to make use of the underlying black earth soils. Eastward, the more humid conditions, from tropical southern Florida to the cool, temperate Northeast, supported deciduous (broadleaf) forest, which gave rise to brown earth soils of moderate to good fertility. North of the deciduous forests and the prairies, a wide band of coniferous (needle-leaf) forest has poor podzol soils and gives way to the tundra along the Arctic Ocean shores.

Today, the eastern mountains have thin soils on steep slopes. Before extensive clear-cutting by loggers in the late 1800s and early 1900s, the eastern mountains supported a profusion of trees containing a much greater variety of species than those in western Europe. The best farming soils formed under forest and grassland in the interior plains, where the old glacial and wind-blown materials and the moderate amounts of precipitation produced rich brown (forest) or black (grassland) soils. In the southeastern United States, the frequent presence of sandy soils with low nutrient content caused some areas to be dominated by pine trees. Along the western coast of the United States, north of San Francisco, huge firs and cedars, growing over 100 m (300 ft.) tall, formed a massive timber resource.

Natural Hazards

A greater range of natural hazards than in any other country in the world marks the natural environments of the United States. The conterminous United States is affected by hurricanes,

Figure 11.16 **North America: natural vegetation.**
Relate this distribution to the climatic environments (see Figure 11.12) and land-use patterns (see Figure 11.22).

severe thunderstorms with accompanying tornadoes, lightning, and hail, earthquakes, volcanoes, floods, blizzards, ice storms, and wildfires. Alaska and Hawaii are affected by various natural hazards including extreme winter weather and earthquakes in Alaska and nearly continuous volcanic activity in Hawaii. All the natural hazards experienced within the borders of the 50 United States have a major impact on the human activities of United States residents.

Each hazard has a geographic area of greatest impact.

- Hurricanes threaten the mid- and southern Atlantic coast and Gulf coast states in late summer and fall, when water temperatures are warmest.

- Tornadoes are most common in the Plains states, the Mississippi Valley, and Florida, where the necessary climate ingredients meet to support their formation. Tornadoes affect Texas in February and move northward to the Great Lakes area by June.

- River floods occur most commonly in the Mississippi River valley following the spring melting of snow on the surrounding hills or heavy summer rains. Flash flooding occurs in all regions of the United States and is particu-

larly dramatic in arid regions in the West after sudden rains fill dry streambeds.

- Lightning is a yearlong event across the southern part of the United States and is especially prevalent in Florida.

- Severe winter storms affect the upper Plains states, the Midwest, the western mountains and the northeast coast of the United States. Ice storms pelt many areas east of the Rocky Mountains.

- Earthquakes and volcanoes occur mainly along the west coast. The meeting of the North American plate and the East Pacific rise causes horizontal displacements along the San Andreas Fault (see Figure 11.11b). Volcanic activity in the Cascades occurs where the Juan de Fuca minor plate plunges beneath part of North American Plate.

Canada is less troubled by most of the natural hazards plaguing the United States, primarily due to its position in the high latitudes of the Northern Hemisphere. The waters off the Canadian coasts are much too cold to support hurricanes. The extreme temperature and moisture contrasts needed for severe thunderstorms and accompanying tornadoes, hail, and lightning are also less common in Canada due to its high latitudinal position. Canada does experience earthquake activity and the threat of tsunamis (tidal waves) along the west coast, and very cold temperatures and high snowfall totals are common across the country during the winter.

Environmental Problems

The emphasis on expansion and economic growth led to many environmental problems in Canada and the United States before the era of concern that developed through the 1900s. The mining of fuels and metallic ores, the outputs of manufacturing industries, and the plowing of land across the Canada–United States region resulted in spectacular examples of environmental degradation.

- By the early part of the 1900s, many fields in the U.S. South that had been cultivated for decades were destroyed by deep gullies and abandoned. Heavy rains had washed out the gullies when soil between row crops such as corn, tobacco, or cotton was kept bare by cultivation. Today, these fields are often covered by kudzu, a Japanese plant introduced to the region.

- The plowing of subhumid grasslands in western Oklahoma, Kansas, and Texas resulted in the **Dust Bowl** disaster of the 1930s, in which winds blew away the dried finer soil particles and left piles of sand behind. Even in the more humid Midwest, the soils thinned as winds and surface runoff removed the topmost layers.

- Farming extended into the arid areas of the southwestern United States, where water for irrigation was available in the underlying rocks or major rivers such as the Colorado, but faced problems from salinization, the construction of engineering works to store and distribute water, and the depletion of groundwater resources.

- Population settlement in arid areas where natural burning cycles frequently occur in the vegetation presents a growing hazard. Wildfires are increasingly affecting populated areas, burning businesses and homes, and placing residents and firefighters alike at risk.

- Large-scale hydroelectricity projects, often linked to flood control, irrigation, or navigation, as in the Colorado River, Columbia River, and Tennessee Valley Authority projects, drowned land and attracted polluting industries (e.g., the Hanford nuclear weapons facility on the Columbia River).

- Mining on a large scale resulted in the digging of huge pits for extracting copper and other metal ores in the West. The strip mining of Appalachian hills for coal after 1950 devastated large areas of eastern Kentucky and West Virginia.

- The increasing scale of industrialization and urbanization in the United States brought problems of storing and getting rid of wastes from factories and homes.

- Cities became plagued by smog generated from the exhaust gases of large numbers of vehicles and thermal power plants.

- Acid rain, derived from power plant emissions, particularly along the Ohio River valley, affected trees and caused rivers and lakes downwind in the northeastern United States and eastern Canada to become more acidic.

Partially due to the enforcement of environmental legislation, the impacts of soil erosion, poor water management, and urban air quality reduced in the late 1900s, despite increases in population and intensity of human activities.

Canada, with a much smaller population in a slightly larger area, has fewer environmental problems than the United States. Industrial pollution largely created in the United States travels through air currents and reaches the ground in parts of Canada in the form of acid rain. Industrial effluents from both Canada and the United States foul the waters of the Great Lakes. Although Canada reduced polluting outputs from smelters on its territory near the British Columbia border and along the northern shore of Lake Superior, more significant reduction is needed from both countries to ensure lasting health of the region's fresh water systems. Canada attempts to diminish industrial effects on the natural environment through federal legislation that imposes heavy financial fines on pollution-creating industries such as the nickel smelters at Sudbury in Ontario.

The increasing size of hydroelectricity projects in Québec and elsewhere led to environmental and Native American protests over the resulting hydrologic and land-use changes. Uncertainties concerning federal and provincial jurisdiction in these areas led to differences of approach among the different parts of Canada. For example, Québec moves ahead with its huge hydroelectric projects against the wishes of local indigenous communities, and Alberta encourages the extraction of coal, oil, and natural gas, while Ontario is more concerned about environmental impacts.

Test Your Understanding 11B

Summary North America contains most of the world's natural environmental types except for the most tropical. Its climates range from tropical to Arctic and its terrains from young folded mountains to wide river plains and ancient shield areas. Natural vegetation and soils reflect climatic differences, with deserts in the southwest, grasslands and black earths west of the Mississippi River, forests and brown earths in the east, and forests with podzols in the north, giving way to tundra in northern Canada and Alaska.

The natural hazards affecting North America rival the richness and variety of the region's natural resources. They include hurricanes, tornadoes, earthquakes, volcanoes, and floods. Other environmental problems result from the intensive uses made of the soils, forests, and mineral resources and from the pollution of air and water.

Questions to Think About

11B.1 What role did glaciers play in shaping the current physical geography of North America?

11B.2 What is the spatial distribution of natural hazards in North America? What role does latitude and proximity to water play in the distribution of climate-related natural hazards in the region?

Key Terms

Dust Bowl

World Roles

The United States and Canada participate in a number of intraregional and interregional diplomatic, economic, and military alliances. Some of the most significant global connections today are directly linked to this region through formally structured coalitions in which the United States and Canada engage with European states and other countries worldwide. Both the United States and Canada, along with Japan, Germany, the United Kingdom, France, Italy, and Russia, are part of the **Group of Eight (G8),** which is an economic discussion forum of the world's eight most materially wealthy countries. The size, military might, and economic expanse of the United States make it a more prominent and powerful player in international relationships than Canada. The United States wields significant influence in the United Nations, the North Atlantic Treaty Organization, the World Bank, the International Monetary Fund, the Organization of American States, and the World Trade Organization, among many others. The UN, World Bank, IMF, and OAS are headquartered in the United States.

The United States and the United Nations

North America plays a unique global role in most areas of the global political economy and international security. The United States enjoys an especially distinctive role as the host country of the United Nations. The international headquarters of the UN is in New York City. The **UN Security Council** is the most powerful branch of the organization. The purpose of the Security Council is the "maintenance of international peace and security." The United States (along with China, France, the United Kingdom, and Russia) is one of five countries granted a permanent seat on the Security Council. The permanent members of the Security Council have the power to veto council decisions. The decisions of the Security Council are binding for the entire UN General Assembly.

The decisions and actions of the United Nations often simultaneously result in positive praise as well as criticism. Domestically, actions of the Security Council produce healthy debate among U.S. citizens concerning the role the UN should play globally and the degree to which the United States should be involved in that role. Internationally, countries and national groups benefiting from actions of the UN in general and Security Council specifically applaud the execution of decisions made by the organization. Other groups, not directly benefiting from specific measures, criticize the organization and often the United States for interfering and question the validity of the unequal distribution of power on the Security Council. Controversy is furthered by the U.S. reluctance to fully participate in the United Nations World Court and to pay its financial dues promptly.

Globalization: The Good, the Bad, and the Ugly

The United States garners the most attention concerning political, economic, and cultural globalization. One of the most common examples given to portray U.S. influence on cultural globalization involves the international growth and distribution of a well-known U.S. fast-food chain. While appropriate as an example of some forms of globalization, this picture is far from complete. Globalization is much more than the residents of Moscow eating a hamburger at a fast-food restaurant owned by a U.S. company. There are countless global connections in communications, the media, transportation, politics and diplomacy, athletic competition, entertainment, recreation, and leisure and business travel. Few aspects of life outside the region are not touched in some way by North American connections, and few aspects of life within the region are not touched by international relationships.

While disagreement and controversy easily bond with many globalization trends, both the United States and Canada enjoy numerous positive relationships within the globalization frame. The Space Shuttle program under the wing of the U.S. National Aeronautics and Space Administration (NASA) is an example of positive global cooperation. The production of the various components of the shuttle and its operation are not solely generated by the United States. Research from Canada, countries in Europe, and Russia plays prominently into the success of the shuttle missions. The flight paths of shuttle missions present one of the more unique and positive aspects of a globalization relationship. During critical takeoff and landing windows as well as during the time the shuttle spends orbiting Earth, the United States has the cooperation of several countries forming a complete ring around the globe to assist in mon-

itoring shuttle flights and, perhaps more important, to ensure safe landing sites, should something go dramatically wrong.

Although the positive aspects of globalization are numerous for the United States and Canada, it seems in recent years that the negative side of globalization relationships garner the greatest attention of the media. Some of the most dramatic manifestations of antiglobalization sentiment are the increasingly violent protests staged at the meeting sites of the G8 or the World Trade Organization, which is the primary international organization governing global trade. Such protests are often aimed directly at the United States because the U.S. government and U.S.-based multinational corporations enjoy considerable influence with the organization. The young organization quickly became the scapegoat for antiglobalization groups. Supporters of the WTO assert the organization opens barriers, creates jobs, provides development opportunities, and infuses capital where there otherwise would be none. Protesters argue the WTO favors more materially wealthy countries, such as the United States and Canada, over more impoverished countries. Protest groups also argue the decisions of the WTO are detrimental to the rights of laborers around the world. The annual meeting of the WTO in 2000 in Seattle, Washington, was engulfed in dramatic protest that resulted in rioting, property destruction, and human injuries. When the G8 met in Genoa, Italy, in 2001, demonstrators clashed with Italian police forces, resulting in the death of a protester.

Global Role of the Regional Economy

The United States had the world's largest economy in 2000, over twice the size of Japan's, the second largest. Canada, with a much smaller population, was eighth. Economic growth and victory in World War II gave the United States political superpower status alongside the Soviet Union. After the 1991 breakup of the Soviet Bloc, the United States remained the sole superpower.

The combination of natural resource wealth, technological and management leadership, and a large internal market quickly delivered the United States to a role as a global economic powerhouse. Before 1950, it dominated the development and output of new products based on automobiles, trucks, airplanes, and consumer goods. U.S. corporations expanded abroad, taking opportunities to do so after World War II during the economic recoveries in Europe and East Asia, and in the context of Cold War competition in developing countries. After 1950, the United States established a huge lead in the initial development and use of computers. Although Western Europe and especially Japan now challenge this economic and technological superiority, no other single country is close to rivaling the total PPP GNI of the United States or the size of its internal market for products (Figures 11.17a and 11.17b).

Canada lagged behind its neighbor through the early 1900s, partly because of restrictions resulting from its colonial ties to Britain and partly because of the smaller size of its home market for goods and protection of its own industries. It gradually became more closely enmeshed with the U.S. economy and dependent on it in many ways. Canada's centers of manufacturing production around Toronto and Vancouver, and to a lesser extent around Montréal and in Alberta, now rival individual centers in the United States in size and diversity of output. Canadians

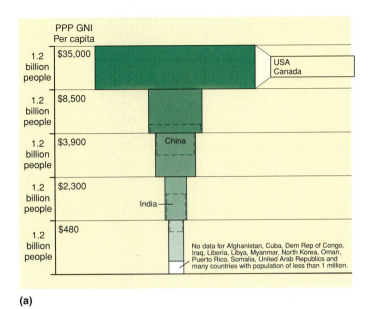

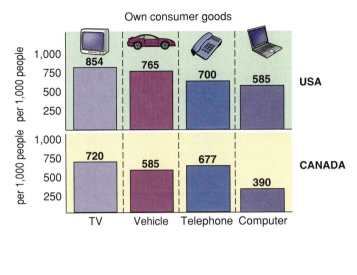

(a)

(b)

Figure 11.17   **North America: country incomes compared.** (a) The countries are listed in the order of their PPP GNI per capita. (b) Ownership of consumer goods is among the highest world levels. Source: (a), (b) Data (for 2000) from *World Development Indicators,* World Bank, 2002.

enjoy material living standards almost equal to those in the United States, and many who live in Canada strongly support the notion that they enjoy an even better quality of life. Some Americans prefer the Canadian way of life and live across the border from Detroit or Buffalo. The 2000 United Nations human development and gender-related development (HDI and GDI) indices, based on 1998 statistics, both placed Canada in first position above European countries, the United States, and Japan.

In 1988, the United States and Canada established the United States and Canada Free Trade Agreement, which led to the North American Free Trade Agreement (NAFTA) in 1994 by adding Mexico to the arrangement. The agreement between the United States and Canada continues to thrive under NAFTA, despite a sequence of disputes over individual items (see the "Point-Counterpoint: North American Free Trade Agreement" box, p. 504).

🌐 Subregions of North America

Regional Population Distribution

The United States of America is a highly urbanized country with over three-quarters of its population living in urban areas and just over half in metropolitan centers of over 1 million people. More than 90 percent of the U.S. population lives within a two-hour drive of a large city of over 300,000 people. Primary population clusters exist along the eastern seaboard in an urban corridor from Washington, D.C., to Boston, Massachusetts (known as the Megalopolis), around the Great Lakes from Chicago to Detroit, in Florida and westward along the Gulf coast

to southeastern Texas, and along the Pacific coast in the west (Figure 11.18). Despite a slight fall in urban population between 1970 and 1980, the rise that began in the early 1800s continued again in the 1980s. Large American cities are the essential feature of the country's geography today. Urban population clusters are more numerous in the eastern half of the contiguous United States. Large areas of the western United States, away from the Pacific coast, are sparsely populated and have very low population densities. Hawaiian population is clustered near Honolulu and resort towns, and Alaska, which is the largest U.S. state in area, has the lowest population density in the country.

Canada is the second-largest country in the world in terms of area. Although the land area of Canada is extensive, the country has an extremely small population equal to approximately only one-tenth of the U.S. population. The size of the country coupled with a very small population produces extremely low population densities. Although the overall population density for the country is low, it is important to consider the distribution of people in Canada to fully understand the spatial patterns of Canadian population. The Canadian people are concentrated in a belt across the southern part of the country nearly parallel to the border with the United States. Thus, portions of this populated belt across southern Canada have high population densities, while most of the remainder of Canada is virtually empty (Figure 11.18).

🌐 The United States

The United States at a Glance

The United States of America dominates the midlatitudes of continental North America with 48 of the country's 50 states situated between the Atlantic and Pacific Oceans to the east

NORTH AMERICAN FREE TRADE AGREEMENT

The formal implementation of the North American Free Trade Agreement (NAFTA) in 1994 created one of the world's largest trading bloc. The United States, Canada, and Mexico moved toward an extensive elimination of trade barriers to increase the economic activity between them and strengthen their economic and political positions at the global scale. The three countries tied by NAFTA are now each other's largest trading partners.

NAFTA has many proponents and critics in each of the three countries. The similar living standards in the United States and Canada contrast sharply with poverty conditions common to millions in Mexico. Proponents of NAFTA claim it creates jobs, strengthens and expands business and industry, diversifies the economies of each of the three country partners, enhances foreign revenues, and fortifies each country's ability to compete in a global trade arena. Those opposing NAFTA assert the agreement results in the exportation of jobs and revenues from Canada and the United States to Mexico, the perpetuation of harsh labor conditions for Mexican workers, the degradation of the natural environment in all countries, and the granting of nearly supreme powers to big business and industry at the expense of workers, consumers, and those living in the shadow of production facilities.

The North American Free Trade Agreement supercedes the U.S.–Canada Free Trade Agreement implemented in 1989. Political leaders from the three NAFTA partners (Mexican President Carlos Salinas de Gortari, 1988–1994; U.S. President George H. W. Bush, 1988–1992, and, later, U.S. President Bill Clinton, 1993–2001; and Canadian President Brian Mulroney, 1984–1993) signed the 1992 trade pact. The respective legislative bodies each then subsequently passed the agreement, which was formally implemented in 1994. The agreement effectively removed most tariffs immediately and established a timeline by which all tariffs would be eliminated within 15 years of implementation. NAFTA eroded barriers, restrictions, and red tape, making it nearly effortless for companies from the three countries to conduct business and sales anywhere in the region.

The United States led efforts to push NAFTA forward in hopes of creating a competitive trade bloc to the European Union. The majority of NAFTA proponents, who pushed through an accelerated legislative process in the United States, were mostly representatives of large corporations. Labor unions, human rights groups, and environmentalists were strongly united against the rapid development of a trade agreement between the three countries and hoped to stop fast-track legislation but argue they were largely not included in the negotiations. Environmental critics of NAFTA claim many of the key negotiators rapidly pushing NAFTA through the U.S. Congress were among the United States' biggest polluters.

Proponents of NAFTA assert that the trade agreement forces the equal treatment of corporations and industry throughout the region. Business leaders claim the package promotes fair and equal business opportunities for all companies in the three countries. Corporate leaders also assert that equal treatment reduces the risks of governmental interference in free market flows. Mexico has a long history of protectionism, or imposing restrictions, taxes, and quotas on goods and services produced outside of Mexico, and nationalization of industry (government takeover and control of business and industry, thus removing private ownership). Business proponents claim protectionist policies would be heavily curtailed under NAFTA. Business leaders firmly believe that capitalism thrives best in NAFTA-created conditions in the region and this will promote and fortify economic and political stability for all three countries.

Strong opposition to NAFTA emanates from labor unions in the United States. Union leaders argue that the agreement exports thousands of jobs from the United States to Mexico. Working conditions in production facilities in Mexico are dramatically different from those in the United States. Mexican wages are, on average, one-eighth to one-tenth of U.S. wages. Benefits and other labor costs are nonexistent or significantly lower in Mexico also. Laws restricting or regulating environmental pollution from production facilities and those concerning labor safety are fewer and less enforced in Mexico. United States-based companies are able to manufacture products in Mexico with significantly lower labor costs and with far fewer environmental or labor restrictions. United States labor leaders say these factors are far too attractive for U.S. companies to forgo, and they claim numerous companies are shifting production to Mexico and therefore replacing U.S. workers with Mexican ones. Unions claim thousands of jobs lost already in the United States and project thousands more as NAFTA progresses and removes any remaining trade barriers.

The Mexican Border Industrial Program, first implemented in 1965, slowly opened the door for foreign firms to establish production facilities in Mexico. The idea of the program was to stimulate growth in northern Mexico, create jobs, and infuse capital into a region that had operated under protectionist and nationalist economic policies for decades. The program created a manufacturing zone across northern Mexico known as the maquila zone (the factories are referred to as maquiladoras). The idea was that foreign firms, such as U.S. companies, could import raw materials or parts, pay Mexican workers low wages to refine or assemble them into finished goods, take the goods back across the border (to the United States for a U.S. firm) relatively tax free, then sell them to consumers. Although maquila enterprise is now permitted virtually anywhere in Mexico, the greatest concentration of maquiladoras remains in the border region in the north.

Environmental conditions in the maquila zone in northern Mexico are among some of the worst in the world. Multidisciplinary research suggests an increasing array of health problems and ecological damage due to the concentration of factories in a country with loose envi-

and west, respectively, Canada to the north, and Mexico to the south (Figure 11.19). Alaska and Hawaii are separated from the conterminous 48 states by Canada and the Pacific Ocean, respectively.

Metropolitan centers of the United States house the majority of the larger manufacturing facilities and financial and business services. Some 85 percent of U.S. real estate value occurs in the 2.5 percent of land occupied by urban areas. Although the contrasts between wealthier and more materially poor groups in American society are most obvious

in the cities, rural areas struggle to maintain their populations as increased mechanization on farms eliminates jobs, land is taken out of agriculture, and low-paying jobs in farming are replaced by low-paying jobs in manufacturing or tourism. Rural areas near large cities show less evidence of such stresses only when they become commuter or second-home hinterlands of the cities.

The United States is the largest market economy in the world, and it has a major influence in spreading this type of economy to the rest of the world. The United States remains ronmental and worker safety laws. The health and environmental issues present in northern Mexico also exist across the border in the southwestern part of the United States. Cancer rates along the Texas-Mexico border are some of the world's highest. When NAFTA discussions began, proponents argued the agreement would ease the problems critics claim were created by maquila enterprise for both Mexico and the United States. Since NAFTA's implementation, critics of the agreement argue that production in the border region of northern Mexico is growing and thus causing the workers' and residents' health problems, as well as the related environmental damage to the region, to increase.

The controversy surrounding NAFTA is far from over. Further trade barriers will be reduced or eliminated in the next several years in accordance with the progression plan of the agreement. Activist groups both for and against NAFTA will continue to lobby the respective governments, especially the U.S. government, to promote their views on the future of the agreement. There are proponents and critics in each of the three countries involved. Depending on the source, convincing statistics touting its value and success, or its detriment to society, the economy, and the health of the region's citizens all make compelling arguments.

SUPPORTERS OF NAFTA	CRITICS OF NAFTA
Opened new markets for the three countries.	Primarily opened new markets for Mexico and Canada.
Competition of lower-priced goods produced in Mexico forces down the price of goods produced in the United States and Canada; thus, U.S. and Canadian consumers "win" with lower-priced goods.	Low labor costs in Mexico and lower-priced goods in the United States and Canada cause U.S. and Canadian companies to move their production operations to Mexico; thus, thousands of jobs are lost ("exported") to Mexico, hurting employment in the United States and Canada.
Strengthens the global economic weight of the three countries and makes them better able to compete with the EU and other trade blocs and countries of the world.	Significant economic disparity exists between the affluent United States and Canada and the relatively materially impoverished Mexico, placing Mexico at a disadvantage and creating more of a service role for Mexico to Canada and especially the United States, rather than truly making it an equal trade partner and stronger international economic player.
Promotes democracy and political stability in Mexico and strengthens the Mexican economy, thus ensuring greater stability for North America.	Perceptions by some Mexicans of heightened economic disparity in their country due to NAFTA result in political instability such as the Zapatista uprising (see Chapter 10).
Creates thousands of jobs in Mexico. Cities and towns in northern Mexico, where the majority of NAFTA-related production (maquila) takes place, enjoy much higher living standards and higher rates of employment than most other parts of the country.	Cultural distinctions are blurred in all three countries. U.S. culture may overpower parts of Canada and especially northern Mexico. Fast food is replacing traditional food; U.S. holiday celebrations are replacing traditional celebrations. Areas in the U.S. Southwest are developing a watered-down culture that is a mixture of U.S. and Mexican elements. The increased use of the Spanish language in parts of the United States, especially areas in the Southwest, increases tension with some English-speaking residents.
Strengthens Mexican environmental conditions through environmental side agreements negotiated along with the primary trade agreement, resulting in a healthier environment for Mexico and especially the border region; reverses environmental damage on the U.S. side of the border.	The side agreements negotiated with NAFTA fall far short of strengthening environmental laws and enforcement in Mexico. U.S. companies are further attracted to relocating in Mexico due to lax enforcement. The wording of NAFTA facilitates environmental abuse by companies in all three countries as NAFTA protects the companies' rights to free trade over the rights of people living in areas polluted by factories and other production facilities.
Forces the equal treatment of corporations in the three countries.	Corporations are too powerful under NAFTA.

the world's entrepreneurial leader, based on flexible labor markets, low taxes, light regulation, and an easier lot for failures than in other countries. In 1996, the United States was first among the countries of the world in terms of raising venture capital. Of the more than $10 billion raised that year in the United States, California's share was $3.7 billion, and Massachusetts raised over $1 billion. Both states were ahead of the U.K., France, and Germany—the next three—and 10 U.S. states raised enough to be ranked in the top 20 world "countries." The United States is a global leader in high technology,

and within its borders are several major centers of technological research, innovation, and production.

The growth of market-based activities within the United States and the increasing geographic complexity and mobility of people and jobs during the 1900s led to the greater involvement of government in regulatory activities and infrastructure provision. With the growth of government funding through income taxes, the expanding role of defense, and increased personal mobility in the 1900s, the federal government became increasingly significant in the country's changing

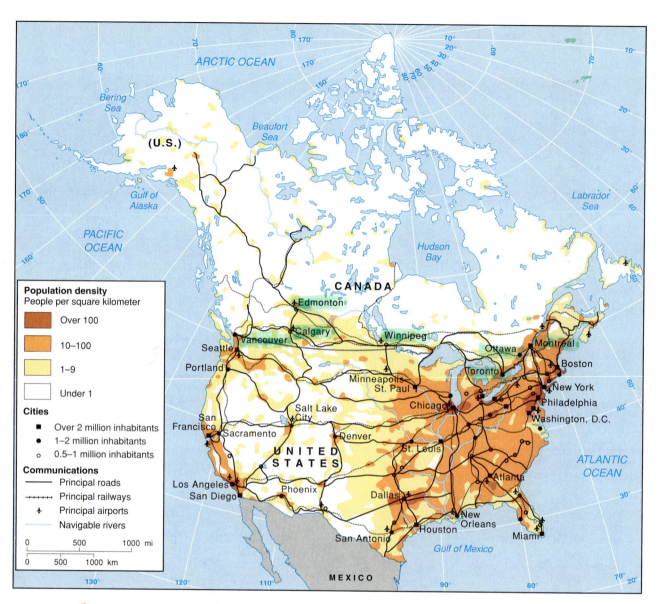

Figure 11.18 **North America: population distribution.** Compare the distribution of population in the eastern and western halves of the United States and note the closeness of most Canadians to the U.S. border. Source: Data from *New Oxford School Atlas*, p. 98, Oxford University Press, UK, 1990.

human geography. It grew in importance as an employer and source of grant aid for infrastructure projects that often opened up impoverished areas to economic development. The state governments retain responsibilities for local physical planning, education, and other services, many of which are delegated to local authorities at the city or county level.

Problems of Affluence

The wealth present in North America is far from reaching all within the borders of Canada and the United States. While educated, two-income families in the United States garner increasing material wealth, those who are located in impoverished regions or areas with few educational opportunities struggle to acquire the basic necessities for nutrition, shelter, and health care. Problems abound from contrasting sets of opportunities, which are often linked to geographic location.

Increasing Gap Between the Materially Wealthy and the Materially Poor

In the world economic system, increasing wealth in a country often combines greater numbers of those with the highest percentages of material wealth with continuing large numbers of materially impoverished peoples. Gaps increase between the wealthier and more economically challenged segments of the populous of a country—a process known as **uneven development.** Such geographic contrasts in material wealth occur within cities and between many metropolitan regions and rural hinterlands.

During the 1970s and 1980s, the gap between the materially wealthy and the materially poor widened in the United States. A large proportion of the increasing inequality is explained by two factors. The first is the lightly regulated labor markets that react to world conditions by depressing wages for unskilled jobs in competition with poorer countries and

Figure 11.19   **The United States of America: the 50 states and major cities.**

increasing salaries for skilled workers as demand from service industries and professions rises. The second factor is the changes in U.S. households that polarized the increasing incomes in a greater number of two-income homes against the growing numbers of single parents who make up 35 percent of the poorest 20 percent of the population. Complicating the disparity is that greater inequality often leads to less efficiency in production, since poor people may be subject to poor health, social stress, and temptations to be involved in crime.

The materially poor areas of many inner cities have lower-quality buildings and services such as education and health because of the inability of small jurisdictions with predominantly poor populations to support such services. Such effects create a downward spiral of living conditions. The contrasts produced by uneven development are also illustrated by the fact that some of the best public secondary education in the country is found in the suburbs of Boston but some of the worst in Boston's inner city—both ends of the spectrum in the same metropolitan area.

Although the poorer groups in American society gain some sympathy, there is less pressure for taxing the materially wealthy further to support the poor than is found in many other wealthy countries. Tax revenues are used for welfare and public health and infrastructure programs, but redistribution of income on a scale of European countries from 1945 to the 1980s does not occur. This is partly because an unusual level of mobility still exists between economic groups. It is increasingly difficult,

however, to climb out of poverty into the next stage of more secure income where most upward mobility begins.

It appears that the increasingly global scale of technological competition leads to increasing income gaps and uneven development throughout the world and has the potential for setting off social unrest or protectionist policies within countries from the wealthiest to the most impoverished.

Congregation and Segregation

The income and material wealth gaps are often linked to perceived ethnic and racial differences. High income-earning people have many choices with respect to where they live. Many choose to congregate with those of similar socioeconomic standing and cultural tastes in trendy inner-city areas or suburban neighborhoods. They may live close to shopping, ethnic restaurants, cinemas, places of worship, or recreation facilities. Less affluent groups in U.S. society have fewer opportunities for choosing where they live and become segregated into communities that more affluent groups avoid. Such segregated groups often occupy inner-city areas that contain high proportions of African American or Hispanic people and have poor access to high-quality education and job opportunities.

Urban and Rural Contrasts

Another set of contrasts intensified across the United States, particularly after 1950, between urban and rural communities.

The growing affluence arising from the expanding manufacturing and service industries was urban-based. It was enhanced by U.S. government investment in interstate highways and airports, which improved transport among and between metropolitan centers. In the 1960s, Americans became conscious of disparities in material wealth between affluent people living in the metropolitan suburbs and the lagging conditions of rural areas such as Appalachia, the Mississippi Delta, the Atlantic coastal plain, the Southwest, and the Upper Great Lakes area. Policy measures passed by governments to help rural areas did not reverse the trend. Throughout the 1900s, farms replaced labor with machines in the most productive farming areas of the Midwest, the high plains and prairies, and central California, and by the 1980s these areas were losing population.

Environmental Impact of Affluence: Disproportionate Consumption

Other problems of affluence result from the environmental impacts of intensive use and extraction of resources. In 1970, the United States, a country whose population represents approximately 4.5 percent of Earth's total, used 40 percent of the world's oil production. The United States' continuing use of large proportions of the world output of fuels, metal ores, soils, and water led to questions about the impacts of continuing economic growth. Such disproportionate consumption often places the United States at the forefront of international protests.

In the beginning of the 1970s, the United States passed stringent environmental legislation to improve its air and water quality. Following the rise of the environmental movement in the United States and related legislative changes, air and water quality standards increased in the 1980s and 1990s. Such regulations, however, sometimes led to the relocation of the most polluting industries to more materially impoverished countries.

The limits of resource usage occasionally became an issue. In 1972, a scare arose over whether world grain supplies could supply the needs of growing populations. In 1973, the oil-exporting countries raised their prices. To many, it seemed that the "American Dream" of access to increasing affluence was coming to an end. The worst, however, did not happen, and new sources of materials, often at cheaper prices, enabled Americans to enjoy increasing affluence.

People

Birth rates in the United States are relatively high, considering the country's economic affluence and social systems. Young immigrant populations within the United States have larger families and higher birth rates than other groups who have been present for generations. The demographic transition for the United States suggests the above-average population increase will maintain population growth for the country in the short term (Figure 11.20). The age and sex diagram for the United States details both a maturing population and new

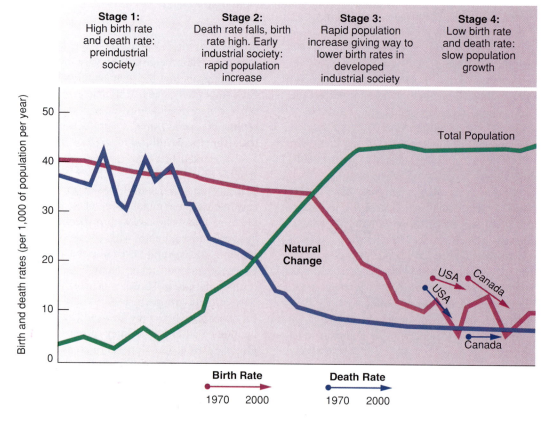

Figure 11.20   **North America: demographic transition.** Both Canada and the United States have relatively high birth rates for affluent countries, a sign of the continuing immigration of young families.

growth in younger age groups due to immigration and higher birth rates among some ethnic populations (Figure 11.21).

Urban Population

The predominantly urbanized nature of the population of the United States resulted in many very large cities and groups of cities. By the 1990s, large and growing metropolitan areas, often consisting of groups of cities that merged around one major center, were increasingly dominant in the U.S. human geography. In 1996, 104 million people lived in 35 metropolitan areas of more than 1 million people. Of these, 32 million lived in the largest cities of the Megalopolis region (mainly Boston, New York, Philadelphia, Baltimore, and Washington, D.C.). A further 23 million people lived in the metropolitan areas of the western Manufacturing Belt and Midwest (Pittsburgh, Cleveland, Detroit, Chicago, St. Louis, and Minneapolis–St. Paul, together with the somewhat smaller Kansas City, Milwaukee, Cincinnati, and Columbus). In the South, 16 million people lived in Miami, Tampa, Atlanta, Houston, and Dallas-Fort Worth, together with Fort Lauderdale, San Antonio, Orlando, and New Orleans. In the West, 31 million people lived in metropolitan areas (Denver, Phoenix, San Diego, Los Angeles, San Francisco, and Seattle, together with San Jose, San Bernardino–Riverside, Sacramento, and Portland).

Economic Development

The United States of America is the world's most developed country in terms of economic prosperity, size, and influence. Although its economy is subject to growth fluctuations, the United States maintains the largest total GNI of any country in the world, although Switzerland, Japan, and some Scandinavian countries exceed its GNI per capita. The United States main-

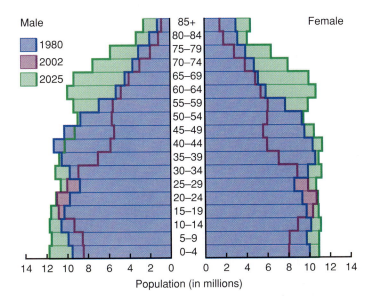

Figure 11.21 **United States: age-sex diagram.** The 2002 graphs show how the baby boom of the 1950s and 1960s produced a bulge in the 20 to 44 age groups. Source: U.S. Census Bureau. International Data Bank.

tains the greatest possible diversity of economic activity and remains at the leading edge of technology in many fields.

Commercial Farming Basis

Commercial farming regions developed as the combination of natural resources (climate, land, soils, water) and economic factors (Figure 11.22) led to areas of specialization. The early farming areas of the East Coast suffered soil erosion, and, unable to compete with more fertile inland areas, much of the land reverted to woodland. The Midwest grain-livestock area, the Mississippi Delta cotton area, the Great Plains feedlots, and the irrigated lands of California became among the world's most productive areas. Drier and higher parts of the West became low-productivity grazing land.

In 1996, the federal government ended its controls on farming. Planted acreages surged in response to rising prices and increased exports to Mexico and, especially, China. This process is likely to exaggerate the move to increased specialization. The government interventions had helped to moderate the swings of farm income. The farmers' problem remains the better management of their output.

In the 1990s, many farmers in the U.S. heartland found that falling or static prices for their output of grain, meat, or milk could not sustain their debts on sophisticated equipment. Some went bankrupt. A few, especially on family-owned farms, returned to lower-cost farming using less equipment and fertilizer and relying on crop rotation to maintain soil fertility. The extra labor previously used on farms before the machinery revolution was not available, so those managing family farms worked harder.

Manufacturing Becomes Central

Until the 1990s, many economists saw the development of the U.S. economy in the 1800s and 1900s as a model for others to follow. Although the conditions necessary for economic development vary significantly depending on time and place, it is worth tracking a history of U.S. economic development to help an understanding of the regional differences within the country.

A marked break occurred between the economies of the traditional cultures of the Native Americans and the settlers who arrived from Europe after 1500. From the outset, most settlers followed commercial practices, seeking valuable minerals and producing crops and goods for sale, often back in Europe. Trade flourished as plantation owners in the South and a wealthy merchant class on the mid-Atlantic coast and in New England emerged among the colonists. Farming and farm produce continued to be the mainstay of the American economy and trade until the mid-1800s.

Manufacturing developed from local crafts (pottery, smithing, weaving) in the early 1800s in southern New England, New Jersey, and eastern Pennsylvania. It was based mainly on water mill power and produced metal goods, textiles, and leather goods. Numerous small mill factories required new transportation facilities to take their products to widespread markets. At first, water transportation was also vital. Local capital, accumulated by merchants, was invested in the early factories, often under family ownership.

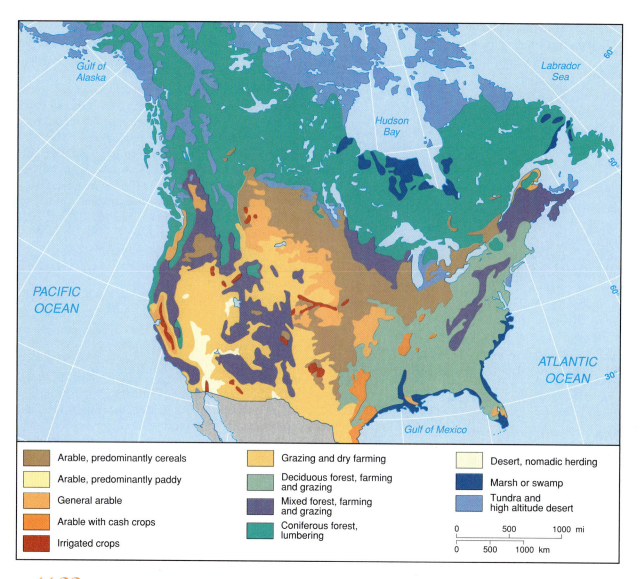

Figure 11.22   **North America: dominant land uses.** The most productive farming regions are in the Midwest of the United States, the Prairie Provinces of Canada, and parts of California, Texas, and Florida. What factors make farming difficult in other areas? Source: Data from *New Oxford School Atlas*, p. 98, Oxford University Press, UK, 1990.

By 1860, railroads reached west to Chicago (Figure 11.23a) from several East Coast cities, and their construction formed the basis of iron industry expansion. In eastern Pennsylvania, the use of charcoal gave way to coal in the iron-smelting process, then diffused westward to metropolitan Pittsburgh.

A major growth in manufacturing occurred after the 1860s with the adoption of steel production and the new possibilities it opened in heavy engineering. Many of the manufacturing industries of this phase were tied closely to their raw materials, such as coal and iron ore. New industrial areas emerged inland from the primary markets on the East Coast. New markets also developed as the interior of the United States was settled.

Major federal investments in transport infrastructure stimulated much of the spread of economic growth. Beginning in the 1860s, the transcontinental railroads were financed by federal and state governments granting lands along the routes to the railroads, which sold them and encouraged homesteading.

In 1860, the value of U.S. manufactured goods exceeded the value of commercial farm products for the first time. Agriculture became industrialized with the increasing use of mechanization and chemicals; markets for crops and livestock products were linked to the growing railroad network (Figure 11.23b).

Until 1950, manufacturing was the primary engine fueling the expansion of the U.S. economy. New products developed in consumer goods and transportation vehicles, including cars, trucks, and airplanes. These industries were less tied to sources of raw materials than the 1800s industries, and many were market-based or assembly industries where the best locations were central to a range of component producers or large markets. The Manufacturing Belt, however, was both the main market area and the location of basic metal industries and continued to be where most of these developments occurred. Factories were built in established market locations, such as the New York City area and around Chicago with its central

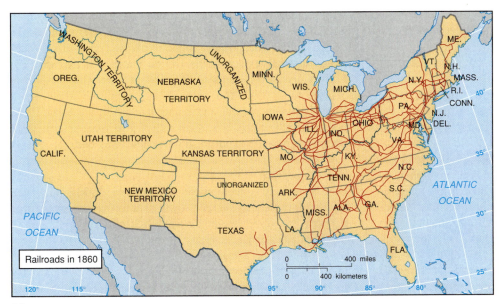

(a)

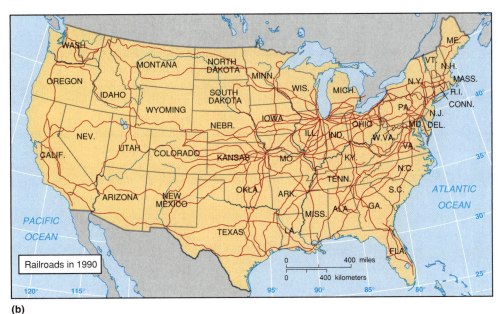

(b)

Figure 11.23 **United States: railroad networks.** (a) 1860. (b) 1990. The network was virtually complete by 1900: the extension across the continent between the late 1860s and 1900 was linked to the settlement of the U.S. West, although the building of transcontinental routes was held back until after the Civil War of 1861–1864. In the second half of the 1900s, many lines in the east were closed as truck transportation increased on the interstate highway system. Source: From P. Guinness and M. Bradshaw, *North America*. Copyright © 1985 Hodder & Stoughton Education, London. Reprinted by permission.

location on the national railroad network, or in new locations, such as in Detroit, on which the auto industry was centered.

After World War II, manufacturing industries expanded along the West Coast and in the southern United States, where they were located during the war. The previous concentration in the Northeast became more widespread, satisfying new markets in what had been the periphery of the country and relating more clearly to growing markets overseas in Latin America and Asia.

Constructing interstate highways, distributing electricity to rural areas, and developing a network of airline routes facilitated the wider geographic diffusion of manufacturing industries. The interstate highway system, begun in 1956, created nearly 80,000 km (50,000 mi.) of limited-access highways across the United States, opening many previously isolated parts of the country to economic development. The construction of airport facilities with federal grants increased the amount of traffic possible. In the 1970s, nearly two-thirds of all

scheduled passenger flights in the world occurred inside the United States, and air freight increased in importance as a result of the combination of widespread airport facilities and developing aircraft technology. The network then spread more widely around the world, with U.S. airlines having a huge initial advantage because of their flying experience on many internal flights. Air transportation of people and freight inside the United States has the potential for continued expansion. Combined with the wide availability of high-quality telecommunications, these developments reduced the costs of distance between rival locations for economic development. They provided a vital infrastructure base for continuing economic growth, but their maintenance is costly. Although the events of September 11, 2001, produced sharp declines in commercial air travel, heightened security and the gradual reestablishment of consumer confidence in the safety of airline service will likely restore the health of the industry.

While products diversified and became more technically sophisticated, production and management techniques developed. American-based multinational corporations took their products to the world and opened factories in many countries in Europe and other continents.

In the later 1900s, U.S. manufacturing continued to be important, although it employed fewer than 20 percent of the work force compared to nearly 40 percent in 1950. Its products were even more diverse, with some heavy industry and producer units moving to countries with lower labor costs. The United States is a world leader in applying high technology to manufacturing and service industries. An increasing range of high-tech goods is produced in the newer industrial areas, including the globally known Silicon Valley of California, metropolitan Boston (Figure 11.24), metropolitan Washington, D.C. (primarily Fairfax County, Virginia), the Research Triangle region in North Carolina, metropolitan Austin, Texas, the Denver-Boulder region in eastern Colorado, and areas of the Pacific Northwest. These all have national as well as regional markets.

As new manufacturing industries developed, older established industries either perished or reinvented themselves. Many of the older industries slimmed their work force, installed new machinery, and became more specialized. Some changed their focus to accommodate new trends. For example, the steel industry, which until the 1950s had been dominated by the huge and inflexible integrated complexes that resulted from economies of scale and vertical integration, lost out in competition with the Japanese in the 1960s. The Japanese had new facilities and technologies, and the American steel mills closed or changed. During the 1980s, the policy of voluntary export restraints for foreign countries selling to the United States, combined with the falling value of the U.S. dollar—

which made foreign products more expensive in the United States—provided an opportunity for the large plants to upgrade their technology. Smaller plants (minimills) established themselves across the United States. For example, the USX Corporation reduced its jobs at Gary, Indiana, from 30,000 to 10,000 and focused its output on sheet steel. Productivity increased, and rising demand in the early 1990s helped much of the U.S. steel industry to meet international competition as the export restraints came to an end.

Service Industries

U.S. economic development in the late 1900s, however, focused increasingly on the growth of information-using occupations, including financial services, publishing, computer services, and design services. Tourism and leisure-based services also became a leading part of many local economies. In 2000, over 50 million tourists from other countries visited the United States, while 51 million went from the United States to other countries, and many millions of Americans were tourists in their own country. Such industries locate economic activity away from the workshop and factory.

Integration with the rest of the world became commonplace, with improved communications and transport easing contacts with people and businesses in other countries. The United States, with its economic and technological lead, has so far been able to take greater advantage of this new economic base than any other country and is investing to maintain the lead.

While comprising a small proportion of U.S. GNI in the 1990s, the primary sector contributed important raw materials and fuels. Mining, forestry, fishing, and agricultural outputs were often linked to the manufacture of metal goods, energy supplies, paper, fertilizer, textiles, and food products. They

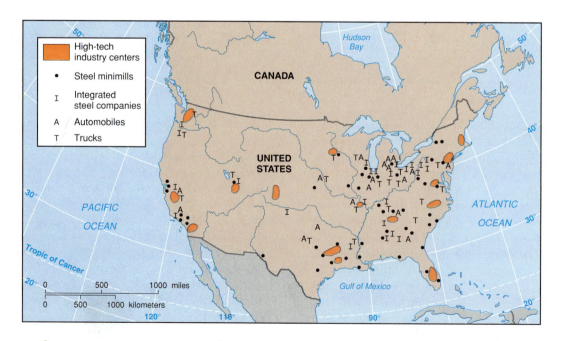

Figure 11.24 **United States: distribution of some major manufacturing industries.** The continuing concentration of steelmaking and vehicle manufacturing in the Manufacturing Belt contrasts with the widespread nature of high-tech industry centers. Steel production in minimills has become more widely distributed since 1970.

were commonly produced by multinational corporations and involved high inputs of capital and technology with declining numbers of employees.

Regional Policies

During the 1900s, rapid growth in some parts of the United States coincided with slower growth and even poverty in others. Uneven development reflected the concentration of capital investments in some regions and cities rather than in others. The favored and less fortunate regions changed over time. Until the Civil War, some of the South was prosperous for a privileged group of planters whose wealth was possible only through the enslavement of people forced to migrate from Africa. After the Civil War, the South became a large and long-lasting region of poverty for the United States. Southern Appalachia was also extremely poor, with hill farmers living on small plots of eroded land. The plight of southern Appalachia was first attacked after 1933 by the establishment of the **Tennessee Valley Authority (TVA),** one of the few federal programs designed to stimulate economic growth in lagging regions. TVA was initially viewed as a possible model for action in other regions, but the dominant political views in the United States were against government-funded regional aid. The huge investments in hydroelectricity and other forms of inexpensive electrical power generation in the Tennessee Valley provided a basis for economic growth and population growth—and delivered some serious environmentally detrimental side effects. For example, the rising demand for inexpensive electricity caused TVA to invest in first thermal and then nuclear power plants that became increasingly costly and dangerous to the environment. In the early 1990s, the TVA nuclear program ended.

By the 1960s, however, it was recognized that a more extensive Appalachian region was the largest of several rural areas that were lagging behind the rest of the country. The main achievement of the **Appalachian Regional Commission (ARC),** established in 1965, was to make available to this region the many federal grant-aid financial packages, particularly in the late 1970s. In the 1980s, the administration of President Ronald Reagan reduced ARC's funding and effectiveness.

The rural problem grew, however, as the metropolitan centers of the United States assumed increasing dominance in the economic and social life of the country. In the 1980s and the early 1990s, the wealthiest farming regions of the U.S. heartland were affected by drought and debt following overinvestment in the 1970s.

Urban Landscapes

Urbanization within the United States produces distinctive landscape forms through the types of building, the differentiation of land uses, and the distribution of groups of people. Human variation in the landscape includes the materially wealthy and the materially poor, African Americans, Hispanic Americans, European or white Americans, Asian Americans, and many others. American cities grew from colonial ports and inland market centers to 1800s industrial and 1900s commercial metropolitan centers. Large cities increasingly dominated the life of the country. Although Americans are often quick to

knock down the old and build new, most cities retain landscape relics of each historic period that combine with the natural environment to give each city a specific character.

Preindustrial U.S. Towns

The colonial ports of New England and the Middle Atlantic were small, with scarcely 10,000 people each and a limited number of functions. Most people lived in the surrounding rural areas. By the mid-1800s, a pattern of market towns and small villages, linked to each other by road, canal, and railroad, covered the farmed plains of Ohio, Indiana, and Illinois. In the Northeast, small factory towns huddled around water mills. South of Chesapeake Bay, plantation agriculture in coastal areas produced individual homes for the planters, surrounded by slave dwellings and the homes of poor white workers. There were few towns, although places such as Charleston, South Carolina, developed as a regional port and important political center (Figure 11.25). Atlanta, Georgia, became the first inland rail junction in the South and began to grow in size before it was destroyed during the Civil War in the 1860s.

Some of the oldest towns grew under the Spanish occupation of the southwestern parts of modern United States, with relics still in Santa Fe and Albuquerque, New Mexico, and San Antonio, Texas. All the early towns comprised relatively small buildings, some in stone or brick, but most in wood. The historic areas composed of such buildings take up smaller areas of modern cities than those in Europe (see Chapter 3).

Industrial and Commercial Cities

In the later 1800s and early 1900s, many cities in the east and center of the United States expanded into large conurbations of housing, factories, shops, and offices with the rapid growth of manufacturing industry and the railroad network. The larger factories and mills required many workers, who were accommodated in surrounding housing units. Areas of worker housing became differentiated from the better housing areas of the managers and owners. New **central business districts (CBDs),** where shops, banks, and other financial services

Figure 11.25 **Historic district: Charleston, South Carolina.** The ornate former homes of cotton plantation owners in the historic district of Charleston, South Carolina. The large homes in the Battery Street neighborhood were built as summer residences to take advantage of coastal sea breezes. Photo: © Kevin Fleming/Corbis.

became concentrated, were easily accessible by streetcar to and from the industrial areas and leafy suburbs.

By 1920, American cities often evolved into a **concentric pattern of urban zones** around the CBD, with poorer housing in inner suburbs and more affluent zones beyond (Figure 11.26a). Some of the inner-city areas housed distinctive segregated groups such as Slavic or Jewish people in areas that became known as **ghettos** (after Jewish quarters in medieval Italian cities). In cities built across a local relief of hills and valleys, the development of railroads, roads, and commercial activities was concentrated along the valleys, giving the city geography a pattern of wedge-shaped urban sectors of different land uses (Figure 11.26b). In some cities, the distinction between land uses was clear, but the distribution was not in such regular patterns, establishing multiple nuclei (Figure 11.26c) in which industrial and commercial activities occurred in several specialized and separated zones, instead of around a single CBD.

During the 1930s and 1940s, American cities grew more slowly than they had in the previous 50 years. The economic depression of the 1930s and World War II made capital either in short supply for house building or directed to military purposes. By the mid-1940s, many families shared cramped inner-city housing units.

Post-1945 Cities

After World War II, an explosion of house and road building opened up huge suburban areas to what had become an ideal of single-family homes for those with rising incomes as the United States dominated the world economy. New federal roads linking major cities followed by the interstate highway system established in 1956 made greater mobility possible and

encouraged suburban housing and commercial development. The outer areas of American cities expanded into formerly rural areas. By the 1990s, for example, metropolitan Chicago occupied 10 times the area its buildings had covered in 1920.

These trends gave rise to a new form of city that developed from the monocentric (single center) before 1940 to the multicentric pattern of today (Figure 11.27). The CBD became more specialized around financial businesses as many of its shopping and industrial functions moved to the suburbs. High land prices led to the construction of tall office buildings with 50 and more floors to accommodate banks and other commercial corporations. Tall buildings formed the skyline of the central area, which sometimes expanded into surrounding warehouse and poor housing districts. New highways, often elevated or in tunnels, circled the CBD.

Beyond this sector of intense commercial activity, the older pre-1940 suburbs, crossed by railroad yards and older highways with commercial strip developments, were often taken over by the expanding African American, Hispanic, or other ethnic groups. Such groups either did not have the resources to move to suburban homes, faced discrimination by financial institutions or home sellers when they tried to do so, or preferred to stay in an area that was familiar to them. The African American ghettos of northern cities did not appear until the influxes of such people from the South and the suburb-bound movements of the white American population in the 1950s.

The post-1945 suburbs were quite different from the inner suburbs built before 1930. They were laid out with more space to allow single-family homes and yards. Such areas often occupy the largest amount of land in American cities. They are interrupted by shopping centers, light manufacturing factories,

Figure 11.26 **United States: urban landscapes, 1900 to 1950.** (a) The concentric zone model, based on a 1922 study of Chicago. At that time, the "ghetto" was not related to African Americans. (b) The sector model, based on 1930s census information, illustrating the impact of railroads and highways on the location of different land uses. (c) The multiple nuclei model of 1945 still has a single center, but commercial and manufacturing districts are not in close proximity to it. Source: From C. D. Harris and E. L. Ullman, "The Nature of Cities" in *The Annals of the American Academy of Political and Social Sciences*, Vol. 242, 1945. Reprinted by permission of the American Academy of Political & Social Studies.

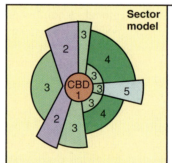

(b)

Sector model

1. Central business district (CBD)
2. Wholesale, light manufacturing
3. Lower-class residential
4. Medium-class residential
5. Higher-class residential

Concentric zone model

1. Central business district (CBD)
2. Wholesale, light manufacturing
3. Lower-class residential
4. Medium-class residential
5. Higher-class residential
6. Residential suburbs

(a)

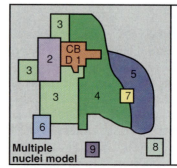

Multiple nuclei model

1. Central business district (CBD)
2. Wholesale, light manufacturing
3. Lower-class residential
4. Medium-class residential
5. Higher-class residential
6. Heavy manufacturing
7. Outlying business district
8. Residential suburb
9. Industrial

(c)

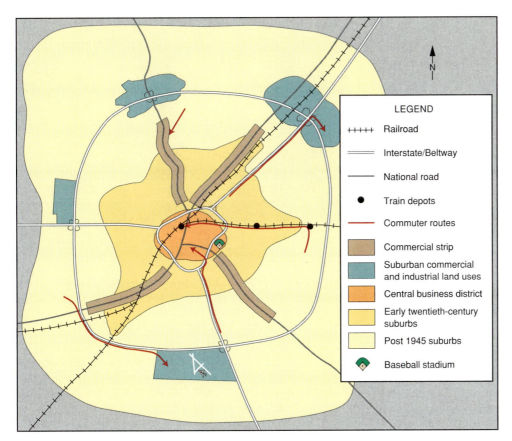

Figure 11.27 **United States: urban landscapes in the late 1900s.**
The monocentric pre-1940 city is now small compared with the post-1945 suburbs in which interstate and other highway links and junctions provide bases for new centers of retail, office, warehouse, and manufacturing employment—the multicentric American city. Some of these centers produce as much as the CBD. The international airport is the center of a distinctive suburban commercial zone. Commuter routes can still be from suburbs to CBD, but many go in other directions. The baseball stadium shown may also be a location for other major sports arenas or the symphony hall.

LEGEND

+++++ Railroad

═══ Interstate/Beltway

─── National road

● Train depots

── Commuter routes

▬ Commercial strip

▬ Suburban commercial and industrial land uses

▬ Central business district

▬ Early twentieth-century suburbs

▬ Post 1945 suburbs

◈ Baseball stadium

and office buildings that associate together and may include more economic activity than in the CBD. Such suburbs are linked by interstate highways and beltways. These types of suburban centers in the larger metropolitan complexes often contain the city's international airport, which itself may be the size of a small city with multiple economic activities surrounding it, from hotels to warehouses and factories.

These major changes in American cities occurred with little or no overall planning apart from designation of land use by zoning. Metropolitan areas in the United States are often a mosaic of local jurisdictions that find it difficult to work together. They take advantage of, or complain about, differences among them instead of looking for possible links. For example, public transit construction forced some jurisdictions to work together, but in other cases, one group of jurisdictions excluded others from the planning process. It is, perhaps, surprising that many regular and common patterns of land use emerge in such a situation. Apart from the absence of green belts around United States cities, the general built environment patterns are similar to what emerged (with a greater degree of planning) in Western Europe (see Chapter 3).

The changing face of U.S. cities since 1950 resulted in many movements of people within the cities. The growth of the suburbs involved people moving out of the older areas to new single-family homes. This trend was labeled "white flight," since it resulted in low-income African Americans and other minority groups being left in the older areas that lost many services and jobs to the growing suburbs. Property values in the older areas declined, sometimes leading to abandonment and dereliction.

The social result in Detroit has been described in graphic terms as a city with an empty center like a donut, or as the "city of death" with high levels of abandonment (Figure 11.28).

Postindustrial Cities

From the late 1970s, expanding American urbanism spread new land uses into former rural areas. The term "edge city" was coined for new exurban developments relying primarily on car and truck transport and secondarily on air travel. Edge cities include large developments of shopping malls, offices, warehouses, and factories, located on the edge of major metropolitan areas (Figure 11.29). Since the 1970s, there were some movements of people from the suburbs farther out from the city into surrounding rural areas and other movements back to the city center.

Gentrification

Gentrification is the movement of higher-income families and individuals into materially poor areas of inner cities, leading to the physical improvement of property conditions and potentially dramatic changes in the quality of life in the districts affected. It was first noticed in British cities in the early 1970s and is now common in urban centers of both the United States and Canada.

Gentrification most commonly involves higher-income men and women typically in the age groups of mid-twenties to mid-thirties and mid-fifties to mid-sixties. They are mostly singles or couples with few children. Demographic trends toward

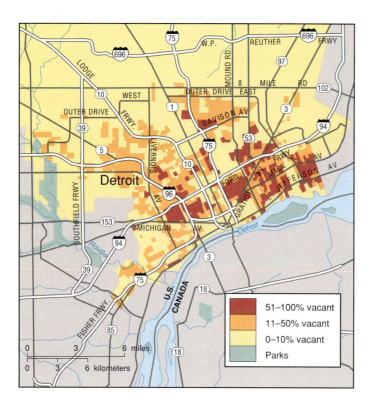

Figure 11.28 **United States: urban landscape in Detroit.** The "city of death" shown in the high levels of housing vacancies in the inner city and older industrial areas, compared to the "fat city" of the suburbs.

Figure 11.29 **An edge city: Bethesda, Maryland.** The characteristics of new blocks of offices and apartments intermingling with parks, highways, and shopping centers are typical of these new urban developments located adjacent to, or just outside large cities. Photo: © Jon Feingersh/Stock, Boston.

later marriage and child-bearing in the younger groups, together with large "empty-nest" homes in the suburbs, rising divorce rates, and widowhood, provide a large affluent group of small households.

With the increasing costs of commuting and suburban housing, many people found it economically worthwhile and convenient with respect to location to move to the inner city. The inner-city properties were on average very low in price in the 1960s and 1970s. In addition to low-cost property and convenience to work, many other pull factors brought wealthier suburbanites back into the city center. Some of the attractive factors include access to leisure-time facilities in theaters, symphony halls, opera houses, museums, restaurants, improved parks and waterfront areas, and major sports facilities. The primary drawbacks consisted of higher crime rates and less funding for public schools. Crime rates have decreased in many inner-city areas resulting from changes in police and city government attitudes as well as changes brought about by the gentrification process itself. Schools have improved in some areas, while in others, it remains common for younger affluent families who begin living in the inner city to move out to suburban locations when their children are of school age.

The movements of people with high incomes into city centers were at first heralded as an "urban renaissance," and much was made of the process in the context of the years of declining and abandoned inner-city areas in the 1950s and 1960s. These cries were muted, however, when it was realized that many

materially poorer people were displaced by the process and had nowhere to go. Quality buildings erected at the end of the 1800s, for instance, often became rooming houses before gentrification turned them into more costly accommodations and displaced the previous occupants. Poorer-quality housing was swept aside and replaced by new apartment blocks. In neither case was provision made for the previous inhabitants.

The extent of gentrification is great in terms of the number of cities affected by it but often modest in relation to the total built environment of each inner city—several blocks rather than larger tracts. In Manhattan, New York, for instance, an area of the Upper West Side between Central Park and the Hudson River and near the Lincoln Center and museums experienced gentrification. Similar areas have been identified in Baltimore, Boston, Philadelphia, Pittsburgh, Washington, D.C., Atlanta, and San Francisco (Figure 11.30).

During the movement to the suburbs in the 1950s and 1960s, large sectors of the older parts of American cities were left behind by middle-class whites and taken over by black and

Figure 11.30 **Gentrification in the United States: San Francisco, California.** Photo: © Emily A. White.

other materially poor minority groups. Housing values dropped and many financial institutions drew lines around areas in which they would not support house purchases with their loans—a process known as "redlining." Redlining led to further deterioration of inner-city areas. Once gentrification began to bring higher-income people back to the city center, the process was reversed ("whitelining"). The financial institutions that had contributed to inner-city decline by refusing loans gained much from the revival of inner-city living, backed by government support in renovation grants. They invested heavily in financing the building of new apartments in the 1970s and 1980s.

Opinion is divided as to whether gentrification is a small-scale process of individual choices over short timescales that affects a few people in each city or whether it is symptomatic of large-scale forces at work in remodeling the cities as part of uneven development. It has not involved the numbers of people that took part in the suburbanization of the 1950s and 1960s, but it has been significant in forming a partial reversal of that trend. It may be the outcome of manipulation by financial institutions, but most people who change their housing locations would not see that as a factor they considered. Certainly gentrification is linked to the continuing changes within urban environments in response to economic shifts, such as the changes in the central business district, demographic changes in family sizes and age

structure, social changes in people's living habits and groupings, and perceived improvements in the environmental conditions of city center areas.

Regions of the United States

The large area of the United States makes it a country of many internal geographic regions that developed from the interaction of people with the natural environments and resources. The historic pattern began on the east coast and spread westward. The geographic distribution of its varied natural resources and human conditions resulted in internal regional units being recognized by groups with widely differing interests, such as geographers, novelists, and government and business administrators.

Many different regional divisions of the United States have been proposed, but the one adopted here is based on the human geographic regions that emerged in the 1800s and early 1900s. Current geographic processes are changing their characters and external linkages, but the regional entities established at that time still form an essential basis for understanding the human geography of the country. The overlapping set of regions in the northeast (Figure 11.31) reflects the outcomes of the initial geographic concentration of European settlements and subsequent industrialization and urbanization.

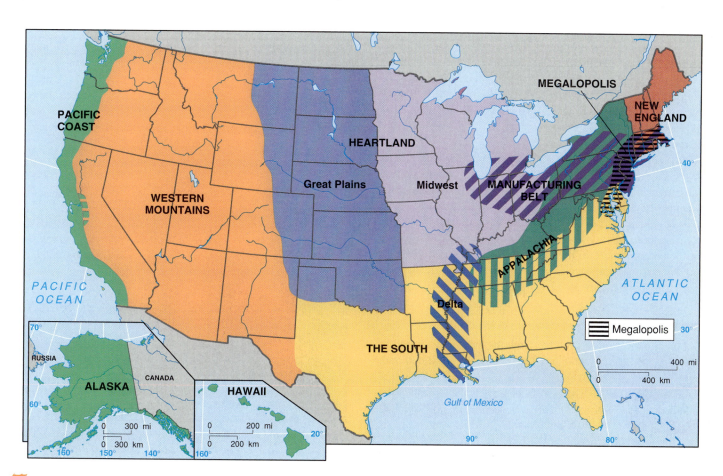

Figure 11.31   **United States: the regions.** They are based on differences in natural environments and on patterns of human occupation. The overlap of regions in the northeast reflects the early concentration of people and changing economic and social emphases.

New England

New England comprises six states in the northeastern corner of the United States. The first European immigrants sailed to the region in 1620 on the *Mayflower* seeking religious freedom. Settlers from Europe first established ports and townships in the southern parts of New England. Secular politics supplanted religious motivations as New Englanders played an important role in the revolution that brought United States independence in 1783. The southern part of the region then became the first part of the United States to be industrialized in the early 1800s, based on water-mill power, local labor, merchant capital, and access to markets by water transport. The early metal, textile, and leather industries grew in scope and factory size through the 1800s. Possessing no coal or oil, the region was at a disadvantage when new industries based on steam and electrical power developed later in the 1800s and early 1900s.

By the mid-1900s, New England's established manufacturing industries were declining. Cities such as Fall River and Lowell, Massachusetts, had many derelict textile mills, high unemployment, and out-migration. During and after World War II, many of the traditional industries were replaced by high-tech computer design and manufacturing, airplane engine-related industry, and others favored again by high levels of defense spending throughout the Cold War period. At this time, the large number of higher education institutions in the region combined research innovations with the local availability of entrepreneurial capital as a basis for a fresh boom of economic development. This is sometimes referred to as the "Massachusetts Miracle" because of the numbers of new jobs created in a declining area. Although periods of struggle and recovery occurred in the mid- and late 1990s for some companies along Boston's growth corridor, it seems the area will remain an important region of high-tech industry concentration.

Another industrial boom may be passing, but many parts of New England retain evidence of wealth as its financial and professional services continue to grow. New England's environment of rocky and sandy coasts, wooded hills, interior hills and mountains, winter sports facilities, and well-preserved early American history, together with the sophisticated provision for entertainment in the cities such as Boston, appeals to many Americans as a place to vacation or as a high-quality living environment (Figure 11.32).

Megalopolis

The Megalopolis region stretches from the greater Washington, D.C., area (by some definitions, as far south as Richmond, Virginia) at the southwest end of its long urban arm to Boston at the northeastern end. The urban chain of the Megalopolis includes the huge metropolitan areas of Baltimore, Philadelphia, and New York, as well as many smaller urban centers, positioned between metropolitan Washington, D.C., and Boston (Figure 11.33). Approximately 50 million people—nearly one-fifth of the total population—of the United States—live in this region. The Megalopolis overlaps the Middle Atlantic and New England regions. In many places, urban land uses follow the

(a)

(b)

Figure 11.32 **New England life.** (a) Coastal New England: waves crash on a rocky coastline. (b) Bostonians enjoying neighborhood festivities on St. Patrick's Day. Photo: (a) © Owaki-Kulla/Corbis; (b) © Ted Spiegel/Corbis.

Figure 11.33 **Megalopolis: spatial distribution.** The Megalopolis region is defined by largely built-over land that has close-spaced metropolitan cities and a predominance of urban and urban-linked activities. Some areas within this definition are rural but linked closely to the urban functions of the region (commuters, truck farming, dairying).

roads and railroads from one city to another without a break. Even the rural areas in this belt are subsidiary to the urban activities in the cities. Farmers produce milk and vegetables for the local urban markets and specialty crops, such as tobacco, for industrial processing. Many rural homes set in wooded lots house commuters, while the coasts and hills provide recreation centers for the urban population.

The Megalopolis phenomenon was first identified in the United States in the late 1950s. The concept is the same as the European conurbation or connected urban chain phenomenon. Such a close-spaced set of cities reflected the historic competition among the closely spaced preindependence colonies and the newly independent states in New England and the Middle Atlantic. Although New York always retained its early advantage as the largest port city, Baltimore grew rapidly in the early 1800s as an exporter of wheat. From 1830, the city fathers of Philadelphia built railroads westward to compete with the New York State Barge (Erie) Canal and New York railroads for access to the newly settled lands in the Midwest. Washington, D.C., emerged as one of the largest U.S. cities when the functions of federal government multiplied in the 1900s. The suburban counties around Washington, D.C., contain some of the most affluent communities in the country.

Despite the migration of some corporation headquarters out of the Megalopolis cities to other parts of the United States, this region still dominates the American economy, federal politics, and the modern media industry. It is the internal core of the United States. Other groups of cities within the United States have been identified as new versions of the megalopolis phenomenon, but none matches this region in its numbers of people or its economic, political, or social leadership.

Manufacturing Belt

Between the 1800s and mid-1900s, an area stretching from Boston on the east coast through New York and Philadelphia westward to Pittsburgh, Cleveland, Detroit, and Chicago became so dominated by manufacturing industries that it was termed the "Manufacturing Belt." The textile and metal goods industries continued to develop in the east as the Pittsburgh-Cleveland area became important in steel and heavy engineering, Detroit became the center of the automobile and truck industries, and the Chicago area produced a range of engineering and farm machinery. The largest cities became the centers of complexes of linked mining and mill towns with high densities of population, high proportions of blue-collar workers, and close-spaced networks of railroad and road links. From Ohio westward, intensive farming produced raw materials for food-processing industries, while other factories made the chemicals and machines needed on the farms. This region produced over two-thirds of U.S. manufactured products until challenged by the West Coast and southern states during and after World War II.

The region remains dominant in manufacturing output within the United States, but older industries are less concentrated here, and many have been replaced by modern consumer-product manufacturing industries taking up sites close to the country's largest markets for such goods. As manufacturing growth slowed from the 1970s, service industries grew in sig-nificance in this region. Chicago became a major center for financial services arising out of its futures trading in agricultural products. Some of the smaller towns became higher education centers, where facilities expanded to accommodate 30,000 and more students on each college campus.

After two decades of slower industrial growth, the western part of the Manufacturing Belt experienced a recovery of its manufacturing industries in the 1990s. Steel output along the southern shores of Lake Michigan rose rapidly to supply demands for sheet steel from expanding auto and consumer goods industries. Unemployment dipped below the national average. Since the late 1970s, the economy of the region diversified within each sector (farming, manufacturing, and services) with rising productivity. In the 1990s, manufacturing employment grew faster there than in the rest of the United States. Autos remained central to the region's factory output, and some Japanese manufacturers sited their factories in the region. The problems of the region changed from high unemployment to a potential shortage of skilled labor.

Pittsburgh, once considered the industrial furnace of the United States, underwent a dramatic transformation during the 1970s and 1980s. Pittsburgh played a prominent role in the growth of the United States through the development of an intense concentration of iron and steel production and related heavy industry. In addition to being known for its significant industrial output, the city also gained a reputation for its pollution. During the city's renaissance of the 1970s and 1980s, many manufacturing jobs were replaced with service positions. Selected factories closed to make way for Fortune 500 companies. The city emerged as an international center for medical research. Pollution-filled air and the glow from the blast furnaces have been replaced with a dramatically modern and vertical skyline (Figure 11.34).

Appalachia

The hilly region of the Appalachian Mountains spans the middle part of the Manufacturing Belt and extends into the U.S. South. Each part of Appalachia has distinctive economic problems.

Eastern Kentucky, West Virginia, western Virginia, and eastern Tennessee form central Appalachia and remain one of the poorest parts of the United States. Coal mining continues to be important but only in areas where the coal has a low sulfur content, making some people materially wealthy from this industry (Figure 11.35). After 1950, the lack of other employment opportunities in central Appalachia caused many younger families to leave it for the prospect of jobs elsewhere. They often found getting jobs difficult because of their poor education and limited skills. The early 1980s boom in coal output when energy prices were high, together with improved infrastructure, health, and education provision, brought people back to central Appalachia, but conditions deteriorated again as coal prices slumped in the mid-1980s.

Although southern Appalachia was one of the poorest parts of the United States in the early 1900s, its economy became diversified with the availability of cheap electrical power through the Tennessee Valley Authority from the late 1930s. Major aluminum companies and federal facilities, such as the Oak Ridge atomic laboratories and the Huntsville rocket center, were joined by many other manufacturing concerns. Cities such as Asheville, Knoxville, Chattanooga, and Huntsville expanded, developing a wide range of service industries. Contrasts of material wealth, infrastructure, health care, and educational opportunities remain between the urban centers and the surrounding isolated rural mountains.

U.S. Heartland: Midwest and Great Plains

The lowland area between the Appalachians and Rockies that is north of the Ohio River and the Ozark Mountains is an essentially agricultural region, including 8 of the top 10 U.S. farming states, and is sometimes called the "breadbasket" of the country because of its production of grain crops. During the period from 1970, this region produced sufficient food to make up for shortages elsewhere in the world, from India to the Soviet Union and parts of Africa. In periods when foreign markets needed less grain, land in the Midwest and Great Plains was idled to prevent overproduction.

The term "Midwest" is mostly applied to the eight states of the Corn Belt and Great Lakes area. The eastern part overlaps with the Manufacturing Belt, so that cities such as Chicago, Detroit, Cleveland, and Cincinnati stand amidst the world's most productive farming region. The Great Plains include North and South Dakota, Nebraska, Kansas, Oklahoma, and parts of Montana, Wyoming, Colorado, and northern Texas. They have a subhumid climate and rise in elevation to over 1,500 m (5,000 ft.) above sea level at the foot of the Rockies.

The settlement of the heartland area in the early and mid-1800s rapidly established the suitability of its environments for different combinations of crop and livestock products. The growing season declines in length toward the north and precipitation toward the west. The brown soils in the eastern states of Ohio and Indiana mostly developed beneath deciduous forest, while the black soils farther west developed beneath grassland; both are fertile and workable.

Commercial farming grew more important in the mid- and late 1800s, as railroad lines linked the region with markets on the east coast and abroad. The new rail lines replaced the longer Ohio and Mississippi River distance as well as the lengthy coastal cargo boat routes that had been commonly traveled. Distance from the main centers such as Chicago and climatic conditions were major factors in the types of farming adopted. A series of agricultural regions emerged, based on a combination of natural advantage and distance costs (Figure 11.36).

- The largest was the Corn Belt, which itself included a mosaic of farming types having varied emphases on corn as a cash crop or—mainly—as feed grain for the fattening of pigs or beef cattle.

- In the cooler conditions north of the Corn Belt, with poorer stony soils on the most recent glacial deposits, dairy cattle were kept and most of the land left in pasture.

- In the drier Great Plains to the west and northwest, wheat became the main grain crop, sown before winter in the south and in spring in the north.

- In favored locations, especially on the eastern lakeshores, fruit growing became a specialty.

- On the hillier southern margins facing the Ohio River and in the Missouri hills, less land was suitable for farming, which was a less profitable occupation.

The patterns of farming established in the 1800s lasted until after 1950, when productivity increased rapidly. Agribusiness—the close commercial linking of inputs to farming, farm activities, and the processing and marketing of farm products—made American agriculture an integral part of a much larger industry. On the farm, agribusiness brought greater mechanization and the use of fertilizers. New varieties of corn became available that would ripen in a shorter growing season, new crops such as soybeans became prominent, new

Figure **11.35** **Appalachia: coal mine in West Virginia.** The Arkwright Mine, with workings that extend 4.8 km (3 mi.) into the hillside. Pickets were posted at the entrance to this mine, near Morgantown, during a miners' strike in the mid 1990s. Many deep mines in the area have been closed as surface mining with its lower costs has increased. This one closed in the late 1990s. Photo: © Michael Bradshaw.

Figure **11.34** **The modern skyline of Pittsburgh, Pennsylvania.** Photo: © Joseph P. Dymond.

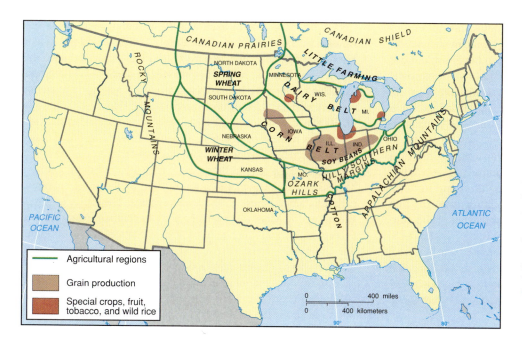

Figure 11.36 **United States Heartland: agricultural regions.** The Corn Belt is the most productive and versatile sector, where corn and soybeans ("grain production" areas on map) are grown for direct sale or for feeding livestock (beef in west, pigs in east). In the cooler north, dairying is important, and in the drier west, wheat and other small grains are dominant, while cattle feedlots dominate toward the south. Farther west, the Great Plains have cattle feedlots and produce wheat in drier conditions. Poor soils, short growing seasons, and mountains mark the boundaries of this productive farming region. High-value crops are grown on costly land close to the main urban areas. Special conditions favor fruit growing on the eastern shores of the lakes where spring comes late, delaying the opening of buds until after killing frosts.

consumer tastes for more vegetables and less dairy foods and meat were expressed, and more rapid road transport made distance costs less significant. World markets for grain opened, especially in the former Soviet Union. By the 1980s, a major shift occurred in the Corn Belt output from grain-fed meat to grain for cash. Soybeans rivaled corn in the southern half of the region, while corn took over from grass in the dairying regions of the north. Many dairy farmers added to their income by growing vegetables and other valuable cash crops.

In the drier areas west of the Missouri River, the development of center-pivot "circle" irrigation led to more corn and other crops, such as sugar beets, being grown in formerly wheat-growing areas. Huge feedlots for cattle were established, with the largest at Sioux Falls, South Dakota. Both irrigation and feedlots required increasing capital inputs at a time of fluctuating prices.

The dependence on groundwater for the expanded use of irrigation raised questions about the limits of water availability. Some streams dried up and the High Plains aquifer—the main source of groundwater—was drawn down. Water levels in the wells of western Kansas fell by 62 percent between 1940 and 1991, raising pumping costs and the specter of using up all the groundwater. Better methods of water use later slowed rates of depletion, and the High Plains aquifer still contains a substantial proportion of the water it contained 50 years ago. While some parts of this area are short of water, others have plenty for the foreseeable future.

Despite the high and increasing farm productivity of this region, major problems arose. Fewer people were able to depend totally for their income on farming and obtained off-farm jobs. Many farmers incurred levels of debt that were too high to sustain so they rented land to other farmers or sold out to banks and other financial businesses. Much land that had resided in family farms became corporately owned. Problems of declining population and few alternative occupations that had plagued other rural regions for decades extended to America's richest farming region during the 1980s.

The Great Plains area, particularly the zone where it meets the Rocky Mountains in Colorado, became a major center of high-tech industries, initially based around the federal facilities in Denver and the University of Colorado at Boulder. The area's population grew rapidly in the early 1990s but slowed late in the decade as air pollution, water shortages, and crowded schools deterred more people from moving in.

The South

The American South extends from the southern Atlantic coastal plain westward to the southern Mississippi River and beyond into eastern Texas. Western Texas has more in common with other states along the Mexican border. The northern panhandle of Texas has some cultural ties to the South but others to the Great Plains and western United States. Florida is physically part of the South, but economically and culturally, it has ties to the southeast and northeast United States, as well as to Latin America.

The region's common cultural identity stems from colonial migration patterns and agricultural practices, as well as membership of the Confederacy states in the Civil War. Most of the South was dominated by the plantation economy, established in colonial times. It received an economic boost with the development of cotton growing from the 1790s. All of the southern states suffered after the Civil War, experiencing decades of poverty at a time when the northern states made huge strides in industrial expansion and garnered significant material wealth. The black slaves were freed but soon became debt slaves, bound to the sharecropped land they cultivated. They did not gain political and social rights for 100 years. The southern Appalachian poor white hill farmers and the small family farmers of much of Texas shared many of the problems of poverty with the African Americans against whom they discriminated.

Economic modernization, including industrialization and the adoption of better farming technology and management, began after World War I. The real changes in the South did not occur until the civil rights of African Americans were acknowledged from the 1960s. This development coincided with the building of the interstate highway system that opened up much of the South to new economic opportunities and the expansion of cities such as Miami, the urbanizing area from Charlotte to Atlanta (Figure 11.37), New Orleans, Houston, Dallas–Fort Worth, Memphis, and Louisville. Service industries, as well as manufacturing, moved into the area, and its changing fortunes were highlighted by local boosters who took up the term "Sun Belt" to refer to the U.S. South with its mild winters and sunnier climate—in contrast to the "Frost Belt" of the U.S. Northeast. Although this term was also applied farther west and much of the economic growth with which the Sun Belt became associated was geographically patchy—more a set of "sun spots" than an evenly prosperous belt—the changes in outlook in the region were real and lasting.

The economic advantages enjoyed by the South since the 1960s include an improved and more focused agriculture, using the better lands for cotton, corn, and peanuts, and leaving the poorer lands to woodland. The demand for cotton increased in the mid-1990s after experiencing a period of market decline.

Figure 11.37 **United States South: Atlanta, Georgia.** The 1996 Olympic stadium is in the foreground, backed by the baseball stadium and the highrise skyline of central Atlanta. The old baseball stadium was demolished after the 1996 Olympics. Atlanta's major league baseball team now plays in the new stadium. Atlanta's growth became a symbol of the economic revival of parts of the U.S. South from the 1960s. Photo: © Marvin E. Newman/The Image Bank/Getty.

Cotton products fought back against artificial fibers while bad weather and insect infestations affected other world producers. The United States' 1994 record cotton harvest was one-fourth of the world total, reflecting increasing yields. While South Carolina's previous highest crop in 1877 was grown on 6.9 million hectares (2.8 million acres), the 1994 crop came from 568,000 hectares (230,000 acres). Exports took half of the U.S. crop. The raising of poultry in factory units is another major capital-intensive farm activity in the region.

Rural areas remain the poorest parts of the South, with the Mississippi Delta lowlands forming one of the most economically challenged areas in the United States. The Delta lowlands extend up the Mississippi River valley to southern Missouri (see Figure 11.31) and were a major area of cotton production until the mid-1900s. Today, the region remains materially poor with relatively less infrastructure, and fewer educational opportunities.

Some new manufacturing jobs have eased the economic deficiencies in parts of the region. Mississippi County, Arkansas, once the leading cotton county in the country, is now one of the largest steel producers. It has two mills making sheet and I-beam steel using scrap metal brought in by river barge. Other industries also moved into this county, but its unemployment is still twice the Arkansas average.

Most Southerners now work in manufacturing and service industries. In the 1920s, textile mills sprang up in a belt northeast of Atlanta on the basis of cheap local land and labor as the cotton farms were mechanized or abandoned. By the 1990s, many of the textile mills had closed as textile industries moved to Latin America and Asia. In the environs of Greenville, South Carolina, such jobs were replaced by those in a variety of new industries, including Michelin tires (from France), BMW cars (from Germany), Lucas automotive electronics (from Britain), and Hitachi electronics (from Japan). Such investments were based on the perception of this region as having skilled labor, often without labor unions, good local support for industry, and pleasant environmental conditions for management. In Texas, the accumulation of oil wealth was paralleled by manufacturing and services development around Houston, Dallas–Fort Worth, and Austin partly related to the National Aeronautics and Space Administration center in Houston and other high-tech research industries.

The service industries in the South include tourism—spreading northward and westward from Florida—plus health care, education, and financial services. Some service industries pay higher wages than semiskilled jobs in manufacturing.

Western Mountains

The western one-third of the United States is a mountainous region. The highest ranges are the Rockies on the east and the Sierra Nevada of California and the Cascades on the west. They are separated by broad plateau areas (see Figure 11.11a) and areas of basin and range. The mountains keep out the rain-bearing winds blowing from the Pacific Ocean, and many parts are in dry, rain shadow areas that result.

Settlement focuses on irrigation farming and towns that grew up as markets, mining centers, or nodes on the transcontinental railroads. A large proportion of the land between such centers is still sparsely settled. Much is owned by the federal government and designated as national parks, national forests, or grazing lands. Even these uses in a sparsely populated area generate conflicts when the interests of ranches and farms interact with environmental conservation or tourism.

The western mountains of the United States are not only high in altitude, but much of the region is arid. Most of the water available falls on the mountains as snow and runs off in swollen streams during the spring season. Many federally funded irrigation projects manage this water supply, with the largest groups of projects along the Colorado River in the south and the Columbia River in the north. Irrigated farming and cheap electricity created a series of productive oases that formed the basis for the growth of cities such as Las Vegas, Nevada, Phoenix, Arizona, and Salt Lake City, Utah. These cities are now service and transportation hubs.

In the southern part of the region, Hispanic people make up a large proportion of the population. Defined in part by the Rio Grande, the border with Mexico is largely a mountainous desert where few people live outside the border towns. The border cities are twinned (Figure 11.38) and hold three-fourths of the Americans and Mexicans living within 80 km (50 mi.) of the border—a population that rose from 9 million to over 15 million people since 1980.

From 1965, the Mexican maquiladora ("mills") program enabled United States and other foreign companies to assemble imported products tax-free in northern Mexico for immediate reexport without duties (see Chapter 10). In the late 1990s, around 2,000 such factories employed half a million Mexicans and brought in one-third of Mexico's foreign exchange. On the U.S. side of the border, many Hispanic people find new homes, making up a high proportion of the local population, which is rising so fast that squatter shantytowns form in southern New Mexico and southern Texas. Mexicans like to buy U.S. goods, spending around half of the wages earned at the maquiladoras (see Chapter 10).

Although the United States–Mexican border areas remain poorer than the U.S. average, people living there have more than twice the incomes of people in Mexico. Environmental conditions in this region are also a concern. The impact of NAFTA on this region is still uncertain, and it will take some years for maquiladora jobs to move to other parts of Mexico (see the "Point-Counterpoint: North American Free Trade Agreement" box, p. 504).

Pacific Coast

From the Seattle area around Puget Sound in the north, to San Diego near the Mexican border in the south, the American Pacific Coast has become a second national core, close in economic importance to that of the region between Boston and Washington, D.C. Settlement increased after the arrival of transcontinental railroads in the 1870s and 1880s. Before World War II, the main products were timber in the north and farm products and the growing motion-picture industry in the south. National defense needs for World War II in the Pacific then placed manufacturing and military centers on the west coast. Puget Sound in the north and San Diego in the south became major naval centers. Seattle, once the corporate headquarters of the Boeing Aircraft Company, grew as the company produced bombers and jet airliners. The Columbia River became a staircase of huge hydroelectricity stations that support 11 aluminum smelters and the Hansford nuclear weapons facility, among other industries—but reduced the salmon catch. Los Angeles in the south also had major aircraft manufacturers, Lockheed and McDonnell-Douglas (now both linked to Boeing), together with a variety of manufacturing and service industries.

After World War II, manufacturing prospered and servicemen returned to live in new communities along the Pacific coast. San Francisco (Figure 11.39) was the only city on the west coast until the early 1900s, when Los Angeles rapidly outstripped its growth. Both cities attracted growing financial service businesses after World War II. Defense cuts in the 1990s led to lost jobs and plant closures, but only a mild recession resulted in California, which was growing again by the mid-1990s as a result of the shift to high-tech jobs, a more diversified range of industries, and increased overseas exports. Over 400,000 new

Figure 11.38
Western Mountains: the United States–Mexico borderlands. The contacts between the two countries led to a series of twinned towns with linked economies. U.S. manufacturers site factories in Mexico to take advantage of cheap labor costs, and less regulation and enforcement of worker safety standards and environmentally polluting effluent.

Figure 11.39 **Pacific Coasts: San Francisco Bay area.** San Francisco occupies the partly cloudy peninsula in the foreground of this space shuttle photo. The Golden Gate inlet is obscured by low clouds forming over the cold water offshore. The line of the San Andreas Fault can be traced from left (north) to right (south) across the foreground. Inland, the urbanized bay shore from Richmond to Berkeley and Oakland, the reclaimed areas of tidal lands, and the complex waterways of the Sacramento–San Joaquin River delta stand out clearly. Photo: NASA.

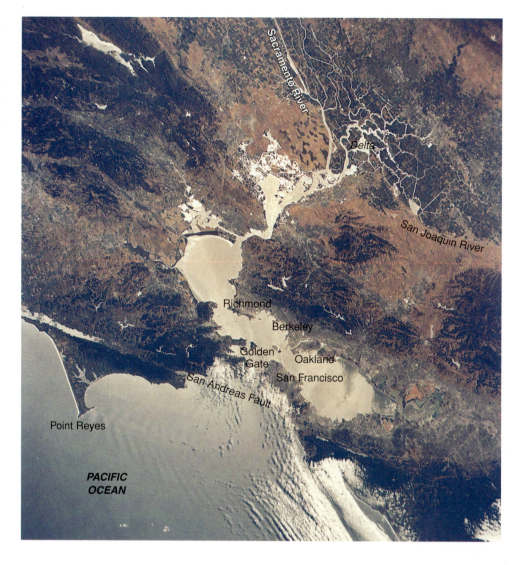

jobs were created in California from mid-1996 to mid-1997, and personal incomes rose. The Los Angeles economy alone is greater than that of South Korea, and the city now contains a reinvigorated motion-picture industry, a growing major port, and more computer software jobs than Silicon Valley, the best-known high-tech center just south of San Francisco. The Seattle area became home to many software firms, including Microsoft.

Important to economic growth in the southern half of California was the supply of water to its largely arid area (Figure 11.40). Federal and state government funds supported huge water storage and distribution projects following the early appropriation of water by the cities of Los Angeles (Owens Valley) and San Francisco (Hetch-Hetchy). The success of these measures is illustrated by the existence of the world's largest desert city in Los Angeles, the world's eighth largest economy in the state of California, and America's most productive farming state.

Years of drought in the 1980s and 1990s stretched limited water supplies. In early 1991, after four consecutive drought years, the major state-level water project reservoirs held less than one-fourth of their capacity, and farmers lost their supplies from that source; the federal project rationed farmers to one-

third of normal amounts. Cities still receive a water supply, but most pay more for it. There are now no further sources of water to be tapped by California. It is likely that farmers, who use 85 percent of the water to produce less than one-tenth of the state's economic output, will have to pay more in the expectation that they will use it more carefully. At present, farmers pay only half the cost of delivering water to them, while city dwellers pay 20 times as much as the farmers pay.

In early 1998, two proposals sought different routes to a middle-term solution of the water shortage problem. They may force a closer study of California's limited water resources and their longer-term usage. The state of California, which used up Colorado River water surpluses from the Nevada and Arizona allocations until they were required in those states, is now forced to return to its original allotment—85 percent of current use. The three states can now exchange water, allowing Arizona to sell enough to Nevada to finance the completion of its own distribution system, the Central Arizona Project. San Diego currently depends on Los Angeles water sources. However, the city is hoping to buy its water from the Imperial Valley to the east, which gained larger Colorado River water concessions than other users in the early 1900s.

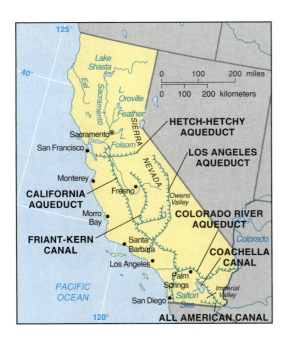

Figure 11.40 **Pacific coasts: water flows in California.** Water flows from the high precipitation areas of the north, where reservoirs store the water, toward the arid south. Water supplies for San Francisco (Hetch-Hetchy) and Los Angeles (Owens Valley) in the early 1900s were later enhanced by federal and state water projects in the Central Valley. The federal Imperial Valley project and Colorado River Aqueduct in the far south took water from the Colorado River to irrigate dry lands and provide water for Los Angeles and San Diego.

The West Coast developed a new importance through an increasing volume of business and personal movements with Asia and other Pacific countries. Seattle and especially Los Angeles are major world ports. From its development as a frontline base for war operations against Japan during World War II and later in Korea and Vietnam, this region is now America's door to the world's most rapidly growing countries. The West Coast attracts most of the growing Asian investment in the United States. Five of California's top 11 banks are Japanese owned; 30 of the smaller ones are backed by Chinese money. Asian capital, particularly Japanese, backs car design centers, media industry corporations, and leisure industries. Half of the 1,400 Taiwanese companies in the United States are based in California, and research-marketing-manufacturing links across the Pacific Ocean are increasing.

The region also has trading links to Europe and Latin America, and close links with British Columbia in Canada to the north and Mexico to the south. In 1990, just under half of California's exports went to Asian countries, around 30 percent went to Europe, and a further 18 percent was shared between Canada and Mexico. Such links attracted Hispanic and Asian people to live and work in the region. Southern California has the highest proportion of Hispanic Americans in the country, with Los Angeles having doubled its middle-class Hispanic population since 1980 to half a million; its Hispanic businesses have doubled since 1993. It is highly unlikely in the short term that the competition between the U.S. Pacific and Atlantic coastal areas is likely to cause the country to splinter as both regions play a role in the country, the hemisphere, and the world.

Alaska and Hawaii

In 1959, Alaska and Hawaii (Figure 11.41) were added to the 48 conterminous states of the mainland United States. They contribute to the extreme variety of environments and resources within the United States but have their own problems.

Alaska is a huge area of northern land that was bought from Russia in 1867. High mountains, icefields, and glaciers mark its southern coast. Inland, the Yukon lowlands have long cold winters, and to the north, the bleak Brooks Range and North Slope leading down to the Arctic Ocean get even colder. No commercial farming is possible. The Yukon lowlands widen westward to the Bering Sea, but there is little mining or manufactured output to trade in that direction. The dramatic mountain scenery and abundant wildlife widespread in the state do support a growing tourist industry.

After purchasing Alaska in 1867, the federal government tried to resist attempts to settle and develop it, but gradually, salmon fishers and gold miners whittled away this policy. Further exploration showed that Alaska contained resources of fuels and metal ores on a large scale, but only a few thousand people lived in the territory. During World War II, when the Japanese invaded the western islands of the Aleutian chain, the strategic importance of Alaska was recognized, and the military presence in Alaska grew. Military activity increased further through the Cold War because the territory was close to the former Soviet Union. Statehood in 1959 gave Alaskans opportunities to select lands for development from the 99 percent owned by the federal government.

The discovery of oil on the North Slope forced the state and federal governments to develop policies for Alaska's native peoples, who live mainly on coastal lands, and for the other settlers, most of whom live in and around Anchorage in the south. The oil from the North Slope made many of Alaska's 500,000 people wealthy in the early 1980s. Residents of the state receive payments from the state government from taxes on oil production. World oil prices fell from the mid-1980s to the late 1990s, with oil production peaking in 1988. The oil discoveries, and especially the oil spill from the Exxon *Valdez* tanker, made Alaska a focus of environmental concern that is now being transferred to other issues in the sparsely settled state. Although oil exploration slowed with the prevention of drilling in the Arctic Wildlife Refuge on the north coast east of Prudhoe Bay, parts of the National Petroleum Reserve to the west of Prudhoe Bay were opened for bids in 1998, looking to a 2010 pumping start.

Hawaii was a long-time hinge-point of U.S. trade in the Pacific, a naval port, and a source of agricultural products such as pineapples and sugar. Involvement in World War II brought its existence home to many more Americans, and it was admitted as the fiftieth state just after Alaska in 1959. Tiny compared to Alaska, but with more people, it has a major international tourist industry. In the late 1990s,

Figure 11.41   **Alaska and Hawaii: the last two states to join the USA.** They contrast in size, climate, and resources. Explain why the smaller Hawaiian Islands have twice the population of Alaska.

Hawaii's economy was in difficulties, with high unemployment. Its tourist industry suffered from cutbacks on travel by the Japanese as the yen lost value. Sugarcane plantations closed as Asian competition and changing U.S. consumption habits reduced the market. Hawaii also suffers from a poor business climate due to many regulations, poor schools, a high union presence, and its geographic isolation from globally connected business systems.

 # Canada

The three thousand miles of distance from Victoria, British Columbia, in Canada's southwest to Halifax, Nova Scotia, on Canada's southeast coast, coupled with the narrow width of the concentrated region of population along the country's southern border results in a series of regionally individualistic groups of people who have to try to interact domestically to make Canada

function as a country. Geographic logic might suggest the likelihood of Canadians interacting more easily with cities and systems across the border in the United States. Although Canadians do interact with urban systems across their border, the people of the country function internally in domestic subregions and on a country level through their federal government. The distinctive subregional groups within Canada's borders include the people of the Atlantic Provinces, the French-speaking peoples of Québec, the multicultural population of Ontario and especially of Toronto, the scattered farming populations and industrial cities of the Prairie Provinces, the people of Vancouver and the British Columbia coast who have an increasing outlook toward Asia, and the indigenous Canadians, or First Nations, of the northlands (Figure 11.42). The Canadian provinces were created a century after the U.S. states, are generally much larger, and have greater internal political control than U.S. states do with respect to the federal government.

In the 1990s, a growing minority of independence-seeking people in Québec attempted to secede from Canada, poten-

The United States continues to be seen as both a friend and an opponent. Canadians enjoy many of the benefits of affluence and national security because of their close integration with the American economy and defenses. Canada, along with the United States, is a member of NATO. Canada has a history of military alliance and cooperation with the United States. Both countries worked together to maintain a line of radar warning stations facing the Soviet Union across the Arctic Ocean during the Cold War. Canada and the United States participate in open free trade and (with Mexico) are each other's most important trade partners. The two countries first entered into formal open trade relations with the United States–Canada Free Trade Agreement of 1987–1988, which eliminated most tariffs and trade barriers. Both countries then joined with Mexico to sign and formally participate in the North American Free Trade Agreement in 1993–1994 (see the "Point-Counterpoint: North American Free Trade Agreement" box, p. 504).

Political and economic ties are not the only areas in which Canadians cooperate with the United States. The two countries have integrated forms of entertainment. Much of Canadian television programming is U.S. produced. The two countries are integrated into many of the same professional sports leagues, including the U.S. National Hockey League and the professional baseball leagues of the United States. Hockey and baseball teams representing the larger cities in Canada participate in the U.S. professional leagues with exactly the same status as teams representing U.S. metropolitan areas.

Although Canada benefits from its geographic proximity to and societal commonalities with the United States, situations result from the same proximity that elicit criticism from the Canadian population toward the United States. Many Canadians object strongly to being the recipients of acid rain from pollution generated by heavy industry in the Ohio River valley of the United States. In the late 1990s, problems arose on the west coast over salmon fishing yields in Alaska, British Columbia, and Washington. Canadians enjoy very low crime rates and have willingly accepted more stringent gun-related legislation from their government. There is a perception in Canada that the United States is exceptionally crime ridden, and many Canadians joke that everyone in the United States is walking around with a gun.

Although living next to an economic giant generates some resentful and even fearful attitudes among Canadians, a few Canadian activities present challenges for the U.S. government. One area of Canadian activity that is of great concern to the United States is the smuggling of high-quality marijuana into the United States from its production area around Vancouver. In addition, Canada's external affairs policies during the Cold War were sometimes at variance with those of the United States. Canada developed relations with some countries in the Soviet camp and some nonaligned, materially poor countries. In the 1990s, Canadian companies took over the development of Cuban mining from the former Soviet groups, despite a U.S. embargo. Nonetheless, the positive aspects of the juxtaposition of the two countries, the shared media and trade, the political cooperation, and the generally friendly attitudes the residents of the two countries have toward one another far outweigh any negative attitudes or friction emanating from either country.

Test Your Understanding 11C

Summary   The United States of America is a city-based country thriving on high-tech manufacturing and growing financial and business services industries. Its federal government structure led to power moving from the states to Washington, D.C., during the 1900s. The variety of regions within the United States reflects the uneven distribution of natural resources and the historic processes of occupation, industrialization, and developing transportation and telecommunications facilities.

The population of the United States is becoming increasingly multicultural as the European groups and African Americans who peopled it until the early 1990s are added to by peoples from Latin America and Asia. Immigration continues to be important in population growth.

The U.S. economy has the largest GNI in the world and is market-oriented. The country's farming is highly productive and commercially organized, its diversified manufacturing output reacts to changes in technology and world demand, and its service industries are now the dominant employers.

Questions to Think About

11C.1 What are some of the ways in which the United States influences global activity?

11C.2 How do the regions within the United States compare and contrast in terms of population distribution, economic activity, physical environment, natural resource availability, and global connections?

11C.3 Which are the two primary regions of dense population, economic, political, and international activity in the United States? How have they changed over time, and what are the key factors in shaping these regions?

Key Terms

Group of Eight (G8)
UN Security Council
uneven development
Tennessee Valley Authority (TVA)

Appalachian Regional Commission (ARC)
central business district (CBD)
concentric pattern of urban zones
ghetto

tially fracturing the Canadian state into at least two or more individual countries. In addition, the closeness of the major centers of population and commercial development to the United States and the beaming of media across the border from the United States have powerful influences on modern Canadian life that could blur the clarity of a cohesive Canadian national identity. The unique population distribution of Canadians, combined with living in the shadow of the globally mighty United States, actually appears to provide strong incentive for, rather than detriment to, the formation of a national identity (the Québec issue aside). Perceptions of people living outside the region, as well as many residents of the United States, that Canada is nothing more than a "satellite" or "subsidiary" of the United States may provide momentum for Canadians to strive to show the world they are separate and different from the United States.

Figure 11.42 **Canada: regions, provinces, territories, and cities.** The Canadian provinces have greater powers than the individual states of the United States. Their geographic distribution in a line north of the U.S. border often makes communications difficult among different parts of the country. The red lines indicate the division of regions within Canada.

People

Population Growth

In the 110 years from 1891 to 2000, the Canadian population increased by approximately 26 million, from 4.8 million to 31 million. Over the same period, the U.S. population increased by over 200 million (63 million in 1890 to 281 million in 2000). Both countries received immigrants in the early 1900s, but two world wars separated by a major economic depression in North America deterred further large-scale increases until after World War II. In the post–World War II era, population

growth in Canada was rapid, doubling from 1941 to 1971 before slowing to a one-fourth increase from 1971 to 1991. Canada's total fertility rate of just under 2 is less than that of the United States, but its population growth rate is similar. Canada has lower death rates than the United States, and its continued immigration rates almost match those of the United States (see Figure 11.20). Canada's age-sex diagram (Figure 11.43) resembles that of the United States.

Regional Changes in Population

Canada's population growth involved regional shifts (Figure 11.44). In 1891, Ontario and Québec had three-fourths of

Canada's people and the Atlantic Provinces just over 18 percent. The rest of Canada held a mere 7 percent. The 1900s growth of the western provinces reduced the dominance of the eastern areas. The Atlantic Provinces declined in their proportion of the Canadian population as the western provinces grew. By 1931, they had less than 10 percent of the total, and although Newfoundland became part of Canada in 1949 and raised the proportion to nearly 12 percent, population growth in these four provinces continued to be slower than in Canada as a whole, and their proportion fell to 8.5 percent in 1991. Ontario and Québec contained over 60 percent of Canada's population between them throughout the 1900s. Québec's pro-portion stayed around 28 percent until 1971, then fell to 25 percent in 1991; Ontario's rose from around 33 percent in 1931 to almost 37 percent in 1991.

The provinces west of Ontario grew rapidly early in the 1900s, and by 1931, nearly 23 percent of Canada's people lived in the Prairie Provinces and a further 7 percent in British Columbia. At this time, Saskatchewan had a higher population than neighboring Manitoba and Alberta. From 1950, however, Alberta increased its industrial capabilities and became the dominant province with a population in excess of the other two combined. The proportion of Canadian people living in the Prairie Provinces declined from 23 percent in 1931 to around 17 percent in 1991, reflecting rural out-migration. On the Pacific coast, British Columbia had around 7 percent of Canada's people in 1931, rising to over 12 percent in 1991.

Canada is experiencing a trend toward metropolitan expansion. The two largest metropolitan areas, Toronto and Montréal, rival many U.S. metropolitan areas in size. Together with Vancouver on the west coast and Ottawa, the capital, these cities contain one-third of Canada's population. Winnipeg, Edmonton, and Calgary are other major Canadian cities. Canadians are less divided by segregation than Americans, but congregation is important among newer immigrant groups in big cities such as Toronto and Vancouver.

Canadian City Landscapes

Of Canadian cities, only Québec has a historic heart with buildings older than the 1800s. Most cities from Toronto westward were built almost entirely in the 1900s. Other major differences between Canadian and American cities include the level of planning involved and the relationships of government units within metropolitan centers.

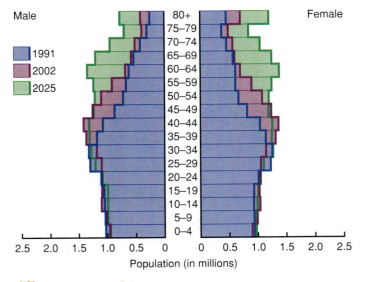

Figure 11.43 **Canada: age-sex diagram.** The 2002 graph details a maturing population with the largest numbers of people in the 35- to 44-years age groups. Source: U.S. Census Bureau. International Data Bank.

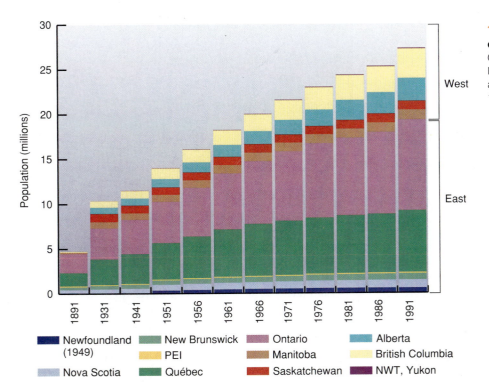

Figure 11.44 **Canada: growth of population by province, 1891 to 1991.** Compare the relative importance of the Atlantic Provinces, Québec, Ontario, the Prairie Provinces, and British Columbia in 1891, 1931, 1961, and 1991.

Newfoundland (1949) | New Brunswick | Ontario | Alberta
Nova Scotia | PEI | Manitoba | British Columbia
 | Québec | Saskatchewan | NWT, Yukon

Toronto is the largest Canadian city and the center of Canadian financial services and manufacturing industries, as well as provincial capital of Ontario, Canada's wealthiest province (Figure 11.45). When Toronto began to extend suburbs into surrounding jurisdictions in the 1950s, the province of Ontario required that these jurisdictions and the city plan cooperatively to make possible the amalgamation of services and regional road construction (Figure 11.46). Further growth in metropolitan Toronto was encouraged around hubs outside downtown, including North York, the area adjacent to the international airport, and a new center with coordinated and concentrated development. A major project was also undertaken to redevelop the waterfront.

Toronto was able to cope with increasing immigrant groups moving into older inner suburbs: Italian, Greek, Portuguese, and Chinese districts form distinctive ethnic enclaves of lively congregation but not segregated ghettos. The city capitalizes on the increasing diversity of its residents. Restaurants, stores, art galleries, and festivals reflecting the vast international heritage of the citizenry continue to multiply throughout the city. Locals and tourists alike crowd ethnic business establishments, helping the diversifying economy of the metropolitan region. Toronto retains a busy downtown and surrounding older suburbs, and it has the lowest homicide rate of large North American cities, uniformly good schools, good public transportation systems, and controlled urban sprawl. Many cities in the United States envy Toronto's development situation, since they have problems of multiple jurisdictions and limited overriding planning controls that affect only a few functions. Some U.S. citizens, however, find Toronto too ordered and even monotonous.

Figure **11.45** **Canada: urban landscape in central Toronto.** The skydome, CN Tower, and downtown commercial buildings as seen from Lake Ontario. Toronto's lake frontage made it a major port with the opening of the St. Lawrence Seaway in 1959. In the late 1900s, many dock areas became derelict, and the waterfront was redeveloped to provide public access to recreation opportunities. Photo: © Wolfgang Kaehler/Corbis.

Economic Development

Canadian economic development generally followed that of the United States but often with a lag of several years until the later 1900s. In the 1990s, Canada combined a continuing emphasis on its natural resource base with being an affluent, high-tech society. It had one of the highest proportions of trade per capita in the world. Canada's GNI per capita in the 1990s rivaled that of many European countries but was still not quite equal to that of the United States. Its relative position worsened during the decade.

Until the mid-1900s, Canada's economy depended mainly on primary products such as grain, timber, and minerals. Canada remains a major world producer of newsprint, wood pulp, and timber, and is one of the world's leading exporters of minerals, wheat, and barley. Canada produces 30 percent of the world's newsprint, which, with wood pulp, is manufactured

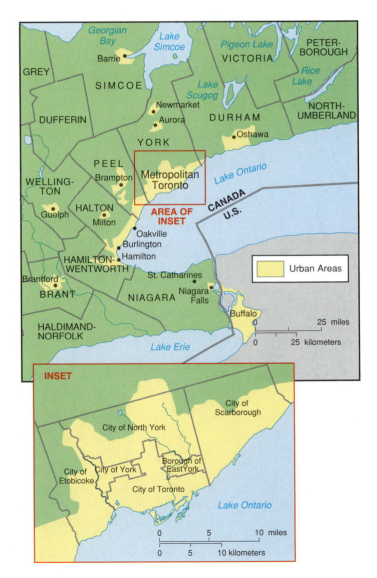

Figure **11.46** **Canada: extent of metropolitan Toronto.** The constituent jurisdictions and surrounding counties in southern Ontario. Toronto's development is controlled by the province of Ontario and has been planned carefully.

mainly along the lower St. Lawrence River valley. Most timber output comes from the west coast forests. The minerals that place Canada in the forefront of world mining countries include coal, oil, and natural gas from Alberta, iron ore from Labrador and Québec, uranium from Ontario and Saskatchewan, nickel from Ontario and Manitoba, and zinc from several places. Agriculture remains significant in the Prairie Provinces, southern Ontario, and the specialized fruit-growing districts of British Columbia, where wine production is gaining international attention (see Figure 11.22).

Industrialization in Canada was a much later phenomenon than in the United States. Beginning before World War II and developing rapidly during the 1940s, the production of aluminum, vehicles, and consumer goods became important as Canada developed import-substitution industries to supply its own markets and avoid total economic domination by the United States. Hamilton at the western end of Lake Ontario became a steelmaking center. Montréal, Toronto, and Vancouver became centers of financial services and a wide range of commercial enterprises. The signing of NAFTA augmented trade between the two countries, resulting in increased U.S. investment in Canadian industries.

Within Canada exists a greater political consciousness of regional disparities and a greater will to do something about it than in the United States. Canadian regional policy enables the federal government to stimulate economic growth in the more materially poor provinces, such as the Atlantic Provinces and the largely agricultural Prairie Provinces by supporting some services and financial investments in infrastructure. Such policies are popular in the recipient provinces, but their impacts on both source and recipient regions are controversial.

Regions of Canada

Canada has a federal government that links the 10 provinces, the Northwest Territories, the Yukon, and Nunavut—the country's main political regions. This arrangement was decided by Britain in the 1867 Act of Confederation. That act united the different parts of what became modern Canada in the face of a perceived military threat from the United States after its Civil War ended in 1865. In 1982, Canada ended its legal ties to Britain, although it remained within the Commonwealth of Nations. Canada was then free to determine the constitutional roles of federal and provincial governments. That was not easy. Several attempts to reconcile the wishes of the people of Québec with those of other provinces highlighted the problems of a federal constitution in which the provinces have, in many ways, greater powers than the central government. This situation contrasts with that in the United States, where the Constitution was generated internally nearly a century earlier, in 1787. The U.S. states are generally smaller in size than Canadian provinces and lost powers to the federal government during the 1900s.

Atlantic Provinces

Canada's political tensions are enhanced by its regional economic and cultural differences. The east coast was settled first and forms the small hilly Atlantic Provinces of Newfoundland

(which has jurisdiction over the almost uninhabited Labrador), Nova Scotia, Prince Edward Island, and New Brunswick. The small-scale economy of these areas based at first on fishing and farming was augmented locally by mining and manufacturing and by the naval base at Halifax. These provinces remain the primary recipients of federal regional aid.

In the 1980s and early 1990s, the region was hit badly by declining fish stocks on the Grand Banks. Some 30,000 fishers and fish plant workers lost their jobs as cod stocks virtually disappeared. Unemployment in Newfoundland rose to over 20 percent. Few new industries were attracted by government efforts. One source of hope could be to switch some fishing capacity to commercial sealing, although environmental concerns have to be assessed against the economic plight. Other prospects, such as pumping oil from the Hibernia field 315 km (200 mi.) offshore in the main iceberg lanes, would be very costly and environmentally risky. The development of nickel and cobalt mining at Voisey Bay, Labrador, scheduled to start production in the early 2000s, will bring wealth to the owners and possibly a few local workers but will have little other local impact. It is significant on a world scale, however, that the cobalt from Voisey Bay replaces the falling output from the previously dominant mines in the Democratic Republic of Congo.

Although the Atlantic Provinces contain some groups of French speakers, particularly in New Brunswick, sympathy for them is eclipsed by the potential outcomes of Québec leaving Canada. An independent Québec would sever direct ground contacts between the Atlantic Provinces and the rest of Canada. The possibility forces the Atlantic Provinces to consider independence for themselves, a seemingly impossible prospect, given the level of federal economic support. If the Québec rift occurred, the Atlantic Provinces might look south: New England's shopping opportunities and events in Boston are closer than those in Toronto and already draw many visits across the border.

Nova Scotia, Prince Edward Island, and, to a lesser degree, New Brunswick are developing tourist industries in an attempt to replace lost components of the economy. Nova Scotia has a well-structured and geographically comprehensive tourist industry covering all parts of the province. New jobs in the region may be found in retail, hotel, and restaurant services.

Québec

The province of Québec was settled shortly after the first Atlantic coast settlements by French people who developed a distinctive type of **long-lot** land settlement along either side of the St. Lawrence River. The French long-lot pattern, which first appeared around Québec City and later diffused upstream to Montréal, was designed to maximize land ownership along the river. Québec became part of British Canada following General James Wolfe's defeat of the French army under Montcalm at Québec City in 1759. The French settlers and their descendant Québecois resented living under their conquerors. The economics of their culture were barely above subsistence, but traders bringing furs from far inland supported a merchant class during the 1800s. The combination of French language and Roman Catholic religion forged a strong loyalty that shifted to political activism in the 1900s as a reaction against

Anglo-Canadian control. The erosion of French speech in the rest of Canada as new immigrants settled the prairies fostered a resolution that this would not happen in Québec.

As a result of such political activities, French was accepted as an equal national language, and Québec gained other concessions from the remainder of Canada despite not acknowledging English as an alternative language within its province. Québec looks beyond Canada to other French-speaking countries and takes a leading role in developing a new global French technical language rather than simply accepting English words.

The province of Québec covers a large area, extending northward to include most of the peninsula east of Hudson Bay (Figure 11.47). The majority of the population lives along the St. Lawrence River estuary. The two largest cities, Québec and Montréal, have a range of manufacturing and service industries that place them among the world's great cities (Figure 11.48). A series of industrial towns use local hydroelectricity to power timber industries, pulp and paper mills, and aluminum refineries along one of the world's main shipping lanes. With global market production of wood pulp extending to tropical areas, the prices—and hence, the well-being of Québec's major industry—fluctuated since the late 1980s. Expansion of global demand, in Asia, Europe, and North America, does not always balance the production output.

North of the St. Lawrence River estuary, the land is bleak, eroded by former ice sheets and covered by coniferous forest, lakes, and tundra. The rocks contain large mineral resources that are gradually being exploited as world markets and transport facilities make this possible. One of the major potential resources is hydroelectricity, and Québec is investing heavily in new facilities to export power to the United States. In doing this, it came up against problems of Native American land rights that are stalling parts of the plan. Overall, Québec produces 25 percent of Canada's manufacturing output and has a strong economy in its own right.

Ontario

The province of Ontario contrasts with Québec. Although it was settled later than the provinces to the east, Ontario became the center of British rule after 1776 and possessed the best farming land. Southern Ontario between Lakes Huron, Ontario, and Erie has a relatively mild climate, considering its interior location on the North American continent. The regional climate is modified by a maritime influence from the very large lakes and from being situated farther south than most other parts of Canada. Soils and climate suitable for pasture, grain crops, and tobacco made this the most attractive part of Canada for immigrant farmers in the 1800s. The city of Toronto arose at that stage, but its main development occurred from the middle of the 1900s, when the opening of the St. Lawrence Seaway turned it into an ocean port, the French-language policies of Québec drove English-speaking businesspeople westward from its former rival, Montréal, and its role as capital of Ontario expanded. Toronto is the largest city of Canada and continues to grow rapidly.

Northern Ontario extends to the shores of Hudson Bay, but its ancient rocks were scraped bare of soil by ice sheets and are covered partially by coniferous forest. Local deposition of clays in meltwater lakes provided usable soils, but growing seasons are short. Most of the settlements in northern Ontario are mining settlements, including the huge complex around the nickel mines of Sudbury. Settlements in the areas west of Sudbury and north of Lakes Huron and Superior are few and far between and constitute a major gap in the linear east-west Canadian settlement pattern.

Prairie Provinces

The Prairie Provinces of Manitoba, Saskatchewan, and Alberta were settled following the building of the Canadian Pacific and Canadian National Railroads in the late 1800s (Figure 11.49). In the early 1900s, they became major wheat-growing areas.

Figure 11.47

Canada: Québec's geography. The major cities, mining areas, industrial centers, and hydroelectricity sites. Nearly all economic activity crowds into the St. Lawrence River lowlands.

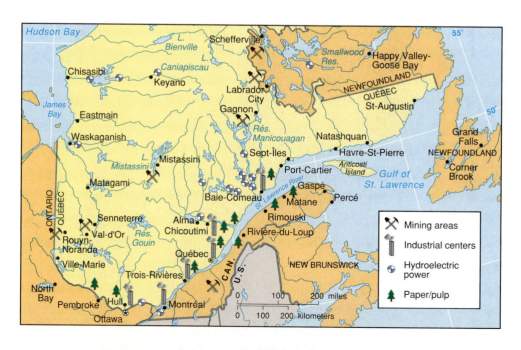

The crop was planted in the spring and ripened in the short growing season, which was as little as 90 days on the northern margin of the plowed area. Toward the west, there was insufficient moisture for this, and cattle ranching took over. The area is essentially a northward extension of the Great Plains in the United States. The ground rises westward in a series of steps, further reducing the growing season and precipitation. Many of the agricultural districts had out-migrations of people since 1950 as a result of the mechanization of farming and the closing of marginal farms.

Winnipeg, Regina, Edmonton, Saskatoon, and Calgary were railroad towns that became grain and meat markets, and three of them became provincial capitals. After 1950, the discovery of coal, oil, and natural gas in Alberta brought extractive and manufacturing industry wealth and more people to that province. The large distances east and west to Canadian ports raised the price of exports and led to some of the main energy markets being southward in the United States.

West Coast

British Columbia (see the "Personal View: Canada" box, p. 534) faces the Pacific Ocean and includes all the mountainous west of Canada apart from a range shared with southwestern Alberta. It is a province of geographically concentrated economic activity, including pockets of mining, fruit growing, and tourism in small lowland areas that are separated by high mountains. Most of the people of the province live in the Vancouver area, extending to Vancouver Island. Attempts to develop the port of Prince Rupert farther north were not so successful.

The people of British Columbia often distance themselves mentally as well as physically from the rest of Canada. Some

Figure 11.48 **Québec City.** The provincial capital of Québec, Québec City is situated at the confluence of the St. Lawrence and St. Charles River. In the foreground, the water front area is referred to as Lower Town. Upper Town looms on a bluff in the background with the imposing Chateau Frontenac hotel dominating the cityscape. Québec City was designated as a World Heritage site by the United Nations. Photo: © Joseph P. Dymond.

of them act more British than the British themselves with overly elaborate rituals of afternoon tea common in Victoria (Figure 11.50), while others emphasize the Pacific links of Vancouver, which has an increasing Asian population and growing economic links with Asia. Asian migrants brought money to metropolitan Vancouver, which facilitated the growth of many metropolitan industries. The growing Asian community further diversifies the culture of the city. In some Vancouver neighborhoods, however, tension developed between some British Canadians and Asian migrants. Many Asians living in Vancouver garnered wealth in business endeavors in Hong Kong, and they tore down traditional dwellings in parts of Vancouver and erected large Asian-style homes in their place, which frustrated long-time residents who preferred the traditional architectural styles of the residential areas of the city.

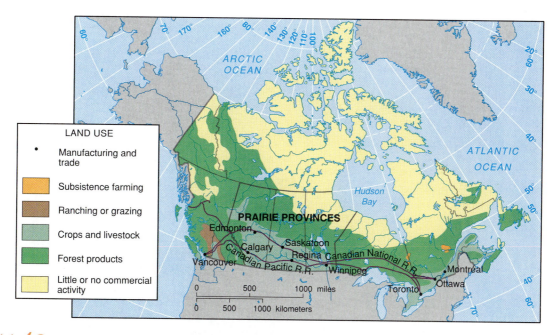

Figure 11.49 **Canada: Prairie Provinces in the context of land use patterns.** The building of the transcontinental railways was linked to the farming development of the prairies and the growth of the major cities west of Toronto.

Personal View

CANADA

Gwen and Jim own and operate a bed and breakfast in Vancouver, British Columbia, Canada. The two are proud to be Canadian, and although they assert the existence of strong ties between the peoples of Canada and the United States, they see a distinction between the cultures of the two exhibited in the respective social structure and systems of each. Gwen and Jim marvel at the vast expanse of their country, the abundant wildlife, the cleanliness, diversity, and safety of Canada's rapidly growing cities, the unique native history and rural livelihood of people thriving in high latitudes on some of the world's least-populated lands, and the seamless fusion of tradition with modernization. When asked about their personal view of Canada, they highlight the diversity of societal roles and cultural traditions as being distinguishing elements of Canadian life and nationality. They discuss the coexistence of traditional fishermen in the Atlantic Provinces of eastern Canada with new service providers for the region's growing tourism industry. They point out the contrasts between the semisubsistence lives of Canada's many native groups, some living very far to the north where summers and winters bring 24 hours of light and 24 hours of darkness, respectively, with Canadians living in large metropolitan areas. They describe Québec and Ontario, the two most populated provinces, as regions where the cities are as cultural and as sophisticated as any in the world. Although they are not French Canadians, they seem to take pride in the remarkable human landscape where Anglo-Canadian and French cultures merge in Québec. They say that in Québec, you could mistake parts of Montréal for Paris and Québec City for a fortified town like St. Malo in northern France. And they will tell you the three Prairie Provinces—Alberta, Saskatchewan and Manitoba—in addition to being a good source region of oil and natural gas for the country, have some of the wealthiest agricultural lands in the world and produce excellent grains.

National pride fully surfaces in Gwen and Jim when they discuss their home province of British Columbia and especially the region surrounding the city of Vancouver (Box Figure 1). The following is their personal view of life in British Columbia.

British Columbia (BC) is a mixture of people from all over the world. The first residents of BC were the First Nations people who arrived from Asia thousands of years ago. The First Nations culture is still an important part of BC life and is exhibited in many forms of art, literature, song, dance, and food. The Spanish and English were the first Europeans to explore the region. The British established fur trading posts that eventually evolved into permanent settlements, together with towns founded near mining, fishing, and logging sites. Compared to the rest of Canada, European settlement in BC is very new.

Most adults living in BC moved here from other places. In the 1950s, 1960s, and 1970s, Europeans were the primary immigrants to the region, while in the 1980s and 1990s, Asians came in large numbers. In the 2000s, Americans are moving to BC in record numbers. These Americans seem to love the region's natural beauty and very low crime rates. The steady growth of the BC economy over the last 50 years created employment opportunities that attracted people from all over the world, resulting in an increasingly plural culture in the region. The depth of multicultural expansion in BC is especially evident in the province's largest city, Vancouver. Metropolitan Vancouver is home to people who trace their ancestral heritage or national origin to more than a hundred different countries. The Chinatown section of the city is the second-largest ethnic Chinese neighborhood in North America. Multilingual signage, varied architecture (especially distinctive cathedrals, mosques, and temples), and dozens of multiethnic festivals celebrating the best aspects of cultures from around the world are some of the many visible elements of Vancouver's diverse cultural landscape. Vancouver's world-class restaurants blend North American, European, Asian, and East Indian flavors. In Canada, these cultures are not melted down into one but are individually preserved, treasured, and celebrated.

One main difference between BC and the rest of Canada is the climate. Most people in BC live in the southwest corner of the province, where the climate is mild and snow is relatively rare (except in the surrounding mountains, where abundant snow supports a thriving winter sports industry). BC people wash their cars on New Year's Day just so they can phone their snow-bound relatives in Edmonton, Winnipeg, Toronto, and Montréal and boast about the mild temperatures. Also on New Year's Day, thousands of Vancouverites go for a swim in English Bay.

Like most Canadians, people in British Columbia enjoy the wilderness. One is never far from the outdoors in BC, and area residents constantly take full advantage of the encompassing wilderness. Downhill and cross-country skiing, hiking, rock climbing, bird watching, and fishing are only minutes away from downtown Vancouver. The mild climate allows the enjoyment of outdoor sports all year. The climate of the city, coupled with the high-elevation mountains nearby, creates conditions in which residents could enjoy boating and snow skiing in the same day. The sheltered waters of the Strait of Georgia make water activities very popular. Fishing, sailing, power boating, and kayaking are enjoyed year-round.

Vancouver is the largest city in British Columbia and one of the best places to live in the world, according to UN surveys. The city is situated on Burrard Inlet with the snow-capped Canadian Rocky Mountains in the background. Visitors describe the natural beauty as "stunning." The city is clean, relatively crime-free, prosperous, and cosmopolitan. Vancouver's

Figure 11.50 **Victoria, British Columbia.** A distinctive British imprint is visible in the cultural landscape of Victoria in the architecture, red double-decker buses, and numerous places to take afternoon tea. Photo: © Joseph P. Dymond.

location makes it a convenient port for goods coming and going to Asia. Vancouver International Airport is the busiest hub in North America for flights to and from Japan, China, and Korea. Recently, Vancouver became a major center for making motion pictures, TV shows, and commercials. Among the numerous recent productions filmed in the area was the popular and long-running *X-Files* science fiction TV series. Production crews are attracted to the varied scenery and unique combination of environments and climate, the availability of a highly skilled and appropriately trained labor force of technicians, stable politics, and low crime, and a low Canadian dollar relative to the U.S. dollar.

Each provincial government in Canada funds its own public education system, creating conditions in which small towns and inner cities have the same quality of schools and teachers as the most materially wealthy suburbs. British Columbia has universities and colleges in all areas of the province, enabling people to study close to home. The Vancouver area has two universities, the University of British Columbia (UBC) and Simon Fraser University. Both offer undergraduate and graduate degrees. UBC is acknowledged as one of the finest research universities in Canada, with many researchers and scholars who are recognized worldwide for their outstanding contributions. Canada's national magazine, *Maclean's,* has consistently ranked Simon Fraser University among the country's top comprehensive universities. Vancouver has also become a world-famous center for learning English as a second language (ESL). Besides having excellent ESL teachers and schools, Vancouver offers a low-crime environment and a beautiful setting for students to enjoy when studying here. There are over 65 ESL schools and colleges in the city. These schools are especially popular with students from Asia and South America. Many of these students love the city so much that they end up living there permanently.

Canadians are proud of the national and provincial health care systems. British Columbia, like the rest of the country, is striving to maintain the standards of its system. Increased costs are challenging these efforts, but Canadians place a high value on the coverage and availability of health care. People from all walks of life in Canada are treated the same way in hospitals and doctors' offices—regardless of their financial status. Canadians envision their system as a national treasure and are willing to work hard to keep it open to everyone and are constantly striving to improve it so future generations will continue to enjoy and value the system.

Canada's proximity to the United States creates a sharp awareness of U.S. political and economic actions on the part of Canadians. Americans are well liked in Canada and are often thought of as extended family. Canadians are often challenged when the U.S. media and political leaders seem to ignore them. George W. Bush's September 20, 2001, speech to Congress is a good example. When he cited a long list of countries rallying to the U.S. side after the attacks of September 11, he did not mention Canada. Canadians had opened their homes and hearts to Americans at this disastrous time and felt quite rebuffed. Many Canadians are also disappointed when U.S. trade barriers are placed to prevent a free flow of trade between Canada and the United States. The stipulations of NAFTA are sometimes ignored by the United States. For example, the lumber tariffs the U.S. government placed on BC's lumber critically damaged the industry and added an unnecessary expense to homeowners in the United States. Canadians love Americans but are sometimes disappointed by U.S. politicians' actions regarding Canada.

Vancouver is a prosperous, clean, friendly city surrounded with spectacular natural scenery. British Columbia is an exciting and breathtakingly beautiful province. Canada is a kind and caring country. These are just a few of the countless reasons why Gwen and Jim have a very fond personal view of living in Canada.

Northern Canada

The northern parts of all the provinces from Saskatchewan to British Columbia end in the barren rock, trees, and tundra that characterize northern Canada. The Northwest and Yukon Territories long formed the largest part of Canada—a federally controlled zone where few people live permanently apart from American Indian and Inuit (Eskimo) groups of Native Americans. In 1999, Nunavut became a new territory (see Figure 11.42).

Mining settlements produce a range of metallic ores, and there are few communications in this wilderness area apart from those linking mines to their markets. In the late 1990s, a new diamond mine near Yellowknife, Northwest Territories, was expected to produce up to 3 percent of the world's diamonds. The diamond mine needs to address local community concerns about caribou herd migration paths and threats to hunting and fishing if it is to establish a lasting positive impact. On a global scale, the mine, together with others outside South Africa, could reduce De Beers's control of diamond markets. Native Americans attracted to the mining settlements are often unable to relate to the contrasting lifestyles and in the past have created a social problem that the Canadian government did not tackle well until the 1980s. The future of this region is not bright as mining activities and federal subsidy incomes decline. A possible alternative income is from tourism.

Test Your Understanding 11D

Summary

Canada is closely tied to the U.S. economy but politically is distinctive—partly because of internal tensions over the future of French-speaking Québec and partly because of its more liberal trading and aid policies. Internal differences are increased by the structure of large provinces that have more powers than do states in the United States.

Canada's regions each have an economic heart close to the U.S. border and huge expanses of sparsely settled northern lands. Canada's economy was largely based on agriculture and forestry until around 1950; since then, Canada has become high-tech and industrialized, and one of the world's major mining countries.

Questions to Think About

11D.1 What is the relationship between Canada and the United States? Do Canadians have a strong sense of national identity? Explain.

11D.2 What issues might cause Canada to fracture into two or more separate countries?

11D.3 What areas of Canada are growing and why? How might growth affect the future cultural landscape in Canada and the Canadian sense of nationalism?

Key Terms

long lot

Online Learning Center

www.mhhe.com/bradshaw

Making Connections

The Online Learning Center accompanying this textbook provides access to a vast range of further information about each chapter and region covered in this text. Go to www.mhhe.com/bradshaw to discover these useful study aids:

- Self-test questions
- Interactive, map-based exercises to identify key places within each region
- PowerWeb readings for further study
- Links to websites relating to topics in this chapter

Chapter 12

Global Connections,
Local Voices

Figure 12.1
Global order or global disorder?

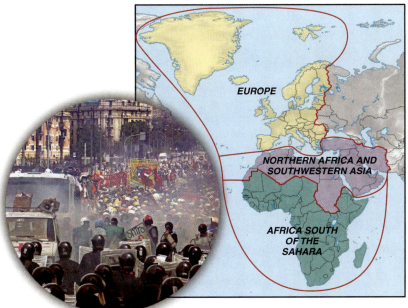

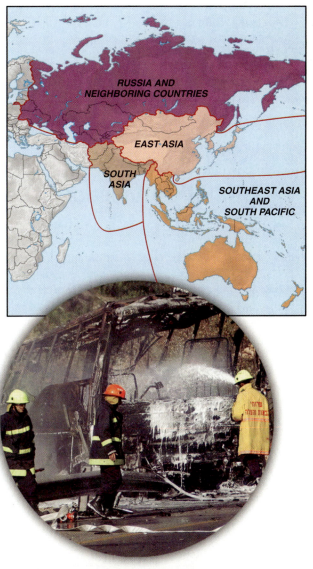

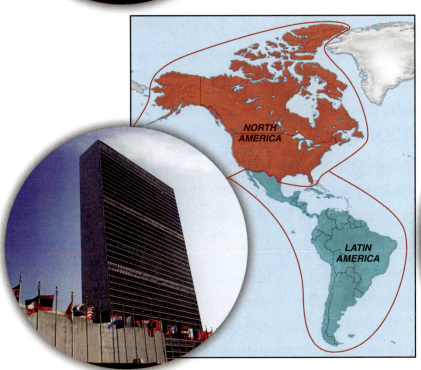

Eyes Opened

The events of Tuesday, September 11, 2001, often referred to as 9/11, made us all more sensitive of the world we live in—one of increasing global connections and of responding local voices (see the "Global Focus: 9/11" box on p. 3). The responses of the United States to those events and its policies related to trade, the natural environment, and intervention in other countries continue to be criticized in many other countries—often despite stated U.S. intentions to help those countries.

Now that you have examined world regions (Figure 12.1), you should have a better understanding of people and countries around the world and a means of assessing the forces of change impacting our world. We can try to assess whether we are moving toward greater global order or disorder. In this chapter we bring together globalization and terrorism, two major areas of concern linked to the September 11, 2001 events. We focus on developments in globalization that need to be understood as the daily news unfolds, and we examine the emergence of local, country-scale, world regional, and global occurrences of terrorism.

 # Global Connections: Trends

Just as New York City is the main world center of the global economy with its many corporations coordinating trade, financial, and stock movements and employing international accountants, lawyers, advertisers, and technologically expert professionals, so Washington, D.C., and particularly the White House and Pentagon, are symbols of U.S. economic, political, and military power. Both the World Trade Center and the Pentagon (and possibly the White House) were symbolic targets for terrorists on September 11, 2001.

The attacks in New York and Washington, D.C., came at a time when the idea of globalization was subject to vocal and sometimes hostile demonstrations, as at the 2000 World Trade Organization (WTO) meetings in Seattle and at the 2001 meeting of the G8 (the world's eight most powerful countries) in Genoa, Italy (Figure 12.2). Demonstrators voiced genuine concerns while extremists indulged in violence. The threat of further disruption at future gatherings of political leaders forced them to retreat behind fortifications and prevented them from listening directly to the clamoring voices.

A Case to Be Answered

Many of the problems highlighted by visible and well-reported groups that oppose globalization focus on the increasing international integration of trade and financial flows. More expansive talk of cultural or political globalization rests on less evidence, but is often more successful in raising emotions. Some of the points made by groups with different agendas need to be heard and unraveled, but their and globalization's defenders' constructive arguments often get lost in oversimplified media rhetoric. The complexities of both good and bad aspects of globalization are difficult to communicate in sound bites. In the following discussion, we separate some issues that antiglobalization critics raise.

- Some critics see international trade as bad: each country would do better to have minimal international links and be concerned about making its own economy self-sufficient. In this view, international institutions such as the WTO, the World Bank, and the International Monetary Fund are often damaging to the places they try to help.

- Another group of protesters and critics concludes that economic growth based on Western experiences of "modernization" destroys cultures and communities, turning them toward materialistic values and actions.

- Others link international trade and globalization to the march of capitalism that, they claim, brings oppression, exploitation, and injustice with it. For them, international terrorism thrives on increasing poverty because the rich-poor gaps get bigger. Nongovernmental organizations supporting some of these critics include many campaigning for justice, human rights, and environmental actions, such as Amnesty International and Greenpeace.

- The above groups are mainly small in numbers, but a wider group of people remains skeptical about the value of globalization, accusing governments of bowing to big business and so losing their own authority and diminishing democracy.

- There is concern that organizations involved in globalization are secretive and undemocratic in their decision-making. This fuels suspicions of a major plot among them aiming to take over the world to the detriment of most people.

Unfair Trading

The issues highlighted raise a number of questions. Is freedom to trade abroad good or bad? Are people's interests sacrificed for the profits of multinational corporations? Should workers in materially poorer countries receive the same wages as those in wealthier countries? Does the movement of businesses to poorer countries result in unemployment in richer countries? Do these questions expect simple answers?

When approaching such important questions, our understanding is helped if we keep the following in mind:

- Human systems and institutions are complex and operated by people making decisions that often favor their personal interests. Each system or institution has its strengths and vulnerabilities. For example, although we may expect economists and politicians to act rationally, straightforwardly, and for the common good, experiences in all countries demonstrate that their actions do not always fulfill such expectations.

- The media have a tendency to oversimplify many complex issues, preferring confrontational positions.
- Critics of systems are not always ready with workable alternatives or, if they are, find them difficult to communicate to nonlistening institutions.

Trading Advantages

The increased trade that is central to the global economic system (Figure 12.3) has some positive outcomes. In the late 1900s, countries that engaged in international trade, such as those in East Asia (see Chapter 5), had faster-rising per capita incomes than those, such as India (see Chapter 7), that were bound by import substitution and attempted self-sufficiency. Countries that reduced foreign trade flows to a minimum had higher rates of corruption through internal bureaucratic control and smuggling; their inward focus stirred up internal political rivalries, as in African countries (see Chapter 9) and Myanmar (see Chapter 6).

Export Pessimism

The "export pessimists" state that if all poor countries had gone for export production in the way many East Asian countries did, the glut of products would have driven down the prices of all exports and everyone would remain impoverished. This is an oversimplified position. It is not established that it could ever happen. Poorer countries are a large group in terms of area and population but are too small economically to have such an effect. Further, many poorer countries have little chance of taking a sizable proportion of the trade that is dominated by the wealthier countries. The exports of the poorest and those with middle incomes make up only 5 percent of total world exports.

Moreover, countries that became materially wealthy as the result of export-encouraging policies did so at different times. For example, Japan did so in the 1970s, other East and Southeast Asian countries in the 1980s and 1990s, and China in the 1990s (see Chapters 5 and 6). In 1955, South Korea was as poor as any country today (US $400 GDP per capita at today's prices) and had poor prospects, with U.S. officials predicting indefinite dependence on international aid. One generation later, South Korea became an international economic power through trading its goods, and internal poverty declined. On the other hand, India pursued a path of resistance to external trade and financial links until the 1990s, and its economy grew slowly compared to East Asia or much of Southeast Asia (see Chapter 7); extensive poverty persisted.

The regional studies in this book make the point that development strategies, such as export-led policies, are time- and place-specific. What worked for South Korea (Chapter 5) is unlikely to work in precisely the same way at a later date for Vietnam (Chapter 6), since Vietnam has to compete with South

Figure 12.2 **Against globalization.** Demonstrators and police face each other in downtown Genoa, Italy, on July 20, 2001, at the time of the G8 (wealthiest 8 countries) summit. Tens of thousands of protesters streamed toward the prohibited zone around the conference venue but were met by police firing tear gas. Photo: © AP/Wide World Photos.

Korea. However, the more materially wealthy countries there are, the more it is likely that addtional countries may take advantage of increasing trade.

Income Gaps Widen

Opening up a country's economy to external trade often goes along with a growing difference between the highest and lowest incomes (see Chapter 6). This is not necessarily a bad thing if linked to an overall rise in incomes, in which the local workers and managers in profitable industries gain first. It is not merely a result of the owners of trading companies or multinational corporation managers gaining at the expense of local people. Evidence exists that extreme poverty is reduced, but the poorest are usually the last in any society to see major changes in their lot.

Differences between poorer and wealthier countries raise the question of what is a "fair wage" paid to workers in industries that produce goods for export. Multinational corporations (MNCs) often move production facilities to countries where labor is cheaper. They justify paying lower wages in some countries because the work force is not as educated, healthy, or productive as in the home country, and technology is not applied so widely that productivity reaches home-country levels. Furthermore,

Figure 12.3 **Global trade.** A huge container ship at the dock, Long Beach, California. United States trade across the Pacific has grown rapidly in the last 50 years—a significant part of globalization. Photo: © Richard Hamilton Smith/Corbis.

MNCs often pay higher rates than local industries and collateral investment in infrastructure (as in highways or electricity generation) often benefits wider groups of people than just those employed directly.

Unemployment in Wealthier Countries

When labor-intensive industries move from wealthier countries to poorer countries, they reduce jobs for low-skilled workers in the wealthier countries. This may devastate families and whole communities in wealthier countries. However, many who lose jobs in the wealthier countries—especially Western Europe (see Chapter 3)—are more likely to have access to financial support and retraining possibilities through social welfare programs. Local enterprise and expertise are likely to attract new opportunities. For example, in Greenville, South Carolina, many textile factories moved out of New England in the 1920s to take advantage of low-wage, nonunionized labor in South Carolina, but were closed in the 1960s. Their owners went out of business or transferred their operations to Latin America. Yet by the early 1990s, Greenville had one of the lowest unemployment rates in the United States as manufacturing corporations such as BMW and Michelin moved in and service industries expanded, bringing higher technology and better wages, encouraged by local conditions that included the existence of a technical college offering training courses.

Profits Before People

When companies including multinational corporations wish to make profits, their leaders do not always consider what is best for their employees. Critics point to the fact that some MNCs buy up local companies and then strip their assets by selling off valuable factories and land. And yet, it is often in the MNCs' interests to pay wages that are somewhat higher than the local norms in poorer countries, to contribute to the education, training, and health care of their workers for reasons of productivity, and to ensure that local infrastructure is adequate to underpin their investments. MNCs work in the context of local conditions, and few poorer countries have rules to ensure worker safety, worker bargaining power, restricted working hours, children going to school and not working in factories, and freedom from physical abuse. However, the resultant low costs of production in poorer countries are seldom passed on to buyers in wealthier countries, nor do higher prices in the latter benefit the poor. For example, baseball caps sold in a U.S. university store for $15 are made in Bangladesh, where workers are paid 6 cents an hour in unsafe conditions and beaten and ostracized for attempting to organize a union.

Reducing Democracy

Are MNCs so powerful that they are replacing country governments? What are the powers that country governments retain?

Reducing the Powers of Country Governments

Some critics suggest that political control is passing from country governments to multinational corporations. Multinational corporations undermine the democratic workings of governments in both poorer and richer countries. It is claimed that MNCs can move production units and so make country borders irrelevant.

Another claim states that 51 of the world's 100 largest economies (including country economies) are business corporations (Figure 12.4). This suggests that MNC boardrooms are becoming more powerful than governments and now make the world's most important decisions.

However, the statement about 51 of the largest 100 economies being MNCs is misleading. It is based only on their total revenues without taking account of costs. These figures should not be compared with government finances that are based on value added (income less costs) across economic sectors. None of the 100 top business corporations can compete with any but the smallest countries in the matter of either income or power. A more recent study by the United Nations Conference on Trade and Development based on value added gives a fairer picture. It found that, although MNC influences are growing, the largest MNC is Exxon, which is behind 45 country economies; the annual Microsoft value added of $20 billion equates with the economy of Uruguay, the world's sixtieth country economy.

However, MNCs do not have many of the powers of country governments. They cannot tax people, conscript them into the military, or deploy police, among other activities. Although, for example, Microsoft acts globally and

Figure 12.4 **Multinational corporations.** The headquarters of MNCs are features of major world cities that are increasingly called global city-regions. The headquarters of the BMW carmaker is in Munich, Germany. Photo: © Adam Woolfitt/Corbis.

takes advantage of differences among places—whereas governments act within their own borders, Microsoft has power only over its own facilities in each country and has to conform to local conditions. It may, however, use its size to garner political influence in the governments of countries where it operates, as Microsoft's 2002 switch in investment from India to China showed. In summary, the statement about MNCs having greater power than most countries is technically incorrect, but MNCs do hold important powers and can hire and fire large groups of workers and move production plants.

Survival of Nation-States

Countries, or nation-states, remain the main political units with far greater powers than the largest corporations. International political integration is far from being complete. Country borders signify jurisdiction and administration rights, and are still resisting globalizing effects to eliminate them. Moreover, for most countries, internal trade is greater than international trade. Trading across borders requires more trust than internal arrangements, and internal country laws take precedence, even where trading agreements are most developed. Even in the European Union (EU) (see Chapter 3), where many cross-border differences have been removed and efforts to harmonize the differences continue, commercial contracts and government rules for using pharmaceutical products that are unique to each country still exist.

Although there are attractions of greater personal and job mobility arising from increasing global connections, people are less mobile than money. Freedom of movement is greater than ever, but most people do not wish to move away from their home country to work in other countries. Many people stay near, or soon return to, their families, friends, and familiar culture. One result is that countries can impose high taxes without worrying that large numbers of people and jobs will take opportunities to move to low-taxation environments. Thus, globalization does not force countries to compete with other countries by lowering taxes. For example, the United Kingdom rejected the EU harmonization of company taxes to preserve the special role of the city of London financial center as a global city-region.

In fact, almost all countries increased taxation of their populations in the 1990s. This refutes the view that MNC activities reduce the scope of country governments to provide services, such as social welfare, from taxes. In 2000, the United States collected 30 percent of gross domestic product in taxes of all types—$3 trillion, or greater than the total GDP of Germany! This would be $30,000 per household if averaged across all U.S. households. In comparison, Sweden's taxes amount to 57 percent of its total GDP. Although some welfare programs in European countries are subject to specific cuts, they come about because of rising technology costs and aging and are not due to the need to be more competitive in the face of MNC pressures. Rather, most European countries continue to increase their overall taxation-sourced inputs to health and education programs (see Chapter 3).

Some of the additional demands global economic interconnections place on country governments may be acceptable in themselves. The breakup of the Soviet Union and its bloc of countries after 1991—partly in response to the failure of central planning in the light of growing economic and political power in the West (see Chapter 4)—reduced internal oppression and allowed freer cross-border people movements.

Country governments may also be curbed in actions that harm human rights of local groups by increasing global connections. The growing international flow of information through the media and Internet draws attention to abuses of people and environments and makes voters more demanding. For example, growing knowledge of the West motivated people in the Eastern European countries and the Soviet Union to press for the end of their Communist governments (see Chapter 4). Groups such as the Zapatistas in southern Mexico and al-Qaeda gained much external support by using such media to make their voices heard.

Foreign Investment and Debt

Some critics raise questions about the major institutions that supply investment funds having hidden agendas. Is the transfer of investment funds to poorer countries merely a plot to get them ensnared in debt? Does the "Washington consensus" of the World Bank, International Monetary Fund, and the U.S. government act against the best interests of poorer countries, mainly to the advantage of the United States? Is the World Trade Organization the most dangerous of all global institutions—unaccountable, unrepresentative, and secretive in judgments that are imposed arbitrarily?

"Washington Consensus"

The "Washington consensus" involves U.S. government support for the World Bank and International Monetary Fund to promote trade and foreign direct investment in poorer countries. It links investments and loans to rules that are also followed widely by nongovernmental organization aid agencies.

Some critics see these institutions as representing the interests of materially wealthier countries, leading to debt and increased poverty in the poorer countries. Greater engagement in world trade and exposure to business cycle fluctuations cause poor countries to take out loans in short-lived better times. They then suffer from high debt charges and slumps in market prices when the financial environment turns against them. Thus, the whole package results in even greater dependency on, and control by, the wealthier countries. Such observations make the "Washington consensus" appear as a plot, but although its institutions have made mistakes, they are often more a case of failure to consider all relevant factors, including local conditions, than of conspiracy.

Types of Foreign Investment

Some types of foreign capital are more effective than others. Foreign direct investment is at the risk of the investor who is, for example, building a power facility. In the longer term, it may cost the host country more than a home-financed project, because profits flow out of the country. However, the gains in more rapid construction and in productivity and technology often make externally funded projects a good value, since in many poorer countries, little local capital is available (see Chapter 9 on Africa).

Other types of foreign capital, however, may bring troubles. Many poorer countries, making efforts to upgrade their infrastructure, as well as borrowing funds for grandiose projects (see Chapter 10), took out recklessly large short-term loans; others were moderate in their plans. Rising interest rates on world markets and defaulting local banks later made the debts impossible to repay in both major and moderate borrowing countries. The wealthier countries must take much blame for this situation, since many governments guaranteed loans by their banks to poorer countries. "Too big to fail" characterized the largest projects that could not be left uncompleted and the loans to the largest poorer countries, such as Mexico, Brazil, Nigeria, and Russia.

Stresses increased when the World Bank and IMF, the main sources of loans to poorer countries, imposed their stringent rules known as structural adjustment (see Chapters 2 and 9, for example). The rules are designed to reorient country economies to be more open to participation in the global economy. For example, debts are to be reduced by cutting government programs and employment, including privatization of publicly owned industries. In this context, some countries reduce spending on public education, health, and food subsidies rather than on defense. Currency devaluations make exports cheaper and imports more expensive. Such moves may make sense in running a business, but a country is more than a business. Governments provide for sectors of society that are adversely affected by market forces. Moreover, each country has its own unique resources and needs, and a single overall set of rules is inappropriate: geographic differences matter.

In general, the actions of these international banks are well intentioned, but the area in which they work is an uncertain world of changing opinions as to what is best, world interest and exchange rate fluctuations, and some local and country governments that do a poor or uneven job of applying and monitoring projects. The World Bank and IMF often receive requests only when a country is past help. If these institutions were closed, the flow of funds to poorer countries would fall, and the wealthier countries in the global economy would probably impose even harsher conditions on loans and local financial systems.

World Trade Organization

When critics of the WTO attack its processes of resolving disputes as antidemocratic, they ignore the full picture. It is sometimes forgotten that the WTO is a meeting of government ministers, rather than an all-powerful institution of world government: all of the 142 countries that are members have to agree to the constitution and any changes in rules.

The WTO bases its advice on the advisability of lowering trade barriers, often with one trading partner compensating the other that accepts its exports. Its secrecy is at the insistence of governments that do not want trade negotiations to be published before agreement is reached.

The WTO acts only as a referee in disputes and does not hand out punishments. For example, the EU and United States have been in a dispute since 1989 over the EU's refusal to import hormone-treated beef. The EU failed to put forward scientific evidence to support its ban and was thus deemed to have violated WTO rules. The WTO might allow the United States to impose retaliatory trade sanctions, but it will not do the punishing itself: it is the United States, backed by the WTO opinion, and not the WTO itself, that does any forcing of the EU to open its markets.

The WTO is never a party in disputes, leaving them to negotiations between country governments. However, this often leaves power in the hands of the wealthier countries, such as the United States, which can ignore decisions they do not like. WTO encouragements to accept more goods from poorer countries are often ignored by the wealthier ones. Nevertheless, if the WTO did not exist, there would probably be more disputes over fair trade at much greater cost.

Reorientation or Revolution?

From 1991, the collapse of Communist governments removed the alternative to capitalism for both the wealthier and poorer countries. The collapse revealed the worst excesses of oppression and bureaucratic centralism that had been hidden under centrally planned Communist governments. The self-sufficiency, central government control, and bureaucratic wastefulness that accompanied Communism have been replaced by free market competition that itself brings hardships to many (see Chapter 4). Remaining Communist governments in China, Vietnam, and Cuba take steps to be more involved in the global economy. So the alternative system has gone. What remains is how different countries shape their policies in response to the impacts of globalization.

It is difficult to see what a total rejection of all current systems linked to globalization, international institutions, and multinational corporations would achieve in the absence of obvious and strong alternatives. At this time, it is very unlikely that there will be a single world government or that countries will cease to be the main political units in the near future. The United Nations (Figure 12.5) does not have the powers or widespread credibility for global government control. It is subject to the votes of individual members

For example, the current resort to protectionism by the United States on behalf of its steel and agricultural industries at a time of economic recession retards the global economy for the poorer countries. Almost all countries follow this trend to some extent. When protectionism ruled in the mid-1900s, nationalism, dictatorships, and central planning replaced an earlier phase of globalization (see Chapter 1). After the 1991 breakup of the Soviet Union (see Chapter 4) and the emergence of China as a major global trader (see Chapter 5), the 1990s became an era of economic growth, freer global trading, and cultural and political exchanges, but still subject to the policies of individual countries.

Yet, there is room for improvement. If the overall human development and human rights aims are to raise the prospects of people in the poorer parts of the world, the recent statements of intent by the United Nations and World Bank need to be taken seriously by both wealthier and poorer countries. None of these suggested actions will be easy to accomplish in the face of powerful invested interests that resist change by countries, MNCs, and labor groups.

- Materially wealthier countries and major MNCs often adopt trading rules that discriminate against poorer countries. This is particularly true for farming and textile products (see Chapters 3, 7, and 9). The wealthier countries spend huge sums on protecting their agriculture—a sector in which poorer countries can compete. The protection of intellectual property in wealthier countries also discriminates unfairly. Poorer country producers need access to wealthier markets and new ideas in ways that avoid their exclusion by protectionists.

- Wealthier countries often persist with internal subsidies that maintain high internal prices or favor wasteful uses of energy and natural resources. Poorer countries can only afford to use methods of less strict conservation. The natural environment needs fuller protection by promoting the more efficient use of natural resources so that long-term sustainable development will be possible.

- Wealthier country governments provide foreign aid to the poorer countries and have begun debt forgiveness. These measures are most effective when the additional funds released are spent on health and education provision and do not go to bankers in wealthier countries or bureaucrats in poorer countries.

- Banks in wealthier countries need to be dissuaded from reckless lending or other forms of mismanagement that bring financial and social crises to poorer countries.

- Good governance in both wealthier and poorer countries reduces MNC pressures on trading policies through accountability and responsibility—instead of deflecting the responsibilities for undesirable outcomes to the World Bank, IMF, or WTO.

What About the People?

This section on aspects of globalization focuses on the economic and political processes because their geographically uneven outcomes are crucial to the lives of people in different regions and countries. Although there are aspects to be improved, global interconnectedness increasingly affects people at local levels—for good or ill. However, the ultimate possibility of "one world" or a "new global order" politically or culturally is still a long way off. Countries and local regions within countries still play the major part in maintaining economic, political, and cultural differences, although international agencies and businesses are slowly increasing their roles.

Throughout this text, the "Personal View" boxes link ways people live in different human and natural environments of

Figure 12.5 **The United Nations building, New York.**
The United Nations has a membership of almost all the world's countries. It carries out a range of programs, such as those intended to improve the well-being of children, and also has links to leading scientific organizations. It also organizes international military support for some countries facing civil strife and was involved in the Korean War of 1950–1953 and the Gulf War of 1990–1991, however, it does not constitute a global government. Photo: © Joseph Sohm/Corbis.

globalization and localization. Re-read about the Bosnian now living in the United States (Chapter 3), the conditions in Irkutsk, Russia (Chapter 4), the Japanese student (Chapter 5), the Malaysian (Chapter 6), the Pakistani banker (Chapter 7), the Omanis (Chapter 8), the Rwandan (Chapter 9), the Bolivian doctor (Chapter 10), and the Canadians (Chapter 11).

🌐 Global Focus: Geography of Terrorism

What Is Terrorism?

Considerable confusion exists about defining the terrorism that has an increasingly global basis. This affects views concerning what to do about it. The war on terrorism declared by U.S. President George W. Bush after the events of September 11, 2001, brought out varied and often paradoxical views and responses. In some countries, terrorism and revolutionary actions ceased or were muted in the face of world opinion; in others, support for freedom fighters was renewed. Where terrorist activities diminished, political leaders, as in Israel, often took advantage of the situation to push ahead with the changes that terrorists resisted, stimulating renewed terrorism. Opposing terrorism is a slippery project.

In essence, terror is about making people frightened and, by extension, attempting to influence the actions of governments by frightening whole populations. Terrorism is the systematic use of terror to coerce others without concern over the impacts on innocent victims. It goes against notions of democratic government in seeking to get results by the bullet instead of the ballot box. However, this definition is apparently not binding, since terrorism may be encouraged by democratic governments in countries they do not agree with. Countries that encourage terrorism in other countries are often called "rogue states."

No total agreement exists on a definition of terrorism because there are so many perspectives about it. One definition that is widely adopted comes from the U.S. Federal Bureau of Investigation: "Terrorism is the unlawful use of force or violence against persons or property to intimidate or coerce the government, the civilian population, or any segment thereof in furtherance of social or political objectives."

Terms such as *terrorist, freedom fighter, revolutionary, guerrilla,* and the older *anarchist* are used to describe those who use terrorism, although each has a separate meaning (Figure 12.6). Freedom fighters are those willing to risk or give up their lives in fighting for a cause they believe will bring better conditions to their country or locality, or give them more power to make changes they want. Revolutionaries are those who wish to introduce major changes in political organization, including the overthrow of a government. Guerillas are those who engage in warfare against a government. Some early terrorists were anarchists, who believed in abolishing all government at the country level to enable everyone to enjoy freedom. All are minority groups.

Today's terrorism is partly a reaction to the globalizing world. Local and minority concerns fight the established government of a country or wider international bodies. The interconnected global world makes it possible for terrorist groups to link with each other and exchange details of practices. Many terrorist groups finance themselves through illegal global trade in drugs, slaves, armaments, and precious metals and stones (see Chapter 9).

A geographic assessment of terrorism begins with the understanding of where issues underlying terrorist actions occur and where the vulnerable places for such actions are. However, the diffusion of terrorist capabilities and global links among revolutionary, anarchist, guerrilla, or freedom fighter groups makes it easier to carry out attacks and more difficult to identify the time and place where they will occur. The attacks on New York and Washington, D.C., on September 11, 2001, were foreseen as a potential and general threat but without any precise knowledge of the location or method. Surprise over the place, timing, and method is often a feature of terrorist attacks on powerful institutions.

PHOTO TAKEN FROM VIDEO OF
AL RAUF BIN AL HABIB BIN YOUSEF AL-JIDDI

PASSPORT TYPE PHOTO OF AL RAUF BIN AL HABIB BIN YOUSEF AL-JIDDI FOUND WITH SUICIDE LETTER

FBI LABORATORY PHOTOGRAPHIC RETOUCH OF AL RAUF BIN AL HABIB BIN YOUSEF AL-JIDDI

PHOTO OF FAKER BOUSSORA
(ASSOCIATE OF AL RAUF BIN AL HABIB BIN YOUSEF AL-JIDDI)

The United States Government is seeking information regarding the identities and whereabouts of the individuals shown above.

IF YOU HAVE ANY INFORMATION CONCERNING THESE PEOPLE, PLEASE CONTACT YOUR LOCAL FBI OFFICE OR THE NEAREST AMERICAN EMBASSY OR CONSULATE.

www.fbi.gov

January 2002

Figure **12.6** **Alleged terrorists.** Two Canadian citizens, identified by the Federal Bureau of Investigation in 2002 as linked to al-Qaeda: Al Rauf Bin Al Habib Bin Yousef Al-Jiddi (top two photos and bottom left, including "retouched" photos to show a Westernized appearance is one person). The other photo is of Faker Boussora, believed to be an associate of Al-Jiddi. Photo: © AP/Wide World Photos.

Terrorism, Global Connections, and Local Voices

Humans have always had an ability to make each other afraid by undermining the security of an established order. Historically, robber bands and pirates extracted booty by threats against those wealthier than themselves, while marauding groups and rival armies fought each other with little interest in what happened to "civilians."

The establishment of modern countries, however, set new targets for dissent and violent action against rulers. Terror became a political instrument, and the word *terrorism* entered European languages as modern countries were defined toward the end of the 1700s. In the emerging United States, the Boston Tea Party set a trend of disobedience and defiance against the British in their American colonies, followed up by violent actions. What the British might have called revolutionaries or terrorists, American patriots saw as freedom fighters. From the 1780s, the French revolutionaries in Paris used terror against the aristocracy with the aim of overturning the imperial order and establishing a democracy of the people.

The link between political ends and violence against people whom most would regard as innocent took another 200 years to develop to the point in 2001 when the blatant events of September 11 caused many countries of all political and religious alignments to attempt to rid the world of terrorism. However, confusion over the nature of terrorism and the motives and ends of terrorists and those opposing them continues. A series of waves or phases of evolving terrorism over the last 200 years with increasing geographic impacts is recognized.

Phase 1: Revolutionary Terrorists, 1800s and Early 1900s

Through the 1800s, revolutionary groups attempted to overturn governments in Europe and Latin America in particular. Latin American countries gained their independence from Spain and Portugal. In France, the revolution of 1789 was followed by another in 1848. In Italy, Giuseppe Garibaldi led 1860s revolution against Austrian occupation and papal obstruction to unite Italian states.

From 1879, after the Russian czars' political and economic reforms had raised hopes that could not be fulfilled quickly, the Narodnaya Volya ("The People's Will") revolutionary group in Russia set out to achieve radical changes in society. Frustration rose as mass uprisings were rapidly suppressed. They saw terrorist methods as a temporary necessity to "raise the consciousness of the masses." They set a pattern for the future by acts of violence perpetrated on symbolic victims. Assassinations used the newly invented dynamite—in which the bomber often died as well as the victim. Terrorism involved learning how to fight and die—by seeking to be martyrs and heroes—and to make the most of criminal trials to embarrass the political regime. The terrorist acts in Russia got attention, created political tensions, and drew government responses. They did not precipitate immediate revolution but appealed to groups such as the anarchists of the late 1800s.

Drives toward universal suffrage, however, attracted anarchists, who feared that their hoped-for future of "free" but disorganized societies would be made impossible in strong and fully democratic countries. They were often opposed by harsh police measures. In the Balkans of southeastern Europe (see Chapter 3), suspicions grew of one country supporting terrorists in another in the context of the failing influence of the Turkish Ottoman Empire and the rising aspirations of people within parts of the Austrian Empire. The Serbian Black Hand and Young Bosnia emerged, and their activities led to the Austrian archduke's 1914 assassination in Sarajevo (Figure 12.7) that set off World War I.

In this first phase, terrorism began to increase tensions between countries. Terrorists on the run gained refuge in and resources from sympathetic countries. Socialists and anarchists without government representation supported revolutionary movements abroad. Russian revolutionaries such as Lenin, for example, had headquarters in Switzerland, launched attacks from Finland, acquired arms through Armenia, and used Japanese funds laundered through American banks.

As one outcome in the 1890s, U.S. President Theodore Roosevelt, on succeeding the assassinated William McKinley, called for an international crusade against terror. In what he called the "golden age of assassins," governments encouraged the development of police information-gathering systems, shared the results with other countries, and controlled their borders more closely.

Figure 12.7 **Terrorist act led to World War I, 1914–1918.** Archduke Franz Ferdinand, heir to the Austrian throne, with his wife, Sophie, on a visit to inspect the military in Sarajevo, Bosnia, approach their car. Moments later, a bomb was thrown, but the blast only wounded one of the officers. Changing their route to visit the officer in the hospital, they survived another assassination attempt when a bomb failed to explode, but the royal couple were then shot dead by Gavrilo Princep, one of the plotters who happened to be on the route. The Serbian terrorists were demonstrating against their own government, but Serbia was blamed and Austrian action against Serbia was supported by Germany; Russia entered the war against Germany and Austria and brought in France, the United Kingdom, and eventually the United States. Photo: © Hulton Archive/Getty.

Phase 2: Colonial Freedom Fighters, Approximately 1920s to 1960s

The 1900s world wars stimulated people in the European colonies of Africa and Asia to press for the freedoms that are linked to self-determination and independence. Achieving the freedoms often meant fighting against sophisticated military forces. Although in some countries this involved traditional warfare (as in Vietnam), many groups adopted terrorism, focusing on colonial police and security forces targets. Atrocities by the security forces in response increased sympathy and support for terrorists. Anticolonial causes appealed to outsiders, but the label *terrorist* had negative connotations and was dropped by many rebel groups. For example, before 1948, Menachim Begin's Irgun group called themselves "Israeli freedom fighters" battling government terror, although they were still labeled "terrorists" by the British administration of Palestine (see Chapter 8). The time between being called a terrorist and being called a statesman can be very short.

Other developments at this time included the city-based cells of close-knit terrorist groups that were difficult for security organizations to penetrate. Semiguerrilla hit-and-run actions were more widely used. However, few groups had high levels of public notoriety or gave warnings of actions.

Although terrorists acknowledged common international revolutionary bonds, cooperation among groups in separate countries was only beginning and heroes were national ones. Thus, the Algerian Liberation Front received aid from other Arabs, including sanctuary for planning attacks on the French colonial government. Greece sponsored a Cypriot uprising against the British. From the early 1920s, Americans supported Irish "anticolonial" efforts in Northern Ireland following the independence of the Republic of Ireland in 1922, but they then abandoned the Irish Republican Army until the 1960s. The Jews in Palestine received funds, weapons, and volunteers from people around the world and at times pressured American foreign policy.

By the 1960s, most colonies had become independent countries and most anticolonial terrorism ended—although many of the lessons learned by terrorists and those who opposed them were taken up in the next phase. Separatist activities continued in Portuguese colonies until 1975 and through the Basque ETA group in Spain, the Armenian Secret Army, and the Irish Republican Army.

Phase 3: International Terrorism by Groups and Countries: Common Training, Airplane Hijackings, and Hostages, Approximately 1950s to 1980s

During the Cold War, a new revolutionary ethos emerged in which Western groups, such as the American Weather Underground (USA), Red Army Faction (Germany), Red Brigades (Italy), and Direct Action (France), often encouraged by the Soviet Union, saw themselves as supporting Third World demands for attention. The Vietnam War demonstrated that Vietcong terror carried out by small numbers with minimal equipment who perceived themselves as fighting an anticolonial war could compete with the American military Goliath and its modern technology.

This phase became one of international terrorism, with common bonds emerging among separate national groups and international targets. After the Vietnam War ended in 1975, the Palestine Liberation Organization (PLO) followed up the 1967 Arab defeat by Israel and became a leading terrorist group—supported by the Soviet Union against the U.S. support for Israel. Lebanon provided a training ground for terrorists from many countries such as Iran, Syria, and Libya, which, in turn, employed terrorists in other countries as part of foreign policy. The PLO became active in Europe; groups in countries with U.S. military installations linked up with others to strike such targets. Combining their European presence and war against Israel, they murdered Israeli athletes in their hostel at the 1972 Olympics in Munich, Germany. Kurdish terrorism became rife in Turkey.

The special tactic of this phase was airline hijacking by terrorists armed with bombs and guns. Over 100 cases occurred in each year of the 1970s. As well as providing spectacular newsworthy hostage takings and airplane destruction, they spread international terrorism. Other hostage takings were also part of terrorist tactics, such as the kidnapping and murder of the Italian prime minister in 1979 and the Sandinistas taking the Nicaragua Congress hostage in 1978.

This was a short phase, but the methods often alienated potential domestic or international support. When the PLO realized that its activities raised opposition in powerful countries, its strategies shifted to negotiations. When some countries sponsored their own terror groups, the outcome often turned against them. Thus, Libya, Syria, and Iran were penalized in the U.K. and France, especially after a bomb exploded in the hold of Pan Am Flight 103 over Scotland (Figure 12.8). Iranian and Syrian attacks in Lebanon in the 1980s caused other countries to withdraw their personnel, abandoning Beirut to destruction.

Countries targeted by terrorism cooperated in counterterrorist efforts. With British support, American bombers attacked Libya and the European Community imposed an arms embargo in 1986. But such cooperation did not lead to an international antiterrorist agency being formed and relied on intercountry action at times of particular risk. Several countries refused applications for the extradition of suspects on the grounds of lack of evidence, the likelihood of convicted people being executed, or sympathy with those they saw as freedom fighters. By the late 1980s, however, revolutionary terrorists were defeated in many places, most airline hijacking was controlled, and Israel had destroyed the PLO training camps in Lebanon. Through the 1980s, terrorist action in western Europe continued during the weakening of Communist country influences. Groups such as the Red Brigade in Italy and the Red Army Faction in West Germany made efforts to raise revolution through terrorism, but such desperate action did not lead to political victories.

Phase 4: Political to Religious Purposes, 1980s and Beyond

The 1979 Islamic Revolution in Iran and the 1989 Soviet Union defeat in Afghanistan both built on the increasing signif-

Figure 12.8 **Terrorist bombing of an airliner.** The nose section of Pan Am Flight 103, which exploded over the small Scottish town of Lockerbie in December 1988. All 270 passengers and crew were killed. Investigators examined a huge area for wreckage, bodies, and clues, and two tiny fragments enabled them to track down the bombers, who were eventually extradited from Libya and convicted at a 2000 trial under Scottish law but held in the Netherlands. Photo: © AP/Wide World Photos.

icance of religious extremists over political dissidents. Furthermore, terrorist groups became more powerful by financing themselves through access to emerging illegal global markets in precious stones, drugs, slaves, and armaments. They increased contacts and exchanged methods among themselves by Internet links and open travel.

When Iranian Shiites took over the government of their country, they held some Americans hostage for more than a year, finally sending them back to the United States when President Ronald Reagan took office in January 1981. They then supported Shiite groups in other countries such as Lebanon (Hizbollah) and Iraq. Self-martyrdom, including suicide bombings, became a part of this process as the previous terrorist strategies were countered by security measures and as organizations found extremists willing to do it. In Sri Lanka, the Tamil Tigers (a secular group fighting for independence) used this tactic more than others, including in the murder of Indian Prime Minister Rajiv Gandhi in 1991.

The new Muslim century that began in 1979 opened up the expectation of a redeemer and motivated new uprisings. A strict Sunni group stormed the Grand Mosque in Mecca. Saudi Arabia and strict Sunnis joined others in a jihad to fight the Soviet army in Afghanistan. In the context of the Cold War, the United States helped this mission with training, arms, and finance. On returning from Afghanistan, such Arab fighters had the will, confidence, training, and experience to initiate terrorist operations at home in such countries as Egypt, Syria, Tunisia, Morocco, Indonesia, and the Philippines.

Non-Muslim religious groups also carried out terrorist acts. Sikhs in Punjab wishing to secede from India occupied the Amritsar Temple and killed Indian Prime Minister Indira Ghandi (see Chapter 7), Jews and Palestinians in Israel and the Palestinian lands terrorized each other (Figure 12.9 and see Chapter 8), Aum Shinrikyo gassed subways in Tokyo, Japan, and extreme Christian groups in the United States bombed abortion clinics. Timothy McVeigh bombed the federal building in Oklahoma City as an expression of antigovernment anger.

But Islamist groups were most active, supported by Iran, Iraq, Libya, Syria, Yemen, Sudan, and Afghanistan. Such groups sought a single and internationally wide-ranging Islamic country to be governed by religious law. To attain the grand ideal of a single Muslim country, a high priority was given to removing Americans from the country of Islam's holiest shrines—Saudi Arabia—and destroying their support for the present countries of that area that are a "residue of colonial rule" (see Chapter 8).

The September 11, 2001 attacks on the United States came at a time of Muslim and Arab impatience with the United States over the rising Israeli-Palestinian tensions and sympathy for poor Iraqis who were still under Saddam Hussein's rule 10 years after the Gulf War ended. The perpetrators of the attacks clearly hoped that these feelings would overcome horror at the event and lead to a unified Muslim attack on the repressive regimes in Muslim countries and the formation of an Islamic country. They miscalculated, but such desperate measures

remain likely to continue despite the history of failure to win wide support. They brought terrorist actions to a new level of massive inputs and impacts and instigated the current "war on terrorism."

While the main aims of many terrorist groups in the 1970s and 1980s were to obtain a place at a bargaining table, many of the new terrorists aim to impede the bargaining process. The latter include some groups in Northern Ireland, Israel, and Palestine.

Conclusion

Terrorism is now rooted in the emerging global connections, partly as a reaction against it. Organizations such as Al-Qaeda train terrorists from many countries—often with improved management and impenetrable organizations across many international borders. Although in the past such organizations could be broken up and police controls improved, new ways of carrying out terrorist activities to avoid security controls are always being devised. The more interconnected terrorist groups can use a small number of people to obtain massive effects.

It is difficult to sustain international agreements to work against terrorists. Disagreements often begin over the meaning and application of the *terrorist* label. In the past, despite huge police efforts within countries such as Cyprus, Jewish Palestine, Algeria, and Ireland, for example, many terrorists remained at large. The costs of combating terrorism increase.

However, leading students of terrorist activities find that although they are increasingly high-tech and highly visible through global media, terrorists have not gained major objectives. Islamic extremism seems to have peaked in its achievements between the 1979 Iranian Islamic Revolution and the 1989 withdrawal of the Soviet army from Afghanistan. Actions in the mid-1990s failed to achieve successes in Bosnia, Algeria, and Egypt, while a more Islamist president in Indonesia was replaced by a secular politician, and extreme Islamists were followed by secular leaders in Sudan and

Figure 12.9 **Suicide bomber in Israel, June 2002.** A Palestinian suicide attacker exploded a bomb in a bus packed with Jewish people in the morning rush hour in northern Israel, killing himself and at least 17 passengers and wounding more than 45 in the flames that engulfed the bus. Photo: © AP/Wide World Photos.

Pakistan. People in many countries are becoming accustomed to terrorist actions as part of life, and governments appear almost immune from their outcomes.

Somalia: Globalization, Localization, and Terrorism

Somalia, a country in Eastern Africa, brings together the major themes of this chapter. It has internal tensions that call out for understanding of the local conditions. Questions arise as to whether its Islamic militants are part of a global network, but the attention of most Somalis is drawn to internal tensions. Both considerations affect Somalia's place in the world and other countries' perceptions of it.

Hopes of Political Security Dashed

Somalia (Figue 12.10) is one of the world's poorest countries. After Somalia was ruled by a corrupt regime from 1969 to 1990 that was supported by the United States in the Cold War, the overturn of that government led to internal strife among factions and clans. As an agent of the United Nations, the United States tried to bring peace but withdrew its forces in 1995. Subsequent efforts to establish a new government failed until meetings in Djibouti in August 2000 established the Transitional National Government (TNG). Mainly run by Mogadishu businessmen, TNG generated hopes of obtaining external aid, although only a few Arab oil-rich countries responded.

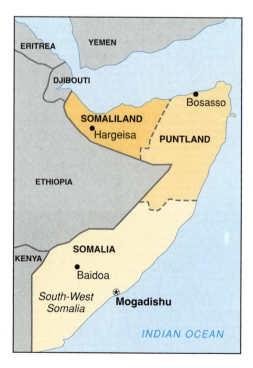

Figure 12.10   **Map of Somalia.** Provinces and main cities.

Even before September 11, 2001 the TNG was in trouble. It controlled only part of Mogadishu and hardly any of the rest of the country and so was unable to make important facilities (ports, airports, police, and judiciary) function. It had no hope of collecting tax revenues. It might have been more successful had it worked closely with Ethiopia, which at first agreed to the process formulated by the conference in Djibouti but then was alarmed by the Somali rush to the Arab and Islamic world for aid. Ethiopia began to back opposition groups inside Somalia in a "proxy war" in which groups supported by Ethiopia clash with groups supported by Arab countries. Somalia remains ungovernable as a "shadow state" in which local groups exploit the political and strategic impulses of foreign interests.

Regional Pressures

A major part of the TNG's problems in trying to govern Somalia stems from the growth of powerful regional interests within the country that were not represented at the Djibouti meeting or in the TNG. Other unrepresented interests include local strongmen, militias, and factions in southern Somalia. The main regional interests are in Somaliland—former British Somaliland that signified a wish to secede in 1991—and Puntland, a nonsecessionist body established in 1998 to create a regional administration in the northeast of Somalia. In 2001, South-West Somalia declared itself autonomous. Although the Intergovernmental Agency for Development, a body that represents surrounding countries (Djibouti, Eritrea, Ethiopia, Sudan, Kenya, and Uganda), tries to reconcile the conflicting groups, its efforts appear to focus solely on how the parliamentary seating "pie" is divided and on the "carrot" of any agreement opening the floodgates of foreign aid. The former goal forgets the most important issues of justice, representation, and taxation, while the latter is wishful thinking.

In fact, the northern parts of Somalia were the most stable parts until 2001. Both continued to export livestock products to Persian Gulf countries, allowing a degree of economic and administrative recovery. But in 2001, Puntland became the battleground for armed conflict between the former president with Ethiopian support and his successor who was loyal to an Islamic group, al-Ittihad al-Islamiya ("Islamic Unity"). In Somaliland, a referendum on the constitution had potential for destabilization as the president proposed a format that would allow him to continue in office instead of rotating it with other leaders.

Economic Collapse

Economic woes made the political tensions worse. Demands for schools and health care facilities were not met, apart from some Islamic schools established by Persian Gulf countries. Even at its best, Somalia's economy was precarious. In the 1990s, most of the country's infrastructure was destroyed in fighting, and low food production caused continuing shortages, malnutrition, and famines.

The main source of income for many is the annual $500 million earned by Somalis working abroad and sent home to their families; for others, there are the exports of livestock from the northern areas, a transit trade of food and consumer goods through southern Somalia and smuggled into Kenya, and telecommunication facilities for money transfers. Since 2000, however, livestock exports to Saudi Arabia ended as the result of cattle fever (although some think the shutdown was encouraged by Saudi ownership of ranches in Australia and New Zealand). This affected the previously stable areas of the north. In Mogadishu, the ruling business faction introduced illegal coins to increase the money supply, but this led to hyperinflation that devastated the purchasing power of poorer households and discredited the TNG government.

Impacts of the 2001 War on Terrorism

Somalia was an early member of a short list of countries thought to be involved in—and likely to be punished for—terrorism. The signs were damning: a collapsed government and an impoverished country that could not control what happened in its territory were considered potential fertile ground for recruiting or harboring terrorists. The indigenous Islamic political movement of al-Ittihad might act like others in Sudan or Yemen. U.S. naval patrols along the coast and aerial surveillance increased intelligence gathering about movements inside the country. The assets of a telecommunications company that dealt with remittances from Somali workers abroad were frozen. Despite such actions, evidence for Somali involvement in terrorism remains slight, and initial reliance on Ethiopian information sources is now questioned in U.S. government departments.

The role of al-Ittihad is the main point at issue. The organization emerged in the late 1980s, when young men educated abroad took steps to oppose the corrupt Somali government through political Islam, a feature of many Arab countries at the time. Involvement in terrorist acts in Ethiopia in 1996 caused the U.S. State Department to label al-Ittihad as a terrorist organization and the Ethiopians to crack down on Islamic groups. Next, al-Ittihad changed its strategy toward greater integration in local communities and decentralization among sympathetic clans with the long-term goal of bringing Islamic rule to Somalia. Its focus on education, the judiciary and media, as well as commerce, is designed to build constituencies and extends contacts and support through social means. This makes the organization a difficult target to evaluate in the war on terrorism, as some of the groups it works through cooperate with Western aid agencies and disavow Islamic links, others keep to themselves, and a few are linked to violent acts. Moreover, while some commentators now dismiss al-Ittihad as discredited or moribund, others still see it as a basis for future terrorist activities. Following September 11, 2001, al-Ittihad's links with al-Qaeda were examined, but no training camps were found in Somalia. Although some Somali businesses borrowed from al-Qaeda-linked Dubai sources, they did not repay enough to fund al-Qaeda excesses. By 2002, it was clear that the link between al-Ittihad and al-Qaeda was not strong and that foreigners of any sort entering the country would stand out from Somalis. The suggestion that al-Ittihad was a silent partner in the TNG with key government posts is regarded as an Ethiopian ploy. It may be that some of the Islamic NGOs that are neither transparent nor accountable in their activities may be greater threats.

Thus, the global war on terrorism may distort the internal political and economic dynamics of countries such as Somalia that experience internal upheavals and do not have much diplomatic support from other countries. From the U.S. point of view, the Department of Defense is more hawkish and ready to act against apparent terrorist-supporting groups; the State Department is less strident. The low level of reliable information makes both attitudes possible and envisaged interventions of dubious value.

Somalia contains many groups of continuing local significance. The impacts of global connections are at times helping to broaden markets for the country's few products and at others reducing them. However, external pressures from Western and Islamic sources that bring economic and religious interventions often enhance local rivalries that make governing difficult and lead to both threats of secession and international interventions inside the country. This country illustrates the continuing significance of country boundaries, although some may be impermanent and subject to change; the significance of local values and power distribution; and the increasing involvement of even the least accessible countries in global affairs.

World Regional Geography: Final Point

World regional geography is always relevant and always changing. The ideas and principles studied in this course should help you to assess world and local events in a context that makes them easier to understand. Of course, our world is not just a changing one but also a very complex set of human and natural environmental processes interacting with each other. Fifty years ago, it was still possible to study a local region or country with little reference to its surroundings, but in the 2000s, the wider world regional and global interconnections are increasingly relevant for us all.

Before World War II, American politicians counseled President Franklin D. Roosevelt to keep out of the war that was erupting in Europe. No one could attack the United States, they asserted, and it would be best for it to remain aloof from the conflict. This ignored the potential threat from the opposite direction, which was realized when Japan attacked Pearl Harbor, Hawaii. Since that time, triumph in World War II and the Cold War encouraged U.S. feelings of supremacy and confidence in its ability to fend off potential problems. The threats of nuclear war from the Soviet Union were diminished by the equal threat of retaliation and subsequently by the breakup of the Soviet

Union. In the late 1990s, some Americans hoped the United States would retreat from some forms of global interaction and decrease its controversial role as a global police force. However, it is geographically impossible for a country with the political, economic, and cultural influence of the United States to shut itself off from the rest of the world. In the increasingly connected global system present today, the actions of all governments, large nongovernmental organizations, and multinational corporations have far-reaching impacts. None may operate in isolation. A better understanding of world regional geography facilitates the ability to connect cause and effect, action and impact, and place and people throughout the world.

Online Learning Center

www.mhhe.com/bradshaw

Making Connections

The Online Learning Center accompanying this textbook provides access to a vast range of further information about each chapter and region covered in this text. Go to www.mhhe.com/bradshaw to discover these useful study aids:

- Self-test questions
- Interactive, map-based exercises to identify key places within each region
- PowerWeb readings for further study
- Links to websites relating to topics in this chapter

Glossary of Key Terms

Each term is defined and linked to the chapter and Test Your Understanding box in which it was a key term.

A

Aborigine (6A). Race of people whose ancestors inhabited Australia before the arrival of Europeans.

absolute location (1B). Location of a place on Earth's surface as defined by latitude and longitude or by distance in km (mi.) from another place.

acid deposition (3B). Dry or wet deposition of acidic material from the atmosphere, often resulting from sulfur and nitrate gases and particles emitted into the air from coal combustion in power plants.

agglomeration economies (3C). The total economies achieved by a production unit because of a large number of related economic activities located in the same area.

agribusiness (3D). The large-scale commercialization of agriculture that places farming within the broader context of inputs of seeds, fertilizer, machinery, and so on, and of outputs of processing, marketing, and distribution.

alluvial fan (7A). A fan- or cone-shaped river deposit, often formed where a stream issues from mountains into a plain.

alluvium (7A). Deposits of sand, clay, and boulders dropped by rivers in the lower parts of their valleys.

altiplano (10B). High plateau.

altitudinal zonation (10B). Significant changes in altitude produce corresponding climate and vegetative changes. Every thousand meters of change produce cooler temperatures, possible variation in soils and precipitation, and distinctive vegetation regimes. Different crops may be cultivated in the changing environmental zones created through increased elevation.

The altitudinal zonation concept is most frequently used in reference to mountainous areas of Latin America, where the combination of elevation change and latitude produce the most dramatic changes between zones. The first zone is *tierra caliente,* where the lowest 1,000 m (3,000 ft.) of elevation has warm-to-hot conditions and tropical forest. The second zone is *tierra templado,* where the next 1,000 m have mild-to-warm temperate conditions and deciduous forest. The third zone is the *tierra fria* between 2,000 m and 3,000 m (6,000 to 10,000 ft.), which has cold-to-

mild conditions and pine forests and some grasslands. The highest zone is the *tierra helado,* in which lies some grassland in the transitional areas but mostly permanent snow and ice cover.

Andean Common Market (10D). A free trade area among the countries of the Northern Andes-Bolivia, Colombia, Ecuador, Peru, and Venezuela.

animism (6A). Traditional religious beliefs based on the worship of natural phenomena and the belief in spirits separable from bodies.

Appalachian Regional Commission (11C). Federal-State agency established in 1963 to enable the poverty-stricken region to access federal grant-aid.

Arabic (8A). Language spoken and written by Arabs, diffused from Arabia with trade, conquest, and religious use. The only language allowed in the Islamic Qu'ran and prayers.

Arab League (8A). Organization created in 1945 to encourage the united action of Arab countries for their mutual benefit.

artesian well (6B). Groundwater that flows naturally out of springs and wells under pressure that is imposed by the geologic arrangement of water-bearing rocks.

Asia-Pacific Economic Cooperation Forum (APEC) (6A). A group of Asian and Pacific countries dedicated to trade liberalization.

Association of Caribbean States (10D). Formed in 1995 by West Indies countries, Mexico, Central American countries, Colombia, and Venezuela as a response to the formation of NAFTA. A possible step to FTAA.

Association of Southeast Asian Nations (ASEAN) (6A). Established in 1967 as a defensive alliance among Indonesia, Malaysia, Philippines, Singapore, and Thailand in response to the advance of Communism in Southeastern Asia. Now increasingly a trading group that admitted Vietnam, a Communist country, as a member in 1995, and Myanmar in 1997.

Aztec (10A). A group of Native Americans who moved from northern Mexico and established control by A.D. 1325 over much of the surrounding region from its base at Tenochtitlán, the site of modern Mexico City.

B

badlands topography (7A). Close-spaced networks of deep gullies, often carved by occasional streams in soft sediments

unprotected by a vegetation cover. These destroy the usefulness of the land.

banana republic (10C). A country that relies on a single export crop (such as bananas), is dependent on world core country markets, and has a poor government that is often a dictatorship.

Benelux (3D). An acronym for **Be**lgium, the **Ne**therlands, and **Lux**embourg. The term was coined in recognition of the close working relationship that these countries have with one another.

Bengali (7C). Language of 98 percent of Bangladesh people; also spoken in eastern India.

Berber (8A). Language of African origin spoken, but little written, in North Africa.

Biharis (7C). People of non-Bengali origin and Urdu-speaking who migrated to East Pakistan (Bangladesh) at the partition of India in 1947.

biome (2C). A world-scale ecosystem type, such as tropical rain forest or savanna grassland.

birth rate (2A). The number of live births per 1,000 of the population in a year.

black earth soils (4B). Highly fertile soil type in which organic matter accumulates near the surface, commonly beneath midlatitude grassland communities.

Black Triangle (3B). A heavily polluted industrial area straddling the Polish, Czech, and German borders.

brain drain (1A). The migration of professional or highly educated people from one country to another for better economic and career prospects.

British East India Company (7A). A British company that traded with and conquered much of the Indian subcontinent. After a 1857 mutiny in India, the British government took over political control.

British Indian Empire (7A). Established after 1857 on the Indian subcontinent, including Ceylon and later extended to Burma. Lasted until independence and partition in 1947.

brown earth soils (4B). Fertile soil type in which plant matter replenishes nutrients in the upper layers, commonly forming beneath midlatitude deciduous forest.

Buddhism (7A). A religion that began in Southern Asia but became the major religion of Eastern Asia. Followers of Buddhism have a greater social openness than do followers of Hinduism and accommodate other philosophies and religions such as Confucianism and Shinto.

buffer state (5A). A country that stands between major world powers and helps to reduce direct conflicts between them.

Burmese (6A). The dominant people in Myanmar (Burma).

C

capital city (2B). The city in which central government functions are concentrated; sometimes the largest city but often one that is specifically designed for the purpose.

capitalism (3A). An economic system in which goods and services are produced and sold by private individuals, corporations, or governments in competitive markets. The means of production are owned by those investing capital, to whom workers sell their labor. Linked to democratic government and increasing trade among places and countries.

Caribbean Community and Common Market (CARICOM) (10D). Formed in 1973 by 13 former British colonies to provide special entry to U.S. markets.

cartel (8A). An organization that coordinates the interests of producing countries (e.g., OPEC).

caste order (7A). A social class system associated with Hinduism that is based on the supremacy of Aryan peoples. The Aryan castes include priests (Brahmans), warriors, and other Aryan people; non-Aryan castes include cultivators, craftspeople, and untouchables.

Celts (3A). European tribes that diffused skills in metal-making (bronze and iron) as they moved from the Alpine area around 1000 B.C. into Spain, France, and Britain.

central business district (11C). The area in the center of a city where shops, offices, and financial services are concentrated, normally to the virtual exclusion of residential facilities.

central planning (4A). The Soviet practice in which the government decided how many goods and services were needed by society, almost without cost considerations

Christianity (8A). A religion that developed out of Judaism, based on the belief that God came to Earth in the form of Jesus Christ. Main religion of the Western World.

city-state (10A). Individually ruled and independently functioning urban centers.

class (2B). Part of a stratification of society that is based on economic or social criteria.

climate (2C). The long-term atmospheric conditions of a place.

Cold War (1A). The period from approximately 1945 to 1990 when Communist countries (Soviet Union and its allies, and China) vied with Western countries for world domination. It ended with the breakup of the Soviet Union.

collectivization (5D). The transformation of rural life in Communist countries such as China and the Soviet Union, in which individual farmers were grouped in cooperatives that took ownership of their land and labor.

colonialism (3A). The system by which one country extends its political control to another territory to economically exploit the human beings and natural resources of the subordinate territory.

command economy (4A). The Soviet practice whereby the government ran the economy, owned all industries, and set quotas, favoring heavy industry over the production of consumer goods.

Commonwealth of Independent States (CIS) (4A). A political and economic organization created in 1991 by 11 republics of the former Soviet Union. Members included Russia, Belarus, Ukraine, Moldova, Armenia, Azerbaijan, Kazakhstan, Kyrgyzstan, Tajikistan, Turkmenistan, and Uzbekistan. Georgia became the twelfth member in 1993. The CIS coordinates relations between member countries, including issues involving economics, foreign policy, and defense matters.

Communauté Financière Africaine (CFA) (9B). Economic links between France and its former colonies in Africa. From independence until 1994, it involved links to the French franc, but this arrangement ended with the devaluation of the CFA franc.

commune (5D). The organization that controls rural life in Communist countries, including agriculture, industry, trade, education, local militia, and family life.

Communism (3A). A system in which the workers govern and collectively own the means of economic production. When spelled with a capital "C," Communism refers not to the system as it was originally envisioned but rather to the totalitarian systems adopted in countries such as the Soviet Union and China, where small elite groups rule or ruled under the guise of communism.

concentration (of agriculture) (3D). Agricultural production carried out on fewer and larger farms and limited to smaller areas of higher productivity.

concentric pattern of urban zones (11C). A pattern of urban geography with a central business district in the middle, surrounded by a hierarchy of residential zones.

Confucius (Kong Fuzi) (5A). A Chinese administrator who established a system of efficient and humane political and social institutions that became the basis of procedures and ways of life in much of Eastern Asia.

continentality (4B). Especially cold winters and hot summers resulting from locations on landmasses that are far from the moder-

ating effects of large water bodies such as oceans.

cottage industry (7C). Manufacturing based on processes that are carried out in dispersed homes instead of in a central factory.

counterurbanization (5B). The movement of people from metropolitan urban areas to take up residence and employment in small towns or rural areas.

country (1A). A self-governing political unit, recognized by other countries.

criollos (10A). Persons living in the Americas during the colonial era whose sole ancestral heritage was tied to Spain or Portugal. Persons of African heritage living in the Caribbean during the colonial era were referred to Black Creoles.

crony capitalism (6C). The economic system in which entrepreneurs gain advantages through close links to government officials and ministers.

cultural fault line (2B). A line or zone between contrasting cultural regions across which conflict is likely to occur.

cultural geography (2B). The geography of the identity of communities at local, country, and world regional scales.

cultural hearth (1B). A small region of the world that acted as a catalyst for developments in technology, religion, or language, and as a base for their diffusion.

cultural hegemony (11A). When a dominant culture group forces another group of people to reject their own culture and accept that of the dominant group.

Cultural Revolution (5D). The attempt by Mao Zedong between 1966 and 1976 to change the basis of Chinese society. The disruption held back the country's economic development.

culture (1A). The ideas, beliefs, and practices that a group of people hold in common, including language, religion, social activities, and the design of products.

D

Daoism (5A). The teachings of Chinese philosopher Laozi, who disliked the organized and hierarchical social system of Confucius and advocated a return to local, village-based communities with little outside interference.

death rate (2A). The number of deaths per 1,000 of the population in a year.

deindustrialization (3C). A rapid fall in manufacturing employment and the abandonment of factories in a once-important industrial region.

democratic centralism (3A). The practice of sole governance by the Communist Party, the political party of the working class, because it is believed that only the Communist Party is the true representative of the people.

demographic transition (2A). The process in which the population of a poor country moves from high birth and death rates toward the low birth and death rates of rich countries, through a period of high birth rates and low death rates—when the population increases rapidly.

demography (2A). The study of human populations in terms of numbers, density, growth or decline, and migrations from place to place.

density (2A). The frequency of a phenomenon within a unit of land area; for example, people per hectare.

deposition (2C). The dropping of particles of rocks carried by rivers, wind, or glaciers when they stop flowing or blowing, or melt, respectively.

desalination plant (8A). A mechanism that extracts fresh water from seawater by evaporation and condensation.

desert biome (2C). Major biome-type characterized by a discontinuous plant cover or none.

desertification (2C). Processes that destroy the productive capacity of an area of land.

deurbanization (6C). The process adopted by the Communist government of Vietnam after 1975 to reduce the size of Ho Chi Minh City and return people to rural areas.

development (2A). The process by which human societies improve their quality of life, including economic, cultural, political, and environmental aspects.

devolution (3C). The process by which local peoples desire less rule from their national governments and seek greater authority in governing themselves.

diaspora (2B). The scattering of a people, such as the Jews and Chinese, to other countries.

direction (1B). The position of one place relative to another, measured in degrees from due north or by the points of the compass.

disease (3A). A condition in which an animal or plant is prevented from carrying out its normal functions by an infection, which may result in death.

distance (1B). The space between places, measured in kilometers (miles), travel time, or travel cost.

distribution (2A). The spatial spread of a phenomenon, as in lines, clusters, or at random.

diversified economy (8C). An economy in which manufactures are more important than primary products and where there is a variety of manufactured products and a growing service sector.

domino theory (6A). The political theory suggesting that the countries of Southeast Asia would fall one by one to Communist pressures. Thailand and Malaysia were "dominos" that did not fall.

Dust Bowl (11B). The area of Oklahoma, Kansas, and Texas, USA, where plowing of subhumid grassland resulted in massive wind erosion in the dry years of the 1930s and since.

Dutch East India Company (6A). The company established in the Netherlands to control trade in the East Indies (approximately equal to Indonesia today).

E

economic geography (2A). The geography of making a living, production, distribution, and consumption.

economic leakage (10D). Foreign ownership and control of industry, such as tourism in many parts of Latin America, where the revenue generated by the industry flows to the foreign owners and foreign employees and largely bypasses local communities and thus "leaks" from the local economy.

economics (1A). The study of the production, distribution, and consumption of goods and resources.

economies of scale (11A). Increased productivity gained by building larger factories and gaining access to larger markets.

ecosystem (2C). The total environment of a community of plants and animals, including heat, light, and nutrient supplies.

ecotourism (6C). Tourism based on viewing natural wonders such as wildlife and rock formations.

ejiditarios (10C). The individual *ejido* farmers who were granted the right to use the land by the Mexican revolutionary government created following the overthrow of dictator Porfirio Díaz. Although the farmers were also granted the theoretical right to pass land use to their offspring, ownership of the land remained with the government.

ejido (10C). Land tenure system created by Mexican revolutionaries who overthrew the Porfirian government in 1911 with the purpose of easing rural poverty. The system created state-owned rural cooperatives, which provide landless peasants with a degree of control over the land they farmed. The *ejido* system resulted in the fragmentation of landholdings as the government took control away from some landowners and gave it to *ejiditarios*. In many cases, the government placed already impoverished farmers on marginal land, exacerbating their situation.

El Niño phenomenon (10B). The process by which the warm equatorial waters push back the cold Peruvian current off the western coast of tropical South America, producing local fish kills and having wider world climatic effects.

encomienda system (10A). The Spanish colonial system employed in Latin America in which lands were allotted to Spaniards who were responsible for exploiting their wealth and had jurisdiction over native peoples.

entrepôt (5D). A port that collects goods from several countries to trade with the wider world and distributes imports to its immediate area or hinterland. Examples include Hong Kong and Singapore.

erosion (2C). The wearing away of rocks at Earth's surface by running water, moving ice, the wind, and the sea to form valleys, cliffs, and other landforms.

ersatz capitalism (6C). The economic system that implies an inferior substitute for locally based economic development when the capital, skills, and management are all imported to produce goods for export.

ethnic cleansing (3F). The process by which a dominant group of people in a country causes another ethnic group to leave a region, often using threats or military force.

ethnic group (2B). A group of people with common racial, national, religious, linguistic, or cultural origins.

ethnic religion (2B). A religion that is linked to a particular ethnic group, such as Judaism and Hinduism.

European Union (EU) (3C). Name adopted by the European Community in 1993, suggesting both an expansion to other European countries following the end of the Cold War and the possibility of a future closer political federation.

Euroregions (3C). Border areas of differing countries within Europe where the people within them work together to make transboundary movement easier.

export-led underdevelopment (10D). The process by which countries export their raw materials while their governments charge taxes on exports to pay for their imports. There is no stimulus to develop manufacturing and so diversify the economy to bring more wealth into it.

extensification (of agriculture) (3D). The production of fewer livestock or crops from the same area.

F

favela (10E). A shantytown in a Brazilian city.

federal government (2B). A division of functions among central government, states or provinces, and local units (counties, cities).

feudalism (1B). A political system based on lord-vassal relationships in which the vassals gave homage and service in return for protection.

First Nation (11A). Indigenous or Native Canadians.

First World (2B). The Western countries, led by the United States.

five-year plans (4A). A series of comprehensive economic plans in the Soviet Union. These plans were followed from 1928 until 1991.

fjord (3E). A river valley that was flooded with ocean water after the sea-level rose in the postglacial age.

Fordism (11A). The application of production lines in the assembly of a wide range of components in various manufacturing industries based on the production system established by Henry Ford for the automobile industry.

forest biome (2C). A major biome-type characterized by close-spaced trees.

formal sector (2A). The economic sector in which workers have recognized or licensed jobs, receive agreed-upon wages, and pay taxes.

Francophone (11A). French-speaking Canadians.

free-market system (2A). The economic system that is based on free competition and pricing of goods determined by the market. It is the basis of capitalism, but market "freedom" is often reduced by government and major producer actions.

free trade zone (10E). An area within a country where components can be imported without tariffs for assembly with a view to exporting the finished goods (*see also* **special economic zones**).

friction of distance (1B). The relative difficulty of moving from one place to another, which increases with kilometers (miles), cost, or travel time.

FUNAI (10E). Amazon Indian Rights Agency (*Fundação Nacional do Indio*), which seeks to provide education, health care, and support for indigenous peoples in the Amazon basin.

G

gender (2B). The cultural implications of being male or female, with particular reference to the inequalities suffered by females in human societies.

genocidal rape (3F). The rape of women of an ethnic group or race in the belief that the women's ethnic group or race will be exterminated after the women are impregnated by the perpetrators' own genetic "seed."

genocide (3A). The systematic extermination of an ethnic group, nation, racial, or religious group.

gentrification (3D). The movement of higher-income groups to occupy and improve residences in older and poorer central parts of Western cities.

geographic inertia (3C). Once capital investments are made in factories and infrastructure that give a region agglomeration economies, production will continue there for a period of years after other areas emerge with lower production costs.

geographic information system (GIS) (2A). The computer-based combination of maps, data, and often satellite images that is a foundation for geographic analysis.

geography (1B). The study of how human beings live in varied ways on different parts of Earth's surface; the study of the environment and space in which humans live.

Germanic peoples (3A). A broadly defined group of peoples from northern Europe who began to move south into Germany and other areas of Europe in 500 B.C. Modern Germans, Austrians, Dutch, and the Scandinavians (e.g., Danes, Norwegians, Swedes) are the most numerous of today's Germanic peoples.

ghetto (11E). An urban area in which a particular group of people such as Jews or African Americans is segregated.

glasnost (4C). The Soviet Union policy of the late 1980s designed to create greater openness and exchange of information.

global city-region (1B). A region that is dominated by one or more cities with major involvements in the global economy.

global choke point (1B). A point at which surface transportation routes converge that is liable to the control by the country in which the choke point occurs; for example, the Suez Canal (Egypt) or the Strait of Hormuz (Iran).

globalization (1A). The growing interconnections of the world's peoples and the integration of economies, technologies, and some aspects of cultures.

global warming (2C). The process by which average temperatures in Earth's atmosphere rise over a period of several decades or centuries, leading to the melting of ice masses and a rising sea level.

grassland biome (2C). A major biome type characterized by grasses, with few, scattered trees.

Great Leap Forward (5D). The attempt by the Chinese Communist government in the late 1950s to increase the pace of industrialization. It failed by ignoring food production at a time when bad weather brought poor harvests and famine.

Greeks (3A). Historically, a group of peoples who formed a foundation of European ideas and established a major empire extending to India around 300 B.C

Green Revolution (6C). The result of introducing high-yielding strains of wheat and rice. The outputs of commercial farms in Southern and Eastern Asia increased by several times, but the costs of seeds, fertilizers, and pesticides were too high for smaller farmers.

gross domestic product (2A). The total value of goods and services produced within a country in a year. Often expressed as GDP per capita, when the total GDP is divided by the country's population.

gross national income (2A). The total value of goods and services produced within a country in a year, together with income from labor and capital working abroad less deductions for payments to those living abroad.

Group of Eight (11C). An economic discussion forum consisting of the world's eight most materially wealthy countries; it includes the United States of America, Canada, Japan, Germany, the United Kingdom, France, Italy, and Russia.

guest worker (3D). A foreigner who has permission to reside in a country to work but is not a citizen of that country. From the German word *"Gastarbeiter."*

gulag (4D). Short for the Russian name *Glavnoe upravlenie ispravitel no-trudovykh lagerei* ("Main Directorate for Corrective Labor Camps"), a collection of prison camps in the Soviet Union where criminals and those who opposed the Communist government were sent to perform hard labor as punishment.

Gulf Cooperation Council (8C). Formed in 1981 under the leadership of Saudi Arabia to focus on political problems raised by the Iran-Iraq War.

H

hacienda (10A). Large production estates developed during the Spanish colonial reign in the Americas; usually owned by a single family with substantial material wealth or by the church.

Han Chinese (5D). The largest group (94%) of people in China. An ethnic grouping based on the administrative culture spread by the Chinese empire in the A.D. 200s and 300s.

heartland (4D). The area of a country that contains a large percentage of the country's population, economic activity, and political influence.

Hebrew (8A). A language related to Arabic that is the official language of Israel.

Hindi (7B). One of the official languages of India but spoken by only 30 percent of the population, mainly in the north.

Hinduism (7A). A religion of Southern Asia, observed mainly in India, that includes the worship of many gods related to varied historic experiences and is associated with the caste system.

hinterland (4D). The areas of a country that lie outside the heartland. The hinterland usually has a relatively small percentage of a country's population, economic activity, and political influence compared to the heartland, though it may be well-endowed in natural resources.

horizontal integration (11A). The combining of producers of the same product in a single corporation to create economies of scale, including the control of prices.

household responsibility system (5D). The replacement for communes in rural China after 1976, returning ownership and decision making to individuals and groups that could sell surpluses in open markets.

human development (2A). A broader view of development that focuses on people rather than economic change, which is regarded as a means to the end of enabling people to enlarge their capabilities so that they can enjoy the richness of being human.

human development index (HDI) (2A). A measure of human development based on income, life expectancy, adult literacy, and infant mortality.

human-environment relations (1B). The study of interactions between human activities and the natural environment.

human geography (1B). The study of geographic aspects of human activities, often with a population, economic, cultural, or social focus.

human poverty index (HPI) (2A). A measure of human poverty, linked to HDI, that indicates levels of personal deprivation.

human rights (2A). The rights that should be a normal part of human experience, including justice, a decent standard of living, personal security, and freedom of thought and speech.

Hutu (9B). The majority ethnic group of Burundi and Rwanda with an agricultural culture.

hyperinflation (10E). The process in which rising prices (and often wages) feed upon themselves and begin to increase rapidly.

I

imperialism (3A). The practice of extending the rule of an empire over foreign lands to expand the lands of the empire.

import substitution (2A). A development policy in which countries develop manufacturing industries to fulfill internal market demands, often protected by high tariffs to exclude foreign competition.

Incas (10A). A group of peoples based in the area of modern southern Peru, South America, that extended their empire northward to modern Ecuador and southward to Chile in the A.D. 1400s.

Indian National Congress Party (7A). Formed by a mainly Hindu elite in 1885, it widened its base to become a major force in the independence movement and then India's main political party.

indigenous people (2B). The first inhabitants of an area or those present when the area is taken over by another group.

Industrial Revolution (3A). The period of the late 1700s and early 1800s when increasingly complicated machines and chemical processes fueled by inanimate power sources such as water and coal replaced traditional ways of making goods by hand with simple tools. The mass production of

goods resulted, as did the need for raw materials. The Industrial Revolution began in England and then spread to other areas of Europe and the world.

industry (2C). A group of business activities, such as the farming industry, manufacturing industry, or tourism industry.

infant mortality (2A). The number of deaths per 1,000 live births in the first year of life.

informal sector (2A). The economic sector in which workers act outside the formal sector.

intensification (of agriculture) (3D). The increased output of crops or livestock per area unit of land.

irredentism (3F). The desire to gain control over lost territories or territories perceived to belong rightfully to a group; associated with nationalism.

Islam (8A). A religion of Northern Africa and Southwestern Asia, and parts of Southern and Eastern Asia, based on the teachings of Muhammad as recorded in the Qu'ran. *Islam* means "reconciliation to God."

isthmus (10C). Naturally occurring land bridge.

J

Judaism (8A). The religion of the Jewish people who worship Yahweh as the creator and lawgiver.

K

Karen (6A). Minority group in Myanmar.

Khmer (6A). A people who moved into the Cambodian area from the north.

kibbutzim (8C). A communal farming village in Israel, the social and spiritual basis of the new Israeli nation after 1948 but now the home of fewer than 5 percent of the Israeli population.

Kurdish (8A). A language akin to Persian spoken by a group of tribes in the hilly lands of eastern Turkey, northern Iraq and Syria, and western Iran. The tribes wish to establish their own country.

L

language (2B). The spoken means of communication among people, mostly also capable of being written down. There are 12 dominant language groups but thousands of individual languages in the world.

Lao (6A). The dominant people in Lao People's Republic.

latitude (1B). The distance of a place north or south of the equator, measured in degrees.

localization (1A). The geographic differentiation of places within countries, together with demands for political autonomy: it may be in response to globalization.

location (2A). A place's location is defined by its position on Earth's surface in terms of

latitude and longitude (absolute location) or by its level of interaction with other places (relative location).

loess (3B). Fine-grained and fertile soils developed from windblown deposits.

longitude (1B). The distance of a place east or west of 0° longitude, measured in degrees.

long lot (11D). The unit of land settlement by French colonists in Anglo America. Each plot of land had a narrow river or road frontage and often extended through different types of land suitable for crops, pasture, or woodland.

M

mallee (6B). A vegetation community occurring in Australia that is composed of drought-resistant eucalyptus shrubs forming impenetrable thickets.

Maori (6A). A people who settled New Zealand from the wider South Pacific around the A.D. 800s and continue to form an important component of the New Zealand population.

map (1B). The representation of the features of Earth's surface on paper at varying scales.

market gardening (3E). The commercial production of high-cash value, specialty fruit and vegetable crops such as table grapes, raisins, oranges, grapefruits, apples, and lettuce.

marsupial (6B). A mammal that raises its young in a pouch instead of a womb. Mainly found in Australia, these include kangaroo and koala.

maquiladora (10C). Mexican government program that encourages foreign-owned factories to be sited in Mexico by not charging import duties on raw materials for assembly.

Maya (10A). A group of Native Americans whose civilization centered on the eastern lowlands of modern southern Mexico, Guatemala, and central Honduras. Main period from A.D. 200s to 900s, with wealth based on a surplus of corn.

medina (8B). The crowded streets of the older sections of Arab towns in Northern Africa and Southwestern Asia.

mediterranean climate (3B). A climatic environment that occurs between midlatitude west coast and tropical arid climatic environments, having wet winters and dry summers.

megalopolis (5B). An expanded urbanized area that includes several metropolitan areas with over a million people and dominates the economy of surrounding areas. First identified in the northeastern United States covering the area between Boston and Washington, D.C.

Melanesian people (6A). "Dark-skinned" peoples who inhabit the South Pacific islands.

meridian of longitude (1B). A line joining places of the same longitude on Earth's surface.

mestizo (10A). A person of mixed Native American and European stock in Latin America.

Micronesian people (6A). Peoples of the South Pacific who inhabit the "small islands" of the northwestern area.

Middle America (10B). Mexico, Central America, and the Caribbean.

midlatitude climates (2C). Climates typical of the zone between the tropics and polar regions.

midlatitude continental interior climate (3B). A climatic environment that is marked by very cold and often dry winters and hot summers with thundery rain.

midlatitude east coast climate environment (5A). A climatic environment with warm-to-hot wet summers and cool-to-cold drier winters.

midlatitude west coast climate (3B). A climatic environment dominated by cyclonic weather systems linked to oceanic heat and moisture sources, interspersed with anticyclones. Typically with cooler summers and milder winters than continental interiors on the same latitude.

migration (2A). The long-term movement of people into (inmigration, immigration) or out of (outmigration, emigration) a place.

Ministry of International Trade and Industry (MITI) (5B). The Japanese government ministry charged with promoting Japanese trade and products abroad.

modernization (2A). The theory of development in which poorer countries attempt to follow the stages through primary and secondary to tertiary sectors that the countries of Western Europe and Anglo America followed in the A.D. 1800s and early 1900s.

Mon (6A). A people who moved into the Cambodian area from the north.

Monroe Doctrine (10C). An 1823 declaration asserting U.S. rights to control activity in the Americas over those of countries outside the region. In the Monroe Doctrine, the United States vowed to resist any intervention from outside countries in Latin American affairs.

monsoon climatic environment (5A). A tropical climatic environment in which there are wind shifts between summer and winter, bringing heavy rains from oceanic air in the summer and dry winds of interior continental air in winter.

Mughal (Mogul) dynasty (7A). Turkish invaders of India from Persia in the A.D. 1500s who conquered most of the region and left a heritage of magnificent buildings such as the Taj Mahal (*see* back cover).

Muhajirs (7C). Muslim people who migrated to Pakistan at the time of partition in 1947. Many still speak Hindi.

multinational corporation (2A). A corporation that makes goods and provides services in several countries but directs operations from headquarters in one country.

Muslim League (7A). Formed in India in 1906 to uphold the position of Muslims in the British Indian Empire.

Muslims (8A). "Those who are reconciled to God." Followers of Islam.

N

nation (2B). An "imagined community" in which a group of people believe that they share common cultural features, often linked to a specific area of land.

nationalism (2B). A desire to make the nation equivalent to a country unit.

nationalize (10A). The assumption of ownership and control of private companies by governments or government institutions.

nation-state (3A). The linking of a separate and distinct people ("nation") and a politically organized territory with its own sovereign government ("state").

nation-state idea (3A). The belief that each people ("nation") must have its own country ("state") in order to be free and govern itself as it desires.

Native American (11A). People who inhabited Anglo America before the European arrival, including Amerinds and Inuits (Eskimos).

natural environment (1A). The world as it might be without humans, including climate, landforms, plants, animals, soils, and oceans.

natural hazard (2C). A natural event, such as a volcanic eruption, earthquake, tornado, hurricane, and flood that interrupts human activities by causing extensive damage and deaths.

natural resource (2C). Materials present in the natural environment and recognized by humans as of practical worth (e.g., minerals, soils, water, building stones, timber).

new rice technology (6C). Improved strains of rice and methods of growing that developed in the Philippines as part of the Green Revolution.

Nile Waters Agreement (8B). Agreement between Egypt and Sudan in 1959 to share Nile River waters, with 70 percent allocated to Egypt.

nongovernmental organization (2A). Groups of people who act outside government and major commercial agencies, mainly in advocacy roles such as delivering aid and lobbying for particular causes.

nonrenewable resource (2C). A natural resource that is used up once it is all extracted; for example, coal or metallic minerals.

North Atlantic Treaty Organization (NATO) (3C). A military alliance of non-Communist European countries and the United States, founded in 1949 to counter the military threat of the Soviet Union. In recent years, former Communist countries have joined the alliance, and Russia has formed a partnership with NATO.

northern coniferous forest (taiga) (4B). A forest composed of coniferous trees (firs, pines, cedars) common in the northern parts of midlatitude continental interior climatic environments.

O

ocean biome (2C). The oceans as ecosystems with energy pathways, nutrient cycles, and food chains.

Organization of Petroleum Exporting Countries (OPEC) (8A). Established in 1960 to further the interests of oil and gas producers throughout the world, often in materially poorer countries and often to resist the overriding power of multinational oil corporations.

Organization of the Islamic Conference (OIC). (8A). Established in 1970 by foreign ministers of Muslim countries throughout the world. It has 45 members but advances individual countries' interests rather than pursuing a common agenda.

overpopulation (6E). A situation in which a country's or region's resources cannot support the population on a subsistence basis. The application of more technology and better management may make it possible for such a region to support a larger population.

P

Pacific Rim (6A). The countries that border the Pacific Ocean, including those in the South Pacific, Eastern Asia, Anglo America, and Latin America.

padi (6C). The main feature of wet-rice cultivation in which fields are inundated with water during the growing season.

Palestine Liberation Organization (PLO) (8A). An organization that promotes the reestablishment of a country of Palestine. It has a secular, left-wing political basis.

PAN (10C). The National Action Party (*Partido de Accion Nacional,* PAN) of Mexico. In July 2000, the PRI's dominance of Mexican politics was overturned with the election of PAN candidate Vicente Fox.

Pan-Arab country (8A). The idea of Arab countries joining to form a single country. Egypt and Syria were joined for a short while in the 1960s, but few attempts have been made to achieve this end since.

parallel of latitude (1B). A circle joining places of the same latitude on Earth's surface.

Parti Québecois (11A). A political party formed with the purpose of achieving

Québec's separation from the Canadian people and political independence from the Canadian federal government.

peninsulare (10A). Used in reference to persons living or working in the New World who were born in Spain or Portugal on the Iberian Peninsula (used more for those born in Spain). *Peninsulares* held the highest offices and received the largest land grants in the New World.

perestroika (4C). The Soviet Union policy of the late 1980s designed to reconstruct the political and economic structure of the country so that it could compete with or in the capitalist world economic system.

permafrost (4B). Permanently frozen ground extending several hundred meters below the surface in Siberia and northern Canada. In summer, water in the surface active layer, 50 to 100 cm deep, melts.

Persian (8A). A language spoken in Iran following its introduction from Eurasia around 3000 B.C. Modern Persian, written in Arabic script, is also known as Farsi.

physical geography (1B). The study of geographic aspects of natural environments.

place (1B). A point or area on Earth's surface having a geographic character defined by what it looks like, what people do there, and how they feel about it.

planned economy (3A). The Communist practice of the government, rather than the free market, deciding what goods and services need to be produced within a country.

plantation agriculture (6C). The large-scale and concentrated commercial cultivation of crops often associated with the tropics and colonial attempts to produce such crops for the home countries.

plateau (9A). An upland area with flat-topped hills or having a tablelike form.

podzol soils (4B). Soils of low fertility in which plant nutrients are removed by water passing through. Commonly develop beneath midlatitude coniferous forest and on sandy soils.

polar biome (2C). Major biome-type where plant growth is inhibited by extreme cold and a short, if any, growing season.

polar climates (2C). Climates typical of the polar regions.

political geography (2B). The study of geographic implications of the ways in which governments control activities within their countries and link with each other.

politics (1A). The art or science of government or influencing government decisions.

Polynesian people (6A). Lighter-skinned peoples of the South Pacific islands who inhabit mainly the eastern islands.

poorer people (1A). People who have below-average possession of material goods.

population density (2A). The numbers of people per given area.

post-Fordist production (1C). Factory production methods based on smaller component inventories and an ability to change rapidly to new products.

poverty (1A). Low income and deprivation of minimally acceptable material requirements; the absence of basic capabilities that enable people to function fully as human beings.

PRI (10C). The Institutional Revolutionary Party (*Partido Revolucionario Institucional*, PRI) arose from the Mexican revolutionary period of 1910–1920. The PRI grew out of the revolutionary movement into the most powerful and lasting political party in the history of Mexico, dominating Mexican politics for more than 70 years. The party controlled the government, nationalized industries, and allegedly fixed many political votes in order to remain in power.

primary sector (2A). The sector of an economy that produces output from natural sources, including mining, forestry, fishing, and farming.

primate city (6C). A city that contains a large proportion of the urban population of a country, often several times the population of the second city.

producer goods (3C). Industrial goods used by other industries to make consumer goods.

producer services (3D). Service industries that are involved in the output of goods and services, including market research, advertising, accountancy, legal, banking, and insurance industries.

production line (11A). The system of manufacturing in which components are made and assembled into the final product in a sequence of factory-based processes.

productive capacity (3C). The total amount of goods a country's industries can produce during a given period of time.

productivity (3D). The measure of the amount of product generated or work completed per hour of labor.

protectionism (10C). Governmental policies that focused on government-controlled industry and the protection of domestic products through tariffs, quotas, and red tape.

Punjabi (7C). A language spoken by two-thirds of Pakistan's population.

purchasing power parity (2A). The measure of GNI or GDP that is based on internal country costs of living rather than external exchange rates related to the U.S. dollar.

Q

Qu'ran (8A). The holy book of Islam.

quaternary sector (2A). The sector of an economy that specializes in producer services, including financial services and information services.

Quechua (10A). The official language of the Inca Empire. Quechua is spoken today by many indigenous groups in Bolivia, Ecuador, and Peru.

R

race (2B). A biologic stock of people with similar physical characteristics or a group of people united by a community of interests.

region (1B). An area of Earth's surface distinguished from others by its physical and human characteristics and interacting with other regions in trade and the exchange of people and ideas.

regional geography (1B). The study of different regions at Earth's surface in their country and global contexts.

relative location (1B). The direction and distance of a place relative to others, often affected by factors that slow or increase contacts among people.

relief (2C). The physical height and slope of the land, as in hills, mountains, and valleys.

religion (2B). An organized system of values and practices, including faith in and worship of a divine being or beings.

renewable resource (2C). A resource that is replaced by natural processes at a rate that is faster than its usage.

revolutionary tourist (10C). Supporters of *Zappatista* and other causes who travel from foreign countries to offer their fighting services to the local cause.

rift valley (9A). A deep valley caused by the rocks of Earth's crust cracking along parallel lines to let down a section of crust to form the valley floor.

Romans (3A). Historically, a group of people who established an empire around the Mediterranean Sea that extended to northwest Europe from the first century B.C.

Russification (4A). Policies directed at making non-Russians into Russians by encouraging or forcing non-Russians to adopt Russian cultural characteristics such as the Russian language.

S

Sahel (9A). The zone immediately to the south of the Sahara desert in Africa that suffers droughts as the arid area expands by natural or human-induced actions.

salinization (8A). The process by which soils become unproductive because of an accumulation of alkaline salts near the surface. Often associated with poorly managed irrigation systems in arid areas.

scale (1B). The relationship of horizontal ground distance to map distance, quoted as a fraction (e.g., 1/10,000) or as a ratio (e.g., 1:10,000) in which one unit on the map represents 10,000 units on the ground.

Second World (2B). The Communist countries, led by the Soviet Union until 1991. Now redundant.

secondary sector (2A). The sector of an economy that changes the raw materials from the primary sector into useful products, thus increasing their value, as in chewing gum or parts for airplanes.

Shan (6A). Minority group in Myanmar.

shantytown (5D). An unplanned residential sector of urban areas in poorer countries. Housing is often built of any materials that come to hand and does not have electricity, water, or waste disposal.

Shia Muslims (8A). Also known as Shiites. Muslims who are partisans of the imam Ali (not acceptable to Sunni Muslims) and look to his return. They make up 90 percent of the Iranian population and are associated with radical movements.

Shinto (5A). The traditional religion of Japan, built on ancient myths and customs that promote the national interests and identity.

Sindhi (7C). A language spoken by 12 percent of Pakistan's population.

Sinhalese (7C). A language spoken by the Buddhist majority of people in Sri Lanka.

Slavs (3A). A broadly defined group of people who migrated from the east, settling primarily in East Central Europe between approximately 400 and 800 A.D. The Slavs developed into three distinct subgroups: Western Slavs (e.g., Poles, Czechs, Slovaks, Sorbs), southern Slavs (e.g., Slovenes, Croats, Bosnians, Serbs, Montenegrins, Macedonians, Bulgarians), and eastern Slavs (*see* Chapter 4) (Russians, Belarussians, Ukrainians).

soil (2C). Weathered rock material that develops by the actions of water, animals, and plants into a basis for plant growth.

Southern Africa Development Conference (SADC) (9A). Established by countries in Southern Africa opposed to South Africa's apartheid policy to organize alternative trade outlets. Now encouraging trade among the constituent countries, including South Africa.

South Pacific Forum (6D). Established in 1971 among 13 countries in the South Pacific, including Australia, New Zealand, and several island countries. Aims to promote regional cooperation and confronts regional problems of external exploitation.

spatial analysis (1B). The study of linkages between places at Earth's surface, often in terms of points, lines, and areas, together with statistical associations.

spatial view (1B). A geographic view that focuses on differences among areas of Earth's surface.

special economic zones (5D). Zones established by China in 1979 to encourage foreign investment and export-oriented manufacturing in the southern coastal provinces. Similar zones are found in other countries (*see* **free trade area**).

specialization (of agriculture) (3D). The concentration on fewer commercial products within a farming region.

state (3A). A country, or division of a country within a federal government.

state socialism (3A). The Communist Party actively running the political, social, and economic activities of the people.

steppe grasslands (4B). Midlatitude grasslands typical of the southern parts of midlatitude continental interior climatic environments.

structural adjustment (2A). The theory of development in which a country becomes more involved in the world economic system by producing export goods, reducing tariffs, encouraging privatization, and developing good government and a balanced budget.

suburbanization (5B). The movement of people and corporations from older central parts of urban areas to new residential and commercial areas on the city outskirts.

Sunni Muslims (8A). Also known as Sunnites. Traditional, conservative followers of Islam, forming the majority in most Muslim countries.

supranationalism (3C). The idea that differing nations can cooperate so closely for their shared mutual benefit that they can share the same government, economy (including currency), social policies, and even military.

sustainable human development (2A). A level of development in which resources are exploited at a rate that is sustainable for future generations.

T

tectonic plate (2C). A large block of Earth's crust and underlying rocks approximately 100 km thick and up to several thousand kilometers across. Earth's interior heat causes plates to move apart and crash together, forming major relief features including ocean basins, mountain systems, and continental areas.

temperature inversion (10B). Naturally occurring periods during the winter months when cold dense air remains "trapped" at the surface under warmer air for several days or even weeks. The cold air is usually trapped by mountains or some related physical barrier. Temperature inversions may produce stagnant air, which may complicate pollution problems in urban areas.

Tennessee Valley Authority (TVA) (11C). Established by the U.S. government in 1933 to stimulate economic growth in southern Appalachia through the damming of rivers to improve transportation, flood control, and electricity costs.

Tenochtitlán (10A). The urban capital of the Aztec Empire located in the Central Valley of Mexico. Modern Mexico City originated on the site of the Aztec capital during the Spanish colonial reign.

Terra Australis (6A). The "Southland" mythical continent of the South Pacific, now Australia, that was discovered by Dutch and British explorers in the 1700s.

tertiary sector (2A). The sector of an economy concerned with the distribution of goods and services, including trade, professions, and government employment.

Thai (6A). The majority people in Thailand.

Third World (2B). The poorer countries, mostly in the Southern Hemisphere, not part of the First or Second Worlds.

toponym (11A). Place name.

total fertility rate (2A). The number of births per woman in her child-bearing years.

transmigration (6C). The name given to the process through which Indonesia attempted to ease overpopulation on Java by moving large numbers of people to the less-inhabited islands.

transnational Chinese economy (5D). Economic and trading links established by Chinese people living outside China, especially in Eastern Asia and Anglo America.

Treaty of Tordesillas (10A). The 1494 agreement that established a demarcation line (approximately 46° W longitude) dividing land rights in the Americas between the Spain and Portugal. The line was created under the authority of the pope. Spain gained control of lands to the west of the line, and Portugal gained control of lands to the east.

tropical climates (2C). Climatic environments typical of the tropical zone.

tundra (4B). Ecosystem type occurring in cold polar environments, consisting of mosses, grasses, and low shrubs.

Turkish (8A). A language family spoken by peoples in Central Asia and extended to Turkey. In Turkey, the language is written in Roman script, but Azeris use Arabic script.

Tutsi (9B). Originally a cattle-herding group that settled in the area of Burundi and Rwanda, assuming a leadership position over the more numerous agricultural Hutu tribe.

U

UN Security Council (11C). One of the principle divisions of the United Nations. The purpose of the Security Council is the maintenance of international peace and security. The United States, China, France, the United Kingdom, and Russia are the five countries granted a permanent seat on the Security Council. The permanent members of the Security Council have the power to veto council decisions. The decisions of the Security Council are binding for the entire UN General Assembly.

uneven development (11C). The increase in the gap between poor and wealthy regions in a country and the shifting locations of economic growth and decline over time, seen by Marxists as a natural outcome of capitalism.

universalizing religion (2B). A religion that seeks to be global in its application, such as Islam and Christianity.

Urdu (7C). The official language of Pakistan, although spoken by only 7 percent of the population.

V

vertical integration (11A). The combining of producers of raw materials, manufacturers that process the materials, and those that assemble the products in a single corporation to achieve economies of scale.

Vietnamese (6A). The dominant people in Vietnam.

Vikings (3A). A group of peoples who spread out from Scandinavia from the A.D. 700s, conquering much of northwest Europe, Iceland, and Greenland and influencing events in Russia.

Virgin Lands Campaign (4D). A Soviet agricultural campaign begun in the 1950s. It promoted farming in lands where it had never taken place before, primarily in lands that were very marginal because the soil was poor or not enough water or heat was present to grow crops. Much of the land was in the semidesert and desert areas of southern Siberia and Central Asia, especially in the Kazak Republic.

visual symbolism (4D). The process by which buildings, statues, and other visual features of the built environment promote a dominant social or political message.

W

Wallace Line (6B). The line between the Australian and Asian plant species drawn by the botanist Alfred Russell Wallace in the mid-1880s.

wealthier people (1A). People who have above-average possession of material goods.

weathering (2C). The action of atmospheric forces (through water circulation and temperature changes) on rocks at Earth's surface that breaks the rocks into fragments, particles, and dissolved chemicals.

White Australia policy (6D). The policy designed to exclude Asian people from Australia. Ended in 1972.

Z

Zappatista (10C). Indigenous revolutionaries existing primarily in the Mexican state of Chiapas. The *Zappatistas* assert a lack of representation for indigenous communities in Mexico. Many seek independence from the Mexican government.

Index

Note: Page numbers followed by f indicate figures.

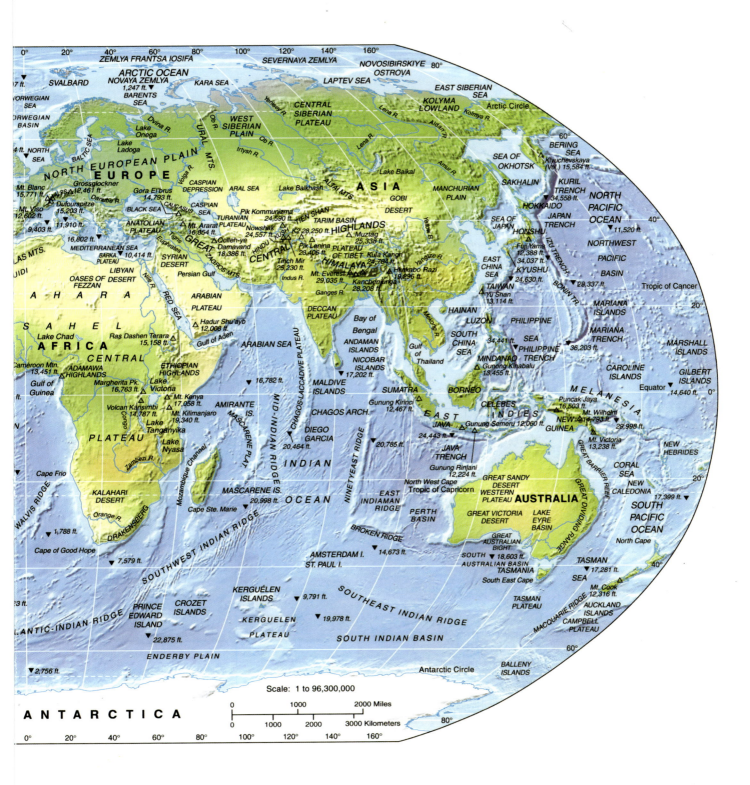

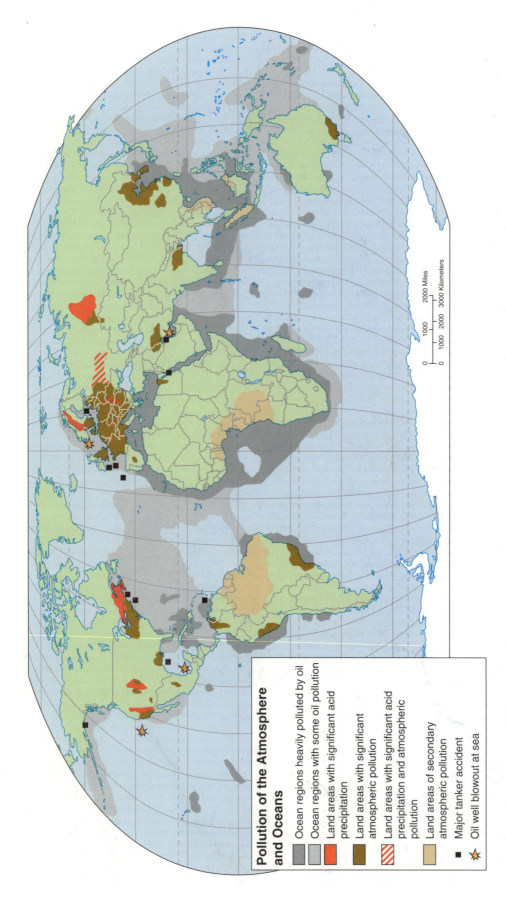

Pollution of the Atmosphere and Oceans

- ▨ Ocean regions heavily polluted by oil
- ▨ Ocean regions with some oil pollution
- ▨ Land areas with significant acid precipitation
- ▨ Land areas with significant atmospheric pollution
- ▨ Land areas with significant acid precipitation and atmospheric pollution
- ▨ Land areas of secondary atmospheric pollution
- ■ Major tanker accident
- ★ Oil well blowout at sea

0   1000   2000 Miles

0   1000   2000   3000 Kilometers

Air and Water Quality

Pollution of the atmosphere and ocean are both vital aspects for future and expanding populations on Earth. The circulations of the atmosphere and the oceans control many features of Earth's natural environments. Land areas downwind of industrial concentrations emitting sulfur and nitrogen gases are subject to acid deposition that damages forest and lake ecosystems; atmospheric pollutants may also be damaging to human health; forest destruction and burning cause "secondary" atmospheric pollution. Oil spills in particular affect the surface plankton basis of life in the oceans. Much of our planet is affected by pollution—a lowering of air or water quality.

Source: Reprinted with permission from *Student Atlas of World Politics*, 4th ed. by John L. Allen. © 2000 The McGraw-Hill Companies, Inc. All rights reserved.